GEOGRAPHICAL
INFORMATION SYSTEMS

GEOGRAPHICAL INFORMATION SYSTEMS

Volume 1
Principles and Technical Issues
Second Edition

Edited by

PAUL A LONGLEY,

MICHAEL F GOODCHILD,

DAVID J MAGUIRE,

and

DAVID W RHIND

JOHN WILEY & SONS, INC.

New York • Chichester • Weinheim • Brisbane • Singapore • Toronto

Volume 1 cover images:

Front cover, clockwise from upper left corner: (1) ESRI—Environmental Systems Research Institute Inc.; (2) JPL/NASA, courtesy of Dr. Anthony Freeman; (3) Volume 2, color plate 51; (4) Canada Center for Remote Sensing, courtesy Lawrence Gray; (Insert) ESRI—Environmental Systems Research Institute Inc.

Back cover, clockwise from upper left corner: (1) Courtesy NASA/JPL; (2) Volume 1 color plate 19, Cambridge University Collection of Air Photos; (3) Courtesy JPL/NASA TOPSAR system; (4) ESRI—Environmental Systems Research Institute, Inc.

This publication is designed to provide accurate and authoritative information in regard to the subject matter covered, It is sold with the understanding that the publisher is not engaged in rendering professional services. If professional advice or other expert assistance is required, the services of a competent professional person should be sought.

Library of Congress Cataloging-In-Publication Data:

ISBN: 0471–33132–5 (Volume 1)
ISBN: 0471–32182–6 (Set)

Printed in the United States of America

10 9 8 7 6 5 4 3 2 1

Dedication

We dedicate this second edition of *Geographical Information Systems* to two pioneers in the field: Professor Terry Coppock and Dr Roger Tomlinson.

Terry Coppock, Professor of Geography at the University of Edinburgh, has devoted a lifetime of research to the subject of humanity's use of the land surface of the Earth. He was one of the very first to recognise the importance of geographical information technologies in collecting, managing, and manipulating the large quantities of data needed to study and understand changing patterns of land-use, and the ways in which new uses compete with traditional ones for the national land resource. He began this work in the late 1950s using the primitive computers of those days. In the mid 1980s he was largely responsible for initiating the *International Journal of Geographical Information Systems* (now the *International Journal of Geographical Information Science*); he became its first editor, and set it on course to its current position as the most influential scientific journal in the field of GIS.

Roger Tomlinson is President of Tomlinson Associates, an Ottawa-based GIS consultancy. He is a past President of the Canadian Association of Geographers, and a recipient of the Royal Geographical Society's Murchison Award and many other honours. In the mid 1960s he recognised that digital computers could be used to analyse the vast quantities of mapped information being created by the Canada Land Inventory. His cost–benefit analyses concluded that computerisation would be the best alternative in spite of the high costs and primitive nature of computers at the time. It was Roger Tomlinson who first coined the term 'geographic information system' for this novel way of using computers, oversaw the extraordinarily innovative creation of the Canada Geographic Information System, and became the world's foremost proponent of the GIS vision over the following decades. He, more than anyone else, is recognised today as the 'father' of GIS.

Dedicating a book is much too modest a way of recognising the magnitude of the contributions of Terry Coppock and Roger Tomlinson – but we hope it draws attention to the high regard in which we hold them.

Contents

Volume 1
PRINCIPLES AND TECHNICAL ISSUES

Dedication — v
Preface — xi
List of contributors — xiii
Acknowledgements — xxiii

1   Introduction · *P A Longley, M F Goodchild, D J Maguire, and D W Rhind* — 1–20

Part 1: Principles

(a)   Space and time in GIS

Introduction · *The Editors* — 23–27
2   Space, time, geography · *H Couclelis* — 29–38
3   Geography and GIS · *R J Johnston* — 39–47
4   Arguments, debates and dialogues: the GIS–social theory debate and the concern for alternatives · *J Pickles* — 49–60
5   Spatial representation: the scientist's perspective · *J F Raper* — 61–70
6   Spatial representation: the social scientist's perspective · *D J Martin* — 71–80
7   Spatial representation: a cognitive view · *D M Mark* — 81–89
8   Time in GIS and geographical databases · *D J Peuquet* — 91–103
9   Representation of terrain · *M F Hutchinson and J C Gallant* — 105–124
10   Generalising spatial data and dealing with multiple representations · *R Weibel and G Dutton* — 125–155
11   Visualising spatial distributions · *M-J Kraak* — 157–173

(b)   Data quality

Introduction · *The Editors* — 175–176
12   Data quality parameters · *H Veregin* — 177–189
13   Models of uncertainty in spatial data · *P F Fisher* — 191–205
14   Propagation of error in spatial modelling with GIS · *G B M Heuvelink* — 207–217
15   Detecting and evaluating errors by graphical methods · *M K Beard and B P Buttenfield* — 219–233

(c)   Spatial analysis

Introduction · *The Editors* — 235–237
16   Spatial statistics · *A Getis* — 239–251
17   Interactive techniques and exploratory spatial data analysis · *L Anselin* — 253–266
18   Applying geocomputation to the analysis of spatial distributions · *S Openshaw and S Alvanides* — 267–282

19   Spatial analysis: retrospect and prospect · *M M Fischer*   283–292

20   Location modelling and GIS · *R L Church*   293–303

Part 2: Technical Issues

(a) GIS architecture issues

   Introduction · *The Editors*   307–308

21   New technology and GIS · *M Batty*   309–316

22   GIS in networked environments · *D J Coleman*   317–329

23   Desktop GIS software · *S Elshaw Thrall and G I Thrall*   331–345

24   GIS interoperability · *M Sondheim, K Gardels, and K Buehler*   347–358

25   GIS customisation · *D J Maguire*   359–369

(b) Spatial databases

   Introduction · *The Editors*   371–372

26   Relational databases and beyond · *M F Worboys*   373–384

27   Spatial access methods · *P van Oosterom*   385–400

28   Interacting with GIS · *M J Egenhofer and W Kuhn*   401–412

29   Principles of spatial database analysis and design · *Y Bédard*   413–424

(c) Technical aspects of GIS data collection

   Introduction · *The Editors*   425–426

30   Spatial referencing and coordinate systems · *H Seeger*   427–436

31   Encoding and validating data from maps and images · *I Dowman*   437–450

32   Digital remotely-sensed data and their characteristics · *M Barnsley*   451–466

33   Using GPS for GIS data capture · *A Lange and C Gilbert*   467–476

(d) Data transformation and linkage

   Introduction · *The Editors*   477–479

34   Spatial interpolation · *L Mitas and H Mitasova*   481–492

35   Multi-criteria evaluation and GIS · *J R Eastman*   493–502

36   Spatial tessellations · *B Boots*   503–526

37   Spatial hydrography and landforms · *L Band*   527–542

38   Intervisibility on terrains · *L De Floriani and P Magillo*   543–556

39   Virtual environments and GIS · *J N Neves and A Câmara*   557–565

40   The future of GIS and spatial analysis · *M F Goodchild and P A Longley*   567–580

Appendix: glossaries and acronyms   xxvii–xxxiii

Author index   xxxiv–lxx

Subject index   lxxi–xcv

Volume 2
MANAGEMENT ISSUES AND APPLICATIONS

List of contributors ix

Acknowledgements xix

Part 3: Management Issues

Introduction · *The Editors* 583–586

(a) Making the GIS efficient, effective, and safe to use

Introduction · *The Editors* 587–588

41 Choosing a GIS · *T Bernhardsen* 589–600

42 Measuring the benefits and costs of GIS · *N J Obermeyer* 601–610

43 Managing an operational GIS · *L J Sugarbaker* 611–620

44 Institutional consequences of the use of GIS · *H J Campbell* 621–631

45 Managing uncertainty in GIS · *G J Hunter* 633–641

46 Liability in the use of GIS and geographical datasets · *H J Onsrud* 643–651

(b) Data as a management issue

Introduction · *The Editors* 653

47 Characteristics and sources of framework data · *N S Smith and D W Rhind* 655–666

48 Characteristics, sources, and management of remotely-sensed data · *J E Estes and T R Loveland* 667–675

49 Metadata and data catalogues · *S C Guptill* 677–692

50 National and international data standards · *F Salgé* 693–706

(c) GIS as a management tool

Introduction · *The Editors* 707–708

51 GIS for business and service planning · *M Birkin, G P Clarke, and M Clarke* 709–722

52 Managing public discourse: towards the augmentation of GIS with multimedia · *M J Shiffer* 723–732

53 Managing a whole economy: the contribution of GIS · *J Smith Patterson and K Siderelis* 733–743

(d) The impact of broad societal issues on GIS

Introduction · *The Editors* 745–746

54 Enabling progress in GIS and education · *P Forer and D Unwin* 747–756

55 Rethinking privacy in a geocoded world · *M R Curry* 757–766

56 National and international geospatial data policies · *D W Rhind* 767–787

Part 4: Applications

Introduction · *The Editors*   791–795

(a) Operational applications

Introduction · *The Editors*   797–799
57   GIS in the utilities · *J Meyers*   801–818
58   GIS in telecommunications · *C Fry*   819–826
59   Transportation GIS: GIS-T · *N Waters*   827–844
60   GIS in emergency management · *T Cova*   845–858
61   GIS in land administration · *P F Dale and R A McLaren*   859–875
62   Urban planning and GIS · *A G-O Yeh*   877–888
63   Military applications of GIS · *D Swann*   889–899
64   Applying GIS in libraries · *P Adler and M Larsgaard*   901–908

(b) Social and environmental applications

Introduction · *The Editors*   909–911
65   The rebuilding of a country: the role of GIS in South Africa · *D R MacDevette,*
     R J Fincham, and G G Forsyth   913–924
66   Health and health care applications · *A Gatrell and M Senior*   925–938
67   GIS and the geography of politics · *M Horn*   939–951
68   Monitoring land cover and land-use for urban and regional planning · *P Bibby and J Shepherd*   953–965
69   GIS and landscape conservation · *R J Aspinall*   967–980
70   Local, national, and global applications of GIS in agriculture · *J P Wilson*   981–998
71   GIS in environmental monitoring and assessment · *L Larsen*   999–1007

72   Epilogue · *P A Longley, M F Goodchild, D J Maguire, and D W Rhind*   1009–1021

Consolidated bibliography   1023–1101
Appendix: glossaries and acronyms   xxiii–xxix
Author index   xxx–lxvi
Subject index   lxvii–xci

Preface

The genealogy of the 'Big Book of GIS' can be traced to the emergent view, just after the 1987 annual meetings of the Association of American Geographers, that the rapidly developing field of GIS had attained sufficient maturity to warrant a large reference compendium. The original editors were appointed in 1988 and *Geographical information systems: principles and applications* appeared as a two-volume boxed set in 1991.

The book fast became the standard reference work for GIS and, despite being joined by an ever increasing number of GIS textbooks and specialised topic volumes, probably remains the most used and most heavily cited work in the field – a fitting testimony to the expertise of the international range of contributors and the quality of their work. Yet any compendium in a fast-developing field has a limited shelf-life, and this edition (initially commissioned, like the first, by Vanessa Lawrence) was commissioned to portray GIS in the late 1990s. The careers of two of the original editors had taken sharp changes in direction since the first edition, which has led them out of academia – David Maguire first became Managing Director of Environment Systems Research Institute (ESRI) UK and then Director of Product Planning at ESRI in California; and David Rhind became Managing Director and Chief Executive of Ordnance Survey (GB) before announcing his departure to be Vice Chancellor of London City University. In order to maintain the academic content and direction of much of the book, Paul Longley was invited to join the original team to co-edit the second edition.

The guiding principles for creating the second edition were fundamentally similar to those used in the first – namely to commission international experts to write benchmark reviews that could be used as a reference against which trends in the field might be assessed. 'History is bunk – but geography isn't' were the watchwords of the early planning stages: the editors decided that none of the original contributions should be retained in any shape or form, that a new list of topics should be drawn up from scratch, and that an entirely new set of

prospective contributors should be assembled without reference to the first edition. The result of this is a complete change in the range of chapters and (the original editors aside) fewer than one in ten of the contributors to this book having contributed to the first edition. This is not intended as a snub to those who were not invited to contribute a second time – all of the original contributions are of enduring relevance – but rather a conscious decision to do all practicable to ensure a complete redefinition, rather than updating, of what we believe epitomises the best in GIS books.

The first edition was divided into two main parts – 'Principles' and 'Applications' – plus an extended overview of GIS detailing definitions, history, and the context to GIS-based analysis. By the late 1990s the developing maturity of the field had made this organising structure less relevant. The material contained in the extended introduction is now much more widely known and taken for granted, while the scale and pace of developments in GIS techniques and management made it desirable to give these issues extended treatments in new and distinct sections. Thus the introduction and epilogue in this book are shorter than in the first edition, and most of the material is arranged into four parts rather than two: 'Principles', 'Technical Issues', 'Management Issues', and 'Applications'. The first edition had 56 numbered chapters, while this edition has 72 – further testimony to both the growth and diversification of the GIS field.

Much is made in this second edition, explicitly and implicitly, about the radical shifts and changes in technology that have taken place since the original 'Big Book' was published; indeed to the editors this is nowhere more apparent than in the ways in which the second edition has been put together. Prospective contributors were first approached in February 1996 and contracts were sent out shortly afterwards: in nearly all cases all significant correspondence and submission between that time and page proofing was electronic. Authors were asked to provide an extended abstract in digital form, and after refereeing by the editors these were all posted on a

WWW site (100 pages and 60 000 words!), along with author contact coordinates (specifically e-mail). The editorial collective gave detailed feedback comments to each of the contributors, particularly with regard to possible overlaps between chapters within the new book structure. Contributors were then encouraged to consult the Web site, in order to ascertain the complementarity of their contributions to others, and to resolve 'low level' problems of overlap and omission with their fellow contributors (again by e-mail). Thus an additional measure of self-regulatory checking was introduced to complement strategic editorial control.

As before, the role of the editors has been to provide focus and direction, and to ensure that the whole of the book is greater than the sum of its individual parts. We feel that, to some extent at least, the 'democratisation' and opening up of dialogue between contributors has made this second edition still more tightly integrated and coordinated than its highly successful forebear. A related point is that all of the contributions were received in digital form and were switched many times between the editors in Bristol, Santa Barbara, Redlands, and Southampton – and innumerable 'field' locations between and beyond. Most of the artwork and colour plates were transmitted to the *GIS World* Web site in Colorado prior to redrawing and sizing. Together, these changes have reduced the lead time between writing and availability of the final product. Whilst it is the hope of the authors that this second edition will prove to be at least as durable as the first, it is also hoped that this reduction in production time will lend the book the freshest possible feel to its first readers.

The successful completion of a major project such as this book requires the cooperation and understanding of many key individuals. We thank firstly our colleagues for their impressive and thoughtful contributions. Vanessa Lawrence originally commissioned this work – her subsequent move to Autodesk represents a considerable loss to the GI publishing industry – and was succeeded by Heather Burkinshaw and Roy Opie at GeoInformation International. Steve Attmore undertook the huge task of overseeing the book's production editing and Rob Garber oversaw the transfer of the project to John Wiley and Sons, Inc. An extended editional meeting was lubricated by Willi Germann's gin and tonics, and Joas made excellent sendwiches. Finally, we thank our wives Mandy, Fiona, Heather, and Christine for putting up with our erratic but intensive work patterns during the project.

Paul A Longley
Michael F Goodchild
David J Maguire
David W Rhind

List of contributors

THE EDITORS

Paul A Longley
Professor at the School of Geographical Sciences in the University of Bristol. Research interests: geographical information systems; fractal geometry; spatial analysis; data integration, especially involving remote sensing and socioeconomic sources; social survey research practice.

School of Geographical Sciences, University of Bristol, University Road, Bristol BS8 1SS, UK
Tel: +44-117-928-7509; Fax: +44-117-928-7878;
E-mail: Paul.Longley@bristol.ac.uk

Michael F Goodchild
Chair of the Executive Committee of the National Center for Geographic Information and Analysis (NCGIA), and Professor of Geography at the University of California, Santa Barbara. Research interests: GIS; environmental modelling; geographical data modelling; spatial analysis; location theory; accuracy of spatial databases; statistical geometry.

Department of Geography, University of California, Santa Barbara, CA 93106-4060, USA
Tel: +1-805-893-8049; Fax: +1-805-893-7095;
E-mail: good@ncgia.ucsb.edu

Dr David J Maguire
Director of Product Planning at the Environmental Systems Research Institute (ESRI) in California. Research interests: spatial databases; GIS customisation; GIS implementation and object-oriented systems.

Environmental Systems Research Institute Inc., 380 New York Street, Redlands, California 92373, USA
Tel: +1-909-793-2853; Fax: +1-909-793-5953;
E-mail: dmaguire@esri.com

David W Rhind
Vice-Chancellor of City University, London and formerly Director General and Chief Executive of the Ordnance Survey of Great Britain. Current research interests include information and data policy issues nationally and internationally; the workings of government; public/private sector interactions.

City University, Northampton Square, London EC1V 0HB, UK
Tel: +44-171-477-8000; Fax: +44-171-477-8560
E-mail: d.rhind@city.ac.uk

THE AUTHORS

Prudence S Adler
Assistant Executive Director at the Association of Research Libraries in Washington, DC. Research interests: information policies; telecommunications; copyright and intellectual property issues.

Association of Research Libraries, 21 Dupont Circle, NW, Washington, DC 20036, USA
Tel: +1-202-296-2296; Fax: +1-202-872-0884;
E-mail: prue@cni.org

Seraphim Alvanides
Research Assistant at the Centre for Computational Geography at the University of Leeds. Research interests: spatial analysis and modelling in GIS environments; scale and aggregation issues; systematic aggregation of areal data.

Centre for Computational Geography, School of Geography, University of Leeds, Leeds LS2 9JT, UK
Tel: +44-113-2431751; Fax: +44-113-333308;
E-mail: s.alvanides@geography.leeds.ac.uk

Luc Anselin
Director, Bruton Center for Development Studies and Professor of Economics, Geography, and Political Economy, University of Texas at Dallas. Research interests: regional economic and demographic analysis; spatial econometrics and spatial statistics; statistical computing; GIS and spatial analysis.

University of Texas at Dallas, PO Box 830688, Richardson, TX 75083–0688, USA
E-mail: lanselin@utdallas.edu

Richard J Aspinall

Director of the Geographic Information and Analysis Center, Montana State University. Research interests: environmental applications of GIS and spatial analysis; data quality issues; integrating socioeconomic and environmental modelling for land-use applications.

Geographic Information and Analysis Center, Montana State University, Bozeman, Montana, MT 59717-0348, USA
Tel: +1-406-994-2374; Fax: +1-406-994-6923;
E-mail: richard@giac.montana.edu

Lawrence E Band

Professor of Geography at the University of Toronto. Research interests include hydrology; geomorphology; GIS and environmental modelling.

University of Toronto, Toronto, Ontario M5S 1A1, Canada
Tel: +1-416-978-3375; Fax: +1-416-978-6729;
E-mail: lband@geog.utoronto.ca

Mike Barnsley

Research Professor of Remote Sensing and GIS at the University of Wales Swansea. Research interests: estimation of land-surface biophysical properties by remote sensing; mapping, monitoring, and analysis of urban areas using very high resolution satellite images, including the development of graph-based spatial analytical tools; scaling and generalisation in remote sensing and GIS.

Department of Geography, University of Wales Swansea, Singleton Park, Swansea SA2 8PP, Wales
Tel: +44-1792-295647; Fax: +44-1792-295955;
E-mail: m.barnsley@swansea.ac.uk

Michael Batty

Professor of Spatial Analysis and Planning at the Centre for Advanced Spatial Analysis, University College London. Research interests include visualisation; urban systems modelling; urban morphology; planning and design processes.

CASA, University College London, 1–19 Torrington Place, London WC1E 6BT, UK
Tel: +44-171-391-1781; Fax: +44-171-813-2843;
E-mail: m.batty@ucl.ac.uk

M Kate Beard

Associate Professor in the Department of Spatial Information Science and Engineering at the University of Maine. Research interests: spatial data quality; metadata; automated generalisation.

NCGIA, University of Maine, Orono, ME 04469, USA
Tel: +1-207-581-2147; Fax: +1-207-581-2206;
E-mail: beard@spatial.maine.edu

Yvan Bédard

Professor in GIS at the Department of Geomatics Sciences, and a member of the Centre for Research in Geomatics, at Laval University. Research interests: spatial database analysis; spatial data warehousing; spatio-temporal reasoning; organisational issues related to the implementation of geomatics technologies.

Department of Geomatics Sciences, Laval University, Québec City G1K 7P4, Canada
Tel: +1-418-656-2131 ext. 3694; Fax: +1-418-656-7411;
E-mail: yvan.bedard@scg.ulaval.ca

Tor Bernhardsen

Senior Consultant at Asplan Viak in Norway. Research interests include implementing GIS and cost/benefit analysis.

Asplan Viak, PO Box 1699, N-4801, Arendal, Norway
Tel: +47-37-035560; Fax: +47-37-023280;
E-mail: tor.bernhardsen@asplanviak.no

Peter R Bibby

Lecturer in Town and Regional Planning at the University of Sheffield. Main research interest is representational systems in urban planning.

Department of Town and Regional Planning, University of Sheffield, Sheffield S10 2TN, UK
Tel: +44-114-222-6181; Fax: +44-114-272-2199;
E-mail: p.r.bibby@sheffield.ac.uk

Mark Birkin

Managing Director, GMAP Ltd and Lecturer in School of Geography, University of Leeds, UK. Main research interests: generation of market intelligence from spatial data; application of geographical models to commercial markets; the use of GIS to improve decision-making within business.

GMAP Ltd, GMAP House, Cromer Terrace, Leeds LS2 9JU, UK
Tel: +44-113-244-6164; Fax: +44-113-234-3173;
E-mail: markb@gmap.leeds.ac.uk

Barry Boots

Professor of Geography and Environmental Studies at Wilfrid Laurier University, Ontario. Research interests: modelling spatial processes; spatial patterns; spatial statistics.

Department of Geography and Environmental Studies, Wilfrid Laurier University, Waterloo, Ontario N2L 3C5, Canada
Tel: +1-519-884-1970; Fax: +1-519-725-1342;
E-mail: bboots@mach1.wlu.ca

Kurt Buehler

Vice President for Technology Development, Open GIS Consortium, Inc. in the USA. Research interests include open systems; geographical data models, standards; open GIS; object-oriented databases; and interoperability.

Open GIS Consortium Inc., 4899 North Old State Road 37, Bloomington, Indiana 47408, USA
Tel: +1-812-334-0601; Fax: +1-812-334-0625;
E-mail: kbuehler@opengis.org

Barbara P Buttenfield

Associate Professor in the Department of Geography at the University of Colorado. Research interests: visualisation; spatial data delivery on the Internet; interface design and evaluation.

Department of Geography, University of Colorado, Boulder, CO 80309, USA
Tel: +1-303-492-3618; Fax: +1-303-492-7501;
E-mail: babs@colorado.edu

Antonio Câmara

Associate Professor in Environmental Systems Analysis at the New University of Lisbon. Research interests: environmental simulation; geographical information systems and multimedia; virtual reality; ecological modelling; water quality modelling.

Department of Environmental Science and Engineering, New University of Lisbon, Monte de Caparica, 2875, Portugal
Tel: +351-1-295-4464 ext. 0104; Fax: +351-1-294-2441;
E-mail: asc@mail.fct.unl.pt

Heather J Campbell

Senior Lecturer in Town and Regional Planning at the University of Sheffield. Research interests: technological innovation; GIS implementation; planning theory.

Department of Town and Regional Planning, University of Sheffield, Western Bank, Sheffield S10 2TN, UK
Tel: +44-114-222-6306; Fax: +44-114-272-2199;
E-mail: h.j.campbell@sheffield.ac.uk

Richard L Church

Professor of Geography at the Department of Geography and the NCGIA, University of California at Santa Barbara. Research interests: location model development; transportation planning and logistics; land management and ecosystems planning; and GIS.

Department of Geography and NCGIA, University of California at Santa Barbara, Santa Barbara, CA 93106 4060, USA
Tel: +1-805-893-4217; Fax: +1-805-839-3146;
E-mail: church@maxz.geog.ucsb.edu

Graham P Clarke

Senior Lecturer in School of Geography, University of Leeds. Research interests: urban geography; retail and marketing geography; GIS and spatial modelling.

School of Geography, University of Leeds, Leeds LS2 9JT, UK
Tel: +44-113-233-3323; Fax: +44-113-233-3308;
E-mail: graham@geog.leeds.ac.uk

Martin Clarke

Chief Executive, GMAP Ltd and Professor of Geographic Modelling at the University of Leeds. Research interests: the development and application of spatial modelling methods including spatial interaction and microsimulation; the development of decision support systems that integrate these methods with GIS.

GMAP Ltd, GMAP House, Cromer Terrace, Leeds LS2 9JU, UK
Tel: +44-113-244-6164; Fax: +44-113-246-0141;
E-mail: martinc@gmap.Leeds.ac.uk

David J Coleman

Associate Professor and Chair of the Department of Geodesy and Geomatics Engineering at the University of New Brunswick. Research interests: system performance determination in network environments and application of computer-supported cooperative work (CSCW) concepts to geomatics production workflow design.

Department of Geodesy and Geomatics Engineering, University of New Brunswick, PO Box 4400, Fredericton, New Brunswick, E3B 5AS, Canada
Tel: +1-506-453-5194; Fax: +1-506-453-4943;
E-mail: dcoleman@unb.ca

Helen Couclelis

Professor in the Department of Geography at the University of California. Research interests: planning; geographical methodology; theories of space and time; cellular automata; geographical data modelling.

Department of Geography, University of California, Santa Barbara, CA 93106, USA
Tel: +1-805-893-2196; Fax: +1-805-893-4146;
E-mail: cook@geog.ucsb.edu

Thomas J Cova

PhD student and Eisenhower Fellow at the University of California, Santa Barbara. Research interests: transportation; hazards; GIS and operations research.

NCGIA/Department of Geography, University of California, Santa Barbara, CA 93106, USA
Tel: +1-805-893-8652; Fax: +1-805-893-8617;
E-mail: cova@geog.ucsb.edu

Michael R Curry

Associate Professor in the Department of Geography, University of California, Los Angeles. Research interests: geographical aspects of technological change; cultural and ethical aspects of geographical technologies; history of geographical ideas.

Department of Geography, University of California, Los Angeles, CA 90095, USA
Tel: +1-310-825-3122; Fax: +1-310-206-5976;
E-mail: curry@geog.ucla.edu

Peter F Dale

Professor in Land Information Management at University College London. Research interests: land, land information and land management; cadastral systems and land registration; professional practice.

Department of Geomatic Engineering, University College London, Gower Street, London WC1E 6BT, UK
Tel: +44-171-387-7050; Fax: +44-171-380-0453;
E-mail: pdale@ge.ucl.ac.uk

Ian J Dowman

Professor of Geomatic Engineering at University College London. Research interests: digital photogrammetry; automation of mapping processes; and mapping from satellite data.

Department of Geomatic Engineering, University College London, Gower Street, London WC1E 6BT, UK
Tel: +44-171-380-7226; Fax: +44-171-380-0453;
E-mail: idowman@ge.ucl.ac.uk

Geoffrey Dutton

Research Associate at the University of Zurich's Department of Geography. Research interests include spatial data modelling; the generalisation of spatial data; and geospatial metadata.

Department of Geography, University of Zurich, Winterthurerstrasse 190, 8057 Zurich, Switzerland
Tel: +41-1-635-52-55; Fax: +41-1-635-68-48;
E-mail: dutton@geo.unizh.ch

J Ronald Eastman

Research interests: the development of decision support routines and methodologies using GIS; incorporation of error and uncertainty in GIS analysis; sustainable implementation of information systems technology; change and time series analysis; and community-based mapping.

The Clark Labs for Cartographic Technology and Geographic Analysis, Clark University, Worcester, MA 01610, USA
Tel: +1-608-793-7526; Fax: +1-508-793-8842;
E-mail: idrisi@vax.clarku.edu

Max J Egenhofer

Associate Professor in Spatial Information Science and Engineering and Director of NCGIA at the University of Maine. Research interests: geographical database systems; spatial reasoning; GIS user interface design.

NCGIA, University of Maine, Orono, ME 04469-5711, USA
Tel: +1-207-581-2114; Fax: +1-207-581-2206;
E-mail: max@spatial.maine.edu

Susan Elshaw Thrall

Professor of Computer Science at Lake City Community College in Florida. Research interests: GIS application programming and programming languages; desktop GIS; geographically enabling data; GIS education.

Business/Industrial Division, Lake City Community College, RT 19, Box 1030, Lake City, Florida 32025, USA
Tel: +1-904-752-1822 ext. 1366; Fax: +1-352-335-7268;
E-mail: links@afn.org

John E Estes

Professor of Geography at the University of California and Director of the Remote Sensing Research Unit. Research interests: interpretation of remotely-sensed data; GIS; regional resources management; global mapping; and environmental forensics.

Department of Geography, University of California, Santa Barbara, CA 93106, USA
Tel: +1-805-893-3649; Fax: +1-805-893-3703;
E-mail: estes@geog.ucsb.edu

Robert J Fincham

Professor and Head of Department of Geography at the University of Natal. Research interests: nutrition surveillance and nutrition information systems development; GIS applications in health and development.

Faculty of Science, University of Natal, Pietermaritzburg, Private Bag X01, Scottsville 3209, South Africa
Tel: +27-331-260-5454; Tel: +27-331-260-5344;
E-mail: fincham@geog.unp.ac.za

Manfred M Fischer

Professor and Chair at the Department of Economic and Social Geography, Wirtschaftsuniversität Wien, Austria, and Director, Institute for Urban and Regional Research, Austrian Academy of Science. Research interests: GIS and spatial analysis; spatial neurocomputing; spatial behaviour and processes, regional labour and housing markets; transportation, communication, and mobility; technological change and regional economic development.

Department of Economic and Social Geography, Wirtschaftsuniversität Wien, Augasse 2-6, A-1090 Wien, Austria; Institute for Urban and Regional Research, Postgasse 7/4, A-1010, Wien, Austria
Tel: +43-1-31336-4836; Fax: +43-1-31336-703;
E-mail: manfred.m.fischer@wu.wien.ac.at

Peter F Fisher

Senior Lecturer in Geographical Information Systems at the University of Leicester. Special research interests include uncertainty modelling and visualisation, and visible area analysis.

Department of Geography, University of Leicester, Leicester LE1 7RH, UK
Tel: +44-116-252-3839; Fax: +44-116-252-3854;
E-mail: pff1@le.ac.uk

Leila De Floriani

Department of Computer and Information Sciences at the University of Genova. Research interests: geometric models and algorithms for GIS; terrain models; structures and algorithms for visibility computations; visualisation.

Department of Computer and Information Sciences, University of Genova, Via Dodecaneso 35, 16146 Genova, Italy
Tel: +39-10-353-6704; Fax: +39-10-353-6699;
E-mail: deflo@disi.unige.it

Pip Forer

Professor of Geography and Geographic Information Studies at the University of Auckland. Pip Forer has symbiotic interests in GIS, individual human geographies, space-time modelling and educational technology. He is currectly engaged in applying GIS to urban structural analysis, tourism planning, Maori economic development, and implementing enhanced learning environments.

Department of Geography, University of Auckland, New Zealand
Tel: +64-9-373-7599 ext. 5183; Fax: +64-9-3737-434;
E-mail: p.forer@auckland.ac.nz

Greg G Forsyth

GIS Project Manager at CSIR in Stellenbosch, South Africa. Research interests: environmental information systems; GIS in natural resource management; GIS in integrated rural development.

CSIR, PO Box 320, Stellenbosch 7599, South Africa
Tel: +27-21-887-5101; Fax: +27-12-886-4659;
E-mail: gforsyth@csir.co.za

Carolyn Fry

Editor of *GIS Europe* and *GeoInformation Africa*. Research interests include science journalism, and geological/environmental applications of GIS.

GeoInformation International, 307 Cambridge Science Park, Milton Road, Cambridge CB4 4ZD, UK
Tel: +44-181-4028181; Fax: +44-181-4028383;
E-mail: carolynf@geoinfo.co.uk

John C Gallant

Postdoctoral Fellow at the Centre for Resource and Environmental Studies at the Australian National University. Research interests: terrain analysis; wavelet and spectral analysis; scale issues in land-surface; and hydrological modelling.

Centre for Resource and Environmental Studies, Australian National University, Canberra, ACT 0200, Australia
Tel: +61-6-249-0666, Fax: +61-6-249-0757;
E-mail: johng@cres.anu.edu.au

Kenn Gardels

Center for Environmental Design Research at the University of California. Research interests: open systems; geographical data models; standards; open GIS; object-oriented databases; and interoperability.

Center for Environmental Design Research, 390 Wurster Hall, University of California, Berkeley, CA 94720-1839, USA
Tel: +1-510-642-9205; Fax: +1-510-643-5571;
E-mail: gardels@ced.berkeley.edu

Anthony C Gatrell

Professor of the Institute of Health Research at Lancaster University, UK. Research interests: geography of health; spatial data analysis; socioeconomic applications of GIS.

Institute of Health Research, Lancaster University, Lancaster LA1 4YB, UK
Tel: +44-1524-593754; Fax: +44-1524-592401;
E-mail: a.gatrell@lancaster.ac.uk

Arthur Getis

Stephen and Mary Birch Professor of Geographical Studies at the Department of Geography, San Diego State University, California. Research interests include the development of spatial statistics, particularly with regard to analysis using large datasets, and the spatial study of disease distributions and urban land-use change.

Department of Geography, San Diego State University, San Diego, CA 92182, USA
Tel: 1-619-594-6639; Fax: 1-619-594-4938;
E-mail: arthurgetis@mail.sdsu.edu

Chuck Gilbert

Technical Services Manager with Trimble Navigation Ltd in California. Special research interest: global positioning systems.

Trimble Navigation Ltd, 645 North Mary Avenue, Sunnyvale, CA 94086, USA
Tel: +1-408-481-2812; Fax: +1-408-481-8699;
E-mail: chuck_gilbert@trimble.com

Stephen C Guptill

Scientific Advisor, US Geological Survey. Research interests: data quality; data structures; GIS design; federated geospatial data systems; GIS policy issues.

US Geological Survey, Reston, VA 20192, USA
Tel: +1-703-648-4520; Fax: +1-703-648-5542;
E-mail: sguptill@usgs.gov

Gerard B M Heuvelink

Reader in Geostatistics at The Netherlands Centre for Geo-ecological Research. Research interests are geostatistics and error propagation in GIS.

Faculty of Environmental Sciences, University of Amsterdam, Nieuwe Prinsengracht 130, 1018 VZ Amsterdam, The Netherlands
Tel: +31-20-5257448; Fax: +31-20-5257431;
E-mail: G.B.M.Heuvelink@frw.uva.nl

Mark E T Horn

Research Engineer at CSIRO Mathematical and Information Sciences in Australia. Research interests: decision support systems; locational analysis; transport planning.

CSIRO Mathematical and Information Sciences, GPO Box 664, Canberra ACT 2601, Australia
Tel: +61-2-6216-7054; Fax: +61-2-6216-7111;
E-mail: mark.horn@cmis.csiro.au

Gary J Hunter

Senior Lecturer in the Department of Geomatics and Deputy Director of the Centre for GIS and Modelling at the University of Melbourne. Research interests: data quality and uncertainty in GIS; spatial data algorithms.

Department of Geomatics, University of Melbourne, Parkville, Victoria 3052, Australia
Tel: +61-3-9344-6806; Fax: +61-3-9347-2916;
E-mail: g.hunter@engineering.unimelb.edu.au

Michael F Hutchinson

Senior Fellow at the Centre for Resource and Environmental Studies at the Australian National University. Research interests: spatial interpolation; digital elevation modelling; spatial and temporal analysis of climate; scale issues in ecological and hydrological modelling.

Centre for Resource and Environmental Studies, Australian National University, Canberra, ACT 2601, Australia
Tel: +61-6-249-4783; Fax: +61-6-249-0757;
E-mail: hutch@cres.anu.edu.au

Ron J Johnston

Professor of Geography at the University of Bristol. Research interests: political geography; electoral geography; political economy of the environment.

Department of Geography, University of Bristol, University Road, Bristol BS8 1SS, UK
Tel: +44-117-9289116; Fax: +44-117-9287878;
E-mail: r.johnston@bristol.ac.uk

Menno-Jan Kraak

Professor of Cartography at ITC in The Netherlands. Research interests: 3-dimensional, temporal, and dynamic visualisation of spatial data.

Department of Geoinformatics, ITC, PO Box 6, 7500 AA Enschede, The Netherlands
Tel: +31-534-784463; Fax: +31-534-874335;
E-mail: kraak@itc.nl

Werner Kuhn

Associate Professor of Geoinformatics and Digital Cartography at the University of Münster, Germany. Research interests: semantics of spatial information; human-computer interaction; workflows with GIS.

Department of Geoinformatics, University of Münster, Robert-Koch-Strasse 26–28, D-48151, Münster, Germany
Tel: +49-251-8334707; Fax: +49-251-8339763;
E-mail: kuhn@ifgi.uni-muenster.de

Art Lange

Product Manager with Trimble Navigation Ltd in California. Special research interest: global positioning systems.

Trimble Navigation Ltd, 645 North Mary Avenue, Sunnyvale, CA 94086, USA
Tel: +1-408-481-2994; Fax: +1-408-481-6074;
E-mail: art_lange@trimble.com

Lars C Larsen

Chief Engineer at the Hydro Informatics Centre in Denmark. Research interests include environmental modelling and information systems.

Hydro Informatics Centre, Danish Hydraulic Institute, Agern Alle 5, DK 2970, Hørsholm, Denmark
Tel: +45-45769555, Fax: +45-45762567;
E-mail: lcl@dhi.dk

Mary L Larsgaard

Map and Imagery Laboratory at the University of California Santa Barbara. Research interests: metadata for georeferenced information and 20th-century topographic and geologic maps.

Map and Imagery Laboratory, University Library, University of California, Santa Barbara, CA 93106, USA
Tel: +1-805-893-4049; Fax: +1-805-893-8799;
E-mail: mary@sdc.ucsb.edu

Thomas R Loveland

US Geological Survey in South Dakota. Research interests: large-area land cover mapping; remote sensing applications; role of geographic data in image classification.

US Geological Survey, EROS Data Center, Sioux Center, SD 57198, USA
Tel: +1-605-594-6066; Fax: +1-605-594-6529;
E-mail: loveland@edcmail.cr.usgs.gov

David R MacDevette

Director of Empowerment for African Sustainable Development (EASD). Research interests: environmental information systems; decision support systems; GIS for education and public information; information for African development.

PO Box 165, Green Point 8051, Cape Town, South Africa
Tel: +27-83-306-0030; Fax: +27-12-841-2689;
E-mail: dmacdev@icon.co.za

Paola Magillo

Department of Computer and Information Sciences at the University of Genova. Research interests: geometric models and algorithms for GIS; terrain models; structures and algorithms for visibility computations; visualisation.

Department of Computer and Information Sciences, University of Genova, Via Dodecaneso 35, 16146 Genova, Italy
Tel: +39-10-353-6704; Fax: +39-10-353-6699;
E-mail: magillo@disi.unige.it

David M Mark

National Center for Geographic Information and Analysis, State University of New York at Buffalo. Research interests: cognitive science; cognitive models of geographical phenomena; critical social history of GIS; languages of spatial relations; qualitative spatial reasoning.

NCGIA/Department of Geography, University of Buffalo, Buffalo, NY 14261, USA
Tel: +1-716-645-2545 ext. 48; Fax: +1-726-645-5957;
E-mail: dmark@geog.buffalo.edu

David Martin

Reader in Geography at the University of Southampton. Research interests: socioeconomic GIS applications; census analysis; medical geography.

Department of Geography, University of Southampton, Southampton SO16 1BJ, UK
Tel: +44-1703-593808; Fax: +44-1703-593295;
E-mail: d.j.martin@soton.ac.uk

Robin A McLaren

Director of Know Edge Ltd in Edinburgh, Scotland. Research interests: business modelling in NLIS; management of change; visualisation; Web-based information services.

Know Edge Ltd, 33 Lockharton Avenue, Edinburgh EH14 1AY, Scotland
Tel: +44-131-443-1872; Fax: +44-131-443-1872;
E-mail: Robin_McLaren@CompuServe.com

Jeffery R Meyers

President of Miner & Miner, Consulting Engineers Inc. in Colorado. Research interests: electrical and gas utilities; GIS design and implementation.

Miner & Miner, Consulting Engineers Inc., 910 27th Avenue, PO Box 548, Greeley, Colorado, USA
Tel: +1-970-352-2707; Fax: +1-970-352-3716

Lubos Mitas

Research Scientist at the National Center for Supercomputing Applications, University of Illinois. Research interests: computational and quantum physics; Monte Carlo methods; spatial interpolations; and modelling of landscape processes.

National Center for Supercomputing Applications, University of Illinois, Urbana, IL 61801, USA
Tel: +1-217-244-1971; Fax: +1-217-244-2909;
E-mail: lmitas@ncsa.uiuc.edu

Helena Mitasova

Research Associate at the Geographic Modeling Systems Laboratory, University of Illinois. Research interests: surface modelling and analysis; multi-dimensional dynamic cartography; modelling of landscape processes; and visualisation.

Department of Geography, University of Illinois, Urbana, IL 61801, USA
Tel: +1-217-333-4735; Fax: +1-217-244-1785;
E-mail: helena@gis.uiuc.edu

Jorge Nelson Neves

PhD student in the Environmental Systems Analysis Group at the New University of Lisbon. Research interests: virtual environments; geographical information systems.

Department of Environmental Science and Engineering, New University of Lisbon, Monte de Caparica, 2875, Portugal
Tel: +351-1-2954464 ext. 0104; Fax: +351-1-2942441;
E-mail: jnn@mail.fct.unl.pt

Nancy J Obermeyer

Associate Professor of Geography at the Indiana State University. Research interests: institutional and societal issues related to the implementation of GIS; political/administrative geography.

Indiana State University, Terre Haute, Indiana 47809, USA
Tel: +1-812-237-4351; Fax: +1-812-237-2567;
E-mail: genancyo@scifac.indstate.edu

Harlan J Onsrud

Associate Professor at the Department of Spatial Information Science and Engineering at the University of Maine. Research interests: legal, policy, and institutional issues surrounding geographic information.

Department of Spatial Information Science and Engineering, University of Maine, Orono, ME 04469-5711, USA
Tel: +1-207-581-2175; Fax: +1-207-581-2206;
E-mail: onsrud@spatial.maine.edu

Peter van Oosterom

Senior Information Manager at Cadastre Netherlands. Research interests: spatial databases; spatial algorithms; map generalisation.

Company Staff, Cadastre Netherlands, PO Box 9046, 7300 GH Apeldoorn, The Netherlands
Tel: +31-55-5285163; Fax: +31-55-3557931;
E-mail: oosterom@kadaster.nl

Stan Openshaw

Professor of Human Geography at the University of Leeds. Research interests: computer modelling and spatial analysis; artificial intelligence, high-performance computing; GIS.

Centre for Computational Geography, School of Geography, University of Leeds, Leeds LS2 9JT, UK
Tel: +44-113-2333320/2431751; Fax: +44 113-2333308;
E-mail: s.openshaw@geography.leeds.ac.uk

Donna J Peuquet

Professor at the Department of Geography at Pennsylvania State University. Research interests: spatial and spatio-temporal representations; spatial cognition; spatial languages; GIS design methodologies; and general issues relating to GIS.

Department of Geography, 302 Walker Building, The Pennsylvania State University, University Park, PA 16802, USA
Tel: +1-814-863-0390; Fax: +1-814-863-7943;
E-mail: peuquet@geog.psu.edu

John Pickles

Professor at the Department of Geography at the University of Kentucky. Research interests: social theory; philosophy of science; political economy of technology and socio-spatial change; and regional development in South Africa and Eastern Europe.

Department of Geography, University of Kentucky, Lexington, KY 40506-0027, USA
Tel: +1-606-257-1362; Fax: +1-606-258-1969;
E-mail: GEG198@ukcc.uky.edu

Jonathan F Raper

Senior Lecturer in Geography at Birkbeck College, University of London. Research interests: philosophy of spatial and temporal representation; 3-dimensional geometric modelling; spatio-temporal modelling; coastal geomorphology.

Department of Geography, Birkbeck College, University of London, 7-15 Gresse Street, London W1P 2LL, UK
Tel: +44-171-631-6470; Fax: +44-171-631-6498;
E-mail: j.raper@geog.bbk.ac.uk

Francois Salgé

Ingénieur en Chef Géographe at the Institut Géographique National in Paris. Research interests: geographic information in all its aspects – production, data management, use and application, economy of the GI sector, quality and quality management, legal and institutional aspects.

Institut Géographique National, 136 bis rue de Grenelle, 75007 Paris, 07SP France
Tel: +33-1-43-98-82-70; Fax: +33-1-43-98-84-00;
E-mail: francois.salge@ign.fr

Hermann Seeger

Director of the Bundesamst für Kartographie und Geodäsie (BFKG) in Germany. Research interests include the theory and practice of international geodetic frameworks.

BFKG, Richard Strauss Allee 11, D 60598, Frankfurt-am-Main, Germany
Tel: +49-69-6333-225; Fax: +49-69-6333-425

Martyn L Senior

Senior Lecturer in the Department of City and Regional Planning at University of Wales Cardiff. Research interests: geography of health; resource allocation; transport planning.

Department of City and Regional Planning, University of Wales Cardiff, PO Box 906, Cardiff CF1 3YN, UK
Tel: +44-1222-874000; Fax: +44-1222-874845;
E-mail: SENIORML@cardiff.ac.uk

John W Shepherd

Professor of Geography at Birkbeck College, University of London. Research interests: urban land-use change; urban and regional planning applications of GIS.

Department of Geography, Birkbeck College, University of London, 7–15 Gresse Street, London W1P 2LL, UK
Tel: +44-171-631-6483; Fax: +44-171-631-6498;
E-mail: j.shepherd@geog.bbk.ac.uk

Michael J Shiffer

Director, Computer Resource Lab, Department of Urban Studies and Planning, Massachusetts Institute of Technology. Research interests: public planning processes; planning support systems; spatial multimedia; urban rail transit.

Department of Urban Studies and Planning, Massachusetts Institute of Technology, 77 Massachusetts Ave., Room 9–514,Cambridge, MA 02139, USA
Tel: +1-617-253-0782; Fax: +1-617-253-3625;
E-mail: mshiffer@mit.edu

Karen Siderelis

Director of North Carolina Center for Geographic Information and Analysis. Research interests: managerial and institutional factors in GIS; national and global spatial data infrastructures.

North Carolina Center for Geographic Information and Analysis, 115 Hillsborough Street, Raleigh, NC 27603, USA
Tel: +1-919-715-0710; Fax: +1-919-715-0725;
E-mail: karen@cgia.state.nc.us

Neil Smith

Senior Consultant at Ordnance Survey. Research interests: information and data policy issues nationally and internationally; interactions between technical, institutional, and managerial issues.

Ordnance Survey, Romsey Road, Southampton SO16 4GU, UK
Tel: +44-1703-792000; Fax: +44-1703-792660;
E-mail: nssmith@ordsvy.gov.uk

Jane Smith Patterson

Advisor to the Governor for Policy, Budget, and Technology in North Carolina and Senior Advisor to the Governor for Science and Technology. Research interests: applications for high-speed networks; bringing technology applications to markets faster; global information; infrastructure development and operations; internetworked applications deployment.

Office of the Governor, 116 West Jones Street, Raleigh, NC 27603, USA
Tel: +1-919-715-0960; Fax: +1-919-715-3775;
E-mail: jpatters@gov.state.nc.us

Mark Sondheim

Head of the Strategic Developments Unit in Geographic Data BC, an agency of the government of British Columbia. Research interests: interoperability; geographic object modelling; large geographic databases; open GIS, object-oriented databases; terrain analysis.

Geographic Data BC, 1802 Douglas Street, Victoria, British Columbia V8T 4K6, Canada
Tel: +1-250-387-9352; Fax: +1-2501-356-7831;
E-mail: sondheim@cmail.gdbc.gov.bc.ca

Larry J Sugarbaker

GIS Manager at the Department of Natural Resources in Washington. Research interests: GIS management; geographical data integration; natural resource applications of GIS.

Department of Natural Resources, PO Box 47020, Olympia, WA 98504-7020, USA
Tel: +1-360-902-1546; Fax: +1-360-902-1790;
E-mail: larry.sugarbaker@wadnr.gov

David Swann

Defense Marketing Manager, ESRI Inc. Main research interest: military applications of GIS, especially communication information.

ESRI Inc., 380 New York Street, Redlands, CA 92373, USA
Tel: +1-909-793-2853; Fax: +1-909-793-5953;
E-mail: dswann@esri.com

Grant I Thrall

Professor of Geography at the University of Florida. Research interests: spatial analysis of urban commercial and residential land markets; commentaries on geographic technology as an emerging business sector.

Department of Geography, 3121 Turlington, University of Florida, Gainesville, FL 32611, USA
Tel: +1-352-392-0494 ; Fax: +1-352-392-8855;
E-mail: thrall@geog.ufl.edu

David Unwin

Professor of Geography, Birkbeck College, University of London. Research interests: visualisation methods; the development and application of local statistics; and the construction of virtual enhancements to field courses, geographical and GISc education.

Department of Geography, Birkbeck College, University of London, 7–15 Gresse Street, London W1P 2LL, UK
Tel: +44-171-631-6485; Fax: +44-171-631-6398;
E-mail: d.unwin@geog.bbk.ac.uk

Howard Veregin

Assistant Professor at the University of Minnesota. Research interests: geospatial data quality assessment; simulation modelling of error; error propagation; metadata analysis; classification accuracy; and the effects of scale and resolution on data quality.

Department of Geography, University of Minnesota, Minneapolis, MN 55455, USA
Tel: +1-612-625-9354 ; Fax: +1-612-624-1044;
E-mail: veregin@atlas.socsci.umn.edu

Nigel M Waters

Professor of Geography at the University of Calgary. Research interests: GIS in transportation; and spatial analytical methods.

Department of Geography, University of Calgary, 2500 University Drive, NW Calgary, Alberta T2N 1N4, Canada
Tel: +1-403-220-6398; Fax: +1-403-282-6561;
E-mail: nwaters@acs.ucalgary.ca

Robert Weibel

Assistant Professor of Spatial Data Handling at the University of Zurich's Department of Geography. Research interests: generalisation of spatial data, digital terrain modelling, and cartographic visualisation.

Department of Geography, University of Zurich, Winterthurerstrasse 190, 8057 Zurich, Switzerland
Tel: +41-1-636-52-55; Fax: +41-1-635-68-48;
E-mail: weibel@geo.unizh.ch

John P Wilson

Professor at the Department of Geography, University of Southern California. Research interests: terrain analysis; environmental modelling; and environmental applications of GIS.

Department of Geography, University of Southern California, Los Angeles, CA 90089-0255, USA
Tel: +1-213-740-1908; Fax: +1-213-740-0056;
E-mail: jpwilson@rcf.usc.edu

Michael F Worboys

Professor of Computer Science at Keele University. Research interests include computational foundations of GIS; geospatial database technology.

Department of Computer Science, Keele University, Keele, Staffordshire ST5 5BG, UK
Tel: +44-1782-583078; Fax: +44-1782-713082;
E-mail: michael@cs.keele.ac.uk

Anthony Gar-On Yeh

Professor and Director at the GIS Research Centre in Hong Kong and Assistant Director of the Centre of Urban Planning and Environmental Management. Research interests: urban and regional applications of GIS; urban development; and planning in China and Asia.

University of Hong Kong, Pokfulam Road, Hong Kong
Tel: +852-2859-2721; Fax: +852-2559-0468;
E-mail: hdxugoy@hkucc.hku.hk

Acknowledgements

Luc Anselin wishes to acknowledge the support of the US National Science Foundation (the research reported on in Chapter 17 was supported in part by Grant SBR-9410612). Special thanks are due to Antony Unwin for providing Figure 1 on Manet, and to Noel Cressie and Jürgen Symanzik for providing Figures 4 and 5 on ArcView-XGobi.

Richard J Aspinall would like to thank Peter Aspinall, Simon Aspinall, David Balharry, Dick Birnie, Marianne Broadgate, Marsailidh Chisholm, Roy Haines-Young, Matt Hare, Rachel Harvey, Ann Humble, Brian Lees, Kim Lowell, David Maguire, Jeff Maxwell, Elaine McAlister, David Miller, Julia Miller, Diane Pearson, Jonathan Raper, Allan Sibbald, Neil Veitch, Paul Walker, and Joanna Wherrett for their many and varied contributions to his thinking on the topics discussed in Chapter 69.

Mike Barnsley wishes to acknowledge the Natural Environment Research Council for data used to construct Plates 21 and 22.

Barry Boots would like to thank Michael Tiefelsdorf who provided the results in Table 1 and who drew Figures 11, 12, and 13 in Chapter 36; and Atsuyuki Okabe and Narushige Shiode who drew Figure 16 using software package PLVOR created by Toshiyuki Imai of the University of Tokyo.

Thomas Cova would like to thank Michael Goodchild for the invitation to contribute to this book and David Maguire for helpful comments on earlier drafts.

Susan Elshaw Thrall and **Grant Ian Thrall** would like to thank Mr Mark McLean of the Department of Geography at University of Florida for his comments on the layout of Table 1 in Chapter 23, and for his comments on the section on 'ready-made maps'.

Manfred Fischer gratefully acknowledges a research grant provided by the Austrian Ministry for Science, Research and Art (EZ 308.937/2 – W/3/95).

Peter Fisher's chapter was completed when the author was Visiting Fellow at the Department of Geomatics, University of Melbourne. The use of facilities, and the kind invitation to visit are both gratefully acknowledged. Figure 3 in Chapter 13 is reproduced with the kind permission of Taylor and Francis. The assistance of Alan Strachan and Paul Longley is also gratefully acknowledged.

Anthony Gatrell and Martyn Senior are grateful to the following for providing, or allowing them to modify, illustrations: Dr Anders Schaerstrom, Dr Danny Dorling, Professor Gerry Rushton, Professor Graham Moon, and Dr Andy Jones.

Art Getis would like to thank Judy Getis, Stuart Phinn, and Serge Rey for reviewing his chapter.

Michael Goodchild acknowledges the support of the National Science Foundation for the National Center for Geographic Information and Analysis (SBR 88–10917 and SBR 96–00465) and the Alexandria Digital Library (IRI 94–11330).

Gerard Heuvelink would like to thank Dr J Bouma and Dr A Stein (Agricultural University Wageningen) for permission to use the Allier dataset.

Mark Horn: acknowledgement is due to Eamonn Clifford and Christine Hansford at the Office of the Surveyor-General of New South Wales, Australia, who produced the illustrations for Chapter 67.

Michael Hutchinson and John Gallant gratefully acknowledge the assistance of Tingbao Xu and Janet Stein in the production of the figures in Chapter 9.

Dave MacDevette, Richard Fincham, and Greg Forsyth would like to thank Adele Wildsehut of the Centre for Rural and Legal Studies, Stellenbosch for permission to reproduce Figure 3 in Chapter 65 from Larry Zietsman's original.

David Mark's paper is a result of research at the US National Center for Geographic Information and Analysis, supported by a grant from the National Science Foundation (SBR-88-10917); support by NSF is gratefully acknowledged.

Robin McLaren wishes to thank the Ministry of Agriculture in Hungary for granting permission to use the cadastral map of Budapest (Figure 5 in Chapter 61).

Jeffery R Meyers wishes to express his gratitude for the invaluable research, editorial, and narrative review assistance provided by Christine M Condit in the preparation of this chapter. Without Ms Condit's efforts, the work would have suffered, and quite possibly not have been completed at all.

Lubos Mitas and Helena Mitasova wish to acknowledge that data for Plates 26 and 27 were supplied by K Auerswald of the Technische Universität München and S Warren of the US Army Construction Engineering Research Laboratories. Data for Plate 28 were supplied by L A K Mertes, Department of Geography, University of California Santa Barbara; data for plate 29 by L Iverson, USDA Forest Service, Delaware, Ohio; and data for Plates 30 and 31 are from US EPA Chesapeake Bay Program Office. The research in GIS applications of spline interpolation methods was supported in part by Strategic Environmental Research and Development Program (SERDP).

Harlan Onsrud's chapter is based upon work partially supported by the National Center for Geographic Information and Analysis (NCGIA) under National Science Foundation grant No. SBR 88-10917. Any opinions, findings, and conclusions are those of the author and do not necessarily reflect the views of the National Science Foundation.

Stan Openshaw and Seraphim Alvanides wish to acknowledge that Cray T3D time was provided by EPSRC under Grant GR/K43933. The 1991 Census data and boundary files are provided by ESRC and JISC and the resulting maps are all Crown Copyright.

Donna Peuquet's work was supported by National Science Foundation grant no. FAW 90-27. Portions of this work was previously published in Donna Peuquet 1988 'Representations of geographic space: toward a conceptual synthesis' in *Annals of the Association of American Geographers* 78: pages 375–94.

John Pickles draws on Chapter 1 of *Ground Truth*, the founding proposal and progress reports of Initiative 19 (I-19) of the National Center for Geographic Information Analysis (NCGIA), as well as his article 'Tool or science? GIS, techno-science, and the theoretical turn' in *Annals of the Association of American Geographers*. In particular, the chapter owes a great deal to the writings of, and discussion with, a small group of colleagues working in the liminal (and at times uncomfortable) spaces between GIS and social theory: Nick Chrisman, Michael Curry, Jon Goss, Carol Hall, Trevor Harris, Ken Hillis, Bob McMaster, David Mark, Patrick McHaffie, Roger Miller, Harlan Onsrud, Eric Sheppard, Paul Schroeder, Dalia Varanka, Dan Weiner; and at the University of Kentucky, Oliver Froehling, Eugene McCann, and Steve Hanna. The chapter draws heavily on the work of this group in I-19, especially the discussions and presentations of the planning group and participants at the Friday Harbor workshop on Geographic Information and Social Theory (1993), the Koinonia Workshop on the Representation of Space, People, and Nature in GIS (1996), the planning group of I-19 (Helen Couclelis, Michael Curry, Trevor Harris, Bob McMaster, David Mark, Eric Sheppard, and Dan Weiner), and the participants in the Critical Social History of GIS Workshop in Santa Barbara (1996) (Michael Curry, Jon Goss, David Mark, Patrick McHaffie, Roger Miller, and Dalia Varanka). Parts of section 4 lean heavily on the founding proposal for I-19 written by the author, Michael Curry, Trevor Harris, Bob McMaster, David Mark, Roger Miller, Eric Sheppard, and Dan Weiner. The summary of GIS-2 was adapted from I-19 discussions presented by Paul Schroeder and Harlan Onsrud. The salient points governing the new systems for a GIS-2 have been abstracted from the results of I-19 and the Public Participation Project, and these can be found at *http://ncgia.maine.edu/pgis/ppgishom.html*. A discussion list for this issue has been set up at *http://ncgia.spatial.maine.edu/webforum.html*. None of the above are responsible for any egregious errors, misinterpretations, or outrageous claims.

Writing the chapter was aided immensely by the opportunity to present these ideas to the national postgraduate programme in geography at the University of Turku in Finland. For this opportunity the author is indebted to Harri Anderson of the Department of Geography at Turku and students in the course.

Jonathan Raper would like to acknowledge that Figure 1 in Chapter 5 was developed by John Walker (*http://www.fourmilab.ch*) – the image is based on the Global Topographic Map from the Marine Geology and Geophysics Division of the National Geophysical Data Center, Boulder, Colorado, USA.

David Rhind wishes to thank Ray Harris and Ian Masser for sight of pre-publication versions of their important books cited in Chapter 56. He also gratefully acknowledges his debt to Nancy Tosta's published work on the US National Spatial Data Infrastructure.

Nigel Waters would like to acknowledge comments, suggestions, ideas and references from Shelley Alexander, Chad Anderson, Robert Arthur, Stefania Bertazzon, Murray Rice, Terry Woods, and Clarence Woudsma (all of the Department of Geography at the University of Calgary); to Tim Nyerges for supplying copies of his seminal papers; and to Harvey Miller for copies of his most recent papers. Finally, he would like to thank Howard Slavin, President, Caliper Corporation, for providing technical documentation on the TransCAD package and for other support.

Robert Weibel and Geoffrey Dutton wish to thank Frank Brazile for helping with the preparation of illustrations. A number of people have generously provided illustrations or helped with the compilation of figures, including Dietmar Grünreich and Brigitte Husen of the University of Hanover, Corinne Plazanet and Anne Ruas of IGN France, and Chris Jones of the University of Glamorgan. Partial support from the Swiss NSF through project 2100-043502.95/1 is gratefully acknowledged.

John Wilson acknowledges the following permissions for reproduction: Plates 60–62 are reprinted with permission from Hutchinson, Nix, McMahon, and Ord *Africa: a topographic and climatic database (version 1)* © 1995 by Australian National University, Canberra, Australia; Plates 63 and 64 are reprinted with permission from Corbett and Carter 'Using GIS to enhance agricultural planning: the example of inter-seasonal rainfall variability in Zimbabwe' *Transactions in GIS 1*: 207–18 © 1997 by GeoInformation International, Cambridge, UK; Figures 1 and 2 in Chapter 70 are reprinted with permission from Bell, Cunningham, and Havens 'Soil drainage class probability using a soil landscape model' *Soil Science Society of America Journal* 58: 464–70 © 1997 by Soil Science Society of America, Madison, Wisconsin; Figure 3 is reprinted with permission from Usery, Pocknee, and Boydell 'Precision farming data management using geographic information systems' *Photogrammetric Engineering and Remote Sensing* 61: 1383–91 © 1995 by American Society for Photogrammetry and Remote Sensing, Falls Church, Virginia.

The editors and contributors are grateful to the following for permission to reproduce copyright figures and tables:

Atsuyaki Okabe and Narushige Shiode for Figure 16 in Chapter 36; *Computing and Statistics* magazine for Figures 4 and 5 in Chapter 17; Garmin (Europe) Ltd for Figures 1 and 2 in Chapter 33; Georgia Tech Virtual GIS project for Figure 2 in Chapter 39; IGN France for Figures 13 and 14 in Chapter 10, courtesy of C Plazenet; Institute of Geography, University of Hanover for Figure 15 in Chapter 10; John Wiley & Sons Inc. for permission to reproduce Figures 1 and 2 in Chapter 15; MEGRIN for Table 2 in Chapter 47; Michael Tiefelsdorf for Figures 11, 12, and 13 in Chapter 36; NASA for permission to reproduce Table 2 in Chapter 45 and Table 2 in Chapter 48; New University of Lisbon for Figure 5 in Chapter 39; Oracle Corporation 1996 for Figure 2 in Chapter 29; Swiss Federal Office of Topography, DHM25©1997, 1263a for Figure 16 in Chapter 10; Tables 1 and 2 in Chapter 43 Courtesy of the State of Washington, Department of Natural Resources; Taylor and Francis, London for Figure 2 in Chapter 8 which appeared in *Time in GIS* by Gail Langran (1992) and for Table 1 in Chapter 44; Trimble Navigation Ltd for Figures 3 and 4 in Chapter 33.

We are grateful to the following for permission to reproduce copyright photographs:

A P Jones for Plate 56; American Society for Photogrammetry and Remote Sensing for permission to reproduce Plate 9; Combined Universities Collection of Air Photographs for Plate 19; Figure 2 in Chapter 29 © Caliper Corporation 1996; Figure 5 in Chapter 61 © Department of Lands and Mapping, Ministry of Agriculture, Hungary; ESRI Inc. for Plate 49; Georgia Tech Virtual GIS project for Plate 37; John Wiley & Sons, Chichester, for Plate 56; Kendall Publishing Co. for Figure 2 in Chapter 54; taken from Morgan J M et al (1996) *Directory of Academic GIS Education*; Longman for Plate 7 which appeared in Kraak and Ormeling *Cartography, visualisation of spatial data*, 1996; Microsoft *Encarta World Atlas* for Plate 8; New University of Lisbon for Plate 40; Office of the Surveyor-General of NSW for Figure 1 in Chapter 67; Plate 57 © NSW Department of Land and Water Conservation 1997; Space Imaging for permission to use Plate 18; Swiss Federal Office of Topography, DHM25©1997, 1263a for Figure 16 in Chapter 10; Taylor and Francis, London, for Figure 4 in Chapter 47; The Caliper Corporation for permission to reproduce Plate 50; Trimble Navigation Ltd for Plates 23 and 24; UCL 3D Image Maker Plate 20.

While every effort has been made to trace the owners of copyright material, in a few cases this has proved impossible and we take this opportunity to offer our apologies to any copyright holders whose rights we may have unwittingly infringed.

1

Introduction

P A LONGLEY, M F GOODCHILD, D J MAGUIRE, AND D W RHIND

Every day in different parts of the world people pose questions just like these:

Politician: 'What is the population of the Sedgefield parliamentary constituency?'

Farmer: 'What are the characteristics of the soils in the Lobley Plantation?'

Retailer: 'Where should I locate my next clothing outlet store?'

Gas engineer: 'Where should I dig up the road to gain access to the gas main?'

Health practitioner: 'How can my authority best respond to the needs of those single parent families with low income and poor housing?'

Climatologist: 'How has the hole in the ozone layer changed in the past 10 years?'

Geologist: 'Are there any trends in the pattern of earthquakes in Italy which could help predict future quakes?'

Planner: 'How has the distribution of urban and rural population changed between the past two censuses?'

Military commander: 'If I deploy my equipment and personnel here who will be able to see me and shoot at me?'

Home delivery service manager: 'What is the shortest route I can use to deliver all these refrigerators to the homes of new customers?'

City accountant: 'What is the total value of the land and property assets which the city has sold in the last 12 months?'

Forester: 'If a fire were to start here on a breezy day, in which direction would it spread and how much timber would be lost?'

Hydrologist: 'A large quantity of a pollutant has been introduced into this well: where will it spread and which customers will be affected?'

All of these questions and many more like them are concerned with geographical patterns and processes on the surface of the Earth. As practitioners of these fields know only too well, answering such questions requires access to geographical information which is characterised by its multidimensional nature (x,y,z coordinates and time), its large volume and high processing cost. To answer apparently simple geographical questions requires that data from several sources be integrated into a consistent form. The art, science, engineering, and technology associated with answering geographical questions is called Geographical Information Systems (GIS). GIS is a generic term denoting the use of computers to create and depict digital representations of the Earth's surface.

From humble beginnings in the 1960s, GIS has developed very rapidly into a major area of application and research, and into an important global business. In 1997 GIS was being taught in over 1500 universities and over 1000 schools, it had over 500 000 regular users (plus innumerable casual map users), and was a global business worth over US $12 billion. It has moved from being an esoteric academic field to being recognised as part of the information technology (IT) mainstream. Today GIS is a vibrant, active and rapidly expanding field which generates considerable public and private interest, debate, and speculation.

1 A BRIEF HISTORY OF GIS

The phenomenon – no other word seems quite as appropriate – now known as 'GIS' has many roots,

1

Table 1 Major GIS textbooks. Note only core text books are included here.

Antenucci J, Brown K, Croswell, Kevany M 1991 *Geographic information systems: a guide to the technology*. New York, Van Nostrand Reinhold

Aronoff S 1989 *Geographic information systems: a management perspective*. Ottawa, WDL Publications

Bernhardsen T 1992 *Geographic information systems*. Arendal, Norway, Viak IT and Norwegian Mapping Authority Cambridge (UK), GeoInformation International

Bonham-Carter G F 1994 *Geographic information systems for geoscientists: modeling with GIS*. New York, Pergamon Press

Burrough P A, McDonnell R A 1997 *Principles of geographical information systems*, 2nd edition. Oxford, Oxford University Press

Cassettari S 1993 *Introduction to integrated geo-information management*. London, Chapman and Hall

Chrisman N R 1997 *Exploring geographic information systems*. New York, John Wiley & Sons Inc.

Clarke K C 1997 *Getting started with geographic information systems* Englewood Cliffs, Prentice-Hall

Dale P F, McLaughlin J D 1989 *Land information management: an introduction*. Oxford, Oxford University Press

Davis B E 1996 *GIS: a visual approach*. Santa Fe, Onword Press

DeMers M N 1996 *Fundamentals of geographic information systems*. New York, John Wiley & Sons Inc.

Huxhold W E 1991 *An introduction to urban geographic information systems*. New York, Oxford University Press

Huxhold W E, Levinsohn A G 1995 *Managing geographic information system projects*. New York, Oxford University Press

Jones C 1997 *Geographical information systems and computer cartography*. Harlow, Longman

Laurini R, Thompson D 1992 *Fundamentals of spatial information systems*. London, Academic Press

Maguire D J, Goodchild M F, Rhind D W 1991 *Geographical information systems: principles and applications*. Harlow, Longman/New York, John Wiley & Sons Inc.

Martin D S 1996 *Geographic information systems: socioeconomic applications*, 2nd edition. London, Routledge

Peuquet D J, Marble D F 1990 *Introductory readings in geographic information systems*. London, Taylor and Francis

Star J L, Estes J E 1990 *Geographic information systems: an introduction*. Englewood Cliffs, Prentice-Hall

Worboys M F 1995 *GIS: a computing perspective*. London, Taylor and Francis

and it is impossible to do justice to all of them in a brief history. The first edition of this 'Big Book of GIS' (Maguire et al 1991) included a full chapter on GIS history; a book on the history of GIS edited by Foresman appeared early in 1998 (Foresman 1998) and many introductory texts include short histories (see Table 1). Rather than attempt to summarise, the emphasis here is on the diversity of GIS's roots, and on updating the story with a brief account of major events and trends since 1991 (when the first edition of this book appeared).

1.1 GIS as data analysis and display tools

The history of GIS is in many (but not all) ways the history of using digital computers to handle and analyse mapped data. Early computers were literally 'number crunchers', not handlers of the complex forms of information found on maps, and were designed to perform a task – the manipulation of numbers – that had no obvious applications in the world of map production and use. Thus it was many years after the development and deployment of the first electronic computers that uses for the new technology for handling maps began to emerge. It is now generally accepted that the British Colossus computer of the early 1940s, used to break the German Enigma codes, was probably the first electronic computer, although an electro-mechanical

one had operated in Harvard a few years earlier. By the 1950s (Rhind 1998), Swedish meteorologists were producing weather maps with the aid of computers. Shortly afterwards, Terry Coppock was geographically analysing agricultural data by computer. At the end of the 1950s, he analysed about half a million records from the Agricultural Census using an early computer in London University. The programmes summarised the data records and classified them ready for mapping by hand. Though the potential value of computer mapping was clearly appreciated at the time, the limitations of machine performance and output devices rendered such automation impossible (Coppock 1962). His work may be the earliest substantive 'GIS-based research'. Working in Canada, Roger Tomlinson (see also section 7 below) is rightly credited with seeing the need for computers to perform certain simple but enormously labour-intensive tasks associated with the Canada Land Inventory in the mid 1960s, and with being the father of the Canada Geographic Information System (CGIS), itself widely acknowledged to be the first real GIS. Tomlinson saw that if a map could be represented in digital form, then it would be easy to make measurements of its basic elements, specifically the areas assigned to various classes of land use. At that time, normal practice involved laborious and tedious hand-measurement of area by counting dots on transparent overlays of known dot density. Tomlinson's cost–benefit analysis showed that computerisation would be cost effective, despite the enormous costs and primitive nature of the computers of the time.

It is, however, important to note that many other pioneers, often working alone, also played a very significant role: for instance, many of the same technical tools were also devised in Australia, while at Northwestern University in the USA, Duane Marble and colleagues became interested in using geographical information technologies to solve transportation and other urban problems.

1.2 GIS as map-making tools

A second and quite distinct history of GIS stems from the benefits of automating the map production process. Once information of any kind is in digital form, it is much easier to manipulate, copy, edit, and transmit. The primary GIS innovator in this context was David Bickmore: at his urging, Ray Boyle invented the 'free pencil' digitiser and, by 1964, Bickmore and Boyle had set up the Oxford system for high quality digital cartography (Rhind 1988). At that time, major mapping agencies – including the US and other military bodies – began the lengthy and often rocky process of automation. The complexity of the issues involved in doing this are confirmed by the fact that even today major map-producing agencies employ a sometimes awkward mix of manual and automated techniques (for a sense of some of the reasons behind this continuing difficulty, see Weibel and Dutton, Chapter 10). Widespread achievement of the benefits of automated cartography had to await the development of suitable mechanisms for input, display, and output of map data, but the necessary devices – map digitiser, interactive graphics display device and plotter, respectively – had become available at reasonable cost by the early to mid 1970s and from then onwards an increasing number of organisations set out to convert all their maps into computerised form.

1.3 Other roots of GIS

A third root of GIS lies in landscape architecture and environmentally sensitive planning. In the 1960s, a view of planning emerged that saw the world as composed of a set of largely independent layers, each representing some component of the environment, and thus some set of environmental concerns. These layers might include groundwater, natural vegetation, or soil. McHarg (1969; 1996) was the foremost proponent of this view, and his group at the University of Pennsylvania applied it in a long series of exemplary studies. Although the initial idea was strictly manual, the computerisation of these ideas in a layer-based raster GIS was a simple step, and many systems owe their origins to McHarg's simple model (e.g. Tomlin 1990).

GIS also has urban and demographic roots. Efforts to automate national population censuses go back to Hollerith and the very early days of office automation, and the mechanical card sorters that predate digital computing. A census is inherently geographical, requiring the tabulation and publication of statistics for a range of geographical units, with complex hierarchical relationships in space (see Martin, Chapter 6). The cost of these aggregations, and the notion that they could be performed automatically from a single representation at the most detailed level, had by the late 1960s driven the US Bureau of the Census to introduce the dual independent map encoding (DIME) system – a primitive GIS representation of the urban street network with simple topology. Interestingly, part of the rationale for the use of this approach to encoding – which initially contained no coordinates – was to permit automated checking of data consistency because the data collection process was spread over many offices. Many of these ideas were reapplied at even more detailed scales in cities in support of such urban functions as infrastructure maintenance, and the Urban and Regional Information Systems Association (URISA) was founded at about this time to foster further development.

Finally, GIS has roots in the stimulus provided by the development of remote sensing, again in the late 1960s and early 1970s, as a potentially cheap and ubiquitous source of Earth observations. While many of the techniques for processing images are highly specialised, more general GIS techniques become important in order to combine information from remote sensing with other information (Star et al 1997). Today, many GIS include extensive functionality for image processing, and all types of remote sensing are increasingly the data source of choice, particularly for detection of landscape change (see Barnsley, Chapter 32; Estes and Loveland, Chapter 48).

1.4 GIS as a coherent, multi-purpose 'thing'

If GIS has so many apparently independent roots, what brought them together, and why has the umbrella term 'GIS' become so widely accepted? First, there are obvious commonalties. For example, the representation of topology invented for the DIME system at the US Bureau of the Census is almost identical to that incorporated in CGIS and in Australian work; the methods of raster processing and storage used in remote sensing systems are almost identical conceptually to those used by systems that have implemented McHarg's multi-layer view of the world. Second, it was easy from the viewpoint of the software engineering paradigms of the 1970s and 1980s to integrate functions around common representations. Once a raster or vector data model had been established, functions that process that data model in different ways were easy to add – thus it was possible, for example, to build large-scale integrations of image processing functions around a common raster representation. By the end of the 1970s, the term 'GIS' had emerged in recognition both of common technical requirements and of the opportunity to build systems that could potentially satisfy all of these applications. It took rather longer for the 'raster GIS' of the McHarg and remote sensing roots to merge with the 'vector GIS' of the CGIS, mapping, urban, and census roots. Debates on whether one or the other was 'better' were commonplace in the 1970s and 1980s, with hybrids like the 'vaster' structure emerging. To some extent this remains a cleavage in GIS to this day, exacerbated by the many variants on the basic raster and vector options (see the various contributions on representational issues in the 'Space and time in GIS' Section of the Principles Part of this volume).

When the first edition of this book was assembled, between 1989 when the project started and 1991 when the book finally appeared, the prevailing view of GIS was this notion of large-scale software integration around a common data model. Since GIS made it possible to store many coverages, software development was seen as providing a large number of functions to operate on those layers, as well as basic housekeeping functions for input, storage, and output. Extending the data model, for example by adding an option to order layers as a temporal sequence, would allow even more functions to be added. Progress in GIS was for a time measured by such additions to the richness of its data models, and associated additions of functionality – all within a monolithic and often proprietary software environment.

This view began to crumble in the early 1990s. First, the demarcation that it implied between geographical and other types of data became less valid. It became possible, for example, to handle an image within a relational database environment or a statistical package; or to make a map from a simple spreadsheet. Second, while such monolithic and expensive packages optimised the overall use of available computer power, this did not necessarily mean that individual GIS operations were performed in the most efficient manner. Third, there was growing resistance in the marketplace to solutions that required all customers to acquire all functions, regardless of need. Finally, customers became increasingly frustrated with the direct and indirect costs of monolithic proprietary solutions.

As we discuss below, today's GIS is in the process of being reinvented. There is much less emphasis on 'system', with all that is implied in that term – a clearly demarcated, monolithic, probably proprietary solution. The 'open GIS' movement, most clearly seen in the Open GIS Consortium (but by no means restricted to it), is driven by a vision of GIS as a collection of interoperable modules, under common standards (Sondheim et al, Chapter 24). The growth of electronic communications networks and associated applications means that it is no longer necessary for the data, the software, and the user to be in the same place at the same time – in the late 1990s vision the activities associated with the term 'GIS' are increasingly distributed (Coleman, Chapter 22). In time these technical innovations are likely to be reflected in institutional changes, as the field moves further from its societal roots. The advent of powerful PCs has provided substantial GIS functionality, shrink-wrapped and relatively stable and easy to use, on the individual desktop. Perhaps most important of all, the advent of the World Wide Web (WWW) has facilitated the routinisation of database linkage (Pleuwe 1997). Since GIS software systems built by many different vendors and running on different hardware in different countries can now be linked routinely together and the data used in combination, the old concepts of GIS are totally dead. This is explored in much greater detail later in this chapter.

In 1980 the GIS collective was dominated by the disciplines that gave it its impetus – landscape architecture, urban and regional planning, geography, cartography, and remote sensing, among others. With the rapid growth of GIS in the 1980s came new alliances, notably with computer science and many of its sub-fields – computer graphics, computational geometry, and database theory. Interest in making GIS easier to use led to alliances with cognitive science and environmental psychology (see Mark, Chapter 7). Increasingly, GIS is seen as a specialised sub-field of information technology and information science, and there are links of growing importance with the library science community (see Adler and Larsgaard, Chapter 64). Perhaps as a result of all this, the large, national and general-purpose GIS conferences popular in the 1980s have begun to lose attendance. They are being replaced in popularity by regional and local general-purpose conferences and by vertical market ones (e.g. GIS appears in utility company conferences).

It is difficult to identify specific individual events in the past seven or so years that have been particularly significant in redirecting GIS. The founding of the Open GIS Consortium may be one, along with the events and trends in the wider information technology arena of 'open systems' that preceded it. Certain moves by GIS vendors – new products, changes of direction, adoption of standards – have also had trend-setting significance, as have various failures, demises, and terminations in the industry. The 1990s marked the final victory of commercial off-the-shelf (COTS) software over the public-sector software development efforts that had characterised earlier decades, and had persisted well into the 1990s in the case of GRASS. It marked very significant moves by major software vendors – Microsoft, Oracle, and Autodesk among them – to establish positions in the geographical information marketplace. It also saw moves by GIS vendors into the consumer software market – an alliance between Intergraph and Egghead, for example, and new consumer GIS products from ESRI (for more on consumer GIS, see Elshaw Thrall and Thrall, Chapter 23). Arguably, however, it is the advent of the WWW that has been the single most important development affecting GIS in the last 20 years.

2 DEFINITION AND CLASSIFICATION OF GIS

Geographical information is information about geography, that is, information tied to some specific set of locations on the Earth's surface (including the zones immediately adjacent to the surface, and thus the sub-surface, oceans, and atmosphere). 'Spatial' is often used synonymously with, or even in preference to, 'geographical' in this context, although in principle it might be taken to include information that is tied to frames other than the Earth's surface, such as the human body (as in medical imaging) or a building (as in architectural drawings). Because of this difficulty, the term 'geospatial' has become popular recently, notably in the context of the US National Spatial Data Infrastructure, the Canadian National Geospatial Infrastructure, and the UK National Geospatial Data Framework. In this book, the terms 'geographical' and 'geospatial' are used interchangeably.

2.1 GIS, GI, and maps

Goodchild (1992a; see also Peuquet, Chapter 8; Gatrell and Senior, Chapter 66) identifies two distinct primitive types of geographical information: field information, in which geography is conceived as a set of spatially continuous functions, each having a unique value everywhere in space; and information about discrete entities, where the world is conceived as populated by geometric objects that litter an otherwise empty space and are characterised by attributes, such that any point in space may lie in any number of discrete entities. The field/object dichotomy underlies many areas of GIS, including its data models, data quality, analysis, and modelling (e.g. Burrough and Frank 1996; see also Raper, Chapter 5; Martin, Chapter 6).

Over the years the vision of a GIS has shifted significantly, but has always included the notion of processing geographical information within an integrated environment. It has been argued that the environment need not be digital, and that the principles of GIS can certainly be taught outside the digital environment, but today's world is increasingly digital and GIS is now almost always associated with digital computing in one form or another. It has also been argued (e.g. Maguire 1991) that the definition of GIS should include much more than the digital environment – in this conception the people who interact with it are also part of the system. Finally, GIS has been defined by its objectives, as in Cowen's definition of a GIS as a spatial decision-support system (Cowen 1988).

Today, the term GIS tends to be applied whenever geographical information in digital form is manipulated, whatever the purpose of that manipulation. Thus using a computer to make a map is as likely to be described as 'GIS' as is using the same computer to analyse geographical information and to make future forecasts using complex models of geographical processes. At the same time, there are significant exceptions. The Earth images collected by remote-sensing satellites are geographical data, but the systems that process them are not likely to be called GIS as long as they remain specialised to this particular form of data – in such cases, 'GIS' tends to be reserved for systems that integrate remotely-sensed data with other types, or process data that have already been cleaned and transformed. Similarly, an atmospheric scientist or oceanographer will tend to associate 'GIS' with systems used more for multidisciplinary work and policy studies, and will use other software environments for modelling and analysis within the confines of his or her own discipline. In short, because GIS implies a generalised software environment that is exclusive to geographical information there is a tendency for it to be most strongly associated with multidisciplinary, integrative work and applications; in more narrowly-defined environments less general solutions may be adequate.

Moreover, there is a persistent – albeit unfortunate and misleading – tendency for 'GIS' to be associated with the digital representation of the kind of geographical information that has traditionally been shown on paper maps, rather than geographical information conceived more generally. While maps may appear to place few restrictions on their compilers and users, in reality they can be highly constrained in the ways they represent the Earth's surface. Traditionally (although with notable and celebrated exceptions) paper map information has typically been:

- static, favouring the representation of fixed aspects of the Earth's surface, because once made, a paper map cannot be changed;
- 2-dimensional, and unable to show many diverse attributes of 3-dimensional socioeconomic

systems such as cities, or physical environments such as the subsurface, oceans, or atmosphere;

- flat, because the curved surface of the Earth must be projected in order for it to be shown on a sheet of paper – or a regular solid like a globe;
- apparently exact, because there have been few applications of cartographic techniques for showing uncertainty in mapped information;
- unconnected to other information that may be available about the same set of places, but cannot be shown on the same map (and possibly cannot even be physically stored in the same place).

Because of its roots in mapping in general, and traditional cartographic practice in particular, much of GIS practice and application has remained similarly shackled to these limitations, unable to move beyond the metaphor of the traditional paper map (but see the Epilogue for a prospective view).

Wright et al (1997) define several different interpretations of what it means, in today's parlance, to be 'doing GIS'. One interpretation might simply be the **application** of a particular class of software, having chosen it from among the classes available today by considering various pros and cons, in order to gain insight, learn more about the world, support some kind of **management** decision-making, etc. In a more general sense, 'doing GIS' might involve applying the **principles** of GIS, including its particular ways of representing the world, and thus operating within a 'GIS paradigm'. Or it might involve furthering GIS **technology** by developing new capabilities. Finally, GIS might provide the medium for studying one or more of the fundamental issues that arise in using digital information technology to examine the surface of the Earth. Wright et al argue that only in the last instance is one necessarily 'doing science' when 'doing GIS'.

This argument, and others related to it, has led to a search for new terms that encompass activities that are less dependent on the particular nature of today's software offerings. Goodchild (1992b) has argued that this can be done by decoding the familiar acronym as geographical information science (GISc), and this idea is reflected in the recent establishment in the USA of the University Consortium for Geographic Information Science (UCGIS), an organisation of the principal GIS research institutions (see *http://www.ucgis.org*). The term geomatics has also gained some popularity, particularly in Europe and Canada and in the surveying engineering and geodetic science communities (see for instance, *http://www.geocan. nrcan.gc.ca*). Geocomputation also has similar connotations, although here the modelling of process may be more important than the modelling of information per se. Forer and Unwin (Chapter 54) have suggested no fewer than three decodings of GIS: GISy for the systems, GISc for the science, and GISt when the focus is on studies of GIS, particularly in the context of society and its institutions.

2.2 Is spatial 'special'?

Ultimately, the continued existence of GIS relies on the belief that there is some value in dealing with geographical information as a special case – that there is 'something special about spatial' (unfortunately there seems to be no available English term to complete the more appropriate 'something . . . about geographical' – 'magical', 'fanatical' don't quite serve the purpose). In the past, the case was argued on several grounds, including:

- the nature of geographical queries, potentially combining topological, geometric, and attribute elements, all with some fuzziness embedded;
- the special data structures, indexing systems, and algorithms needed for efficient processing of geographical information;
- the multi-dimensional nature of geographical information ($x, y, z, n . . .$);
- the voluminous nature of much geographical information;
- the fundamental inability to create a perfect representation of the Earth's surface, forcing users of GIS to deal with problems of data quality, accuracy, and uncertainty;
- the isolated nature of traditional production arrangements for geographical data, including the existence of public sector mapping agencies in most countries;
- the need for special standards for geographical information;
- the combination of distinct legal and economic contexts of geographical information, including copyright laws, liability, privacy protection, freedom of information laws, and costs of acquisition, that vary markedly from one country to another.

Recently, however, much of this basis for demarcation has diminished, if not disappeared altogether. In today's software environments, the special structures needed for handling geographical data are largely invisible to the user. The size of a single remotely-sensed image from a sensor like Landsat no longer seems formidable when personal computers often include gigabytes of storage. And debates about the legal and economic contexts of GIS are increasingly embedded within much broader debates about information policy and practice in general. Moreover, several recent technical developments have reduced the need to maintain distinctions within today's computing environments. Open standards like Microsoft's Object Linking and Embedding/Component Object Model (OLE/COM) and Object Management Group's Common Object Request Broker Architecture (CORBA) allow information of different types to be passed between environments, suitably enclosed in 'wrappers' (interfaces) that describe the type to the host. Thus it is increasingly possible to hold geographical information within an environment designed for processing text – that is, a familiar word processor. In effect, these technologies decouple the handling of a container of information from the nature of its contents, treating all information as 'bags of bits'. Structured Query Language (SQL) and other query languages have been extended recently to handle the special cases of geographical information and geographical queries, and extensions like Oracle's SDO increasingly allow geographical information to be handled within the frameworks of mainstream database management systems.

2.3 Geographical Information is special

Unlike GIS software, geographical information *is* special in many ways, but some of the more fundamental of these have little to do with its manipulation in digital systems. Anselin (1989) has argued that 'spatial is special' in two crucial respects. The first is expressed in Tobler's famous 'First Law of Geography' (Tobler 1970): 'all things are related but nearby things are more related than distant things'. This property of spatial dependence, or at least autocorrelation, is endemic to geographical data, violates the principle of independence that underlies much of classical statistics, and is the basis on which any representation of the infinite complexity of the Earth's surface is even approximately possible.

Anselin's second special characteristic is spatial heterogeneity, the propensity of geographical data to 'drift' such that conditions at one place are not the same as conditions elsewhere. Statistically, this concept corresponds to non-stationarity, and is well-known in geostatistics (e.g. Isaaks and Srivastava 1989). Practically, it means that the results of any analysis are always dependent on how the boundaries of the study are drawn – whereas it is often (erroneously) assumed that a geographical study area is analogous to a sample in statistics, drawn from the set of all possible study areas by some random process, and thus that the choice of study area has minimal effect on the results. Many of the arguments that emerge from this point can be found in the fractal literature (e.g. Mandelbrot 1982). More recently, Fotheringham (1997), Getis and Ord (1992), and others have argued for a new approach to geographical analysis based on the need to determine the local characteristics of places, rather than universal generalities (see also Getis, Chapter 16).

To these two might be added a third, which is particularly apposite in the context of GIS. The idea of expressing geography as a series of layers suggests that each layer captures something unique to it; statistically, that each layer makes an independent contribution to the total picture of geographical variability. In practice, however, geographical layers are almost always highly (if variably) correlated. It is very difficult to imagine that two layers representing different aspects of the same geographical area would not somehow reveal that fact through similar patterns. For example, a map of rainfall and a map of population density would often clearly have *some* similarities: population could be dependent on agricultural production and thus rainfall (or irrigation!), or might tend to avoid steep slopes and high elevations where rainfall was also highest. Of course, these correlations are often indirect, with other controlling variables and cultural features and inertia playing important roles.

These special characteristics of geographical data are undoubtedly important, but often not unique. Dependence is also endemic in time series; non-stationarity occurs in many contexts. While there is every reason for users of GIS to be aware of the ecological fallacy (Robinson 1950) and the Modifiable Areal Unit Problem (Openshaw 1984) – and these themes are explored at greater length in the chapters on spatial analysis later in this volume

(e.g. Openshaw and Alvanides, Chapter 18) – it is difficult to argue that they justify the demarcation of GIS from other types of software.

One final characteristic is worth discussion, because it appears to be of increasing significance as the information society moves to reliance on a world of distributed computing. Society's arrangements for production, storage, and use of information depend critically on how interest in that information is determined. In the case of detailed geographical information, interest tends to be highly localised – interest in a street map of Manchester is clearly of greater importance to users located in Manchester than it is to users in Paris. Traditionally, this has been reflected in the pattern of availability of that information in libraries, bookshops, etc. In a world in which information is distributed over a myriad of servers accessible through tools such as the Web it is of critical importance to know where a particular set of information can be found. That issue is resolved in the case of textual information through the existence of search engines, which use Web crawlers to find and catalogue text by key word. But no comparable mechanism yet exists for geographical information though embryonic Web-based geographical services already exist. In developing new geographical data search engines, the new world of distributed computing is likely to find new ways in which 'spatial is special'.

3 CURRENT TRENDS IN GIS

3.1 The evolving GIS environment

GIS is a young area of technological innovation and application. It is also a very rapidly changing one. Without doubt, developments in computer technology have been a major contributor to the rapid advances of GIS. Thus in exploring the world of GIS it is appropriate to begin by charting the main relevant technological advances of recent years and seeking to gauge their impact on GIS.

Perhaps the root cause of all technological advances, as far as GIS is concerned, is improvement in computer hardware. Twenty years ago Gordon Moore, co-founder of the microprocessor company Intel, suggested that computer hardware performance would double and price would halve every 18 months. In the intervening years this prediction, subsequently dubbed 'Moore's Law', has held true and it appears that for the foreseeable future hardware will continue to improve at this rate. In mid 1997, however, after many years of close adherence to Moore's Law, announcements by IBM and Intel predicted that the rate of growth of processor speed would be even faster in the next few years. IBM announced a technique to replace aluminium connections on microprocessors with copper (which has greater conductivity), and Intel announced 'flash' technology, which allows two or even more bits to be processed by each processor element instead of one.

As a result of these developments, not only have hardware systems become faster and cheaper, but their physical size has also decreased. Notebook and field portable computers, for example, are now very commonly used in GIS applications. Yet the full implications of improvements in computer processor speed have yet to be fully recognised in GIS applications. Perhaps inevitably, hardware bottlenecks do remain in today's computers, notably with respect to the internal communication bus and the speed of disk access. Some of the hardware performance increases have been soaked up by the development of ever more sophisticated graphical user interfaces (GUIs), while the emphasis in spatial analysis has been to use enhanced hardware performance to support visualisation and data exploration rather than data modelling as more traditionally conceived.

Only a few years ago, the engineering workstation with its UNIX operating system was the dominant platform for delivering GIS. Since then, there has been the shift towards the personal computer, the innovation of desktop computing, and the gradual domination of Microsoft (the Windows operating system) and Intel's microprocessors (the 'Wintel' combination). By 1997 the Wintel combination had become the system of choice for GIS applications on the desktop. For server machines and specialist applications, UNIX remains a credible and important alternative. But Windows has become so widely adopted in GIS applications because of its widespread use in general applications, its (comparative) ease of use, its ability to run both GIS and non-GIS applications, and its low cost. As a consequence, the major GIS software systems have a remarkably similar 'look and feel'.

As we saw in the opening paragraphs of this introduction, one of the fundamental characteristics of GIS applications has been their use of large and very large quantities of multi-dimensional data

(i.e. x,y,z coordinates) and the need for multi-user access to spatially continuous databases. The early GIS software systems used binary flat files to store data and specialist data management routines for data organisation and access. Fairly quickly, with the rapid growth of relational database management system (RDBMS) technology, many software developers began to manage non-geometric data using RDBMS. Today, the issues of performance, multi-user access, and data compression have largely been resolved and it is the norm for GIS software systems to store both geometric and non-geometric data in an RDBMS. With the development of Object-Relational DBMS and their capability for extension so that they can manage complex data types, like spatial, these are expected quickly to become the standard.

Most early GIS were individual isolated islands of technology. Since then, the rise in importance of network technology has had a profound impact on GIS. The words of Scott McNealy, President of Sun Microsystems, 'the computer is the network, the network is the computer', clearly state the importance of networks. In the late 1980s there was a move to connect machines together using local area network technology. More recently, wide area network (WAN) technology has been of interest to users. None of these can really compare, however, to the growth in interest and rapid uptake of the Internet as network-based technology.

The Internet is the world's largest public network. It is a multi-faceted mosaic of computer servers supplying information upon request to multiple clients. The Internet is unified by common use of the Internet Protocol (IP). This communication standard allows heterogeneous hardware to communicate in a simple, but effective, fashion. The WWW is a popular application which operates over the Internet. The Web is a distributed collection of sites (servers) composed of multimedia documents. These are linked together using the hypertext transmission protocol (http) and are spatially referenced using a uniform resource locator (URL). Web use has increased at a truly incredible rate in recent years, establishing new standards for many types of GIS application. Those focusing on data publishing, simple display, and query have been most successfully implemented.

While the Internet is almost certainly the technological innovation that is exerting the greatest external influence upon GIS at the present time, its impacts are all the more far-reaching because of contemporaneous developments within GIS. Central to these developments has been the establishment of the Open GIS Consortium (OGC) in August 1994. This is an international consortium of more than 100 corporations, government agencies, and universities. The OGC has put considerable effort into the development of 'interoperable' software using OpenGIS (Open Geodata Interoperability Specification) to build links between different proprietary systems (Sondheim et al, Chapter 24). Allied with the development of the Internet, open object standards and object brokers have been used to support distributed computing. The CORBA and OLE/COM standards allow 'objects', or packages of digital information, to be passed freely between different software environments, and make the contents of objects understandable to systems. More recently, the Java language has provided a means for sending program modules over the Internet as well as data, allowing one system to send a process for another system to execute. Other fragments of programs known as 'applets', 'plug-ins' and 'add-ons' are now routinely distributed from one system to another. Each of these developments is contributing to a new Internet-based computing environment in which it is as common to distribute the ability to process as it is to distribute the subject of processing – that is, the data. This increasing fragmentation of programs is extending the GIS environment ever further beyond its self-contained, monolithic roots.

The combined effect of the application of these technologies is that GIS software is breaking up into reusable 'plug-and-play' modules, which can be assembled and used through the Internet. It is also leading to the development of packages of software modules and data for use as so-called 'desktop GIS' (Elshaw Thrall and Thrall, Chapter 23): some observers view this as a transitory phase on the way towards use of the Internet as the principal platform for GIS.

Each of these advances in technology has, of course, been designed to improve the ability to store, manage, manipulate, display, and query geographical data. Together they have also profoundly changed the way that computing is carried out, as the practice of a user interacting with a file server becomes supplemented by 'peer-to-peer' computing in which every user is potentially both a client and a server – both a source and a destination for computation.

3.2 Our digital world

There have also been a number of significant changes in the way data are used and disseminated which have additionally influenced GIS applications. Spatial referencing is by definition essential to any GIS application, yet application-specific thematic layers alone rarely create a readily-recognisable view of the world – as anyone who has been presented with a choropleth map of an unfamiliar area will testify. Important developments are taking place in the provision of digital 'framework data' for GIS (Rhind 1997b). Framework data provide information pertaining to the location of topographic and other key features in the natural, built, or cultural landscape, which may be used as a backcloth to application-specific thematic data. Since the first edition of this book, such data have been created by a number of national mapping, cadastral, and census agencies and these present officially sanctioned views of the surface of the Earth, to a range of emergent data standards (Salgé, Chapter 50). 'Unofficial' sources of framework data also exist in the form of classified high-resolution satellite images, obtained from the new generation of high-resolution remote sensing satellites or from the new radar sources (which are less limited by cloud).

Each of these sources of framework data has become increasingly commercialised during the 1990s – on the one hand, national mapping and census agencies in many parts of the world are developing commercial datasets in order to meet their cost recovery targets; while, on the other, the break up of the former Soviet Union and the launch of new commercial satellites has done much to multiply the number of sources of remote sensing imagery. The latter commercial developments have become of wider import to GIS given recent technical developments in softcopy photogrammetry and pattern recognition. These are leading to the widespread creation of new products such as digital orthophoto maps and elevation models (DEMs) at much lower cost than has previously been the case.

With the general proliferation of digital datasets it has become increasingly difficult for the GIS user to know what datasets exist, what quality they are, and how they might be obtained. Allied to the development of the Internet, an important current development is the creation of on-line metadata – data about data – services, a number of which are designed for use with geographical location as a primary search criterion. An interesting development in 1997 was the creation of comparatively low cost intelligent data products containing functionality and metadata which allow fast direct access by GIS software packages. More generally, the development of whole digital libraries of geographical information is becoming feasible, and there is growing interest in using the metaphor of libraries to support geographical information management and data sharing (Adler and Larsgaard, Chapter 64).

Just as it is becoming easier for GIS users to find out exactly which digital data exist, so it is also becoming easier for them to collect their own digital data. Although many of the bottlenecks of digitising data from old hardcopy sources remain, much new data are now collected using the global positioning system (GPS) technology that has developed rapidly during the 1990s (Lange and Gilbert, Chapter 33). Low cost hand-held or mounted GPS receivers are suitable for many (but by no means all) field data collection purposes, and record geographical location routinely to quite high levels of precision (40–100 metres for civilian 'selective availability' applications and 10–32 metres for military applications) by reference to the US NAVigation Satellite Timing And Ranging Global Positioning System (NAVSTAR GPS) or its Russian equivalent (GLONASS). Much higher resolutions are obtainable using differential GPS and post-processing. This technology has revolutionised data collection for a wide swath of applications, particularly as receivers have been developed which also permit input of aspatial attribute data during the data collection phase.

Even in 1991 it was clear that information in general and geographical information in particular were becoming both a tradable commodity and a strategic resource. Nowhere in GIS has this continuing trend become more apparent than in business applications of GIS, where a huge value added reseller (VAR) and consultancy industry has developed to service business client needs. The data for most business applications have hitherto largely been obtained by combining census variables into composite 'geodemographic' indicators, which experience has shown bear an identifiable correspondence with observed consumer behaviour. More recently, the proliferation of digital customer records, allied to the collection of data from new customer loyalty programmes, is leading to the creation of more and more 'lifestyles' databases.

These are not as geographically comprehensive as conventional geodemographics, but are much more frequently updateable and contain data which might be judged more pertinent to prediction of customer behaviour than those from conventional censuses.

3.3 Scientific trends and research directions

Elsewhere in this book we will explore the broader scientific trends in GIS: the current emphasis on the big questions of geographic information science (GISc) over the small technical questions; the growth of interest in human cognition that should make GIS easier to use (Mark, Chapter 7); the shift in emphasis towards data modelling and ontological issues (Raper, Chapter 5; Martin, Chapter 6); and the development of new strong links to mainstream computer science (e.g. Worboys, Chapter 26; Oosterom, Chapter 27). These and many other interesting developments and research directions are discussed at length throughout the book, and particularly in the first two sections.

4 WHAT WAS WRONG LAST TIME

The message of all of this is that GIS continues to be a vibrant and fast-changing area of business, application development, and research. From its origins in the 1970s, through its rapid growth phase in the 1980s, GIS has rapidly expanded and matured into a general-purpose information technology that is capable of solving the widest range of problems in a geographical context. Although its disciplinary heart lies in academic geography (Couclelis, Chapter 2; Johnston, Chapter 3), its continued growth and vitality is much more broadly-based than this – GIS is at least as much grounded in people's enduring fascination with maps, and the ease of spatial expression and reasoning that maps allow, as in any particular disciplinary matrix.

The first edition of this book (Maguire et al 1991) attempted not just to set out the whole panoply of GIS circa 1991, but also to anticipate the directions in which its inherent dynamism would move it. If book sales and patterns of academic citations are anything to go by, the first edition certainly provided an accessible and comprehensive snapshot of the state of GIS at the time of its publication, but it is only now with the benefit of hindsight that we can identify the respects in which it failed to anticipate the direction and strength of change.

Perhaps the most glaring omission is the complete failure of the book to anticipate the growth of the Internet and the World Wide Web into a massive global computer. It follows that there was far too little discussion of the technologies required to support distributed databases, distributed processing, and above all distributed users, together with the emergent role of the Internet in supporting vast numbers of servers and clients.

Second, in retrospect, there is the sense throughout the book that the most important technical problems had all been solved and that the big remaining ones concerned GIS management and institutional usage. While there is undoubtedly truth in the latter, it is clear in hindsight that very big technical issues still remain, whilst in the related area of methodology the emergence of GISc and geocomputation suggests that spatial analytical elements may not have been afforded sufficient prominence last time.

Third, there was a sense in the first edition of a quest for the Holy Grail of an 'all-singing, not all-dancing' GIS which would permit the fullest range of analytical operations to be performed. Even from the brief discussion of current trends contained in the previous section, it should be clear that a strong counter-trend has been the break-up of GIS software into packaged components, and that data components are often of similar importance to analytical functions in such systems. The Internet has had the opposite effect in allowing software to converge across different domains, and as a result users have been able to assemble task-oriented systems at will and as needs dictate – particularly given that the drive towards interoperability has meant that component software modules need not all originate from a single source. Neither trend has fostered the development of a single integrated GIS software system. Indeed the emphasis upon the development of analytical functions proved to be a distraction from the under-played information management functions of GIS, development of which has subsequently been key to the wider dissemination and adoption of GIS.

Fourth, passages of the first edition are redolent of a rather more technocentric view of the world – a sentiment which also characterises most of the first generation of GIS textbooks. This sense of mechanistic manipulation has subsequently dissipated somewhat, with the advent of social critiques of GIS and the wider realisation that GIS can be as much an empowering technology as it is a technology of control. The reasons for this emphasis in the first edition probably lie in the then prohibitively high cost of GIS software systems (at a time prior to licensing deals for higher education and government usage, for example) and a fascination with the implications of plummeting costs of computation for analytical functionality rather than the far wider distribution of PC and networked computer technology. The technocentric view is epitomised by the amount of space devoted to the promise of artificial intelligence – a theme which requires surrender of power to the machine rather than encouraging user empowerment, and which subsequent experience suggests cannot deliver much of its early promise.

Finally, there is a recurring sense throughout the first edition that because 'spatial is special' the GIS industry would continue to comprise a set of isolated, proprietary, specialised vendors. Most of those have subsequently disappeared, although two of the early market leaders (ESRI and Intergraph) retain large market shares. The new entrants to the industry are the IT heavyweights Microsoft, Autodesk, and Oracle – as we will discuss further in the next section.

5 THE WORLD OF GIS

There are several encouraging signs that in recent years GIS has reached new levels of popularity, respectability and maturity, and here we will provide something of the flavour of the state of GIS in the late 1990s. It is impossible to be comprehensive in summarising the state of GIS. Quite apart from anything else, space – even in a book at large as this – does not permit it. Rather the approach we will take is to review some of the major strands of development and current interest.

A key sign of the maturity of any discipline or business area is the development of coordinating bodies and academic and professional societies. GIS now has these in abundance. In the USA, the best known include: ACSM (American Congress on Surveying and Mapping), the GIS speciality group of the AAG (Association of American Geographers), AM/FM (Automated Mapping and Facilities Management: also in Europe), ASPRS (American Society of Photogrammetry and Remote Sensing), UCGIS (University Consortium for Geographic

Information Science), and URISA (Urban and Regional Information Systems Association). In other parts of the world comparable organisations include: AGI (the UK Association for Geographic Information), EUROGI (the European GI organisation), AGILE (Association of Geographic Information Laboratories in Europe), CPGIS (Chinese Professionals in GIS), GISRUK (GIS Research – UK) and UDMS (the Urban Data Management Society in Europe). These and many other bodies regularly organise society meetings featuring conferences and exhibitions. Together with a parallel set of meetings organised by private companies and public agencies (notably under the auspices of the OGC, discussed in section 3.1 above), GIS events often feature several thousand participants and provide close interaction between vendors, users, consultants, and researchers.

OGC, through OpenGIS, has brought forward standards for the interoperability of GIS software. The initial standard is based on the straightforward exchange of simple features (points, lines, and polygons) between commercial systems. Comparable international standards bodies that are focusing effort on developing *de jure* standards for GIS include ISO (the International Standards Organisation) and CEN (Comité Européen de Normalisation: Salgé, Chapter 50). ISO is an international body with representatives in many countries and CEN is a European umbrella organisation. These and other organisations are seeking to standardise almost all aspects of GIS, from metadata to database interfaces. If these standards are complementary and are widely adopted then they should further stimulate the growth of GIS.

One of the interesting aspects of GIS is the close involvement of software vendors in the continued evolution. Two of the earliest and most successful vendors – Environmental Systems Research Institute Inc. (ESRI) and Intergraph Corporation – remain the GIS market leaders. However, the increasing use of GIS on the desktop has led to new market entrants such as Mapinfo Corporation, while the movement of GIS to the Web and the ever closer relationships between computer-aided design (CAD) and GIS software has brought firms like Autodesk and Bentley into the GIS market. At the same time, IBM Corporation, Informix Corporation and Oracle Corporation have extended their respective DBMS

to incorporate spatial data. In late 1997 the value of the global software market was estimated to be worth between US$627 and $904 million, depending upon whether a narrow or broad definition of GIS was used, with ESRI and Intergraph having market shares of about 33 per cent each (using the narrow definition) or 20 per cent each (using the broad definition) (Crockett 1997). Each of the market leaders is diversifying into emergent market niches and data-related products. Smallworld Systems maintains a strong position in utilities. After a period of rationalisation (because of takeovers and bankruptcies) GIS has become dominated by just a handful of vendors. By 1997, the GIS software market was probably worth about $1 billion worldwide.

Overall, expenditures on GIS are much higher than simply those on software. The US Office of Management and Budget (OMB) found in 1993 that total expenditures on digital geographical information in Federal agencies amounted to over US$4 billion. Adding the effects of activities at the state and local levels, and the activities of the private sector and non-governmental organisations leads to estimates of between $10 billion and $14 billion for the total value of the digital geographical information industry in the USA, although this is almost certainly an underestimate. Precise estimates of the total number of GIS users are similarly difficult to ascertain. A conservative estimate is that there are about 100 000 highly technical or professional GIS users in the world. When the 500 000 desktop users and one million casual viewers are added, the total becomes about 1.6 million. This is well in excess of the 250 000 or so predicted by the editors of the first edition of this book (Maguire et al 1991). At the current rate of expansion there could be eight million GIS users worldwide by the year 2000.

Just as the number of users has grown, so has the interest and involvement of academics. Education in GIS began in the universities, but has spread over the years to include significant efforts in training colleges and vocational programs, secondary schools, and even elementary schools. These are largely complementary to the training programs offered by major GIS vendors. Recently there has been much interest in distance learning, to address what is perceived to be a lack of educational opportunities for professionals in mid-career, and the UNIGIS consortium now offers distance learning through a network of institutions in several countries. University-based research has been stimulated in many countries by major funding for centres. In the USA the National Center for Geographic Information and Analysis (NCGIA) was established in 1988, with funding from the National Science Foundation, as a consortium of three institutions. In the UK, the Regional Research Laboratories stimulated the development of a network of universities committed to GIS-based research, funded by the Economic and Social Research Council between 1987 and 1991. Similar national research programmes exist in Korea, the Netherlands, France, Japan, and many other countries. The University Consortium for Geographic Information Science (UCGIS) was established in the USA in 1995 as a network of major research universities, and now has nearly 50 members. The European Science Foundation's GISDATA program coordinated and stimulated GIS research in a network of European countries between 1993 and 1997.

6 GIS: PRINCIPLES, TECHNIQUES, MANAGEMENT, AND APPLICATIONS

Just about the only thing that has not changed about GIS during the 1990s is its inherent dynamism. It is seven years since the first edition of this 'Big Book of GIS' appeared, and the editors of this second edition find themselves dealing with a subject which has developed and expanded enormously – not least in the range of geographical realities that GIS used to represent and the wider range of media through which digital representations of that reality may be constructed. Since the first edition was published the scale and pace of human interactions with computers has accelerated, and the provision and use of digital geographical information has provided one means of navigating through a geographical reality that we understand to be ever more detailed and complex. What, in the face of these remarkable upheavals, are the prospects for recreating a GIS reference work that is as relevant in terms of content and coverage as its forebear?

It is perhaps best to begin with a view of what this book is not. First, in these two volumes we have not sought to revisit all of the principles expounded in the first edition, since much of this material has

completed the transition from application-led research and practice to standard textbook material. Table 1 on page 2 lists some of the general GIS textbooks that are available. Even in a work of this length, it is impossible to cover everything in GIS from first principles, given the vast expansion of the field since the first edition. Second, neither is it possible to cover the entire range of GIS applications, and our aim here has been to review those applications from operational and strategic GIS practice which we judge to be of key importance in understanding the breadth of the field. Applications of GIS are truly legion and the detail of practice is as fast-changing as the field of GIS itself. For this reason, readers with particular application interests should instead consult any of the range of GIS journals and professional magazines, listed in Table 2, which contain periodic reports of the experience of a wide range of GIS applications – many of these are targeted at national or supranational markets, which adds further specificity to the experience that is reported. Third, it is not just an extended guide to the latest research in GIS by academics – various monographs (notably the GISDATA and Innovations in GIS series, and the books arising out of the NCGIA initiatives) exist to document these rapid developments and changes.

Table 2 Major GIS journals and magazines

(a) Journals

Cartography and Geographic Information Systems
Computers and Geosciences
Computers, Environment, and Urban Systems
Earth Observation Science
Geographical Analysis
Geoinformatica
International Journal of Geographical Information Science
Journal of the Urban and Regional Information Systems Association
Photogrammetric Engineering and Remote Sensing
Transactions in GIS

(b) Magazines

Geo Info Systems
GIM International: Geomatics Info Magazine
GIS Africa
GIS Asia Pacific
GIS Europe
GIS World
Mapping Awareness

Instead we have attempted to produce a work which is focused towards 'frontiers in GIS' and which discusses and explains the issues and practices important to everybody who comes into contact with GIS. Thus we have tried to summarise existing state-of-the-art knowledge and best practice, to explain recent developments, and to anticipate possible future ones. We have sought to cross-reference related themes and to provide pointers to other textbooks, research papers, and consultancy reports wherever appropriate. We hope that readers will find this new edition at least as comprehensive, readable and well-illustrated, and as thoroughly up-to-date as the first edition. In short, we have attempted to create a hybrid of relevant pedagogy and research and development, produced by the leading writers in the GIS field. The result looks very different to the first edition, but this is only fitting given the transformation of GIS itself over the last seven years.

In producing a second edition of what we hope will remain the definitive GIS reference book ('Big Book Two') we began essentially from scratch. At an early stage in our deliberations we recognised that we should separate our discussion of *technical issues* from underlying *principles* in order to reflect different interests among our readership. Due recognition of the wider *management* functions that GIS now has would require that a separate section be devoted to such issues. Finally, a new range of *applications* would be used in order to illustrate the ways in which theory, technique, and management map into a representative range of operational and strategic situations in practice. **Principles** and **Technical Issues** are discussed in the first volume of this set, and **Management Issues** and **Applications** in the second.

Of course it is not just the world of GIS that has changed so profoundly during the 1990s, but also those many aspects of the real world that GIS seeks to abstract and to model. At its simplest, if we recognise that the world is not the same as it was, then we should not be surprised if the ways in which we order it are not the same either. Science is also changing, as many of the old certainties are breaking down in response to the challenges of relativism. We thus begin the wholly rewritten **Principles** Part of this book with a review and reappraisal of the central role of GIS in structuring our geographical understanding of the world, including the arguments, debates, and dialogues that have developed since the first edition was published. New chapters also chart developments in the representation and visualisation of spatial phenomena. Data quality, error, and uncertainty are also given new and extended treatments, and an expanded group of contributions on spatial analysis present a contemporary view of the usefulness of GIS in analysing spatial distributions.

As we have seen, the technological setting to GIS has been transformed since the publication of the first edition – so our new **Technical Issues** Part traces the emergence of new technologies such as the development of networked and 'open' GIS and the introduction of GIS for the desktop. New techniques of spatial database management receive extensive attention, as does data capture through the latest remote sensing and GPS technologies. Finally in this section, a range of techniques for transforming and linking geographical data are discussed, notably in the context of terrain modelling, hydrographical analysis, and the creation of virtual GIS environments.

As GIS comes to play an important role in an ever-wider range of organisations, so management issues such as the choice between different commercial GIS, data availability and operational management become of importance to increasing numbers of people. These issues are addressed in the all-new **Management Issues** Part of the book. Information managers also need to be aware of legal liability issues in the provision and use of GIS, as well as data pricing and availability, and issues of privacy and confidentiality. This Part provides comprehensive introductions to these important emergent topics in GIS usage.

In many respects applications are the most important aspect of GIS since the only real point of working with GIS is to solve substantive real-world problems. Diverse though the range of GIS applications is, many nevertheless share common themes. In the **Applications** Part of this book we have selected a range of operational ('nitty gritty') and more strategic social and environmental applications. The former generally focus on practical issues such as cost effectiveness, service provision, system performance, competitive advantage, and database creation/access/use; while the latter are often more concerned with model sophistication, the social and environmental consequences of results, and the precision and accuracy of the findings.

In the Epilogue the editors draw some conclusions and indulge in some speculation as to what the future holds for GIS. We hope that readers will judge the end result to be an authoritative, comprehensive, and up-to-date statement of all that is relevant and interesting about GIS.

7 SOME INDEPENDENT VIEWS ON THE STATE, RELEVANCE, VALUE, OR FUTURE OF GIS

The act of producing a book, even one as large and diverse as this, is liable to force some degree of homogeneity on the contributions. Each author is honour-bound to report the latest trends or research findings in his or her field and assess these in a rational way; the editors need to ensure balance and provide cross-links between chapters. We considered this and agreed that a small number of iconoclastic, individual and personal views could add materially to the book. This would be especially true if they were written by individuals known to be incapable of being seduced by editorial or other blandishments

and who had worked in the furnace at the centre of some major GIS developments.

As a consequence, we invited five contributions from well-known figures, with use of the first person to emphasise this personal viewpoint. Their brief was to write about the state, relevance, value, or future of GIS. We suggested that they might use 'major historical events', 'GIS in a societal context', 'future trends', 'how has GIS changed the way we live today?', 'a personal story about becoming involved in GIS' or 'what are the remaining challenges to GIS?' as the basis for their contributions, but no restrictions were placed on comments.

What follows represents some of the wider strands of thinking about GIS worldwide.

GIS as the national Majlis

by Sheik Ahmed Bin Hamad Al-Thani
Centre for Geographic Information Systems, Doha, Qatar

The Majlis, an informal village meeting to discuss community issues and resolve differences, is an ancient tradition known throughout the Middle East. Even as a child, I wondered at the ease with which this simple, open forum prompted inquiry, discussion, analysis, and resolution.

As a member of the Qatari government I faced, with others, the challenge of establishing methods of master planning and the redevelopment of our cities in a systematic way that would rectify the make-or-break construction projects of the past and provide a definitive guide for future development.

In the late 1980s I saw, by chance, my first demonstration of GIS. It was as if a beacon, or guiding light, was suddenly sighted and I realised that this technology was the key that would provide the framework for developing an information infrastructure for our entire country.

As with all computer-based technologies, compatibility was the central issue. If we were to implement a successful national GIS, standardisation would be critical. With the authority of the senior

members of our government, I was able to establish a National GIS Steering Committee responsible for developing and maintaining national standards and the Centre for GIS which was tasked with implementing these standards. Today Qatar enjoys a unique, nationwide GIS in which all participating government agencies are connected by a high-speed optic fibre network. Each agency can access the data of all others but the responsibility for maintaining the data rests with the individual data custodians, the different agencies. As a result of all this, Qatar now has a GIS that will facilitate intragovernmental cooperation and coordination for many generations to come.

It is clear to me that, for successful implementation of a national GIS, those in the highest levels of government must understand the benefits of the technology and must actively support its implementation. GIS provides an easy method of standardising and sharing a wide variety of information amongst all levels of government. Like the Majlis, it fosters cooperation, interaction, analysis, and well-considered decisions, solving real problems in real time – from which a society can only benefit.

Technology changes everything

by John O'Callaghan
Cooperative Research Centre for Advanced Computational Systems,
The Australian National University, Canberra, Australia

I think the opportunities for GIS in the current age of 'convergence' are really exciting. We have now entered the age where the integration of computing, communications, and content is providing an information infrastructure which is fuelling the widespread use of GIS by government, industry, and the community.

GIS have built on the rapid advances in information technology and, since the 1960s, have exhibited typical stages of growth towards maturity: the experimentation with GIS technologies, the demonstration of GIS on practical applications, the consolidation of the geographical data infrastructure, and the realisation of benefits from operational GIS.

My own country – Australia – has been an early adopter of information technology and this, coupled with our coordinated approach to land ownership, our large geographical size and our dependence on natural resources, has resulted in Australia playing a leading role in the development and application of GIS.

Today, the most obvious demonstration of 'convergence' is the Internet, which is revolutionising the way we access data, interact with systems, and communicate with people. For GIS, the Internet is enabling the rapid deployment and widespread dissemination of geographical information services.

My group's research is now focused on enriching the user interfaces to these kinds of services: on-line navigation and analysis of large and distributed geographical databases; 3-dimensional modelling and visualisation of geographical data using 'immersive' display and haptic devices; and cooperative working on geographically-based simulations at several locations. We expect the results of this research to be adopted rapidly through the information infrastructure of the Internet and to contribute to the huge opportunities for GIS in this age of convergence.

How it all began and the importance of bright people

by Roger F Tomlinson
Tomlinson Associates, Ottawa, Ontario, Canada

The Canadian contribution to the development of GIS centres around the idea of using *computers to ask questions of maps*. This idea stemmed from the need for multiple map overlay and analysis facing Spartan Air Services, an Ottawa company working in Kenya in 1960. Later, in 1962, the approach was proposed by Spartan Air Services to the federal government of Canada, who adopted it for the Canada Land Inventory then planning to generate thousands of new maps to describe current and potential land use in Canada. This very successful federal-provincial programme funded the development of

GIS in Canada for the next decade. From the basic idea came the concept that many maps in digital form could be linked across Canada to form a continent-wide map database to be permanently available for analysis, and further, that these digital maps could be linked intelligently to digital databases of statistics (particularly the Census of Canada) so that a wide range of spatial questions could be answered.

I directed the development of the Canada Geographic Information System from its conception until 1969. During that time over 40 people were involved in the

17

work and there are many who deserve great credit. Lee Pratt was the young head of the Canada Land Inventory who, as a civil servant, took the entire risk of funding the new ideas. D R Thompson of IBM designed and built the first 48 x 48 cartographic scanner for primary map input. A R Boyle, then working for Dobbie McInnes (Electronics) Ltd in Scotland, designed and built the first 48 x 48 high precision free cursor digitising tables used to input point data. Guy Morton designed the continent-wide data structure incorporating a brilliant tessellation schema (the Morton Matrix) that allowed many maps to be handled by the tiny (in terms of speed and capacity) computers of the time. Don Lever was central to most of the logic of converting scanner data to topologically coded map format. It was the first use of the arc-node concept of line encoding incorporated in a GIS. Bruce Sparks and Peter Bédard made major contributions to the automatic map sheet edge match capability, which topologically matched polygons and contents seamlessly over a continent. Art Benjamin played a major part in designing the automatic topological map error recognition capability and in designing the links between map data and statistical data. Bob Kerneny developed the essential map data compaction methods using eight-directional codes originated by Galton and later called Freeman codes. Frank Jankaluk devised the reference coordinate system and made the calculations of error in calculation algorithms. Bob Whittaker designed the system for error correction and updating. Also incorporated in the system were map projection change, rubber sheet stretch, scale change, line smoothing and generalisation, automatic gap closing, area measurement, dissolve and merge, circle generation and new polygon generation, all operating in the topological domain.

The computer command language that recognised geographical analysis terms used to pose spatial questions, and that could be understood by a wide range of potential users, was a very important part of the system. Peter Kingston was responsible for the overall design of this data retrieval system and particularly for the efficient polygon-on-polygon overlay process. He also designed the command language, together with Ken Ward, Bruce Ferrier, Mike Doyle, John Sacker, Frank Jankaluk, Harry Knight, and Peter Hatfield.

Our most useful links to the academic world were through Waldo Tobler and Duane Marble in the USA, and Terry Coppock in the United Kingdom. In Canada the principal initiatives came from within private industry and government rather than academia. The links to work in the UK were through David Bickmore of the Oxford Cartographic System who, in the early 1960s, was responsible for many of the ideas for using computers to make maps. We disagreed on almost everything in the early days, but eventually our paths converged and we became firm friends.

The 1960s in Canada were exciting years, and I am happy to have been part of that excitement. While we all worked extremely hard, there was a spirit of adventure and the feeling that if you could imagine it you could make it. In those days, a few key individuals – many of them mentioned above – really counted. In the process I described, the first GIS was born and the field was named. We still call them the Champagne years.

GIS, politics, and technology

by Nancy Tosta
Director of Forecasting and Growth Strategy, Puget Sound Regional Council, Seattle, Washington, USA

In 1978, I tried to convince the Director of the California Department of Forestry that pixels were good for him and his agency. In those days, appointed and elected officials were highly suspicious of any form of geospatial technology. Their fears were justified. The price tags were huge and no one had proved that spending all those dollars to digitise data would pay off. I remember him asking why there were all those little squares on the map/image. Why didn't it look like the maps he usually used? How could the data be used? Now, writing in the early months of 1997, I would be hard pressed to find an elected official who does not know the meaning of GIS and who does not have a story to tell about how GIS was used to clarify or solve a problem. I knew that we had crossed a watershed in political acceptance of the technology in 1994 when President Clinton signed Executive Order 12906: 'Coordinating Geographic Data Acquisition and Access: The National Spatial Data Infrastructure'. While labouring in the preparation of that order, I was astounded at the lack of questions from the White House and others about the technology. The assumption was that GIS was valuable and that data should be coordinated and shared to use the technology more effectively. Other nations have used Clinton's Order to generate political support for their GIS data efforts. The local elected officials I interact with today may not know about Federal Executive Orders, or exactly how much has been expended to develop their GIS, or what the software does, but they accept that the technology works. What more do we need to make a difference?

It's all about money, stupid!

by Joe Lobley

Lobley Associates, Santatol, Southlands, USA

Much rubbish has been talked about the special value of GIS. Even more rubbish has been heard about the essential contributions of academic research and the role of government in GIS. These two groups have made almost no contribution to the evolution of GIS to date nor will they greatly influence its future. Government talks a lot, produces lots of paper, and consumes our taxes. Other than spasmodic politically correct initiatives to 'modernise' itself, government is as moribund as ever it was (and will be). Academics are supposed to exist to question what is taken for granted but when did we ever see anything really critical or new come out of the geographers at least? Technically, it was probably in the mid 1960s. Since then we have spent loads of money on fancy research centres to little effect except airline revenues. Maybe some social geographers have hit something interesting in this ethics business but their posturing and soul-bearing seems a mite contrived to me (and has no real effect other than to cause more trees to be felled for their precious publications, read only by themselves).

No, the mainspring of everything important that has happened in GIS is business and the profit motive.

Nothing of any significance started until the first commercial GIS became available. The growth in use of GIS has been fuelled by the decrease in cost of technology, driven in turn by commercial competition and salesmanship. Unlike most academics, some government data producers have a potentially important role simply because they hold valuable data assets. It's just a pity that they are typically complacent and act on geological timescales; the only way to jolt them out of all this is to contract out many of their activities. So far as access to software, hardware, and data are concerned – if people won't pay for software, data, and services, they don't really need them. If we pay for software and hardware from the commercial sector, why should we not pay for data from it – and why should government be involved at all?

The moral is obvious. Official history is created by those with the luxury of time to write and claim the credit. But the real achievers are those who have put their money on the line and built a business worldwide. I don't expect this situation to change much in future and I don't really care. But don't forget who really makes GIS happen!

References

Anselin L 1989 *What is special about spatial data? Alternative perspectives on spacial data analysis.* Technical paper 89-4. Santa Barbara, NCGIA

Burrough P A, Frank A U (eds) 1996 *Geographic objects with indeterminate boundaries.* London, Taylor and Francis

Collins M, Rhind J 1997 Developing global environmental databases: lessons learned about framework information. In Rhind D W (ed.) *Framework for the world.* Cambridge (UK), GeoInformation International: 120–9

Coppock J T 1962 Electronic data processing in geographical research. *Professional Geographer* 14: 1–4

Cowen D J 1988 GIS versus CAD versus DBMS: what are the differences? *Photogrammetric Engineering and Remote Sensing* 54: 1551–4

Crockett M 1997 GIS companies race for market share. *GIS World* 10 (4): 54–7

Foresman T W (ed.) 1998 *The history of geographic information systems: perspectives from the pioneers.* Upper Saddle River, Prentice-Hall

Fotheringham A S 1997 Trends in quantitative methods 1: stressing the local. *Progress in Human Geography* 21: 88–96

Getis A, Ord J K 1992 The analysis of spatial association by use of distance statistics. *Geographical Analysis* 24: 189–206

Goodchild M F 1992a Geographic data modeling. *Computers and Geosciences* 18: 401–8

Goodchild M F 1992b Geographical information science. *International Journal of Geographical Information Systems* 6: 31–46

Isaaks E H, Srivastava R M 1989 *Applied geostatistics.* New York, Oxford University Press

Maguire D J 1991 An overview and definition of GIS. In Maguire D J, Goodchild M F, Rhind D W (eds) *Geographical information systems: principles and applications.* Harlow, Longman/New York, John Wiley & Sons Inc. Vol. 1: 9–20

Maguire D J, Goodchild M F, Rhind D W 1991 *Geographical information systems: principles and applications.* Harlow, Longman/New York, John Wiley & Sons Inc.

Mandelbrot B B 1982 *The fractal geometry of nature.* San Francisco, Freeman

McHarg I L 1969 *Design with nature.* New York, Natural History Press

McHarg I L 1996 *A quest for life.* New York, John Wiley & Sons Inc.

Openshaw S 1984 *The modifiable areal unit problem.* Concepts and Techniques in Modern Geography Vol. 38. Norwich, GeoBooks

Pleuwe B 1997 *GIS online: information retrieval, mapping and the Internet.* Santa Fe, Onword Press

Rhind D W 1988 Personality as a factor in the development of a discipline. *American Cartographer* 15: 3277–89

Rhind D W 1997b *Framework for the world.* Cambridge (UK), GeoInformation International

Rhind D J 1998 The incubation of GIS in Europe. In Foresman T W (ed.) *The history of geographic information systems.* Upper Saddle River, Prentice-Hall: 293–306

Robinson G K 1950 Ecological correlation and the behavior of individuals. *American Sociological Review* 15: 351–7

Star J L, Estes J E, McGwire K C 1997 *Integration of geographic information systems and remote sensing.* New York, Cambridge University Press

Tobler W R 1970 A computer movie: simulation of population change in the Detroit Region. *Economic Geography* 46: 234–40

Tomlin C D 1990 *Geographic information systems and cartographic modeling.* Englewood Cliffs, Prentice-Hall

Wright D J, Goodchild M F, Proctor J D 1997 Demystifying the persistent ambiguity of GIS as 'tool' versus 'science'. *Annals of the Association of American Geographers* 87: 34–62

PART 1

PRINCIPLES

Contents

(a) Space and time in GIS

Introduction   23–27
The Editors

2. Space, time, geography   29–38
    H Couclelis

3. Geography and GIS   39–47
    R J Johnston

4. Arguments, debates and dialogues: 49–60
    the GIS–social theory debate and the
    concern for alternatives
    J Pickles

5. Spatial representation: the   61–70
    scientist's perspective
    J F Raper

6. Spatial representation: the social   71–80
    scientist's perspective
    D Martin

7. Spatial representation: a cognitive view   81–89
    D M Mark

8. Time in GIS and geographical   91–103
    databases
    D J Peuquet

9. Representation of terrain   105–124
    M F Hutchinson and J C Gallant

10. Generalising spatial data and dealing   125–155
    with multiple representations
    R Weibel and G Dutton

11. Visualising spatial distributions   157–173
    M-J Kraak

(b) Data quality

Introduction   175–176
The Editors

12. Data quality parameters   177–189
    H Veregin

13. Models of uncertainty in spatial data   191–205
    P F Fisher

14. Propagation of error in spatial   207–217
    modelling with GIS
    G B M Heuvelink

15. Detecting and evaluating errors by   219–233
    graphical methods
    M K Beard and B P Buttenfield

(c) Spatial analysis

Introduction   235–237
The Editors

16. Spatial statistics   239–251
    A Getis

17. Interactive techniques and exploratory   253–266
    spatial data analysis
    L Anselin

18. Applying geocomputation to the   267–282
    analysis of spatial distributions
    S Openshaw and S Alvanides

19. Spatial analysis: retrospect and prospect   283–292
    M M Fischer

20. Location modelling and GIS   293–303
    R L Church

Introduction

THE EDITORS

The term GIS is fundamentally about the use of digital data to represent space and time, and few of the readers of this book will be unfamiliar with the standard sequence of operations that GIS invoke to create such representations – data input, storage, manipulation, and output. For many users of GIS, this simple chronology of operations has provided an adequate framework for understanding what GIS is about. Yet reality is infinitely complex in its totality, and our digital representations are inevitably simplifications or 'models' of it. With experience, and perhaps the demands of wider domain and strategic applications, many GIS users will begin to get a feel (from the 'bottom up') for the nature of the simplifying assumptions, or 'transformations' (Martin 1996) which are inherent in reducing the myriad complexities of geographical reality to digital computer records. From a quite different perspective, the fundamental ('top down') views of social science and science held by some academics bring into question the very validity of GIS-based representations of the real world. The opening five chapters of this book seek to set out the context to GIS, as a contribution towards reconciling philosophy and science with practice, concepts with application, analytical capability with social context. As such, and although avowedly academic in emphasis, they contain material of relevance to everyone who has considered using GIS to formulate and analyse problems in the real world. Successive chapters begin to translate these abstract notions and ideas into firmer guiding principles of GIS, in order that principles in turn might coalesce into operational guidelines for implementation.

In the opening chapter to this Section, Helen Couclelis traces the disciplinary origins of interest in representing space and time to the disciplines of geography, mathematics, philosophy, and physics. The traditional paper map subsequently emerged as the dominant paradigm of spatial representation, with its goal of depicting spatial phenomena using established and recognisable schemes of representation. The more recent innovation of GIS has sought to develop and enhance such analogue models of the world using computer hardware, software, and digital data. More detailed and sophisticated than paper maps they may be, but most GIS-based maps remain similarly constrained – they must present a world that has been projected onto a flat plane; they must be static and 2-dimensional; they depict the world as if it were known perfectly, or at least as accurately as the scale of the map allows; and they must present the world at a uniform scale or level of geographical detail. These are all examples of simplifications of reality, yet the GIS medium is fundamentally more capable of relaxing these assumptions, constraints, and conventions than paper mapping. Thus it is with some confidence that Couclelis sees GIS as rising to the challenge of achieving 'the seamless integration of space and time, the representation of relative and non-metric spaces, the representation of inexact geographical entities and phenomena, and the accommodation of multiple spatio-temporal perspectives to meet a variety of user purposes and needs'.

Of the different disciplines that have sought to represent space and time, it has been geography that has identified itself most closely with the innovation of GIS – although (as a number of the contributors to this section note) geography has not been central to its technological development. Ron Johnston (Chapter 3) uses the debates that have developed within geography to explore the implications of the fundamentally empiricist view of the world that GIS provides – that is, a view founded upon the philosophical belief that there is a separate objective world that is outside and independent of any individual observer. Empirical scientific approaches have come to be viewed with disdain by some academics working in human geography, yet they remain the predominant *modus operandi* in physical geography. Johnston concedes that a pragmatic application-led science, couched in the world of appearances, is not universally attractive to all geographers (although it has undoubtedly enhanced the status of their discipline), yet his closing remarks suggest that the very richness of digital media no longer need constrain GIS in this way. If, as Couclelis asserts, GIS has already come a long way since the era of early computer cartography, then the rapidity and pace of current developments should in turn now begin to suggest ways in which GIS might inform other, non-empiricist, approaches to social science.

In Chapter 4, John Pickles develops the critique of GIS from a more functionalist perspective: that is, how its approach to science has impacted upon its technological capacities and social uses. The early 1990s critique of GIS within academic geography was fundamentally one of empiricism – that is, the approach to science in which, in Johnston's words, 'facts speak for themselves'. As such, it was to some extent a re-run of the critiques of quantitative geography that had developed in the 1970s and 1980s. This time, however, the detailed critique developed on two quite different fronts. First, what was different for some this time was the power of the technology, and the drive towards data-rich depictions of geographical reality capable of eroding privacy and increasing (social, political, military . . .) control. Second, empiricist approaches (in contrast to other social science approaches such as social and critical theory) were deemed most unlikely to shed light on questions of valid and intrinsic academic interest, and thus infusion of GIS into the discipline of geography would never create more than a diversion and irrelevant distraction. Although perhaps contradictory (if a technology truly is capable of eroding privacy, then it surely is a worthy focus of academic concern), these two perspectives shared the common sentiment that GIS has introduced a technological distraction to legitimate academic discourse and thus has reinvigorated an approach to social science in geography which by the 1990s many had thought discredited. Some of the later chapters in the Management Issues and Applications sections air these issues in much greater detail. Pickles' important contribution, here as well as elsewhere (Pickles 1995), has been to open up these issues to constructive dialogue between the 'top down' views of the best practice of science and the 'bottom up' empirical experience of GIS users. Such dialogue is likely to lead to GIS applications breaking free from notions of 'objective' reality and

(echoing the views of Johnston) may also lead to supplementation of quantifiable attributes and characteristics of geographical reality with measures of local knowledge, place-based information and other qualitative considerations.

There are lessons here for even the most unequivocal advocate of GIS – namely the need for cognisance of the philosophical background and context to analysis, and the inherent subjectivity of even the most apparently 'objective' models of reality that are abstracted within GIS. In Jonathan Raper's view this relativist honing of GIS to context is not restricted to social science applications that embrace human agency, as suggested in his review of scientific representation (Chapter 5). Physical science lies much more uncontroversially in the empiricist domain than social science, and hence it might be taken as axiomatic that there is a strong correspondence between increased richness of digital information and the accurate and orderly depiction of real-world systems. Raper formalises the empiricist conception of GIS in natural scientific applications as representing a 'bridge' between scientific theory and the real world. Yet even within natural science the way in which this structure is fashioned is profoundly influenced by our information sources and scientific conventions. 'Scientific conventions' are the ways in which we define and give significance to geographical phenomena, and the ways that we identify phenomena in space and time within GIS (when is a sand dune not a sand dune? what are its boundaries? what sort of time increments should be used to represent and model its dynamics?).

Raper's view is that the spatial and temporal context to natural science representations in GIS should be specified inductively rather than deductively, in a context-sensitive manner and in a spirit of humility rather than conviction – a view developed later in this section with respect to digital terrain models by Hutchinson and Gallant. As such, empirical science should refocus more of its efforts on the ways in which simplified representations of an infinitely complex reality are developed within GIS, and greater detail and volume of information is seen as but one ingredient of an enhanced approach to model-building. Spatial representation is thus an intrinsic component of scientific method, and must be related to theory about the way the world works and how geographical reality is structured. This mapping of reality into model is seen as being accomplished through geographic information science (Goodchild 1992) – that is, the development of formal conventions and rules for the appropriate representation of phenomena within conventional and unconventional (3-dimensional models, video and multimedia) GIS representations.

The issues and problems that Raper identifies are at least as problematic as many in the socioeconomic realm, yet here there are additional problems arising out of the strictures of confidentiality (and consequent areal aggregation of data), the ways in which boundaries are imposed around continuous spatial features, and the ways in which time is discretised during data collection. These and other problems create some differences in the definition and handling of geographical objects of study between natural and social science applications. David Martin reviews these differences here (Chapter 6). Spatial boundaries and temporal intervals pose more than an analytical inconvenience, since they lie outside of the control of the GIS analyst and cannot be changed. Moreover, there are conceptual difficulties in assigning precise geographical coordinates to human individuals and describing their activity patterns. Martin describes how these provide additional challenges to effective representation, analysis, and display within socioeconomic GIS, and discusses how representational strategies may be used to contain ecological fallacy and modifiable areal unit effects.

Taken together, the contribution of these chapters is to steer GIS users towards a more relativist conception of reality and its representation within GIS, and thence to identify how GIS data structures and architectures can be developed to accommodate the widest range of information sources. This provides a general framework for enhancing GIS-based representations of reality which are tailored to the perceptions and needs of the many. In short, it presents a broad canvas to GIS applications: the remaining chapters in this section investigate a range of topics that can further improve representation within GIS.

One socially significant facet of this critique is that if GIS has been rendered accessible only to the scientific community, then successful users have been placed in the role of experts. As a consequence, GIS is intolerant of diversity of viewpoint. Cognitive interest in GIS stems from a desire to make it easier to use, by making its user interfaces and representations more compatible with the ways people naturally think and reason about the world around them. David Mark's chapter (7) reviews the current state of cognitive research in GIS, and discusses some of the issues that are raised by this line of reasoning. He begins by appraising the correspondence between the ways in which humans perceive real-world phenomena and the ways they are represented within GIS. His approach is avowedly empiricist in approach, and sets out to examine the possible mismatches between objective measurement and cognitive models of reality. He shares some of Johnston's optimism that GIS may provide a suitable medium through which realist models of geographical reality might be built. The detail of his empirical analysis of cognitive categories substantiates the views of Raper, Martin, Pickles, and Coucelis that boundaries are *de facto* often indistinct, fuzzy, and graduated. He also describes how distance, direction, reference frames, and topology are all subjectively manipulated in common parlance, and how variations may be compounded by natural language differences (when is a lake not a lake but a pond? why are 'lodge' water bodies apparently confined to northern England?).

As Raper has already intimated, our inability to deal with time, and thus to represent the dynamic elements of the geographical world, is perhaps the most compelling of the inadequacies of maps and traditional GIS. A technology that is forced to represent the world as static inevitably favours the static aspects of the world. Thus (as Martin describes) our maps show the locations of buildings and roads rather than people, and the same biases have been inherited by GIS. The chapter by Donna Peuquet (8) reviews the state of the art in the representation of time in GIS, and specifically the problems that arise out of the inadequate definition and representation of events and timeframes within GIS. She assesses the merits of different data structures for representing time in GIS – for example, as raster coverage-based snapshots, variable length (raster) pixels, entity-based (vector) representations and geographical objects. Peuquet then reviews modes of exploring and visualising space-time interactions using GIS. It is clear from this that query languages for identifying temporal change are much better developed for aspatial, rather than spatial, database management systems (DBMS), because of the reduced dimensionality of the queries. Nevertheless, in an upbeat conclusion, she anticipates considerable improvements in the representational power and analytical capabilities of GIS in this regard.

The remaining chapters share a pragmatic emphasis on the ways in which data models may be used to fulfil a variety of end-uses. Much of the conceptual debate surrounding GIS has arisen simply because the user of today's GIS is faced with many more options in representation. Choices must be made between different scales, or levels of geographical detail; between raster and vector options; and between various approaches to representing change. Nowhere are these choices more apparent than in the representation of topography, or the form of the Earth's surface. Data are available at various scales, and in three major representational schemes: the meshes of triangles known as triangulated irregular networks (TINs); grids of regularly spaced sample elevations (digital elevation models or DEMs); or digitised contour lines. But many more complex and subtle issues exist in finding accurate and useful representations.

Michael Hutchinson and John Gallant's main concern (Chapter 9) is with identification of the guiding principles for generating digital terrain models, and an area in which (firmly in the empiricist tradition) spatial analysis of form is frequently used to draw inference about environmental process. Digital terrain models are also used in the conceptualisation (cf. Raper) and display. Accuracy and extent of spatial coverage are of importance here, of course, but there is also a sense of the recursive relationship between the way that relevant phenomena are identified and defined (ontology) and the ways in which they are subsequently analysed. Additionally, the representation of 3-dimensional structures creates a potentially vast increase in the amount of data that might be stored within GIS, many of which are likely to be redundant: clear thinking, coupled with appropriate choice of analytical technique are thus required in order to create realistic yet manipulable models of real-world 3-dimensional structures. This raises a wide range of considerations in making the choice of data model: how to anticipate/manage errors associated with GPS data capture (Lange and Gilbert, Chapter 33), whether priorities favour capturing surface variability through use of variable point densities across a surface or by representing local properties of curvature, whether and how grids may be adapted to local terrain structure, the range

of scales relevant to the end-user, and the ways in which features are defined and parameterised. As the previous chapters in this section imply, the model-building process does not then terminate with a single pass through the data, and Hutchinson and Gallant describe further recursive stages of data quality assessment and model interpretation.

Robert Weibel and Geoffrey Dutton (Chapter 10) broaden this analysis, looking at the generalisation of geographical objects and cartographic features. They develop a typology of motivations for generalisation within GIS, ranging from data storage, through improved data robustness to optimising visual communication. All of the preceding contributions have emphasised that good environmental and social science data models are sensitive to context in what, through abstraction, they retain and discard. Yet such reflection is clearly not practicable where a multitude of routine decisions must be made, or where the outcome of data modelling is a cartographic product for visual display. In such circumstances, sensitivity to context may nevertheless be achieved using a range of automated and semi-automated knowledge-based methods, such as generalisation algorithms and methods for structure/shape recognition, and further methods for evaluating the 'quality' of generalisations. The principles underpinning such methods are seen as an automated development of traditional map-making conventions, in which the cartographer was always to some extent the arbiter, even architect, of cartographic form. That said, progress towards automated generalisation of digital maps has apparently been rather slower than was anticipated in the first edition of this book: however, Weibel and Dutton provide evidence of encouraging prospects in this regard, suggesting that automated generalisation may not be a 'holy grail'.

The final chapter in this section (11), by Menno-Jan Kraak, extends the discussion of the theme of scientific visualisation – that is, the presentation, analysis, and exploration of geographical phenomena. Although we naturally tend to think of visual displays of GIS databases as the digital equivalent of making paper maps, there are significant differences and opportunities. The design of a paper map is permanent, but visual displays can be manipulated and transformed freely. Scale, for example, takes on a different and more interesting meaning in a world of zoomable displays.

GIS displays can be animated, raising a host of new issues for the user's ability to perceive and understand geography through visual display. Although computer display screens are approximately flat, their use allows us to re-examine the significance of map projections, and to ask whether they are actually necessary in a digital geographical world, since there are no flat surfaces in a digital computer that are as constraining as the inevitable flatness of paper. There are also echoes here of many of the previous contributions in the description of the 'overlay model' as a simplified, error-prone depiction of reality, and a review of the effective use of symbolisation and other cartographic conventions to present spatial distributions – as well as to interpret data reliability and quality. As such, visualisation is considered an important adjunct to explanation, which helps through query, re-expression, multiple views, dynamics, animation and changes in dimensionality. Kraak's view (cf. Raper) is that GIS provides a bridge between the map and the database 'text', and he anticipates some of the ways in which video and other multimedia are set to develop and enhance the links between digital models of reality and their visual front ends.

Other issues of representation are addressed elsewhere in this book. The representation of uncertainty, another missing element in traditional maps, is taken up in the next section; and object-oriented issues in representation are discussed later as Technical Issues. But research on data modelling is proceeding at such a pace as to make it impossible to achieve a complete coverage in the space available here. This is why we have presented an extended overview of guiding principles rather than fast-changing practices. The interested reader is referred to the references in each chapter, to recent collections (for example, Molenaar and Hoop 1994), and to the continual stream of new research papers appearing in the journals of the field.

References

Goodchild M F 1992b Geographical information science. *International Journal of Geographical Information Systems* 6: 31–45

Martin D J 1996 *Geographic information systems: socioeconomic applications,* 2nd edition. London, Routledge

Molenaar M, Hoop S de 1994 *Advanced geographic data modelling.* Publications on Geodesy, New Series No. 40. Delft, Netherlands Geodetic Commission

Pickles J (ed.) 1995a *Ground truth: the social implications of geographic information systems.* New York, Guilford Press

2

Space, time, geography

H COUCLELIS

Specific notions of space and often also time underlie every GIS application. This chapter reviews the conceptual roots of space and time representations in the four traditional disciplines of geography, mathematics, philosophy, and physics, recently augmented by cognitive and sociocultural perspectives. It then assesses the place of GIS as information technology at the intersection of several different views of space and time, and addresses the tension between that plurality and GIS's strong basis in a single view (the 'map' view). Finally, it outlines four challenges for GIS research in the domain of spatio-temporal representation: the seamless integration of space and time, the representation of relative and non-metric spaces, the representation of inexact geographical entities and phenomena, and the accommodation of multiple spatio-temporal perspectives to meet a variety of user purposes and needs.

1 INTRODUCTION: WHY 'GEOGRAPHICAL' INFORMATION SYSTEMS?

Of the many millions of users of GIS, only a small fraction have any formal ties with the discipline of geography. Planners, foresters, natural and social scientists, utilities managers, marketing consultants, transportation engineers, and many others now use these systems on a daily basis without giving too much thought to what the 'G' in GIS might stand for. Obviously 'geographical' refers to something of very broad import that far transcends the bounds of a particular discipline. The 'geo' in 'geography' is in fact a great common denominator for all of us living on the surface of the Earth, as we are all more or less familiar with the same basic things that populate our planet. There are, in particular, two large categories of geographical concepts with which most people are acquainted either through their professional activities or simply as part of everyday life: geographical entities and phenomena, and the spatial and temporal properties and relations characterising these. Geographical information systems derive their name from the fact that they are designed around both these categories of concepts:

they are not just about the things listed in geographical atlases, nor are they just 'spatial' information systems.

The first class of widely shared geographical concepts are thus the entities and phenomena of the world at geographical scales, and their changes over time. These entities can be as small as a village square or as large as the planet itself: this is the range that the notion of geographical scale covers. Typical geographical entities are mountains, rivers, valleys, and coastlines, but artificial features such as cities and roads are also among them. Phenomena are the things that happen, rather than those which are on the landscape: brush fires, weather systems, floods, droughts, erosion, land reapportionment, urban growth. Often the most useful applications of GIS have to do with the complex interactions between relatively static geographical entities and the dynamic phenomena through which these entities themselves evolve.

The second category of universally shared geographical concepts concerns the notions of space and time applicable at geographical scales, and in particular the spatial and temporal relations among geographical entities and phenomena. Where something *is* in geographical space is still the quintessential geographical question, though both the question and its possible answers are usually less simple than might appear at first sight. 'Where' may mean on which continent as well as at what precise coordinates or address, or in what direction, how far, next to what, where else, in which part of a region. A useful answer to the 'where' question may be given in latitude/longitude terms, or be something like 'near the lake but not too close to the forest'. Similarly, questions regarding the temporal dimensions of geographical entities and phenomena go well beyond simple 'when' inquiries about clock time and date: what changes since, how fast, what could have caused this, what else happened at about the same time, what came first. As an example of this latter kind of question, GIS is already being used to help arbitrate in debates arising from charges of 'environmental racism', where a critical question is often whether the noxious land-use or the affected minority population was there first. But at what point in time is a land-use or a population 'there'? Even disregarding the difficulty of pinning down the spatial component in this question, we are clearly dealing with two possibly interconnected spatio-temporal processes neither of which can be neatly time-stamped.

Geography and a number of related disciplines have developed an array of methods and tools to help answer these kinds of questions through the spatial and temporal analysis of data about geographical entities and phenomena. Many of these have been incorporated in GIS, and their underlying assumptions about space and time are reflected in the systems' data models, functions and graphic user interfaces. Thus, while few users of GIS may be concerned with space and time *per se*, they all have to live with the consequences of how a particular GIS implicitly treats these notions, and deal with the problems of spatio-temporal representations that may be mutually conflicting or inappropriate for the task at hand.

The purpose of this chapter is threefold: first, it will review the conceptual roots of space and time representations generally, and particularly in the case of GIS. Second, it will assess the place of GIS as information technology at the intersection of several different perspectives on space and time. Third, it will examine the challenges GIS faces in striving to embody appropriate conceptualisations of space and time to meet increasingly complex and sophisticated user needs.

2 DISCIPLINARY ROOTS OF SPATIO-TEMPORAL PERSPECTIVES

2.1 A brief history of space and time

Though invisible and nonsensible, space and time have preoccupied people since antiquity (Jammer 1964). Systematic thinking about space in particular has its roots in four traditional disciplines: mathematics, physics, philosophy, and geography. These represent, respectively, the formal, theoretical, conceptual, and empirical perspectives on the subject. Each of these comprises a large number of different fields or views. For example, mathematics includes geometry, topology, and trigonometry; philosophy comprises epistemology and the philosophy of science; theoretical physics includes classical and relativistic mechanics and quantum theory; and geography is subdivided into human and physical. All these fields, and several others, have developed their own perspectives on space and time (indeed, each of them may encompass a number of substantially different such perspectives). The multiple overlaps among these four major disciplines, and the particular fields and subfields within them, have given rise to additional insights and ways of thinking. This is illustrated in Figure 1. Time has often, though not always, been considered along with space, either as an extension of space or in analogy with it.

Surely the oldest of the four, the geographical way of looking at the world represents the empirical perspective on the subject of space and time at geographical scales. Throughout their history as a

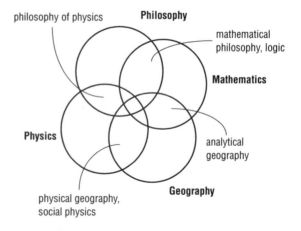

Fig 1. Historical roots of spatio-temporal perspectives.

species humans have always had to deal with rivers, mountains, lakes, oceans, bogs, forests, weather systems, and eventually also roads, cities, and dams. Over the millennia people have evolved a very sophisticated practical knowledge of the spatial and other properties of these entities, the spatial relations possible among them, the range of their variations from place to place, and the changes these may undergo at different timescales (daily, seasonal, or longer term). That knowledge has been recorded by geographers since ancient times, has been codified in maps and nautical charts, has been made increasingly more precise through advancing surveying and positioning technology, and has become to a large extent quantitative and analytical in more recent years.

In their attempts to describe accurately, explain and solve problems relating to the geographical environment, people from early on have turned to mathematics. By some accounts geometry, 'the language of space' (Harvey 1969), originated in ancient Egypt where land surveyors needed to re-establish property boundaries annually following the seasonal flooding of the Nile. The Greek mathematicians, Euclid and the Pythagoreans in particular, brought the science of geometry to a level of perfection that remained unsurpassed for two millennia, while the seventeenth-century work of Newton on the calculus also provided a language for time. Mathematics represents the formal perspective on space and time, bringing its formidable deductive power to the representation, manipulation, and analysis of these elusive concepts.

Of the many kinds of space represented in mathematics, only a few appear to be naturally applicable to geographical-scale entities and phenomena and are thus of direct interest to GIS (Worboys 1995). Euclidean space, the space described by Euclid's five axioms, is an abstraction of people's experience with the spatial properties of the local to medium-scale environment. The basic elements it deals with – points, lines, areas, and volumes – have intuitive interpretations in the geographical world. Euclidean space is also an instance of a metric space, that is, a space in which the notion of distance between two points and its properties are axiomatically defined and quantifiable. The Euclidean distance metric is defined as:

$$d_{ij} = \sqrt{[(x_i - x_j)^2 + (y_i - y_j)^2]}$$

where d is the distance between two points i and j with coordinates (x_i, y_i) and (x_j, y_j). There exist other metrics defining other geometries. The 'Manhattan' or 'taxicab' metric works well in gridded spaces where distance measurement must follow the grid lines (e.g. the gridiron road network in Manhattan). Taxicab distance is defined as follows:

$$d_{ij} = |x_i - x_j| + |y_i - y_j|$$

It behaves differently from Euclidean distance but shares with it the properties of all metrics: it is symmetric (the distance from i to j is the same as that from j to i), and it obeys the 'triangle inequality', meaning that for any three points, the distance between any two of them is never greater than the sum of the distances from these points to the third one. These conditions are easily violated in real environments: distances are usually not symmetric in areas with one way streets, and, if measured in terms of travel time rather than miles, the shortest route between two points is often not the direct route. Variable-metric spaces in which the variation is not systematic are very difficult to represent mathematically.

Genuine non-metric spaces are more general and very powerful. Topological spaces are those dealing with the properties of figures that remain invariant under continuous transformations (e.g. stretching, twisting, squeezing, folding, but not cutting or puncturing). More formally, topological spaces are sets of arbitrary elements (called 'points' of the space) in which a concept of continuity, based on the existence of local (neighbourhood) relations, is defined: it is precisely these relations which are preserved in a continuous mapping from one figure onto another (Alexandroff 1961). Familiar concepts such as inside and out, right and left, touching and overlapping, being connected with, and so on, also express topological relations because they do not depend on metric properties such as shape, size, and distance. Connectivity in particular is a central topological property and is at the basis of the definition of relative spaces, briefly discussed below. What is known as topology in vector GIS is thus a very restricted view of a much broader and more fundamental notion. Topology is a popular area of inquiry among a number of GIS researchers who rightly see it as a rich source of formal insights about how geographical entities may relate to each other in space (Egenhofer and Franzosa 1995; Egenhofer and Mark 1995; Worboys 1995).

The most recent of the four traditions as a distinct discipline, physics has its roots in mathematics and philosophy. Indeed, well into the nineteenth century physics was synonymous with either natural philosophy or applied mathematics. Through physics people were gradually able to organise their formal and conceptual understanding of the world (consisting in large part of the geographical world) into a systematic framework connecting the different pieces of knowledge together. Prominent in that edifice, though not always explicitly so, were the notions of space and time, upon which physics lent its distinct theoretical perspective. Newton's work on classical mechanics could not have been developed in the absence of an underlying model of space and time. Much of the modern understanding of these concepts is attributed to the work of Newton in the seventeenth century, even though the essence of the Newtonian space–time concept was already contained in Aristotle's *Physics*.

Newton's mechanics presupposes a space conceptualised as a neutral container of things and events. Newton himself called this absolute space, in contradistinction to relative space which came to be associated with Newton's contemporary and rival, Leibnitz. Relative space emerges out of the relations among things and events: contrary to absolute space, there can be no such thing as empty relative space. Absolute space is endowed with a 3-dimensional Cartesian frame of reference, to which time may be added as a fourth orthogonal axis. Relative space by contrast does not depend on any frame of reference extrinsic to the spatio-temporal relations represented, and its dimensionality and general properties can vary widely with the geometry entailed by these relations. The triumph of classical mechanics ensured that the notion of absolute space became orthodoxy for three full centuries. It is only in recent decades, following the formulation of alternative notions of space-time in both general relativity and quantum mechanics, that interest in relative space has been revived. Thus for Gatrell (1991) 'space is taken to mean "a relation defined on a set of objects"'. Gatrell goes on to argue for the relevance of that view of space for GIS, which thus far has been based almost exclusively on the absolute-space model. However, the contributions of general relativity and quantum mechanics to our understanding of space and time go well beyond the absolute–relative controversy. It is an open question

whether these new conceptions (some of which are downright bizarre), developed for the immensely large and the vanishingly small, have any relevance for space and time at geographical scales and for GIS in particular.

Finally, philosophy is another ancient tradition representing the conceptual perspective on the issues of space and time. From Pythagoras to Russell, Poincaré and Heisenberg, the best philosophers of space and time have often been the great physicists and mathematicians striving to clarify the implications of their own discoveries for our conceptual understanding of the world. Of the debates that took place for over two millennia, a few are directly relevant to GIS. Prominent among these is the question of whether things or properties are the world's primary ingredients (Hooker 1973). This is the fundamental controversy between the 'atomic' and 'plenum' ontologies, allowing two conflicting hypotheses to be formulated (for a discussion of the implications of these hypotheses for GIS, see Couclelis 1992):

- There exist things in time and space which have (known and unknown) attributes;
- The spatio-temporal clusters of known attributes are the things.

The first hypothesis leads to an ontology of objects, the second one to an ontology of fields. Both are in principle compatible with either a relative or an absolute view of space-time, though an advanced exploration of the plenum ontology is likely to lead to a relative view whereby the properties of the space itself come to depend on the properties of the field. According to Einstein (1920: 155):

'There is no such thing as empty space, i.e. space without field. Space-time does not claim existence on its own, but only as a structural quality of the field.'

Another old philosophical debate recently found to be of great relevance to GIS is that regarding the ontological status of space and time: are these objective properties of the world, or are they constructs of human understanding? The latter, less popular position was taken by Kant in his *Critique of Pure Reason* (see Friedrich 1977), who argued that space is a 'synthetic *a priori*': something that appears to be the way it is because human minds are such as they are. In recent years a neo-Kantian view of space has been adopted by many geographers and GIS researchers exploring the cognitive dimensions of our understanding of space.

2.2 Spatio-temporal conceptions in the age of GIS

Newton and his contemporaries and followers set the tone for the modern intellectual tradition which was marked by the search for objective knowledge independent of any observer. With the decline of that tradition in the second half of the twentieth century and the advent of postmodernity, two new perspectives on space and time were added to the traditional four: the cognitive and the sociocultural. Both are based on the premise that there is no single objective reality that is the same for all, but that different realities exist for different minds or for different sociocultural identities. This implies that the world as described by mathematics and physics is not the only world there is, and that in fact the world so described may be of little relevance to people's thinking and activities. On the cognitive side of the argument, the experiential perspective in particular, propounded primarily by Lakoff and Johnson (1980) and Lakoff (1987), has attracted a lot of attention among a number of GIS researchers (Mark, Chapter 7; Mark and Frank 1996), while the multiple realities viewed from the standpoint of different sociocultural perspectives have been the subject of investigation by a growing number of critical theorists and cultural geographers (see Pickles, Chapter 4). Thus we may view the four historical 'objective' approaches to space and time as being embedded in the intersubjectivity of the cognitive individual on the one hand and the sociocultural group on the other. This is illustrated in Figure 2. This means that, far from being resolved, the question of space and time has become more complex over the centuries. It is this growing conceptual quagmire that GIS is being called to address in practical terms.

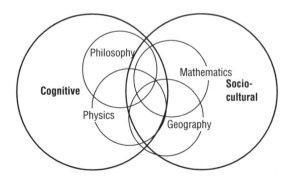

Fig 2. Spatio-temporal perspectives in the age of GIS.

For historical reasons the current generation of GIS embodies the spatial views of a small number of applied disciplines: cartography, computer aided design, landscape architecture, remote sensing. From these it has inherited a strong basis in Euclidean, analytical and computational geometry and a dual spatial ontology of fields (the remote-sensing legacy) and objects (the landscape-architecture legacy), while the temporal aspect, which was mostly absent in the parent disciplines, has been largely neglected until recently and is still often treated almost as an afterthought.

Of the several contributing disciplines cartography has surely had the strongest and most lasting impact on GIS. In fact, GIS may be described (or criticised) as presenting the 'map' view of the world, tied to the notion of an absolute space equipped with a Cartesian or other system of 2-dimensional coordinates. The representation and manipulation of geographical coordinate systems under different geometric projections, and the association of attribute information with specific (x, y) coordinates (geocoding), is as central to GIS as it is to cartography. Much of the power of GIS derives from its strong roots in that ancient discipline which over the centuries has evolved a formidable arsenal of methods for recording, measuring, and representing the surface of the Earth. However, that strength is also the source of several of GIS's weaknesses, as the map view of the world can have serious limitations if stretched beyond its intended purposes: maps are static, flat, 2-dimensional, precise, and not well suited for conveying the fact that the level of knowledge or certainty over their range is often far from uniform (Goodchild 1996). Section 4 below includes a brief discussion on how these limitations are currently being addressed, and what research challenges remain for the future.

3 AN INFORMATION SCIENCE FOR SPATIO-TEMPORAL PHENOMENA

As an information technology the purpose of GIS is not to add another perspective or view on space and time to the many already available, but rather to help convey to the users spatio-temporal information in a form suitable for the task at hand. This simple-sounding requirement is in fact very complex because of the multiple spatio-temporal views simultaneously present in a GIS. There are indeed four critical aspects or players here (see also Goodchild and Longley, Chapter 40):

1 the builder of the database, who is driven by an empirical understanding of the geographical entities and phenomena being measured;
2 the data model on which the database is mapped, which has to conform to the spatio-temporal 'understanding' of the digital computer;
3 the user of the database, who needs to extract the information necessary for a given task from the primarily graphical representations presented by the system;
4 the sociocultural (including disciplinary) context of the task, which determines, among other things, what kinds of questions are to be asked, and what forms of answers are acceptable.

There are thus four qualitatively different spatio-temporal perspectives involved in this process: an empirical one, attempting to capture the spatio-temporal and other properties of cities, lakes, forests, rivers, and so on as accurately as possible; a formal one, based on the properties of points, lines, areas, or pixels, and on the constraints of digital representations; an experiential one, using spatial metaphors and other cognitive devices to convert graphics and other computer-generated signs back into expert geographical understanding; and a social one, focusing inquiry and determining what the ontologies of interest should be. These views are partially conflicting. For example, the point–line–area data model view of vector GIS is ill adapted to the need to represent fuzziness and uncertainty in geographical entities and phenomena as apprehended from either the empirical or the experiential or the social perspectives (Burrough and Frank 1996); on the other hand, the discrete field view represented in raster data models contradicts two basic intuitive notions prominent in the experiential perspective: that geographical space is continuous, and that it is populated with individual things (Couclelis 1992). As another example of such internal conflicts, the temporal aspect is implicitly present in the experiential perspective, explicitly absent or superficially added on in the formal view represented by most current data models, and either absent or present, as the case may be, in the empirical and social views.

Clearly there are issues here that far transcend the technical. Geographical information science has

developed out of the maturing GIS technology to address just these kinds of questions that cannot be resolved merely through smarter software and better system design (Goodchild 1992). Elsewhere I have proposed a framework for geographical information science anchored on four vertices representing the above four perspectives: the empirical, the formal, the experiential, and the social (Couclelis 1997). The edges and faces of the resulting tetrahedron represent particular research perspectives in geographical information science, while the core questions, partaking of all four perspectives, are represented by the tetrahedron's interior (Figure 3). Prominent in that scheme is the base triangle defined by the empirical–formal–experiential triad of vertices, which represents the map view of the world, critically augmented by the temporal consciousness and intersubjectivity inherent in the experiential perspective. The 'social' vertex is a more recent addition, marking the growth of the GIS field from a computer-aided technology to a discipline capable of reflecting on the multiple two-way connections between that technology and its social, political, cultural, and philosophical context.

4 CHALLENGES FOR GIS

Geographical information science is a 'meta' science: it is not about the geographical world, it is about information about the geographical world. Contrary to some common misconceptions, information is not a thing – i.e. a bunch of bits – but a relation between a sign and an intentionality: the sign(s) being, in this case, the various graphic and other

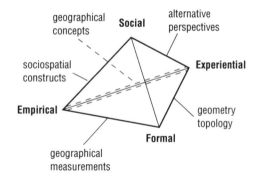

Fig 3. A framework for geographic information science: dimensions of time and space representations for GIS.

forms of GIS output, and the intentionality, the purposeful human intelligence giving meaning to these signs (Couclelis 1997). This implies that the right way to represent geographical information is a function of who is looking at it, and for what purpose: there can be no single right way. What is true of all information is complicated fourfold by the fact that in GIS there are always the four concurrent perspectives (empirical, formal, experiential, and social), each of them with its own preferred views of time and space. In the domain of spatio-temporal representation, the greatest challenge for GIS is thus to move beyond its traditional quasi-exclusive identification with a single view – the map view – useful though this may be for so many purposes, and to permit the simultaneous accommodation of the multiple views required in each case.

Looking back at the variety of approaches to space and time outlined in section 2, it is clear that the map view of the geographical world is only one of many possible. Whether in the 'fields' or 'objects' version, the map view is rooted in absolute Newtonian space and Euclidean geometry. The former is ill matched with the explicit representation and treatment of relations (witness the intractable problem posed in Newtonian physics by the 'three-body problem'); the latter deals with discrete figures, volumes and surfaces defined through infinitely small points and infinitely thin and crisp lines and surfaces. Both presuppose a homogeneous, isotropic space that is a neutral container, and neither is integrated with time (indeed, time cannot even be defined within Euclidean geometry). These properties contradict many aspects of real-world geographical entities and phenomena, which are strongly time dependent, not precisely bounded in either space or time, inhomogeneous and anisotropic as to their attributes and dynamic properties, not properly representable either as geometric figures or as fields, and have complex relations in both space and time with other entities and phenomena.

Researchers have long recognised the limitations of the map view and have proposed several useful extensions of standard GIS that try to address one or the other of these problems, as many of the chapters in this Section demonstrate. Some of these limitations, such as the difficulty of representing more than two dimensions in GIS, are primarily technical. Other efforts focus on more fundamental problems. These may be discussed under the following four headings:

- Integration of space and time
- Representation of relative and non-metric spaces (and times)
- Representation of inexact spaces (and times)
- Representation of commonsense views of space and time.

4.1 Integration of space and time

The static quality of the map has been the primary reason why the integration of the temporal perspective in GIS (and the representation of dynamic phenomena and changing features) continues to be so difficult. Efforts to do justice to the temporal essence of geographical phenomena are relatively recent (Peuquet, Chapter 8; Langran 1992; Langran and Chrisman 1988; Peuquet 1994; Worboys 1995). Standard approaches to representing change within the map view are mostly variations of the 'timeslice' model, consisting of producing a sequence of time-stamped maps corresponding to different time points within a given time interval. The resulting sequence may be represented in GIS either as an ordered set of independent maps, or as a space–time composite layer, or as a 3-dimensional spatio-temporal structure. While sufficient for many purposes, this kinematic (as opposed to dynamic) representation breaks up the continuity of phenomena, may miss temporal orderings indicating causal connections between events, and leaves open the question of what may have happened in the intervals between timeslices.

Advances in temporal GIS involve various departures from the notion of time as a single extra axis added to a Cartesian spatial frame. Two-dimensional time defined on both a real-world time dimension and a database time dimension, and nonlinear time (in the form of forward or backward branching time) have been successfully implemented by several researchers (Snodgrass 1992). Even more advanced notions of time as defined through events, change, motion, and process have also been proposed, though most of these remain at the conceptual level (Clifford and Tuzhilin 1995; Kelmelis 1991). Thus, while great progress has been made in developing data models for GIS that go beyond the timeslice approach, the creation of a truly spatio-temporal GIS remains an unmet challenge (see also Peuquet, Chapter 8).

4.2 Representation of relative and non-metric spaces (and times)

Many geographical phenomena are defined in whole or in part through relations holding among relevant entities. These relations may be material exchanges such as fluid flows or human or animal migration flows between places, functional connections of influence, communication, accessibility, potential interaction, and so on, or cognitive properties of ordering, classification, association, or differentiation. In all but the simplest cases these relations are best seen as defining a relative space, i.e. a space whose properties depend on the configuration of the relevant relations. Handling relative space well will become increasingly important for GIS as cyberspace, the space of electronic connections, continues to expand its hold on every aspect of society.

There are two problems here for GIS. First, rooted as it is in absolute space, GIS does not represent relations well. This is because in absolute space geocoded locations are bound to *a priori* existing relations of geometry and topology among the corresponding points in the space, whereas in relative space the definition of a set of arbitrary relations comes first and the geometry and topology follow. Thus even relations that can be represented on the plane, such as those defined by communication or movement over physical networks, are confounded by the underlying Euclidean metric. For example, a relation of proximity among a set of places cannot be properly represented by the transportation network connecting these places if proximity is defined in terms of travel time rather than distance. While it is often possible to represent well-behaved time distances by subjecting the original map to an appropriate geometric transformation, in other cases the resulting space is non-metric or non-planar and cannot be so represented. More generally, the conflicting properties of absolute and relative space prevent the satisfactory representation of relations in GIS. This is also largely the reason why the proper integration of GIS with geographical models, especially those describing social phenomena, continues to be so difficult. Takeyama and Couclelis (1997) present a partial solution to this problem by formalising the notion of a relational space combining properties from both absolute and relative. Points in relational space behave as in absolute space but are also linked to information on their functional neighbourhoods, i.e. their place in the relative space(s)

of which they are part. Elements of this idea are also contained in Tomlin's (1992) map algebra.

A further problem is that for the most part relative spaces are n-dimensional, where n can be any arbitrary integer. Such spaces defy not only map-based GIS but any analogue (visual or material) representation medium. However, formal and digital representations of n-dimensional spaces abound, along with several very useful analysis techniques (e.g. multidimensional scaling, cluster analysis, Q-analysis). There is no reason why these could not be part of GIS data models through which users could derive appropriate partial views linking relative and absolute spaces. An illustration of this possibility is given by Portugali and Sonis (1991), where a 7-dimensional space of labour relations is sequentially projected on an ordinary map of Israel. A more general approach to this problem is known as spatialisation, whereby arbitrary n-dimensional spaces (not necessarily derived from geographical phenomena) are transformed into and analysed as familiar (often geographical) spatio-temporal representations.

4.3 Representation of inexact spaces (and times)

In contrast to the more general challenges of space-time and relative-space representation, this one is of GIS's own making. Euclidean geometry, georeferencing and the map view together conspire in forcing GIS into one or the other end of a representational spectrum ranging from crisply delineated, internally homogeneous objects to continuously varying attribute fields. Most geographical entities and phenomena or their most useful representations do not fall neatly into either category (Burrough and Frank 1996). In the literature this problem has usually been treated in terms of fuzziness and uncertainty. The distinction made is that between the geometrical properties of the entities and phenomena themselves, which may or may not be crisply delineated (fuzziness), and the state of our knowledge about these properties, which may or may not be accurate (uncertainty). The latter aspect is being attacked with the tools of probability theory (Goodchild and Gopal 1989), while a growing number of researchers are applying fuzzy set theory to address the many cases where the lack of clear boundaries is an intrinsic property of the entities studied (Burrough 1996). The graphic representation of fuzziness and uncertainty, especially where both aspects coexist, is currently an active area of investigation.

The fuzziness/uncertainty perspective on the issue of inexact spaces decomposes the problem into an objective and a subjective component, whereby a clear distinction is drawn between how things really are out there in the empirical world, and how they are known to be by imperfect human observers. The simplicity of that perspective has fostered much robust research, but it seems likely that alternative views will soon be required in order to deal with more sophisticated demands on future GIS. Thus Couclelis (1996) distinguishes 480 different potential cases of geographical entities with ill-defined boundaries, based on different possible combinations of empirical characteristics, observation mode, and user purpose. While that number in itself is not significant, there are two points worth noting: first, only a few of these cases are fully accounted for through the fuzziness/uncertainty perspective; and second, the variety of cases calls for a corresponding variety of different models of space and time, not all of which can be supported by GIS based on the map perspective.

4.4 Representation of commonsense views of space and time

Introducing purpose and the user perspective into the picture calls for some radically different approaches to the treatment of space and time in GIS. First, this requires the ability to present multiple views of the same information so as to meet different individual interests, skill levels, and needs. Second, the construction of these alternative views must be thoroughly informed by the active ongoing research on the cognitive and social dimensions of space and time. Third, the presentation of the information must be flexible enough to conform to how people use graphic representations of spatio-temporal configurations and phenomena (from sketch maps and flow diagrams to photographs and animations) in reasoning, problem-solving, collaborative work, public debate, teaching, and communication. Work on space and time in GIS is thus expanding well beyond the traditional domain of geometry and topology, investigating issues of language, culture, semantics, metaphor, cognitive configurations, and social constructions (Kuhn 1995; Mark and Frank 1991, 1996; Pickles 1995). Translating these multiple and sometimes conflicting insights into data models that work is no mean task. One of the major research frontiers in GIS lies in this area.

5 CONCLUSION

GIS has come a long way since it was little more than the latest in computer mapping software. With the rapidly increasing technical and conceptual sophistication of the technology came increasingly complex demands and expectations from an ever expanding and diverse user community. Directly or indirectly, the representation of space and time has been central to the pressures for better GIS. Advanced applications require better integration of space and time and ways to represent the entities of interest that go beyond the objects/fields dichotomy; the introduction of the social perspective demands more attention to the issues of subjective perceptions and multiple views of spatio-temporal entities and phenomena; and the establishment of the information age challenges absolute physical space as the sole, undisputed framework for representing geographical reality. We are learning that the geographical is not just the mappable, the spatial is not just the visible, the temporal is not an independent domain, and not all users see the world through the same eyes. The purpose of this chapter has been to review the field of spatio-temporal representation and highlight the unresolved issues, which are many. Research in GIS is moving so rapidly that the next edition of this book is bound to report on some spectacular progress – as well as on the next set of challenges.

References

Alexandroff P 1961 *Elementary concepts of topology*. New York, Dover

Burrough P A 1996 Natural objects with indeterminate boundaries. In Burrough P A, Frank A U (eds) *Geographic objects with indeterminate boundaries*. London, Taylor and Francis: 3–28

Burrough P A, Frank A U (eds) 1996 *Geographic objects with indeterminate boundaries*. London, Taylor and Francis

Clifford J, Tuzhilin A 1995 Recent advances in temporal databases. *Proceedings, International Workshop on Temporal Databases, Zurich, Switzerland*. Berlin, Springer

Couclelis 1992 People manipulate objects (but cultivate fields): beyond the raster-vector debate in GIS. In Frank A U, Campari I, Formentini U (eds) *Theories and methods of spatio-temporal reasoning in geographic space*. Berlin, Springer: 65–77

Couclelis H 1996 Towards an operational typology of geographic entities with ill-defined boundaries. In Burrough P A, Frank A U (eds) *Geographic objects with indeterminate boundaries*. London, Taylor and Francis: 45–55

Couclelis H 1997 GIS without computers: building geographic information science from the ground up. In Kemp Z (ed.) *Innovations in GIS 4: selected papers from the Fourth National Conference on GIS Research UK*. London, Taylor and Francis: 219–26

Egenhofer M J, Franzosa R D 1995 On the equivalence of topological relations. *International Journal of Geographical Information Systems* 9: 133–52

Egenhofer M J, Mark D M 1995 Modelling conceptual neighbourhoods of topological line–region relations. *International Journal of Geographical Information Systems* 9: 555–65

Einstein A 1920 *Relativity*. London, Methuen

Friedrich C J 1977 *The philosophy of Kant*. New York, The Modern Library

Gatrell A C 1991 Concepts of space and geographical data. In Maguire D J, Goodchild M F, Rhind D W (eds) *Geographical information systems: principles and applications*. Harlow, Longman/New York, John Wiley & Sons Inc. Vol. 1: 119–34

Goodchild M F 1992b Geographical information science. *International Journal of Geographical Information Systems* 6: 31–46

Goodchild M F 1996 Multiple roles for GIS in global change research. Paper presented at the second NSF/ESF Young Scholars International Institute in Geographic Information, Berlin, July 1996

Harvey D 1969 *Explanation in geography*. London, Edward Arnold

Hooker C A 1973 Metaphysics and modern physics: a prolegomenon to the understanding of quantum theory. In Hooker C A (ed.) *Contemporary research in the foundations and philosophy of quantum theory*. Dordrecht, Reidel

Jammer M 1964 The concept of space in antiquity. In Smart J J (ed.) *Problems of space and time*. New York, Macmillan

Kelmelis J A 1991 'Time and space in geographic information: towards a 4-dimensional spatio-temporal data model.' Unpublished PhD dissertation, The Pennsylvania State University

Kuhn W 1995a *Semantics of geographic information*. GeoInfo, Department of Geoinformation, Technical University of Vienna

Lakoff G 1987 *Women, fire, and dangerous things*. Chicago, University of Chicago Press

Lakoff G, Johnson M E 1980 *Metaphors we live by*. Chicago, University of Chicago Press

Langran G 1992 *Time in geographic information systems*. London, Taylor and Francis

Langran G, Chrisman N R 1988 A framework for temporal geographic information. *Cartographica* 25: 1–14

Mark D M, Frank A U (eds) 1991 *Cognitive and linguistic aspects of geographic space*. NATO ASI Series 63. Dordrecht, Kluwer

Mark D M, Frank A U 1996 Experiential and formal models of geographic space. *Environment and Planning B: Planning and Design* 23: 3–24

Peuquet D J 1994 It's about time: a conceptual framework for the representation of temporal dynamics in geographic information systems. *Annals of the Association of American Geographers* 84: 441–61

Pickles J 1995 *Ground truth: the social implications of geographic information systems*. New York, Guilford Press

Portugali J, Sonis M 1991 Palestinian national identity and the Israeli labor market: Q-analysis. *The Professional Geographer* 43: 265–79

Snodgrass T R 1992 Temporal databases. In Frank A U, Campari I, Formentini U (eds) *Theories of spatio-temporal reasoning in geographic space*. Berlin, Springer: 22–64

Takeyama M, Couclelis H 1997 Map dynamics: integrating cellular automata and GIS through Geo-Algebra. *International Journal of Geographical Information Science* 11: 73–91

Tomlin C D 1992 *Geographic information systems and cartographic modeling (revised edition)*. Englewood Cliffs, Prentice-Hall

Worboys M F 1995 *GIS: a computing perspective*. London, Taylor and Francis

3

Geography and GIS

R J JOHNSTON

Geographers study three major concepts – environment, space, and place – but their approaches to these have varied considerably over recent decades, incorporating three very different conceptions of science and views of the world: all of these can be encapsulated into 'geography as spatial science' and 'geography as social theory'. As reviewed here, most applications of GIS to date have fallen within the first of these categories: their utility in the latter is also discussed.

1 INTRODUCTION

There are very strong links between GIS and the academic discipline of geography, which extend well beyond the commonality of titles. The academic discipline has been the 'home' for much of the research, development, and training for GIS practitioners, and some see GIS as a major element in the discipline's 'survival package' for the foreseeable future, in a period of considerable pressures on public sector funding for academic work (see Gober et al 1995, on employment prospects for geographers in the USA). Nevertheless, the discipline embraces a great deal more than GIS and, as Openshaw (WWW 1996) has argued, much of what is done with GIS has very little to do with that discipline as generally practised: over 90 per cent of all applications, he claims, are 'of no significant consequence to people and society. They ... are concerned with the management of the physical infrastructure ... [and] involve little more than a digital replacement for various large-scale paper map-making, recording and handling industries'. This chapter looks at the context within which much GIS work has been, and continues to be, nurtured, while recognising that the 'child' has grown immensely in stature and influence beyond the 'academic nest' in the last two decades.

Significantly, different views on the practice of geography have been advanced on several occasions over the last half-century, but core beliefs regarding the discipline's *raison d'être* have remained constant. There is little difference, for example, between Hartshorne's (1939) classic definitional statements:

'... geography is a science that interprets the realities of areal differentiation of the world as they are found, not only in terms of the differences in certain things from place to place, but also in terms of the total combination of phenomena in each place, different from those in every other place'

so that '... geography is concerned to provide accurate, orderly and rational description and interpretation of the variable character of the Earth's surface' which

'... seeks to acquire a complete knowledge of the areal differentiation of the world, and therefore discriminates among the phenomena that vary in different parts of the world only in terms of their geographic significance – i.e. their relation to the total differentiation of areas. Phenomena significant to areal differentiation have areal expression – not necessarily in terms of physical extent over the ground, but as a characteristic of an area of more or less definite extent'.

through its updating (Hartshorne 1959):

'... geography is that discipline that seeks to describe and interpret the variable character from place to place of the earth as the world of man' and '... geography is primarily concerned to describe... the variable character of areas as formed by existing features in interrelationships'.

and McDowell's (1995) modern rendering, that geography is a discipline '... whose *raison d'être* is the explanation of difference and diversity' (1995: 280) with geographers needing '...theoretical perspectives that not only permit the elucidation of the main outlines of difference and diversity, of the contradictory patterns of spatial differentiation in an increasingly complex world, where ever-tighter global interconnections coexist with extreme differences between localities; but perspectives that also allow us to say something about the significance of these differences.' Geography remains the study of differences across the Earth's surface.

Throughout this period, geographical research and writing have focused on three main concepts:

1 *Environment* – or nature
2 *Space*
3 *Place*

although the relative stress placed on each, and on their interactions, has varied somewhat. Geographers are concerned with the where, how, and why of the physical environment on which material life is based, the spatial organisational structures erected and operated by human societies to sustain and promote their material well-being, and the nature of the places which they have created within those structures. In pursuing those interests, they have engaged with a variety of approaches to science, and applied their findings in a number of separate ways.

2 APPROACHES TO SCIENCE

Discussions of the nature of science reflect not only the changing worlds in which scientists live and work but also their own varying conceptions of what comprises knowledge, how it can be obtained, and to what purposes it should be used. This literature is simplified into a typology containing just three categories, each of which has been used and argued about within geography in recent decades.

The first type (that most associated with the term 'science') is the *empirical*, which contains several sub-categories. Knowledge is acquired through direct experience, especially visual: its production involves accurate observation, and its propagation calls for unambiguous reportage. Within empirical science, therefore, facts speak for themselves. The scientist, having decided what to study, is a neutral observer-reporter, presenting material using accepted languages and categories (and, if necessary, proposing extensions to them). The acid test of the validity of a piece of science is that when replicated it produces the same outcome.

Many empirical scientists seek not only to describe accurately but also to explain: geographers aim both to show what is where and to indicate why. Positivist empirical science explains through generalisation: an event is accounted for as an occurrence of the operation of one or more general laws, whose identification is the scientists' goal. (Logical positivist science claims that there is no other route to knowledge.) Scientific laws can be used to predict, and that power is associated with the power to change, applying scientific findings to achieve certain ends.

The empirical sciences are founded on a belief in a separate world outside and independent of any individual observer, and some claim that its methods are equally valid in the social as in the natural sciences – in human as well as physical geography. The *hermeneutic* sciences challenge that: in them, nothing exists outside the observer, because any perception of an empirical event involves interpretation, using human constructs (especially language) to give meaning to what is being recorded. Individuals ascribe meanings to the worlds they live in, and act according to those meanings, so understanding an action involves appreciation of the meanings on which it was based. Natural scientists operate in this way as well as social scientists, but whereas the latter study the creation and transmission of meanings by human actors the natural scientist's subject matter comprises phenomena lacking the distinguishing features of 'humanness' – memory and reason.

The hermeneutic social sciences study people continuously interacting with their natural and human environments – individuals learning and using their powers of reasoning to interpret their observations, in often-changing situations. Explanation through generalisation is not possible, therefore, since humans cannot be equated with machines which always respond in the same way to given stimuli: they may react differently because of their learning processes (which may involve reinterpretation of the stimulus itself), or because the stimuli and the contexts in which they are encountered are changed. Instead of explanation, therefore, hermeneutic science advances understanding, appreciation of why people acted in particular ways: this promotes awareness of the past and the present, and provides a guide to the future, but no more.

The final type of science – the *critical* – promotes explanation, but not necessarily through generalisation. Underpinning these sciences is a belief that observation is insufficient to appreciate the world, so that theories are needed which can account for the hidden structures involved in creating what is recorded by the senses – the theories must be consistent with the outcomes of those hidden processes. The 'law of gravity' exemplifies this. Gravity cannot be observed: all that can be recorded is behaviour which is consistent with its assumed operation. The law is accepted as valid because of its successful predictive power in repeated experiments under controlled conditions; if it should fail, however, then its validity in similar circumstances must be reconsidered.

Application of the rules of the hidden structures in the subject matter studied by social scientists involves human interpretation and action, as argued in the hermeneutic sciences. Biological drives demand food and liquid sustenance in humans, for example, but what foods and liquids is not determined: people individually and collectively decide how to meet those needs, not by making *ad hoc* decisions every time that they are thirsty and hungry but by developing strategies to meet their requirements on a regular basis. Those strategies vary over time and space, reflecting the conditions within which they are developed (such as the environmental context), the processes of learning, and the lessons which are passed on (as part of a society's cultural inheritance). Their outcomes may be inscribed in the landscape – as with land-use patterns which reflect food-raising strategies. The strategies may occasionally be reconsidered – either as a matter of urgency (because of a crop failure following an environmental disaster, for example) or more slowly (as population pressure on environmental resources builds).

The critical sciences identify three 'domains of interest':

1 The *real*, comprising the underlying, unobservable, mechanisms;
2 The *actual*, involving the operation of those mechanisms;
3 The *empirical*, which is the outcome of those operations.

In the natural sciences, the law of gravity occupies the first level: it is put into operation at the second level when rain falls, for example; and the empirical outcome is the rain's impact on the landscape. In the social sciences, both the biological drives necessary to human survival and the mechanisms put in place to sustain them socially occupy the domain of the *real* – as with the capitalist and socialist modes of production which have evolved to ensure individual and collective well-being, and improving material standards. The domain of the *actual* involves implementing those mechanisms, the myriad decisions made by 'knowing individuals' applying the fruits of their own learning and acquired meanings within the 'rules'; and the domain of the *empirical* contains the outcomes – which become part of the context within which future decisions on operating the mechanisms are made.

In both natural and social science it is possible to observe the outcomes and the procedures by which they are generated, but not the underlying mechanisms – one can no more 'see' capitalism than one can 'see' the law of gravity, so that the nature of each has to be inferred (or theorised) and the validity of that thought process evaluated against reality. Beyond this similarity there is a major difference between the two, however. The law of gravity is assumed to be invariant, and nothing that humans can do will remove it: it is an enduring fact. The mechanisms studied in the social sciences are not invariant: not only are they human creations which can be destroyed by human action but they are also continually being changed, as societies respond to situations in the light of previous decisions and their consequences. Hence, whereas a critical realist approach to the natural sciences underpins the search for explanation via generalisation (since the mechanisms are invariant and all that can alter is how they are put into operation) such a strategy is not viable in the social sciences, since the same conditions may never be repeated: events in the natural sciences may be unique; in the social sciences they may be singular (Johnston 1989).

The three types of science – empirical, hermeneutic, and critical – represent different world-views, different conceptions of how the world can be understood. They are incommensurate and cannot be evaluated for their validity against some external criterion: choosing a world-view involves determining one's conception of what is involved in achieving understanding. Each world-view contains a number of separate approaches (sometimes termed disciplinary matrices), different strategies accepted by scientists for achieving their desired ends. In turn, each of those strategies has its own detailed exemplars, paradigm instances of how the science should be done. When becoming a scientist, therefore, an individual adopts a world-view, a general orientation for her or his work. The next decision is which community to join, whose strategy to adopt. Finally, which detailed exemplar(s) to follow has to be determined. (These three decisions and scales are readily equated with Kuhn's paradigm model of scientific practice: Kuhn 1962 – and for its application to geography: Johnston 1997.) These decisions are not taken in isolation, of course: they are made during the process of becoming a scientist, of being socialised into its culture and accepting its norms, the would-be researcher already having decided his or her scientific subject-matter – what discipline to join. And they can be revisited, with different exemplars being adopted as the models to follow, different strategies being entertained, and even different world-views canvassed.

Those decisions are not irrevocable. Change is most likely at the level of the exemplar, or *modus operandi*: as science progresses so new ways of achieving its goals are identified and adopted by practitioners. It may occasionally occur at the intermediate level, when a scientist decides that one community's strategy is better than another's. Change is rare at the highest level, however, since this involves altering one's world-view which cannot be done 'objectively' against predetermined criteria: it involves deciding that one's approach to science to date has been entirely wrong, and that an alternative offers a 'better' path to understanding – and that decision is at root a subjective one, a 'leap of faith' from one cosmology to another. A number of geographers have made such leaps in recent decades, and many more have switched their preferred disciplinary matrix.

3 GEOGRAPHY AND THE SCIENCES

Over recent decades geographers have encountered, experimented with, and debated the relative merits of a range of approaches to science, at each scale. At the largest, whereas physical geographers have largely remained within the empirical world-view (Gregory 1985; see also Aspinall, Chapter 69), human geographers have embraced all three – with some individuals shifting conceptions (see Sheppard 1995) and many more being socialised into the one proving particularly attractive to their generation. Within each world-view, different strategies have attracted varying support – as exemplified among physical geographers by debates over the relative merits of logical positivism and Popperian critical rationalism (Haines-Young and Petch 1985) and among human geographers by the debates on the relevance of the 'economic man' model (see Barnes 1996; Johnston 1997). Within each strategy, the importance of different exemplars has waxed and waned: technical developments in the analysis of spatial data generated a major shift in methodological orientation between the two editions of Haggett's *Locational analysis in human geography*, for example (Haggett 1965; Haggett et al 1977). A third edition, prepared in the wake of developments in GIS since, would undoubtedly be a very different book yet again.

The history of geography in recent decades has been charted in a number of volumes and will not be repeated here (Gregory 1985; Johnston 1997). Instead, focus is on broad categorisations of how geographers approach their discipline – and why. Some have simply divided geographers into two camps, with separate world-views – the spatial analysts and the social theorists (Sheppard 1995): all physical geographers fall into the former, but human geographers are divided between the two (see Bennett 1989). In slightly more detail, Buttimer (1993) has identified four 'root metaphors' at the heart of different approaches to the discipline and four separate rationales for its study.

3.1 The root metaphors

Metaphor is basic to many approaches to science – it provides a way of moving from the known to the unknown, via analogy (Barnes and Duncan 1992; Barnes 1996). Root metaphors provide analogies for use in a wide range of circumstances.

Buttimer's (1993) first metaphor – *the world as a mosaic of forms* – involves the 'recognition of forms and patterns, similarities and differences, among people, places and events'. These are described – verbally or through other representational forms, such as maps in geography – using 'language' which stresses their similarity (as with the centre/periphery model applied in a variety of contexts within human geography and the analogies between river systems and settlement patterns identified by Woldenberg and Berry 1967; see also Haggett and Chorley 1969, on networks). The approach provides a way of summarising the vast volumes of geoinformation now available: its main concern is description, saying what is where rather than why it is.

The second metaphor – *the world as a mechanical system* – emphasises the mechanisms that provide an explanatory base for the study of forms, as in developments in physical geography from the late 1960s when a 'systems approach' offered an integration of studies of processes and forms (Chorley and Kennedy 1971; and Bennett and Chorley 1978 promoted its application in human as well as physical geography).

Buttimer's third metaphor – *the world as organic whole* – focuses on the interrelationships among society and nature, using organic analogies to argue that the whole is greater than the sum of its parts. Whereas the mechanistic analogy promotes the study of order, rationality, and certainty, the organic concentrates on human individuality, identity, and ingenuity. The organic metaphor was initially applied to the identification of regions, territorial units in which society and milieu interacted to form identifiable niches in the spatial organisation of urban as well as rural worlds (on which see Gregory 1994; and Livingstone 1992).

Finally, *the world as arena of events* focuses on 'Events, in all their complexity, possible uniqueness, and contingency...; one looks for holistic understanding of particular events rather than ways of fitting them into some *a priori* schema of form, process, or organic whole.' (Buttimer 1993: 187). In its search for explanation this has similarities with the mechanistic vision (rather than the hermeneutic goal of the organic), but it rejects the search for universal truths and order underpinning the Enlightenment project that informs the mechanistic approach (Barnes 1996). It stresses pragmatic decision-making in the critical science mould, drawing conclusions within an embracing context of interpretations, as in the Marxist and realist approaches promoted by Harvey (1982) and Sayer (1992) respectively. This metaphor embraces a wide range of disciplinary matrices, including those such as postmodernism which promotes a relativistic

form of knowledge and denies the existence of universal truths; all accounts are incommensurate (having been produced within different contexts and from different positions) and thus are of equal value.

3.2 The metaphors in use

These four metaphors have varied in their application within the practice of geography. In physical geography, for example, the promotion of studying form rather than process led to its relative demise within the discipline in the United States for several decades (see Marcus 1979) – a situation not repeated in the UK (Johnston and Gregory 1994); the mechanical system metaphor now dominates. Human geography, on the other hand, has experienced considerable, continuing debate over their validity and significant variations in their relative importance (as chronicled in Johnston 1997). These variations have reflected changing emphases on the role of science and social science within society (on which see Taylor 1985). Buttimer (1993) identified four roles, or vocational meanings:

1 *Poesis* encourages curiosity about the relationships among individuals, society and nature by evoking geographical awareness;
2 *Paideia* promotes self-education through reflective practice regarding life and landscape;
3 *Logos* emphasises the search for order and generalisation via analytical rigour;
4 *Ergon* involves a focus on appropriate behaviour (applied geography) in the face of contemporary societal and environmental problems.

As a broad generalisation, the first two are associated with the organic and arena metaphors whereas the others are linked with form and mechanism.

Human geographers have long debated the validity of these metaphors and roles, characterised by Sheppard (1995) as a contest between spatial science and social theory. While that has continued, discussions have taken place within each camp regarding which disciplinary matrix to adopt and which exemplars to follow. Progress within each matrix has led to the replacement of some exemplars by others. With regard to GIS, this has occurred almost entirely within the spatial science world-view, involving the mechanism and form metaphors (and the virtual rejection of organism and arena) and the *logos* and *ergon* roles.

43

4 GIS AND GEOGRAPHY AS SPATIAL SCIENCE

Progress within geography as spatial science has been substantially linked to technical advances in the collection, collation, display, and analysis of data; computers with greater size, speed, and power have enabled the expansion of existing approaches and the introduction of new.

Several topics illustrate how GIS has enabled analyses which were previously technically extremely time-consuming, if not impossible.

- In the analysis of spatial form, maps of different phenomena are overlaid to identify correlations. This may involve the straightforward linking of point and area patterns, as in Ravenhill's (1955) investigations of the pattern of Celtic settlements in Cornwall. It may, however, involve attempts to link data from incommensurate sources, such as the classic paper by Robinson et al (1961) on areal correlations which related point (rainfall amounts) to area (population density) data through interpolation processes involving manually-created hexagonal tesselations.[1]
- In analyses of spatial diffusion using simulation techniques, time-consuming methods were employed to generate changing patterns over time using mean information fields (as in Hägerstrand's 1968 classic and works by his followers, such as Morrill 1965a, 1965b) – so much so that some simulations were based on a single iteration only (Robinson 1981).
- Although the ultimate focus of much research was on individual decision-making and behaviour, a great deal of the available data related to spatially-defined aggregates. Their analysis involved not only the ecological fallacy of inferring from a population (usually defined as those living in a particular area) to its individual members but also the aggregation and scale problems which exacerbated the difficulties of drawing valid conclusions. The simulation exercises involving what became known as the modifiable areal unit problem (MAUP: Openshaw 1977; Openshaw and Alvanides, Chapter 18) were further examples of time-consuming procedures which significantly constrained evaluation not only of the technical issues but also of the underlying substantive concerns (Wrigley 1995).[2]

In each of these, GIS technology has increased the efficiency of research endeavours many-fold,

illustrated for the first topic by Openshaw's proto-GIS work on the location of nuclear facilities (Openshaw 1986) and his 'geographical analysis machine' for the evaluation of significant clusters of disease outbreaks (Openshaw et al 1987). Their development does not eliminate decision-making regarding technical issues, as illustrated by the concerns of GIS researchers regarding how to integrate datasets compiled on incommensurate spatial frameworks (indeed in some ways they may extend them), but they make experiments much more easy to conduct.

In addition to improving the efficiency and effectiveness of existing research approaches, technological advances open opportunities for work in previously-closed areas. This is exemplified by 3-dimensional modelling of landscapes (Hutchinson and Gallant, Chapter 9) – for which the proto-technology (the Harvard mapping package) was extremely limited and precluded interactive experiments involving, for example, altering perspective and changing various components of a map format.

The introduction of GIS to geographical research thus provided new exemplars for the conduct of original investigations and assisted progress in spatial science approaches to both human and physical geography: use of the form and machine metaphors was significantly enhanced. These changes occurred during a period of very considerable debate over the use to which science should be put and the priorities over investment in its enhancement. In the early–mid 1980s, for example, the funding of new staff positions in British Universities under a competitive 'new blood' scheme focused substantially on investments in 'enabling technologies': of the 11.5 posts allocated to Departments of Geography, five were for work on remote sensing and digital mapping and three for mathematical modelling (see Smith 1985). More recently, investigations of the nature of posts vacated and filled in US university geography departments indicate a net growth in GIS specialists (Gober et al 1995; Miyares and McGlade 1994).

Of the various roles for geography-as-spatial-science, *ergon* has been promoted over *logos*, increasingly so in an economic environment which stresses the material gains to be derived from educational investment (Johnston 1995), on the grounds that successful technical applications are more likely to bring political (and thus financial) recognition for the discipline than are improvements in explanatory power. Thus established location–allocation models are used in 'spatial decision-making systems' applied to questions regarding the optimal location of service facilities (Birkin et al, Chapter 51; Clarke and Clarke 1995). Some argue that the vast amounts of data available to geographers should be dredged in an empiricist fashion rather than addressed in a theoretically-informed context (Openshaw 1989: 73, refers to 'data-driven computer modelling in an information economy'; and see Openshaw and Alvanides, Chapter 18): their expertise with spatially-referenced data allows them to 'add value' to them, thereby making that expertise more saleable (see also Rhind 1989), as exemplified by the growing field of geodemographics which uses computer-generated classifications of ecological data to identify target markets for 'niche products' (Batey and Brown 1995; Birkin 1995). Openshaw (1994: 202) has argued that a combination of three recent developments – in large-scale parallel computing, the use of artificial intelligence, and what he terms the 'GIS revolution' – is stimulating the emergence of an 'IT State' characterised by:

'large scale, benign, universal data capture covering most aspects of modern life, the computerisation of virtually all of the management and control systems on which societies and economies depend, the linkage of separate computer systems, the dissolution of technical obstacles to systems integration by the emphasis on open systems, and the increased reliance on computers at all levels and all scales for the continuation of life on Earth'.

This offers the potential of a new mode of 'computational geography ... sufficiently broadly defined, generic in its technology, and flexible in philosophical outlook so as to encompass most, if not all, areas of human and physical geography' (208–9; see also the discussions in Openshaw and Alvanides, Chapter 18 and Longley et al, Chapter 72). Despite that claim, however, Openshaw has yet to convince many (mainly human) geographers committed to organism and arena, poesis, and paideia that GIS has much to offer them and their research agenda.

5 GIS AND GEOGRAPHY AS SOCIAL THEORY

The division within geography between spatial scientists and social theorists, the association of GIS with the former, and the intensity of some of the debate between the two camps (including over GIS: see Johnston 1997) have largely blocked substantial exploration of the potential for GIS (and the 'IT State' more generally) within the latter camp. Gilbert (1995) argued that:

'The use of computers in geography has been stunted by identification with ... GIS, by both its proponents and critics. By and large geographers have come to regard computers as analytical machines, and have ignored their growing potential as a distinctively new mode of expression.'

Computers are much more than ever-faster and more-powerful calculating machines, but within geography the continuity of personnel from spatial science into GIS has created a barrier to this realisation among social theorists, despite the increasingly widespread use of multi- and hyper-media within the humanities.

This has come about despite the early use of a proto-GIS in one of the pioneer hypermedia educational tools – the BBC Domesday project of 1986 (Openshaw et al 1986). This integrated textual and visual material in novel ways, since very significantly enhanced. Spatial scientists increasingly employ hypermedia in decision-support systems, whereby a range of digitally-coded spatial data (remotely-sensed images, maps, other images, video etc.) is combined with textual and other material to aid a range of projects (Shiffer, Chapter 52; Raper and Livingstone 1995). Such usages have been subject to substantial critiques (as in several of the contributions to Pickles 1995) and the role of GIS in creating new images of the world is increasingly appreciated (Roberts and Schein 1995),[3] but the technology's positive potential has been submerged under the weight of this (usually valid) assessment of likely negative impacts.

A central argument of geographers attracted to the postmodernist disciplinary matrix is its emphasis on variety in both space and time that cannot be readily captured by conventional literary forms and styles – contemporaneity is extremely difficult to express (see the attempt by Pile and Rose 1992). Hypermedia make this much more possible, allowing what Gregory (1989) refers to as alternative textual strategies 'which do not attempt to reduce differences and fragmentations to a single overarching account, particularly where there is a concern to convey the spatial simultaneity of different experiences' (Gilbert 1995: 8) which allow 'changes in perspective, jump-cuts and cross-cuts between scenes, dislocations of chronology and composition, commentaries on the construction of the text by author and reader, and so on'. (See also Crang 1992, on polyphony.) The ability to integrate textual and other materials (including moving images with sound) would enable exploration of many of the issues raised by adherents to the postmodernity approach (as well as, more prosaically, allowing readers to refer to illustrative material alongside the relevant text, and to access cited material without relying on the – potentially biased – selection made by another author; they need not depend on the world as represented to them by someone else, but can access the world directly themselves).

6 CONCLUSION

There have been many changes in the practice of geography over recent decades, but the discipline has sustained its core concern with spatial variations in the nature of and the interactions among environment, space, and place. Four separate world-views, each with a number of associated disciplinary matrices and their paradigm exemplars, have been called on to address these concerns – with four separate 'applied goals'. These have been reduced in this brief discussion to just two major world-views – spatial science and social theory – with their associated applications.

Over the last decade, GIS has been associated with the former world-view – largely because of its perceived strong links with the technical apparatus and empiricist/positivist aims of spatial science. It has both increased the efficiency and efficacy of established research directions therein and opened up other possibilities. By contrast, applications within geography-as-social-theory have been few, and its potential within hypermedia and other applications largely ignored; the challenges thus remain massive and should be extremely fruitful.

Endnotes

1 The hexagons were not regular, because of an error by one of the researchers (personnel information).

2 As in several other areas, Stan Openshaw's creative abilities as computer programmer partially overcame these (Openshaw and Taylor 1979).

3 As exemplified by the NCGIA's Initiative 19 – 'GIS and society: the social implications of how people, space and environment and represented in GIS'. This addressing the

forms of representation adopted and rejected in GIS applications, the level of empowerment which user groups obtain, the ethical and regulatory issues raised by GIS applications, and the potential for using GIS in democratic resolutions of social and environmental conflicts.

References

Barnes T J 1996 *Logics of dislocation: models, metaphors, and meanings of economic space.* New York, Guilford Press

Barnes T J, Duncan J S 1992 Introduction: writing worlds. In Barnes T J, Duncan J S (eds) *Writing worlds: discourse, text & metaphor in the representation of landscape.* London, Routledge: 1–17

Batey P, Brown P 1995. In Longley P A, Clarke G (eds) *GIS for business and service planning.* Cambridge (UK), GeoInformation International

Bennett R J 1989 Whither models and geography in a post-welfarist world. In MacMillan W (ed.) *Remodelling geography.* Oxford, Basil Blackwell: 373–90

Bennett R J, Chorley R J 1978 *Environmental systems: philosophy, analysis and control.* London, Methuen

Birkin M 1995 Customer targeting, geodemographics, and life-style approaches. In Longley P A, Clarke G (eds) *GIS for business and service planning.* Cambridge (UK), GeoInformation International

Buttimer A 1993 *Geography and the human spirit.* Baltimore, Johns Hopkins University Press

Chorley R J, Kennedy B A 1971 *Physical geography: a systems approach.* Englewood Cliffs, Prentice-Hall

Clarke M, Clarke G 1995 The development and benefits of customised spatial decision support systems. In Longley P A, Clarke G (eds) *GIS for business and service planning.* Cambridge (UK), GeoInformation International: 227–46

Crang P 1992 The politics of polyphony: reconfigurations in geographical authority. *Environment and Planning D: Society and Space* 10: 527–49

Gilbert D 1995 Between two cultures: geography, computing, and the humanities. *Ecumene* 2: 1–13

Gober P A, Glasmeier A K, Goodman J M, Plane D A, Stafford H A, Wood J S 1995 Employment trends in geography. *The Professional Geographer* 47: 317–46

Gregory D 1989 Areal differentiation and postmodern human geography. In Gregory D, Walford R (eds) *Horizons in human geography.* London, Macmillan: 67–96

Gregory D 1994 *Geographical imaginations.* Oxford, Blackwell

Gregory K J 1985 *The nature of physical geography.* London, Edward Arnold

Hägerstrand T 1968 *Innovation diffusion as a spatial process.* Chicago, University of Chicago Press

Haggett P, Cliff A D, Frey A E 1977 *Locational analysis in human geography,* 2nd edition. London, Edward Arnold

Haggett P 1965 *Locational analysis in human geography.* London, Edward Arnold

Haggett P, Chorley R J 1969 *Network models in geography*. London, Edward Arnold

Haines-Young R, Petch J R 1985 *Physical geography: its nature and methods*. London, Harper and Row

Hartshorne R 1939 *The nature of geography: a critical survey of current thought in the light of the past*. Association of American Geographers: 21, 462–3

Hartshorne R 1959 *Perspective on the nature of geography*. Chicago, Rand McNally/London, John Murray

Harvey D 1982 *The limits to capital*. Oxford, Basil Blackwell

Johnston R J 1989 Philosophy, ideology, and geography. In Gregory D, Walford R (eds) *Horizons in human geography*. London, Macmillan: 48–66

Johnston R J 1995 The business of British geography. In Cliff A D, Gould P R, Hoare A G, Thrift N J (eds) *Diffusing geography: essays for Peter Haggett*. Oxford, Blackwell: 317–41

Johnston R J 1997 *Geography and geographers: Anglo-American human geography since 1945*, 5th edition. London, Edward Arnold

Johnston R J, Gregory S 1994 The United Kingdom. In Johnston R J, Claval P (eds) *Geography since the Second World War, an international survey*. London, Croom Helm: 107–31

Kuhn T S 1962 *The structure of scientific revolutions*. Chicago, University of Chicago Press.

Livingstone D N 1992 *The geographical tradition: episodes in the history of a contested enterprise*. Oxford, Blackwell

Marcus M G 1979 Coming full circle: physical geography in the twentieth century. *Annals of the Association of American Geographers* 69: 521–32

McDowell L 1995 Understanding diversity: the problem of/for theory. In Johnston R J, Taylor P J, Watts M J (eds) *Geographies of global change: remapping the world in the late twentieth century*. Oxford, Blackwell: 280–94

Miyares I M, McGlade M S 1994 Specialisation in 'Jobs in Geography' 1980–1993. *The Professional Geographer* 46: 170–7

Morrill R L 1965a *Migration and the growth of urban settlement*. Lund, C W K Gleerup

Morrill R L 1965b The negro ghetto. *The Geographical Review* 55: 339–61

Openshaw S 1977 A geographical study of scale and aggregation problems in region-building, partitioning, and spatial modelling. *Transactions of the Institute of British Geographers* NS2: 459–72

Openshaw S 1986 *Nuclear power: siting and safety*. London, Routledge

Openshaw S 1989 Computer modelling in human geography. In MacMillan W (ed.) *Remodelling geography*. Oxford, Blackwell: 70–88

Openshaw S 1994 Computational human geography: exploring the geocyberspace. *The Leeds Review* 37: 201–20

Openshaw S 1996b GIS and society: a lot of fuss about very little that matters and not enough about that which does! In Harris T, Weiner D (eds) *GIS and society: the social*

implications of how people, space and environment are represented in GIS. Scientific report for the Initiative–19 Specialist Meeting, 2–5 March 1996, Koinia Research Center, South Haven, NCGIA Technical Report 96–7: D54–D58. Santa Barbara, NCGIA

Openshaw S, Taylor P J 1979 A million or so correlation coefficients: three experiments on the modifiable areal unit problem. In Bennett R J, Thrift N J, Wrigley N (eds) *Statistical applications in the spatial sciences*. London, Pion

Openshaw S, Wymer C, Charlton M 1986 A geographical information and mapping system for the BBC Domesday optical discs. *Transaction of the Institute of British Geographers* 11: 296–304

Openshaw S, Charlton M, Wymer C, Craft A W 1987 A Mark I geographical analysis machine for the automated analysis of point datasets. *International Journal of Geographical Information Systems* 1: 335–58

Pickles J (ed) 1995 *Ground truth: the social implications of geographic information systems*. New York, Guilford Press

Pile S, Rose G 1992 All or nothing? Politics and critique in the modernism–postmodernism debate. *Environment and Planning D: Society and Space* 10: 123–36

Raper J, Livingstone D 1995 The development of a spatial data explorer for an environmental hyperdocument. *Environment and Planning B: Planning and Design* 22: 679–87

Ravenhill W L D 1955 The settlement of Cornwall during the Celtic period. *Geography* 40: 237–48

Rhind D W 1989 Computing, academic geography, and the world outside. In MacMillan W (ed.) *Remodelling geography*. Oxford, Blackwell: 177–90

Roberts S M, Schein R H 1995 Earth shattering: global imagery and GIS. In Pickles J (ed.) *Ground truth: the social implications of geographic information systems*. New York, Guilford Press: 171–95

Robinson A H, Lindberg J B, Brinkmann L W 1961 A correlation and regression analysis applied to rural farm densities in the Great Plains. *Annals of the Association of American Geographers* 51: 211–21

Robinson V B 1981. In Jackson P, Smith S J (eds) *Social interaction and ethnic segregation*. London, Academic Press

Sayer A 1992 *Method in social science: a realist approach*, 2nd edition. London, Routledge

Sheppard E S 1995 Dissenting from spatial analysis. *Urban Geography* 16: 283–303

Smith D M 1985 The 'new blood' scheme and its application to geography. *Area* 17: 237–43

Taylor P J 1985 The value of a geographical perspective. In Johnston R J (ed.) *The future of geography*. London, Methuen: 92–110

Woldenberg M J, Berry B J L 1967 Rivers and central places: analogous systems? *Journal of Regional Science* 7: 129–40

Wrigley N 1995 Revisiting the modifiable areal unit problem and the ecological fallacy. In Cliff A D, Gould P R, Hoare A G, Thrift N J (eds) *Diffusing geography: essays for Peter Haggett*. Oxford, Blackwell: 49–71

4

Arguments, debates, and dialogues: the GIS–social theory debate and the concern for alternatives

J PICKLES

The chapter provides an historical overview of the emergence of the debate in geography about the disciplinary and social implications of GIS: its theory of science, technological capacities, and social uses. It describes the epistemological, methodological and at times political positions that have guided both practitioners and theorists of GIS on the one hand, and those who have expressed concern about these positions and the social implications on the other. The chapter charts these debates from opposition arguments through lively debates to recent critical engagements and joint projects between scholars of GIS and social theory. The chapter ends with discussion of Research Initiative 19 of the US National Center for Geographic Information and Analysis (NCGIA) on 'GIS and Society' and some of its contributions.

1 GIS AND SOCIETY

In recent years GIS practitioners have begun to argue for the importance of building a more flexible, open, and theoretical science of geographic information systems and geographic information – a geographic information science (Goodchild 1992, 1993, 1995; Openshaw 1991, 1992, 1996; Wright et al 1997). This theoretical turn has emerged as GIS itself has changed from an enterprise involving the development and testing of software and hardware, to the application of GIS and the study of data structures and visualisation techniques, to a field that has become so generalised in everyday life and in academic research that the specific role of any single discipline – especially one with a special relationship to GIS (geography) has to be rethought (Pickles 1997; Wright et al 1997). This chapter maps out the parallel evolution of responses to these phases of GIS development in geography, and geographers' attempts to come to grips with the changing possibilities and problems that GIS has brought to the discipline and the wider society (see also Forer and Unwin, Chapter 54; Martin, Chapter 6).

Specifically, the chapter seeks to locate the GIS social theory debate in geography (and their respective claims to method, science, and

knowledge) in terms of a decade of changing technological and institutional ensembles, discourses, and practices which have brought about different responses and forms of engagement. We seek to capture something of the dynamism in the debate that occurred in the transition from the mid 1980s to the mid 1990s. This debate ranged from GIS as a research tool and scholarly practice (and the epistemological grounds on which these battles were fought), to debate about its fundamental assumptions and transformative capacities, to dialogue about alternative pathways for a technology that is increasingly realising both its utopian and dystopian possibilities.

The chapter outlines briefly the nature of the opposition arguments that emerged as a result of the disciplinary impacts wrought by GIS in the 1980s. It then shows how these opposition arguments – while they still continued in some quarters – gradually began to take the form of a constructive debate about the real material and intellectual effects of GIS. We go on to show how this debate is currently leading to experiments in dialogue among individuals and groups with quite distinct goals and perhaps different conceptions of GIS as technology, practice, and body of ideas. A different understanding of the possibilities and constraints of GIS as a tool, and of the study of GIS as a social practice, emerges from these engagements (Gilbert 1995).

The primary goals of this kind of work should be spelled out briefly, given the suspicion about the critique that has emerged within the field:

- to contribute to a theory of GIS which is neither technical nor instrumental, but locates GIS as an object, set of institutions, discourses, and practices that have disciplinary and societal effects;
- to show how these disciplinary and societal effects operate;
- to push against the limits of GIS and its unacknowledged conditions and unintended consequences of development and practice (e.g. corporate influence, epistemological assumptions, and understanding of appropriate applications);
- to ask whether GIS could have been different, or in what ways it may be made different in the future.

2 GIS AND GEOGRAPHY: NEW SCIENCE OR OLD WINE?

GIS is not only big business, it is becoming bigger and bigger business with every passing year. In the 1980s and 1990s GIS and related spatial data handling and imaging systems became central elements in the restructuring of economic activity, the modernisation of the state, and the administration of social life by public and private organisations (Cowen 1995). In the 1960s most geographers would probably have welcomed such changes and lauded as progressive the rationalisation of planning. In the 1990s these matters have given rise to deep divisions within the discipline about the role and function of social engineering and the information revolution that makes it possible in new forms. Although a substantial part of the discipline cannot understand why the geographic profession displays such distrust of the developments in GIS and why it remains sceptical about motives, potential value, and political consequences of adoption, another part of the discipline cannot understand why these questions have not yet been asked within the GIS community, how practitioners cannot see the problems raised by corporate control, proprietary systems, limitations on available data, and the uses to which GIS has been put in recent years. For some the revolution in spatial data processing and digital

imaging systems offers new opportunities for constructing 'informed' societies and pursuing rational and efficient social planning; for others the new systems of knowledge engineering and social engineering raise serious questions about freedom, civil society, and democratic practice (for further discussion see Curry 1994, 1995, 1996, 1997; Goss 1995a, 1995b; Harris et al 1995; Lake 1993; Miller 1995; Pickles 1991, 1992, 1993, 1995, 1997; Sheppard 1995; Sui 1994). Thus, as GIS has become a more significant element in restructuring public and private life, it becomes crucial to ask what impacts these technologies and applications have on the ways in which people interact with one another.

Until recently, discussions of the social impacts of GIS have been limited mainly to an internal analysis of technique and methodology: improving accuracy, extending capabilities, and widening the scope of applications that are possible. Little attention has been given to the broader discussions in geography about the interests that influence scientific research, the socially constituted nature of objects, categories, and concepts, the gendering of science, or the differing commitments of empiricist, hermeneutic, and critical epistemologies (Johnston, Chapter 2; Gregory 1978, 1994). Instead, much of the discussion has taken the form of a theoretical advocacy and an almost evangelical celebration of the possibilities offered by GIS to save geography – from its marginal economic position in universities, from its weak professional status in areas of public policy, from its underdeveloped technical capacities in applied fields, and from its humpty-dumpty like fragmentation in the discipline (Abler 1993; Openshaw 1991, 1992). In each of these domains GIS, it is claimed, offers rigorous science, useful technique, and universal possibilities for application. An objectivist epistemology and a pragmatic politics combine to reject any broader theorising of the consequences of this form of knowledge production and management.

Other geographers sometimes disagreed. Jordan's 1988 Presidential column in the newsletter of the Association of American Geographers (Jordan 1988) signalled the first reaction on the part of the old guard in the discipline to what was perceived as the pretension of GIS and its claim to intellectual standing: GIS was, in his view, merely a technical field without intellectual vigour or promise. Moreover, the need for large investments in capital equipment, personnel, training, and recurring costs for maintenance and upgrade was not matched initially by quality output and clear results. Indeed, the points of contact with GIS for most geographers in the 1980s were requests for budget and faculty lines on the one hand, and faint, Cubist- and Futurist-like map images on the other hand. As a result, many Realists in the discipline greeted the emergence of GIS with quiet resistance and the knowing scepticism of the bourgeois critic, comfortable in the assurance that the fad would pass.

One unfortunate and unnecessary side-effect of such opposition positions has been the tendency of one side or the other to reject as 'unreasonable' the arguments of the other. The result has been a closing down of constructive and open debate on both sides, and the emergence of 'cultures of indifference' on both sides. Since the personal, institutional, and social stakes are high this is not unexpected, but what was lost in this opposition was any serious debate with some important issues on each side. Where a fuller engagement with the ideas and claims of each has occurred, the result has been an 'energising' of both communities and an opening of new avenues of research.

3 EPISTEMOLOGICAL CRITIQUE: DEBATING THE ASSUMPTIONS

The first serious engagements between GIS and social theory occurred over issues related to the politics of knowledge and the social impacts of use (Lake 1993; Miller 1995; Pickles 1991, 1995; Sheppard 1995; Sui 1994). In his trenchant critique of GIS as the new imperialist geography, Taylor (1990) suggested that GIS emerged as a two-part strategy on the part of unreconstructed 'quantifiers' who had 'bypassed' the critiques levied against the empiricism of spatial analysis, and at the same time captured the rhetorical ground of a progressive modernism by readily accepting the switch from knowledge to information:

'Knowledge is about ideas, about putting ideas together into integrated systems of thought we call disciplines. Information is about facts, about separating out a particular feature of a situation and recording it as an autonomous observation . . . The positivist's revenge has been to retreat to information and leave their knowledge problems – and their opponents – stranded on a foreign shore. But the result has been a return of the very worst sort of positivism, a most naive empiricism.' (Taylor 1990: 211–212)

In this (re)turn the geographical is defined as the study of anything that is spatial:

'GIS is a technological package that can treat any systematic collection of facts that are individually identified spatially. These facts may be medical statistics, remote-sensing images, crime files, land-use data, population registers or whatever. In terms of the package, spatial patterns can be produced irrespective of what the information is about . . . Such quantifiers can produce a maverick geography dealing with crime one week, bronchitis the next, and so on.' (Taylor 1990: 212)

The colonising aspirations of such claims are, for Taylor, transparent. But many practitioners of GIS saw these claims as exaggerated at best and false at worst, or, as Openshaw (1991) argued they represent reductionist assertions and derogatory and confrontational language; 'knockabout stuff' that emerges from a reactionary desire to protect a particular system of order and power. Thus, for Openshaw the crisis to which Taylor points is redefined as 'contrived' and should be replaced by a notion of 'creative tensions' between at times complementary, at times competing, but equally productive intellectual projects. In place of any narrow delimitation of the possibilities of GIS, Openshaw (1991) offered an expansive vision of emerging GIS practice:

'A geographer of the impending new order may well be able to analyse river networks on Mars on Monday, study cancer in Bristol on Tuesday, map the underclass of London on Wednesday, analyse groundwater flow in the Amazon basin on Friday. What of it? Indeed, this is only the beginning.' (Openshaw 1991: 624)

This new order geography needs GIS in order to 'put the pieces of geography back together again to form a coherent scientific discipline':

'It would appear then that GIS can provide an information system domain within which virtually all of geography can be performed. GIS would emphasise an holistic view of geography that is broad enough to encompass nearly all geographers and all of geography. At the same time it would offer a means of creating a new scientific look to geography, and confer upon the subject a degree of currency and relevancy that has, arguably, been missing.' (Openshaw 1991: 626)

In this view, GIS has an overreaching technology and approach broad enough to allow any geographer to pursue his or her research questions: GIS offers the epistemological and methodological flexibility to the creative researcher to be adapted to any practical circumstance.

The divide is not, in this sense, between GIS and social theory, but between a social theory and notion of science rooted in empiricism (in which theory is that which accounts for the outcome of model testing) and social theory in which theory is the precondition for any understanding and analysis in the first place.

For these reasons, several commentators have argued strongly against the particular view of the discipline, of science, and of research practice and application that ties the development of GIS to the 'resurrection' of a rational model of planning and a positivist epistemology (Lake 1993; Sui 1994):

'... the unrelenting embrace of the rational model by planning and applied geography is not adequately described merely in terms of the tenacity and inertia of convenient and familiar practices. The rational model has been actively resurrected and rehabilitated by the ascendance of GIS to a position near to or at the core of both planning and geography.' (Lake 1993: 404)

In the 1980s, human geography developed strong critiques of the reductionist ontology of spatialism and turned to questions of contextual knowledge, contingency and necessity, society, space, and Nature, the (social/political/gendered) construction of space, and the production of scale, each of which in various ways problematised aspects of Cartesian science and the ontology of spatial analysis. These approaches questioned the overemphasis on pattern, challenged geographers to rethink the meaning of space, problematised the dominance of natural science method in the study of social phenomena, and raised questions about the underlying ontology of objects, location, and application on which spatial analysis was predicated. Yet, in his 1993 review of the field, Lake found few publications on the part of GIS proponents which consider these epistemological, political and ethical critiques of positivism, or any serious engagement with what he terms the 'fundamental disjuncture growing at the core of the disciplines'.

By the decade of the 1990s, social theorists within the discipline began to take aim at what they saw as the transformative capacities of GIS, both in disciplinary and broader social terms. The author's own 1991 essay on 'The Surveillant Society', Gregory's (1994) claim that GIS positivists represented the 'new Victorians', and Smith's (1992) charge that the war against Iraq – the Gulf War – represented the first GIS war, incensed many practitioners and theorists of GIS. How could these neophytes and outsiders levy such charges, particularly against the only part of the discipline that really exercised rigour in its work and power in regard to other disciplines and funding agencies?

Such concern turned to outrage as more GIS practitioners interpreted claims about GIS and its origins in surveillance and battlefield logistical needs in the military (Pickles 1991) as a direct attack on their own credibility and commitments, and 'GIS über Alles' (the title of the first section in *Real wars, theory wars*: Smith 1992) and the purposeful ambiguity in the first two sentences: 'The war against Iraq in 1990–91 was the first full-scale GIS war. It put geography on the public agenda in a palpable if unpalatable way as it claimed an estimated 200 000 Iraqi lives' as suggesting that GIS, fascism, and imperialistic warmongering were somehow synonymous.

There was a double irony here. First, in the *declarations of war* against social theory and the expressions of the need to *mobilise in defence* of GIS against this onslaught that ensued in various gatherings of geographers. Second, that these responses occurred at the very time when, for example, Dangermond was bringing Ralph Nader to speak to the ESRI Users' Conference to argue for a 'vigorous GIS' – that is, one that recognised its current embeddedness within the institutions of government, military, and corporate interests, and instead sought to foster democratic access and public participation. This at a time when Openshaw (1991, 1992) was calling for a more open, *flexible* GIS, and when Goodchild (1992) was arguing for the need for a geographical information science that would address the impacts, as well as the possibilities, of the use of GIS. In one sense, social theorists and theorists of GIS had reached similar conclusions but by different paths.

In essence, the speed at which the technology was changing, the breadth of adoption and use, and the depth of the impacts of contemporary GIS had changed the terrain on which the discussion would occur. The opposition logics of the 1980s were no longer practically helpful. User-friendly software had increased the number of GIS users. GIS had grown institutionally strong and – with its own conferences, journals, and funding sources – no longer took the arguments of disciplinary theorists seriously. At the same time, few critics followed sufficiently closely the emerging capacities and the new applications to understand the changes they wrought. In particular, few understood that while the instrumental logics and positivist justifications they attacked were being ever more deeply ingrained, they were also being fundamentally challenged by new practices and notions of space, object, and science that did not fit within such positivist frameworks: GIS itself was beginning to experience contradictions in its own claims and practices.

Despite last ditch efforts on the part of the traditionalists (Jordan 1988), GIS could not be wished away, nor could the hard resource decisions be avoided by departments and individuals in their research and teaching. When the Chancellor of the University and Manager of the Office of Facilities Planning both pull up GIS for the day-to-day management of their campus, when city planners are digitising every street in the city, when city engineers are GPSing every waterline and powerline they manage, when new forms of red-lining using GIS maps have become second nature to insurance companies, and when the US Department of Defense solves complex peace negotiations over delimiting territorial borders in Bosnia with digital terrain models and repeated flyovers for negotiators, there can be no question that geographers must take GIS seriously as a set of tools, institutions, ideas, and practices that are shaping our lives and landscapes, and that are transforming the possibilities for certain types of research in the discipline. How to ask these questions was the crucial issue.

4 GIS AND SOCIETY: DIALOGUE AND ENGAGEMENT

In response to the sterile binaries of uncritical support and outright denial, Brian Harley and the author decided (following two sessions of 'GIS and Society' held at the Annual Conference of the Association of American Geographers in 1991) to edit a book of essays that attempted to theorise the impacts of GIS in the discipline and in the wider society as a means of stimulating students to begin to think about alternatives to the rather sterile pro and con positions that dominated discussions at the time. To our

surprise, the 'idea of the book' *Ground truth* (Pickles 1995) achieved some of these goals prior to publication. This 'idea of the book' began to circulate on list servers like GIS-L, and concerns were expressed that such a book could undermine the growing position of GIS in the field: the book was somehow to be thought of as a dangerous attack on GIS.

One outgrowth of these discussions was an NCGIA-sponsored workshop 'GIS and Society'. The workshop addressed the kinds of questions that needed to be asked to understand the growing influence and social implications of GIS development and use, to consider how and in what ways such questioning might be sustained, and to investigate the possibilities for future critical engagements among GIS and social theory (Poiker 1993).

Perhaps the single issue that causes confusion in geography over the possibilities and limits of GIS use is what generally is represented as the debate about positivism, a term that has served geographers as a recurring moment for mobilisation or vilification. The concept itself stands as a signifier for something broader and it is here that the problem needs to be located. The apparent incommensurability between GIS and social theory critiques has its origin, perhaps, in how one understands the appropriate scope for inquiry (see also Johnston, Chapter 2). Most discussion of GIS operates within a very circumscribed understanding of the appropriate domain of inquiry, and this bounding of the field has variously been criticised as technicist, instrumentalist, and positivist. Social theorists have gradually broadened their own understanding of the appropriate scope within which inquiry must be situated, and currently any single social theoretic critique might operate at any one scale ranging from theories of geography, and science, to theories of society and technology (including the role of commerce, planning, and strategic thinking), to theories of modernity (including political theories of liberalism and critiques of masculinism, imperialism, and observer epistemologies) to Enlightenment thought itself. For each of these domains distinct literatures and languages have been carefully developed to enable critical thought. *Ground truth* aimed to locate discussions of GIS in a variety of these possible interpretative frameworks, and thereby to provide illustrations that might lead others to deepen the analysis of the intellectual and practical commitments and impacts of GIS.

This was also the goal for the 1995 special issue of *Cartography and Geographic Information Systems* – 'GIS and Society' – edited by Sheppard (1995). In his introduction Sheppard argued that the opposition nature of the debates occasioned by the emergence of GIS was full of heroic images and cruel caricatures, and that supporters and critics of GIS could learn from each other. Sheppard demonstrated how the origins of GIS affect the ways of thinking that can be employed. First, the dependence of GIS on digital computing (as opposed to analog computers, for example) constrains GIS by the structure and logic of the Turing machine, which employs deductive, Aristotelian logic. Second, the link between GIS and computers means that GIS is embedded in a broader set of social relations within which the computer is deployed:

'A major theme of the post-war era, in both the first and second worlds, has been extending the ability of both public and private institutions to control and organise the production and delivery of goods and services effectively. The principles of operations research as a methodology for optimally achieving well-defined goals, so effectively demonstrated in the armed services during the second world war, have been promoted as facilitating the rationality of both private enterprise operating in a free market and of public planning in a welfare or socialist state.' (Sheppard 1995: 8)

The result is that 'large institutional actors favour, and finance, those developments meeting their needs' (Sheppard 1995), and thus influence the development of computing and the directions taken by applications such as GIS. Since these large institutional actors have primarily been corporate, military, or public administration institutions, it should be no surprise that applications that favour surveillance, private sector interests, and control functions have been more common than those favouring public participation, data access, and community-defined goals. Such biases may be unproblematic for some, but for others they present a serious challenge to the possibility of a critical and rigorous science. Either way, GIS is a product of such technological and social constraints and its capacities have been influenced and delimited by these constraints.

5 GIS AND SOCIETY – NCGIA INITIATIVE 19 (I-19)

Following the Friday Harbor workshop, and in part stimulated by it, a group of interested individuals proposed that the US National Center for Geographic Information and Analysis dedicate one of its research initiatives to the issues now known as 'GIS and society'. A proposal was submitted to the NCGIA Board of Directors and approved, and a meeting of specialists convened in early 1996 (Harris and Weiner 1996). This section examines the continuing work of the initiative in some detail.

What marks the Friday Harbour and I-19 workshops as unique and important in the emerging theory of GIS is that Friday Harbour ended with, and I-19 began with, a set of assumptions that have been absent from debates about GIS until recently. Questions of origins, epistemology, data selection and data access, forms of representation, and the politics and ethics of information have generally been seen as marginal to the more technical questions of systems development and application (Martin, Chapter 6; see also Raper, Chapter 5). At these meetings they were seen as essential for any discussion of GIS and society. GIS is thus seen as a set of institutionalised systems of data handling and imaging technologies and practices situated within particular economic, political, cultural, and legal structures. They can thus be thought of as spatial data institutions (Curry 1995) and sociotechnological ensembles (Latour 1993). Understanding GIS as both a set of social practices and institutions embedded in a particular discourse is, perhaps, unique in the history of the engagement between GIS and social theory. Certainly, such social constructionist, genealogical or post-positivist theoretical frameworks have been virtually absent until recently in the debates over GIS.

Deploying such frameworks has been an important part of an emerging theory of GIS and society in which description (of the development of particular logics, systems, and uses of GIS), analysis (of the limits of access, range of diffusion, and effects of use) and critique (focused on the epistemological assumptions embedded in systems and use, conceptions of language in use, and logics and representations) are all present.

5.1 Critical social history of GIS

The written history of GIS is quite limited and few detailed case studies have appeared in print (Coppock and Rhind 1991; Goodchild 1988; Petchenik 1988). But it is vital to any critical field of inquiry that its practitioners know about the origins of the choices made and those rejected in defining and delimiting the field. In particular, it is vital that the technical, logical, and epistemological constraints on what GIS does, and the ways in which particular logics and visualisation techniques, values systems, forms of reasoning, and ways of understanding the world have been incorporated into existing GIS techniques are understood. It is equally important that practitioners and theorists understand the ways in which alternative forms of representation have been filtered out.

In the first instance, this has to do with the development paths taken within GIS and the possible alternatives that were not chosen but were available.

- Accepting that scientific knowledge is socially produced and rejecting any linear path of technical development, what were the debates and decisions leading to certain system choices and foundational logics rather than others within GIS over the past 30 years?
- Second, if alternatives were not pursued or accepted at the time, what were these and what were the conditions under which they were rejected or not pursued?
- Third, if there are always choices being made in the design and implementation of any technology and research tool, can alternative cultural and social conceptions of objects (property, land, resource relational values, historical meaning) be incorporated within GIS, and what are the actual possibilities for extending GIS to incorporate new ways of understanding the world?
- Fourth, since system and procedural choices have already been made and are now rooted in place through technical, financial, and practical inertia, what are the limits on what present-day GIS can do and what any reformed GIS might achieve?

There is a broader context that is also relevant here. This has to do with the issue of historical antecedents. GIS does not spring full blown or completely new into our world (Coppock and Rhind 1991; Goodchild 1988; Petchenik 1988). It emerges out of systems of land surveying, mapping, and data collection each with long heritages, and each having been centrally placed in the systematising and formalising of social life under capitalism. It is a constant surprise to social theorists in geography that the published histories of GIS tend to be what Livingstone (1992) referred to as 'internalist' and 'hagiographic', and do not deal with these historical antecedents, the ways in which GIS developed and diffused (who funded development, what options were considered and rejected, what institutional and intellectual linkages were forged in the development of GIS, etc.) and the patterns of production, marketing, and use that emerge in different cultures and settings. This would seem to be vitally important for any area of science in assessing the effectiveness, value, and limitation of its own technical and theoretical practices. Moreover, such questions locate the study of GIS at the heart of contemporary geographical issues (Wright et al 1997).

Recognising that GI comprises a series of institutions, discourses, and practices (as well as a set of tools) means that any theory of GIS must account for its origins and effects. In other words, GIS as a socially embedded and historically produced set of practices must account for its own history. It is to this question that the Critical History of GIS (CHGIS) Group, an activity initiated under I-19, has recently turned its attention.

Attempting to write a history of GIS that is not internalist or hagiographic, the CHGIS Project aims to bring a variety of theoretical perspectives from contemporary social theory to bear on the question of GIS as social practice. It also attempts to contextualise GIS in its social, political, and economic context, to locate GIS in terms of a broader history of science and technology than heretofore – and specifically to do so through an engagement with the systems and logics that were developed, the paths that were not taken, and the institutional linkages that provided the context for that which emerged.

5.2 Marginalised groups and the politics of access, exclusion, and control

In recent years, new technological capacities and an expansion of the scope of their application in many areas of social life have made it increasingly important to think about the ways in which the logics, systems, and representations deployed by contemporary GIS support particular types of social practice and inhibit others. What effects are GIS having? If GIS has been influenced by the demands of their developers and funders, many of them tied to large institutional and corporate interests and high-cost applications, what forms of access to information do these systems promote and deny? Specifically, how has the proliferation and dissemination of databases associated with GIS, as well as differential access to these databases, influenced the ability of different social groups to gain access to and use this information for their own purposes (see Rhind, Chapter 56)? Second, what types of knowledge and forms of reasoning are not well represented within GIS and what are the consequences of their exclusion (Onsrud 1992a, 1992b)?

A theory of GIS and society must address the impacts of these limits and impediments on groups or individuals where unequal access to software, hardware, and technical skills present real barriers to use, and seriously affect the types of outcome that result from the use of GIS in making decisions.

Differential access to databases is, clearly, becoming one of the central issues facing scholars and users of GIS and all forms of electronic data. As spatial data handling capabilities increase in power, the social impacts become more important. Geodemographic spatial data handling, for example, is already raising serious questions about privacy and access to databases (Curry, Chapter 55; 1997; Goss 1995a, 1995b). Until very recently, the primary sites at which GIS have been developed have been at national and local (in the USA, the state) level. In Britain, GIS has been used for land-use applications related to zoning, long-term planning, and the like. But the increasing availability and ease of use of GIS, accelerated by the development and deployment of global positioning systems and remote sensing systems, now constitute a powerful means of systematically tracking a wide range of natural and social phenomena, and in particular of developing monitoring systems for tracking populations (Graham 1997; Pickles 1991). The development of these systems raises a wide range of questions about the types of assumption, data, and representation that are incorporated in any GIS. Who decides which data are to be collected? Who decides how those data are collected, which categories (of race, gender, species and so on) are to be used? How will the accuracy and validity of those data be measured and guaranteed, not in the technical sense of data error, but in a political sense of data appropriateness? Finally, because state agencies are both users and regulators of software, hardware, and data, questions arise concerning the ways in which these agencies adjudicate their sometimes competing responsibilities of protecting citizens and promoting use. (See Goodchild and Longley, Chapter 40, for a discussion of the technical implications of some of these issues.) In summary, how is the balance between rights to access and rights to privacy currently being struck (Curry 1995)?

The emergence of geodemographic information systems (GDIS) as targeted marketing strategies has already pointed to the emergent dangers of the use of GIS to further the commodification of everyday life (Curry 1997; Goss 1995a, 1995b). In the case of GDIS the issues go beyond the increasing efficiency of marketing agencies to target consumers with particular tastes and purchasing habits. They involve questions about the constitution of identity. GDIS consumer profiles, are aggregate profiles based on neighbourhood level data from which individual profiles are constructed. The targeting of commercial, political, and public service materials to individuals based on neighbourhood-derived profiles in turn 'produces' new identities (in that it channels and restricts the information individuals in that neighbourhood receive). Thus, even beyond questions of access and privacy, GDIS raises fundamental questions about the ethics of using information systems in ways that presuppose (and in turn contribute to the development of) socially homogeneous neighbourhoods.

There is a basic paradox in using GIS to address issues of land-use planning of any sort. On the one hand, conflicts over the use of space typically involve competing sets of values, assumptions, and interests. Not unexpectedly, the representations incorporated in GIS models of landuse conflicts tend to reflect the views, values and interests of dominant sectors of society. Ethnic, racial, and sexual minorities whose values and interests differ from those of culturally or economically dominant groups may be doubly disadvantaged when attempts to resolve conflict involve a significant GIS component (see Fisher, Chapter 13). Not only are their interests not intrinsic to the models on which technical solutions to complex problems are based, but they may lack access to the tools used by planners and politicians in making their decisions (Aitken and Michel 1995; Lake 1993; Miller 1992; Yapa 1991).

5.3 Ways of knowing

Beyond questions of access and exclusion is a related set of issues having to do with the ways in which knowledge and information are represented. An interesting change in the thinking of geographers seems to have occurred as GIS has been applied to more and more questions of this sort. Geographical information is increasingly assumed to refer to that which is captured or could be captured by GIS. Since GIS typically assume a universal set of objectifiable and 'self-evident' components of the processes they model (Sheppard 1995), GIS representations are often based on the assumption that there is a single version of reality to be modelled, and that land-use planning and conflict resolution principally involve the discovery of the most efficient solution to this objectifiable location problem. The use of GIS in locational conflict resolution has, in one important sense, poorly served the interests of those whose viewpoints and values differ from those incorporated in GIS models. Other forms of geographical information: place-based information, local knowledge, historical memory of land-use struggles, past events etc., are being marginalised as subjective information, *doxa*, or opinion (Curry 1996, 1997; but see Fisher, Chapter 13, and Veregin, Chapter 12, for discussions of uncertainty and data quality, respectively).

One example already addressed in literature is the case of the use of GIS to revisit claims of North Americans whose lands were ceded to the government in the nineteenth century, and whose abrogated treaty rights are now a basis for re-evaluation of that land alienation process. A basic problem emerges in the fact that GIS is far better at incorporating certain types of variable than others (Fischer, Chapter 13; Poiker 1993). Clearly, the variables incorporated in GIS representations are

not always tangible: for instance, both physical forest resources and conceptual property boundaries are included in GIS databases used in adjudicating land disputes. However, intangible factors related to competing value systems are not usually present in such analyses. How factors such as emotional attachments and the sacredness of place, the role of place in creating and maintaining community, use rights versus property ownership rights, and alternative views of nature are incorporated adequately into the GIS analysis of such conflicts has a huge impact on the types of claims and decisions that can be made (Rundstrom 1991). Rundstrom (1995) has even gone so far as to ask whether decisions should be based on GIS analysis at all in cases where such calculi are not amenable to incorporation into GIS models.

It is not yet clear how any technical systems can deal with alternative knowledge systems in cross-cultural settings. Some ways of knowing are privileged in existing GIS approaches, but it is not clear how different types of knowledge and information can be included. Nor is it clear whether the apparent technical and epistemological limitations of present systems could incorporate different ways of knowing without reducing one to the other, or whether new, different system logics, configuration, and practices need to be developed. The possibilities and the difficulties involved in these efforts are well documented by Harris et al (1995) and Weiner et al (1995).

With the inclusion of locationally fuzzy knowledge many issues arise as to how the multi-objective goals, based on multiple criteria, and using spatially imprecise and possibly conflicting data might actually achieve what is assumed to be consensus decision-making. Perhaps one reason why GIS has achieved such astounding 'success' to date in decision-making support roles is that it is based on only one seemingly non-contradictory perception of reality. Collaborative spatial decision-making is a complex issue even among participants with similar world views and knowledge. In the absence of this commonality the difficulties are qualitatively greater. But these difficulties are also opportunities; they arise as such partly because of new technical capacities for handling large datasets and displaying and disseminating spatial images. What a 'pluralistic GIS' (one containing multiple views of resource value, potentially fuzzy, and conflicting information) would look like and what it would imply for the ways in which GIS can be used in collaborative decision-making remain open questions.

5.4 Public participation and GIS-2

If it is the case that the systems and logics that underpin much GIS emerged in response to the requirements and influence of large institutional supporters (be they public or private), then issues of surveillance, ownership, and control raise questions about the possibility of access, participation, and community-based involvement in GIS. This is even more pressing if one is not willing to reduce such issues of access and participation to the logics already present in existing systems. If GIS has emerged in its present form as a result of influences from a variety of financial and institutional interests, and if it does operate (through its technical demands, cost structure, types of data, and differential access) as a top-down technology and practice, can it be democratised? The democratisation of GIS means that the emerging possibilities of the technology must be considered. What must also be considered is how the types of systems and logics emerged within contemporary GIS and whether they can be changed.

If these forms of embeddedness do function as real constraints on public participation, can alternative social relations, ways of knowing, and marginalised groups be represented or given access in ways that do not reduce their own positions and logics to those of current GIS practice? How can the knowledge, needs, desires, and hopes of marginalised social groups be represented adequately as input to a decision-making process, and what are the possibilities and limitations of GIS as a way of encoding and using such representations?

If contemporary GIS can be thought of as predicated on the computerisation of the cartographic industry (GIS-1), can alternatives (GIS-2) be thought of which might range from 'knowledge creation environments' (Goodchild 1995) to public access centres and which address these issues? Also, how should people, space, and nature be represented? Who should have the right to speak on the nature of the representations that are created (Latour 1993)? What criteria might govern the emergence of such a GIS-2?

This question was raised and discussed at Friday Harbor and has become a central focus of I-19 research. The issue of system design is being addressed in the public participation project at the University of Maine, headed by Schroeder and Onsrud. Questions of legal and ethical conditions that enable and prevent intrusion are being addressed in a joint project between UCLA and the University of Minnesota – specifically by Curry, Sheppard, and Miller. The nature of geographical information in situations involving social conflict, and its relationship to the present capacities of GIS, is being addressed variously in projects in Minnesota and UCLA, and at the University of Kentucky and West Virginia University.

These efforts are aimed at asking what GIS-2 might look like. It would certainly have to be cheaper, more accessible, and sufficiently flexible to be of use to a wider range of users. But it would also have to address public concerns about privacy and access to information. Such a public GIS would have to guard against the reduction of multiple ways of knowing to a single logic and the premature resolution of differences. Instead, it will have to develop ways to represent different conceptions of space or Nature, and preserve contradiction, inconsistency, and disputes. Finally, a more flexible and accessible GIS-2 needs to be capable of integrating all data components, such as WWW, data archives, parallel and counter texts in diverse media, standard maps and datasets, and sketch map and field notes, all from one interface (Harris and Weiner 1996; *http://ncgia.spatial.maine.edu/ppgis/ppgishom.html*).

6 CONCLUSIONS

What are the results of the engagements described above? In the first place, these are early days in each of these projects and concrete research results are limited. Several conceptual advances have, however, been made.

- The relationship between the speed of developments and depth of the impacts of GIS technology, theory, and practice can now be seen in the context of a field that has been reticent to acknowledge the conditions of its own production, that has been lax in building its own archive, and that has by and large failed to develop sustained and detailed critical reflection upon its own practices. The discussion around GIS until the late 1980s remained focused largely on technical issues, unreflective in nature, and theoretical only insofar as theory referred to either empirical findings or internal technical concerns (but see, for example, Chrisman 1987a, 1991a; Coppock and Rhind 1991; and Goodchild 1995).

- The debate thus far has broadened discussion of GIS practice so that it now encompasses the social impacts of GIS. This is particularly important as new cyberspaces emerge and new forms of geographical information are finding a home through which important reconfigurations of material life are being affected.
- The need to think of GIS as a social object with its own institutional contexts, discourses, and practices has been demonstrated. This is not, however, an argument for a form of exceptionalism or professionalising of GIS as a discipline. Instead it calls for the necessity of locating those institutions, discourses, and practices in terms of broader debates in social theory about science/technology/society, theories of science, and the political economy of informatics on the one hand, and the recontextualising of GIS practice within the broader debates about geography on the other hand.
- The engagement has rectified one important absence within GIS communities (the legitimacy of 'GIS and society' questions and the availability of sites and groups among whom such discussion can continue).

The emergence of critical dialogue between GIS and social theory offers great promise for the emergence of a critical GIS aware of its own effects and striving to open its capacities to the needs, questions, and ways of knowing of broader and different 'publics'.

References

Abler R F 1993 Everything in its place: GPS, GIS, and geography in the 1990s. *The Professional Geographer* 45: 131–9

Aitken S C, Michel S M 1995 Who contrives the 'real' in GIS? Geographic information, planning, and critical theory. *Cartography and Geographic Information Systems* 22: 17–29

Chrisman N R 1987a Directions for research in GIS. *Proceedings, IGIS Symposium* 1: 101–12

Chrisman N R 1991a A geography of geographic information: placing GIS in cultural and historical context. *Mimeo*

Coppock J T, Rhind D W 1991 The history of GIS. In Maguire D J, Goodchild M F, Rhind D W (eds) *Geographical information systems: principles and applications.* Harlow, Longman/New York, John Wiley & Sons Inc. Vol 1: 21–43

Cowen D J 1995 The importance of GIS for the average person. *GIS in government: the federal perspective 1994.* Proceedings First Federal Geographic Technology Conference, 26–28 September 1994. Washington DC, GIS World Inc: 7–11

Curry M R 1994 Images, practice, and the hidden impacts of GIS. *Progress in Human Geography* 18: 441–59

Curry M R 1995 Rethinking rights and responsibilities in GIS: beyond the power of imagery. *Cartography and Geographic Information Systems* 22: 58–69

Curry M R 1996a Digital people, digital places: rethinking privacy in a world of geographic information. Paper presented at the conference on Technological Assaults on Privacy, Rochester Institute of Technology, Rochester (USA), 18–19 April

Curry M R 1997a Geodemographics and the end of the private realm. *Annals of the Association of American Geographers* 87: 681–99

Gilbert D 1995 Between two cultures: geography, computing, and the humanities. *Ecumene* 2: 1–13

Goodchild M F 1988 Stepping over the line: technological constraints and the new cartography. *The American Cartographer* (special issue on the history of GIS) 15: 311–22

Goodchild M F 1992 Geographical information science. *International Journal of Geographical Information Systems* 6: 31–45

Goodchild M F 1993 Ten years ahead: Dobson's automated geography in 1993. *The Professional Geographer* 45: 444–5

Goodchild M F 1995a GIS and geographical research. In Pickles J (ed.) *Ground truth: the social implications of GIS.* New York, Guilford Press: 31–50

Goss J 1995a Marketing the new marketing: the strategic discourse of geodemographic information systems. In Pickles J (ed.) *Ground truth: the social implications of GIS.* New York, Guilford Press: 130–70

Goss J 1995b 'We know who you are and we know where you live': the instrumental rationality of geodemographic information systems. *Economic Geography* 71: 171–98

Graham S 1997 Surveillant simulation and the city: telematics and the new urban control revolution. *Environment and Planning D: Society and Space*

Gregory D 1978 *Ideology, science, and human geography.* London, Hutchinson

Gregory D 1994 *Geographical imaginations.* Cambridge (USA), Blackwell

Harris T M, Weiner D 1996 *GIS and society: the social implications of how people, space, and environment are represented in GIS.* Scientific report for the Initiative-19 Specialist Meeting, 2–5 March, Koininia Retreat Center, South Haven. NCGIA Technical Report 96–7. Santa Barbara, NCGIA

Harris T M, Weiner D, Warner T A, Levin R 1995 Pursuing social goals through participatory GIS. In Pickles J (ed.) *Ground truth: the social implications of GIS.* New York, Guilford Press: 196–222

Jordan T 1988 The intellectual core: President's column. *AAG Newsletter* 23: 1

Lake R W 1993 Planning and applied geography: positivism, ethics, and GIS. *Progress in Human Geography* 17: 404–13

Latour B 1993 *We have never been modern.* Cambridge (USA), Harvard University Press

Livingstone D N 1992a *The geographical imagination*. Cambridge (USA), Blackwell

Miller B 1992 Collective action and rational choice: place, community, and the limits of individual self-interest. *Economic Geography* 68: 22–42

Miller B 1995 Beyond method, beyond ethics: integrating social theory into GIS and GIS into social theory. *Cartography and Geographic Information Systems* 22: 98–103

Onsrud H J 1992a In support of open access for publicly held geographic information. *GIS Law* 1: 3–6

Onsrud H J 1992b In support of cost recovery for publicly held geographic information. *GIS Law* 1: 1–7

Openshaw S 1991a A view on the GIS crisis in geography, or using GIS to put Humpty Dumpty back together again. *Environment and Planning A* 23: 621–8

Openshaw S 1992 Further thoughts on geography and GIS: a reply. *Environment and Planning A* 24: 463–6

Openshaw S 1996b GIS and society: a lot of fuss about very little that matters and not enough about that which does! In Harris T, Weiner D (eds) *GIS and society: the social implications of how people, space, and environment are represented in GIS*. Scientific report for the Initiative-19 Specialist Meeting, 2–5 March 1996, Koininia Retreat Center, South Haven. NCGIA Technical Report 96–7: D.54–D.58. Santa Barbara, NCGIA

Petchenik B B (ed.) 1988 Special issue on the history of GIS. *The American Cartographer* 15: 249–322

Pickles J 1991 Geography, GIS, and the surveillant society. *Papers and Proceedings of Applied Geography Conferences* 14: 80–91

Pickles J 1992 Review of D Martin 'GIS and their socio-economic applications'. *Environment and Planning D: Society and Space* 10: 597–606

Pickles J 1993 Discourse on method and the history of discipline: reflections on Jerome Dobson's 1983 'Automated geography'. *The Professional Geographer* 45: 451–5

Pickles J (ed.) 1995 *Ground truth: the social implications of geographic information systems*. New York, Guilford Press

Pickles J 1997 Tool or science? GIS, technoscience, and the theoretical turn. *Annals of the Association of American Cartographers* 87: 363–72

Poiker T (ed.) 1993 *Proceedings 'Geographic Information and Society' workshop, Friday Harbor, 11–14 November*. Santa Barbara, NCGIA

Rundstrom R 1991 Mapping, postmodernism, indigenous people, and the changing direction of North American cartography. *Cartographica* 28: 1–12

Rundstrom R 1995 GIS, indigenous peoples, and epistemological diversity. *Cartography and Geographic Information Systems* 22: 45–57

Sheppard E (ed.) 1995 Special issue: GIS and society. *Cartography and GIS* 22 (1)

Smith N 1992 Real wars, theory wars. *Progress in Human Geography* 16: 257–71

Sui D Z 1994 GIS and urban studies: positivism, post-positivism, and beyond. *Urban Geography* 15: 258–78

Taylor P 1990 Editorial comment: GKS. *Political Geography Quarterly* 9: 211–12

Weiner D, Warner T A, Harris T M, Levin R M 1995 Apartheid representations in a digital landscape: GIS, remote sensing and local knowledge in Kierpersol, South Africa. *Cartography and Geographic Information Systems* 22: 30–44

Wright D J, Goodchild M F, Proctor J D 1977 Demystifying the persistent ambiguity of GIS as 'tool' versus 'science'. *Annals of the Association of American Geographers* 87: 34–62

Yapa L 1991 Is GIS appropriate technology? *International Journal of Geographic Information Systems* 5: 41–58

5

Spatial representation: the scientist's perspective

J F RAPER

In this chapter, spatial representation is seen as providing a bridge between scientific theories and the 'real world', a bridge which is fundamentally important yet is also inherently bound by scientific conventions. Such conventions effectively prescribe the correspondence between geographical space and the constrained physics of forces and mass in the world. This much has emerged from a range of recent reappraisals of the general scientific world view, as well as from the more specific experience of generalising and modelling real-world phenomena using GIS. Having reviewed this work, we go on to explore some of the metaphysical, ontological, and epistemological considerations that underpin spatial and temporal representation. This is done in part through discussion of the general and specific conventions that characterise current GIS practice, but is also viewed more prospectively in relation to the emergence of 'geographic information science'. All of this suggests that GIS-based spatial representations should become richer than they are at present, and that they should be more firmly grounded in method. If spatial representation is to remain central to our theorising about the external world, it follows that the challenge for those who wish to create and use spatial representations is to employ existing GIS critically and to look for new ways to enlarge their scope and expressiveness. In this way, spatial representation will continue to open up new ways of exploring structure, relationships, and causality in the world.

1 INTRODUCTION

Spatial representation is essential to science. It provides science with a means to establish correspondences between theories and the world. If a theory states that the sun sets in the west every day, then spatial representation is required to establish the meaning of 'west' and to locate a viewpoint. However, science itself depends on a theory of knowledge that governs how spatial representations are made and used to reason about the world. For example, the map of the areas of darkness and light on the Earth's surface shown in Plate 1 permits the testing of the 'sun setting in the west' theory suggested above and poses a number of challenges to the assertion. However, this scientific form of spatial

representation can only be related to observations of the world by understanding the scope of the theory in question and the nature of the observations which are made to test it. As a consequence, users of scientific spatial representations such as maps or models must know something of science and its conventions to relate such representations effectively to the external world (Raper 1996a).

A key aspect of scientific spatial representation is that it assumes a correspondence between geographical space and the physics of forces and mass in the world. There are different ways to express this: for example, the energy required to move an object from one place to another is related to the distance between them. The correspondence between geographical space and physics has been used to provide a foundation for many geographical models such as the theories of regionalised variables (Journel and Huibregts 1978) and spatial autocorrelation (Cliff and Ord 1973) in which distance is used as an explanation of spatial variation and association. Geographical space also creates physical constraints such as those upon interaction (e.g. at crossings, meeting points), on searching processes, and on topological conditions (such as junctions). The substitution of space for time (ergodicity) in process is also dependent on this correspondence. In the scientific world-view the explanatory power of spatial representations derives from these correspondences between the forces of physics and the dimensions of geographical space.

In the last three decades the scientific world-view has been challenged on the grounds that it cannot offer a universal theory of knowledge (for example, by Foucault 1972; Habermas 1978; Giddens 1979) and that its methods are flawed (Kuhn 1962; Feyerabend 1975; Lakatos 1976). Other theories of knowledge (epistemologies) have been defined which provide alternative frameworks for research. Johnston (Chapter 3) describes the three methodological world-views identified by Habermas (1978): the 'empirical–analytic' world-view (corresponding roughly to explanation in natural science); the 'historical hermeneutic' world-view (concerned with the understanding of meanings), and the 'critical' world-view (concerned with the underlying structures creating what we observe). In the 'historical–hermeneutic' world-view the correspondence between space and physics is denied *per se*, as it is considered that the world is entirely created within discourse and does not exist independently of an observer. In the 'critical' world-view, largely based on the critical realist philosophy of Bhaskar (1978), there is an external 'real' world but it is only observable via the empirical outcomes of mechanisms which structure the real world. This view implies that spatial representations do not necessarily have privileged explanatory status through generalisation, as causal relations are created by the local and particular interaction of the generative mechanisms.

Yet amongst these competing world-views science is intensively scrutinised and remains the pre-eminent methodology focused on explanation or causation, particularly in the natural sciences (Musgrave 1993). By contrast the 'historical–hermeneutic' world-view is essentially reflective,

promoting understanding, ironically evaluating a world itself transformed by science (Collier 1994). Accordingly, scientific spatial representations remain widely used and are often highly successful when used in conjunction with modelling (for example, tidal predictions). In some cases the use of scientific spatial representation produces insights which have been found essential to human life such as hurricane warnings or quarantine measures in the control of disease. This is the essence of the process described by Hacking (1983: 31): 'we represent and we intervene'. The key issues in the appropriate use of scientific spatial representation are, therefore, whether the scientific world-view is appropriate for a particular use, and, if so, how spatial representation fits within the theory of knowledge used by science. This chapter will explore the nature of scientific spatial representation in more detail and then survey the implications for the emerging field of geographic information science.

2 THE NATURE OF SCIENTIFIC SPATIAL REPRESENTATION

The contemporary critique of the science world-view has renewed the importance of making an explicit justification for the use of spatial representations and showing how they lead to the production of new knowledge. Such a justification should involve a review of the phenomena and processes whose spatial dimensions are to be represented in spatial form in order to establish:

- the approach to conceptualisation of the world (metaphysics) employed;
- the methodology by which specific phenomena and processes that the world contains are given significance (ontology);
- the approach by which knowledge of the world is established and tested (epistemology).

Few researchers carry out such a justification explicitly, preferring rather to rely upon conventions already established in their application field. These conventions range from the informal ones associated with institutions (e.g. mapping agencies: see Smith and Rhind, Chapter 47) to the formal ones required by a profession or government (e.g. international standards: see Salgé, Chapter 50).

However, spatial representation has been made newly accessible by digital technology – opening the way for its much wider use (e.g. see Goodchild and Longley, Chapter 40). The development of a geographic information science (GISc: Goodchild 1992) as a counterpart to the use and application of geographical information systems (GIS) has as a consequence generated an interest in the proper methodological context for spatial representation. The goals of GISc are, therefore, to question the foundations of the conventions established in the application of GIS. This project will help to avoid the generation of dangerous 'spatial' fallacies with newly powerful computer tools. It is also perhaps the way to identify new avenues for research given the central role of spatial representation in the ontology of the physical world.

2.1 The scientific context of spatial representation

Most spatial representations are made within the scientific world-view since users of other epistemologies often do not regard spatial representation as especially significant in understanding the world. Spatial representation is essentially a product of the science world-view. Hence, monitoring the use of spatial representations involves critically examining the scientific context within which the conventions have been developed.

The conventions associated with spatial representation can be divided into the high level conventions of the science world-view and the low-level conventions associated with the implementation of representations. These low-level conventions can be identified with the 'disciplinary matrix' and 'paradigms' of GISc in the terms used by Kuhn (1962). Yet the high and low levels are thoroughly interwoven: the development of high-level conventions in 'general science' can generate a false sense of security for the implementors of spatial representations by suggesting that certain assumptions can be made quite generally. While 'general' science might suggest that judgements about the energy output of a chemical reaction can be made free of ethical considerations, the same assumption is not likely to be valid in the case of the regionalisation of a city for the determination of voting district boundaries. Implementors and users of spatial representations must, therefore, satisfy themselves that the high level implicit assumptions of general science are appropriate in the low-level implementation and use of spatial representation.

Consequently, the following sections provide short but explicit summaries of the issues that scientific users of GIS should reflect upon when employing these tools. If such assumptions and conventions are explored and spelled out publicly (i.e. in publication) it will go some way to exposing the methodological foundations of spatial representation to the critical examination suggested by Pickles (Chapter 4).

2.2 The metaphysics of spatial representation

Metaphysics may seem remote from the interests of most users of spatial representations. The way the world is conceptualised is usually considered to be subject to a wide and uncontroversial consensus; expressions such as 'real-world conditions' or 'ground truth' assume this. While in certain respects this assumption is robust (e.g. solar energy output is enduring and relatively unchanging), in most other ways it proves less universal both culturally and theoretically. In fact there are considered to be two distinct metaphysical positions (Musgrave 1993):

- *Idealism* argues that nothing exists outside the mind to create perceptions. In this view appearance and reality are one and external realities are merely ideas based on sensory data. Some social theoretic approaches to geographical research are associated with this view since they argue that social structures and beliefs control the way the world is conceptualised.
- *Materialism* argues that phenomena exist in the world independently of minds and that it is their true existence that causes sensory perceptions. This is the view which is associated with science.

These metaphysical positions illustrate that widely different views can be put forward for the cause of our sensory perceptions and the nature of the 'real' world. Since materialism is associated with science this is the commonly assumed position in the creation of spatial representations. A typical process of reasoning would be that if the 'real' world exists and can be observed, the goal of spatial representation should be to capture some particular essence of the world as in, for example, an environmental model. If this is satisfactorily achieved then the spatial representation can be studied and conclusions drawn from it that will prove applicable to the world (Collier 1994). However, in many circumstances this approach does not lead to satisfactory explanation of the world, making it difficult to decide whether the metaphysical position or the process of making the spatial representation is flawed. If it is the former case then science is inadequate; if it is the latter case then the conduct of science is at fault. Such conundrums illustrate the importance of understanding the metaphysical foundations on which spatial representations are constructed.

2.3 Ontological perspectives on spatial representation

The process by which phenomena are defined and given significance is the study of ontology; essentially, it is the study of what phenomena exist. Since the dimensions of space and time are pervasive in perception they are generally regarded as being critical to ontology (Heller 1990). There is, however, a range of distinct approaches to the understanding of space and time, each of which implies a different ontology. Users of spatial representations need to examine the possible ontologies to avoid making a commitment by default to an ontology that is not appropriate.

One ontological divide is between those who regard space and time as a universal physical reference framework and those who regard them as simply a set of relations between phenomena. The first view (absolute space and time) arises out of Newtonian physics and implies that phenomena can be defined in themselves by where and when they are found. Spatial representation simply requires the bounding or sampling of the phenomena. The second view (relative space and time) was first elaborated by Leibniz and suggests that space can be defined as the set of all possible relations between phenomena. By analogy time can be defined as the order of succession of phenomena. Here spatial representation must be fully spatio-temporal in nature: it is the (causal) connections and dependencies between phenomena which have ontological importance in this scheme.

Space and time can be seen as a kind of index to phenomena (first view) or as a domain through which phenomena can be causally interconnected (second view). Although the distinction between these two is highly theoretical, use of the second approach has offered some potentially new solutions to spatio-temporal representation for complex environmental problems. For example Raper and Livingstone (1995) argue that in some dynamic

environments such as the coast it is not meaningful to attempt an *a priori* space and time partition where lines are drawn to separate distinct entities such as channels, beaches, and dunes. Rather, investigation should map all forms and measure representative processes over time in order to create a database of system states. This then allows the researcher to explore the range of movement in space and the evolution of identity through time for all the phenomena of interest, without the constraints of identifying spatial identity (for example by capturing boundaries into a GIS) at the outset. This approach means that an ontology is generated *from* the behaviour of the phenomena rather than imposed *onto* it through a space and time framework.

The ontological significance of space and time in representation is also not scale invariant since different ontologies apply to domains outside the direct experience of human beings such as the microscopic (the scale of subatomic particles) or the macroscopic (the scale of galaxies). Most spatial representation relates to the 'mesoscopic' or geographical domain which ranges in extension approximately from the size of a human being to the dimensions of the Earth. In this domain it has been suggested that concepts of space may be driven by cognitive processes scaled to the dimensions of the human body. Zubin (1989) suggests that different concepts of space are developed for domains which are directly manipulable by human beings (e.g. within rooms) compared to those developed for domains which can be viewed but not manipulated (e.g. landscapes). Do users of spatial representations in science preferentially scale the entities they decide to measure according to these built-in rules? Few implemented spatial representations reflect this cognitive variation in ontology or explicitly document their assumptions. Spatial representations also suffer from the modifiable areal unit problem (MAUP: Openshaw and Alvanides, Chapter 18; Openshaw 1984) in which spatial variation can be seen to differ spatially depending on the scale at which geographical units of aggregation for observations are set. This raises the question of which scale level has ontological primacy (see Openshaw and Alvanides, Chapter 18, for an empirical perspective upon this issue).

2.4 The epistemology of spatial representation

The contemporary world-view of science is associated with an epistemology (largely derived from Popper's 'critical rationalism') governing the way science is conducted. Popper (1959) argued that it is reasonable to believe deductively generated hypotheses that have withstood severe criticism (non-falsifiability) through empirical observation. However, important constraints on the creation and use of spatial representations are implied by the use of the critical rationalist epistemology which need to be incorporated in their use.

First, spatial representations are forms of observation. Yet observation is heavily contested epistemologically: the relativist critique suggests that observation cannot be either rational or theoretically neutral (Feyerabend 1975; Kuhn 1962; Lakatos 1976); critics of science have suggested that observation must inevitably be a product of the world-view within which it is defined (Foucault 1972; Habermas 1978); and critics of positivism have suggested that observation is fixated with external expression rather than internal meaning (Gregory 1994). The implication of this critique is that all observation is defined by and applicable to a certain domain: in other words it is acutely contextual. This is a conclusion that applies equally to observation when used to make spatial representations such as maps. Hence, the 'features' that public topographic maps capture were largely decided by eighteenth century army generals and nineteenth century civil engineers and emphasise engineering plant, street furniture, building outlines, and administrative boundaries at the expense of all the other features that could be included. Such observations rapidly become a 'standard' form of observation which may be commodified and used beyond its domain of original applicability.

Second, spatial representations provide an operational method by which correspondences between theories and the world can be established in science. However, if spatial representations are seen as catalogues of observation then their analysis may involve the fallacy of induction i.e. reasoning about one area of geographical space or time interval from evidence gathered elsewhere in space or time. Spatial representations may also reflect the ruling paradigm of observational theory (as for example in the sampling interval adopted when recording temporal change). If spatial representations are regarded as theories of process and implemented as models then their form must be stated in such a way that they can be tested and, if necessary, falsified (Popper 1959). The nature of the testing of spatial representation is

of importance: maps as spatial representation can be characterised as 'normal' science, although the potential of digital spatial representation may facilitate the development of a critical science by the creation of emancipatory spatial representations. João (1994) analysed the generalisation effects embedded in British and Portuguese topographic maps, finding that they differed significantly across the scales of maps published; in one sense this study was an important attempt to falsify maps as a form of knowledge (see also Weibel and Dutton, Chapter 10).

2.5 Status of spatial representation

In summary, making and using spatial representations is a key part of the methodology of the scientific world-view. However, it is not merely a process of descriptive observation as critics of GIS have charged (Pickles 1995; see also Pickles, Chapter 4). The need to make metaphysical, ontological, and epistemological commitments when forming spatial representations requires theoretical work by users. Of these three, ontology now poses the greatest challenges and has stimulated work in several related fields outside geography, e.g. Smith (1995) on the philosophy of boundary drawing or Cohn (1995) on computable qualitative spatial reasoning. In developing the tools for the development of new ontologies of space, geographical information science has opened the way for the discovery of new spatio-temporal structures in the mesoscale geographic world.

3 THE IMPLICATIONS FOR THE DEVELOPMENT OF GEOGRAPHIC INFORMATION SCIENCE

Lying at the overlap of geography, geoscience, computer science, cognitive science and cartography, geographic information science has developed a new interdisciplinary focus on conceptual and computable aspects of spatial representation. However, questions such as 'whose real world' (metaphysics), 'what are the objects and at which scale' (ontology), and 'how are geographical conjectures tested' (epistemology) pose substantive challenges to geographic information science. If geographic information science is to contribute to (for example) the study of the risks of natural hazards, the epidemiology of disease, or prediction of the evolution of environmental systems then spatial representations must be richer than at present and firmly grounded in method (Raper 1996a).

Concrete steps towards these goals can be taken into two areas: first, in the connection of the 'high level' conventions of general science with the 'low level' conventions of geographic information science, specifically those being built into the next generation of GIS software systems; second, in the enrichment of spatial representation from the dominant 2-dimensional model in order to make them multidimensional.

3.1 Connecting the conventions of general science with those of geographic information science

The conventions of general science have counterparts in geographic information science (see also Forer and Unwin, Chapter 54). The process of conceptualising the external world in order to create a spatial representation is generally referred to as 'data and process modelling' in geographic information science, as in computer science generally (Herring 1991). This form of modelling is concerned with the way correspondences are established between geographical aspects of the 'real world' and the elements of the spatial representation. A distinction can be drawn between data and process modelling.

Data modelling in geographic information science has largely developed from techniques employed in database management systems: hence Tsitchizris and Lochovsky (1977: 21) defined a data model as 'a set of guidelines for the representation of the logical organisation of the data in a database . . . [consisting] of named logical units of data and the relationships between them'. Frank (1992: 410) put this in the context of geographic information science when he characterised a data model as 'a set of objects with the appropriate operations and integrity rules defined formally'. These definitions establish data modelling as concerned with essentially static interrelated and discrete entities. In geographic information science these entities have spatial extent and therefore the 'operations' and 'integrity' rules referred to by Frank (1992) must also apply spatially.

Data modelling procedures for natural science applications of GIS are generally ad hoc exercises (since there are few prototypical situations) starting with heterogeneous collections of data. Typical collections of data include direct observations of position, form, or behaviour for an entity of interest (such as a risk zone for incidence of a disease). To form a spatial representation in a GIS it needs to be georeferenced with respect to a global datum (e.g. using a Global Positioning System: see Lange and Gilbert, Chapter 33), a national datum (using the positions defined by National Surveys) or a local datum (specific to a particular project). The key issue in the handling of such data is how the entities identified should be represented by the available geometric primitives, viz. 0-dimensional points; 1-dimensional lines, 2-dimensional areas, and 3-dimensional volumes (note that many GIS have introduced their own terminology for these geometric primitives). *Vector* approaches to spatial data modelling involve using these points, lines, areas, and volumes to make representations, while *raster* approaches involve using tessellations of equal size cells in a grid of a fixed resolution, usually to represent areas.

An early convention developed in spatial data modelling was that entities must be described separately in terms of their nonspatial and spatial identity. This convention originated in the software architectures of the early GIS which integrated a geometry engine with an alphanumeric database management system. More recently, object-oriented languages (see Worboys, Chapter 26) have been used to build GIS with the result that spatial data models can now be constructed in most GIS without starting the data modelling from geometric concepts. This is done by identifying real-world entities (such as rivers and buildings) or conceptual entities (such as samples or boundaries) and specifying both their attributes (whether alphanumeric, spatial, or multimedia in form) and their interrelationships with other entities (see also Hutchinson and Gallant, Chapter 9). It is still rare to find any explicitly temporal concepts in the tools provided for spatial data modelling.

The user must now define how the dimensionality of the 'real' entity is represented by the selected geometric primitive and in which GIS will the result be stored. Hence, points in a spatially non-regular sample (e.g. rainfall observations) can be generalised to space-exhausting areas by creating Thiessen polygons (defined as that polygon bounding the space closer to the selected point than to any other point: see Boots, Chapter 36) from the point locations of the samples. Alternatively, entities extending over areas could be reduced to points by taking the geometric centroid of the area (a function available in many GIS if the boundary is stored), or the areas could be stored using tessellations of grid cells where each cell in the grid has an associated value. Often the spatial data are supplied to the user in a form that dictates the nature of the spatial representation to be used in the GIS. For example, satellite imagery is produced by a scanner generating raster data while new roads are surveyed along a centreline implying vector spatial representation. Some recent work has focused on the forms of spatial representation appropriate for entities with indeterminate boundaries (Burrough and Frank 1996; see also Fisher, Chapter 13).

The other key source of data beside direct observation is mapping. Cartographers have developed sophisticated conventions for the representation of a consensual view of the natural and built environment by using a well documented form of cartographic communication. Most national surveys now offer versions of their paper maps in scanned raster or digitised vector form. Paper maps employ a form of symbology considered appropriate to each case at each scale. However, in digital representations the form of paper map features must be transformed into georeferenced geometric data. Note that maps normally present a governmental or commercial perspective because of the high cost of creating maps for individuals or even for corporations. Hence, mapping has itself become both a reflection of public policy and a powerful influence on perceptions of space as the symbology of topographic maps is taught in schools and the terminology of the map is employed in law and professional practice.

Process modelling has generally developed outside geographic information science in fields such as environmental and ecological modelling (Kemp 1992). The aim of process modelling is to represent the behaviour of continuous 'real' physical systems and their change in status over both time and space (Ziegler 1976; Casti 1989). This is generally achieved by establishing a finite process model whose behaviour is distributed over space and time. Typically the spatial representation is a raster while temporal representation is based on application-specific intervals (Jørgensen 1990). There is a greater concern in process modelling than in data modelling about the establishment of correspondences between the 'real world' and the modelling environment since designing a simulation model is inherently conjectural in nature.

Spatial representations employed by process models are generally divided into finite difference models (based on raster grids) and finite element models (triangulated irregular networks). These 'finite' forms of spatial representation specify a limited number of topologically closed areas for which states of a physical system can be computed and stored. These process models are limited by the number of interactions permitted across the boundaries between grid cells or triangles and by the fixed scale of the spatial representation. Despite these limitations such models have been used successfully in a range of fields (Goodchild et al 1993; Goodchild et al 1996). Casti (1997) suggests that such models can be populated by autonomous elements called agents which inhabit the simulated environment and take decisions by reasoning about the knowledge they have of the system. This knowledge is implicitly and explicitly 'local' in nature since no agents have access to the global state of the system. Casti (1997) argues that such systems exhibit 'emergent' behaviour whereby outcomes emerge from the aggregate behaviour of the agents.

The wide availability of GIS and process models has led many users to adopt the representational tools of current systems as a basis for the expression of their world-view rather than the other way around. A set of well defined conventions has developed around GIS and process models in many disciplines and many scientists employ them. However, prospective scientific users of GIS and process models should distinguish between the mass of available systems and the rather richer prototypes which are available at the sophisticated end of the market and those emerging from research laboratories.

3.2 Enriching the spatial representations in current GIS

At present most GIS implement a continuous 2-dimensional geometric form of spatial representation mimicking that used on maps. However, 2-dimensional geometry is a rather limited representation of real surfaces such as terrains which frequently contain overfolds and holes. In the natural environment most physical processes operate in true 3-dimensional domains such as the solid earth, oceans, or atmosphere, making 2-dimensional geometry even more limited (Raper 1989). A variety of 3-dimensional GIS has developed to offer representational tools suitable for visualisation and analysis of volume property variation, structural reconstructions, and volume interpolation. New

3-dimensional data structures have evolved from solid modelling techniques (Mortenson 1985) which extend geometric representations from 2-dimensional raster and vector geometry to their 3-dimensional analogues (Raper 1992).

Three-dimensional raster representations are generally constructed from a block of volume elements (voxels). The values of these voxels may be directly generated from source data as in geophysical reconstruction, rasterised from vector data such as cross-sections, or interpolated by distance weighting, Kriging or spline-based approaches (Houlding 1994). The voxels may be visualised as cells in a 3-dimensional matrix or as isosurfaces (surfaces joining points of equal value) as in the model in Plate 2 where the interpolated value is mean particle size in the phi scale (fine grain is blue, coarse grain is orange). Three-dimensional raster representations are ideal for the calculation of volume and are easily logically partitioned, especially if indexed using octrees (Gargantini 1992).

There are fewer implemented examples of 3-dimensional vector representations as the data structures are more complex. Vector geometry in three dimensions requires an addition to the basic Euclidean point, line, and polygon primitives to include solids (polyhedra sometimes referred to as 'volgons'). While interactive reconstruction can give good results in some circumstances, the most widely used approaches build solids from points distributed non-regularly in 3-dimensional space. This can be done using a 3-dimensional triangulation (Mallet 1992) or by constructing tetrahedra as in Plate 3 (Lattuada and Raper 1996). Plates 2 and 3 are visualisations of the same cross-sectional block of a spit landform based on the same data showing how the interpolated raster approach gives superior visualisation whereas the exact-fit tetrahedron approach is more precise.

The concepts of time employed in spatial representations are also limited in the widely used systems (see also Veregin, Chapter 12). Time is usually assumed to be absolute, operating as a frame of reference where events partition a single universal timeline. By convention, representations of space in two or three dimensions are realised at an instant in time creating 'timeslices/time volumes' which can be operationalised as layers or volumes. Change can then be defined as geometric differences between 'timeslices/time volumes' (Langran 1992). There has been a variety of attempts to develop more

sophisticated spatio-temporal data structures which can be queried and analysed in more complex ways: O'Conaill et al (1993) extended linear quadtrees to represent phenomena in a 4-dimensional space-time cube; Peuquet and Duan (1995) developed the event-based spatio-temporal data model (ESTDM) to handle forestry change; Wachowicz and Healey (1994) proposed a design in which 'events affecting spatial objects' create 'versioned spatial objects' such that temporally different versions of the same object can exist; Ramachandran et al (1994) propose a design called TCObject in which objects with geometric and non-geometric attributes are given past, present, and future states depending on the dates of birth and death for the object; and Raper and Livingstone (1995) proposed a system which assigned spatio-temporal references to all instances of all variables, thereby avoiding the use of a single timeline.

Spatial representations used by GIS and associated tools are generally made of geometric primitives. These basic primitives are treated as uncontroversially isomorphic with selected 'real world' objects to create a spatial representation; for example, a road is treated as a vector line, or a lake is treated as a connected set of grid cells. However, the process of establishing a geometric isomorphism between world and GIS reveals the extremely limited expressive power that this process currently gives. Although both vector and raster forms of representation link to certain fundamental concepts (vectors are associated with the cognitive importance of entification, rasters are associated with the visual field), both are limited to a sense of physical extension. There are other modalities in the human senses, notably motion parallax, sound, smell, and touch which are critical in the formation of spatial representations cognitively.

Digital video and sound offer new ways to make spatial representations which address these other senses. Yet these new forms of representation have been adopted in an extremely limited way in GIS where at present video and sound can generally be associated with just a single vector or raster primitive. Video and sound representational primitives are usually assumed to be spatially and temporally dimensionless as their internal times do not relate to 'world' time but to 'playback' time. No current GIS can operate on the fields of view or audible ranges of these primitives, neither can they be temporally related to any implemented 'timeline' in the system. Neither could a GIS reconcile multiple views of the same entity from different directions or deal with different sounds made simultaneously in different places. Such issues (even the latter) are however being addressed in multimedia information systems, often in an explicitly spatial and temporal framework (Raper 1996b; see also Shiffer, Chapter 52).

4 CONCLUSIONS

Spatial representation is at the heart of much theorising about the external world. However, the recent rapid growth in geographic information science has created new demands for GIS which often still use geometric concepts that are manifestly limited in scope. The challenge for those who wish to create and use spatial representations is to employ the existing systems critically and to look for new ways to enlarge their scope and expressiveness. Only then will spatial representation offer new means to explore conjectures about structure, relationships, and causality in the world.

References

Bhaskar R 1978 *A realist theory of science*. Brighton, Harvester

Burrough P A, Frank A U (eds) 1996 *Geographic objects with indeterminate boundaries*. London, Taylor and Francis

Casti J L 1989 *Alternative realities: mathematical models of nature and man*. Chichester, John Wiley & Sons

Casti J 1997 *Would-be worlds*. Chichester, John Wiley & Sons

Cliff A D, Ord J K 1973 *Spatial autocorrelation*. London, Pion

Cohn A G 1995 The challenge of qualitative spatial reasoning. *Computing Surveys* 273: 323–7

Collier A 1994 *Critical realism*. London, Verso

Feyerabend P K 1975 *Against method*. London, Verso

Foucault M 1972 *The archaeology of knowledge*. London, Tavistock

Frank A U 1992b Spatial concept, geometric data models, and geometric data structures. *Computers and Geosciences* 18: 409–17

Gargantini I 1992 Modelling natural objects via octrees. In Turner K (ed.) *Three-dimensional modelling with geoscientific information systems*. Dordrecht, Kluwer: 145–58

Giddens A 1979 *Central problems in social theory: action structure and contradiction in social analysis*. London, Macmillan

Goodchild M F 1992 Geographical information science. *International Journal of Geographical Information Systems* 6: 31–45

Goodchild M F, Parks B O, Steyaert L T 1993 *Environmental modelling with GIS*. Oxford, Oxford University Press

Goodchild M F, Steyaert L, Parks B, Johnston C, Maidment D, Crane M, Glendinning S 1996 *GIS and environmental modelling*. Fort Collins, GIS World Inc.

Gregory D 1994 *Geographical imaginations*. Cambridge (USA), Blackwell

Habermas J 1978 *Knowledge and human interests*. London, Heinemann

Hacking I 1983 *Representing and intervening*. Cambridge (UK), Cambridge University Press

Heller M 1990 *The ontology of physical objects*. Cambridge (UK), Cambridge University Press

Herring J 1991 The mathematical modelling of spatial and nonspatial information in GIS. In Mark D M, Frank A U (eds) *Cognitive and linguistic aspects of geographic space*. NATO ASID Dordrecht, Kluwer: 313–50

Houlding S 1994 *Three-dimensional geoscience modelling*. Berlin, Springer

João E 1994 'Causes and consequences of generalisation'. PhD thesis, Birkbeck College, University of London

Jørgensen S E 1990 Modelling concepts. In Jørgensen S E (ed.) *Modelling in ecotoxicology*. Developments in environmental modelling 16. Amsterdam, Elsevier Science Publishers

Journel A G, Huibregts C J 1978 *Mining geostatistics*. New York, Academic Press

Kemp K K 1992 *Environmental modelling with GIS: a strategy for dealing with spatial continuity*. National Center for Geographic Information and Analysis Technical Report 93–3. Santa Barbara, NCGIA

Kuhn T S 1962 *The structure of scientific revolutions*. Chicago, University of Chicago Press

Lakatos I 1976 *Proofs and refutations*. Cambridge (UK), Cambridge University Press

Langran G 1992 *Time in geographical information systems*. London, Taylor and Francis

Lattuada R, Raper J F 1996 Applications of 3D Delaunay triangulation algorithms in geoscientific modelling. In *Proceedings Third International Conference/Workshop on Integrating GIS and Environmental Modeling*. Santa Fe, 21–26 January. Santa Barbara, NCGIA. CD

Mallet J-L 1992 *GOCAD: a computer-aided design application for geological applications*. University of Nancy

McHaffie P H 1994 Manufacturing metaphors: public cartography, the market, and democracy. In Pickles J (ed.) *Ground truth: the social implications of geographic information systems*. New York, Guilford Press: 113–29

Mortenson M E 1985 *Geometric modelling*. New York, John Wiley & Sons Inc.

Musgrave A 1993 *Common sense science and scepticism*. Cambridge (UK), Cambridge University Press

O'Conaill M A, Mason D C, Bell S B M 1993 Spatio-temporal GIS techniques for environmental modelling. In Mather P M (ed.) *Geographical information handling: research and applications*. Chichester, John Wiley & Sons: 103–112

Openshaw S 1984 The modifiable areal unit problem. *Concepts and Techniques in Modern Geography* 38. Norwich, Geo-Books

Peuquet D J, Duan N 1995 An event-based spatio-temporal data model (ESTDM) for temporal analysis of geographical data. *International Journal of Geographical Information Systems* 9: 7–24

Pickles J (ed.) 1995 *Ground truth: the social implications of geographic information systems*. New York, Guilford Press

Popper K 1959 *The logic of scientific discovery*. London, Hutchinson

Ramachandran B, MacLeod F, Dowers S 1994 Modelling temporal changes in a GIS using an object-oriented approach. In Waugh T C , Healey R G (eds) *Advances in GIS research: Proceedings Sixth International Symposium on Spatial Data Handling, Edinburgh, 5–9 September*: 518–37

Raper J F (ed.) 1989 *Three-dimensional applications in geographical information systems*. London, Taylor and Francis

Raper J F 1992b Key 3D modelling concepts for geoscientific analysis. In Turner K (ed) *Three-dimensional modelling with geoscientific information systems*. Dordrecht, Kluwer: 215–32

Raper J F 1996a Unsolved problems of spatial representation. In Kraak M-J, Molenaar M (eds) *Advances in GIS research 2: Seventh International Symposium on Spatial Data Handling, Delft, 12–16 August*. International Geographical Union: 14.1–11

Raper J F 1996b Progress towards spatial multimedia. In Craglia M, Couclelis H (eds) *Geographic information research: bridging the Atlantic*. London, Taylor and Francis: 512–30

Raper J F, Livingstone D N 1995 Development of a geomorphological data model using object-oriented design. *International Journal of Geographical Information Systems* 9: 359–83

Smith B 1995 More things in heaven and earth. *Grazer Philosophische Studien* 50: 187–200

Tsitchizris T C, Lochovsky F H 1977 *Database management systems*. New York, Academic Press

Wachowicz M, Healey R G 1994 Towards temporality in GIS. In Worboys M (ed.) *Innovations in GIS 1*. London, Taylor and Francis, 105–15

Ziegler B 1976 *Theory of modelling and simulation*. New York, John Wiley & Sons Inc.

Zubin D 1989 Untitled. In Mark D et al (eds) *Languages of spatial relations: Initiative 2 Specialist Meeting Report*. Technical Paper 89-2. Santa Barbara, NCGIA

6

Spatial representation: the social scientist's perspective

D J MARTIN

This chapter focuses on the representation of socioeconomic phenomena within GIS. Such phenomena include the distribution and characteristics of population, economic activity, and aspects of the built environment. The discussion aims to demonstrate that the association of many such phenomena with precise geographical coordinates is problematic, and there is considerable diversity among the available approaches at both conceptual and technical levels. GIS offer extensive tools for the manipulation and modelling of socioeconomic data, but each option has its own advantages and disadvantages: there are no fundamentally 'right' solutions. There is a need to consider carefully the potential effects of each representation strategy on data output and display.

1 INTRODUCTION

GIS are powerful tools for the manipulation of spatial objects. The application of GIS technology to socioeconomic uses has continued to be a major international growth area, with the population censuses of the 1990s providing an added impetus. Specialised systems for service planning, neighbourhood profiling, and market analysis have been vigorously developed to utilise new digital data (for examples, see Birkin et al, Chapter 51; Longley and Clarke 1995). In many countries, although notably not the United Kingdom, cadastral applications have been some of the major implementations and the utilities are important for GIS users worldwide (see Bibby and Shepherd, Chapter 68). This chapter explores the conceptualisation of socioeconomic phenomena as spatial objects, without assuming any particular application such as marketing or resource allocation. Spatial objects in this sense are entities having both spatial location and spatially independent attribute characteristics (Gatrell 1991). The treatment here of representation as a primarily technical problem permits integration with much other GIS literature,

but in so doing it is important to recognise that we do not adequately investigate the very important issues of exactly what is to be represented, who makes the selection, nor the purposes to which it may be put. These issues, particularly relevant to the application of GIS to social phenomena, have been the subject of considerable debate (see for example Pickles, Chapter 4; 1995a), and are also addressed briefly here. All GIS representations are essentially abstract models of selected aspects of the real world, and there will be important questions which cannot be addressed adequately within this framework.

In Chapter 5, Raper considered the discretisation of physical and environmental phenomena for representation in GIS, thereby implementing widely used 'consensus' approaches to conceptualisation. Some phenomena of interest to the social scientist, particularly those relating to the built environment such as property ownership and values, may be treated in much the same way as these physical characteristics. However, the association between many socioeconomic phenomena such as unemployment, population density or ill health and spatial coordinates can be even more problematic than for their physical counterparts. Intuitively, we understand that socioeconomic phenomena vary across geographical space, but their values cannot usually be measured unambiguously at any given location, and it is not clear what (if any) are the fundamental spatial units to which they relate (see also Veregin, Chapter 12).

The rest of this chapter is divided into four sections. The following section reviews the range of potentially spatial socioeconomic data, and considers in more detail the ways in which these may be georeferenced, that is, associated with specific locations. Section 3 extends this discussion to examine the ways in which the basic spatial objects representing socioeconomic phenomena may be represented within a GIS: representation strategies may be divided into those which actually transform the data in order to create different types of spatial object, and those primarily concerned with display. Section 4 then addresses some further considerations relevant to spatial representation in this context, but which do not relate directly to technical processes internal to GIS.

2 GEOREFERENCING SOCIOECONOMIC PHENOMENA

This section considers the links between the enormous variety of socioeconomic phenomena and the relatively limited range of objects which can be used for their spatial representation in computer databases. One of the most basic socioeconomic tasks is the measurement of population. Rhind (1991) divides the sources of large-scale population data into three groups: the conduct of censuses, the maintenance of population registers, and the estimation of population size by indirect means such as the interpretation of remotely-sensed imagery (see also Smith and Rhind, Chapter 47). Additional data sources providing detailed characterisation of the population and its activities include the wide range of social data collected and published by government departments and statistical organisations (statistics pertaining to unemployment, health etc.), and the ever-growing range of information about individuals maintained by commercial organisations as part of their business activities. These include data about flows of many types, such as commuters, migrants, freight, and information. Initially, most of these data do not contain explicit geographical locations in the form of map coordinates, but are associated with the

addresses, place names, or regions used in the organisation of service delivery or political representation. In developing GIS applications which attempt to deal with these population-related phenomena, it is thus always necessary to use some form of indirect spatial referencing, frequently via the geographies of the administrative or built environments. In these applications, the GIS user is almost always using secondary data, and should be particularly conscious of data quality and fitness-for-purpose. These socioeconomic data, as collected, can be georeferenced by one of three types of spatial object: areas, lines, or points. Line referencing occurs when only a street or route location is given, providing referencing to a linear object or flow of some kind, but this is actually quite uncommon. The endpoints of such lines are generally points such as street intersections or areal units such as local government areas, and we shall therefore focus on the use of points and areas, in theory and in practice.

2.1 Theory

Population data are most commonly related to geographical locations by reference to areal geographies such as census zones, electoral constituencies, local government areas, or regular grid squares. These are frequently the only areal units for which socioeconomic data are reported, and information derived initially from individuals (either as a sample, or from the whole population), is therefore aggregated to provide summary values for each areal unit. The difficulty with these areal units as a method for georeferencing is that they are essentially 'imposed' rather than 'natural' units (Unwin 1981). This means that the locations of boundaries may be arbitrarily related to the phenomena which are being measured. This has two aspects: first, a large region may be subdivided into smaller areas at many different scales. For example, the United Kingdom may be divided into around 70 counties, 460 districts, or 10 000 wards. Second, at a given scale, there are different ways of configuring the boundaries of these areal units, each of which results in a different aggregation of the individual-level data. Each of the last three censuses in the United Kingdom has had a different configuration of approximately 10 000 wards. Each reconfiguration would produce a different distribution of zone characteristics, even if there were no change in the underlying population. This is known as the *modifiable areal unit problem* (MAUP), and is discussed more fully by Openshaw and Taylor (1981) and Openshaw (1984). Its effects on spatial analysis have been the subject of continued debate (Fotheringham and Wong 1991), and awareness of the problem is of particular importance when designing zonal systems, as discussed by Openshaw and Alvanides (Chapter 18). The use of areal units with irregular boundaries also presents difficulties for the representation of socioeconomic phenomena on a map, as large, sparsely populated areas will tend to dominate the visual image (but see Elshaw Thrall and Thrall, Chapter 23, for some different cartographic representations). Most zoning schemes of this kind are designed to cover the entire land surface and therefore include extensive areas of unpopulated land, leading to wide variations in population density between areas. The cities in which most people live are represented by small zones covering a very small proportion of the mapped area. A further, and related, difficulty with all types of aggregate data is known as the *ecological fallacy* (Blalock 1964): relationships between variables which are observed at one level of aggregation (e.g. a correlation between household income and educational achievement at county level) may not hold at the individual, or any other, level of aggregation.

An alternative way of thinking about population is as data relating to points. The difficulty here is that it is hard to determine precisely the location of an individual or household. This is usually performed by reference to the home address, for which a location may be derived from a street segment, postal code, or individual property location. Data of this type are most commonly derived from consumer surveys, customer (patient, visitor, etc.) lists, and other address-based lists, including the population registers common in some European countries (Redfern 1989; Ottoson and Rystedt 1991). Individual-level data are hard to visualise in GIS, and many concepts such as an unemployment 'rate' cannot be measured for an individual (who will be registered employed or not), but only have meaning in relation to aggregate data. For some purposes (e.g. employment mapping), the home address may not be the most appropriate spatial reference, but is usually the only option. A full database of individual locations with associated social characteristics would however offer an ideal basis for purpose-specific areal aggregation in which the size and shape of areal units could be redesigned as required.

The modifiable areal unit problem and ecological fallacy, together with the visual constraints associated with zone and point mapping are not restricted to GIS, but also apply to traditional cartographic methods using these types of data. The additional significance of these difficulties in a GIS context is that their effects may be propagated in complex ways through many subsequent operations in which the initial data are transformed for analysis or visualisation, making it difficult to unravel their impact on the final output from the system.

2.2 Practice

In addition to these theoretical considerations, a brief survey of georeferencing in practice reveals the enormous increase in the resolution and range of products available for attaching socioeconomic data to specific locations, illustrated for the case of the United Kingdom in Figure 1. Early census mapping was dependent on rather large and poorly defined spatial units, but over time there has been a massive improvement in the resolution and quality of such data sources. The dual independent map encoding (DIME) system was developed for the 1970 US census, and included records for each street segment, with grid references for street intersections and information about the blocks falling on each side of the street segments. The DIME data structure was an important development (Peucker and Chrisman 1975), but the database covered only major metropolitan areas, and was fundamentally tied to the geography of the built environment (Barr 1996). The original motivation for such databases in the USA was the actual organisation of the census enumeration process, whereas in the United Kingdom mapping products were developed primarily for data display and analysis. For the 1971 UK census, no digital boundary data were available, but census data were published for 1-km grid squares nationally and 100-m squares in urban areas, providing a direct method for the production of maps such as those in the census atlas *People in Britain* (CRU/OPCS/GRO(S) 1980). A single (subjectively determined) centroid location was also provided for each enumeration district (ED) – the smallest areal units for which census data are published. By the early 1980s DIME had been extended to cover many more areas and in the UK a set of digital boundaries was produced at the ward level (each ward typically contains around 13 EDs) in addition to the ED centroids. With the massive growth in GIS in the mid 1980s, a number of organisations began to produce additional digital data for socioeconomic zones in local areas, such as the production of ED boundaries. In the UK, where there is no direct correspondence between census and postal geographies, there was also a marked increase in interest in postal geography as a georeferencing system (Raper et al 1992), and a national directory called the Central Postcode Directory (CPD) came into widespread use, containing a 100-m grid reference for each unit postcode in the country, typically covering only 15 addresses, compared to 169 in the average ED.

The early 1990s have seen another major increase in the completeness and resolution of basic georeferencing products. In the USA, the Topologically Integrated Geographic Encoding and Referencing (TIGER) system now provides national coverage of DIME-type information, with extensive topology, information about the shape of street segments and relationships with a variety of other statistical and administrative zoning systems (Broome and Meixler 1990). The TIGER database has been

maintained and made widely available, ensuring continual enhancement, as data suppliers take the basic files and 'add value' by integration with other datasets. In the United Kingdom, two commercially produced digital boundary sets were produced for 1991 census EDs, together with a directory indicating the intersections between EDs and unit postcodes. 1996 saw the completion of ADDRESS-POINT, a national database containing 0.1-m grid references and unique property reference numbers (UPRNs) for all properties, which will provide the basis for further derived products in the future (Smith and Rhind, Chapter 47; Martin and Higgs 1996). Similar enhancements to the resolution of georeferencing products have occurred elsewhere, with an increasing focus on individual addresses (e.g. Lind and Christensen 1996). It should be noted, however, that the process of associating socioeconomic data with appropriate georeferences is frequently error-prone, because of the inherent uncertainty involved in the integration of datasets created by different organisations at different times, yet purporting to describe the same entities. Recent initiatives to develop national standards for address referencing in the UK, for example, are still a long way from widespread implementation (Cushnie 1994).

The preceding discussion illustrates that there have been increases in both the number of products available, and in spatial resolution over the last two decades. It is tempting to assume that georeferencing is all that is involved in the representation of socioeconomic phenomena, yet despite the improvements noted above, none of these products provides fundamentally the 'correct' spatial object for representation in GIS. As explored in the following section, many alternative representations of these same data sources may be constructed using the manipulation tools available within GIS software. This can be done either by transforming the input data to another type of spatial object, such as a continuous surface model, or by transforming geographical space for the purposes of visualisation, for example in cartogram construction.

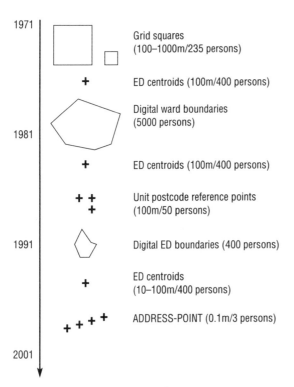

Fig 1. Increases in georeferencing resolution in the UK, 1971–present.

3 SPATIAL REPRESENTATION OF SOCIOECONOMIC PHENOMENA

The previous section has provided a discussion of the spatially-referenced data available to describe various phenomena of interest to the social scientist. Although there may be many thousands of data series in existence which describe aspects of the socioeconomic world, there are relatively few options available for actually attaching these measurements to specific geographical locations. It is important to understand the relationship between these geographical data series, usually produced externally to the GIS, and the different representation strategies which may be adopted within a GIS.

This discussion is best introduced by a simple example: to attach disease incidence to household addresses and thus to point grid references assigns a particular type of representational model to that disease. This will tend to promote a different way of thinking about the disease to that which might have been adopted if cases were treated as flows, assigned to census zones or modelled as some kind of continuously varying surface phenomenon or field (Figure 2). Different analytical techniques may be used on the different representational models, and different answers may result from questions concerning the degree of clustering in disease incidence, or its association with the geographical distribution of some other variable (see also Gatrell and Senior, Chapter 66). In the following chapter, Mark (Chapter 7) discusses the issue of deciding what is to be represented in more detail, making the distinction between 'entities' which exist in the real world and 'objects' which are part of the digital representation. An important feature of GIS is the ability to remodel data from one spatial object type to another, including the generation of complex objects from simple or primitive ones, as identified for example by Gatrell (1991; see also Fisher, Chapter 13).

It is useful to consider representation not as a single operation, but as a process, whereby selected

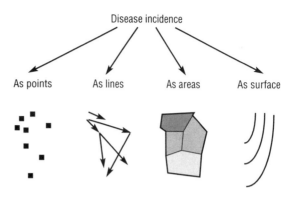

Fig 2. Different spatial conceptualisations of disease incidence.

information about a real-world entity passes through a number of stages along the road to visualisation or some other form of data output, each of which may affect the way in which its spatial characteristics are stored and accessed. This discussion is organised around the examples given in Table 1. Much has been made in GIS literature of the distinction between vector and raster data structures (e.g. Maffini 1987; see also Batty, Chapter 21). Although important to the operation of the GIS at a technical level, these issues are largely independent of the issues being considered here, and the different approaches to representation may be implemented with relative ease using each of the available data structures.

The columns of Table 1 illustrate four 'classes' of geographical phenomena, namely points, lines, areas, and surfaces. As will be seen, this division based on spatial dimensionality is not without its difficulties, but it will serve here to illustrate some of the many

Table 1 Examples of the representation of socioeconomic phenomena in GIS.

	Point	Line	Area	Surface
Real-world entity	individual disease case?	journey to work	property ownership	population density?
Digital object	property or postcode coordinate	street segment coordinates	land parcel, census zone boundary	TIN or elevation model
Manipulation techniques	nearest neighbour analysis, boundary generation, surface generation	network functions, topological analysis	areal interpolation, centroid generation, surface generation	slope analysis, TIN creation, DEM creation, analysis of surface form
Visualisation techniques	point mapping, multivariate glyphs, convert to 3D	line mapping, line cartograms, convert to 3D	choropleth mapping, area cartograms, convert to 3D	isoline mapping, TIN mapping, grid mapping, convert to 3D

possible approaches to representation. The rows of the table illustrate four stages in the representation process, which are further illustrated in Figure 3. Geographical entities are considered to exist in the real world; georeferencing provides the link with digital objects which may be used to represent the locations of such entities; GIS provide manipulation tools for the creation of new objects, moving both within and between object classes; and finally, visualisation techniques may be applied in each case. Between these stages are: (1) data collection and entry, (2) data manipulation, (3) data output transformations discussed in Martin (1996), which are analogous to the transformation stages in the traditional cartographic process described by Clarke (1995). In traditional cartography the printed map embodies a single representation of each real-world entity, but in GIS there is a variety of representation strategies available. Many forms of GIS analysis may not actually result in visualisation, but will produce some other kind of non-graphical output (query results, statistical summaries, etc.) directly by manipulation of the digital objects. Another important parallel with cartography is that the relationship between the 'real' world and representation at each stage is not simply determined by the accuracy of coordinates and attributes, but is influenced by operator decisions about which phenomena are to be included, and how they are to be measured, classified, and symbolised.

In the first row of Table 1 are examples of different real-world entities which are of interest to the social scientist. For some such entities there is general consensus regarding their object class, for example a

journey to work is conceptualised as a linear feature or flow, and legal land ownership by definition relates to a specified parcel of land, the ownership documents often being accompanied by some form of map or description of the area enclosed. The other two examples are more ambiguous: as in the example above, most people would probably think of disease incidence in terms of a pattern of individual points over space, but others might argue that these too are more correctly considered as flows or even as a continuously varying surface. Population density is tentatively included as an example of a real-world entity which may actually be a surface, although this is again subject to some debate and is further addressed below. In many cases, the 'true' class depends on the scale at which the object is being considered, thus population density which at very large mapping scales may be considered as a point pattern is more convincingly a surface at smaller scales.

The second row of the table contains examples of digital objects falling into each of the classes. These are the kinds of spatial 'data' which have been discussed above, and which are used for the georeferencing of many socioeconomic phenomena which do not have any precisely measurable locational characteristics. The coordinates of a property will frequently be used as a proxy for the location of an individual member of the population, perhaps with a high degree of spatial precision, or perhaps less precisely via the grid reference of the corresponding postcode. Encoding of a journey to work, or other route information, is most commonly accomplished by reference to line segments in the spatial database or perhaps to a computed line

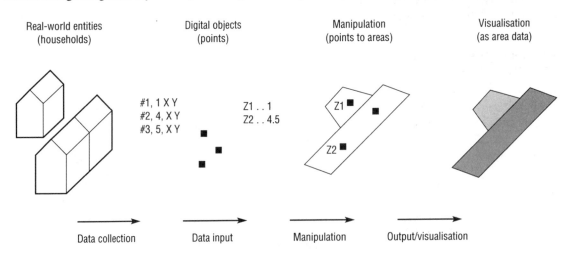

Fig 3. **Representation in GIS viewed as a process.**

between two address locations. In this context, it is appropriate to mention non-Cartesian concepts of space, in which relations such as time or cost may replace Euclidean distance in defining the separation between locations (see Couclelis, Chapter 2). Such spaces may indeed be the most appropriate frameworks within which to explore certain socioeconomic phenomena, but they are not easily handled by contemporary GIS data structures, and there has been little development work in this area. Areal phenomena are readily encoded by a series of coordinates defining an enclosed area. It should be noted, however, that the use of areal objects to describe such phenomena as population results in a change of object class from point or surface (depending on the user's point of view) to area. It is not possible to record directly any social phenomena in surface form, although various representation strategies exist for the storage of surface models – some of which are discussed below.

The third row of Table 1 contains examples of a range of manipulation functions which may operate on spatial objects within a GIS. These include both techniques for the analysis of spatial form and tools for the transformation of objects from one class to another. Spatial analysis techniques are dealt with more fully in later chapters in this volume by Openshaw and Alvanides (Chapter 18), Getis (Chapter 16), and Fischer (Chapter 19). Spatial analytical tools are heavily influenced by the object class of the models on which they operate, and the decision to represent phenomena using different models will therefore have a significant impact on the types of analysis which may be performed (Bailey and Gatrell 1995). The transformation of objects between classes includes, for example, the generation of boundaries around point data by the creation of Thiessen polygons (Boots, Chapter 36; 1986). This allows both data associated with the original points to be mapped as areal data, and areal manipulation and analysis functions to be applied. An extensive series of functions exists for the analysis of network topology and routing problems, which work on suitably structured line data. Although precise distance calculations are possible along such networks, it is frequently difficult to define the endpoints of flows or journeys with sufficient detail, and single origin and destination points are frequently used to represent the flows between zones. The definition of such centroids is another transformation operation, permitting values

originally relating to zones to be applied to points and thus mapped and analysed as point patterns. The acceptability of the assumptions involved in each of these types of estimation will vary according to the specific application. A frequently encountered problem with socioeconomic data is the need to relate phenomena recorded for two incompatible sets of areal units. A variety of approaches has been suggested, including the use of ancillary variables to aid in the interpolation of data values between the two sets of boundaries, and the intermediate estimation of values in an underlying surface (Flowerdew and Green 1991; Goodchild et al 1993).

This last approach indicates that it is also possible to remodel data from each of the other source object classes into surface form. The common practice of georeferencing population-related phenomena by reference to discrete locations such as address points or census zones has tended to reinforce the view that these phenomena are indeed discrete, but there are good grounds for reconceptualising many such phenomena as continuously varying over space, and attempting to represent them in this way. Nordbeck and Rystedt (1970) describe phenomena such as population density as 'reference interval functions' which cannot be measured at a single point, but only have meaning in relation to some reference interval, such as 200 persons per hectare. Such functions may be treated as surfaces with validity. None of the existing data models used by GIS is able to represent fully continuous variation over space, but structures such as altitude matrices and Triangulated Irregular Network (TIN) models can represent surface form adequately if used at an appropriate scale. Tobler (1979) presents a volume-preserving approach to the construction of socioeconomic surfaces from areal data, and Martin (1989) illustrates surface construction from zone centroids, a technique developed further by Bracken and Martin (1995), for example.

The final row of the table illustrates the kinds of display and visualisation technique which may be applied to spatial data representing socioeconomic phenomena. Visualisation may involve the direct display of data values using a visual representation of the same object class, or may involve further transformation of the data for display which does not directly affect the form of the digital objects in the database. Traditional cartographic representation makes use of points, lines, and areas together with particular conventions for symbolisation, projection, and scale (Elshaw Thrall and Thrall, Chapter 23; Kraak, Chapter 11). These techniques may be directly reproduced using GIS technology, but a broader range of possibilities also becomes available. Gatrell (1994) explores the visualisation of point patterns, which may be applied to the types of postal and address information discussed above, but the use of choropleth (shaded area) maps for areal data is still the most widespread type of socioeconomic data mapping, embodying all the difficulties associated with modifiable areal units. Flow data are conventionally represented by mapped lines, while surfaces are variously depicted by isolines (lines of equal value) such as contours, sample points, or quasi-continuous shading of small areas – such as that which may be output from an elevation model comprising a grid of small cells. Less conventional, but potentially more powerful, representations of each object class may be achieved by the use of mapping tools in which strict geographical relationships are relaxed in order to represent other features of the data. Dorling (1994) presents a range of such approaches, including multidimensional glyphs for point data and area cartograms in which the geographical area of zones is replaced by a non-spatial attribute such as population size in determining the size of the mapped symbol.

There is increasing interest in 3-dimensional visualisation tools in GIS, both in order to recreate 'realistic' scenes and as an aid to the understanding of patterns in data. In architectural and planning applications, there is increasing integration between GIS and 3-dimensional tools for the exploration of urban scenes which may be affected by redevelopment, for example (Levy 1993; Liggett and Jepson 1995). Data visualisation in three dimensions is less well developed in socioeconomic applications, but there is considerable scope for the use of 3-dimensional scenes in which some of the visual parameters – such as the colour and texture of objects – are used to symbolise otherwise unobservable pattern in socioeconomic phenomena. An interesting example is provided by Wood et al (1996), in which 3-dimensional models are used to explore features of the population structure of Greater London.

4 DISCUSSION

In the light of these many different possibilities, is there a single 'right' way to represent socioeconomic phenomena within GIS? The answer to this question is almost certainly 'no', the best strategy being highly dependent on the specific application. The implication of this is that considerable onus is placed on users to fully understand the implications of the representation strategy which they choose to adopt. In addition to the technical and conceptual considerations already addressed, a number of further relevant issues are addressed here, although fuller discussions of some of these will be found in later chapters. (e.g. see Cova, Chapter 60, for a discussion of representational issues in emergency management; Larsen, Chapter 71, for issues in environmental monitoring and assessment applications).

In concentrating on spatial representation, this discussion has largely assumed that the spatially independent attribute characteristics of objects are unproblematic, but there will be many situations in which this is not the case. Attribute values will be subject to broadly similar operator selectivity and measurement error to their spatial counterparts, and the extent of these influences may vary over space. Data input to GIS, and therefore the range of issues which can reasonably be addressed, is usually constrained by the questions which were asked in some previously existing survey or census (see Goodchild and Longley, Chapter 40). The 1991 UK Census was subject to differing degrees of underenumeration, which is generally understood to have been highest among the young adult male populations concentrated in large metropolitan centres. In the analysis of the resulting data, there may therefore be complex interactions between the spatial and attribute characteristics. The use of GIS to optimise zoning schemes for socioeconomic data, as illustrated by Openshaw and Alvanides (Chapter 18) and by Openshaw and Rao (1995), potentially offers important advances in the control of attribute variation by spatial manipulation.

Other GIS developments which potentially offer benefits to the representation of socioeconomic phenomena include the incorporation of fuzzy concepts, and non-Euclidean spaces (see Couclelis, Chapter 2, and Fisher, Chapter 13). Openshaw (1989) uses the term 'fuzzy geodemographics' to refer to an approach in which the inherent uncertainty in spatial referencing and attribute values for socioeconomic GIS data are acknowledged. By contrast, in some current applications, where all the information to be processed relates to the same set of zones, or the task is essentially one of list matching (for example, between entries in a directory of post/zip codes), then it may not be necessary to use GIS at all (Barr 1993).

A longstanding issue when dealing with information concerning individuals in computer-readable form has been that of confidentiality, and the need to protect individuals from the inadvertent disclosure of personal information. This has usually been a guiding principle in the publication of data from censuses by government, typically resulting in the imposition of a population threshold size below which no data are published. However, this deliberate aggregation of data runs counter to many business applications, in which an objective is to identify individuals or households with particular characteristics as precisely as possible, and there is considerable commercial interest in the association of different data series in order to build up detailed individual-level information. The 'representation' of individuals in this way raises ethical issues which have not really been addressed by most GIS users, but which are considered more fully by Curry (Chapter 55). The use of GIS in this area has been the subject of broad-ranging critique, such as that found in Pickles (Chapter 4; 1995b). There is not space here to develop a response to these arguments, but it is certainly true that maps have always been tools associated with the exercise of power, representing both the physical and socioeconomic worlds selectively, for the purposes of particular groups and individuals (Wood 1992). These possibilities are multiplied rather than reduced by GIS, with its multiple representational strategies. There is still a considerable gap between the academic critics and proponents of GIS technology, but at present many of these issues would probably not even be recognised by many non-academic GIS users who routinely handle socioeconomic data. As with the technical choices concerning representation, there is a heavy responsibility placed on GIS users to be fully aware of the implications of their actions.

5 CONCLUSION

This chapter has considered the representation of phenomena likely to be of interest to social scientists within GIS. GIS offer a wide range of tools for the manipulation and analysis of models of geographical reality stored as digital objects, but the primary difficulty in socioeconomic applications concerns the most appropriate way in which to measure and model the phenomena of interest.

There are many sources of computer-readable data about population and economic activity, but these are not usually associated with directly measurable geographical locations, and some form of indirect spatial referencing is required. The association of population-related phenomena with discrete geographical locations poses technical, conceptual, and ethical difficulties. There have been improvements in the number and resolution of available datasets for georeferencing, but this has not made the conceptualisation of phenomena such as unemployment or population density any easier. There remains disagreement over the precise spatial form of these phenomena, and therefore the most appropriate technical solutions which should be adopted. Initial decisions about georeferencing affect all subsequent transformations of the data, both for the creation of new digital objects and for visualisation and output. GIS users in this area also need to address more carefully some of the broader questions about representation which have perhaps to some degree been provoked by uncritical use of the available data sources.

References

Bailey T C, Gatrell A C 1995 *Interactive spatial data analysis.* Harlow, Longman/New York, John Wiley & Sons Inc.

Barr R 1993 Mapping and spatial analysis. In Dale A, Marsh C (eds) *The 1991 Census user's guide.* London, HMSO: 248–68

Barr R 1996 A comparison of aspects of the US and UK censuses of population. *Transactions in GIS* 1: 49–60

Blalock M 1964 *Causal inferences in nonexperimental research.* Chapel Hill, University of North Carolina

Boots B N 1986 *Voroni (Thiessen) polygons* Concepts and Techniques in Modern Geography 45. Norwich, Geo-Books

Bracken I, Martin D 1995 Linkage of the 1981 and 1991 censuses using surface modelling concepts. *Environment and Planning A* 27: 379–90

Broome F R, Meixler D B 1990 The TIGER database structure. In Marx R W (ed.) *The Census Bureau's TIGER system.* Bethesda, ACSM: 39–47

Clarke K C 1995 *Analytic and computer cartography*, 2nd edition. Englewood Cliffs, Prentice-Hall

CRU/OPCS/GRO(S) 1980 *People in Britain: a census atlas.* London, Her Majesty's Stationery Office

Cushnie J 1994 A British Standard is published. *Mapping Awareness* 8: 40–43

Dorling D 1994a Cartograms for visualising human geography. In Hearnshaw H M, Unwin D J (eds) *Visualisation in geographical information systems.* Chichester, John Wiley & Sons: 85–102

Flowerdew R, Green M 1991 Data integration: statistical methods for transferring data between zonal systems. In Masser I, Blakemore M (eds) *Handling geographic information: methodology and potential applications*. Harlow, Longman/New York, John Wiley & Sons Inc.: 38–54

Fotheringham A S, Wong D W S 1991 The modifiable areal unit problem in multivariate statistical analysis. *Environment and Planning A* 23: 1025–34

Gatrell A C 1991 Concepts of space and geographical data. In Maguire D J, Goodchild M F, Rhind D W (eds) *Geographical information systems: principles and applications*. Harlow, Longman/New York, John Wiley & Sons Inc. Vol. 1: 119–34

Gatrell A C 1994 Density estimation and the visualisation of point patterns. In Hearnshaw H M, Unwin D J (eds) *Visualisation in geographical information systems*. Chichester, John Wiley & Sons: 65–75

Goodchild M F, Anselin L, Deichmann U 1993 A framework for the areal interpolation of socioeconomic data. *Environment and Planning A* 25: 383–97

Levy R M 1995 Visualisation of urban alternatives. *Environment and Planning B: Planning and Design* 22: 343–58

Liggett R S, Jepson W H 1995 An integrated environment for urban simulation. *Environment and Planning B: Planning and Design* 22: 291–302

Lind M, Christensen T 1996 The address as key to GIS. *Proceedings of the second joint European conference and exhibition on geographical information*. Barcelona Vol. 2: 868–77

Longley P, Clarke G 1995a *GIS for business and service planning*. Cambridge (UK), GeoInformation International

Maffini G 1987 Raster versus vector data encoding and handling: a commentary. *Photogrammetric Engineering and Remote Sensing* 53: 1397–8

Martin D 1989 Mapping population data from zone centroid locations. *Transactions of the Institute of British Geographers* NS 14: 90–7

Martin D 1996 *Geographic information systems: socioeconomic applications*, 2nd edition. London, Routledge

Martin D, Higgs G 1996 Georeferencing people and places: a comparison of detailed datasets. In Parker D (ed.) *Innovations in GIS 3*. London, Taylor and Francis: 37–47

Nordbeck S, Rystedt B 1970 Isarithmic maps and the continuity of reference interval functions. *Geografiska Annaler* 52B: 92–123

Openshaw S 1984 *The modifiable areal unit problem*. Concepts and Techniques in Modern Geography 38. Norwich, Geo-Books

Openshaw S 1989b Learning to live with error in spatial databases. In Goodchild M F, Gopal S (eds) *Accuracy of spatial databases*. London, Taylor and Francis: 263–76

Openshaw S, Rao L 1995 Algorithms for reengineering 1991 Census geography. *Environment and Planning A* 27: 425–46

Openshaw S, Taylor P J 1981 The modifiable areal unit problem. In Wrigley N, Bennett R J (eds) *Quantitative geography: a British view*. London, Routledge: 60–70

Ottoson L, Rystedt B 1991 National GIS programmes in Sweden. In Maguire D J, Goodchild M F, Rhind D W (eds) *Geographical information systems: principles and applications*. Harlow, Longman/New York, John Wiley & Sons Inc.Vol. 2: 39–46

Peucker T K, Chrisman N 1975 Cartographic data structures. *The American Cartographer* 2: 55–69

Pickles J (ed.) 1995a *Ground truth: the social implications of geographic information systems*. London, Guilford Press

Pickles J 1995b Representations in an electronic age: geography GIS and democracy. In Pickles J (ed.) *Ground truth: the social implications of geographic information systems*. London, Guilford Press: 1–30

Raper J F, Rhind D W, Shepherd J W 1992 *Postcodes: the new geography*. Harlow, Longman

Redfern P 1989 Population registers: some administrative and statistical pros and cons. *Journal of the Royal Statistical Society* A 152: 1–28

Rhind D W 1991 Counting the people. In Maguire D J, Goodchild M F, Rhind D W (eds) *Geographical information systems: principles and applications*. Harlow, Longman/ New York, John Wiley & Sons Inc. Vol. 2: 127–37

Tobler W R 1979 Smooth pycnophylactic interpolation for geographical regions. *Journal of the American Statistical Association* 74: 519–36

Unwin D J 1981 *Introductory spatial analysis*. London, Methuen

Wood D 1992 *The power of maps*. London, Routledge/New York, Guilford Press

Wood J, Unwin D, Stynes K, Fisher P, Dykes J 1996 Using the landscape metaphor to understand population data. *http://www.geog.le.ac.uk/argus/poplandscape/index.html*

7

Spatial representation: a cognitive view

D M MARK

Approaches to the representation and modelling of geographical phenomena in digital form have expanded recently with the introduction of new perspectives, notably from cognitive science. The chapter reviews the roots of this cognitive perspective, and examines the impacts it is having on GIS. A formal model of topological relations between geographical objects is introduced, and used as the basis of a simple experiment with human subjects to show how its concepts are mapped into human cognition. Such experiments show promise for improved designs for easier-to-use systems.

1 INTRODUCTION

Most computer programs gain their meaning from the relationships between their internal formalisms and some aspect of the external world. Digital objects are used to represent entities in the real world, and computational procedures represent real-world processes. In the science of artificial intelligence, a representation has been defined as 'a set of conventions about how to describe a set of things' (Winston 1984: 21). Winston goes on: 'A description makes use of the conventions of a representation to describe some particular things'. The fidelity of these representations of entities and processes of the world is a key factor in the potential utility and usability of almost any computational system. Representation is a key concept in any application that uses computers.

This chapter provides a cognitive perspective on representations of geographical space, entities, and relationships that underpin geographical information systems. Representations of things geographical that have been developed in the cognitive sciences for use in those sciences will not be a focus here, but will be reviewed for what they may contain that is relevant to GIS. The review will concentrate on how a cognitive science perspective can inform the definition and design of digital objects to represent geography. It will also concentrate on graphic symbols that can represent those internal digital representations for presentation to and interaction with GIS users.

2 REALISM AND EXPERIENTIALISM

Many of the characteristics of the geographical world can be determined through objective methods such as measurement although human perceptions of the geographical world are also established through such cognitive functions as perception, reasoning, and memory. These two worlds, of measurement and cognition, are clearly compatible in many aspects. Even if there is no functional connection between them (as a phenomenologist might assert), experiences provide feedbacks that tune internal mental representations to match reality. This is a restatement of experiential realism, a philosophical stance outlined and named by Lakoff (1987). Experiential realism appears to be a good base for geographical representation in GIS (Couclelis 1988; Frank and Mark 1991; Mark and Frank 1996). Also, if mental representations exist, and are shaped both by the nature of human brains, bodies, and senses and by the nature of the world, then the same representations should be appropriate both to model spatial cognition and for scientific models of geographical phenomena. Thus the representations discussed in this chapter should not differ systematically from those in other chapters in this section; rather, different motivations and sources of evidence should lead to similar representational systems.

3 PERCEPTION, BEHAVIOUR, LANGUAGE, AND COGNITION

'Cognition' and 'cognitive' are words that refer to conscious thinking, including memory, reasoning, and perception. 'Perception' normally has a more narrow meaning, being used to refer to mental sensations and processes that relate to the senses and which occur in the direct presence of sensory stimuli (this psychological definition of perception differs from the common (mis)use of this term in the geographical literature to refer to any mental or cognitive phenomenon, as in 'environmental perception'). Some brain functions, and some brain controls of the body, are outside consciousness – motor skills are a good example. Such non-conscious processes are usually excluded from 'cognition'. Since human behaviour is a combination of perceptual inputs, conscious decisions, and motor acts, cognition is just one aspect influencing human behaviour. Since people's reasoning and decisions cannot be better than their knowledge of situations, cognitive factors often influence behaviour more than objective properties of the world such as distances or directions.

Language can be considered to be a very specialised form of behaviour, but it also appears to provide a unique window on mental processes. The relation between language and thought is, however, controversial. Decades ago, amateur linguist Benjamin Lee Whorf proposed an extreme level of interdependence between language and thought – that people think in language, and that native speakers of different languages think (reason) differently (Whorf 1940). In recent times, many linguists have attacked this conjecture, sometimes known as the Sapir-Whorf Hypothesis, claiming that it is false, and even ridiculous (cf. Pinker 1994). Whether or not language influences thought, it seems obvious that thought influences language, and that studying both the structure and the use of natural language describing spatial situations is a good way to study how people think about such situations. More specifically, distinctions made by language, and distinctions in spatial situations which must be made in order to account for linguistic differences, may reveal fundamentals of spatial representation that may be useful in GIS as well.

4 ENTITIES, FIELDS, AND PHENOMENA

The first problem in representation is to decide what is to be represented. What things exist? To put this more formally, what is the ontology of geographical space? This chapter will follow the terminology of the US Spatial Data Transfer Standard (SDTS). 'Entities' are things in the real world; 'objects' are things in the digital world. Digital objects and associated attributes and values represent geographical entities. The distinction between entity and object makes explicit the difference between things and their representations in a formal system.

4.1 Entities versus fields

Many geographical phenomena are best described scientifically as fields. Good examples are topographic elevations, air temperatures, and soil moisture content. A 2-dimensional field may be defined as any single-valued function of location in a 2-dimensional space. Fields can be classified according to the levels of measurement (data types) of the dependent variable. Two of the most important types are continuous fields, in which the dependent variable is measured on an interval or ratio scale, and discrete fields, with nominal dependent variables. It appears that any geographical phenomena can be represented either as a field or as a collection of digital objects. However, in many cases one approach is a much better basis for efficient computation; also, one may be a better model of people's mental representations of the phenomena. For example, a set of states or provinces within a country would commonly be represented in GIS as a set of areal objects, or as a set of linear objects that form their boundaries. However, the same phenomenon could also be represented as a discrete 2-dimensional field where each point in space is mapped into the nominal category indicating which state it falls in. Fields can be digitally represented by vector approaches, but often are represented by raster data structures.

Whether the entity or the field model is more appropriate is particularly interesting for topographic elevations. Topographic data normally are represented in GIS as fields, either through gridded digital elevation models (DEMs) or as triangular tessellations (TINs: see Hutchinson and Gallant, Chapter 9). Although TINs are often placed in the vector GIS class, they still represent a field. However, in discourse, people normally communicate about terrain in terms of features such as hills, ridges, valleys, gullies, canyons, etc. Thus, a commonsense GIS, capable of being used by untrained people, should be able to handle queries about 'this hill' or 'that canyon'. However, this is not just an issue for easy-to-use GIS: a lot of field data from the pre-Global Positioning System (GPS) days – such as the labels on biological specimens in museums – have location specified only by feature names and descriptions, such as 'on the east ridge of Sumas Mountain' (see also Seeger, Chapter 30; Shiffer, Chapter 52; and Smith and Rhind, Chapter 47). A system to geocode such museum locality data automatically would have to be able to recognise ridges and mountains. Technically, either an objects or a fields model could be used internally in a GIS; the other model could then be presented to users as a 'view'.

4.2 Dimensionality: points, lines, and areas

It has become standard practice to classify geographical entities according to their dimensionality as points, lines, or areas. This typology is especially prevalent in cartography. For the conventional wisdom of US cartography there is arguably no better place to look than the various editions of *Elements of Cartography* by Arthur H Robinson and his various colleagues at the University of Wisconsin. In the second edition, Robinson (1958: 137) wrote about four kinds of geographical quantities (point, line, area, and volume) and three kinds of cartographic symbols (point, line, and area). Robinson discusses 2-dimensional data in a chapter on 'Mapping quantitative point, line, and area data', and separates volume data under the title 'Mapping 3-dimensional data'. Campbell's textbook (1984) follows an almost identical chapter subdivision, with chapters on 'Mapping spatial variations: points, lines, and areas' and on 'Mapping spatial variations: surfaces'. It is not surprising that GIS follows the same basic classification, using dimensionality as the highest level of subdivision for geographical objects (cf. Fegeas et al 1992).

It is interesting that cognitive and linguistic works also use such a classification, apparently without knowledge of the cartography and GIS literatures. For example, in Herskovits' (1986) study of the prepositions in English, she uses exactly the same classification based on dimensionality; as does Talmy's classic work on the relation between language and the structure of space (Talmy 1983). Evidently, although entities must be 3-dimensional to have a real physical existence in a 3-dimensional world, it is common if not universal for people to conceptualise some geographical entities as points, others as lines, and still others as areas (regions).

4.3 Entities with uncertain boundaries

Whereas some geographical entities have distinct or crisp boundaries, many lack these and instead are 'bounded' by transition zones (Fisher, Chapter 13). The frequency of such indistinct boundaries is one of the most distinctive things about geographical entities, compared with manipulable (table-top) entities. The frequency of geographical entities with indistinct boundaries has been known for some time, yet vector GIS is tuned to represent entities with crisp boundaries, whereas raster GIS does not represent entity boundaries at all. Thus, formal methods for the representation of geographical entities with uncertain or graded boundaries is an important new research topic in GIS (cf. Burrough and Frank 1996). Fuzzy set theory represents a possible approach to modelling entities with graded boundaries, but it has problems (see Fisher, Chapter 13). Fuzzy membership functions can be the dependent variables in membership fields, perhaps represented by rasters, but it is not clear whether a full range of GIS functions can be based on such data or whether results of such implementations would be either cognitively acceptable or scientifically valid. Much work in implementation, testing, and human subjects' evaluation lies ahead.

4.4 Entity types

Classification is a fundamental cognitive process, and it is widely held that categories lie near the heart of cognition (Lakoff 1987; Rosch 1973, 1978). Categories carry a great deal of generic, or default, characteristics of entities and allow people to 'know' some aspects of novel situations. Geographical entities are no exception, and classification of geographical entities into categories is a well-known process both in everyday thinking and in scientific work. Various subfields of geography have developed elaborate classifications for kinds of landforms, vegetation assemblages, settlements etc.

Mark (1993) described a cognitive theory of categories and showed how such a theory could form a sound basis for entity types in geographical data interchange standards such as the US SDTS and other data standards. Using an example of inland water bodies, Mark showed that the conceptual boundaries between adjacent categories of water bodies in two closely related European languages (lake, pond, lagoon in English compared with lac, étang, lagune in French) did not match up. The category of water bodies classified as étangs in French might be considered to be lakes or ponds or lagoons in English. In standard English, lakes are distinguished from ponds mainly by size and lagoons are distinguished mainly by their position relative to the sea. On the other hand, the distinction between *étang* and *lac* in French seems to be mainly one of water quality with étangs having stagnant water. A weakness of Mark's analysis is that it was based on dictionaries and examples, rather than on human subjects' data. Clearly, further research is required. The example points to the relation of geographical categories to cognition, and to the potential of cultural differences in geographical entity type definition, which could be an impediment to cross-cultural geographical information exchange.

5 SPATIAL RELATIONS

Spatial relations are what distinguish spatial information from other information. Spatial relations are often encoded in human natural language by closed-class linguistic elements, typically prepositions in western European languages. 'Closed-class' means that there is a relatively small and fixed set of words (lexicon), and thus a limited number of categories that can be distinguished. Spatial relations also can be expressed through verbs that describe trajectories or other spatial actions.

Mathematically, in a metric case, there is an infinite continuum of possible spatial relations between any two entities. Topological characteristics of a situation are invariant under certain transformations. There is evidence that many of the cognitively important spatial relations are entirely or predominantly topological. Because of combinatorial principles, even

a small number of topological distinctions can lead to a large number of mathematically-distinct topological spatial relations. Most work on formalising spatial relations, both in GIS and in cognitive science, has relied on researchers' intuitions as to which relations are worthy of distinction and which are not. Some cognitive principles for motivating formal models of spatial relations are presented in this section (5.1–5.4).

5.1 Topology and metrics

Topology may be defined as 'those properties of geometrical figures that are invariant under continuous deformation' (McDonnell and Kemp 1995). Many spatial relations between objects are topological in nature, including adjacency, containment, and overlap. It has long been known that topological spatial relationships are learned by humans at a very early age, well under one year (Piaget and Inhelder 1956). Metric information such as size, shape, distance, or direction can also be very important cognitively, but is often used to identify entity types. Metric aspects of spatial relations often refine, rather than define, spatial relations. The relative roles of topological and metric properties in defining spatial relations in reasoning or language is complex. Some terms indicate relations which are purely topological, and in those cases metric properties may be irrelevant – 'within' and 'enters' are probably examples of this. In other cases, metric properties such as distance or direction, expressed either quantitatively or qualitatively, may determine the meanings of various terms, for example 'north of' or 'near' both normally refine the 'disjoint' topological relation, and are ill defined for non-disjoint entities.

5.2 Spatial relations between disjoint objects

It seems that spatial relations between disjoint entities, which neither touch nor overlap, are characterised by a system of distinctions that is essentially independent of the system used to describe and classify spatial relations for non-disjoint entities. This section (5.2) deals with disjoint entities, and the following section will deal with the other case.

5.2.1 Distance
Distance may be pure Euclidean distance. In natural language, 'hedge' words such as 'about' are often associated with approximate numerical distance.

Even more commonly, distance may be given in qualitative rather than metric terms, dividing distances into perhaps just three categories: 'at', 'near', and 'far'. There may be gender differences in the tendency to use or rely on metric distances, compared with landmarks. Concepts such as 'near' are ill defined, and Robinson and his colleagues (Robinson and Lundberg 1987; Robinson and Wong 1987) worked on calibrating the meaning of 'near' using fuzzy set theory.

5.2.2 Direction

Direction also may be either qualitative or quantitative. Direction is an orientation specified relative to some reference frame (see section 5.2.3). Again, in everyday speech, directions seem normally to be specified qualitatively, typically in either four or eight directions. For science and navigation, more precise metric measures of directions are needed, normally specified in degrees from some arbitrary direction. Frank (1992) has discussed qualitative reasoning about cardinal directions, and has shown that eight directions are an adequate basis for most spatial reasoning.

Ideally, directional relations are thought of as being between points. Directions are not so straightforward between spatially-extended entities, since a large range of directions may exist, between any point in one entity and any point in the other. Peuquet and Zhan (1987) discussed this problem and provided heuristic rules for computing the direction relation between extended entities. However, there were apparently no human subjects tests to evaluate whether the algorithm's conclusions match human intuitions or the usage of directional terms. The approach advocated in this chapter requires eventual cognitive testing with human subjects before the heuristics developed by Peuquet and Zhan can be considered to be effective and workable representations (see also Peuquet, Chapter 8).

5.2.3 Reference frames

Various reference frames are used in discourse and spatial reasoning. Geographically, in many cultures, a reference frame based on cardinal directions seems dominant for outdoor ('geographical') spaces, whereas viewer-centred or object-centred reference frames often dominate for bodily or tabletop ('manipulable') spaces and entities (Mark et al 1987). But even in speech communities that typically use cardinal directions in geographical space, it is common to describe location as being, say, 'in front of

the library', which invokes a directional reference frame centred on and oriented to the library. Systems to use and understand many spatial terms, especially directional, will have to accommodate multiple reference frames and switching contexts. Individuals may need to switch among these reference frames during discourse or spatial reasoning. Preferences among multiple available reference frames vary with situation, scale, and culture, with a few cultures typically using cardinal directions even indoors and for body parts, and with others failing to have orthogonal geographical coordinate schemes equivalent to cardinal directions (Pederson 1993).

5.3 Spatial relations between non-disjoint objects

This section discusses topological spatial relations between entities using a particular formal model termed the 9-intersection (Egenhofer and Kuhn, this volume; Egenhofer and Herring 1994). This model is based on very simple views of spatial entities: each entity is defined to have an interior, a boundary, and an exterior. The exterior fills the 'universe', except for the parts of that universe occupied by the entity itself and its boundary. Uses of the 9-intersection, and the discussions of it in this chapter, have been restricted to entities in a 2-dimensional space, although it can also be applied to spaces of fewer or more dimensions.

The simple form of the 9-intersection model tests each 'part' (interior, boundary, exterior) of one spatial entity against each such part of the other and simply records which of the nine possible intersections (3×3) are empty and which are not. With two possible 'states' for each of nine possible intersections, this model could distinguish 2^9, or 512, spatial relationships. However, if continuity constraints are placed on the entities involved, the number of possible spatial relations is greatly reduced. The 9-intersection model offers no improvement in explanatory or descriptive power for region–region relations, compared with previous models. For a pair of regions, eight spatial relations can be distinguished, as shown in Figure 1. For two simple unbranched line segments, however, the 9-intersection model distinguishes 33 different topological spatial relations. It seems unlikely that people distinguish that many relations and we advocate human subjects testing to sort out how these 33 relations are organised in cognitive and linguistic systems.

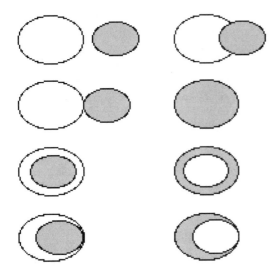

Fig 1. The eight distinct spatial relations between two simply connected 2-dimensional objects in a 2-dimensional space, according to the 9-intersection model.

Line–region relations offer an interesting intermediate level of complexity and resolution. Egenhofer and Herring (1994) showed that 19 spatial relations are possible between a line and a region. These relations have formed the basis for a program of cognitive evaluation of spatial relations that Mark and Egenhofer have been conducting over the last several years. This work will be reviewed next as a case study.

5.4 Case study: spatial relations between lines and regions

As noted above, the simple (Boolean) version of the 9-intersection model distinguishes 19 spatial relations between a simple, unbranched line and a simply connected region without holes. Intuition would suggest that people would rarely distinguish that many spatial relations. However, such intuitions may be wrong, and human subjects testing is essential to determine what distinctions are actually made in language or spatial reasoning. Tests of users' reactions to spatial relations in queries or reasoning situations apparently have not been performed. However, tests using language have been done by Mark, Egenhofer and colleagues, and this work will be summarised here to demonstrate the value of subjects testing combined with formal models.

Mark and Egenhofer first explored the cognitive validity of the 9-intersection relations using a

grouping task (Mark and Egenhofer 1994b). All of their experiments so far have been performed with a geographical context of a road and a park. A park outline shape was kept constant for all experiments, and roads were placed in 38 different positions, two for each 9-intersection relation. Subjects were asked to put drawings into groups so that the same phrase or sentence could be used to describe all drawings in the group. Subjects were native speakers of English, Mandarin, or German. The experiment produced two main findings: first, topologically identical drawings (according to the 9-intersection) were almost always grouped together by any subjects; but second, subjects grouped topological relations in a very wide variety of ways (Mark and Egenhofer 1994b). Beyond the 9-intersection classes, individual differences dominated the results, and cultural or linguistic differences, if present, could not be detected given the small sample sizes and the within-group variance. Tendencies for relations different in only one of the nine intersections to be grouped together were used by Egenhofer and Mark (1995) as part of an evaluation of conceptual neighbourhoods for spatial relations. However, different experimental protocols were needed to get at more subtle differences in cognitive representations of spatial relations.

Mark and Egenhofer designed an agreement task and a drawing task. In the drawing task, subjects were presented with blank outlines of a polygon said to be a park, with a sentence under each drawing. They were asked to draw a line to represent a road that had the spatial relation indicated in the sentence. This test was applied to 32 English subjects, each of whom was asked to draw 64 sentences, and 19 Spanish-speaking subjects who drew 43 sentences (Mark and Egenhofer 1995). Such a protocol tends to identify prototypical configurations. Respondents produced a total of more than 2800 drawings; 88 per cent of which fell into just five spatial relations, roughly equivalent to 'inside', 'outside' (disjoint), 'enters', 'crosses', and 'goes to'. Sharif (1996) analysed the geometry of the drawings by the English language subjects and found that there were many empirical regularities in the spatial relations drawn.

In the agreement task, subjects were presented with a road–park map and a sentence (such as 'the road crosses the park') and were asked to evaluate the degree to which the sentence described the spatial relation between the road and park shown in the map (Mark and Egenhofer 1994a). This task was much more controlled, and produced useful results. It has been applied to 12 sentences in English (see Mark et al 1995 for a summary of most of this work) and one sentence in Spanish; Abrahamson (1994) also used the same stimuli and experimental protocol for five sentences in Norwegian. A total of more than 550 subjects have been tested for English sentences, and about 90 subjects for the Spanish sentence. When responses are averaged for each map–sentence pair, across all subjects within a language, and rescaled to a 0–1 scale, the values can be interpreted as the 'membership' of that map in the fuzzy set of all configurations described by, for example, 'the road crosses the park'. Correlations between mean agreements to different sentences, across all maps, can be interpreted as a measure of the similarity in meanings of the sentences tested. The mean agreement values can also be related to other parameters describing the configurations, in efforts to explain subjects' responses and perhaps to characterise the meanings of the sentences themselves.

The most solid data are for the English sentence 'the road crosses the park' (about 150 subjects) and the Spanish 'la carretera cruza el parque' (about 90 subjects). These are plotted in Figure 2, and the simple correlations between the means by diagram is 0.985 (97 per cent of variance in common). Not only are the means related to topology in about the same way, but geometric variations also reduce agreement for the English-speaking and Spanish-speaking subjects in almost exactly the same way.

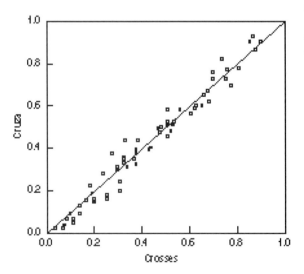

Fig 2. Mean agreement values for 'cruza' against 'crosses'.

The results of all this testing has been a general confirmation of the validity of the 9-intersection model. However, the relations it distinguishes must in most cases be aggregated in order to make intuitive spatial relations. Previous work on formal models of spatial relations has generally assumed, either explicitly or implicitly, that the relations distinguished form a uniform set of equally distinct spatial relations. In the new model, the same 'primitive' spatial relation might be an element in two or more cognitive spatial relations. This result requires both a powerful formal model of spatial relations, and human subjects testing, using several protocols, to confirm or refine the formal models.

6 SUMMARY AND PROSPECTS

Experiential realism appears to bridge the gap between naive realism and developmental models of cognition. Cognition is related to perception, language, and behaviour. The exact relation of language to cognition is controversial, with some believing that language shapes cognition, whereas others see cognition as being independent of language. Whatever the relation, language is certainly related to culture, and makes an excellent site for the study of certain aspects of cognition.

Both entities and fields exist in cognitive models. Entities are typically conceptualised as being organised by dimensionality: points, lines, areas, volumes. Entities are often thought of as having indistinct boundaries, a fact which is at odds with typical GIS representation schemes. Entities are also categorised, and since many aspects of nature form a continuum, categories may be relatively arbitrary and thus subject to cultural differences. Spatial relations, on the other hand, seem to be very similar in disparate cultures and languages. Cognitive spatial relations are predominantly topological but metric factors such as distance and direction often refine the relations and characterise prototypical relations. A case study, that of spatial relations between lines and regions, was used to describe the value of research approaches combining formal models and human subjects testing.

There is a very real sense in which all representations are cognitive. Mathematics is, after all, a formalisation of how at least some people think. The cognitive view of spatial relations, however, emphasises the importance of human subjects testing, preferably under laboratory controlled conditions, in defining the nature of the spatial representations that are needed for geographical information systems and spatial analysis.

References

Abrahamson F 1994 'Uklarhet og noyaktighet i geografiske informasjonssystemer (fuzziness and accuracy in geographical information systems)'. Unpublished Master's thesis, Division of Surveying and Mapping, University of Trondheim (Norway) and Norwegian Institute of Technology

Burrough P A, Frank A U (eds) 1996 *Geographic objects with indeterminate boundaries*. London, Taylor and Francis

Campbell J 1984 *Introductory cartography*. Englewood Cliffs, Prentice-Hall

Couclelis H 1988 The truth seekers: geographers in search of the human world. In Golledge R, Couclelis H, Gould P (eds) *A ground for common search*. Sanata Barbara, The Santa Barbara Geographical Press: 148–155

Egenhofer M J, Herring J 1994 Categorising topological spatial relations between point, line, and area objects. In Egenhofer M J, Mark D M, Herring J R (eds) *The 9-Intersection: formalism and its use for natural-language spatial predicates*. Technical Report 94–1. Santa Barbara, NCGIA

Egenhofer M J, Mark D M 1995 Modelling conceptual neighbourhoods of topological line–region relations. *International Journal of Geographical Information Systems* 9: 555–65

Fegeas R G, Cascio J L, Lazar R A 1992 An overview of FIPS 173, the Spatial Data Transfer Standard. *Cartography and Geographic Information Systems* 19: 278–93

Frank A U, Mark D M 1991 Language issues for GIS. In Maguire D J, Goodchild M F, Rhind D W (eds) *Geographical information systems: principles and applications*. London, Longman/New York, John Wiley & Sons Inc. Vol. 1: 147–63

Frank A U 1992a Qualitative spatial reasoning about distances and directions in geographic space. *Journal of Visual Languages and Computing* 3: 343–71

Herskovits A 1986 *Language and spatial cognition: an interdisciplinary study of the prepositions in English*. Cambridge (UK), Cambridge University Press

Lakoff G 1987 *Women, fire, and dangerous things: what categories reveal about the mind*. Chicago, University of Chicago Press

Mark D M 1993 Toward a theoretical framework for geographic entity types. In Frank A U, Campari I (eds) *Spatial information theory: a theoretical basis for GIS*. Lecture Notes in Computer Sciences No. 716. Berlin, Springer: 270–83

Mark D M, Comas D, Egenhofer M J, Freundschuh S M, Gould M D, Nunes J 1995 Evaluating and refining computational models of spatial relations through cross-linguistic human-subjects testing. In Frank A U, Kuhn W (eds) *Spatial information theory: a theoretical basis for GIS*. Lecture Notes in Computer Science No. 988. Berlin, Springer: 553–68

Mark D M, Egenhofer M J 1994a Calibrating the meanings of spatial predicates from natural language: line–region relations. *Proceedings, Spatial Data Handling 1994* 1: 538–53

Mark D M, Egenhofer M J 1994b Modelling spatial relations between lines and regions: combining formal mathematical models and human subjects testing. *Cartography and Geographic Information Systems* 21: 195–212

Mark D M, Egenhofer M J 1995 Topology of prototypical spatial relations between lines and regions in English and Spanish. *Proceedings, AutoCarto 12, Charlotte, North Carolina, March 1995*: 245–54

Mark D M, Frank A U 1996 Experiential and formal models of geographic space. *Environment and Planning B: Planning and Design* 23: 3–24

Mark D M, Svorou S, Zubin D 1987 Spatial terms and spatial concepts: geographic, cognitive, and linguistic perspectives. *Proceedings, International Symposium on Geographic Information Systems: The Research Agenda, November.* Crystal City, Virginia 2: 101–12

McDonnell R, Kemp K 1995 *International GIS dictionary*. Cambridge (UK), GeoInformation International

Pederson E 1993 Geographic and manipulable space in two Tamil linguistic systems. In Frank A U, Campari I (eds) *Spatial information theory: a theoretical basis for GIS*. Lecture Notes in Computer Science No. 716. Berlin, Springer: 294–311

Peuquet D J, Zhan C-X 1987 An algorithm to determine the directional relationship between arbitrarily-shaped polygons in the plane. *Pattern Recognition* 20: 65–74

Piaget J, Inhelder B 1956 *The child's conception of space*. London, Routledge

Pinker S 1994 *The language instinct*. New York, William Morrow & Co. Inc.

Robinson A H 1958 *Elements of cartography,* 2nd edition. New York, John Wiley & Sons Inc.

Robinson V B, Lundberg G 1987 Organisation and knowledge base considerations for the design of distributed geographic information systems – lessons from semantic modeling. *Proceedings, International Symposium on Geographic Information Systems: The Research Agenda, November 1987.* Crystal City, Virginia 2: 245–55

Robinson V B, Wong R 1987 Acquiring approximate representation of some spatial relations. *Proceedings, Eighth International Symposium on Computer-Assisted Cartography*: 604–22

Rosch E 1973 On the internal structure of perceptual and semantic categories. In Moore T E (ed.) *Cognitive development and the acquisition of language*. New York, Academic Press

Rosch E 1978 Principles of categorisation. In Rosch E, Lloyd B B (eds) *Cognition and categorisation*. Hillsdale, Erlbaum

Sharif A R B M 1996 'Natural-language spatial relations: metric refinements of topological properties'. Unpublished

Doctoral dissertation, Department of Spatial Information Science and Engineering, University of Maine

Talmy L 1983 How language structures space. In Pick H, Acredolo L (eds) *Spatial orientation: theory, research, and application.* New York, Plenum Press

Whorf B L 1940 Science and linguistics. *Technology Review (MIT):* 42. Reprinted in Carroll J B (ed.) 1956 *Language, thought, and reality: selected writings of Benjamin Lee Whorf.* Cambridge (USA), The MIT Press: 207–19

Winston P H 1984 *Artificial intelligence,* 2nd edition. Reading (USA), Addison-Wesley

8

Time in GIS and geographical databases

D J PEUQUET

Although GIS and geographical databases have existed for over 30 years, it has only been within the past few that the addition of the temporal dimension has gained a significant amount of attention. This has been driven by the need to analyse how spatial patterns change over time (in order to better understand large-scale Earth processes) and by the availability of the data and computing power required by that space–time analysis. This chapter reviews the basic representational and analytical approaches currently being investigated.

1 INTRODUCTION

Representations used historically within GIS assume a world that exists only in the present. Information contained within a spatial database may be added to or modified over time, but a sense of change or dynamics through time is not maintained. This limitation of current GIS capabilities has been receiving substantial attention recently, and the impetus for this attention has occurred on both a theoretical and a practical level.

On a theoretical level GIS are intended to provide an integrated and flexible tool for investigating geographical phenomena. The world never stands still. This means that GIS should be able to represent these phenomena in both space and time (see also Veregin, Chapter 12). If GIS are to fulfil their envisioned role as decision-making tools, they will need to represent information in a manner that more closely approximates human representation of geographical space. An important element of human representation of the world around us is the retention of information relating to past events. Cognitive science has shown that the retention and accumulation of such information is essential to deriving more generalised concepts, and of learning in general (Mark, Chapter 7).

On a practical level, as a result of widespread use of GIS over the past 15 years and increasing reliance on GIS in everyday applications, users are

increasingly encountering the issue of how to keep a geographical database current without overwriting outdated information. This problem has come to be known as 'the agony of delete' (Copeland 1982; Langran 1992). The rapidly decreasing cost of memory and the availability of larger memory capacities of all types is also eliminating the need to throw away information as a practical necessity. In addition, the need within governmental policy-making organisations (and subsequently in science) to understand the effects of human activities better on the natural environment at all geographical scales is now viewed with increasing urgency. The emphasis is shifting in natural resource management within the developed world from inventory and exploitation toward maintaining the long-term productivity of the environment. This task requires integrated and broad-scale process analysis in order to understand natural and human processes better and how they are interrelated. Global Circulation Models (GCMs) are currently being used to study climate dynamics, ocean dynamics, and global warming (Simmons and Bengtsson 1988).

The need for a more detailed examination and understanding of the dynamics of human–environment interactions at urban and regional scales is also a continuing priority (Hunter and Williamson 1990; Vrana 1989). Diffusion theory (Hägerstrand 1970) has been applied to a diverse range of topics including agricultural innovation, the spread of political unrest, and the spread of AIDS (Gould 1993a; Parkes and Thrift 1980).

All of these require the analysis of change through time and of patterns of change through time, with the goal of gaining insights about cause-and-effect relationships. The advent of remotely-sensed satellite data in addition to the accumulation of other spatio-temporal observational data has made the empirical study of large-scale, complex spatio-temporal processes possible – further increasing the demand for integrated computer-based tools for this task (see Barnsley, Chapter 32; Estes and Loveland, Chapter 48). The extension of GIS and geographical database capabilities in order to provide efficient and flexible data storage and access has therefore attracted much attention within the GIS research community within the past few years.

2 ISSUES OF REPRESENTATION

In order to understand and compare approaches to space–time representations for GIS and geographical databases in general, a number of preliminary concepts and definitions are necessary.

2.1 Form vs function in space–time representation

There are two basic types of questions that can be asked using a space–time model: those concerning the 'world state' at a given time and those concerned with change in the properties of locations or spatial entities over time. These can be interpreted respectively as relating to static and dynamic views in time. Based on these two types of questions, the following generalised types of queries can be defined:

1 *World state*; what was/is/will be the spatial distribution of a given phenomenon at a given time? (e.g. where were the locations devoted to recreational land use in 1993? What was the spatial configuration of the 42nd Congressional District in the last election?)

2 *Change*; which elements changed/are changing/will change during a given time span? (e.g. where has growth in recreational land-use occurred between 1988 and 1998? Which congressional districts have shown an increase in unemployment over the past four years?)

The capability to ask change-related questions requires a dynamic view of the relevant phenomena. With this in mind, work toward providing temporal capabilities in a GIS context must include the development of methods for representing spatial change as it occurs through time in an explicit way, as well as 'states' at given times. Similarly, there must also be methods allowing direct manipulation and comparison of simulated and observational data in the temporal as well as the spatial dimension.

Any representational scheme is inextricably linked with specific types of questions or uses. This is why a strip map or route map is more easily used for travelling from one place to another than an overall areal map, whereas a route map is virtually useless for showing the overall distribution of various geographical entities within a given area. Thus, if change-related questions are to be addressed, it is necessary to utilise a type of representation that is specifically suited to that type of application.

2.2 Representing time and change

Conceptually, the basic objective of any temporal database is to record or portray change over time. Change is normally described as an event or collection of events. Perhaps the most encompassing definition of an event is 'something of significance that happens' (Mackaness 1993). For the purpose of space–time modelling a better definition might be 'a change in state of one or more locations, entities, or both'. For example, a change in the dominant species within a forest, a forest fire, change of ownership of the land, or building of a road would all be events. Change, and therefore also events, can be distinguished in terms of their temporal pattern into four types:

- *continuous* – going on throughout some interval of time
- *majorative* – going on most of the time
- *sporadic* – occurring some of the time
- *unique* – occurring only once.

This means that duration and frequency become important characteristics in describing temporal pattern.

These patterns can be very complex: just as a spatial distribution can be random, uniform, or clustered a temporal distribution can be chaotic, steady state, or cyclic. Similarity of states of locations or entities through time can also be converging, diverging, or combinations such as a dampened oscillation. Individual events can be characterised as clustered, forming *episodes*, that perhaps can be further grouped into cycles. Perfectly cyclical distributions are an important form of steady-state behaviour over longer temporal intervals. Thus a series of dry years in California can be grouped as a drought episode. This drought episode may in turn be part of the El Niño cycle. Variations in the length of El Niño cycles may be seen as chaotic, as may governmental response to various economic and social repercussions of such climatic cycles.

Change relating to entities or locations can be sudden or gradual. Entities appear, progress through various changes, then disappear over time. They may also change in complex ways. A critical issue is maintaining or changing the identity of entities: how are various states of the same entity, such as a stand of trees or a town, maintained as it changes through time? What kinds of change denote a change in the identity of an entity? Spatially, an entity may move, expand, shrink, change shape, split in two, or merge with an adjoining entity.

Change of some sort is always occurring. Change also occurs at different rates for various entities and locations. Since it is impossible to go back to a location in time as one can go back to a location in space, some events will inevitably be observed after-the-fact (or after the onset), with the exact time, duration, and nature of the events inferred. Some events will simply go unrecorded or be obliterated within a series of unobserved events. This translates to some level of inherent error and incompleteness in space–time data.

Although space and time are continuous, they are conventionally broken into discrete units of uniform or variable length for purposes of objective measurement. Time is divided into units that are necessarily different from those of space (we cannot measure time in metres or feet). Temporal units can be seconds, minutes, days, seasons, political administrations, or other units that may be convenient. Whether a single temporal scale or a hierarchy of scales is used, the smallest unit of recorded time is called a *chronon* (Jensen et al 1993).

An important capability for geographical modelling applications is to be able to represent alternative versions of the same reality. The idea of multiple realities over time is called *branching* (Langran 1993; Lester 1990). The idea of branching time is that of independent, yet synchronous states. This allows various model simulation results to be compared or to compare simulation results to observed data. As such, branching time is an integral notion as part of process model calibration and validation – regardless of whether the model is of mesoscale climate change or urban growth.

Given this general review of the nature of time and the representational issues that must be taken into account our focus now turns to the more concrete issue of present cartographic and GIS approaches for representing space–time information.

3 APPROACHES FOR REPRESENTING SPATIO-TEMPORAL DATA IN GIS

3.1 Location-based representations for spatio-temporal data

The only data model available within existing GIS that can be viewed as a spatio-temporal representation is a temporal series of spatially-registered 'snapshots', as shown graphically in Figure 1. This is not an intended representation but is rather a convenient redefinition within a standard GIS database organisation of what is stored in the

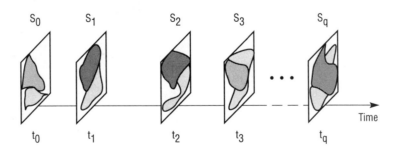

Fig 1. The 'snapshot' approach for representing spatio-temporal data: each 'snapshot', S_i, represents the state for a given point in time, t_i.
Source: Peuquet and Duan 1995

individual thematic layers. The 'snapshot' approach for space–time representation usually employs a grid data model, although a vector model can also be used (Pigot and Hazelton 1992). Instead of storing all information relating to a given thematic domain (e.g. elevation or land-use) within a single layer, a layer holds information relating to a single thematic domain at a single known time. Data are thus recorded over a series of discrete temporal intervals. The distinguishing feature of the snapshot representation is that a 'world state map' S_i at each given point in time t_i is stored as a complete image or snapshot. Everything is included regardless of what has or has not changed since the previous snapshot, and the temporal distance between snapshots is not necessarily uniform.

With this conceptually straightforward approach the state of any location or entity at a given time can be easily retrieved. This is also an obvious representation for data that are similarly collected as exhaustive coverages at discrete intervals, such as the decennial Census. There are, however, three drawbacks inherent in this approach.

1 The data volume increases enormously when the number of snapshots increases since each snapshot is a complete map of the entire region. This necessitates storage of a significant amount of redundant data since in most cases the spatial changes in two consecutive snapshots are only a small portion of the total data volume.

2 The changes of spatial entities that accumulate between two points in time are stored implicitly in the snapshots and can only be retrieved via a cell-by-cell (or vector-by-vector) comparison of adjacent snapshots. This process can be very time consuming. More importantly, however, some critical yet short-lived change at some location may occur between two consecutive snapshots and thus may not be represented.

3 Exactly when any individual change occurred cannot be determined. Chrisman warned against the use of snapshots on the basis of these last two characteristics alone since volume problems can be overcome with greater hardware storage capacity (Chrisman 1994).

A modification of the grid model that allows the time and place of individual changes (i.e. events) to be recorded was proposed within a GIS context by Langran (1990) and has been implemented on a prototype basis (Peuquet and Qian 1996). This model is also used in electronic circuitry design analysis (Fujimoto 1990). Instead of recording only a single value for a single pixel a variable-length list is associated with each pixel. Each entry in the list records a change at that specific location denoted by the new value and the time at which the change occurred. This is shown in Figure 2 in which each new change for a given location is added to the beginning of the list for that location. The result is a set of variable-length lists referenced to grid cells. Each list represents the event history for that cell location sorted in temporal order. The present (i.e. most recently recorded) world state for the entire area is easily retrieved as the first value stored in all of the locationally-referenced lists. In contrast to the snapshot representation, this representation stores only the changes related to specific locations and avoids storing redundant information (i.e. values for locations which remain unchanged).

3.2 Entity-based representations for spatio-temporal data

Several spatio-temporal models have also been proposed that explicitly record spatial changes through time as they relate to specific geographical entities instead of locations (Hazelton 1991; Kelmelis 1991; Langran 1992). On a broad conceptual level, all of these proposed models represent extensions of the topological vector approach. As such, they track changes in the geometry of entities through time. These spatio-temporal models rely on the concept of

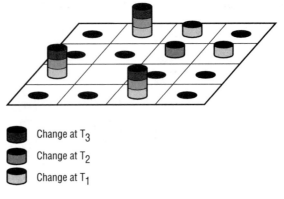

Change at T_3

Change at T_2

Change at T_1

Fig 2. The temporal grid approach for representing spatio-temporal data.
Source: Langran 1992

amendments, where any changes subsequent to some initial point in time in the configuration of polygonal or linear entities are incrementally recorded. The first of these models was proposed by Langran (1989b) and relies on what she describes as 'amendment vectors'. As a simple graphic example Figure 3 shows the historical sequence for a small portion of the roadways in a growing urbanised area. The thin black line shows the original configuration of a road at time t_1. At some later time, t_2, the route of the original road was straightened. Note that this modification required cutting the original line at two points, designating the piece of the original route between those two points as obsolete, and inserting a new line segment between the same two points to represent the new portion of the road. This results in four line segments where there was only one before the update. At some still later time, t_3, a new road is built and entered into the database which has an intersection point along the realigned segment of the first road. The time, t_n, when the change occurred is recorded as an attribute of each vector. This organisation allows the integrity of individual entities (e.g. lakes, roads, etc.), components of those

entities (e.g. boundary lines), and the vector topology to be explicitly maintained over time.

Hazelton utilised this basic idea within a 4-dimensional Cartesian space-time and proposed an extended hierarchy comprised of nodes, lines, polygons, polyhedra, polytopes, and polytope families as the conceptual organisational basis (Hazelton et al 1990). Kelmelis proposed a similar hierarchy of nodes, lines, surfaces, and volumes within a 4-dimensional Cartesian space–time coordinate space as an extension of the DLG-3 data model (Guptill 1990; Kelmelis 1991).

Besides explicitly maintaining the integrity of individual entities and their changing topology through time, the amendment vector approach also has the advantage of being able to represent asynchronous changes to entity geometries. This capability, however, comes at significant cost: as time progresses and the number of amendment vectors accumulate, the space–time topology of these vectors becomes increasingly complex.

There are many aspatial entity attributes that can also change over time. To track the history of aspatial attributes a separate relational database representation can also be used where an entity is represented as a tuple, composed of a unique identifier and a sequence of attribute values. This is an extension of the separation of spatial/non-spatial attribute storage commonly used in current GIS. However if the components that make up the spatial and aspatial aspects of any given entity are changing at different times and at different rates, as would commonly happen, maintaining the identity of individual entities in such a disjoint representation becomes difficult. Accurately maintaining identities becomes particularly complex when entities split or merge through time: which components carry which identifiers?

The object-oriented approach to representation has been proposed by a number of researchers as a means of handling this problem (Kucera and Sondheim 1992; Ramachandran et al 1994; Roshannejad and Kainz 1995; Worboys 1994; see also Worboys, Chapter 26). The object-oriented approach was originally developed as a generally applicable data representation method for software design and implementation (Booch 1994). The key to the object-oriented approach is the idea of storing, as an integral unit, all components that define a particular 'thing', as a *concept* (e.g. a single bank transaction, geographical entity, etc.). This is known as *encapsulation*. Another basic element of

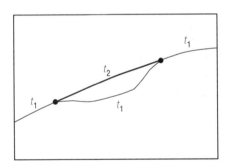

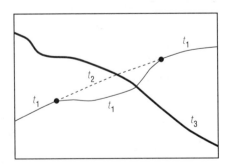

Fig 3. The 'amendment vector' approach.

the object-oriented approach is *inheritance*: the explicit linkage of objects to a taxonomic hierarchy identifying object types. For the case of geographical entities Lake Erie *is a* lake would be an example of an inheritance relation. Figure 4 gives a conceptual schematic of how an entity might be represented using this approach. This conceptual example includes aspatial, spatial, temporal, and taxonomic components. The object-oriented approach thus provides a cohesive representation that allows the identity of objects, as well as complex interrelationships, to be maintained through time. Rules for determining how to split or merge entities can be stored as part of the entity and entity class definitions. Perhaps the most mature implementation so far of an object-oriented spatio-temporal model is the Spatial Archive and Interchange Format (SAIF). Version 1 of this format was adopted in 1991 by the Canadian General Standards Board Committee on Geomatics as the Canadian standard for the exchange of geographical data (Kucera and Sondheim 1992).

3.3 Time-based representations for spatio-temporal data

Spatio-temporal representations that use time as the organisational basis have also been proposed recently (Peuquet and Duan 1995; Peuquet and Wentz 1994). These also suggest maintaining the explicit storage of temporal topology as an adjunct to location- and entity-based representations in a temporal GIS. In the time-based representation proposed by Peuquet and Duan, shown diagrammatically in Figure 5, all changes are stored as a sequence of events through time. The time associated with each change is stored in increasing temporal order from an initial, stored 'world state' (see Figure 5a). Differences between stored times denote the temporal intervals between successive events. Changes stored within this timeline or 'temporal vector' can relate to locations, entities, or to both (see Figure 5b). Such a timeline, then, represents an ordered progression through time of known changes from some known starting date or moment (t_0) to some other known, later point in time (t_n). Each location in time along the timeline (with its temporal location noted as $t_0, t_1,..., t_n$) can have associated with it a particular set of locations and entities in space-time that changed (or were observed as having changed) at that particular time and a notation of the specific changes.

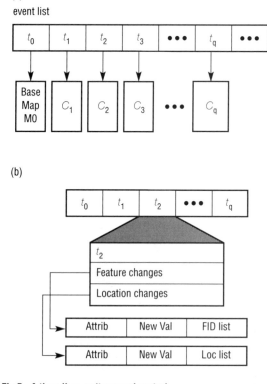

Fig 5. A time-line, or 'temporal vector'.
Sources: Peuquet and Duan 1995; Peuquet and Qian 1996

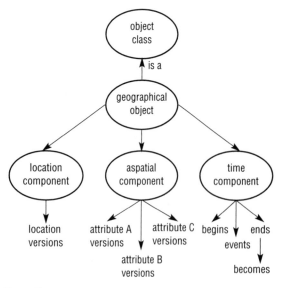

Fig 4. The object-oriented approach.

Besides sudden change as might be caused by some catastrophic event such as a forest fire or industrial plant closing, change can also be gradual such as the amount of rainfall or income level associated with a particular location, drainage basin, county, etc. For such instances of gradual change a change 'event' is recorded at the time when the amount of accumulated change since the last recorded change is considered to be significant or by some other domain-specific rule. It would also be possible to extend this basic model to denote explicitly and separately the start and end of gradual or longer-term changes as specific event types. Mackaness (1993) has called these longer-term changes *episodes* in order to distinguish them from shorter-term 'events'.

With this type of time-based representation, the changes relating to times are explicitly stored. This type of representation has the unique advantage of facilitating time-based queries (e.g. retrieve all events that occurred between January 1 and March 30, 1995). Adding new events as time progresses is also straightforward; they are simply added to the end of the timeline.

Similar timeline approaches have been proposed by Lester (1990) and by Frank (1994). In these approaches the timeline is strictly an ordinal model where the precise dates of events are unknown. A strictly ordinal timeline facilitates the representation of *branching* time in order to represent alternative or parallel sequences of events resulting from specific occurrences. Limitation to the temporal ordering of events is encountered in application domains such as archaeology, urban history, and ecology.

3.4 A combined approach for spatio-temporal representation

Associating additional temporal information with individual entities provides a means of recording entity histories, and thereby allows histories of entities and types of entities to be easily traced and compared. Similarly, associating temporal information with locations allows the history of individual locations and sets of locations to be traced and compared. Locational overlay operations can also be used in a location-based spatio-temporal representation to characterise locations on the basis or colocation of multiple changes or types of change. Since these different types of representation provide differing perspectives of the data, each facilitates a

different class of queries. This is precisely the reason why the current trend to provide both location-based and entity-based representations within current GIS is being extended to temporal GIS as well (Peuquet 1988, 1994; Yuan 1994).

4 TEMPORAL RELATIONS

In using a spatio-temporal GIS to analyse processes it is essential to be able to examine change on the basis of time, retrieving locations and entities on the basis of the temporal relationships of a specified event and, moreover, to be able to examine overall patterns of temporal relationships. Within an automated database context relationships act as *operators* on stored observational values. These relational operators may be implemented algorithmically or stored as explicit links. As operators, they enable the analyst to select and manipulate stored data in an ad hoc manner. Three basic types of temporal relationships have been defined (Peuquet 1994):

- association between elements within a given temporal distribution at a given temporal scale;
- combination of elements from different temporal distributions;
- transformations between temporal scales.

Selective retrieval and manipulation of data on the basis of these basic temporal relationships is particularly important from an analytical perspective because it allows the examination and derivation of cause and effect as well as overall temporal patterns (i.e. temporal *cycles* and rhythms).

The first type of basic temporal relationship includes metric and topological relations. Given that time is 1-dimensional there is only a single temporal metric, namely temporal distance. Temporal distance is the quantitative measurement of the interval between any two given points in time. Temporal distance can be used to denote the length of an event (i.e. its duration), the duration between two events, or the duration of some continuous state.

Temporal topology is the expression of relationships between events strictly in an ordering sense. In this type of relation the timeline itself can be viewed as elastic with events recorded upon it as knots or singularities. In terms of analysis this also allows for examination of different sequences of events that may occur at varying time scales.

Temporal topology has been studied in a number of fields including philosophy, mathematics, linguistics, and artificial intelligence (Allen 1984; Davis 1990; Shoham and McDermott 1988). These relationships are shown graphically in Figure 6. Note that six of the seven basic relationships have inverses so there are 13 possible temporal topological relationships in all.

The second basic type of temporal relationship includes those that combine different types of temporal distributions. These are the temporal overlay relationships that act as Boolean set operators (intersection, union, negation). This type of relationship allows temporal co-occurrence (or non co-occurrence) of different states or events over specific temporal intervals to be examined.

The third basic type of temporal relationship, temporal scale change, includes generalisation and extrapolation over a specific temporal distribution. This type of relationship involves combining events, episodes, and states in the case of generalisation. In the case of extrapolation events, episodes, and states may be inferred in order to fill in an expanded temporal sequence. The method of temporal generalisation (or extrapolation) used is dependent upon the given temporal measurement used (e.g. four seasons, 12 months, or 365 days (usually) each equal one year). The method may also vary from scale to scale.

In addition to these basic temporal relationships there are other temporal relationships that also relate directly to specific locations or to specific entities. As such they operate respectively within location-based or entity-based temporal representations.

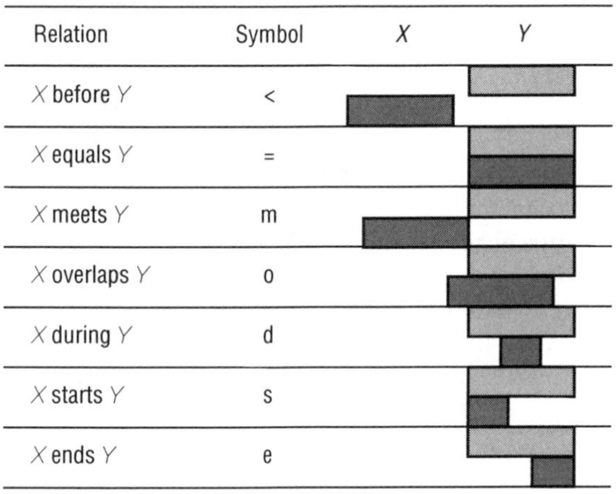

The most apparent locationally-based temporal relationship is essentially a variation of spatial overlay but in this case is used to compare change or similarity of state at a specific location or set of locations at differing times. This is the temporal operation that can already be performed in existing GIS utilising temporal snapshots. This differs from the temporally-based overlay relationships described above in that the comparison is tied to specific locations over specific temporal intervals. In contrast, the basic temporal overlay relationship previously described is tied to specific times but not to specific locations.

Cause-and-effect is an entity-based temporal relationship relation. This relationship is certainly the most important but also the most complex of the temporal relationships to determine. Establishing a cause-and-effect relationship is usually the final goal of most spatio-temporal analyses. One-to-one cause-and-effect relationships may be derived from *a priori* knowledge or direct observation of before/after states. As such these are most often explicitly stored as causes/caused-by or becomes/precedes links within an entity-based temporal representation.

5 METHODS OF SPATIO-TEMPORAL ANALYSIS

The ability to retrieve data in an ad hoc manner is a very basic and fundamental form of qualitative analysis. For the full analytical power of any modern GIS, temporal or otherwise, to be realised, all three types of analysis should be present: qualitative, quantitative (numerical), and visual.

One method of spatio-temporal analysis within a GIS is the use of the basic temporal relationships and space–time overlay in conjunction with visualisation techniques in an exploratory mode. This is a powerful technique that employs the power of the human visual system to detect pattern in an intuitive and ad hoc manner and has been proposed by a number of researchers within the cartographic community (Dorling 1992; Kraak and MacEachren 1994; Monmonier 1990).

As reported by Kucera (1996), experiments with visualisation of space–time dynamics of geographical phenomena have led to some startling results in uncovering patterns that are not otherwise evident. Relatively early investigation by Dorling revealed a space–time association between pub closing times, football matches, and telephone calls

Relation	Symbol	X	Y
X before Y	<		
X equals Y	=		
X meets Y	m		
X overlaps Y	o		
X during Y	d		
X starts Y	s		
X ends Y	e		

Fig 6. Allen's temporal relationships.
Source: Allen 1984

to emergency services (Dorling 1992). Dorling and Openshaw (1992) and Gould (1993a) have applied dynamic visualisation to the understanding of how disease spreads. Others have applied visualisation techniques to the space–time analysis of traffic patterns (Ganter and Cashwell 1994); other applications include climate change, regional groundwater pollution analysis, and crop prediction (Mitasova et al 1995).

Dorling and Openshaw (1992) have also demonstrated that direct user manipulation of temporal scale and other graphic variables for dynamic display can enhance the discriminatory power of visualisation. MacEachren has elaborated upon the potential of independently manipulating spatial and temporal scales in order to 'bring the phenomena to human scale (both spatially and temporally) so that we can examine them using a sensory system particularly adapted to this scale. To be able to see (and hear, and manipulate) "objects" as if they were a flower, or rock, or bird in our hand...' (MacEachren 1995: 159). Kraak and MacEachren (1994) also reviewed some general issues that need to be addressed towards the development of a different kind of cartography that is more attuned to dynamic display. Issues include ordering the display on the basis of variables other than time, synchronisation of two or more series, and duration of display time as it corresponds to both real-world time and to human perception.

A variety of quantitative methods for analysis of space–time data has been developed. Some of these are deterministic models, while others are purely descriptive. Perhaps the most generally known are the studies of diffusion which originated with Hägerstrand (1953). Methods for modelling diffusion processes have included Markovian probability sequences and gravity models. More recently, Gould (1993b) has utilised spatial adaptive filtering to study diffusion processes. With this method, the goal is to derive structure from a given space–time distribution instead of simulating the process as it unfolds from some starting point. Gould (1994) has also experimented with neural nets on temporal data. Layers of 'neurones' are interconnected using trial and error and then weighted by using training data (Fischer, Chapter 19).

Another approach that has been proposed is the use of cellular automata. An individual sequential state 'machine' is located within each cell of a regular grid. Each cell is in some individual but known state. Depending upon which protocol is used, changes in these states can be triggered by either (a) discrete temporal increments, or by (b) the occurrence of discrete events. At each temporal increment (or event occurrence), each cell determines its new state using some explicit transition rule. This rule may include the states of the neighbouring cells as well as global state changes as factors. Couclelis (1985) and Hogweg (1988) proposed the use of cellular automata for modelling environmental processes. Itami and Clark (1992) implemented a prototype version of this approach within the cell-based Map Analysis Package (MAP) originally developed by Tomlin (1983). Nevertheless, this approach has not progressed beyond these tentative and very preliminary proposals.

As theories of frequency, cycle, and causality are developed there is also need for a method to express these patterns. Examples abound within various disciplines in the use of self-contained numerical models for space–time modelling. These include a large number of air and water flow models, such as stormwater runoff models, global climate models, and ocean circulation models. Some of these use gridded representations and simulate varying states at these grid points depending upon surrounding states. Others use a link–node representation. All of these are fluid dynamics models and as such are based upon the fundamental physical continuity equations for the conservation of mass, momentum, and energy (Anthes et al 1987; Orlob 1983; Sklar and Costanza 1991). Ecological models that are space–time specific focus on simulating the distribution of some phenomenon over a landscape-scale surface. These are typically specific to a given geographical context. Examples include oceanic ecosystem models, pest infestation models, and wildfire models. The model developed by Show (1979) to describe the distribution of plankton in the Gulf of Mexico is one example of an oceanic ecosystem model. Perhaps the most widely used of this type of model are wildfire models, such as the one developed by Kessel (1975). These ecological models divide space either into a rectangular grid or into contiguous polygonal units and simulate successive states within these units.

The basic difficulty of modelling natural and social processes is that they are usually very complex, involving the interplay of a wide variety of variables that differentially change over time and space. Processes are also interlinked, and at varying spatial and temporal scales. Regardless of the application domain of such models, a limiting factor in their continuing refinement and validation has historically been lack of data and lack of sufficient computing capacity to handle available data. Certainly both are being rapidly overcome with modern computing advances and the availability of large, integrated government databases. The primary issue is now becoming the lack of integration with GIS. The powerful data access and manipulation capabilities of GIS need to be linked with numerical simulation models in order to refine these models (and our general understanding of the phenomena represented) in a more interactive and incremental manner, comparing before and after states over a sequence of space-time and comparing alternative simulations.

Sophisticated qualitative approaches have been proposed as an alternative for expressing the individual elements of a process in a manner that is cognitively more straightforward than numerical models. Davis (1990) and Worboys (1990) suggested the use of rule-based temporal logics. This approach is based upon formalised rules and axioms using relationships as operators. However, temporal logics, regardless of the application context, are at a very early stage of development.

6 TIME IN ASPATIAL DATABASES

The integration of time into database management systems (DBMS) has been an active field of research for a number of years. Although DBMS technology historically emphasises the storage and handling of aspatial data, the prospect of temporal DBMS will have both a direct and indirect impact on handling time in GIS. DBMS have been used for a long time as an adjunct within GIS for handling the non-coordinate, or attribute, data within a GIS. DBMS with temporal capabilities will perform an equivalent complementary role in handling temporal attribute data within a temporal GIS. The availability of temporal DBMS will thus have a direct impact on the capabilities of temporal GIS. The development of concepts and methodologies for integrating time into DBMS will also have an indirect effect by providing insight into equivalent representational and processing issues within the spatial context. These less tangible, indirect effects are discussed in more detail below. A more thorough review of temporal DBMS research and how specific areas could be applied to GIS has been given by Langran (1989a).

Much more progress has been made in representing time within DBMS as compared to spatial databases. Within the typically aspatial context of DBMS the issues involved are not as complex because of the reduced dimensionality of the data. Perhaps at least in part because of this, progress on temporal DBMS is rapidly moving this area out of a purely research realm. Representations used for DBMS have been developed using extensions to traditional relational as well as object-oriented approaches (Ozsoyoglu and Snodgrass 1995; Snodgrass 1995). Some of this work has also been applied to the GIS context, particularly by Stonebraker and the Sequoia 2000 project (Guptill and Stonebraker 1992).

A number of query languages have also been proposed (Egenhofer and Kuhn, Chapter 28). A standard language for temporal DBMS – SQL2 – was recently adopted as the ISO standard. This language represents an extension of SQL. There are now efforts toward a further extension of SQL in a new standard (SQL/Temporal) that will also incorporate features of SQL/MM. SQL/MM was designed as an extension of SQL to address the specific needs of multimedia applications. Most importantly, in order to deal with imagery and video, SQL/MM incorporates spatial as well as temporal concepts (Kucera 1996).

Two issues uniquely related to the temporal dimension have received much attention in DBMS and seem relevant to the spatio-temporal context also. First, it is for practical purposes often impossible to enter data into a database at the moment the relevant event occurs in the real world. There is a distinction between when a state or condition is current or valid in the real world and when that state or condition was entered into the database (i.e. 'valid time' vs 'transaction time'). These are also known as 'world time' and 'database time', respectively (Jensen et al 1993). This distinction has been used for a long time in banking transaction records; for example, when a bank teller received a deposit and when that deposit was actually credited to the account. The difference between these two times is used to detect potential bookkeeping errors and to guard against fraud (e.g. to allow time for the cheque to clear before allowing withdrawals on this amount). How to effectively represent both types of information within the same database, therefore, is an important issue. This becomes an important distinction in certain GIS applications such as cadastral systems (Al-Taha 1992; Hunter and Williamson 1990; Worboys 1994).

A second major area of interest within the temporal DBMS arena is how to perform retrospective updates (i.e. inserting new information concerning past conditions or events). This issue involves how to determine which other information already in the database is affected by the new information and must therefore also be changed in order to maintain the integrity of the data. Pioneering research on this issue within the temporal DBMS community involves the use of a variety of artificial intelligence techniques and formal logics for temporal reasoning (Kowalski 1992; Maiocchi et al 1992).

Two classical and unresolvable issues in traditional GIS and cartography, which also carry over into the temporal GIS context, are being discovered anew by the temporal DBMS research community. These are temporal generalisation and temporal resolution. Temporal resolution is known as the 'granularity' of the database (Jensen et al 1993) within temporal DBMS. Many of the problems associated with resolution and discretising values for continuous phenomena as they have become known in the spatial realm are also being encountered in the aspatial realm.

7 THE FUTURE?

Much remains to be done before a true temporal GIS can be realised. As was the case in current spatial database systems, the introduction of representational power and analytical capabilities in commercial systems will itself be a temporal process. However, progress should be more rapid given the current state of knowledge and technology. This brief discussion on the related field of DBMS also provides a glimpse of how much there is to be gained from interdisciplinary research on temporal databases and temporal representation in general.

References

Allen J F 1984 Towards a general theory of action and time. *Artificial Intelligence* 23: 123–54

Al-Taha K K 1992 'Temporal reasoning in cadastral systems'. Unpublished PhD dissertation, Department of Surveying Engineering, University of Maine

Anthes R A, Hsie E Y, Kuo Y H 1987 *Description of the Penn State/NCAR Mesoscale Model version 4 (MM4)*. Report No. NCAR/TN-2282+STR. Boulder, National Center for Atmospheric Research

Booch G 1994 *Object-oriented analysis and design*, 2nd edition. Redwood City, Benjamin Cummings

Chrisman N R 1994 'Beyond the snapshot: changing the approach to change, error, and process'. Unpublished manuscript

Copeland G 1982 What if mass storage were free? *Computer* 15: 27–35

Couclelis H 1985 Cellular worlds: a framework for modeling micro-macro dynamics. *Environment and Planning A* 17: 585–96

Davis E 1990 *Representations of commonsense knowledge*. San Mateo, Morgan Kaufmann

Dorling D 1992 Visualising people in time and space. *Environment and Planning B: Planning and Design* 19: 613–37

Dorling D, Openshaw S 1992 Using computer animation to visualise space–time patterns. *Environment and Planning B: Planning and Design* 19: 639–50

Frank A U 1994 Qualitative temporal reasoning in GIS – ordered timescales. *Sixth International Symposium on Spatial Data Handling, Edinburgh, Scotland*. International Geographical Union: 410–30

Fujimoto R M 1990 Parallel discrete event simulation. *Communications of the ACM* 33: 30–53

Ganter J, Cashwell J W 1994 Display techniques for dynamic network data in transportation GIS. *Seventh Symposium on GIS for Transportation, Norfolk, Virginia*: 42–53

Gould P 1993a *The slow plague: a geography of the AIDS pandemic*. Cambridge (USA), Blackwell

Gould P 1993b Why not? The search for spatio-temporal structure. *Environment and Planning A* (Anniversary Issue): 48–55

Gould P 1994 Neural computing and the AIDS pandemic: the case of Ohio. In Hewitson B, Crane R (eds) *Neural nets, applications in geography*. Boston (USA), Kluwer Academic

Guptill S C 1990 *An enhanced digital line graph design*. United States Geological Survey Circular No. 1048

Guptill S C, Stonebraker M 1992 The Sequoia 2000 approach to managing large spatial object databases. *Proceedings, Fifth International Symposium on Spatial Data Handling, Charleston, South Carolina*. International Geographical Union: 642–51

Hägerstrand T 1953 *Innovation diffusion as a spatial process*. Chicago, The University of Chicago Press

Hägerstrand T 1970 What about people in regional science? *Papers of the Regional Science Association* 14: 7–21

Hazelton N W J 1991 'Integrating time, dynamic modelling, and geographical information systems: development of 4-dimensional GIS'. Unpublished PhD dissertation, Dept of Surveying and Land Information, University of Melbourne

Hazelton N W J, Leahy F J, Williamson I P 1990 On the design of temporally-referenced, 3-dimensional geographical information systems: development of 4-dimensional GIS. *GIS/LIS 90 Proceedings*: 357–72

Hogweg P 1988 Cellular automata as a paradigm for ecological modeling. *Mathematics and Computation* 27: 81–100

Hunter G J, Williamson I P 1990 The development of a historical digital cadastral database. *International Journal of Geographical Information Systems* 4(2): 169–79

Itami R M, Clark J D 1992 Spatial dynamic simulations using discrete time and discrete event theory in cell-based GIS systems. *Proceedings, Fifth International Symposium on Spatial Data Handling, Charleston, South Carolina*. International Geographical Union: 702–12

Jensen C S, Clifford J, Gadia S K, Segev A, Snodgrass T R 1993 A glossary of temporal database concepts. *Proceedings, International Workshop on an Infrastructure for Temporal Databases, Arlington, Texas*: A25–A29

Kelmelis J 1991 'Time and space in geographic information: toward a 4-dimensional spatio-temporal data model'. Unpublished PhD dissertation, Department of Geography, Pennsylvania State University

Kessel S R 1975 The Glacier National Park basic resources and fire ecology model. *Bulletin of the Ecological Society of America* 56: 49

Kowalski R 1992 Database updates in the event calculus. *Journal of Logic Programming* 12: 121–46

Kraak M-J, MacEachren A M 1994b Visualisation of the temporal component of spatial data. *Sixth International Symposium on Spatial Data Handling, Edinburgh, Scotland*. International Geographical Union: 391–409

Kucera G L 1996 *Temporal extensions to spatial data models*. Final Report to Intergraph Corp. under contract entitled 'Spatial-temporal modeling for environmental decision support'. Victoria (Canada), Mercator Systems Ltd

Kucera H A, Sondheim M 1992 SAIF – conquering space and time. *GIS 92 Symposium*. Vancouver, Canada

Langran G 1989a A review of temporal database research and its use in GIS applications. *International Journal of Geographical Information Systems* 3(3): 215–32

Langran G 1989b 'Time in geographic information systems'. PhD dissertation, University of Washington

Langran G 1990 Temporal GIS design tradeoffs. *Journal of the Urban and Regional Information Systems Association* 2: 16–25

Langran G 1992 *Time in geographic information systems*. London, Taylor and Francis

Langran G 1993 One GIS, many realities. *GIS 93, Vancouver*

Lester M 1990 Tracking the temporal polygon: a conceptual model of multidimensional time for geographic information systems. Presented at Temporal Workshop (NCGIA), Orono, Maine

MacEachren A M 1995 *How maps work: representation, visualisation, and design*. New York, Guilford Press

Mackaness W 1993 *Events and episodes – just patterns in time*. ICA Conference Proceedings, Cologne

Maiocchi R, Pernici B, Barbic F 1992 Automatic deduction of temporal information. *ACM Transactions of Database Systems* 17: 647–88

Mitasova H, Mitas L, Brown W M, Gerdes D P, Kosinovsky I, Baker T 1995 Modelling spatially and temporally distributed phenomena: new methods and tools for GRASS GIS. *International Journal of Geographical Information Systems* 9: 433–46

Monmonier M 1990 Strategies for the visualization of geographic time-series data. *Cartographica* 27: 30–45

Orlob G T (ed.) 1983 *Mathematical modeling of water quality: streams, lakes and reservoirs*. New York, John Wiley & Sons Inc.

Ozsoyoglu G, Snodgrass T R 1995 Temporal and real-time databases: a survey. *IEEE Knowledge and Data Engineering* 7: 513–32

Parkes D, Thrift N 1980 *Times, spaces, and places: a chronological perspective*. New York, John Wiley & Sons Inc.

Peuquet D J 1988 Representations of geographic space: toward a conceptual synthesis. *Annals of the Association of American Geographers* 78: 375–94

Peuquet D J 1994 It's about time: a conceptual framework for the representation of temporal dynamics in geographic information systems. *Annals of the Association of American Geographers* 84: 441–61

Peuquet D J, Duan N 1995 An event-based spatio-temporal data model (ESTDM) for temporal analysis of geographic data. *International Journal of Geographical Information Systems* 9: 2–24

Peuquet D J, Qian L 1996 An integrated database design for temporal GIS. *Seventh International Symposium on Spatial Data Handling, Delft, Holland*. International Geographical Union: 2–1 to 2–11

Peuquet D J, Wentz E 1994 An approach for time-based analysis of spatio-temporal data. *Sixth International Symposium on Spatial Data Handling, Edinburgh, Scotland*. International Geographical Union: 489–504

Pigot S, Hazelton B 1992 The fundamentals of a topological model for a 4-dimensional GIS. *Proceedings, Fifth International Symposium on Spatial Data Handling, Charleston, South Carolina*. International Geographical Union: 580–91

Ramachandran B, MacLeod F, Dowers S 1994 Modelling temporal changes in a GIS using an object-oriented approach. *Sixth International Symposium on Spatial Data Handling, Edinburgh, Scotland, 5–9 September*: 518–37

Roshannejad A A, Kainz W 1995 Handling identities in spatio-temporal databases. *AutoCarto 12, Charlotte, North Carolina*: 119–26

Sack R 1980 *Conceptions of space in social thought: a geographic perspective*. Minneapolis, University of Minnesota Press

Shoham Y, McDermott D 1988 Problems in formal temporal reasoning. *Artificial Intelligence* 36: 49–61

Show I T 1979 Plankton community and physical environment simulation for the Gulf of Mexico region. *The 1979 Summer Computer Simulation Conference, San Diego*. Society for Computer Simulation

Simmons A J, Bengtsson L 1988 Atmospheric general circulation models: their design and use for climate studies. In Schlesinger M (ed.) *Physically-based modelling and simulation of climate and climatic change*. Boston (USA), Kluwer Academic Vol. 2: 627–52

Sklar F H, Costanza R 1991 The development of dynamic spatial models for landscape ecology: a review and prognosis. In Turner M G, Gardner R H (eds) *Quantitative methods in landscape ecology: the analysis and interpretation of landscape heterogeneity*. New York, Springer: 239–88

Snodgrass T R 1995 Temporal object-oriented databases: a critical comparison. In Kim W (ed.) *Modern database systems*. Reading (USA), Addison-Wesley: 386–408

Tomlin C D 1983 'Digital cartographic modeling techniques in environmental planning'. Unpublished PhD dissertation, Yale University

Vrana R 1989 Historical data as an explicit component of land information systems. *International Journal of Geographical Information Systems* 3(1): 33–49

Worboys M F 1990 Reasoning about GIS using temporal and dynamic logics. Temporal Workshop (NCGIA), Orono, Maine

Worboys M F 1994a A unified model for spatial and temporal information. *The Computer Journal* 37: 26–33

Yuan M 1994 Wildfire conceptual modeling for building GIS space–time models. *GIS/LIS 94, Phoenix, Arizona*: 860–869

9

Representation of terrain

M F HUTCHINSON AND J C GALLANT

This chapter demonstrates the central role played by representations of terrain in environmental modelling and landscape visualisation. Current trends in digital terrain modelling are discussed. Topographical data sources and digital elevation model (DEM) interpolation and filtering methods are described in relation to the requirements of environmental models. Accurate representation of surface shape and drainage structure is a common requirement, and is facilitated by the development of locally-adaptive, process-based DEM interpolation techniques. The role of traditional contour data sources and remotely-sensed data sources is also examined. Methods for interpreting terrain include terrain parameters, as simplifications of key environmental processes, and a range of terrain features associated with secondary terrain structures. The issue of spatial scale is discussed in relation to the multi-scale requirements of environmental modelling and the identification of scaling properties of DEMs and associated terrain parameters. Multi-scale terrain feature analysis permits the incorporation of terrain structure into analyses of scale.

1 INTRODUCTION

Terrain plays a fundamental role in modulating Earth surface and atmospheric processes. So strong is this linkage that understanding of the nature of terrain can confer understanding of the nature of these processes directly, in both subjective and analytical terms. Thus, analyses and representations of terrain have provided cardinal examples for many activities in GIS and environmental modelling. They have stimulated directly the development of new methods for obtaining digital environmental data (Barnsley, Chapter 32; Dowman, Chapter 31; Lange and Gilbert, Chapter 33), new spatial interpolation methods (Mitas and Mitasova, Chapter 34; Hutchinson 1996), and new methods for assessing data quality (see below). Since 3-dimensional representations of terrain form natural backgrounds for the display of spatially distributed quantities and entities, representations of terrain have also played a prominent role in the development of methods for conceptualisation (Raper, Chapter 5; Weibel and Dutton, Chapter 10) and visualisation (Neves and Câmara, Chapter 39; Kraak, Chapter 11) of 3-dimensional data.

Of central importance for the assessment and management of natural resources is the accuracy and spatial coverage that can be achieved in environmental modelling by incorporating appropriate dependencies on terrain. This particularly applies to improved representations of surface climate (Hutchinson 1995; Running and Thornton 1996) and hydrology (Moore and Grayson 1991) which are key factors in geomorphological and biological applications. This has led to the consideration of the underlying physical processes, and the spatial scales at which they operate, coupled with an increasing focus on explicit mathematical analysis, leading to the development of new methods for representing and interpreting terrain data (Gallant and Hutchinson 1996; see also Heuvelink, Chapter 14). These developments are consistent with key conclusions of the survey of digital terrain modelling by Weibel and Heller (1991), who emphasised a need to combine mathematical and algorithmic approaches with environmental and geomorphological understanding.

Weibel and Heller also asserted that digital terrain modelling had satisfied a number of goals, and that future developments would concentrate on refining current techniques and enlarging their scope. This chapter discusses current developments in terrain modelling in the light of their review, with particular emphasis on methods for the generation and interpretation of DEMs. These are the two areas of terrain representation which are directly related to the modelling of Earth surface processes. Their relationship to the overall context of digital terrain modelling is shown in Figure 1, which is a revised version of Figure 19.1 of Weibel and Heller.

Figure 1 clarifies the main functional connections between the tasks, particularly the interaction between DEM generation and DEM interpretation, and the overriding context provided by a wide range of applications. The issue of spatial scale arises at various points in this scheme. The scale of source data should guide the choice of resolution of generated DEMs and the scales of DEM interpretations should be matched to the natural scales of terrain-dependent applications. Recent developments in terrain features derived from DEMs are seen as having the potential to address issues of both scale and structure in digital terrain analysis.

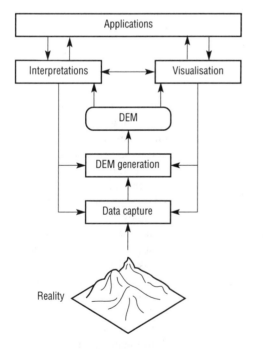

Fig 1. The main tasks associated with digital terrain modelling.

2 CURRENT TRENDS IN DIGITAL TERRAIN MODELLING

The rationale for the revised digital terrain modelling scheme shown in Figure 1 is discussed in relation to current trends in digital terrain modelling and the issues to be addressed in further detail by this chapter.

2.1 Elevation data capture

The role of digital elevation data capture has been enhanced, reflecting recent developments in airborne and spaceborne remote sensing, such as laser and synthetic aperture radar systems, and the development of the Global Positioning System (GPS) for ground data survey (see Lange and Gilbert, Chapter 33). Analysis of the errors associated with these data sources is an essential part of DEM generation.

Elevation contours continue to be the principal data source for the interpolation of DEMs, as well as being useful representations of terrain in their own right. They are widely available from existing topographic maps and, despite inherent sampling biases, can accurately reflect surface structure, particularly if they are coupled with a high-quality interpolation technique.

2.2 DEM generation

The development of methods for interpolation and filtering of DEM data continues to be a central area of digital terrain analysis, but the methods are now applied to a wider variety of data sources. These include traditional data sources such as points, profiles, contours, stream-lines, and break-lines, for which specific interpolation techniques have been developed, and remotely-sensed elevation data, for which various filtering procedures are required. Included in the task of DEM generation is a variety of associated DEM manipulation tasks such as DEM editing, DEM resampling, and data structure conversion between regular grids and triangulated irregular networks (TINs), the two dominant forms of terrain representation.

DEM interpolation methods based on triangulations have been seen as attractive because they can be adapted to various terrain structures and to varying data densities. However, it has been difficult to constrain triangulations to greatest advantage, and TINs can have deficiencies in representing terrain shape parameters such as slope and curvature. On the other hand, techniques for interpolating and analysing regular grids tend to be relatively straightforward, and recent developments in locally adaptive gridding techniques have enhanced the sensitivity of interpolated regular grids to terrain structure, including ridges and stream-lines. Interpolation using local spline surface patches can also achieve a degree of local adaptivity to terrain structure.

TINs have seen most use as a data reduction tool, particularly useful in visualisation applications (De Floriani and Magillo, Chapter 38), while regular-grid DEMs have become the dominant vehicle for environmental modelling and natural resources assessment (Band, Chapter 37). Regular-grid DEMs can be readily integrated with remotely-sensed environmental data sources and gridding methods can be adapted to the filtering of noisy remotely-sensed elevation data.

2.3 DEM interpretation

Interpretation of DEMs includes scale analyses, terrain parameters, and a variety of terrain features that can be constructed from DEMs. Many DEM interpretations have been evolved to support hydrological analyses of DEMs, which are also discussed by Band (Chapter 37).

2.3.1 Scale
Scale and resolution enter into terrain analysis in several ways. The most fundamental is the choice of scale or grid resolution, which is analogous to the choice of map scale in cartography (see also Veregin, Chapter 12). The choice is usually a compromise between achieving fidelity to the true surface, and respecting practical limits on the density and accuracy of the source data. Grid resolution can be used as an index of information content. This has important consequences for the construction of meaningful linkages of DEMs with other data sources (Goodchild and Longley, Chapter 40).

The identification of characteristic scales in terrain, and the degree to which surface form changes with scale, is important for deciding on the scales or resolutions required to model terrain-dependent processes. As yet no satisfactory model of the changes of surface form with scale has been developed. The fractal model has been found to be too simplistic for most applications, since a single scaling law does not apply across all scales of interest. The fractal model also does not recognise important structural features such as drainage networks.

2.3.2 Terrain parameters
Terrain parameters, or topographic indices, are descriptions of surface form that can be computed directly at every point on a DEM. A substantial collection of such parameters has been developed to facilitate analyses of surface hydrological and ecological processes (Moore et al 1991). Most terrain parameters depend on the DEM having an accurate representation of surface shape (see also Mitas and Mitasova, Chapter 34). They exhibit scale dependencies which have yet to be fully understood and quantified.

2.3.3 Terrain features
A variety of terrain features have been constructed from DEMs to support terrain-dependent analyses. They are usually associated with secondary terrain structures defined in terms of surface shape and drainage structure. Many of these coincide with common conceptions of landscape features, such as mountain ranges, ridges, catchments, rivers, and valleys.

Dissection of the DEMs into catchments and sub-catchments is an established procedure using the technique of Jenson and Domingue (1988). Terrain can also be dissected into a set of stream tubes bounded by contour lines and flow-lines (Moore et al 1988), particularly suited for hydrological applications. A multi-scale representation of terrain as a collection of overlapping topographic features at different scales has been recently developed by Gallant and Hutchinson (1996).

2.4 DEM interpretation and DEM visualisation

Visualisation techniques may be applied directly to DEMs, as well as to various interpretations of DEMs. Visualisation of DEMs can provide subjective assessments, such as perspective views and intervisibility analyses for various planning and monitoring applications. Intervisibility analyses of DEMs, represented as TINs or as regular grids, are discussed in detail by De Floriani and Magillo (Chapter 38). Visualisations of DEMs draped with various textures can also provide valuable insight into the nature of the processes being represented. They are an essential component of many virtual environment systems (Neves and Câmara, Chapter 39).

Interpretation and visualisation of DEMs can provide assessments of DEM quality which have direct implications for DEM generation and data capture, as indicated in Figure 1. Automated graphical techniques for detection of errors in source data are particularly important since most source topographic datasets are large and contain errors. Non-classical measures of data quality based on visualisation methods offer rare opportunities for confirmatory data analysis (CDA).

2.5 DEM applications

The overriding influence of applications on terrain representation and analysis is indicated in Figure 1. Applications may be found across a wide range of spatial scales, in civil engineering, planning and resource management, Earth sciences, and military studies. The general trend in representations of terrain for environmental modelling has been to move from broader continental and regional scales, closely allied to the representation of major drainage divisions (Jenson 1991), mesoscale representations of surface climate (Hutchinson 1991) and associated flora and fauna (Nix 1986), to finer scales suited to the modelling of surface hydrology, vegetation, and soil properties (Gessler et al 1996; Mackey 1996; Moore and Grayson 1991; Quinn et al 1991). This general trend has been accompanied by improvements in methods for representing fine-scale shape and structure in DEMs, supported by the steady increase in the speed and storage capacity of computing platforms.

This has brought into sharper focus the issue of determining appropriate spatial scales for modelling Earth-surface processes (Steyaert 1993) and for hydrological modelling in particular (Blöschl and Sivaplan 1995). More recently there has been a renewed appreciation of the utility of broader-scale DEMs, with a spatial resolution of about one kilometre, for the purposes of environmental modelling at global level. This has been accompanied by the development of new broader-scale remote-sensing instruments (Barnsley, Chapter 32; Dowman, Chapter 31) and the compilation of global coverages of terrain data (Verdin and Jenson 1996) and Earth-surface data commensurate with this resolution (Steyaert 1996).

3 DEM GENERATION

DEM-generation procedures need to be guided by both the nature of the source data and the intended applications of the generated DEM. For most applications, accurate representation of surface shape and drainage structure is more important than absolute elevation accuracy, particularly in areas with low relief.

3.1 Sources of elevation data

Three main classes of source elevation data may be recognised, for which different DEM-generation techniques are applicable.

3.1.1 Surface-specific point elevation data
Surface-specific point elevations, including high and low points, saddle points, and points on streams and ridges make up the skeleton of terrain (Clarke 1990). They are an ideal data source for most interpolation techniques, including triangulation methods and specially adapted gridding methods. These data may be obtained by ground survey and by manually assisted photogrammetric stereo models (Makarovic 1984). They can also be obtained from grid DEMs to construct TIN models (Heller 1990; Lee 1991). The advent of the GPS has enhanced the availability of accurate ground-surveyed data (Lange and Gilbert, Chapter 33; Dixon 1991), but such data are available only for relatively small areas.

3.1.2 Contour and stream-line data
Contour data are still the most common terrain data source for large areas. Many of these data have been digitised from existing topographic maps which are the only source of elevation data for some parts of the world. The conversion of contour maps to digital form is a major activity of mapping organisations worldwide (Hobbs 1995). Contours can also be generated automatically from photogrammetric stereo models (Lemmens 1988), although these methods are subject to error. A sample contour and stream-line dataset is shown in Figure 2, with some additional point data. Contours implicitly encode a number of terrain features, including points on stream-lines and ridges. The main disadvantage of contour data is that they can significantly under-sample the areas between contour lines, especially in areas of low relief, such as the lower right hand portion of Figure 2. This has led most investigators to prefer contour-specific algorithms over general-purpose algorithms when interpolating contour data (Clarke et al 1982; Mark 1986).

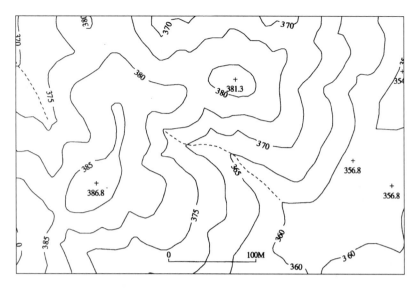

Fig 2. Contour, stream, and point elevation data.

Contour data differ from other elevation data sources in that they imply a degree of smoothness of the underlying terrain. When contours are obtained by manually assisted photogrammetric techniques, the operator can remove the effects of obstructions such as vegetation cover and buildings. Contour data, when coupled with a suitable interpolation technique, can in fact be a superior data source in low-relief areas (Garbrecht and Starks 1995), where moderate elevation errors in remotely-sensed data can effectively preclude accurate determination of surface shape and drainage.

Stream-lines are also widely available from topographic maps and provide important structural information about the landscape. However, few interpolation techniques are able to make use of stream-line data without associated elevation values. The method developed by Hutchinson (1988, 1989) can use such stream-line data, provided that the stream-lines are digitised in the downhill direction. This imposes a significant editing task, which can be achieved by using a GIS with network capabilities.

3.1.3 Remotely-sensed elevation data

Gridded DEMs may be calculated directly by stereoscopic interpretation of data collected by airborne and satellite sensors (Dowman, Chapter 31). The traditional source of these data is aerial photography (Kelly et al 1977) which, in the absence of vegetation cover, can deliver elevations to sub-metre accuracy (Ackermann 1978; Lemmens

1988). Stereoscopic methods have been applied to SPOT imagery (Day and Muller 1988; Konecny et al 1987), and more recently to airborne and spaceborne synthetic aperture radar (SAR). Spaceborne lasers can also provide elevation data in narrow swathes (Harding et al 1994). A major impetus for these developments is the yet unrealised goal of generating high-resolution DEMs with global coverage (Dixon 1995; Zebker et al 1994).

Remote-sensing methods can provide broad spatial coverage, but have a number of generic limitations. None of the sensors can measure the ground elevations underneath vegetation cover reliably. Even in the absence of ground cover, all methods measure elevations with significant random

Fig 3. Shaded relief view of a 10-m-resolution DEM obtained from airborne SAR in an area with low relief.

errors, which depend on the inherent limitations of the observing instruments, as well as surface slope and roughness (Dixon 1995; Harding et al 1994). The methods also require accurately located ground control points to minimise systematic error. These points are not always easy to locate, especially in remote regions. Best-possible standard elevation errors with spaceborne systems currently range between 1 and 10 metres, but elevation errors can be much larger, up to 100 metres, under unfavourable conditions (Harding et al 1994; Lanari et al 1997; Sasowsky et al 1992; Zebker et al 1994). Averaging of data obtained from multiple passes of the sensor can reduce these errors.

Airborne SAR data are available for areas of limited extent. Standard elevation errors for DEMs derived from these data can be as small as 1 to 3 metres (Dixon 1995). Figure 3 shows a shaded relief view of a DEM derived from airborne SAR in an area with low relief. The figure shows occasional large errors, evidenced as high points and holes, random elevation errors across the whole DEM, and significant anomalies in the form of spurious ridges along tree-lined watercourses. Careful filtering of such data is required to derive a useful representation of surface shape and drainage structure.

3.2 Interpolation methods

Interpolation (Mitas and Mitasova, Chapter 34) is required to generate DEMs from surface-specific points and from contour and stream-line data. Since datasets are usually very large, high-quality global interpolation methods, such as thin plate splines, in which every interpolated point depends explicitly on every data point, are computationally impracticable. Such methods cannot be adapted easily to the strong anisotropy evidenced by real terrain surfaces. On the other hand, local interpolation methods, such as inverse distance weighting, local Kriging, and unconstrained triangulation methods, achieve computational efficiency at the expense of somewhat arbitrary restrictions on the form of the fitted surface. Three classes of interpolation methods are in use. All achieve a degree of local adaptivity to anisotropic terrain structure.

3.2.1 Triangulation

Interpolation based on triangulation is achieved by constructing a triangulation of the data points, which form the vertices of the triangles, and then fitting local polynomial functions across each triangle (Weibel and

Heller 1991; see also Weibel and Dutton, Chapter 10, for a broader discussion of generalisation). Linear interpolation is the simplest case, but a variety of higher-order interpolations have been devised to ensure that the interpolated surface has continuous first derivatives (Akima 1978; Auerbach and Schaeben 1990; Sambridge et al 1995; Sibson 1981; Watson and Philip 1984). Considerable attention has been directed towards methods for constructing the triangulation. The Delaunay triangulation is the most popular method and several efficient algorithms have been devised (e.g. Aurenhammer 1991; Heller 1990; Tsai 1993). The dual of the Delaunay triangulation is the Dirichlet tessellation. Both structures have been used to assess neighbourhood relationships of point data in 2- and 3-dimensional space (Boots, Chapter 36).

Triangulation methods have been seen as attractive because they can be adapted to various terrain structures, such as ridge-lines and streams, using a minimal number of data points (McCullagh 1988). However, these methods are sensitive to the positions of the data points and the triangulation needs to be constrained to produce optimal results (Pries 1995; Weibel and Heller 1991). Triangulation methods are known to have difficulties interpolating contour data, which generate many flat triangles unless additional structural data points along streams and ridges can be provided (Clarke 1990: 204–37). Such data may be obtained by detailed ground or photogrammetric survey, but have not been readily obtained from existing contour maps.

Figure 4(a) shows surface-specific data points selected from corners in the contour data shown in Figure 2 and from stream-lines and ridges inferred from these data by the locally adaptive gridding method described below. The corresponding TIN is shown in Figure 4(b). Using Akima's method to interpolate across the triangles, the TIN is contoured in Figure 4(c) at half the elevation spacing of the data contours. A shaded-relief view is shown in Figure 4(d). This triangulation accurately represents the broad structure of the terrain, but the contour and shaded-relief views reveal minor deficiencies in surface shape. These are typically associated with small narrow triangles which are difficult to avoid. The outstanding feature of this representation is its numerical efficiency, with the number of vertices in the TIN less than one per cent of the number of nodes in the grid DEM shown in Figure 5. Examples of TIN generation from a gridded DEM are shown by Lee (1991), by Weibel and Heller (1991), and by Neves and Câmara (Chapter 39).

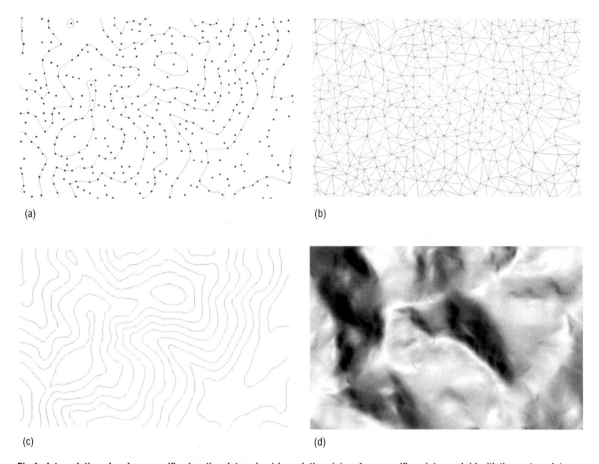

(a)

(b)

(c)

(d)

Fig 4. Interpolation of surface-specific elevation data using triangulation: (a) surface-specific points overlaid with the contour data from Figure 2; (b) TIN derived from the surface-specific points; (c) contours interpolated from the TIN; and (d) shaded relief view of the surface interpolated from the TIN.

3.2.2 Local surface patches

Interpolation by local surface patches is achieved by applying a global interpolation method to overlapping regions, usually rectangular in shape, and then smoothly blending the overlapping surfaces. Franke (1982) and Mitasova and Mitas (1993) have used respectively thin plate splines and regularised splines in tension in this way: see also Mitas and Mitasova (Chapter 34). These methods overcome the computational problems posed by large datasets and permit a degree of local anisotropy. They can also perform data smoothing when the data have elevation errors. There are some difficulties in defining patches when data are very irregularly spaced and anisotropy is limited to one direction across each surface patch. Nevertheless, Mitasova and Mitas (1993) have obtained good

performance on sparsely-distributed contour data. An advantage of this method for applications is that topographic parameters such as slope and curvature, as well as flow-lines and catchment areas, can be calculated directly from the fitted surface patches which have continuous first and second derivatives (Mitasova et al 1996). Local surface patches can also be readily converted into regular grids.

3.2.3 Locally adaptive gridding

Direct gridding or finite-difference methods can provide a computationally efficient means of applying high-quality interpolation methods to large elevation datasets. Iterative methods which fit discretised splines in tension have been described by Hutchinson (1989) and Smith and Wessel (1990). Both are based on the method developed by

Briggs (1974). Computational efficiency is achieved by using a simple multi-grid strategy. The use of splines in tension is indicated by the statistical nature of actual terrain surfaces (Frederiksen et al 1985; Goodchild and Mark 1987). It overcomes the tendency of minimum-curvature splines to generate spurious surface oscillations in complex areas and has been similarly applied to interpolation of elevation by local surface patches.

Former limitations in the ability of general gridding methods to adapt to strong anisotropic structure in actual terrain surfaces, as noted by Ebner et al (1988), have been largely overcome by applying a series of locally-adaptive constraints to the basic gridding procedure. These constraints can be applied between each pair of adjacent grid points, allowing maximum flexibility. Constraints which have direct relevance for hydrological applications are those imposed by the drainage enforcement algorithm devised by Hutchinson (1989). This algorithm removes spurious depressions in the fitted DEM, in recognition of the fact that sinks are usually quite rare in nature (Band 1986; Goodchild and Mark 1987). This can significantly improve the drainage quality and overall structure of the fitted DEM, especially in data-sparse areas.

A related locally-adaptive feature is an algorithm which automatically calculates ridge- and stream-lines from points of locally maximum curvature on contour lines (Hutchinson 1988). This permits interpolation of the fine structure in contours across the area between the contour lines in a more reliable fashion than methods which use linear or cubic interpolation along straight lines in a limited number of directions (Clarke et al 1982; Cole et al 1990;

Legates and Willmott 1986; Oswald and Raetzsch 1984). A partly similar approach, combining triangulation and grid structures, has been described by Aumann et al (1992). The result of applying the locally-adaptive gridding procedure to the contour and stream-line data in Figure 2 is shown in Figure 5. The inferred stream-lines and ridges are curvilinear, particularly in the data-sparse, low-relief portion of the figure, and there are no spurious depressions. The derived contours closely match the data contours and the shaded-relief view confirms that the surface has no fine-scale artefacts. The locally-adaptive method has overcome problems formerly encountered by gridding methods in accurately representing drainage structure in low-relief areas (Carter 1988; Douglas 1986).

The procedure also yields a systematic classification of the landscape into simple, connected, approximately-planar terrain elements, bounded by contour segments and flow-line segments. These are similar to the elements calculated by Moore et al (1988), but are determined in a more stable manner which incorporates both uphill and downhill searches, depending on the shape of the terrain.

Recent developments in this locally-adaptive gridding method include a locally-adaptive data-smoothing algorithm, which allows for the local slope-dependent errors associated with the finite-difference representation of terrain, and a locally-adaptive surface roughness penalty, which minimises profile curvature (Hutchinson 1996). The smoothing method has yielded useful error estimates for grid DEMs and a criterion for matching grid resolution to the information content of source data.

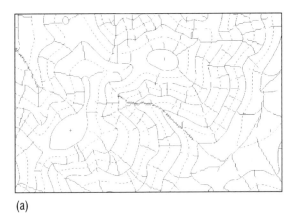

(a)

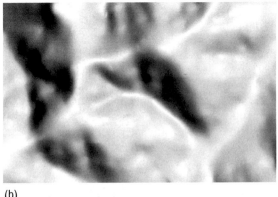

(b)

Fig 5. Locally-adaptive gridding of the contour and stream-line data shown in Figure 2: (a) structure lines (ridges and stream-lines) generated by the gridding method and contours derived from the fitted DEM; and (b) shaded relief view of the fitted DEM.

3.3 Filtering of remotely-sensed grid DEMs

Filtering of remotely-sensed grid DEMs is required to remove surface noise, which can have both random and systematic components. Filtering is usually associated with a coarsening of the DEM resolution. Methods include simple nearest-neighbour sub-sampling techniques and standard filtering techniques, including median and moving-average filtering in the spatial domain, and low-pass filtering in the frequency domain. Several authors have recognised the desirability of filtering remotely-sensed DEMs to improve the representation of surface shape.

Sasowsky et al (1992) and Bolstad and Stowe (1994) used the nearest-neighbour method to sub-sample SPOT DEMs, with a spatial resolution of 10 metres, to DEMs with spatial resolutions ranging from 20 to 70 metres. This generally enhanced the representation of surface shape, although significant errors remained. Giles and Franklin (1996) applied median and moving-average filtering methods to a 20-m-resolution SPOT DEM. This similarly improved representation of slope and solar-incidence angles, although elevation errors were as large as 80 metres and no effective representation of profile curvature could be obtained.

Hutchinson et al (1997) removed the large outliers from the airborne SAR data shown in Figure 3 and applied moving-average smoothing to generate a 50-metre-resolution DEM with accurate representation of surface aspect, except in those areas affected by vegetation cover. Lanari et al (1997) have applied a Kalman (moving-average) filter to spaceborne SAR data obtained on three different wavelengths. Standard errors ranged between about 5 and 80 metres, depending on land surface conditions.

The data in Figure 3 indicate that standard filtering techniques are not sufficient adequately to reduce error. Points associated with random large data errors and systematic errors attributable to surface cover need to be detected and replaced by interpolation. This process would be assisted by making use of techniques which enforce appropriate drainage conditions on the filtered DEM.

3.4 DEM quality assessment

The quality of a derived DEM can vary greatly depending on the data source and the interpolation technique. The desired quality depends on the application for which the DEM is to be used, but a DEM created for one application is often used for other purposes. Any DEM should therefore be created with care, using the best available data sources and processing techniques. Efficient detection of spurious features in DEMs can lead to improvements in DEM generation techniques, as well as detection of errors in source data.

Since most applications of DEMs depend on representations of surface shape and drainage structure, absolute measures of elevation error do not provide a complete assessment of DEM quality. A number of graphical techniques for assessing data quality have been developed. These are non-classical measures of data quality which offer means of confirmatory data analysis without the use of an accurate reference DEM. Assessment of DEMs in terms of their representation of surface aspect has been examined by Wise (1997).

Spurious sinks or local depressions in DEMs are frequently encountered and are a significant source of problems in hydrological applications (Band, Chapter 37). Sinks may be caused by incorrect or insufficient data, or by an interpolation technique that does not enforce surface drainage. They are easily detected by comparing elevations with surrounding neighbours. Hutchinson and Dowling (1991) noted the sensitivity of this method in detecting elevation errors as small as 20 metres in source data used to interpolate a continent-wide DEM with a horizontal resolution of 2.5 kilometres. More subtle drainage artefacts in a DEM can be detected by performing a full drainage analysis to derive catchment boundaries and stream-line networks, using the technique of Jenson and Domingue (1988).

Computing shaded relief allows a rapid visual inspection of the DEM for local anomalies that show up as bright or dark spots. It can indicate both random and systematic errors, as shown in Figures 3 and 4(d). It can identify problems with insufficient vertical resolution, since low-relief areas will show as highly visible steps between flat areas. It can also detect edge-matching problems (Hunter and Goodchild 1995). Shaded relief is a graphical way of checking the representation of slopes and aspects in the DEM. These can also be checked by standard statistical analysis if there is an accurate reference DEM or accurately surveyed ground data (Bolstad and Stowe 1994; Giles and Franklin 1996; Sasowsky et al 1992).

113

Contours derived from a DEM provide a sensitive check on terrain structure since their position, aspect, and curvature depend directly on the elevation, aspect, and plan curvature respectively of the DEM. Derived contours are a particularly useful diagnostic tool because of their sensitivity to elevation errors in source data. Subtle errors in labelling source-data contours digitised from topographic maps are common, particularly for small contour isolations which may have no label on the printed map. An example is shown in Figure 6, which also shows the utility of plotting sinks. The contours in Figure 6(b) and (c) were derived from a DEM calculated by the locally-adaptive gridding procedure described above.

Other deficiencies in the quality of a DEM can be detected by examining frequency histograms of elevation and aspect. DEMs derived from contour data usually show an increased frequency of contour elevations in the elevation histogram. The severity of this bias depends on the interpolation algorithm. The frequency histogram of aspect can be biased towards multiples of 45 and 90 degrees by interpolation algorithms that restrict searching to a few specific directions between pairs of data points.

4 DEM INTERPRETATION

4.1 Scale

4.1.1 Matching the resolution of grid DEMs to source data

Determination of the DEM resolution which matches the information content of the source data is desirable for several reasons. It directly facilitates efficient data inventory. It also permits interpretation of the horizontal resolution of the DEM as an index of information content. This is an important consideration when linking DEMs to other grid datasets and when filtering remotely-sensed DEMs. Moreover, it can facilitate the assessment of scale dependencies in terrain-dependent applications.

A simple method for matching DEM resolution to source data information content has been developed as part of the locally-adaptive gridding technique of Hutchinson (1996). The method monitors the root-mean-square slope of all DEM points associated with elevation data. The optimum DEM grid spacing is determined by refining the DEM spacing until further refinements produce no significant increase in the root-mean-square DEM

slopes. The method is particularly appropriate when source data have been obtained in a spatially-uniform manner, such as elevation contours from topographic maps at a fixed scale, or from remotely-sensed gridded elevation data.

4.1.2 Spectral and fractal analyses of scale

Understanding of the scaling characteristics of land-surface elevation is useful for identifying characteristic scales and predicting how sensitive the surface is to changes in resolution. This scaling behaviour can be studied using measures that are sensitive to the magnitude of variation at different spatial scales, such as the variogram (Oliver and Webster 1986) and the Fourier power spectrum (Mulla 1988; Pike and Rozema 1975). The power spectrum can discriminate degrees of smoothness that are indistinguishable using variograms (Gallant et al 1994).

The fractal model of scaling asserts that variance changes with scale according to a power-law function. This translates to a straight line in the logarithmic plot of the power spectrum with the magnitude of the slope between 1 and 3. A single scaling exponent across all scales is acknowledged to be unrealistic (Burrough 1981; Mandelbrot 1977; Mark and Aronson 1984) and several straight segments with different slopes are considered to satisfy the fractal model, provided the slopes are in the allowable range.

Figure 7 shows the power spectra of two DEMs in an area with moderate relief, one at 5-m resolution from 1:10 000-scale contours and stream-lines and the other at 20-m resolution from 1:25 000-scale data. These are the optimum resolutions for the source data, as determined by the procedure described above. Multiple straight lines are apparent in both spectra but, apart from the broadest-scale segment, the spectral slopes are too steep to be interpreted as fractal surfaces. The steep spectral slope indicates low fine-scale variance relative to coarse-scale variance.

Figure 7 also demonstrates that the spectral slope at fine scale is sensitive to both DEM resolution and the scale of the source data. Coarser-scale source data and coarser DEM resolution result in a smoother surface and steeper spectral slopes. However, the common increase in spectral slope at about 200-m wavelength for both curves is likely to be a function of the actual topography, related to hill-slope length and drainage density.

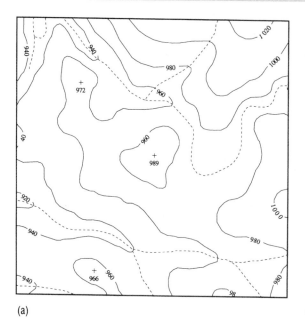

(a)

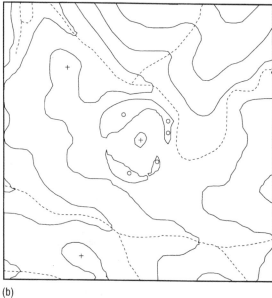

(b)

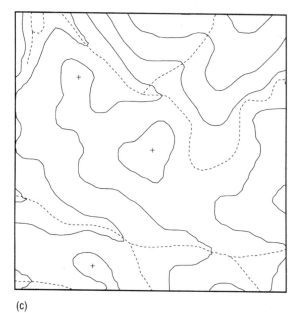

(c)

Fig 6. Use of drainage artefacts and derived contours to detect errors in source data: (a) contour and stream-line data with one contour label in error by one contour interval; (b) contours and spurious sinks, denoted by small circles, derived from a DEM fitted to the erroneous data in (a); and (c) contours derived from a DEM fitted to corrected data.

4.2 Terrain parameters

Terrain parameters, or topographic indices or attributes, are descriptive parameters of land-surface form that can be measured on the real surface and computed from a DEM. They have been developed as simplifications of specific Earth-surface processes in order to characterise the spatial variability of these processes across the landscape (Moore et al 1991; Speight 1980; Zevenbergen and Thorne 1987). Some parameters, such as slope and curvature, are defined in terms of local surface shape while others, such as specific catchment area, topographic wetness index, and flow length, are dependent on the shape of the surface some distance away from the reference location.

The more commonly used terrain parameters and their hydrological applications are described by Moore et al (1991, 1993) and Wilson and Gallant (1997). Slope and aspect modulate solar insolation, evaporation, and surface-water flow rates. Plan curvature and specific catchment area are parameters that describe the accumulation of surface water, closely related to the formation of streams and to the processes of soil erosion and soil aggradation. A 3-dimensional perspective view of specific catchment area derived from the DEM in Figure 5 is shown in Plate 4.

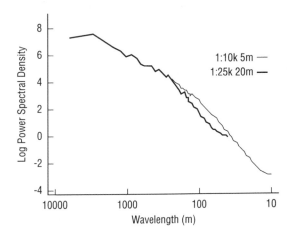

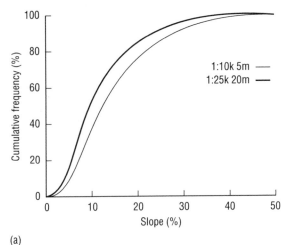

(a)

Fig 7. Terrain power spectra for DEMs at 5-m and 20-m horizontal resolutions.

The sensitivity of terrain parameters to DEM resolution has been demonstrated by Moore et al (1993). As DEM resolution becomes coarser, surface detail is lost, leading to reduced slopes and curvatures, and an increasingly simplified drainage network. This behaviour must be considered when using terrain parameters to represent landscapes in applications. In view of the absence of a satisfactory theory of scale dependence, comparisons between different landscapes are feasible only when the terrain surfaces are represented at the same resolution. The effects of changing DEM resolution and source-data scale on the cumulative distributions of slope and specific catchment area are shown in Figure 8.

4.3 Features

Three methods of dissecting the landscape into area features are described. Further classifications of landscapes into line and area features related to surface hydrology are described by Band (Chapter 37).

4.3.1 Catchments and sub-catchments

Catchments and sub-catchments form natural hierarchical dissections of landscapes. They can be readily calculated from DEMs, across a wide range of spatial scales, provided the DEMs represent surface drainage accurately. In contrast to many terrain parameters, they are robustly defined with respect to DEM resolution. Figure 9 shows a catchment determined from two coarse-scale continent-wide DEMs, with resolutions of 1/40th and 1/20th degree

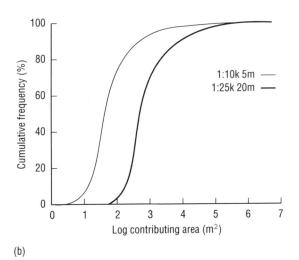

(b)

Fig 8. Cumulative distributions of terrain parameters derived from DEMs with horizontal resolutions of 5 and 20 metres: (a) slope; and (b) specific catchment area.

(approximately 2.5 and 5 kilometres). The two boundaries are in close agreement with the boundary determined from the 250-m-resolution DEM used to produce Figure 10.

Sub-catchments also form a natural unit for modelling and characterising biological and hydrological activity. Grouping environmental attributes across sub-catchments can greatly reduce model complexity, typically by around two orders of magnitude (Lewis et al 1991). The sub-catchments in Figure 10 were calculated using the technique of Jenson and Domingue (1988) to contain a minimum of 200 grid cells.

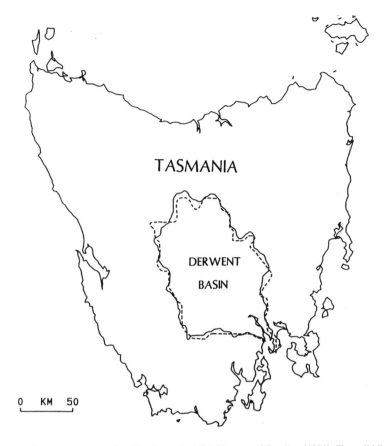

Fig 9. The Derwent River catchment of Tasmania, after Figure 2 of Hutchinson and Dowling (1991). The solid line denotes the boundary calculated from a 1/40th-degree DEM. The dashed line denotes the boundary calculated from a 1/20th-degree DEM.

4.3.2 Contour-flowline networks

Terrain surfaces can be dissected into small essentially planar elements bounded by contour lines and flow-lines (Moore et al 1988). This produces a natural discretisation of the landscape that reflects the convergence and divergence of surface water flow. This structure simplifies hydrological analyses, which become essentially 1-dimensional on each flow element. An efficient distributed-parameter dynamic hydrological model based on this structure has been developed by Grayson et al (1995).

Figure 11 shows a network of elements derived from contours using the TAPES-C program (Moore and Grayson 1991). The TOPOG program (Vertessy et al 1993) operates in a similar fashion. The elements are constructed by taking fixed-size steps along the lowest contour and successively connecting flow-lines to the next-highest contours.

The flow-lines are ideally orthogonal to every contour and follow the line of steepest descent across the landscape. In practice, straight line segments are used so the segments tend not to be orthogonal to contours at both ends, particularly where the contours are sharply curved. Hilltops and saddle points must be carefully specified to provide the connectivity required by the model (Dawes and Short 1994).

Current methods for defining the elements are stable for divergent topography, where flow-lines converge in the uphill direction, but unstable for convergent topography where flow-lines converge downhill. The construction of flow-lines in valley bottoms is therefore difficult and frequently produces large and uneven elements. Contour data often need to be augmented with intermediate contours to produce satisfactory elements.

117

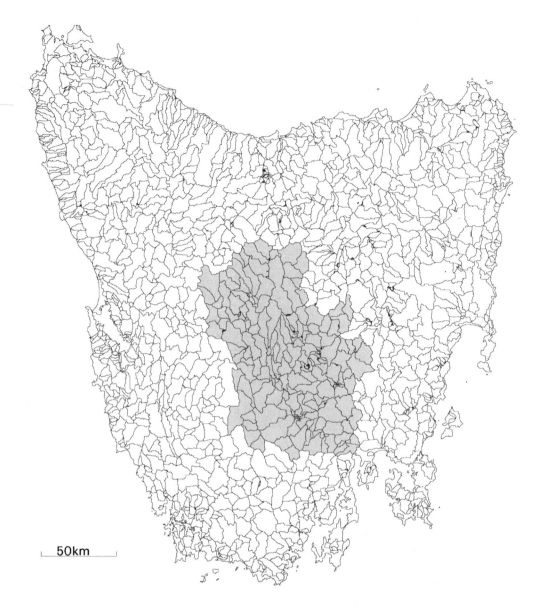

Fig 10. Sub-catchments calculated from a 250-m-resolution DEM for Tasmania overlaid with the Derwent catchment shown in Figure 9.

Because of these difficulties the method is usually restricted to well-defined catchments with limited interior complexity.

4.3.3 Multi-scale feature model
The extraction of particular topographic features from surface representations, usually grid DEMs or contour maps, has received attention from many authors, including Band (1986), Graff and Usery (1993), O'Callaghan and Mark (1984), Speight (1974), and Tribe (1992). However, there have been few attempts to automate an explicit representation of terrain using such features. A method inspired by wavelet methods and the shortcomings of fractal analysis has been developed by Gallant and Hutchinson (1996). It represents terrain as a

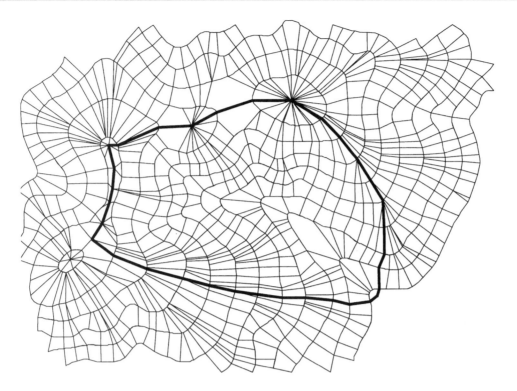

Fig 11. Finite elements bounded by refined contours and flow-lines for the area shown in Figure 2.

collection of features at different scales. These features differ from sub-catchment and contour-based features in having overlapping elliptical coverages. They have a bell-shaped profile form that blends smoothly to zero at the feature boundary.

Each feature is specified by six parameters describing spatial location, length, width as a fraction of length, orientation, and height, which may be negative. A grid DEM can be decomposed into these features using an iterative technique that repeatedly detects features, using the wavelet-based correlation-detection algorithm developed by Watson and Jones (1993), to remove them from the surface and detect new features from the residual surface.

Figure 12 illustrates this technique with a progressively-refined representation of the catchment shown in Figure 5. The four broadest-scale features define a ridge and valley within which the catchment is embedded. Adding further features improves the representation of topographic structure until detailed catchment structure is represented using just 34 features. This representation uses fewer parameters than the TIN shown in Figure 4, but provides a more accurate representation of surface shape.

The particular value of this representation is that the length and height parameters capture scale directly, facilitating study of the scaling properties of terrain and of the connections between scale and shape. The representation permits generalisation of the surface by removal of fine-scale features, and refinement of particular areas by addition of new features based on additional site data. These features may also be used to obtain information about shape and orientation of the terrain surface.

5 SCALES OF APPLICATIONS OF DEMS IN ENVIRONMENTAL MODELLING

Steyaert (1993) and others have recognised the need to identify appropriate scales for modelling various Earth-surface processes and the need for effective methods to integrate data and analyses across different scales. Accordingly, applications of grid DEMs in environmental modelling are best described in relation to their spatial resolution.

The finest DEM spatial resolutions, from 5 to 50 metres, are typically used for spatially-distributed

119

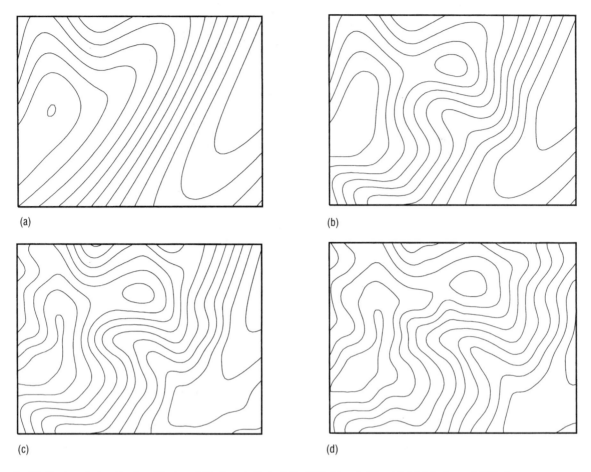

(a)

(b)

(c)

(d)

Fig 12. Progressively refined positive wavelet feature analysis of the DEM shown in Figure 5: (a) 4 features; (b) 7 features; (c) 18 features; and (d) 34 features.

hydrological modelling (Binley and Beven 1992; Zhang and Montgomery 1994) and analysis of soil properties (Gessler et al 1996). The determination of appropriate spatial scales for hydrological modelling is an active research issue (Blöschl and Sivaplan 1995). DEMs at this scale can also be used to make aspect-based corrections to remotely-sensed data (Ekstrund 1996; Hinton 1996). These applications are distinguished by their dependence on accurate representation of terrain shape.

Fine mesoscale or 'toposcale' DEMs, with spatial resolutions from 50 to 200 metres, are used to model aspect-related microclimatic variations, particularly in solar radiation, evaporation, and associated vegetation patterns (Mackey 1996; Wigmosta et al 1994). This scale is appropriate for broader-scale distributed-parameter hydrological models which incorporate

remotely-sensed land-cover data (Kite 1995). It is also appropriate for defining sub-catchment units for lumped-parameter hydrological models and assessments of biodiversity (Lewis et al 1991).

Mesoscale DEMs, with spatial resolutions from 200 metres to 5 kilometres, are appropriate for topographically-dependent representations of surface temperature and rainfall, key determinants of biological activity. For these variables elevation is more important than surface shape, giving rise to temperature and precipitation elevation lapse rates, so that the spatial distributions of these variables are truly 3-dimensional. Precipitation is best described by a model which permits spatially-varying elevation lapse rates (Hutchinson 1995), as illustrated in Plate 5. There are secondary aspect effects related to prevailing wind directions (Daly et al 1994), and local relief at this scale can be used to assist the interpolation of surface windspeeds (Weiringa 1986). This scale is also useful for determining continental-scale drainage structure (Hutchinson and Dowling 1991; Hutchinson et al 1996; Jenson 1991), and for providing fundamental terrain and climatic constraints on agricultural productivity (as discussed by Wilson in Chapter 70).

Macro scales with spatial resolutions from 50 to 500 kilometres are used for broad-scale atmospheric modelling. The DEMs used in these applications are very generalised and accuracy is not critical. Terrain shape is still significant in terms of defining major orographic barriers. DEMs at much finer resolutions are required to distribute the outputs of these broad-scale models spatially (Steyaert 1996).

6 CONCLUSION

This chapter has demonstrated the central role played by representations of terrain in environmental modelling and landscape visualisation. An important theme for providers of source topographic data and DEM interpolation methods is the need by most applications for accurate representations of terrain shape and drainage structure.

This has prompted the development of locally-adaptive process-based interpolation methods and a renewed interest in contour and stream-line data sources which represent surface shape explicitly. Remotely-sensed elevation data sources hold the promise of providing DEMs with global coverage, but filtering methods which respect surface structure and drainage need to be developed to reduce the inherent errors in these data, particularly in areas with low relief.

Spatial scale has become an important issue. The need for multi-scale representations of Earth-surface processes is now recognised, as is the need for representations of terrain to have spatial scales consistent with these processes. The scaling properties of DEMs and various associated terrain parameters have yet to be determined satisfactorily. Spectral analyses of terrain reveal some information about terrain structure, but associated fractal models of scale have been found to have shortcomings. A multi-scale feature model shows promise in incorporating relevant aspects of shape and drainage structure into terrain scale analyses.

References

Ackermann F 1978 Experimental investigation into the accuracy of contouring from DTM. *Photogrammetric Engineering and Remote Sensing* 44: 1537–48

Akima H 1978 A method of bivariate interpolation and smooth surface fitting for irregularly distributed data points. *ACM Transactions on Mathematical Software* 4: 148–59

Auerbach S, Schaeben H 1990 Surface representation reproducing given digitised contour lines. *Mathematical Geology* 22: 723–42

Aumann G, Ebner H, Tang L 1992 Automatic derivation of skeleton lines from digitised contours. *ISPRS Journal of Photogrammetry and Remote Sensing* 46: 259–68

Aurenhammer F 1991 Voronoi diagrams – a survey of fundamental geometric data structure. *ACM Computing Surveys* 23: 345–405

Band L E 1986b Topographic partition of watersheds with digital elevation models. *Water Resources Research* 22: 15–24

Binley A, Beven K 1992 Three-dimensional modelling of hill-slope hydrology. *Hydrological Processes* 6: 347–59

Blöschl G, Sivaplan M 1995 Scale issues in hydrological modelling: a review. *Hydrological Processes* 9: 313–30

Bolstad P V, Stowe T 1994 An evaluation of DEM accuracy: elevation, slope, and aspect. *Photogrammetric Engineering and Remote Sensing* 60: 1327–32

Briggs I C 1974 Machine contouring using minimum curvature. *Geophysics* 39: 39–48

Burrough P A 1981 Fractal dimensions of landscapes and other environmental data. *Nature* 294: 240–2

Carter J R 1988 Digital representations of topographic surfaces. *Photogrammetric Engineering and Remote Sensing* 54: 1577–80

Clarke A L, Grün A, Loon J C 1982 The application of contour data for generating high fidelity grid digital elevation models. *Proceedings AutoCarto 5*. Falls Church, ASPRS: 213–22

Clarke K C 1990 *Analytical and computer cartography*. Englewood Cliffs, Prentice-Hall

Cole G, MacInnes S, Miller J 1990 Conversion of contoured topography to digital terrain data. *Computers and Geosciences* 16: 101–9

Daly C, Neilson R P, Phillips D L 1994 A statistical-topography model for mapping climatological precipitation over mountainous terrain. *Journal of Applied Meteorology* 33: 140–58

Dawes W, Short D 1994 The significance of topology for modelling the surface hydrology of fluvial landscapes. *Water Resources Research* 30: 1045–55

Day T, Muller J-P 1988 Quality assessment of digital elevation models produced by automatic stereo matchers from SPOT image pairs. *International Archives of Photogrammetry and Remote Sensing* 27: 148–59

Dixon T H 1991 An introduction to the global positioning system and some geological applications. *Reviews of Geophysics* 29: 249–76

Dixon T H (ed.) 1995 *SAR interferometry and surface change detection*. RSMAS Technical Report TR 95-003, University of Miami

Douglas D H 1986 Experiments to locate ridges and channels to create a new type of digital elevation model. *Cartographica* 23: 29–61

Ebner H, Reinhardt W, Hössler R 1988 Generation, management and utilisation of high-fidelity digital terrain models. *International Archives of Photogrammetry and Remote Sensing* 27: 556–65

Ekstrund S 1996 Landsat TM-based forest damage assessment: correction for topographic effects. *Photogrammetric Engineering and Remote Sensing* 62: 151–61

Franke R 1982b Smooth interpolation of scattered data by local thin plate splines. *Computers and Mathematics with Applications* 8: 273–81

Frederiksen P, Jacobi O, Kubik K 1985 A review of current trends in terrain modelling. *ITC Journal* 1985: 101–6

Gallant J C, Hutchinson M F, Moore I D, Gessler P E 1994 Estimating fractal dimensions: an empirical approach. *Mathematical Geology* 26: 455–81

Gallant J C, Hutchinson M F 1996 Towards an understanding of landscape scale and structure. *Proceedings, Third International Conference/Workshop on Integrating GIS and Environmental Modelling*. Santa Barbara, NCGIA. CD

Garbrecht J, Starks P 1995 Note on the use of USGS level 1 7.5-minute DEM coverages for landscape drainage analyses. *Photogrammetric Engineering and Remote Sensing* 61: 519–22

Gessler P E, McKenzie N J, Hutchinson M F 1996 Progress in soil-landscape modelling and spatial prediction of soil attributes for environmental models. *Proceedings, Third International Conference/Workshop on Integrating GIS and Environmental Modelling*. Santa Barbara, NCGIA. CD

Giles P T, Franklin S E 1996 Comparison of derivative topographic surfaces of a DEM generated from stereographic SPOT images with field measurements. *Photogrammetric Engineering and Remote Sensing* 62: 1165–71

Goodchild M F, Mark D M 1987 The fractal nature of geographic phenomena. *Annals of the Association of American Geographers* 77: 265–78

Graff L H, Usery E L 1993 Automated classification of generic terrain features in digital elevation models. *Photogrammetric Engineering and Remote Sensing* 59: 1409–17

Grayson R B, Blöschl G, Moore I D 1995 Distributed parameter hydrologic modelling using vector elevation data: Thales and TAPES-C. In Singh V P (ed.) *Computer models of watershed hydrology*. Highland Ranch, Water Resources Publications: 669–96

Harding D J, Bufton J L, Frawley J 1994 Satellite laser altimetry of terrestrial topography: vertical accuracy as a function of surface slope, roughness, and cloud cover. *IEEE Transactions on Geoscience and Remote Sensing* 32: 329–39

Heller M 1990 Triangulation algorithms for adaptive terrain modelling. *Proceedings, Fourth International Symposium on Spatial Data Handling*. Columbus, International Geographical Union: 163–74

Hinton J C 1996 GIS and remote sensing integration for environmental applications. *International Journal of Geographical Information Systems* 10: 877–90

Hobbs F 1995 The rendering of relief images from digital contour data. *The Cartographic Journal* 32: 111–16

Hunter G J, Goodchild M F 1995b Dealing with error in spatial databases: a simple case study. *Photogrammetric Engineering and Remote Sensing* 61: 529–37

Hutchinson M F 1988 Calculation of hydrologically sound digital elevation models. *Proceedings, Third International Symposium on Spatial Data Handling*. Columbus, International Geographical Union: 117–33

Hutchinson M F 1989 A new procedure for gridding elevation and streamline data with automatic removal of spurious pits. *Journal of Hydrology* 106: 211–32

Hutchinson M F 1991 The application of thin plate smoothing splines to continent-wide data assimilation. In Jasper J D (ed.) *Data assimilation systems*. BMRC Research Report No. 27. Melbourne, Bureau of Meteorology: 104–13

Hutchinson M F 1995b Stochastic space-time weather models from ground-based data. *Agricultural and Forest Meteorology* 73: 237–64

Hutchinson M F 1996 A locally adaptive approach to the interpolation of digital elevation models. *Proceedings, Third International Conference/Workshop on Integrating GIS and Environmental Modelling*. Santa Barbara, NCGIA. CD

Hutchinson M F, Dowling T I 1991 A continental hydrological assessment of a new grid-based digital elevation model of Australia. *Hydrological Processes* 5: 45–58

Hutchinson M F, Gessler P E, Xu T, Gallant J C 1997 Filtering Wagga TOPSAR data to improve drainage accuracy. In Milne A K (ed.) *Proceedings, International Workshop on Radar Image Processing and Applications*. Canberra, CSIRO: 80–3

Hutchinson M F, Nix H A, McMahon J P, Ord K D 1996 The development of a topographic and climate database for Africa. *Proceedings, Third International Conference/Workshop on Integrating GIS and Environmental Modelling*. Santa Barbara, NCGIA. CD

Jenson S K 1991 Applications of hydrologic information automatically extracted from digital elevation models. *Hydrological Processes* 5: 31–44

Jenson S K, Domingue J O 1988 Extracting topographic structure from digital elevation data for geographic information system analysis. *Photogrammetric Engineering and Remote Sensing* 54: 1593–1600

Kelly R E, McConnell P R H, Mildenberger S J 1977 The Gestalt photomapping system. *Photogrammetric Engineering and Remote Sensing* 43: 1407–17

Kite G W 1995 The SLURP model. In Singh V P (ed.) *Computer models of watershed hydrology*. Highland Ranch, Water Resources Publications: 521–62

Konecny G, Lohmann P, Engel H, Kruck E 1987 Evaluation of SPOT imagery on analytical instruments. *Photogrammetric Engineering and Remote Sensing* 53: 1223–30

Lanari R, Fornaro G, Riccio D, Migliaccio M, Papathanassiou K, Moreira J, Schwabisch M, Dutra L, Puglisi G, Franceschetti G, Coltelli M 1997 Generation of digital elevation models by using SIR-C/X-SAR multifrequency two-pass interferometry: the Etna case study. *IEEE Transactions on Geoscience and Remote Sensing* 34: 1097–114

Lee J 1991 Comparison of existing methods for building triangular irregular network models of terrain from grid digital elevation models. *International Journal of Geographical Information Systems* 5: 267–85

Legates D R, Willmott C J 1986 Interpolation of point values from isoline maps. *The American Cartographer* 13: 308–23

Lemmens M J P M 1988 A survey on stereo matching techniques. *International Archives of Photogrammetry and Remote Sensing* 27: V11–V23

Lewis A, Stein J L, Stein J A, Nix H A, Mackey B G, Bowyer J K 1991 *An assessment of regional conservation adequacy: Tasmania*. Forestry and Timber Inquiry Consultancy Series No. FTC91/17. Canberra, Resource Assessment Commission

Mackey B G 1996 The role of GIS and environmental modelling in the conservation of biodiversity. *Proceedings, Third International Conference/Workshop on Integrating GIS and Environmental Modelling*. Santa Barbara, NCGIA. CD

Makarovic B 1984 Structures for geo-information and their application in selective sampling for digital terrain models. *ITC Journal* 1984: 285–95

Mandelbrot B B 1977 *Fractals: form, chance, and dimension*. San Francisco, Freeman

Mark D M 1986 Knowledge-based approaches for contour-to-grid interpolation on desert pediments and similar surfaces of low relief. *Second International Symposium on Spatial Data Handling*. Columbus, International Geographical Union: 225–34

Mark D M, Aronson P B 1984 Scale-dependent fractal dimensions of topographic surfaces: an empirical investigation with application to geomorphology and computer mapping. *Mathematical Geology* 16: 671–83

McCullagh M J 1988 Terrain and surface modelling systems: theory and practice. *Photogrammetric Record* 12: 747–79

Mitasova H, Hofierka J, Zlocha M, Iverson L 1996 Modelling topographic potential for erosion and deposition using GIS. *International Journal of Geographical Information Systems* 10: 629–41

Mitasova H, Mitas L 1993 Interpolation by regularised spline with tension: I. Theory and implementation. *Mathematical Geology* 25: 641–55

Moore I D, Grayson R B 1991 Terrain-based catchment partitioning and runoff prediction using vector elevation data. *Water Resources Research* 27: 1177–91

Moore I D, Grayson R B, Ladson A R 1991. Digital terrain modelling: a review of hydrological, geomorphological, and biological application. *Hydrological Processes* 5: 3–30

Moore I D, Lewis A, Gallant J C 1993 Terrain attributes: estimation of scale effects. In Jakeman A J, Beck M B, McAleer M (eds) *Modelling change in environmental systems*. New York, John Wiley & Sons Inc.: 189–214

Moore I D, O'Loughlin E M, Burch G J 1988 A contour-based topographic model for hydrological and ecological applications. *Earth Surface Processes and Landforms* 13: 305–20

Mulla D J 1988 Using geostatistics and spectral analysis to study spatial patterns in the topography of southeastern Washington State, USA. *Earth Surface Processes and Landforms* 13: 389–405

Nix H A 1986 A biogeographic analysis of Australian elapid snakes. In Longmore R (ed.) *Atlas of elapid snakes of Australia*. Australian Flora and Fauna Series No.7. Canberra, Australian Government: 4–15

O'Callaghan J F, Mark D M 1984 The extraction of drainage networks from digital elevation data. *Computer Vision, Graphics, and Image Processing* 28: 323–44

Oliver M A, Webster R 1986 Semi-variograms for modelling the spatial pattern of landform and soil properties. *Earth Surface Processes and Landforms* 11: 491–504

Oswald H, Raetzsch H 1984 A system for generation and display of digital elevation models. *Geo-Processing* 2: 197–218

Pike R J, Rozema W J 1975 Spectral analysis of landforms. *Annals of the Association of American Geographers* 65: 499–516

Pries R A 1995 A system for large-scale image mapping and GIS data collection. *Photogrammetric Engineering and Remote Sensing* 61: 503–11

Quinn P F, Beven K J, Chevallier P, Planchon O 1991 The prediction of hill-slope flow paths for distributed hydrological modelling using digital terrain models. *Hydrological Processes* 5: 59–79

Running S W, Thornton P E 1996 Generating daily surfaces of temperature and precipitation over complex topography. In Goodchild M F, Steyaert L T, Parks B O, Johnston C, Maidment D, Crane M, Glendinning S (eds) *GIS and environmental modelling: progress and research issues*. Fort Collins, GIS World Books: 93–8

Sambridge M, Braun J, McQueen H 1995 Geophysical parameterisation and interpolation of irregular data using natural neighbours. *Geophysical Journal International* 122: 837–57

Sasowsky K C, Peterson G W, Evans B M 1992 Accuracy of SPOT digital elevation model and derivatives: utility for Alaska's North Slope. *Photogrammetric Engineering and Remote Sensing* 60: 1327–32

Sibson R 1981 A brief description of natural neighbour interpolation. In Barnett V (ed.) *Interpreting multivariate data*. Chichester, John Wiley & Sons: 21–36

Smith W H F, Wessel P 1990 Gridding with continuous curvature. *Geophysics* 55: 293–305

Speight J G 1974 A parametric approach to landform regions. *Special Publication Institute of British Geographers* 7: 213–20

Speight J G 1980 The role of topography in controlling throughflow generation: a discussion. *Earth Surface Processes and Landforms* 5: 187–91

Steyaert L T 1993 A perspective on the state of environmental simulation modelling. In Goodchild M F, Parks B O, Steyaert L T (eds) *Environmental modelling with GIS*. New York, Oxford University Press: 16–30

Steyaert L T 1996 Status of land data for environmental modelling and challenges for geographic information systems in land characterisation. In Goodchild M F, Steyaert L T, Parks B O, Johnston C, Maidment D, Crane M, Glendinning S (eds) *GIS and environmental modelling: progress and research issues*. Fort Collins, GIS World Inc.: 17–27

Tribe A 1992 Problems in automated recognition of valley features from digital elevation models and a new method toward their resolution. *Earth Surface Processes and Landforms* 17: 437–54

Tsai V 1993 Delaunay triangulations in TIN creation: an overview and linear-time algorithm. *International Journal of Geographical Information Systems* 7: 501–24

Verdin K L, Jenson S K 1996 Development of continental scale digital elevation models and extraction of hydrographic features. *Proceedings, Third International Conference/ Workshop on Integrating GIS and Environmental Modelling*. Santa Barbara, NCGIA. CD

Vertessy R A, Hatton T J, O'Shaugnessy P J, Jayasuroya M D A 1993 Predicting water yield from a mountain ash forest catchment using a terrain analysis based catchment model. *Journal of Hydrology* 150: 665–700

Watson D F, Philip G M 1984 Triangle based interpolation. *Mathematical Geology* 16: 779–95

Watson G H, Jones J G 1993 Positive wavelet representation of fractal signals and images. In Crilly A J, Earnshaw R A, Jones H (eds) *Applications of fractals and chaos*. Berlin, Springer-Verlag: 117–35

Weibel R, Heller M 1991 Digital terrain modelling. In Maguire D J, Goodchild M F, Rhind D W (eds) *Geographical information systems: principles and applications*. Harlow, Longman/New York, John Wiley & Sons Inc. Vol. 1: 269–97

Weiringa J 1986 Roughness-dependent geographical interpolation of surface wind speed averages. *Quarterly Journal of the Royal Meteorological Society* 112: 867–89

Wigmosta M S, Vail L W, Lettenmaier D P 1994 A distributed hydrology-vegetation model for complex terrain. *Water Resources Research* 30: 1665–79

Wilson J P, Gallant J C 1997 Terrain-based approaches to environmental resource evaluation. In Lowe S N, Richards K S, Chandler J H (eds) *Landform monitoring, modelling and analysis*. Chichester, John Wiley & Sons: 219–40

Wise S 1997 The effect of GIS interpolation errors on the use of digital elevation models in geomorphology. In Lowe S N, Richards K S, Chandler J H (eds) *Landform monitoring, modelling, and analysis*. Chichester, John Wiley & Sons: Chapter 7

Zebker H A, Werner C, Rosen P A, Hensley S 1994 Accuracy of topographic maps derived from ERS-1 interferometric radar. *IEEE Transactions on Geoscience and Remote Sensing* 32: 823–36

Zevenbergen L W, Thorne C R 1987 Quantitative analysis of land surface topography. *Earth Surface Processes and Landforms* 12: 47–56

Zhang W, Montgomery D R 1994 Digital elevation model grid size, landscape representation, and hydrologic simulation. *Water Resources Research* 30: 1019–28

10

Generalising spatial data and dealing with multiple representations

R WEIBEL AND G DUTTON

Functions for generalising spatial data are of fundamental importance in GIS because of a variety of requirements for scale-changing as well as thematic reduction and emphasis. Following a brief introduction, section 2 discusses what generalisation is, what its objectives are, and why it is important. In section 3, the distinction is developed between process-oriented and representation-oriented approaches to generalisation, first discussing the latter in the context of multiple representations and multi-scale databases. Section 4 provides an introduction to the range of conceptual models of process-oriented generalisation, outlining the nature and relationships of data models, operators, objectives, and controls. Section 5 sketches the requirements for a successful digital generalisation system, including elements of data modelling, structure and shape analysis, generalisation algorithms, knowledge-based methods, human-computer interaction, and quality evaluation. The chapter concludes with a brief look at some operational generalisation systems, recently developed and still evolving.

1 INTRODUCTION

'Wherever there is life, there is twist and mess: the frizz of an arctic lichen, the tangle of brush along a bank, the dogleg of a dog's leg, the way a line has got to curve, split or knob. The planet is characterised by its very jaggedness, its random heaps of mountains, its frayed fringes of shore . . . Think of a contour globe, whose mountain ranges cast shadows, whose continents rise in bas-relief above the oceans. But then: think of how it really is. These heights aren't just suggested; they're there . . . What if you had an enormous globe in relief that was so huge it showed roads and houses – a geological survey globe, a quarter of a mile to an inch – of the whole world, and the ocean floor! Looking at it, you would know what had to be left out: the free-standing sculptural arrangement of furniture in rooms, the jumble of broken rocks in a creek bed, tools in a box, labyrinthine ocean liners, the shape of snapdragons, walrus . . . The relief globe couldn't begin to show trees, between whose overlapping boughs birds raise broods, or the furrows in bark, where whole creatures, creatures easily visible, live out their lives and call it world enough.' (Dillard 1974: 141–3)

A Highway Department studies how to widen and straighten a road that runs through towns and villages in a river valley. A property owner decides to develop land that has restrictions by the state on construction within 100 metres of waterways. A planning board attempts to rationalise a community's haphazard zoning code to minimise future conflicts between adjacent land uses. A farmer buys a new, larger tractor and needs to revise how he or she ploughs and intercrops in hilly terrain.

All of these are real-world analogues of what cartographers call map generalisation. Although they concern physical forms and processes on the landscape, these decisions also involve abstractions familiar to users of GIS, such as categorical coverages, curve geometries, proximity buffers, and

125

spatial autocorrelation. Were the people described above to make use of GIS to help solve their problems, they would quickly see that in making and manipulating 2-dimensional representations of their project worlds they would be making cartographic decisions, and furthermore that their systems would not be of very much help in the process.

This chapter describes what generalisation of spatial data is, why it is necessary, some techniques for performing it in the digital domain, and what tools currently exist to support it, providing pointers to some of the recent research literature. More detailed surveys of generalisation techniques have been published, notably by McMaster and Shea (1992) and Weibel (1997), who focus on algorithmic methods. This topic is discussed from the perspective of data quality by Veregin (Chapter 12). Compilations of recent research are provided by Buttenfield and McMaster (1991), McMaster (1989), Molenaar (1996a), Müller et al (1995), and Weibel (1995a). For discussions of generalisation in the context of particular applications see, for example, Larsen (Chapter 71), Meyers (Chapter 57), Wilson (Chapter 70), and Yeh (Chapter 62).

2 DESCRIBING GENERALISATION

Our world presents us with an infinite regress of detail having neither beginning nor end. Digital computers, being primitive machines with limited memory for and no real understanding of facts, must be coaxed and prodded artfully to do anything useful with data describing the planet, its regions and phenomena. As Dillard thoughtfully observes, if we try to model the world it is impossible not to generalise spatial data, whether we intend doing so or not.

2.1 Generalisation in conventional cartography

In conventional cartography, map generalisation is responsible for reducing complexity in a map in a scale reduction process, emphasising the essential while suppressing the unimportant, maintaining logical and unambiguous relations between map objects, and preserving aesthetic quality. The main objective then is to create maps of high graphical clarity, so that the map image can be easily perceived and the message the map intends to deliver can be readily understood. This position is expressed by the concise definition which equates map generalisation to 'the selection and simplified representation of detail appropriate to the scale and/or the purpose of a map' (ICA 1973: 173).

Scale reduction from a source map to a target map leads to a competition for space among map features caused by two cumulative effects: at a reduced scale, less space is available on the map to place symbols representing map features, while at the same time, symbol size increases relative to the ground it covers in order to maintain size relations and legibility. Figure 1 illustrates the spatial conflicts that arise from reduction of available map space and enlargement of symbol sizes. These can be resolved by simplifying symbolism, by selecting only a subset of features to depict, and by displacing some features away from others (e.g. moving buildings away from streets).

However, note that map scale is not the only factor that influences generalisation. Map purpose is equally (and perhaps even more) important. A good map should focus on the information that is essential to its intended audience. Thus, a map for cyclists will emphasise a different selection of roads than a map targeted at car drivers. Map purpose also influences

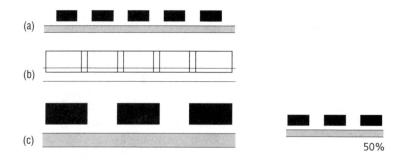

Fig 1. Competition for space among map features as a consequence of scale reduction.

directly the selection of the appropriate map scale, as spatial phenomena and processes should be studied at the level of scale at which they are most relevant (Dikau 1990). Other factors that control traditional map generalisation are the quality of the source material, the symbol specifications (e.g. the width and colour of line symbols to depict roads, political boundaries, etc.), and technical reproduction capabilities (SSC 1977). The combination of these factors is termed the 'controls of generalisation'.

2.2 Generalisation in digital systems

In digital cartographic systems and GIS, generalisation has gradually assumed an even wider meaning. It can be understood as a process which realises transitions between different models representing a portion of the real world at decreasing detail, while maximising information content with respect to a given application. Figure 2 shows how transitions take place in three different areas along the database and map production workflow; the terminology used here was originally developed for the German Amtliches Topographisch–Kartographisches Informations system (ATKIS) project (Grünreich 1992), but has since been adopted by other authors. Generalisation takes place:

- as part of building a primary model of the real world (a so-called digital landscape model or DLM) – also known as object generalisation;
- as part of the derivation of special-purpose secondary models of reduced contents and/or resolution from the primary model – also known

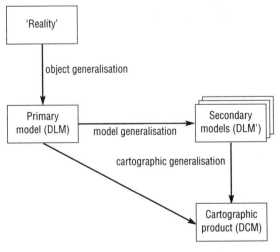

Fig 2. Generalisation as a sequence of modelling operations (after Grünreich 1985).

as model generalisation (also termed 'model-oriented', or statistical (database) generalisation by different authors; cf. Weibel 1995b);
- as part of the derivation of cartographic visualisations (digital cartographic models or DCMs) from either primary or secondary models – commonly known as cartographic generalisation.

The next section takes a closer look at the scope and the objectives of these three generalisation types.

2.2.1 Object generalisation
This process takes place at the time of defining and building the original database, called the 'primary model' in Figure 2. Since databases are abstract representations of a portion of the real world, a certain degree of generalisation (in the sense of abstraction, selection, and reduction) must take place, as only the subset of information relevant for the intended use(s) is represented in this database. Although seen from the perspective of generalisation here, this operation is sufficiently explained by methods of semantic and geometric data modelling (which define the relevant object classes and their attributes and relations), as well as sampling methods (which define the sampling strategy and desired resolution and accuracy), combined with human interpretation skills (e.g. if photogrammetric data capture is used: see Dowman, Chapter 31).

2.2.2 Model generalisation
While the process of object generalisation had to be carried out in much the same way when preparing data for a traditional map, model generalisation is new and specific to the digital domain. In digital systems, generalisation can affect directly not only the map graphics, but also the map data. The main objective of model generalisation is controlled data reduction for various purposes. Data reduction may be desirable to save storage and increase the computational efficiency of analytical functions. It also speeds data transfer via communication networks. It may further serve the purpose of deriving datasets of reduced accuracy and/or resolution. This capability is particularly useful in the integration of datasets of differing resolution and accuracy as well as in the context of multi-resolution databases (Goodchild and Longley, Chapter 40). While model generalisation may also be used as a preprocessing step to cartographic generalisation, it is important to note that it is not oriented towards graphical depiction, and thus involves no artistic, intuitive components. Instead, it encompasses processes which can be modelled

completely formally (Weibel 1995b); these may, however, have aesthetic consequences for subsequent cartographic generalisation.

2.2.3 Cartographic generalisation

This is the term commonly used to describe the generalisation of spatial data for cartographic visualisation. It is the process most people typically think of when they hear the term 'generalisation'. The difference between this and model generalisation is that it is aimed at generating visualisations, and brings about graphical symbolisation of data objects. Therefore, cartographic generalisation must also encompass operations to deal with problems created by symbology, such as feature displacement (cf. Figure 1), which model generalisation does not. The objectives of digital cartographic generalisation remain basically the same as in conventional cartography (see above). However, technological change has also brought along new tasks with new requirements (Kraak, Chapter 11) such as interactive zooming, visualisation for exploratory data analysis, or progressively adapting the level of detail of 3-dimensional perspective views to the viewing depth. The concept of cartographic generalisation thus needs to be extended. On the other hand, typical maps generated in GIS are no longer complex multi-purpose maps with a multitude of feature classes involved, but rather single-purpose maps consisting of a small number of layers. Furthermore, maps and other forms of visualisations are often presented by means of a series of different partial views in a multi-window arrangement, particularly in exploratory data analysis (Anselin, Chapter 17). Together with the capabilities of interactive direct manipulation these new forms of cartographic presentations partially alleviate (but by no means eliminate) some of the generalisation problems, or at least make them less salient for many GIS users.

2.3 Motivations for generalisation

The discussion above has already alluded to some of the reasons for generalisation. Extending Müller's (1991) discussion of requirements for generalisation, it is possible to develop a more detailed list of motivations.

1 *Develop a primary database*: build a digital model of the real world, with the resolution and content appropriate to the intended application(s), and populate it (object generalisation):
   - select objects;
   - approximate objects.

2 *Use resources economically*: minimise use of computing resources by filtering and selection within tolerable (and controllable) accuracy limits:
   - save storage space;
   - save processing time.

3 *Increase/ensure data robustness*: build clean, lean, and consistent spatial databases by reducing spurious and/or unnecessary detail:
   - suppress unneeded high-frequency detail;
   - detect and suppress errors and random variations of data capture;
   - homogenise (standardise) resolution and accuracy of heterogeneous data for data integration.

4 *Derive data and maps for a range of purposes*: from a detailed multi-purpose database, derive data and map products according to specific requirements:
   - derive secondary scale and/or theme-specific datasets;
   - compose special-purpose maps (i.e. all new maps);
   - avoid redundancy, increase consistency.

5 *Optimise visual communication*: develop meaningful and legible visualisations:
   - maintain legibility of cartographic visualisations of a database;
   - convey an unambiguous message by focusing on main theme;
   - adapt to properties of varying output media.

Examination of the above list reveals that classical cartographic generalisation mainly relates to task 5 (visual communication) and to a lesser extent also to task 4, while tasks 1 to 3 are more specific to the digital domain (object generalisation, model generalisation). In task 5, an aspect of cartographic generalisation germane to a GIS environment is that output may be generated for media of varying specifications, such as high-resolution plotted maps or low-resolution CRT (cathode ray tube) views, requiring consideration of the resolution of the output media when composing maps for display (Spiess 1995).

3 GENERALISATION AND MULTIPLE REPRESENTATIONS

3.1 Generalisation: process-oriented vs representation-oriented view

The above discussion has implicitly taken a process-oriented view of generalisation, understanding it as the process of transforming a detailed database into a database or map of reduced complexity at arbitrary scale. As was already mentioned, generalisation is a complex process, and indeed, satisfactory implementations of all the transformation operations (and their interactions) necessary to achieve comprehensive automated generalisation largely remain to be developed.

An alternative – or complementary – approach is to develop multi-scale databases. We term this approach the 'representation-oriented view', because it attempts to develop databases that integrate single representations at different fixed scales into a consistent multiple representation. An example may help to illustrate the concept. Consider a set of four maps of the same region, drawn at scales 1:1000, 1:25 000, 1:100 000, and 1:250 000. In a particular settlement, dwellings may be represented by their detailed footprints at 1:1000. At 1:25 000 buildings are now represented by simplified rectangular shapes; some building polygons may even have been aggregated. At the next smaller scale, 1:100 000, built-up areas are depicted as tinted blocks defined by the street network and urban boundaries, thus are transformed into larger, less regular polygons. Finally, the 1:250 000 scale map shows only aggregations of blocks, tinting entire cities as one polygonal feature, and transforming smaller towns into point symbols. At still smaller scales, all settlements might be depicted as point symbols. Were all of these maps to be digitised into a GIS, some way would be needed to encode and associate these transformations, in which features group together, acquiring different topologies as well as new symbolism. There are still many open problems regarding how such representational issues should be handled. This section reviews some of the pertinent research efforts, but starts by examining the concept of multiple representations.

Analogous to the concept of views in tabular databases, the term 'multiple representations' is sometimes encountered in discussions of spatial databases (Buttenfield and DeLotto 1989; Devogele et al 1997; Kidner and Jones 1994). It is used to describe a number of different things,

including alternative graphical depictions, scale-filtered versions of digital data, changes in database schema, and hierarchical data structures. Multiple representations are often associated with specific display scales, but also may tailor data to serve particular thematic or analytic purposes. In both respects they are intimately related to generalisation.

The most common form of multiple (scale) representations are topographic map and navigational chart series. National mapping agencies and private map producers normally publish maps at different scales, whereby each scale serves as a basis for the compilation of the next smaller one, forming a map series. When map series are updated, usually the large-scale representations are modified first, then the changes are propagated manually through the other scales. In describing the difficulties this entails, Charles Schwarz of the US National Ocean Service wrote:

'The problem of multiple representations is that it is difficult to maintain consistency. One of the most important concepts in database management is that it is preferable to keep only a single copy of a data item, so that consistency is automatically insured. The alternative is controlled redundancy, where one attempts to exercise control procedurally. This is difficult to enforce.' (quoted in Buttenfield and DeLotto 1989: 77)

3.2 Implicit multiple representations: generalisation by preprocessing

Several methods to deal with 'controlled redundancy' – that is, to enable consistent matching of features between different levels of scales – have been proposed. One possible approach is to preprocess the spatial data and compute and store the results of generalisations from a single database for subsequent interactive on-the-fly retrieval across a range of scales. A series of implicit ('latent') multiple representations is thus stored and the need for explicit representations of scale transitions avoided. Oosterom (1993) and Oosterom and Schenkelaars (1995) describe a set of 'reactive data structures' to accomplish this using the object-oriented DBMS Postgres to store spatial data, indices, and procedures (see also Oosterom, Chapter 27). One data structure, the binary line generalisation (BLG) tree, encodes results from Douglas-Peucker line simplification (Douglas and Peucker 1973) to allow retrieval of linear objects at any level of precomputed detail. Another structure, the Reactive tree (based on R-trees), stores collections of points, polylines, and polygons indexed by minimum bounding rectangles (MBR) and by some measure of their 'importance' such as perimeter or area. When zooming in or out, a roughly constant number of objects is selected for display according to their importance. The third data structure, the GAP-tree, is called upon to fill the gaps between regions caused by omitting less important areas. The GAP-tree contains (precomputed) alternative topologies for the omitted polygons by merging the areas they cover with a similar or dominant (most important) neighbour. The importance of an area can, for instance, be expressed as a function of its size and an application-specific weighting factor (e.g. an urban area may be more important than a grassland area). Note that this approach requires complete topology for all areal features to be known in order to build the GAP-tree, as well as extensive analysis and preprocessing of all geometric data. For further details, see Oosterom (Chapter 27).

3.3 Explicit multiple representations: multi-scale databases

Rather than deriving implicit multiple representations by preprocessing, one might wish to build multiple representations by integrating existing mono-scale representations and by modelling explicitly the transitions between scales. According to Devogele et al (1997), the design of such multi-scale databases entails three types of problem:

1 *Correspondence between abstractions*: database schemata translate phenomena of the real world into abstracted instances of databases, by focusing only on relevant parts of these phenomena; integration of abstractions thus requires methods for schema integration on the semantic level.

2 *Correspondence between objects of different representations*: data models are required to describe the links between corresponding objects of the different representations.

3 *Defining the matching process between objects*: in order to identify corresponding (homologous) objects, two sets of geographical data must be searched for objects that represent the same real-world objects; methods for this purpose are subsumed under the term 'data matching'.

Devogele et al (1997) concentrate on the first and last problems, schema integration and data matching. In their research, they aim at developing methods for building a multi-scale database from two road databases available at the French Institut Géographique National (IGN), BDCarto and GéoRoute, for purposes of road navigation. While the first database relates to a mapping scale of 1:100 000, the second contains more detailed road data for urban areas. Since the two databases employ different definitions of feature classes and their attributes, schema integration is required in a first step to arrive at a common schema that allows the two individual schemata to be related and the feature classes matched. In a second step, the individual objects of the two databases are matched, involving the matching of entire roads, as well as individual crossroads and sections. Road matching is achieved by comparing semantic information (attributes such as road number), and crossroad matching by a combination of topological and metric criteria. Section matching is carried out in two steps: first by semantic criteria (sections belonging to the same road are identified), and second by a metric search using the Hausdorff distance (for a definition of Hausdorff distance see, for example, Huttenlocher et al 1992).

A significant part of the research on multi-scale databases has focused on the second of the above problems, the design of data models and data structures to encode the correspondence relations between the individual representations. Most of the work was inspired by the largely hierarchical nature of transitions between scales. Extending a classification proposed by Beard (1991), such hierarchical relations can be found in four fundamental data domains: the domains of spatial primitives (geometric components of entity abstractions, such as points, lines, and polygons); features (real-world referents, such as buildings, rivers, and political units); attributes; and the spatial domain.

In the *primitive domain*, emphasis of hierarchical multiple representations is on the manipulation of detail of spatial primitives. Examples include the BLG-tree described above (Oosterom 1993) and the equivalent 'simplification tree' of Cromley (1991), both of which precompute the order of disappearance of vertices on a line using the Douglas-Peucker line simplification algorithm. In the *feature domain*, a great variety of hierarchies exist which lend themselves to multiple representations.

Sets of features can nest within one another, for example political territories or census geography. Network data, particularly hydrography and roads, may also be classified hierarchically, by ordering stream segments (Rusak Mazur and Castner 1990) and designating routes (Ruas 1995a). A related technique is to identify containment relations between smaller and larger features. For instance, a set of buildings can be represented by a centroid of a block or other feature that contains them. Features can also be grouped into 'placeholders' (e.g. a group of buildings that is turned into a single building or a polygon representing a built-up area at a smaller scale). Timpf and Frank (1995) have proposed the use of a directed acyclic graph to represent such transitions. In the attribute domain, the classical example of hierarchies is categorical data which may have inherent hierarchy, such as land-use or soil classifications. Such inherent relations can be formalised for storage and retrieval as hierarchies (Molenaar 1996b; Richardson 1994). Finally, in the spatial domain, a number of space-primary data structures can be used to represent spatial objects at varying levels of resolution, including quadtrees (Samet 1990), pyramids, and spherical quadtrees (Dutton 1989, 1997; Fekete 1990; Goodchild and Yang 1992). The use of such data structures in generalisation is discussed in more detail in section 5.2.

Clearly, the hierarchy levels used to represent spatial objects in the four domains must be in harmony to achieve a good generalisation. For instance, when feature and attribute hierarchies are used to reduce detail by reducing the number of objects, the primitives that make up these objects must also be simplified. None of the existing techniques for representing multi-scale relations has undertaken to address the hierarchies in all four domains comprehensively, yet examples of a combined treatment of hierarchies exist. The combination of the BLG- and GAP-trees documented by Oosterom and Schenkelaars (1995) allows hierarchies to be linked in the primitive and feature domains (cf. section 3.2). A prototypical GIS that links the hierarchies in the primitive, feature, and attribute domains has been described by Kidner and Jones (1994). This testbed employs an object-oriented database (OODB), as well as object-oriented programming (OOP) techniques, to construct class hierarchies of objects and methods that allow variants of cartographic elements to be stored, manipulated, and accessed for query and display. The system is intended to handle multi-source, multi-scale, and multi-temporal versions of spatial data, incorporating processing histories and other metadata specific to point sets, polylines, polygons, triangulated irregular networks (TINs), and raster images managed by the GIS. An enhancement of this system, now named GEODYSSEY (Jones et al 1996), relies more heavily on metadata and assertions about spatial relations to match multiply-represented features, both exactly and probabilistically. The system can determine if two representations are similar enough to delete one, and will invoke simplification procedures to satisfy queries to which the feature database provides no suitable match. Topological relations and assertions are respected, and can be derived from geometry if necessary.

Finally, while it is easy to find evidence of hierarchical relationships between multiple scales, it is also true that features can change their shape gradually between scales, making it unfeasible to model such transitions in a purely hierarchical fashion. To deal with smooth shape modifications and displacements, Monmonier (1991) has described methods for interpolating (linear) features from two or more representations digitised at different scales, conflated to match corresponding critical points along them. Recently, some GIS vendors have added rubber sheeting tools to their systems; these are primarily designed for map conflation (e.g. for building transportation databases), but may also have applications in the realm of map generalisation.

3.4 Multiple representations of terrain

Terrain surfaces are an important component of many GIS applications, and the need often arises to simplify their structures for analysis or display. Until recently, most digital terrain models (DTMs) were elevation grids having fixed resolution. Many environmental modelling applications utilise such data, which may be resampled to suit the scales and purposes of inquiry. But as Dikau (1990) observes, different types of geomorphological complexity and processes exist at different scales, and all cannot easily be captured in a single DTM, and may be obscured due to resampling operations. Most non-raster-based GIS model terrain using triangular irregular networks (TINs). The requirement to construct TINs at different resolutions is as common as it is for handling planimetric data. A number of solutions have been proposed for the analytic construction of multi-resolution TINs (De Floriani and Magillo, Chapter 38). Early methods for building hierarchical triangulations (De Floriani et al 1984) started with a coarse triangulation of highly significant points such as local extremes, to which less significant points are progressively inserted to yield finer levels of resolution, subdividing the initial triangles while maintaining their edges. This approach allows the hierarchical TIN to be represented as a tree structure, but a problem is that skinny triangles may be formed: this may be alleviated to some extent by using more sophisticated splitting rules (Scarlatos and Pavlidis 1991), or by employing the Delaunay criterion (Boots, Chapter 36) when subdividing coarser triangles (De Floriani and Puppo 1995). As triangle edges of coarser levels of the hierarchy remain unchanged, however, skinny triangles may still form and show up as distinct artefacts on surface displays (De Floriani and Puppo 1995). Such adverse effects can only be avoided effectively by optimising the triangulation independently at each level of resolution (e.g. by constructing a full Delaunay triangulation at each level). This implies a departure from the simple tree structure and building more complex directed acyclic graphs (Berg and Dobrindt 1995), but the surface will be approximated more consistently at each individual level.

One problem that hierarchical TINs must overcome is the elimination of vertical discontinuities that occur when new vertices are added along edges of a coarser triangulation. This necessitates retriangulation in the vicinity of the added points and complicates data management. An alternative approach, called 'implicit TINs' (Jones et al 1994), attacks this problem by not storing any triangles at all; instead, each surface-specific point is labelled with a detail level. This parameter may be derived in a number of ways, most commonly via a full triangulation from which the least significant vertical deviations are identified and these vertices successively removed. When a surface having a certain degree of detail is required for use, all points with greater or equal importance to the criterion are retrieved and triangulated, on the fly, by a Delaunay triangulation which takes into account linear constraints such as surface breaklines (constrained Delaunay triangulation). As constrained Delaunay triangulation is a relatively fast operation (especially when points are spatially indexed), this just-in-time approach to generating multiple terrain representations is more flexible than

hierarchical TINs and may prove workable in a number of applications. Similar implicit or online techniques have been presented by Puppo (1996) and Misund (1997) to build TINs of variable resolution. Variable resolution TINs are needed, for instance, in flight simulation where resolution decreases with increasing viewing distance or in other applications where some regions may be in the focus of interest while others are not (Misund 1997).

4 CONCEPTUAL MODELS OF GENERALISATION

Following the discussion of multiple representations in the previous section, the remainder of this chapter will adopt again the process-oriented view of generalisation. To that end, it first examines the models that have been developed in the literature to describe conceptually the processes necessary to derive generalised datasets or visualisations from detailed databases.

4.1 Conceptual frameworks of the generalisation process

In order to understand, much less to render, a complex and holistic process such as generalisation amenable to automation, conceptual frameworks need to be developed. Such theoretical models must be capable of describing the overall process and must at the same time identify essential process components and steps. McMaster and Shea (1992) review several conceptual frameworks proposed in the literature, and then go on to present a comprehensive model of digital generalisation which extends a similar framework proposed by Brassel and Weibel (1988), specifying details for model components which were previously defined only in general terms. The model of McMaster and Shea summarised in Figure 3 decomposes the overall process into three operational areas: first, consideration of the philosophical objectives of why to generalise; second, cartometric evaluation of the conditions which indicate when to generalise; and third, selection of appropriate spatial and attribute transformations which provide techniques on how to generalise. Resolving generalisation is then seen as an attempt to answer each of the three questions in turn, whereby each one forms a prerequisite for the subsequent one.

Digital generalisation

Philosophical objectives (Why to generalise)	Cartometric evaluation (When to generalise)	Spatial and attribute transformations (How to generalise)
Theoretical elements	**Geometrical conditions**	**Spatial transformations**
reducing complexity	congestion	simplification
maintaining spatial accuracy	coalescence	smoothing
maintaining attribute accuracy	conflict	aggregation
maintaining aesthetic quality	complication	amalgamation
maintaining a logical hierarchy	inconsistency	merging
consistently applying rules	imperceptibility	collapse
		refinement
Application-specific elements	**Spatial and holistic measures**	exaggeration
map purpose and intended audience	density measurements	enhancement
appropriateness of scale	distribution measurements	displacement
retention of clarity	length and sinuosity measures	
	shape measures	**Attribute transformations**
Computational elements	distance measures	classification
cost effective algorithms	Gestalt measures	symbolisation
maximum data reduction	abstract measures	
minimum memory/storage usage		
	Transformation controls	
	generalisation operator selection	
	algorithm selection	
	parameter selection	

Fig 3. The conceptual framework of digital generalisation by McMaster and Shea (1992).

The discussion of the philosophical objectives (why to generalise) of McMaster and Shea (1992) uses similar arguments to the ones raised in our review of motivations for generalisation above.

The second area of the McMaster and Shea model, cartometric evaluation (when to generalise) is essentially equivalent in scope to the key steps in the framework of Brassel and Weibel (1988), called structure recognition and process recognition. *Spatial and holistic measures* are employed to characterise the shape and structure of the source data by quantifying the density of feature clustering, the spatial distribution of features, the length, sinuosity, and shape of features, and more. These measures then serve as parameters in evaluating whether critical *geometrical conditions* are reached which trigger generalisation, such as congestion (crowding) of map objects, coalescence of adjacent objects, conflicts (e.g. overlap), imperceptible objects (e.g. objects that are too small to be clearly visible),

etc. Process recognition, as specified in the Brassel and Weibel model, is served by *transformation controls* to help select appropriate operators, algorithms, and parameters to resolve the critical geometrical conditions. Points not explicitly mentioned in the McMaster and Shea model, but which are nonetheless of utmost importance to generalisation, are the identification of topological, semantic, and proximity relations, as well as the establishment of priority orderings among features. Examples of cartometric evaluation and structure recognition are provided in section 5.3.

Finally in the third area, spatial and attribute transformations (how to generalise) consisting of a list of 12 *generalisation operators* are identified (originally proposed in an earlier paper by Shea and McMaster in 1989), subdivided into ten operators performing spatial transformations – simplification, smoothing, aggregation, amalgamation, merging, collapse, refinement, exaggeration, enhancement, and displacement – and two operators for attribute transformations – classification and symbolisation. The definition of a useful set of operators is of particular interest in the conceptual modelling of generalisation, and deserves further discussion in the following two sections.

4.2 Generalisation operators

The overall process of generalisation is often decomposed into individual sub-processes (Hake 1975). Depending on the author, the term 'operator' may be used, or other terms such as 'operation' or 'process'. Cartographers have traditionally used terms such as 'selection', 'simplification', 'combination' and 'displacement' to describe the various facets of generalisation, an example of which is the definition of generalisation by the ICA (1973) given in section 2.1. A detailed list of terms occurring in traditional cartography has been provided by Steward (1974). In the digital context, a functional breakdown into operators has obviously become even more important, as it clarifies identification of constituents of generalisation and informs the development of specific solutions to implement these sub-problems. Figure 4 illustrates some of the generalisation operators used in the discussion of section 4.3 (Table 1) for a simple map example. Three levels of scale are shown, each one at 100, 50 and 25 per cent, respectively. The appropriate reduction is highlighted by a double frame. Naturally, given the holistic nature of the generalisation process, this reductionist approach is too simple, as the whole can be expected to be more than just the sum of its parts, but it provides a useful starting point for understanding a complex of diffuse and challenging problems.

Shea and McMaster's (1989) typology is the first detailed one which also attempts to accommodate the requirements of digital generalisation, and spans a variety of data types including point, line, area, and volume data. Still, closer inspection of this set of operators reveals that some fundamental operators are missing (e.g. selection/elimination) and that the definitions of some operators are perhaps not sufficiently clear (e.g. refinement) or overlapping (aggregation, amalgamation, merging). Even worse, cartographers may use different definitions for the same term or use different terms for the same definition, as a recent study by Rieger and Coulson (1993) has shown. This has led other authors (e.g. Plazanet 1996; Ruas and Lagrange 1995) to extend

this classification by adding operators and by refining definitions of existing ones. The composition of a comprehensive set of generalisation operators is still the subject of an ongoing debate; it is hoped that having it would assist the development of adequate generalisation algorithms as well as their integration into comprehensive workflows.

No matter what set of operators is defined, however, the relationship between generalisation *operators* and generalisation *algorithms* is hierarchical. An operator defines the transformation that is to be achieved; a generalisation algorithm is then used to implement the particular transformation. This also implies that operators are independent of a particular data model (e.g. vector or raster). Algorithms are linked to a specific representation, usually the one that is best suited to implement an operator for a given purpose. For most operators, a number of algorithms with different characteristics have been developed. In particular, a wide range of different algorithms exists for line simplification in vector mode. Note also that operators often are phenomenon-specific, as Figure 5 shows. Generalisation algorithms may be phenomenon-specific and must take into account the particular shape properties and semantics of the real-world features in producing a generalised version. Also, the selection of operators will vary depending on the feature classes and scales.

4.3 Relations between operators, data, and generalisation objectives

Available conceptual models of cartographic generalisation such as the ones referred to above tend to stop short of describing how different operators come into play for map elements and specific purposes. This section attempts to synthesise some of these concepts, illustrating how generalisation operators (procedures), operands (data), and objectives relate to one another.

We have attempted to integrate the diverse and often conflicting operator sets proposed in the literature into a combined list (Table 1). Note that this typology is solely meant to support our discussion and is not intended as yet another proposal of the ultimate set of generalisation operators. There is, for instance, some debate over whether a distinction should be made between simplification (by weeding redundant points from a

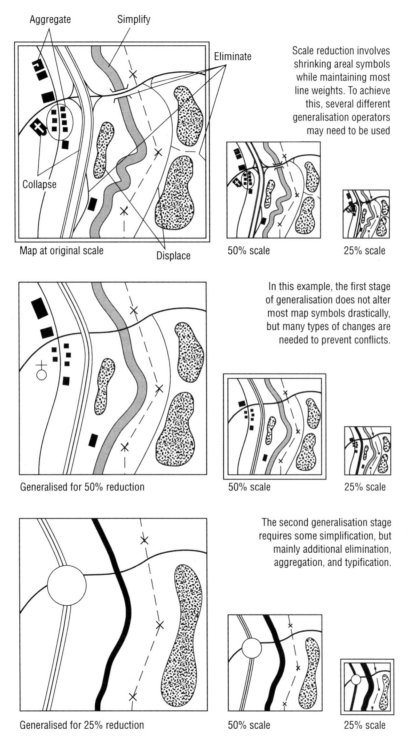

Aggregate   Simplify

Eliminate

Scale reduction involves shrinking areal symbols while maintaining most line weights. To achieve this, several different generalisation operators may need to be used

Collapse

Map at original scale   Displace

50% scale   25% scale

In this example, the first stage of generalisation does not alter most map symbols drastically, but many types of changes are needed to prevent conflicts.

Generalised for 50% reduction   50% scale   25% scale

The second generalisation stage requires some simplification, but mainly additional elimination, aggregation, and typification.

Generalised for 25% reduction   50% scale   25% scale

Fig 4. Application of map generalisation operators (cf. discussion in section 4.3, Table 1).

Table 1 Relations of generalisation operators and data (operands).

OPERATOR	TYPE OF SPATIAL DISTRIBUTION				
	Points	Lines	Areas	Fields	Hierarchies
Eliminate/ Select	*Weed based on attributes or priorities*	*Eliminate minor branches*	<u>Eliminate small areas or sub-polygons</u>	*Recode less significant values to null*	Ignore nodes with few children
Simplify	Weed to min. neighbour distance	Eliminate minor segments or inflections	Remove <u>islands</u> and concavities	*Collapse category definitions*	Shift to lower level of detail
Smooth	Make distribution more uniform	Reduce angularity	Soften concavities and crenulations	*Average or convolve variations*	Average at nodes of tessellation
Aggregate/ Amalgamate/Merge	*Combine similar neighbours*	<u>Simplify intersections</u>	<u>Delete edges between similar features</u>	Interpolate to larger cell size	Derive lower levels from higher ones
Collapse	<u>Replace by area symbol or convex hull</u>	<u>Combine nearly parallel features</u>	<u>Shrink to point or medial axis</u>	*Merge similar categories*	Redefine or reorganise hierarchy
Displace	Disperse from each other and larger objects	Increase separation of parallel lines	Move away from linear elements	(not normally attempted)	<u>Move data to less occupied neighbours</u>
Enhance/ Exaggerate	Impute, randomise, or densify distributions	Impute or emphasise changes in direction	Complicate boundaries, <u>impute inclusions</u>	*Emphasise differences, equalise histogram*	Extrapolate to levels of higher resolution

Note: Table entries are suggestive only and are not intended to be exhaustive.
<u>Underlined entries may require/result in topological transformations</u>.
Italic entries require reference to attributes or change their domain.

line) and smoothing (by modifying coordinates on a line to plane away small irregularities), as cartographers normally do not make a conscious distinction between the two operations when simplifying the shape of a map object (Plazanet 1996; Ruas 1995b; Weibel 1997). One of the criteria that guided our choice of operators was that each operator should possibly be valid for all common data types including points, lines, areas, fields (usually raster data), and hierarchies (trees and recursive tessellations). This contrasts with Shea and McMaster's (1989) typology, where some operators are only applicable to certain data types. In Table 1, basic examples are given for each operator, and note is made of whether the operator may alter topological relations or affects the attribute domain. For some of the operators alternative terms found in the literature are also given.

Table 1 focuses on the 'how to' aspect of generalisation, categorising some functions that might exist in a process library (Brassel and Weibel 1988). Yet selection of such operators is also shaped by the objectives or 'whys' of generalisation. To achieve these objectives, each situation should be evaluated cartometrically to determine if communication goals are being met (Brassel and Weibel 1988; McMaster and Shea 1992). Table 2 describes this process in terms of simple rules that may be applicable depending on the generalisation objective(s). Terminology for 'cartometric criteria' is taken from a larger set described by McMaster and Shea (1992: 42–51), and can be described as:

1 *Crowding*: excessive feature density attributable to scale reduction;
2 *Conflict*: symbolism for features which overlap or cannot be distinguished;
3 *Consistency*: uniformity of symbolism and value classification across a map;
4 *Perceptibility*: maintaining legibility when features or symbols are shrunk.

Lastly, generalisation operators can be related to the fundamental data domains introduced in section 3.3, by discriminating between those that: (1) select or modify spatial primitives; (2) select or modify features; (3) transform the attribute domain; or (4) transform the spatial domain. This is useful because it helps to

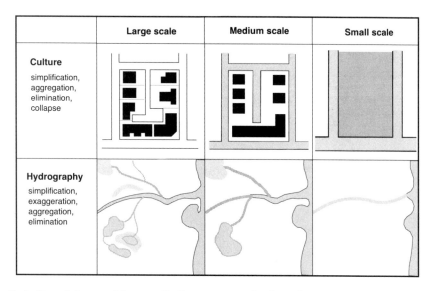

	Large scale	Medium scale	Small scale
Culture simplification, aggregation, elimination, collapse			
Hydrography simplification, exaggeration, aggregation, elimination			

Fig 5. The set of appropriate generalisation operators varies depending on the feature classes and scales.

identify the point at which operators get applied, and identifies what types of side-effect they may have.

Knowing which data domains are targeted by a given procedure helps users (and applications) to select procedures from a process library (algorithms available to implement operators). The content and structure of the database must be appropriate for a given operator to work, and the manner in which an operator functions is highly dependent on what hooks data models and data structures provide for handling cartographic primitives, features, attributes, and space itself, as well as what types of constraint generalisation must satisfy (cf. sections 5.2 and 5.4). Clearly, it may be necessary to take several of these actions together, even if one approach is primary. In practice, operations from more than one domain will be invoked to generalise a given digital map for a stated purpose. Operations in different domains have many close relationships, and can be combined in a number of natural and effective ways. As a simple example, aggregation of class values in the attribute domain of a categorical land-use map to form superclasses – e.g. aggregating 'deciduous forest' and 'coniferous forest' to 'forest', or 'residential area' and 'industrial area' to 'built-up area' – will have an effect on the geometry of the primitive domain (some polygon boundaries will disappear) and on the feature domain (new and larger regions will be formed).

5 ELEMENTS OF A GENERALISATION SYSTEM

Whether generalisation functionality is implemented as an ad hoc set of tools made available in a toolbox, or represents a sub-system of a larger GIS or cartography system, or forms a stand-alone generalisation system is not in itself important. (See Openshaw and Alvanides, Chapter 18, for a similar view with regard to the implementation of spatial analytic functions.) What matters is that the necessary elements required to solve a given class or several classes of generalisation problems are made available in a way which enables the system or user to invoke the right actions. This section attempts to identify the elements that need to be present in an idealised (not yet available) comprehensive generalisation system. Discussion of these elements will serve as a review of selected existing techniques and will highlight critical issues requiring additional research. The systems-oriented approach helps to identify the interrelations between the various elements.

5.1 An idealised generalisation system

The following discussion is based on two constraining assumptions. First, the discussion here will be restricted to functions for generalisation in the narrow sense. That is, we assume that basic graphics functions needed for map production

Table 2 Rules for achieving map design objectives.

GENERALISATION OBJECTIVES	CARTOMETRIC CRITERIA			
	Crowding	Conflict	Consistency	Perceptibility
Reduce/Maintain Graphic Complexity	Enforce radical law where practical	Displace or eliminate overlapping symbols	Apply tolerances and thresholds uniformly	Weed detail to perceptual limit
Maintain/Standardise Spatial Accuracy	Respect map accuracy standards	Displace less accurate or critical features	Generalise uniformly within a given feature class	Do not represent features that cannot be distinguished
Maintain/Standardise Attribute Accuracy	Streamline attribute classification	Do not construct overlapping symbol classes	Always use the same symbol for a given attribute within a map	Limit no. of value classes more as features shrink
Maintain/Standardise Aesthetic Quality	Use minimum appropriate symbol sizes	Maintain figure/ground rels., use compatible colours, textures	Use distinct but related colours and line styles	Be judicious when using multi-variate symbolism
Reduce/Maintain Attribute Hierarchy	Combine related feature classes	Eliminate minor features by size or attribute	Use same value classification everywhere for a feature class	Ensure that the classification conveys enough information

Note: Table entries are suggestive only and are not intended to be exhaustive.

(cartographic symbolisation, projections, zooming, etc.) will be available to the envisaged ideal system, or will be supplied by a carrier system (e.g. a GIS). Second, we assume that a human (user) will be involved in the generalisation process in some capacity, interacting with the system. User involvement may range from close to zero (invoking 'batch' modules) to a constant stream of interactions (fully interactive mode of operation, with no built-in machine 'intelligence'). In all cases, users will always be involved, even if the involvement is restricted to visual evaluation of results. That is, the more the system relies on user interaction, the more responsibility is put on the user. Cartographic experts will obtain better results than novices.

The next question we have to ask is what software paradigm is used to build the system. A look at prior research in digital cartographic generalisation reveals that neither purely algorithmic methods (Leberl 1986; Lichtner 1979) nor knowledge-based techniques such as expert systems (Fisher and Mackaness 1987; Nickerson 1988) have been capable of solving the problem comprehensively. While the former suffer from a lack of flexibility (since they are usually designed to perform a certain task) and from weak definition of objectives, the development of the latter was impeded by the scarcity of formalised cartographic knowledge and the problems encountered in acquiring it (Weibel et al 1995).

More recent research has therefore concentrated on systems that attempt to integrate different paradigms into a single coherent approach. Workbench systems designed to support research on more complex, contextual generalisation operators such as aggregation and displacement today use a combination of algorithmic (deterministic) and knowledge-based techniques commonly implemented on the basis of object-oriented technology. Examples of this class of research systems are MAGE (Bundy et al 1995; Jones et al 1995) and Stratège (Ruas 1995a; Ruas and Plazanet 1997). Another paradigm from artificial intelligence research is that of autonomous agents, which is just starting to be applied to generalisation (Baeijs et al 1996). If we consider how a system could be used in a production environment to solve actual generalisation problems, we find that the decision support system (DSS) paradigm, a strategy often used to solve ill-defined problems, may be an appropriate approach to take. A particular approach in this vein has been termed *amplified intelligence* (Weibel 1991). As visualisation and generalisation are essentially regarded as creative design processes, the human is kept in the loop: key decisions default explicitly to the user, who initiates and controls a range of algorithms that automatically carry out generalisation tasks (Figure 6). Algorithms are embedded in an interactive environment and are

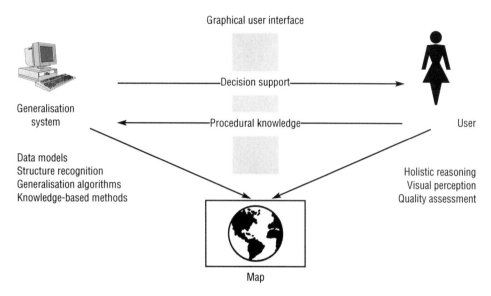

Fig 6. The concept of amplified intelligence for map generalisation.

complemented by various tools for structure and shape recognition giving cartometric information on object properties and clustering, spatial conflicts and overlaps, and providing decision support to the user as well as to knowledge-based components. Ideally, interactive control by the user reduces to zero for tasks which have been adequately formalised and for which automated solutions could be developed.

In a system such as Figure 6 depicts, algorithms serve the purpose of implementing tasks for which sufficiently accurate objectives can be defined. This includes cartographic generalisation operators such as simplification, aggregation, or displacement; functions for structure and shape recognition, including shape measures, density measures, and detection of spatial conflicts; and model generalisation functions.

Knowledge-based methods can be used to extend the range of applicability of algorithms and code expert knowledge into the system. This initially builds on methods for knowledge acquisition; for instance, machine learning may help to establish a set of parameter values that controls the selection and operation of particular algorithms in a given generalisation situation (Weibel et al 1995). Second, procedural knowledge and control strategies are needed: once the expert knowledge is formalised, it can be used to select an appropriate set and sequence of operators and algorithms and to

establish a strategy to solve a particular generalisation problem.

An ideal system builds on a hierarchy of control levels. The human expert makes high-level design decisions and evaluates system output. Knowledge-based methods operate at an intermediate level and are responsible for selecting appropriate operators and algorithms and for conflict-resolution strategies. Finally, algorithms are the workhorses at the lowest level, forming the foundation of everything else.

In summary, the following main areas – which will be discussed in the remainder of this section – need to be addressed to implement a comprehensive generalisation system based on the above assumptions:

- data representations and data models;
- structure and shape recognition;
- generalisation algorithms, including model generalisation;
- knowledge-based methods;
- human–computer interaction;
- generalisation quality assessment.

5.2 Data representations and data models

Most generalisation algorithms available today are related to operators such as selection, simplification, or smoothing which are context-independent, treating map objects largely independently of their context. Few solutions are available for context-dependent operators such as aggregation or displacement which the spatial context (spatial relations of objects, object density, etc.) triggers and guides to completion. Various authors have argued that the scarcity of available context-dependent generalisation algorithms is caused by the fact that commonly used spatial data models are unable to provide adequate support of such complex functions, in particular those requiring a representation of proximity relations between disjoint objects (Bundy et al 1995; Dutton 1984; Ruas and Lagrange 1995; Weibel 1997). In recent years, research has therefore started to exploit alternative data models. The problem needs to be addressed at two levels, involving representations for geometric primitives as well as complex data models.

5.2.1 Representations for geometric primitives

Adequate data structures must be available for representing geometric primitives (points, lines, areas), including methods such as polygonal chains (or polylines), mathematical curves, and rasters. These primitive representations must be capable of capturing the shape of the modelled features accurately and in a compact and expressive way.

In vector mode generalisation, polylines are by far the most commonly used representation scheme for geometric primitives. Regardless of its popularity, the polyline representation also imposes impediments on the development of generalisation algorithms (Fritsch and Lagrange 1995; Werschlein 1996), essentially restricting the design options to simplification (vertex weeding) and smoothing (vertex modification). The fact that a polyline is simply a sequence of points implies that it is difficult to model entire shapes such as a bend of a road properly or compactly. Complementary representations to polylines are therefore being investigated. Work by Affholder (reported by Plazanet et al 1995) on geometric modelling of road data is an example of fitting the representation scheme more closely to the object that needs to be represented. Affholder models roads by a series of cubic arcs, leading to a more compact and also more realistic representation of these manmade features which offers potential for the development of new algorithms.

Parametric curves based on curvature can be usefully exploited for shape analysis as critical points such as inflection points show up as extremes (Werschlein 1996). The magnitude of these extremes exhibits the size of the shape that is associated with the critical point and thus allows prioritisation.

Wavelets (Chui 1992) are a relatively untried but promising approach to generalisation of both surfaces (Schröder and Sweldens 1995) and lines (Fritsch and Lagrange 1995; Plazanet et al 1995; Werschlein 1996). They either require a raster representation or, for vector data, that its geometry be first transformed into a function (e.g. a parametric curve). A geometric basis function, the 'mother wavelet', is then applied to fit the representation over doublings of resolution. Wavelet coefficients can be analysed to determine critical points and shapes, and they can also be filtered to yield generalised versions of the original feature. It is also possible to eliminate entire shapes selectively by setting the coefficients of the wavelets supporting a particular shape to zero (Werschlein 1996).

5.2.2 Complex data models

Complex data models let one integrate primitives into a common model (e.g. a topological vector data model) and record their spatial and semantic relations. While the search for alternative primitive representations is mainly guided by the requirements of representing and analysing shapes, research into complex data models is driven by the need to support context-dependent generalisation. Improved complex data models must: allow representation of relevant metric (proximity), topological, and semantic object relations within and across feature classes; enable object modelling (including differentiation between primitives and features, complex objects, and shared primitives); and permit the integration of auxiliary data structures such as triangulations, uniform grids, or hierarchical tessellations for computing and representing proximity relations.

As a consequence of these requirements the main data model should be an object-oriented extension of the basic topological vector model (as opposed to layer-based). Data models of this kind are now beginning to appear in some commercial GIS. Integrated auxiliary data structures for proximity relations are not yet available in commercial systems, but research is under way in that direction.

Data structures for proximity relations are commonly based on tessellations (see Boots, Chapter 36 for a comprehensive review of tessellations in GIS). Space-primary tessellations use regular subdivision of space and can be used as a simple mechanism to assess spatial conflict within

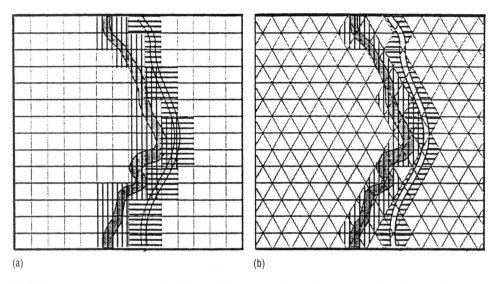

Fig 7. Space-primary assessment of spatial conflict between a river and a road using alternative tessellations: (a) rasterising to a rectangular grid; (b) rasterising to a triangular grid.

a fixed resolution, usually relating to the resolution of the target map (Figure 7). They can also be turned easily into hierarchical data structures. By discretising space in a hierarchical fashion, vertices and entire primitives can be coalesced or eliminated when several are found to occupy the same location (a 'chunk' of space). Encoded features can have addresses that indicate where and how big they are, allowing access both by location and resolution. Generalisation often proceeds in such methods by aliasing the positions of points or grid cells to some median location (see Figure 8). Alternatively, detected overlaps or coalescing primitives can be resolved by displacement. Quadtrees (Samet 1990) and pyramids are commonly used to partition map space hierarchically; conventional approaches index planar coordinates using rectangular subdivision. Their use in geoinformation processing, however, has been principally limited to multi-resolution image reconstruction. Spherical quadtrees (Dutton 1989; Fekete 1990; Goodchild and Yang 1992) enable planetary indexing of global geographical coordinates by partitioning the facets of regular polyhedra into forests of triangular quadtrees. A line generalisation method via hierarchical coarsening using a triangular quadtree structure – the quaternary triangular mesh (QTM) – has been presented by Dutton and Buttenfield (1993) and

Dutton (1997). This is illustrated in Figure 8 in which the 'level' of generalisation denotes the resolution (doubling of scale) of each hierarchical encoding: level 16 corresponds to 150 m, level 12 to 2.5 km resolution.

Most approaches to represent proximity relations between spatial objects accurately, however, have concentrated on the use of object-primary tessellations including Delaunay triangulations or Voronoi diagrams (Jones et al 1995; Ruas 1995a; Ruas and Plazanet 1997; Ware et al 1995; Ware and Jones 1996; Yang and Gold 1997). The data models used by Ruas (1995a) and by Bundy et al (1995) are both based on Delaunay triangulations. Both approaches concentrate on the support of methods for detecting and resolving spatial conflicts (e.g. feature overlap or coalescence), and both use the space subdivision scheme as a means to compute proximity relations, compute displacement vectors (if needed), and keep track of displacements. Beyond these similarities, however, the two data models take a different approach.

Ruas (1995a; see also Ruas and Plazanet 1997) subdivides map space according to the hierarchy of the road network. Within each of these irregular tiles, conflict detection and resolution again takes place, starting at level 1 and proceeding to finer levels. A local, temporary Delaunay triangulation is

141

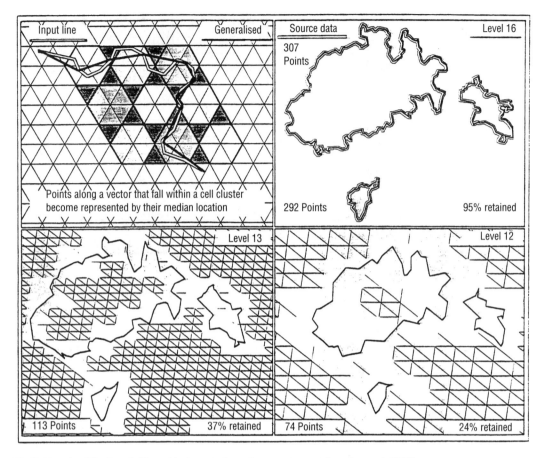

Fig 8. Line simplification via hierarchical coarsening using a quaternary triangular mesh (QTM).

Fig 9. Local Delaunay triangulation between buildings and adjacent roads (after Ruas 1995a).

then built within each partition to negotiate spatial conflicts there. The triangulation connects the centroids of the small area objects and point objects falling within the tile, as well as projection points on the surrounding roads forming the tile boundary. The edges of the triangulation are classified according to the types of object they connect. Thus in Figure 9, edge e1 denotes an edge connecting two buildings; e2 connects two vertices on a road; and e3 connects a building and a road. If the shape of a bounding road is changed or buildings are enlarged or moved, the triangulation is used to determine any conflicts that might have arisen. Displacement vectors are then computed from the distances between objects and displacement propagation is activated using distance decay functions. This is illustrated in Figure 10, in which a road (a) is modified, leading to overlap (b). Displacement vectors are calculated (c) and buildings are rotated

142

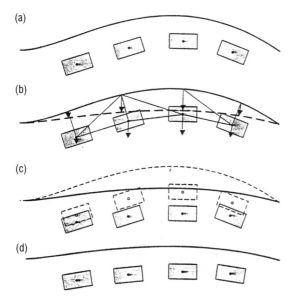

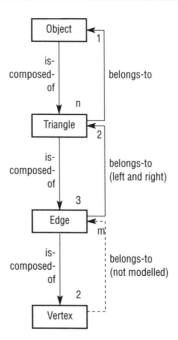

Fig 10. Displacement of buildings after simplification of a road (after Ruas 1995a).

Fig 12. Entity relationships in the simplicial data structure (after Jones et al 1995).

and realigned with the modified road (d). (For the sake of clarity only the road centreline is shown in this figure; the symbol width of the road would be taken into account when computing the displacement vectors.)

In the triangulated data model developed by researchers at the University of Glamorgan (Bundy et al 1995; see also Jones et al 1995; Ware et al 1995; Ware and Jones 1997) the triangulation forms the core of the data model. Rather than connecting centroids of map objects, a constrained Delaunay triangulation of all the vertices of all map objects is built (Figure 11). The resulting simplicial data structure (SDS) is

represented through a set of relations which are stored by pointers corresponding to the entity relationships illustrated in Figure 12. Since the SDS comprises all the geometric information of the original objects, all generalisation operations are carried out directly on the SDS. The target application for the MAGE system built around the SDS is the generalisation of large-scale topographic map data of the Ordnance Survey of Great Britain. To that end, a palette of generalisation operators has been developed including object exaggeration (enlargement), object collapse (constructing the centreline of road casings), operators for areal object amalgamation, and building simplification using corner flipping of triangles.

5.3 Structure and shape recognition

As discussed in section 4.1, structure and shape recognition (cartometric evaluation) are logically prior to the application of generalisation operators (Brassel and Weibel 1988; McMaster and Shea 1992). They can determine when and where to generalise and inform the selection, sequencing, and parameterisation of a set of generalisation operators for a given problem. Because cartographic data tend not to be richly structured, parts of features (e.g. a

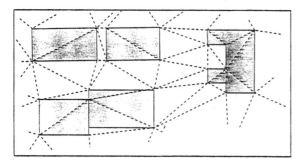

Fig 11. A sample section of the constrained Delaunay triangulation forming the simplicial data structure (after Ware et al 1995).

hairpin bend on a road or an annex of a building) are rarely coded explicitly. Likewise, little information is normally stored on shape properties of map features. Structure and shape recognition therefore aims at enriching the semantics of source map data, and deriving secondary metric, topologic, and semantic properties including shape characteristics, object density and distribution, object partitioning, proximity relations, relative importance (priority) of map objects, and logical relations between objects.

In recent years, research in this area has intensified. First attempts at cartographic line characterisation were made by Buttenfield (1985). Recently, an approach has been presented by Plazanet (Plazanet 1995, 1996; Plazanet et al 1995) which generates a hierarchical segmentation of cartographic lines according to a *homogeneity* criterion. The resulting tree structure is called the 'descriptive tree'. Figure 13 illustrates such a tree; the homogeneity of the individual sections is intuitively apparent. The homogeneity definition used to split up the line is based on the variation of the distances between consecutive inflection points which have been previously extracted (Figure 14). While the objective of segmentation is mainly to obtain sections of the line that are geometrically sufficiently homogeneous to be tractable by the same generalisation algorithm and parameter values, further information is added to the descriptive tree that characterises the *sinuosity* (and thus the prevailing geometric character) of each

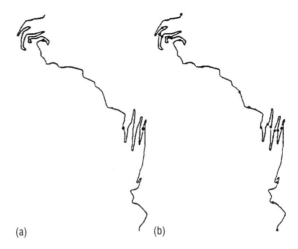

Fig 14. (a) Detected inflection points; (b) critical points retained automatically for segmentation.
Source: C Plazanet, IGN France.

line section. A variety of measures can be obtained from the deviation of the cartographic line from a trend line formed by the base line connecting consecutive inflection points. These measures are then used to classify line sections according to their degree of sinuosity.

Analysis of complex situations involving disjoint objects such as buildings and roads in built-up areas can obviously benefit from data models such as those discussed in section 5.2.2. Proximity and adjacency relations can be assessed and possible overlaps detected by direct analysis of the data model. More complex relations involving many objects, however, require further processing. An example of such a procedure has been presented by Regnauld (1997), who proposed the use of the minimum spanning tree (MST) to detect clusters of buildings of like shape and alignment in order to form candidate sets for aggregation and typification operations.

In the area of terrain generalisation, Weibel (1992) has reported on the use of procedures for geomorphometric analysis to drive the selection of appropriate generalisation methods. Geomorphometric analysis is first applied at the global level, segmenting an entire DTM into homogeneous regions amenable for a particular generalisation method in an approach similar in nature to Plazanet's (1995) technique for line segmentation. Next, structure lines (drainage channels and ridges) are automatically extracted from the DTM to form the so-called

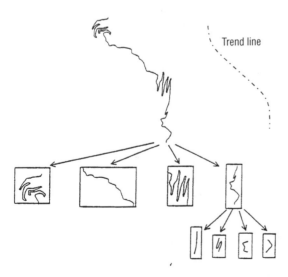

Fig 13. An example of line segmentation into a descriptive tree.
Source: C Plazanet, IGN France.

'structure line model' (SLM) which is seen as a 3-dimensional skeletal representation of the terrain surface and used as a basis for the generalisation method termed 'heuristic generalisation'. This procedure generalises the SLM by modifying links in the networks of channels and ridges and interpolates a generalised DTM from the modified SLM (Figure 16 (b) shows a result of this procedure).

5.4 Generalisation algorithms

In section 4.2, it was noted that generalisation algorithms implement generalisation operators which in turn define the spatial transformations necessary to achieve generalisation. Generalisation algorithms are thus at the heart of the generalisation system, 'making it happen'. As Figure 5 illustrates, however, generalisation algorithms tend to be phenomenon-specific because they must take into account the particular shape properties and semantics of the real-world features they aim to depict in a generalised version. Only careful analysis of the structure and shape of map objects – exploiting the resources of a rich spatial data model – can give generalisation algorithms the guidance they need.

Countless generalisation algorithms have been developed over the past three decades; this discussion is restricted to algorithms for vector data. Raster-based algorithms are reviewed by Schylberg (1993); they are less common, but naturally lend themselves to implementation of context-dependent operators, as Figure 7 indicates. Most vector algorithms are intended to generalise point and line primitives using context-independent operators such as selection, simplification, and smoothing (see McMaster and Shea 1992 for descriptions). Research in more recent years has been characterised by increasing interest in more complex context-dependent operators such as aggregation and displacement. Displacement algorithms have profited from data models such as the ones discussed in section 5.2, but also from research in displacement propagation functions (Mackaness 1994). Examples of aggregation algorithms include procedures for area patch generalisation described by Müller and Wang (1992) as well as the research at the University of Glamorgan by Jones et al (1995).

Attention in research on generalisation algorithms has also turned to the development of algorithms which are constrained by the particular requirements and characteristics of specific feature classes. Instead of generalising line or polygon primitives, the focus is on the feature classes that these primitives purport to represent. That is, specific methods for generalising building outlines, traffic networks, or polygonal land-use maps are developed. Such techniques may be based on more primitive algorithms, integrating them to build fine-tuned methods. An example of a system that has employed specialised algorithms at an early stage to develop functionality for large-scale topographic map generalisation is the CHANGE system which integrates two decades of research at the University of Hanover (Grünreich et al 1992). This system includes specific algorithms for generalisation of roads (centreline generation, simplification, symbolisation), building generalisation (outline simplification, aggregation), and identification and editing of spatial conflicts. A sample output is shown in Figure 15. It represents the result of a fully automated procedure; interactive editing is usually necessary to clean up residual problems.

Other research has addressed the generalisation of road networks (Mackaness and Beard 1993; Thompson and Richardson 1995). Owing to the topologic constraints of networks, these methods are based on an analysis of the network structure using graph-theoretic algorithms. Techniques for terrain generalisation have been proposed by Weibel (1992), who also summarises other approaches. A strategy is used that employs three types of generalisation methods, two of which apply filtering techniques, while the third is based on an extraction of the networks of drainage channels and ridges (cf. section 5.3; see also Hutchinson and Gallant, Chapter 9). Figure 16 is based on an application of the third method and shows: (a) an original surface (7.75 × 5.5 km; 25 m resolution); (b) a generalised surface, resulting from the extraction and generalisation of the network of topographic structure lines (channels and ridges); and (c) a further modification of the automatically generalised surface through interactive retouching to enhance prominent landforms using a DTM editor developed by Bär (1995). This editor offers a range of tools, most of which are implemented as local convolution filters in the spatial domain and which can be controlled interactively.

Developing algorithms which can meet the needs of particular feature classes, scale ranges, and map

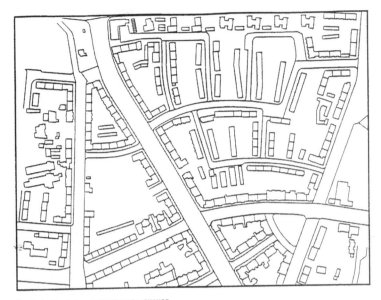

Result of automated generalisation using CHANGE

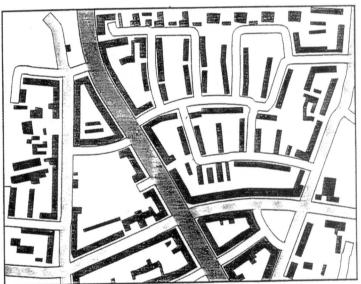

Scale 1:5000

Scale 1:10 000

Fig 15. An example of large-scale topographic map generalisation using the CHANGE system.
Source: Grünreich et al 1992: courtesy of the Institute of Cartography, University of Hanover. Note that scale is only **approximate.**

types requires that the constraints which govern a particular generalisation task are defined accurately (Beard 1991; Weibel 1997). A good example of a constraint-based algorithm is the procedure proposed by Berg et al (1995) for the simplification of polygonal maps. Their algorithm satisfies four constraints:

1 all points on the simplified polygon boundary (chain) are within a prespecified error distance from the input boundary;
2 the simplified chain has no self-intersections;
3 the simplified chain may not intersect other chains of the polygonal map;
4 all points of an additional point set lie to the same side of the resulting polygon boundaries as before simplification.

Finally, the algorithms toolbox must also include methods for model generalisation. Relatively little attention has so far been paid to this segment of generalisation, however. The volume edited by Molenaar (1996a) contains a selection of papers on the topic. Weibel (1995b) discusses the requirements of model generalisation and their differences from those of cartographic generalisation. Heller (1990) presents a method for filtering a grid or TIN DTM which can be used for model generalisation of terrain models (see also section 3.4).

5.5 Knowledge-based methods

If a generalisation system were based solely on the data models and algorithms discussed above, much of the higher-level reasoning and decision strategies would be lacking, making it necessary to rely entirely on the user for the provision of this missing knowledge. Knowledge-based methods have been proposed as a way to overcome this reliance on the individual user and to build more completely automated solutions. Knowledge-based methods with a potential applicability to generalisation encompass expert systems (or knowledge-based systems) as well as machine-learning techniques including methods such as inductive learning, case-based reasoning, genetic algorithms, or artificial neural networks (Carbonell 1990). The potential of these methods lies in two areas: in acquiring and representing human knowledge *explicitly* (e.g. inductive learning for knowledge acquisition and rules of an expert system for

knowledge representation), and in complementing or replacing algorithmic techniques by use of *implicitly* encoded knowledge (e.g. knowledge which is latently contained in large sets of examples) as well as computational learning strategies (e.g. genetic algorithms or neural networks: see Fischer, Chapter 19).

Use of knowledge-based methods for the latter purpose has been restricted to a few isolated attempts to use genetic algorithms, neural networks, or case-based reasoning in generalisation, and is still in an undeveloped state (see Weibel et al 1995 for a review). The former area, however, has received more attention. Whatever the intended use of knowledge-based methods may be, knowledge acquisition (KA) is the key to success. This is particularly true for expert systems, as they derive their power from the knowledge they contain and not from the particular knowledge representation schemes and inference mechanisms they employ. Because of the scarcity of available formalised cartographic knowledge, success of expert generalisation systems to date has been limited (with a few exceptions, such as that described by Nickerson 1988), a situation which has sometimes been termed the 'knowledge acquisition bottleneck'.

According to Armstrong (1991) cartographic knowledge takes three different forms. Geometrical knowledge describes the geometry (locations), shape, and distribution of cartographic objects. Structural knowledge represents the structure of cartographic features in terms of their geomorphological, economic, or cultural meaning, and thus relates to the term 'semantic knowledge' used by other authors (e.g. Chang and McMaster 1993). Finally, procedural knowledge is used to select the appropriate generalisation operators, algorithms, and parameter settings required to perform a generalisation task. It is the knowledge that is needed to control the flow of operations. While geometric knowledge is largely contributed by methods for structure and shape recognition (section 5.3), and structural knowledge can be feature-coded into the database (or possibly extracted by shape analysis), the acquisition of procedural knowledge is largely an open problem. Its main task is to find rules which relate generalisation operators, algorithms, and parameter values to map scale, map purpose, feature classes, and shape properties. That is, procedures are to be linked to structure and semantics.

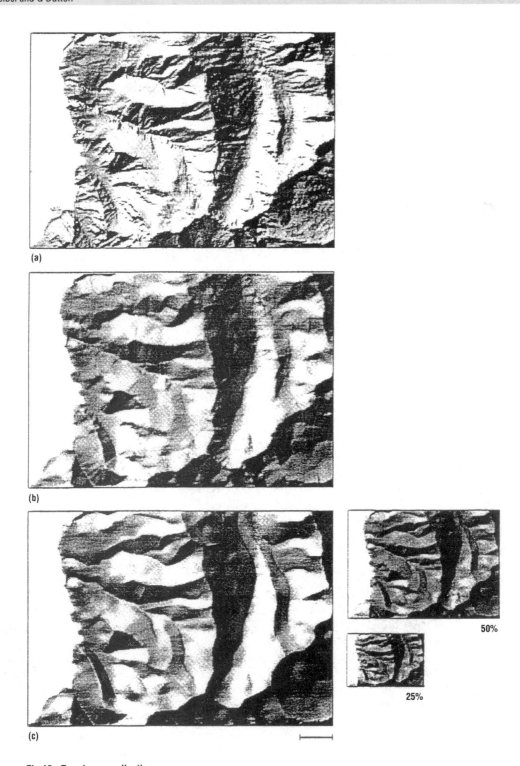

(a)

(b)

(c)

50%

25%

Fig 16. Terrain generalisation.
Source: DTM data courtesy of Swiss Federal Office of Topography.

Cartographic knowledge is different from other knowledge types (e.g. the knowledge needed in medical diagnosis) in that it is essentially graphical and therefore hard to verbalise and formalise. It may be acquired from different sources, necessitating different KA methods, which can be exploited in combination. Weibel (1995b) and Weibel et al (1995) discuss the potential of various KA methods and knowledge sources. The obvious source of knowledge is the *human expert* (cartographer). According to McGraw and Harbison-Briggs (1989), methods for eliciting knowledge from experts include interviews, learning by being told, and learning by observation. In generalisation, direct knowledge elicitation from experts has been restricted to projects linked to national mapping agencies (NMAs), such as those described by Nickerson (1991) or Plazanet (1996). *Text documents* form a second possible knowledge source. Apart from textbooks, written guidelines are available at NMAs, for example coding guides for digitising hardcopy maps (USGS 1994). These textual descriptions, often including positive and negative sample illustrations, may be used as a basis to develop formal rules. In summarising an attempt to cast generalisation guidelines in use at Ordnance Survey of Great Britain into rules, Robinson (1995) observes that the resulting rules revealed that the written guidelines were incomplete and often vaguely specified. *Maps* as a third knowledge source embody cartographic knowledge in graphical form. It may be hoped that a study of the evolution of features across the scales of a map series can reveal the procedural knowledge that was used to create the generalisation. Practical experiments using this type of 'reverse engineering' procedure have been useful in establishing quantitative relations between scales such as the percentage of objects retained for specific feature classes (Leitner and Buttenfield 1995), but they have also demonstrated the difficulty of reliably identifying more complex procedural knowledge such as the operators used to produce a generalisation (Weibel 1995b). It is often impossible to determine the operations that led to the result from the result alone. Finally, *process tracing* in interactive systems offers a method that could complement KA from experts. Instead of eliciting knowledge directly from experts, interactions of the expert with an interactive system are logged and later analysed to extract rules. Thus, the system acts as a mediator which already achieves a first step of translating human knowledge into the 'language' of generalisation systems. Use of this approach has been proposed by several authors, including Weibel (1991) and McMaster and Mark (1991). An implementation of this method to determine appropriate parameter values for line generalisation algorithms is described by Weibel et al (1995) and Reichenbacher (1995). The trace of interactions is analysed using inductive machine-learning algorithms in order to extract rules. Further experiments using inductive learning have been carried out by Plazanet (1996).

5.6 Human–computer interaction

Assuming that any comprehensive generalisation system is likely to take at least a partially interactive approach, consideration must be given to the design of user interfaces for generalisation. Human–computer interaction (HCI) in GIS and cartographic systems has profited greatly from the widespread availability of graphical user interfaces (GUI), but the specific requirements of generalisation may necessitate the development of optimised HCI mechanisms. Specific HCI mechanisms are needed during several phases of the development and use of generalisation systems: during algorithm development and testing, for selection and fine-tuning of generalisation algorithms and control parameters, for the evaluation of results, for retouching, and more.

Beard and Mackaness (1991) discuss the problem of how the 'cognitive responsibility' of the design process is best shared between the generalisation system and the user, and how the general workflow could be captured in the user interface. According to the general user interface requirements for generalisation systems proposed by McMaster and Mark (1991), the user interface should provide a broad set of generalisation operators, tools for identifying map features, assistance in selecting parameters for algorithms, warnings about inconsistencies, and traces of generalisation actions. User feedback should include hypermedia-based documentation and diagrams, graphical and numeric measures of success, and highlighting of features or regions in need of generalisation. Chang and McMaster (1993) report on a prototype system that implements parts of these elements, and develops specific HCI mechanisms to support them. The system was mainly intended for experimentation with the different line generalisation algorithms described by McMaster (1987). Results of up to four algorithms can be displayed, and tolerance parameters of the algorithms can be controlled via a slider bar, effecting real-time generalisation. The slider bar interaction technique was developed concurrently in the MGE Map Generalizer system described by Lee (1995). Schlegel and Weibel (1995) developed a prototype system which was intended to illustrate some of the HCI requirements of a generalisation system useful for production. Elements that were added to the design included multiple windows – a working window and two windows to display the map at the source and target scale, respectively – together with extensive feature selection functions (e.g. by selection via a histogram of shape measures), and full symbolisation with correct symbol size at target scale (necessary to assess the need for displacement). Finally, a novel interaction technique termed *generalisation by example* was proposed by Keller (1995). To specify tolerance parameters for a line generalisation algorithm, the user draws a representative sample line. The system picks up this sample and finds the parameter value(s) which can best reproduce the example with the specified generalisation algorithm; the optimisation procedure uses a genetic algorithm.

5.7 Generalisation quality assessment

As is true for digital cartography in general, assessment of the quality of generalisation results has received relatively little attention in research so far. The availability of a comprehensive palette of evaluation methods and strategies, however, is indispensable for progress in generalisation research, making it possible to assist such diverse tasks as the comparative appraisal of algorithms or entire software systems, the development of built-in evaluation functions for genetic algorithms, or the selection of positive examples for the training of neural networks (Weibel 1995b).

To date, evaluation has largely relied on visual assessment of the results obtained from a particular procedure or system, for instance by comparing digital results with a manually produced solution. This approach, however, is neither rigorous nor always possible, as the manual reference map may not be available. More rigorous, intersubjective, and repeatable methods are needed in digital systems. Quantitative assessment techniques can largely draw from structure and shape recognition methods

(cf. section 5.3), as *a priori* structure recognition is similar to *a posteriori* assessment of results. The former analyses the structure of the objects in a source database, while the latter analyses the difference between a source and a derived database.

In order to assess the geometric performance of line simplification algorithms, McMaster developed a variety of geometric measures to analyse shape distortion in terms of the difference between the original and the simplified line (McMaster 1987). While this method is valid for entire lines, Mustière (1995) has proposed a series of measures which analyse a cartographic line for possible conflicts which might arise due to symbol enlargement, suggesting, for instance, elimination or enlargement of a particular bend. Checks for topological consistency such as tests for self-intersection, intersections between neighbouring lines, or point containment violations caused by line simplification are relatively easy to develop (Berg et al 1995). In general, quantitative assessment methods are weak when multiple objects are involved or the quality of an entire map needs to be characterised. This area still requires additional research.

As generalisation involves subjective decisions and aesthetic considerations, one cannot evaluate its results completely quantitatively, nor can quality assessment simply consist of a series of atomic tests and measures. An integrated approach is needed to capture the more holistic elements of generalisation. A methodology to integrate quantitative measures with qualitative judgements by cartographic experts in a consistent way has been proposed by Ehrliholzer (1995, 1996). Using it, an assessment process starts by carefully defining the relevant criteria to be used in describing the quality of a particular generalisation application. These criteria may be measurable, such as the minimum size of small areas, or qualitative and holistic, such as 'maintenance of the overall character of the map'. Application of these criteria will yield a 'quality description' with interval/ratio values for quantitative measures and symbolic descriptions or keywords for qualitative criteria. The two sets of results may then be integrated by transforming both into scores or ranks on a rating scale.

5.8 Digital generalisation in practice

Most commercial GIS today offer some generalisation functionality. Usually, however, generalisation capabilities in GIS are restricted to a few functions (e.g. line simplification, polygon aggregation, or various filters for raster data) which must be applied through independent commands in a toolbox approach (Morehouse 1995) or integrated to build more complex functions (Schlegel and Weibel 1995). Ruas (1995b) reviews the use of a general-purpose GIS (ARC/INFO) for the production of a database at scale 1:1 000 000 from a database at 1:100 000 at IGN France. Brandenberger (1995) reports on an experiment of producing a 1:10 000 scale map from data captured at 1:1000 using another major commercial system (Intergraph MGE).

Special-purpose commercial generalisation systems exist. The CHANGE system (Grünreich et al 1992), mentioned in section 5.4, builds on a suite of batch modules controlled by parameter sets. Figure 16 shows a sample run, in which the (approximate scale) maps represent the results of a fully automated procedure; following this first step, interactive editing is usually necessary to clean up remaining problem cases. Intergraph's MGE Map Generalizer product offers a palette of operators and algorithms embedded in an interactive environment under full user control (Lee 1995). Both systems have recently been evaluated in a software test as part of the activities of the Working Group on Generalisation of the OEEPE (Baella et al 1994; Rousseau et al 1995; Weibel and Ehrliholzer 1995).

Commercial GIS and cartography systems can be and are indeed used for the production of generalised maps today, which certainly is an improvement over the situation a decade ago when this was largely impossible. However, the generalisation workflow in practice still involves a great deal of interactive guidance and retouching (and essentially results in multiple representations if generalisation output is committed to one GIS database), owing to limitations of available generalisation capabilities. While generalisation functionality of current systems still needs considerable extension with respect to all of the elements of generalisation discussed above, perhaps the most limiting factors to date are the ones that form the foundations of a successful generalisation system: data representations and data models, generalisation algorithms, and structure and shape recognition (Schlegel and Weibel 1995).

Despite these limitations, however, the present situation in the commercial sector gives rise to substantial hope for future improvement. Since the first release of Intergraph's MGE Map Generalizer product in 1992 (Lee 1995), a growing number of GIS vendors including, among others, ESRI and Laser-Scan have become aware of the relevance of generalisation capabilities and have begun to extend the range of tools for generalisation and to articulate intentions to perform more long-term research and development (ESRI 1996; Hardy 1996; Woodsford 1995).

6 CONCLUSIONS

Generalisation is a highly complex process for which no simple solutions exist. Yet, after a period of relative stagnation during the late 1970s and the 1980s, generalisation has again attracted significant interest in the GIS community and beyond. The topic is well represented at key GIS conferences and pursued by several international working groups. The International Cartographic Association (ICA), the European Organisation for Experimental Photogrammetric Research (OEEPE), and the International Society of Photogrammetry and Remote Sensing (ISPRS) have all formed commissions and working groups to coordinate international research efforts. Similar research initiatives are being pursued at the national level in most of the larger industrialised countries. Additionally, the commercial sector is investing more effort in research and development in generalisation (see also Salgé, Chapter 50; Smith and Rhind, Chapter 47).

This situation has been caused by a confluence of three factors. First, there is an *increasing demand* for generalisation functionality by many types of GIS users, ranging from major national mapping agencies to specialised application builders. Apart from the 'classical' requirements of map production, the importance of generalisation in digital cartography and GIS is accentuated by continuing rapid growth of the number and volume of spatial databases and by the need to produce data meeting specific requirements and share them among different user groups (see Rhind, Chapter 56). Second, as a consequence of the first factor these forces have turned generalisation from a problem considered as too hard to tackle into *emergent solutions and market opportunities*, as technology suppliers (academic research and GIS vendors) react to these new demands. The research community is finding a renewed interest in these issues, addressing them with new conceptual approaches and more modern software architectures. GIS vendors are responding with apparent commitments to extend and improve their systems' generalisation capabilities. Third, the *technological setting has matured*; more powerful enabling technologies have become available, reliable, and affordable. This last factor is of utmost importance as it has finally created a situation where the processing power, networking capabilities, software engineering technologies, and graphics and analytical functions are available to approach realistically the task of developing complex generalisation functionality (see Batty, Chapter 21; Coleman, Chapter 22). The next few years will show whether the current high level of activity represents a persistent evolution or whether it merely marks a passing enthusiasm, to be tempered by the difficulty of the challenges facing the development of generalisation technology. Given the fundamental importance of generalisation in the context of flexible and distributed spatial data handling, however, we expect that a lasting coalition of users, researchers, and vendors will form with a strong interest in and commitment to developing workable solutions.

References

Armstrong M P 1991 Knowledge classification and organization. In Buttenfield B P, McMaster R B (eds) *Map generalisation: making rules for knowledge representation*. Harlow, Longman: 86–102

Baeijs C, Demazeau Y, Alvares L 1996 SIGMA: application of multi-agent systems to cartographic generalisation. *Seventh European Workshop on Modelling Autonomous Agents in a Multi-Agent World (MAAMAW 96), Eindhoven, The Netherlands*. Lecture Notes in Artificial Intelligence 1038. Springer

Baella B, Colomer J L, Pla M 1994 CHANGE. Technical report. Barcelona, Institut Cartogràfic de Catalunya: 1–16

Bär H R 1995 Interaktive Bearbeitung von Geländeoberflächen – Konzepte, Methoden, Versuche. (PhD dissertation) Geoprocessing series. Department of Geography, University of Zurich. Vol. 25: 1–140

Beard M K 1991 Constraints on rule formation. In Buttenfield B P, McMaster R B (eds) *Map generalisation: making rules for knowledge representation*. Harlow, Longman: 121–35

Beard M K, Mackaness W A 1991 Generalisation operators and supporting structures. *Technical Papers, ACSM/ASPRS Annual Convention 6 (Proceedings AutoCarto 10)*: 29–45

Berg M de, Kreveld M van, Schirra S 1995 A new approach to sub-division simplification. *ACSM/ASPRS Annual Convention and Exposition 4 (Proceedings AutoCarto 12)*: 79–88

Berg M de, Dobrindt K T G 1995 On levels of detail in terrains. *Proceedings, Eleventh Annual ACM Symposium on Computational Geometry*: C26–C27

Brandenberger C 1995 From digital cadastral data to large-scale topographic maps. *Cartography crossing borders (Proceedings, Seventeenth International Cartographic Conference)*. Barcelona, Institut Cartogràfic de Catalunya: 1771–80

Brassel K, Weibel R 1988 A review and framework of automated map generalisation. *International Journal of Geographical Information Systems* 2: 229–44

Bundy G L, Jones C B, Furse E 1995 Holistic generalisation of large-scale cartographic data. In Müller J-C, Lagrange J-P, Weibel R (eds) *GIS and generalisation: methodological and practical issues*. London, Taylor and Francis: 106–19

Buttenfield B P 1985 Treatment of the cartographic line. *Cartographica* 22: 1–26

Buttenfield B P, DeLotto J S 1989 Multiple representations: scientific report for the specialist meeting. *NCGIA Technical Report 89–3*. Santa Barbara, NCGIA

Buttenfield B P, McMaster R B (eds) 1991 *Map generalisation: making rules for knowledge representation*. Harlow, Longman

Carbonell J G 1990 Introduction: paradigms for machine learning. In Carbonell J (ed.) *Machine learning: paradigms and methods*. Cambridge (USA), MIT Press

Chang H, McMaster R B 1993 Interface design and knowledge acquisition for cartographic generalisation. *Proceedings AutoCarto 9*. Bethesda, ACSM/ASPRS: 187–96

Chui C K 1992 *An introduction to wavelets*. San Diego, Academic Press

Cromley R G 1991 Hierarchical methods of line simplification. *Cartography and Geographic Information Systems* 18: 125–31

De Floriani L, Falcidieno B, Nagy G, Pienovi C 1984 A hierarchical structure for surface approximation. *Computer Graphics (Proceedings SIGGRAPH 84)* 18: 183–93

De Floriani L, Puppo E 1995 Hierarchical triangulation for multiresolution surface description. *ACM Transactions on Graphics* 14: 363–411

Devogele T, Trevisan J, Raynal L 1997 Building a multi-scale database with scale-transition relationships. In Kraak M-J, Molenaar M (eds) *Advances in GIS research II (Proceedings Seventh International Symposium on Spatial Data Handling)*. London, Taylor and Francis: 337–52

Dikau R 1990 Geomorphic landform modelling based on hierarchy theory. *Proceedings, Fourth International Symposium on Spatial Data Handling, Zurich*: 230–39

Dillard A 1974 *Pilgrim at Tinker Creek: a mystical excursion into the natural world*. New York, Bantam

Douglas D H, Peucker T K 1973 Algorithms for the reduction of the number of points required to represent a digitised line or its caricature. *The Canadian Cartographer* 10: 112–22

Dutton G 1984 Truth and its consequences in digital cartography. *Proceedings ASP/ACSM 44th Annual Meeting*. Falls Church, ACSM: 273–81

Dutton G 1989 Modelling locational uncertainty via hierarchical tessellation. In Goodchild M F, Gopal S (eds) *Accuracy of spatial databases*. London, Taylor and Francis: 125–40

Dutton G 1997 Encoding and handling geospatial data with hierarchical triangular meshes. In Kraak M-J, Molenaar M (eds) *Advances in GIS research II (Proceedings Seventh International Symposium on Spatial Data Handling)*. London, Taylor and Francis: 505–18

Dutton G, Buttenfield B P 1993 Scale change via hierarchical coarsening: cartographic properties of quaternary triangular meshes. *Proceedings, Sixteenth International Cartographic Conference*. Bielefeld, German Society of Cartography: 847–62

Ehrliholzer R 1995 Quality assessment in generalisation: integrating quantitative and qualitative methods. *Cartography crossing borders (Proceedings Seventeenth International Cartographic Conference)*. Barcelona, Institut Cartogràfic de Catalunya: 2241–50

Ehrliholzer R 1996 Methoden für die Bewertung der Qualität von Generalisierungslösungen. Unpublished MSc thesis. Department of Geography, University of Zurich

ESRI 1996b Automation of map generalisation – the cutting-edge technology. White paper. Redlands, ESRI Inc.

Fekete G 1990 Rendering and managing spherical data with sphere quadtree. *Proceedings Visualisation 90 (first IEEE conference on visualisation, San Francisco, October 23–26 1990)*. Los Alamitos, IEEE Computer Society Press

Fisher P F, Mackaness W A 1987 Are cartographic expert systems possible? *Proceedings AutoCarto 8 (Eighth International Symposium on Computer-Assisted Cartography)*. Falls Church, American Congress on Surveying and Mapping: 530–34

Fritsch E, Lagrange J-P 1995 Spectral representations of linear features for generalisation. In Frank A U, Kuhn W (eds) *Spatial information theory – a theoretical basis for GIS (Proceedings COSIT 95)*. Lecture Notes in Computer Science 988. Berlin, Springer: 157–71

Goodchild M F, Yang S 1992 A hierarchical data structure for global geographic information systems. *Computer Vision, Graphics, and Image Processing*: 54: 31–44

Grünreich D 1985 Computer-assisted generalisation. *Papers CERCO Cartography Course*. Frankfurt am Main, Institut für Angewandte Geodäsie

Grünreich D 1992 ATKIS – a topographic information system as a basis for GIS and digital cartography in Germany. In Vinken R (ed.) From digital map series in geosciences to geo-information systems. *Geologisches Jahrbuch Reihe A*, Heft 122. Hanover, Federal Institute of Geosciences and Resources: 207–16

Grünreich D, Powitz B M, Schmidt C 1992 Research and development in computer-assisted generalisation of topographic information at the Institute of Cartography, Hannover University. *Proceedings EGIS 92 (Third European Conference and Exhibition on Geographical Information Systems)*. Utrecht, EGIS Foundation: 532–41

Hake G 1975 Zum Begriffssystem der Generalisierung. *Nachrichten aus dem Karten- und Vermessungswesen* Sonderheft zum 65 Geburtstag von Prof Knorr: 53–62

Hardy P 1996 Map generalisation – the Laser-Scan way. Technical paper. Cambridge (UK), Laser-Scan Ltd

Heller M 1990 Triangulation algorithms for adaptive terrain modeling. *Proceedings, Fourth International Symposium on Spatial Data Handling, Zurich*: 163–74

Huttenlocher D P, Klanderman G A, Rucklidge W J 1993 Comparing images using the Hausdorff distance. *IEEE Transactions on Pattern Analysis and Machine Intelligence* 15: 850–63

ICA (International Cartographic Association) 1973 *Multilingual dictionary of technical terms in cartography*. Wiesbaden, Franz Steiner

Jones C B, Bundy G L, Ware J M 1995 Map generalisation with a triangulated data structure. *Cartography and Geographic Information Systems* 22: 317–31

Jones C B, Kidner D B, Ware J M 1994 The implicit triangular irregular network and multiscale databases. *The Computer Journal* 37: 43–57

Jones C B, Kidner D B, Luo L Q, Bundy G L, Ware J M 1996 Database design for a multi-scale spatial information system. *International Journal of Geographical Information Systems* 10: 901–20

Keller S F 1995 'Generalisation by example': interactive parameter control in line generalisation using genetic algorithms. *Cartography crossing borders (Proceedings Seventeenth International Cartographic Conference)*. Barcelona, Institut Cartogràfic de Catalunya: 1974–83

Kidner D B, Jones C B 1994 A deductive object-oriented GIS for handling multiple representations. In Waugh T C, Healey R G (eds) *Advances in GIS research (Proceedings Sixth International Symposium on Spatial Data Handling)*: 882–900

Leberl F W 1986 ASTRA – a system for automated scale transition. *Photogrammetric Engineering and Remote Sensing* 52: 251–8

Lee D 1995 Experiment on formalising the generalisation process. In Müller J-C, Lagrange J-P, Weibel R (eds) *GIS and generalisation: methodological and practical issues*. London, Taylor and Francis: 219–34

Leitner M, Buttenfield B P 1995 Acquisition of procedural cartographic knowledge by reverse engineering. *Cartography and Geographic Information Systems* 22: 232–41

153

Lichtner W 1979 Computer-assisted processes of cartographic generalisation in topographic maps. *Geo-Processing* 1: 183–99

Mackaness, W A 1994 An algorithm for conflict identification and feature displacement in automated map generalisation. *Cartography and Geographic Information Systems* 21: 219–32

Mackaness W A, Beard M K 1993 Use of graph theory to support map generalisation. *Cartography and Geographic Information Systems* 20: 210–21

McGraw K L, Harbison-Briggs K 1989 *Knowledge acquisition: principles and guidelines.* Englewood Cliffs, Prentice-Hall

McMaster R B 1987 Automated line generalisation. *Cartographica* 24: 74–111

McMaster R B (ed.) 1989 Numerical generalisation in cartography. *Cartographica* 26

McMaster R B, Mark D M 1991 The design of a graphical user interface for knowledge acquisition in cartographic generalisation. *Proceedings GIS/LIS 91.* Bethesda, ACSM/ASPRS: 311–20

McMaster R B, Shea K S 1992 *Generalisation in digital cartography.* Washington DC, Association of American Geographers

Misund G 1997 Varioscale TIN based surfaces. In Kraak M-J, Molenaar M (eds) *Advances in GIS research II (Proceedings Seventh International Symposium on Spatial Data Handling).* London, Taylor and Francis: 353–64

Molenaar M (ed.) 1996a *Methods for the generalisation of geo-databases. Publications on Geodesy No. 43.* Delft, Netherlands Geodetic Commission

Molenaar M 1996b The role of topologic and hierarchical spatial object models in database generalisation. In Molenaar M (ed.) *Methods for the generalisation of geo-databases. Publications on Geodesy.* Delft, Netherlands Geodetic Commission 43: 13–36

Monmonier M 1991b Role of interpolation in feature displacement. In Buttenfield B P, McMaster R B (eds) *Map generalisation: making rules for knowledge representation.* Harlow, Longman: 189–203

Morehouse S 1995 GIS-based map compilation and generalisation. In Müller J-C, Lagrange J-P, Weibel R (eds) *GIS and generalisation: methodological and practical issues.* London, Taylor and Francis: 21–30

Müller J-C 1991 Generalisation of spatial databases. In Maguire D J, Goodchild M F, Rhind D W (eds) *Geographical information systems: principles and applications.* Harlow, Longman/New York, John Wiley & Sons Inc. 1: 457–75

Müller J-C, Lagrange J-P, Weibel R 1995 (eds) *GIS and generalisation: methodological and practical issues.* London, Taylor and Francis

Müller J-C, Wang Z 1992 Area patch generalisation: a competitive approach. *The Cartographic Journal* 29: 137–44

Mustière S 1995 Mesures de la qualité de la généralisation du linéaire. Unpublished MSc thesis, St-Mandé (F), Ecole Nationale des Sciences Géographiques

Nickerson B G 1988 Automated cartographic generalisation for linear features. *Cartographica* 25: 15–66

Nickerson B G 1991 Knowledge engineering for generalisation. In Buttenfield B P, McMaster R B (eds) *Map generalisation: making rules for knowledge representation.* Harlow, Longman: 40–56

Oosterom P van 1993 *Reactive data structures for geographic information systems.* Oxford, Oxford University Press

Oosterom P van, Schenkelaars V 1995 The development of a multi-scale GIS. *International Journal of Geographical Information Systems* 9: 489–508

Plazanet C 1995 Measurements, characterisation, and classification for automated line feature generalisation. *ACSM/ASPRS Annual Convention and Exposition 4 (Proceedings AutoCarto 12)*: 59–68

Plazanet C 1996 Analyse de la géométrie des objets linéaires pour l'enrichissement des bases de données. Intégration dans le processus de généralisation cartographique des routes. PhD Thesis, Université Marne la Vallée

Plazanet C, Affholder J-G, Fritsch E 1995 The importance of geometric modeling in linear feature generalisation. *Cartography and Geographic Information Systems* 22: 291–305

Puppo E 1996 Variable resolution terrain surfaces. *Proceedings, Canadian Conference on Computational Geometry*: 202–10

Regnauld N 1997 Recognition of building clusters for generalisation. In Kraak M-J, Molenaar M (eds) *Advances in GIS research II (Proceedings Seventh International Symposium on Spatial Data Handling).* London, Taylor and Francis: 185–98

Reichenbacher T 1995 Knowledge acquisition in map generalisation using interactive systems and machine learning. *Cartography crossing borders (Proceedings 17th International Cartographic Conference).* Barcelona, Institut Cartogràfic de Catalunya: 2221–30

Richardson D E 1994 Generalisation of spatial and thematic data using inheritance and classification and aggregation hierarchies. In Waugh T C, Healey R G (eds) *Advances in GIS research (Proceedings Sixth International Symposium on Spatial Data Handling)*: 901–20

Rieger M, Coulson M 1993 Consensus or confusion: cartographers' knowledge of generalisation. *Cartographica* 30: 69–80

Robinson G J 1995 A hierarchical top-down bottom-up approach to topographical map generalisation. In Buttenfield B P, McMaster R B (eds) *Map generalisation: making rules for knowledge representation.* Harlow, Longman: 235–45

Rousseau D, Rousseau T, Lecordix F 1995 An evaluation of map generaliser, an Intergraph interactive generalisation software package. Technical report. St-Mandé (F), Institut Géographique National, Laboratoire COGIT

Ruas A 1995a Multiple paradigms for automating map generalisation: geometry, topology, hierarchical partitioning and local triangulation. *ACSM/ASPRS Annual Convention and Exposition 4 (Proceedings AutoCarto 12)*: 69–78

Ruas A 1995b Multiple representations and generalisation. Lecture notes for Nordic Cartography Seminar 95. *ftp://sturm.ign.fr*

Ruas A, Lagrange J-P 1995 Data and knowledge modelling for generalisation. In Müller J-C, Lagrange J-P, Weibel R (eds) *GIS and generalisation: methodological and practical issues.* London, Taylor and Francis: 73–90

Ruas A, Plazanet C 1997 Strategies for automated generalisation. In Kraak M-J, Molenaar M (eds) *Advances in GIS research II (Proceedings Seventh International Symposium on Spatial Data Handling).* London, Taylor and Francis 319–36

Rusak Mazur E, Castner H W 1990 Horton's ordering scheme and the generalisation of river networks. *The Cartographic Journal* 27: 104–12

Samet H 1990 *The design and analysis of spatial data structures.* Reading (USA), Addison-Wesley

Scarlatos L, Pavlidis T 1991 Adaptive hierarchical triangulation. *Proceedings AutoCarto 10.* Bethesda, ACSM/ASPRS: 234–46

Schlegel A, Weibel R 1995 Extending a general-purpose GIS for computer-assisted generalisation. *Cartography crossing borders (Proceedings Seventeenth International Cartographic Conference).* Barcelona, Institut Cartogràfic de Catalunya: 2211–20

Schröder P, Sweldens W 1995 *Spherical wavelets: efficiently representing functions on the sphere.* Research Report, Department of Mathematics, University of South Carolina

Schylberg L 1993 *Computational methods for generalisation of cartographic data in a raster environment.* Fotogrammetric Reports No. 60. Stockholm, Royal Institute of Technology, Department of Geodesy and Photogrammetry

Shea K S, McMaster R B 1989 Cartographic generalisation in a digital environment: when and how to generalise. *Proceedings AutoCarto 9.* Bethesda ASPRS/ACSM: 56–67

Spiess E 1995 The need for generalisation in a GIS environment. In Müller J-C, Lagrange J-P, Weibel R (eds) *GIS and generalisation: methodological and practical issues.* London, Taylor and Francis: 31–46

SSC (Swiss Society of Cartography) 1977 *Cartographic generalisation – topographic maps.* Cartographic Publication Series, Vol. 2. Zurich, Swiss Society of Cartography

Steward H J 1974 Cartographic generalisation: some concepts and explanations. *Cartographica Monographs* 10

Thompson R C, Richardson D E 1995 A graph theory approach to road network generalisation. *Cartography crossing borders (Proceedings Seventeenth International Cartographic Conference).* Barcelona, Institut Cartogràfic de Catalunya: 1871–80

Timpf S, Frank A U 1995 A multi-scale DAG for cartographic objects. *ACSM/ASPRS Annual Convention and Exposition. (Proceedings AutoCarto 12)* 4: 157–63

USGS 1994 *Standards for digital line graphs.* Reston, US Geological Survey: 1–107

Ware J M, Jones C B 1997 A spatial model for detecting (and resolving) conflict caused by scale reduction. In Kraak M-J, Molenaar M (eds) *Advances in GIS research II (Proceedings Seventh International Symposium on Spatial Data Handling).* London, Taylor and Francis: 547–58

Ware J M, Jones C B, Bundy G L 1995 A triangulated spatial model for cartographic generalisation of areal objects. In Frank A U, Kuhn W (eds) *Spatial information theory – a theoretical basis for GIS (Proceedings COSIT '95).* Lecture Notes in Computer Science 988. Berlin, Springer: 173–92

Weibel R 1991 Amplified intelligence and rule-based systems. In Buttenfield B P, McMaster R B (eds) *Map generalisation: making rules for knowledge representation.* Harlow, Longman: 172–86

Weibel R 1992 Models and experiments for adaptive computer-assisted terrain generalisation. *Cartography and Geographic Information Systems* 19: 133–53

Weibel R 1995a Map generalisation. *Cartography and Geographic Information Systems* (special issue) 22

Weibel R 1995b Three essential building blocks for automated generalisation. In Müller J-C, Lagrange J-P, Weibel R (eds) *GIS and generalisation: methodological and practical issues.* London, Taylor and Francis: 56–69

Weibel R 1997a A typology of constraints to line simplification. In Kraak M-J, Molenaar M (eds) *Advances in GIS research II (Proceedings Seventh International Symposium on Spatial Data Handling).* London, Taylor and Francis: 533–46

Weibel R, Ehrliholzer R 1995 An evaluation of MGE Map Generalizer – interim report. Technical report, Department of Geography, University of Zurich

Weibel R, Keller S, Reichenbacher T 1995 Overcoming the knowledge acquisition bottleneck in map generalisation: the role of interactive systems and computational intelligence. In Frank A U, Kuhn W (eds) *Spatial information theory – a theoretical basis for GIS (Proceedings COSIT 95).* Lecture Notes in Computer Science 988. Berlin, Springer: 139–56

Weibel R 1997 Generalization of spatial data – principles and selected algorithms. In Kreveld M van, Nievergelt J, Roos T, Widmayer P (eds) *Algorithmic foundations of geographical information systems.* Lecture Notes in Computer Science. Berlin, Springer: 99–152

Werschlein T 1996 Frequenzbasierte Linienrepräsentationen für die Kartographische Generalisierung. MSc Thesis, Department of Geography, University of Zurich

Woodsford P A 1995 Object-orientation, cartographic generalisation, and multi-product databases. *Cartography crossing borders (Proceedings Seventeenth International Cartographic Conference).* Barcelona, Institut Cartogràfic de Catalunya: 1054–8

Yang W, Gold C 1996 Managing spatial objects with the VMO tree. In Kraak M-J, Molenaar M (eds) *Advances in GIS research II (Proceedings Seventh International Symposium on Spatial Data Handling).* London, Taylor and Francis

11

Visualising spatial distributions

M-J KRAAK

Maps are an integral part of the process of spatial data handling. They are used to visualise spatial data, to reveal and understand spatial distributions and relations. Recent developments such as scientific visualisation and exploratory data analysis have had a great impact. In contemporary cartography three roles for visualisation can be recognised. First, visualisation may be used to present spatial information where one needs function to create well-designed maps. Second, visualisation may be used to analyse. Here functions are required to access individual map components to extract information, and functions to process, manipulate, or summarise that information. Third, visualisation may be used to explore. Functions are required to allow the user to explore the spatial data visually, for instance by animation or linked views.

1 INTRODUCTION

Developments in spatial data handling have been considerable during the past few decades and it seems likely that this will continue. GIS have introduced the integration of spatial data from different kinds of sources, such as remote sensing, statistical databases, and recycled paper maps. Their functionality offers the ability to manipulate, analyse, and visualise the combined data. Their users can link application-based models to them and try to find answers to (spatial) questions. The purpose of most GIS is to function as decision support systems in the specific environment of an organisation.

Maps are important tools in this process. They are used to visualise spatial data, to reveal and understand spatial distributions and relations. However, maps are no longer the final products they used to be. Maps are now an integral part of the process of spatial data handling. The growth of GIS has changed their use, and as such has changed the world of those involved in cartography and those working with spatial data in general. This is caused by many factors which can be grouped in three main categories. First, technological developments in fields such as databases, computer graphics,

multimedia, and virtual reality have boosted interest in graphics and stimulated sophisticated (spatial) data presentation. From this perspective it appears that there are almost no barriers left. Second, user-oriented developments, often as an explicit reaction to technological developments, have stimulated scientific visualisation and exploratory data analysis (Anselin, Chapter 17). Also, the cartographic discipline has reacted to these changes. New concepts such as dynamic variables, digital landscape models, and digital cartographic models have been introduced. Map-based multimedia and cartographic animation, as well as the visualisation of quality aspects of spatial data, are core topics in contemporary cartographic research.

Tomorrow's users of GIS will require a direct and interactive interface to the geographical and other (multimedia) data. This will allow them to search spatial patterns, steered by the knowledge of the phenomena and processes being represented by the interface. One of the reasons for this is the switch from a data-poor to a data-rich environment, but it is also because of the intensified link between GIS and application-based models. As a result, an increase in the demand for more advanced and sophisticated visualisation techniques can be seen.

The developments described here have led to cartographers redefining the word 'visualisation' (Taylor 1994; Wood 1994). In cartography, 'to visualise' used to mean just 'to make visible', and as such incorporated all cartographic products. According to the newly established Commission on Visualisation of the International Cartographic Association it reflects '. . . modern technology that offers the opportunity for real-time interactive visualisation'. The key concepts here are interaction and dynamics.

The main drive behind these changes has been the development in science and engineering of the field known as 'visualisation in scientific computing' (ViSC), also known as scientific visualisation. During the last decade this was stimulated by the availability of advanced hardware and software. In their prominent report McCormick et al (1987) describe it as the study of 'those mechanisms in humans and computers which allow them in concert to perceive, use and communicate visual information'. In GIS, especially when exploring data, users can work with the highly interactive tools and techniques from scientific visualisation. DiBiase (1990) was among the first to realise this. He introduced a model with two components: 'private visual thinking' and 'public visual communication'. Private visual thinking refers to situations where Earth scientists explore their own data, for example. Cartographers and their well-designed maps provide an example of public visual communication. The first can be described as geographical or map-based scientific visualisation (Fisher et al 1993; MacEachren and Monmonier 1992). In this interactive 'brainstorming' environment the raw data can be georeferenced resulting in maps and diagrams, while other data can result in images and text. By the publication of two books, *Visualization in geographical information systems* (Hearnshaw and Unwin 1994) and *Visualization in modern cartography* (MacEachren and Taylor 1994) the spatial data handling community clearly demonstrated their understanding of the impact and importance of scientific visualisation on their discipline. Both publications address many aspects of the relationships between the fields of cartography and GIS on the one hand, and scientific visualisation on the other. According to Taylor (1994) this trend of visualisation should be seen as an independent development that will have a major influence on cartography. In his view the basic

aspects of cognition (analysis and applications), communication (new display techniques), and formalism (new computer technologies) are linked by interactive visualisation (Figure 1).

Three roles for visualisation may be recognised:

- First, visualisation may be used to present spatial information. The results of spatial analysis operations can be displayed in well-designed maps easily understood by a wide audience. Questions such as 'what is?', or 'where is?', and 'what belongs together?' can be answered. The cartographic discipline offers design rules to help answer such questions through functions which create proper well-designed maps (Kraak and Ormeling 1996; MacEachren 1994a; Robinson et al 1994).

- Second, visualisation may be used to analyse, for instance in order to manipulate known data. In a planning environment the nature of two separate datasets can be fully understood, but not their relationship. A spatial analysis operation, such as (visual) overlay, combines both datasets to determine their possible spatial relationship. Questions like 'what is the best site?' or 'what is the shortest route?' can be answered. What is required are functions to access individual map components to extract information and functions to process, manipulate, or summarise that information (Bonham-Carter 1994).

- Third, visualisation may be used to explore, for instance in order to play with unknown and often

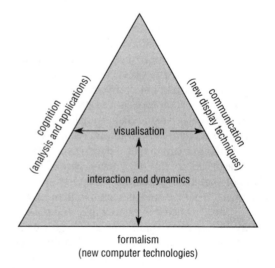

Fig 1. Cartographic visualisation (Taylor 1994).

raw data. In several applications, such as those dealing with remote sensing data, there are abundant (temporal) data available. Questions like 'what is the nature of the dataset?', or 'which of those datasets reveal patterns related to the current problem studied?', and 'what if . . .?' have to be answered before the data can actually be used in a spatial analysis operation. Functions are required which allow the user to explore the spatial data visually (for instance by animation or by linked views – MacEachren 1995; Peterson 1995).

These three strategies can be positioned in the map use cube defined by MacEachren (1994b). As shown in Figure 2, the axes of the cube represent the nature of the data (from known to unknown), the audience (from a wide audience to a private person) and the interactivity (from low to high). The spheres representing the visualisation strategies can be positioned along the diagonal from the lower left front corner (present: low interactivity, known data, and wide audience) to the upper right back corner (explore: high interactivity, unknown data, private person). Locating cartographic publications within the cube would reveal a concentration in the lower left front corner. However, colouring the dots to differentiate the publications according to their age would show many recent publications outside this corner and along the diagonal.

The functionality needed for these three strategies will shape this chapter. Each of them requires its own visualisation approach, described in turn in the

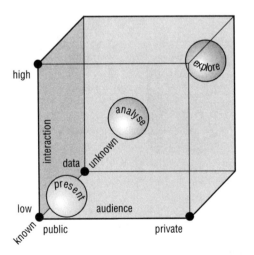

Fig 2. The three visualisation strategies plotted in MacEachren's (1994) map use cube.

following three sections. The first section provides some map basics. It will briefly explain cartographic grammar, its rules and conventions. Depending on the nature of a spatial distribution, it will suggest particular mapping solutions. This strategy has the most developed tools available to create effective maps to communicate the characteristics of spatial distributions. When discussing the second strategy, visualisations to support analysis, it will be demonstrated how the map can work in this environment, and how information critical for decision-making can be visualised. In a data exploration environment, the third strategy, it is likely that the user is unfamiliar with the exact nature of the data. It is obvious that, compared to both other strategies, more appropriate visualisation methods will have to be applied. Specific visual exploration tools in close relation to 'new' mapping methods such as animation and hypermaps (multimedia) will be discussed. It is this strategy that will benefit most from developments in scientific visualisation.

2 PRESENTING SPATIAL DISTRIBUTIONS

Maps are uniquely powerful tools for the transfer of spatial information. Using a map one can locate geographical objects, while the shapes and colours of its signs and symbols inform us about the characteristics of the objects represented. Maps reveal spatial relations and patterns, and offer the user insight into the distribution of particular phenomena. Board (1993) defines the map as 'a representation or abstraction of geographical reality' and 'a tool for presenting geographical information in a way that is visual, digital or tactile'. Traditionally cartographers have concentrated most of their research efforts on enhancing the transfer of spatial data. This knowledge is very valuable, although some additional new concepts need to be introduced as illustrated in Figure 3. The traditional paper map functioned not only as an analogue database but also as an information transfer medium. Today a clear distinction is made between the database and presentation functions of the map, known respectively as the Digital Landscape Model (DLM) and Digital Cartographic Model (DCM). A DLM can be considered as a model of reality, based on a selection process. Depending on the purpose of the database, particular geographical objects have been selected from reality, and are represented in the database by a data structure (see Dowman, Chapter 31; Martin, Chapter 6). Multiple DCMs can be generated from the same landscape model, depending on the output medium or map design. To visualise data in the form of a paper map requires a different approach to an onscreen visualisation, and a road map for a vehicle navigation system will look different from a map designed for a casual tourist. Both, however, can be derived from the same DLM.

Next in importance to its contents, the usefulness of a map depends on its scale. For certain GIS applications one needs very detailed large-scale maps, while others require small-scale maps. Figure 4 shows a small-scale map (on the left) and a large-scale map. Traditionally maps have been divided into topographic and thematic types. Topographic maps portray the Earth's surface as accurately as possible subject to the limitations of the map scale. Topographic maps may include such features as houses, roads, vegetation, relief, geographical names, and a reference grid. Thematic maps represent the distribution of a particular phenomenon. In Plate 6 the upper map shows the topography of the peak of Mount Kilimanjaro in Africa. The lower thematic map shows the geology of the same area. As can be noted, the thematic map contains information also found in the topographic map, since to be able to understand the theme represented one needs to be able to locate it as well. The amount of topographic information required depends on the map theme. A geological map will need more topographic data than a population density map, which normally only needs administrative boundaries. The digital environment has diminished the distinction between the two map types. Often both the topographic and the thematic maps are stored in layers, and the user is able to switch layers on or off at will.

The design of topographic maps is mostly based on conventions, of which some date back to the nineteenth century. Examples are representing water in blue (see MacDevette et al, Chapter 65), forests in green, major roads in red, urban areas in black, etc. The design of thematic maps, however, is based on a set of cartographic rules, also called cartographic grammar. The application of the rules can be translated in the question 'how do I say what to whom?'. 'What' refers to spatial data and its characteristics – for instance whether they are of a qualitative or quantitative nature. 'Whom' refers to the map audience and the purpose of the map – a map for scientists requires a different approach to a map on the same topic aimed at children. 'How' refers to the design rules themselves.

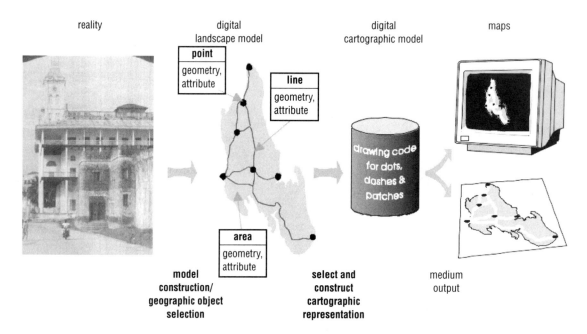

Fig 3. Spatial data characteristics: from reality to the map via a digital landscape model and a digital cartographic model.

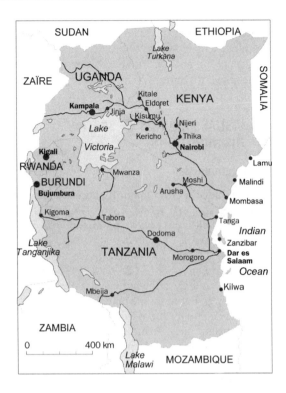 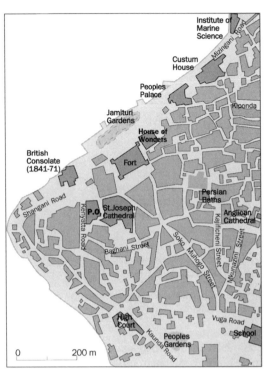

Fig 4. A small-scale map of East Africa, and a large-scale map of Stone Town (Zanzibar).

To identify the proper symbology for a map one has to conduct cartographic data analysis. The objective of such analysis is to access the characteristics of the data components in order to find out how they can be visualised. The first step in the analysis process is to find a common denominator for all of the data. This common denominator will then be used as the title of the map. Next the individual component(s) should be accessed and their nature described. This can be done by determining the measurement scale, which can be nominal, ordinal, interval, or ratio (see Martin 1996 for a discussion of geographical counterparts to these). Qualitative data such as land-use categories are measured on a nominal scale, while quantitative data are measured on the remaining scales. Qualitative data are classified according to disciplinary convention, such as a soil classification system, while quantitative data are grouped together by mathematical method.

When all the information is available the data components should be linked with the graphic sign system. Bertin (1983) created the base of this system. He distinguished six graphical variables: size, value, texture (grain), colour, orientation, and shape (Plate 7). Together with the location of the symbols in use these are known as visual variables. Graphical variables stimulate a certain perceptual behaviour with the map user. Shape, orientation, and colour allow differentiation between qualitative data values. Size is a good variable to use when the purpose of the map is to show the distribution amounts, while value functions well in mapping data measured on an interval scale. The design process results in thematic maps that are instantly understandable (for example newspaper maps and simple maps such as the one in Figure 5), and maps which may take some time to study (for example road maps or topographic maps – Plate 6(a)). A final category includes maps which require additional interpretative skills on the part of the user (for example geological or soil maps – Plate 6(b)).

Figure 6 presents an overview of some possible thematic maps. They represent different mapping

161

Fig 5. Zanzibar Town: appealing map design by visual hierarchy and the use of fonts.

methods, many of which are found in the cartographic component of GIS software. In addition to the measurement scale, it is also important to take into account the distribution of the phenomenon, whether continuous or discontinuous, whether boundaries are smooth or not, and whether the data refer to point, line, area, or volume objects. The maps in Figure 6 are ordered in a matrix with the (dis)continuous nature along one side and qualitative/quantitative nature along the other side. From the above it will be clear that each spatial distribution requires a unique mapping solution depending on its character (see also Elshaw Thrall and Thrall, Chapter 23).

However, if all rules are applied mechanically the result can still be quite sterile and uninteresting. There is an additional need for a design that is appealing as well. Figure 5 provides an example of good design. Here information is ordered according to importance and is translated into a visual hierarchy. The urban area of Zanzibar Town is the first item on the map that will catch the eye of the map user. The map also shows some other important ingredients needed, such as an indication of the map scale and its orientation. Placement and style of text can be seen to play a prominent role too. Text can be used to convey information additional to that represented by the symbols alone, and the graphical

font used for the wording of 'Zanzibar Town' has been chosen to express its oriental Arabic atmosphere. However, to be effective the text must be placed in an appropriate position with respect to the relevant symbols.

3 VISUAL ANALYSIS OF SPATIAL DISTRIBUTIONS

3.1 Introduction

Since one of a GIS's major functions is to act as a decision support system, it seems logical that the map as such should play a prominent role. With the map in this role one can even speak of visual decision support. The maps provide a direct and interactive interface to GIS data. They can be used as visual indices to the individual objects represented in the map. Based on the map, users will get answers to more complex questions such as 'what relationship exists?' This ability to work with maps and to analyse and interpret them correctly is one very important aspect of GIS use. However, to get the right answers the user should adhere to proper map use strategies.

Figure 7 demonstrates that this is not easy at all. The map displayed shows the northern part of the Netherlands. It is a result of a GIS analysis executed by an insurance company which wanted to know if it would make sense to initiate a regional operation. A first look at the map, which shows the number of traffic accidents for each municipality, would indeed suggest so. The eastern region seems to have worse drivers than the western region. However, a closer look at the map should make one less sure. First, the geographical units in the western part are much larger than the average units in the east; because each unit has a symbol the map looks much denser in the east. Second, when looking at the legend it can be seen that the small squares can represent from 1 to 99 accidents; the map shows some small squares representing only one accident, while others represent over 92. The west could therefore still have far more accidents then the east. The example illustrates not only that care is required when interpreting maps, but also that access to the map's single objects and the database behind the objects is a necessity. Additional relevant information such as the number of cars and the length of road should be available as well.

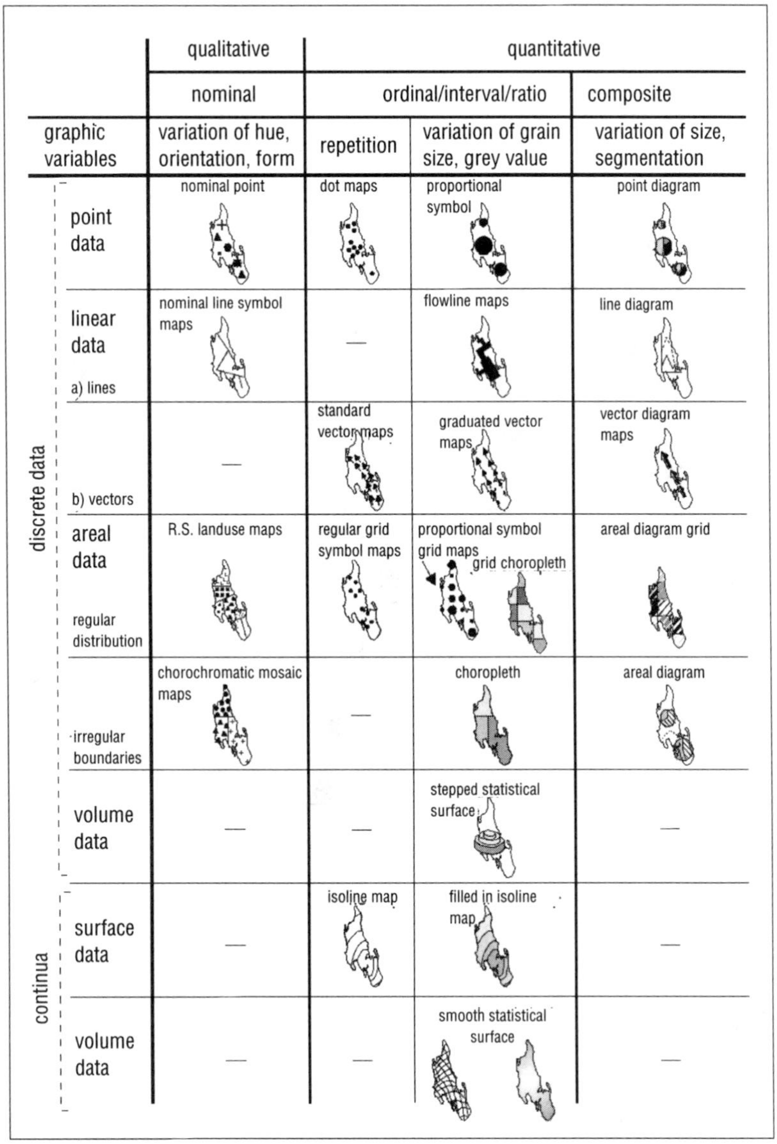

graphic variables	qualitative	quantitative		
	nominal	ordinal/interval/ratio		composite
	variation of hue, orientation, form	repetition	variation of grain size, grey value	variation of size, segmentation
discrete data — point data	nominal point	dot maps	proportional symbol	point diagram
linear data a) lines	nominal line symbol maps	—	flowline maps	line diagram
b) vectors	—	standard vector maps	graduated vector maps	vector diagram maps
areal data — regular distribution	R.S. landuse maps	regular grid symbol maps	proportional symbol grid maps / grid choropleth	areal diagram grid
irregular boundaries	chorochromatic mosaic maps	—	choropleth	areal diagram
volume data	—	—	stepped statistical surface	—
continua — surface data	—	isoline map	filled in isoline map	—
volume data	—	—	smooth statistical surface	—

Fig 6. A subdivision of thematic map types, based on the nature of the data (after Kraak and Ormeling 1996).

How can map tools help with the visual analysis of spatial distributions (Armstrong et al 1992)? Little is known about how people make decisions on the basis of map study and analysis. Giffin (1983) found that the strategies followed by individuals vary widely in relation to map type and complexity as well as according to individual characteristics. From the example above, it becomes clear that the user needs to have access to the appropriate spatial data in order to solve spatial tasks. Compared to the mapping activities in the previous section, the link between map and database (DCM and DLM) as well as access to the tools to describe and manipulate the data are of major importance. A key word here is *interaction*.

In order to make justifiable decisions based on spatial information, its nature and its quality (or reliability) must be known (Beard and Buttenfield,

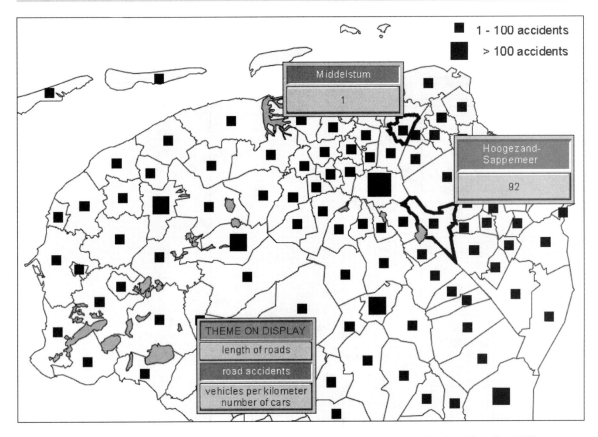

Fig 7. Maps and decision-making: traffic accident in the Netherlands and insurance policy (from Kraak and Ormeling 1996).

Chapter 15; Fisher, Chapter 13; Heuvelink, Chapter 14; Veregin, Chapter 12; Buttenfield and Beard 1994). Whether the data are fit for use is a complex matter, especially where combinations of different datasets are used. Visual decision support tools can help the user to make sensible spatial decisions based on maps. This is the most efficient way of communicating information about spatial reliability. This requires formalisation which can be done by providing functionality for data integration, standardisation (e.g. exchange formats), documentation (e.g. metadata), and modelling (e.g. generalisation and classification). This will lead to insights into the quality of the data on which the user will base spatial decisions. This is necessary because GIS is very good at combining datasets; notwithstanding the fact that these datasets might refer to different survey dates, different degrees of spatial resolution, or might even be conceptually unfit for combination: the software will not mind, but instead will happily combine them and present the results. The schema shown in Figure 8, and described extensively by Kraak et al (1995), summarises this approach.

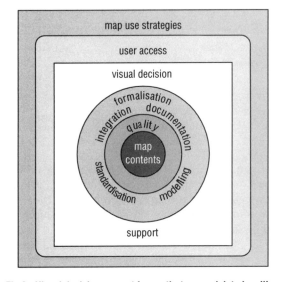

Fig 8. Visual decision support for spatio-temporal data handling (from Kraak et al 1995).

While working with spatial data in a GIS environment one commonly has to deal with 'where?', 'what?', and 'when?' queries. In a spatial analysis operation the queries will result in the manipulation of geometric, attribute, or temporal data components, separately or in combination. However, just looking at a map that displays the data already allows an evaluation of how certain phenomena vary in quantity or quality over the mapped area. Often one is not just interested in a single phenomenon but in multiple phenomena. For some aspects analytical operations are required, but sometimes a visual comparison will reveal interesting patterns for further study. Spatial, thematic, and temporal comparisons can be distinguished (Kraak and Ormeling 1996).

3.2 Comparing spatial data's geometric component

Comparing two areas seems to be relatively easy while focusing on a single theme – for example, hydrology, relief, settlements, or road networks. However, to make a sensible comparison the maps under study should have been compiled according to the same methods. They should have the same scale and the same level of generalisation or adhere to the same classification methods. For instance, if one is comparing the hydrological patterns in two river basins the individual rivers should be represented at the same level of detail in respect to generalisation and order of branches.

Figure 9 shows a comparison of the islands of Zanzibar and Pemba. They have been isolated from their original location and positioned next to each other. The coastline, reefs, road network, and villages are displayed, all derived from the Digital Chart of the World. It can be seen that Pemba, the island on the right, has a typical north–south settlement pattern, while Zanzibar, slightly larger, has a more evenly spread settlement with a larger urban area on the west coast (see Openshaw and Alvanides, Chapter 18, for a discussion of the analysis of geographically averaged data).

3.3 Comparing the attribute components of spatial data

If two or more themes related to a particular area are mapped according to the same method, it is possible to compare the maps and judge similarities or differences. However, not all mapping methods

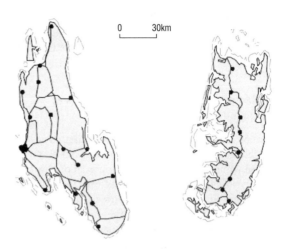

Fig 9. Comparing location: Zanzibar (Unguja) and Pemba.

are easy to compare. Choropleth maps are the simplest to compare, at least as long as the administrative units are the same in both maps. Isoline maps can be compared by measuring values in each map at the same locations.

Figure 10 compares a chorochromatic map (a soil map, right) with an isoline map (precipitation, left). At first sight it appears that low precipitation corresponds with a soil type that dominates the eastern part of the island, and that high precipitation results in a wider diversity of soils. Those familiar with Earth science in general will know that there is no necessary link between the two topics, but the above visual map analysis could be

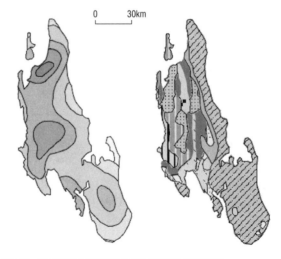

Fig 10. Comparing attributes: precipitation and soils.

true. It shows that only the expert can do the real analytical work, but comparing or overlaying two datasets can be done by anybody – but whether the operation makes sense remains unanswered.

3.4 Comparing the temporal components of spatial data

Users of GIS are no longer satisfied with analysis of snapshot data but would like to understand and analyse whole processes. A common goal of this type of analysis is to identify typical patterns in space-time. Change can be visually represented in a single map. Understanding the temporal phenomena from a single map will depend on the cartographic skills of both the map maker and map user, since these maps tend to be relatively complex. An alternative is the use of a series of single maps each representing a moment in time. Comparing these maps will give the user an idea of change. The number of maps is limited since it is difficult to follow long series of images. Another, relatively new alternative is the use of *dynamic* displays or animation (Kraak and MacEachren 1994). Change in the display over time provides a more direct impression of change in the phenomenon represented.

Figure 11 visualises the growth of the population of Zanzibar. From the maps it becomes clear that there is growth, and that growth in the urban district is faster than in the other parts of the island.

4 VISUAL EXPLORATION OF SPATIAL DISTRIBUTIONS

4.1 Introduction

Keller and Keller (1992) identify three steps in the visualisation process: first, to identify the visualisation goal; second, to remove mental roadblocks; and third, to design the display in detail. In cartography the first step is summarised by the phrase 'how do I say what to whom?', which was addressed in section 2. In the second step, the authors suggest removing oneself some distance from the discipline in order to reduce the effects of traditional constraints and conventional wisdom. Why not choose an alternative mapping method? For instance, one might use an animation instead of a set of single maps to display change over time; show a video of the landscape next to a topographic map; or change the dimension of the map from 2 dimensions to 3 dimensions. New, fresh, creative graphics could be the result, would probably have a greater and longer lasting impact than traditional mapping methods, and might also offer different insight. During the third step, which is particularly applicable in an exploratory environment, one has to decide between mapping data or visualising phenomena. An example of the mapping of the amount of rainfall may be used to clarify this distinction (Figure 12). Experts exploring rainfall patterns would like to distinguish between different

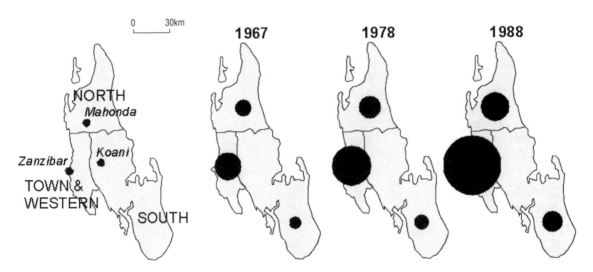

Fig 11. Comparing time: population growth.

precipitation classes, by using different colours for each class, such as blue, red, yellow, and green. A wider television audience might prefer a map showing areas with high and low precipitation. This can be realised using one colour, for instance blue, in different tints for all classes. Making dark tints correspond with high rainfall and light tints with low rainfall would result in an instantly understandable map. When exploring, data visualisation might be favoured; while presenting, phenomena visualisation may be preferred.

This approach to visualisation requires that a flexible and extensive functionality be available. The keywords 'interaction' and 'dynamics' were mentioned before. Compared with the presentation and analytical visualisation strategies these are clearly the extras. However, options to visualise the third dimension as well as temporal datasets should also be available. When exploring their data, users can work with the highly interactive tools and techniques from scientific visualisation. How are those tools implemented in a geographical exploratory visualisation environment?

Work is currently underway to develop tools for this exploratory environment (DiBiase et al 1992; Fisher et al 1993; Kraak 1994; Monmonier 1992; Slocum et al 1994). In 1990 Monmonier introduced the term 'brushing', as illustrated in Figure 13. It is

about the direct relationship between the map and other graphics related to the mapped phenomenon, like diagrams and scatter plots. The selection of an object in the map will automatically highlight the corresponding elements in the other graphics. Depending on the view in which the object is selected, the options are with geographical brushing (clicking in the map), attribute brushing (clicking in the diagram), and temporal brushing (clicking on the time line). Similar experiments on classification of choropleth maps have been made by Egbert and Slocum (1992). MacDougall (1992) followed a similar approach, while Haslett et al (1990) developed the Regard package as an interactive graphic approach to visualising statistical data. Other applications are discussed by DiBiase et al (1992), and Anselin (Chapter 17). Dykes (1995) has built a prototype of what he calls a cartographic data visualiser (CDV) which has much exploratory functionality. The system consists of a set of linked widgets, such as slide bars, buttons, and labels.

The illustrations in Figure 14 show some of the important functions that should be available to execute an exploratory visualisation strategy. The following functions are discussed in the works referred to above:

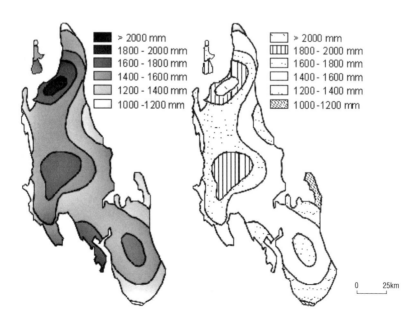

Fig 12. Visualising the classification: phenomena (left) or data (right).

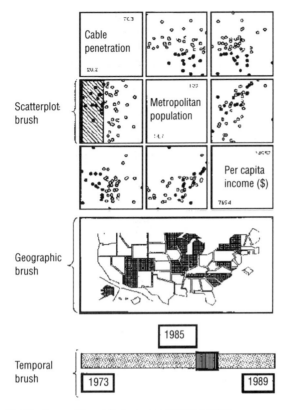

Cable penetration

Scatterplot brush

Metropolitan population

Per capita income ($)

Geographic brush

Temporal brush

1985

1973   1989

Fig 13. Geographic, attribute, and temporal brushing (Monmonier 1990).

- *Query*: an elementary function, that should always be available whatever the strategy. The user can query the map by clicking a symbol, which will activate the database. Electronic atlases incorporate this functionality (Figure 14(a): see also Elshaw Thrall and Thrall, Chapter 23).
- *Re-expression*: this function allows the same data, or part of the data, to be visualised in different ways. A time series of earthquakes could be reordered by the Richter scale instead, which could reveal interesting spatial patterns; or the classification method followed could be changed and the grey tints inverted as well – as can be seen in Figure 14(b).
- *Multiple views*: this approach could be described as interactive cartography. The same data could be displayed according to different mapping methods. Population statistics could be visualised as a dot map, a proportional circle map or a diagram map as shown in Figure 14(c).
- *Linked views*: this option is related to Monmonier's brushing principle. Selecting a

geographical object in one map will automatically highlight the same object in other views. For instance clicking a geographical unit in a cartogram would change the colour of the same unit in a geometrically correct map. In Figure 14(d), clicking the diagram showing clove production reveals a photograph of a clove plant and a map with the distribution of clove plantations in that particular year. This type of functionality allows one to introduce the multimedia components which will be discussed later in this section.

- *Animation*: the dynamic display of (temporal) processes is best done by animation. As will be explained in the next section interaction is a necessary add-on to animation (Figure 14(e)).
- *Dimensionality*: to view 3-dimensional spatial data one should be able to position the map in 3-dimensional space with respect to the map's purpose and the phenomena mapped (see Hutchinson and Gallant, Chapter 19). This means that all kinds of interactive geometric transformation functions to scale, translate, rotate, and zoom should be available, because it may be that the features of interest are located behind other features in the image (Figure 15).

4.2 Animation

Maps often represent complex processes which can be explained expressively by animation. To present the structure of a city, for example, animations can be used to show subsequent map layers which explain the logic of this structure (first relief, followed by hydrography, infrastructure, and land-use, etc.). Animation is also an excellent way to introduce the temporal component of spatial data, as in the evolution of a river delta, the history of the Dutch coastline, or the weather conditions of last week. An interesting example is ClockWork's Centennia (previously Millennium: ClockWork 1995; *http://www.clockwk. com*), a historical electronic atlas which presents an interactive animation of Europe's boundary changes between the years 1000 and 1995. This type of product can be used to explore or analyse the history of Europe.

The need in the GIS environment to deal with processes as a whole, and no longer with single time-slices, also influences visualisation. It is no longer

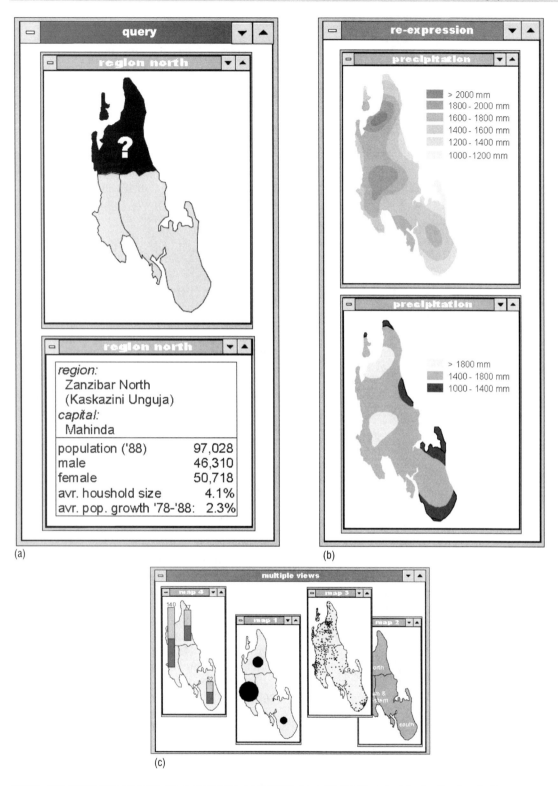

(a)

(b)

(c)

Fig 14. Data exploration: (a) query; (b) re-expression; (c) multiple views; (d) linked views; (e) animation (overleaf).

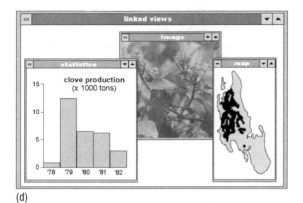

(d)

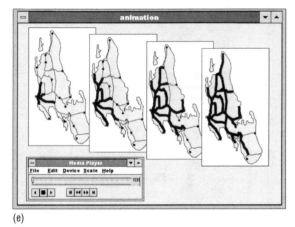

(e)

Fig 14. (cont.)

efficient to visualise models or planning operations using static paper maps. However, the onscreen map does offer opportunities to work with moving and blinking symbols, and is very suitable for animation. Such maps provide a strong method of visual

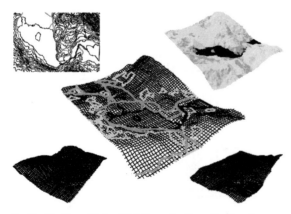

Fig 15. Working with the third dimension (from Kraak and Ormeling 1996).

communication, especially as they can incorporate real data, as well as abstract and conceptual data. Animations not only tell a story or explain a process, but also have the capability to reveal patterns or relationships which would not be evident if one looked at individual maps.

Attempts to apply animation to visualise spatial distributions date from the 1960s (see, for example, Thrower 1961; Tobler 1970) although only non-digital cartoons were possible initially. During the 1980s technological developments gave a second impulse to cartographic animation (see Moellering 1980). A third wave of interest in animation has developed, driven by interest in GIS (DiBiase et al 1992; Langran 1992; Monmonier 1990; Weber and Buttenfield 1993). Historic overviews are given by Campbell and Egbert (1990) and Peterson (1995).

The field of (cartographic) animation is about to change. Peterson (1995) expresses this as 'what happens between each frame is more important then what exists on each frame'. This should worry cartographers since their tools were developed mainly for the design of static maps. How can we deal with this new phenomenon? Is it possible to provide the producers of cartographic animation with sets of tools and rules to create 'good' animation, in the form 'If your data are . . ., and your aim is . . ., then use the variables . . .'? To be able to do so, and to take advantage of knowledge of computer graphics developments and the 'Hollywood' scene, the nature and characteristics of cartographic animations have to be understood. However, the problem is that 'understanding' animations alone will not be of much help, since the environment where they are used, the purpose of their use, and the users themselves will greatly influence 'performance'.

How can an animation be designed to make sure the viewer indeed understands the trend or phenomenon? The traditional graphic variables, as explained earlier, are used to represent the spatial data in each individual frame. Bertin, the first to write on graphic variables, had a negative approach to dynamic maps. He stated in his work (1967): '. . . however, movement only introduces one additional variable, it will be dominant, it will distract all attention from the other (graphic) variables'. Recent research, however, has demonstrated that this is not the case. Here we should remember that technological opportunities offered at the end of the 1960s were limited compared with those of today. Koussoulakou and Kraak (1992) found that the viewer of an animation would not necessarily get a better or worse understanding of the contents of the animation when compared with static maps. DiBiase et al (1992) found that movement would give the traditional variable new energy.

In this context DiBiase introduced three so-called dynamic variables: duration, order, and rate of change. MacEachren (1994b) added frequency, display time and synchronisation to the list:

- Display time – the time at which some display change is initiated.
- Duration – the length of time during which nothing in the display changes.
- Frequency – the same as duration: either can be defined in terms of the other.
- Order – the sequence of frames or scenes.
- Rate of change – the difference in magnitude of change per unit time for each of a sequence of frames or scenes.
- Synchronisation – (phase correspondence) refers to the temporal correspondence of two or more time series.

In the animation literature the so-called animation variables have surfaced (Hayward, 1984). They include size, position, orientation, speed of scene, colour, texture, perspective (viewpoint), shot (distance), and sound. The last of these is not considered here but can have an important impact (see Krygier 1994). These variables are shown in Figure 16 in relation to the graphic and dynamic variables. From this figure it can be seen that Bertin's graphic variables each have a match with one of the animation variables. From the dynamic variables only order and duration have a match, but they are the strongest in telling a story. Research is currently under way to validate and elaborate the new dynamic variables.

4.3 Maps and multimedia components

This section presents a cartographic perspective on multimedia. The relationship between the map and the individual multimedia components in relation to visual exploration, analysis, and presentation will be discussed (see Plate 8).

Maps supported with sound to present spatial information are often less interactive than those created to analyse or explore. In some electronic atlases pointing to a country on a world map plays the national anthem of the country (Electronic World Atlas; Electromap 1994). In this category one can also find the application of sound as background music to enhance a

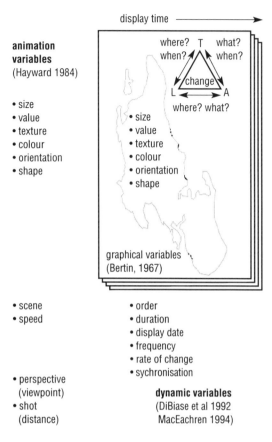

Fig 16. Cartographic animation and variable types.

mapped phenomenon, such as industry, infrastructure, or history. Experiments with maps in relation to sound are known on topics such as noise nuisance and map accuracy (Fisher 1994; Krygier 1994). In both cases the location of a pointing device in the map defines the volume of the noise. Moving the pointer to a less accurate region increases the noise level. The same approach could be used to explore a country's language – moving the mouse would start a short sentence in a region's dialect.

GIS is probably the best representation of the link between a map and text (the GIS database). As shown in Plate 8 the user can point to a geographical unit to reveal the data behind the map. Electronic atlases often have various kinds of encyclopaedic information linked to the map as a whole or to individual map elements. It is possible to analyse or explore this information. Country statistics can be compared. However, multimedia has more to offer. Now scanned text documents, such as those that describe the ownership of parcels,

171

can be included. Text in the format of hypertext can be used as a lead to other textual information or other multimedia components.

Maps are models of reality. Linking video or photographs to the map will offer the user a different perspective on reality. Topographic maps present the landscape, but it is also possible to present, next to this map, a non-interpreted satellite image or aerial photograph to help the user in his or her understanding of the landscape. The analysis of a geological map can be enhanced by showing landscape views (video or photographs) from characteristic spots in the area. A real estate agent could use the map as an index to explore all houses for sale on company file. Pointing at a specific house would show a photograph of the house, the construction drawings, and a video would start showing the house's interior. New opportunities in the framework are offered by the application of virtual reality in GIS.

References

Armstrong M P, Densham P J, Lolonis P, Rushton G 1992 Cartographic displays to support locational decision making. *Cartography and Geographic Information Systems* 19: 154–64

Bertin J 1983 *Semiology of graphics*. Madison, University of Wisconsin Press (original in French, 1967)

Board C 1993 Spatial processes. In Kanakubo T (ed.) *The selected main theoretical issues facing cartography: report of the ICA Working Group to Define the Main Theoretical Issues in Cartography*. Cologne, International Cartographic Association: 21–4

Bonham-Carter G F 1994 *Geographical information systems for geo-scientists: modelling with GIS*. New York, Pergamon Press

Buttenfield B P, Beard M K 1994 Graphical and geographical components of data quality. In Hearnshaw H, Unwin D J (eds) *Visualisation in geographic information systems*. Chichester, John Wiley & Sons: 150–7

Campbell C S, Egbert S L 1990 Animated cartography: thirty years of scratching the surface. *Cartographica* 27: 24–46

ClockWork Software 1995 *Centennia*. PO Box 148036, Chicago 60614, USA

DiBiase D 1990 Visualization in earth sciences. *Earth and Mineral Sciences*, Bulletin of the College of Earth and Mineral Sciences, Pennsylvania State University 59: 13–18

DiBiase D, MacEachren A M, Krygier J B, Reeves C 1992 Animation and the role of map design in scientific visualisation. *Cartography and Geographic Information Systems* 19: 201–14

Dykes J 1995 Cartographic visualisation for spatial analysis. *Proceedings, Seventeenth International Cartographic Conference, Barcelona*: 1365–70

Egbert S L, Slocum T A 1992 EXPLOREMAP: an exploration system for choropleth maps. *Annals of the Association of American Geographers* 82: 275–88

Fisher P F 1994a Randomization and sound for the visualization of uncertain spatial information. In Hearnshaw H, Unwin D J (eds) *Visualization in geographic information systems*. Chichester, John Wiley & Sons: 181–5

Fisher P, Dykes J, Wood J 1993 Map design and visualisation. *The Cartographic Journal* 30: 136–42

Giffin T L C 1983 Problem-solving on maps – the importance of user strategies. *The Cartographic Journal* 20: 101–109

Haslett J, Willis G, Unwin A 1990 SPIDER: an interactive statistical tool for the analysis of spatially distributed data. *International Journal of Geographical Information Systems* 4: 285–96

Hayward S 1984 *Computers for animation*. Norwich, Page Bros

Hearnshaw H M, Unwin D J (eds) 1994 *Visualization in geographical information systems*. Chichester, John Wiley & Sons

Keller P R, Keller M M 1992 *Visual cues, practical data visualization*. Piscataway, IEEE Press

Koussoulakou A, Kraak M J 1992 Spatio-temporal maps and cartographic communication. *The Cartographic Journal* 29:101–8

Kraak M J 1994 Interactive modelling environment for 3-D maps, functionality and interface issues. In MacEachren A M, Taylor D R F (eds) *Visualization in modern cartography*. Oxford, Pergamon: 269–86

Kraak M J, MacEachren A M 1994b Visualization of spatial data's temporal component. In Waugh T C, Healey R G (eds) *Advances in GIS research – Proceedings Fifth Spatial Data Handling Conference*. London, Taylor and Francis: 391–409

Kraak M-J, Ormeling F J 1996 *Cartography, visualization of spatial data*. Harrow, Longman

Kraak M-J, Ormeling F J, Müller J-C 1995 GIS-cartography: visual decision support for spatio-temporal data handling. *International Journal of Geographical Information Systems* 9: 637–45

Krygier J 1994 Sound and cartographic visualization. In MacEachren A M, Taylor D R F (eds) *Visualization in modern cartography*. Oxford, Pergamon: 149–66

Langran G 1992 *Time in geographical information systems*. London, Taylor and Francis

MacDougall E B 1992 Exploratory analysis, dynamic statistical visualization, and geographic information systems. *Cartography and Geographic Information Systems* 19: 237–46

MacEachren A M 1994a *Some truth with maps: a primer on design and symbolization*. Washington DC, Association of American Geographers

MacEachren A M 1994b Visualization in modern cartography: setting the agenda. In MacEachren A M, Taylor D R F (eds) *Visualization in modern cartography*. Oxford, Pergamon: 1–12

MacEachren A M 1995 *How maps work*. New York, Guilford Press

MacEachren A M, Monmonier M 1992 Geographic visualization: introduction. *Cartography and Geographic Information Systems* 19: 197–200

MacEachren A M, Taylor D R F (eds) 1994 *Visualization in modern cartography*. Oxford, Pergamon

Martin D J 1996 *Geographic information systems: socioeconomic applications*. London, Routledge

McCormick B, DeFanti T A, Brown M D 1987 Visualization in scientific computing. *ACM SIGGRAPH Computer Graphics* 21 special issue.

Moellering H 1980 The real-time animation of 3-dimensional maps. *The American Cartographer* 7: 67–75

Monmonier M 1990 Strategies for the visualization of geographic time-series data. *Cartographica* 27: 30–45

Monmonier M 1992 Authoring graphic scripts: experiences and principles. *Cartography and Geographic Information Systems* 19: 247–60

Peterson M P 1995 *Interactive and animated cartography*. Englewood Cliffs, Prentice-Hall

Robinson A H, Morrison J L, Muehrcke P C, Kimerling A J, Guptill S C 1994 *Elements of cartography*, 6th edition. New York, John Wiley & Sons Inc.

Slocum T A , Egbert S, Weber C, Bishop I, Dungan J, Armstrong M, Ruggles A, Demetrius-Kleanthis D, Rhyne T, Knapp L, Carron J, Okazaki D 1994 Visualization software tools. In MacEachren A M, Taylor D R F (eds) *Visualization in modern cartography*. Oxford, Pergamon: 91–122

Taylor D R F 1994 Perspectives on visualization and modern cartography. In MacEachren A M, Taylor D R F (eds) *Visualization in modern cartography*. Oxford, Pergamon: 333–42

Thrower N 1961 Animated cartography in the United States. *International Yearbook of Cartography*: 20–8

Tobler W R 1970 A computer movie: simulation of population change in the Detroit region. *Economic Geography* 46: 234–40

Weber R, Buttenfield B P 1993 A cartographic animation of average yearly surface temperatures for the 48 contiguous United States: 1897–1986. *Cartography and Geographic Information Systems* 20: 141–50

Wood M 1994 Visualization in a historical context. In MacEachren A M, Taylor D R F (eds) *Visualization in modern cartography*. Oxford, Pergamon: 13–26

Plate 1
Spatial representation
used to map the areas of
darkness and light on the
Earth's surface.

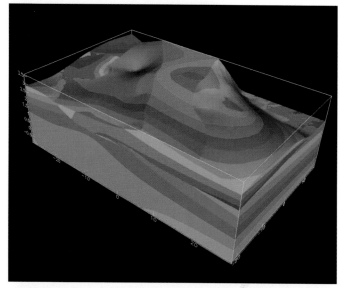

Plate 2
Raster (voxel)
representation in a
3-dimensional matrix.

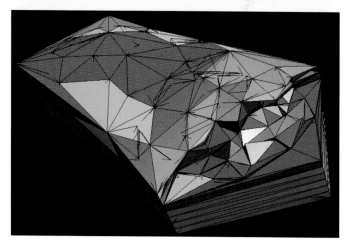

Plate 3
Vector representation of
solids using tetrahedra.

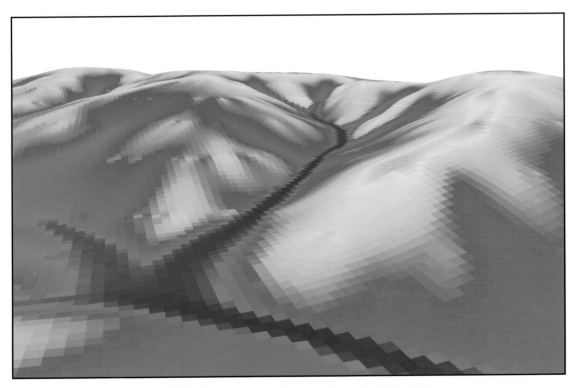

Plate 4 Three-dimensional view of the DEM in Figure 5 of Chapter 9 overlaid with specific catchment area.

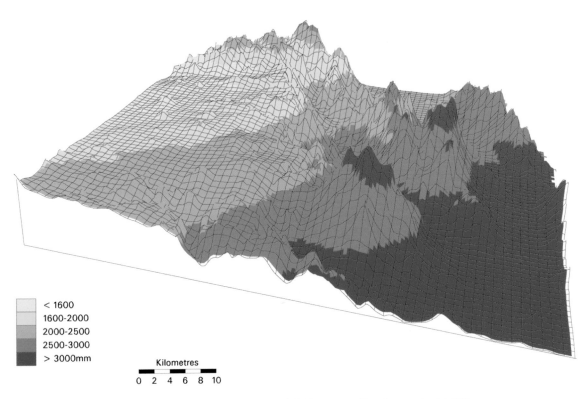

< 1600
1600-2000
2000-2500
2500-3000
> 3000mm

Kilometres
0 2 4 6 8 10

Plate 5 Annual mean precipitation overlaid on a 250-m resolution DEM for an area of length approximately 25 km.

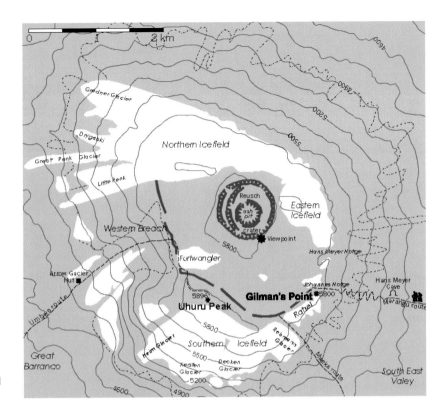

Plate 6(a)
Mount Kilimanjaro: a topographical view.

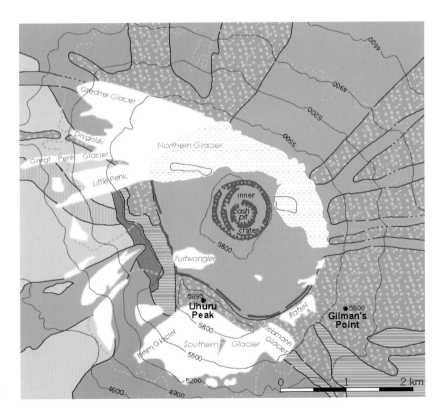

Plate 6(b)
Mount Kilimanjaro: a thematic view.

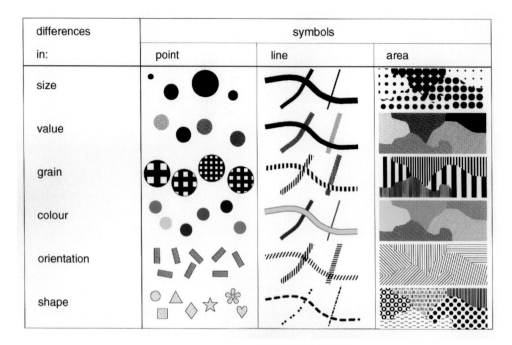

Plate 7
Representing spatial information: Bertin's (1993)
graphical variables: size, value, texture (grain),
orientation, and shape expressed in point, line, and
area symbols.

(*Source*: Kraak and Ormeling 1996)

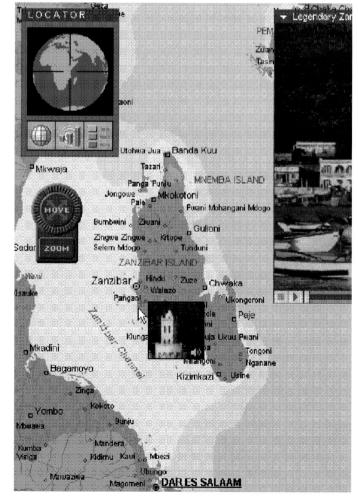

Plate 8
Detail from Microsoft's Encarta World Atlas.

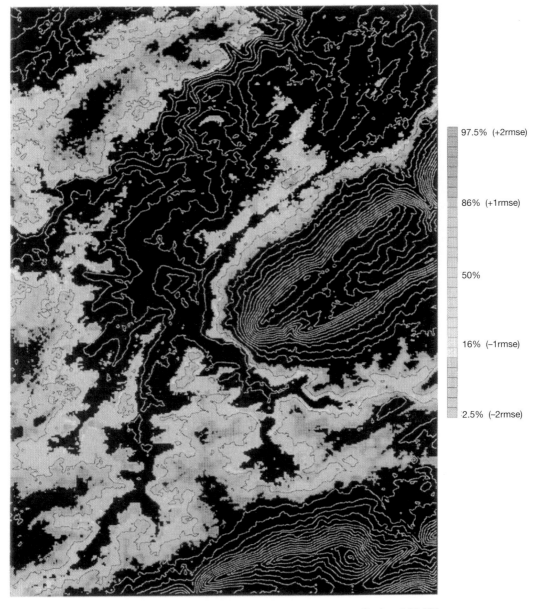

97.5% (+2rmse)

86% (+1rmse)

50%

16% (−1rmse)

2.5% (−2rmse)

Scale   1:60 000

Plate 9 Graphic depiction of the error in a single-elevation value.
(*Source*: Hunter and Goodchild 1995a)

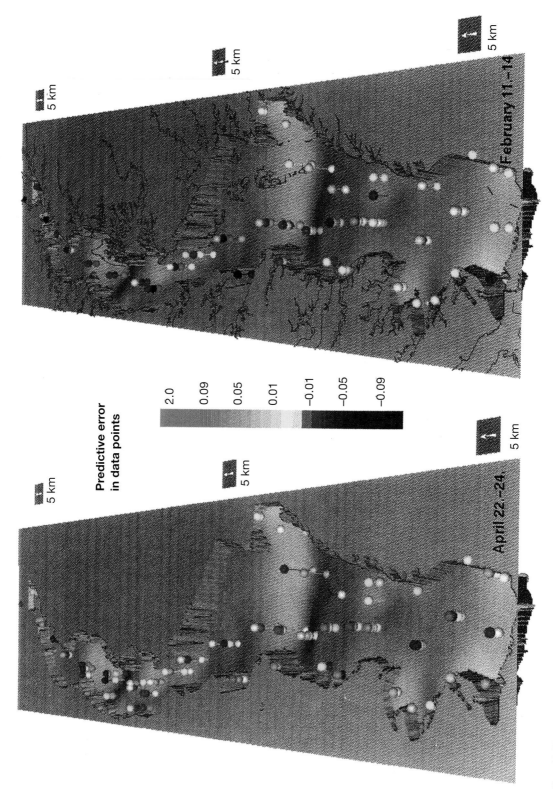

Plate 10 The cross-validation error shown separately from the data in a side-by-side display for two different periods. The cross-validation error is displayed as glyphs (coloured balls on pins) with the colour of each ball representing the error at sampled depths.

(*Source:* Mitasova et al 1995)

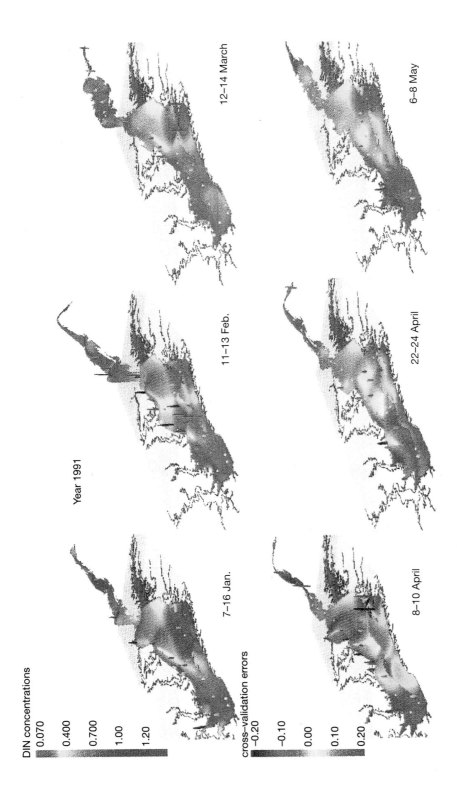

Plate 11
Images showing estimated surfaces of dissolved inorganic nitrogen for two time periods with cross-validation error at each observation station. Cross-validation error is displayed as a glyph. The height and colour of the glyph indicate the value of the cross-validation error.

(*Source*: Mitasova et al 1995)

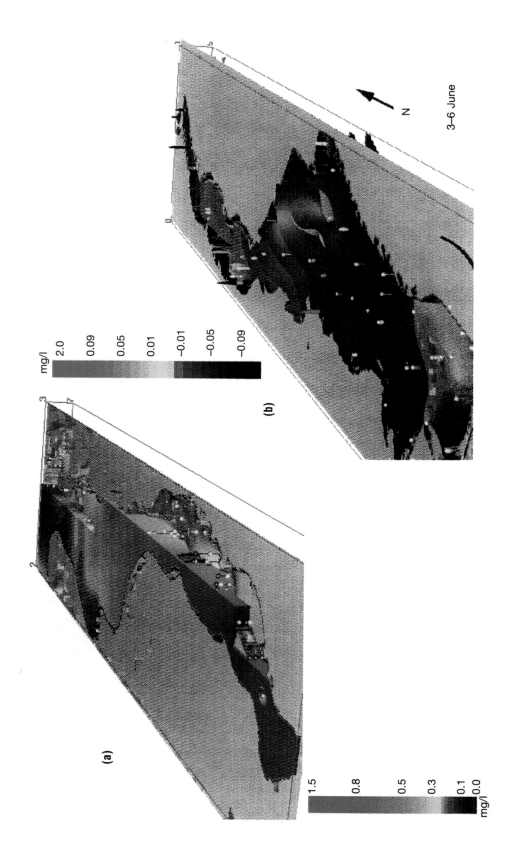

Plate 12

Images showing combinations of data display with error for three dimensions: (a) vertical cuts through the data to display predicted values along the cut in combination with glyphs showing prediction error; (b) isosurfaces in combination with the glyphs showing cross-validation error.

(*Source:* Mitasova et al 1995)

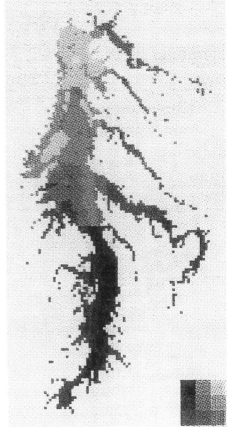

Plate 13
This image shows a bivariate mapping scheme from RVIS that uses saturation to represent uncertainty and lightness to represent the data.
(*Source*: MacEachran et al 1993)

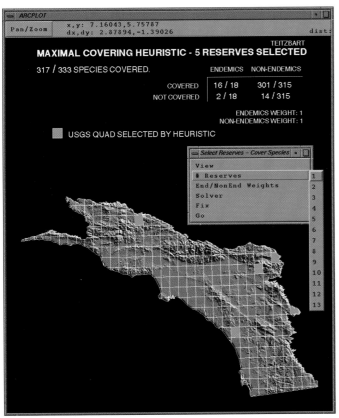

Plate 14
An example of reserve site selection for biodiversity in southern California.

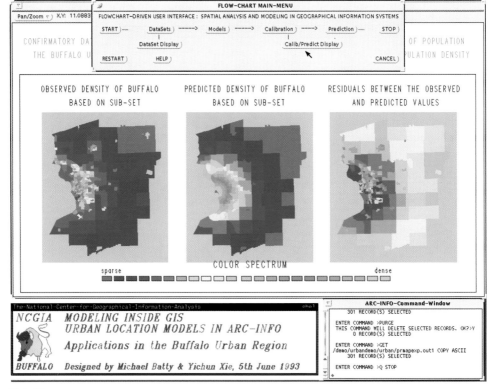

Plate 15

A typical interface from workstation GIS.

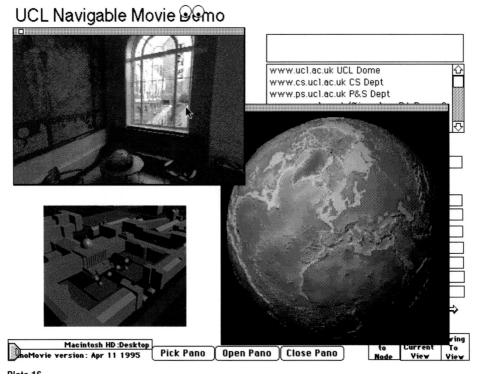

Plate 16

Visual interfaces to GIS incorporating animation and 3D.

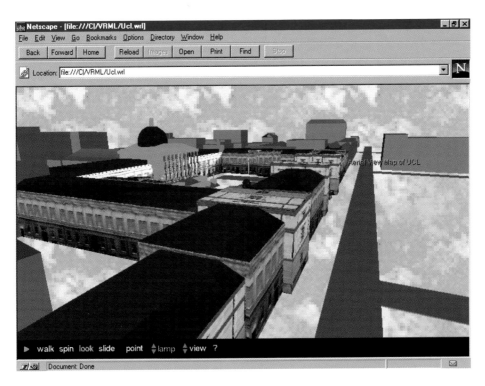

Plate 17
A VRML interface to GIS on the WWW.

Plate 18
High resolution satellite data of
Mountain View, California.

(*Source*: Space Imaging 1996)

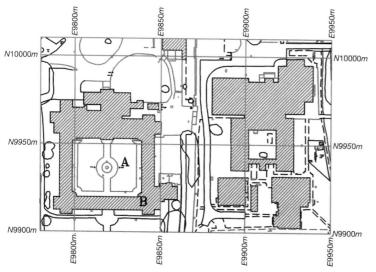

Attribute	Pt A	Comment	Pt B	Comment
Layer	99	Lamppost	122	Building
Mode	3	Symbol	1	Line
Rotation	0.0	Orientation of text		
Width				
Group	18	Post		Solid
Pen	1	Colour		Colour
Easting	9826.20		9837.01	
Northing	9936.40		9919.90	
Height	122.66		*130.05	

* The building height is measured on the roof line.

Plate 19
A pair of aerial photographs with a plot and example of the data record.
(*Source*: Cambridge University Collection of Air Photographs)

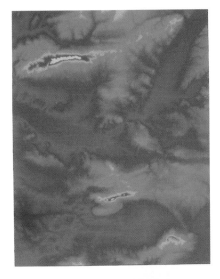
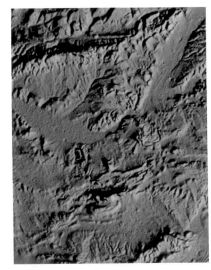

Plate 20
An example of a DEM: (*left*) colour-coded DEM of area of SE France; (*right*) a vertical hill shaded view of same area.
(*Source*: UCL 3D Image Maker)

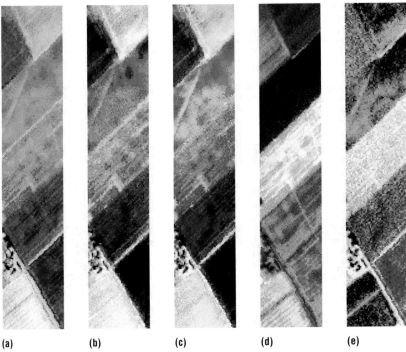

(**a**)   (**b**)   (**c**)   (**d**)   (**e**)

Plate 21 (above)
Comparison between a standard (i.e. multi-spectral) false-colour composite covering an area of arable farmland (**a**), and four single-band, multiple view angle (MVA) false-colour composite images of the same site (**b–e**). The MVA composites have each been constructed using data acquired in a single spectral waveband – (**b**) green, (**c**) red, (**d**) near-infrared, (**e**) middle-infrared – but at three different sensor view angles (two opposing oblique angles, plus nadir). The images show the potential value of directional (angular) reflectance data for distinguishing Earth surface materials.
(*Source*: Barnsley et al 1997a)

Plate 22 (right)
Multi-spectral false-colour composite image comprising data from near-infrared, red and green wavebands. These data were acquired by an airborne scanner over Orpington in the London Borough of Bromley. The spatial resolution of these data is approximately 2 metres. The figure illustrates the type of data that will be available from the new generation of very high spatial resolution, commercial satellite sensors.

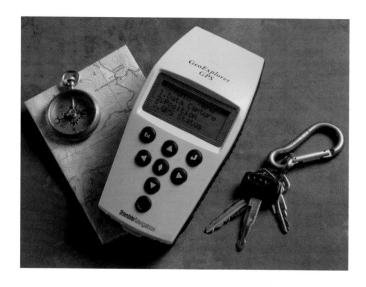

Plate 23
A hand-held 12-channel GPS receiver.

Plate 24
In-vehicle use of GPS.

Plate 25
Use of a hand-held GPS receiver.

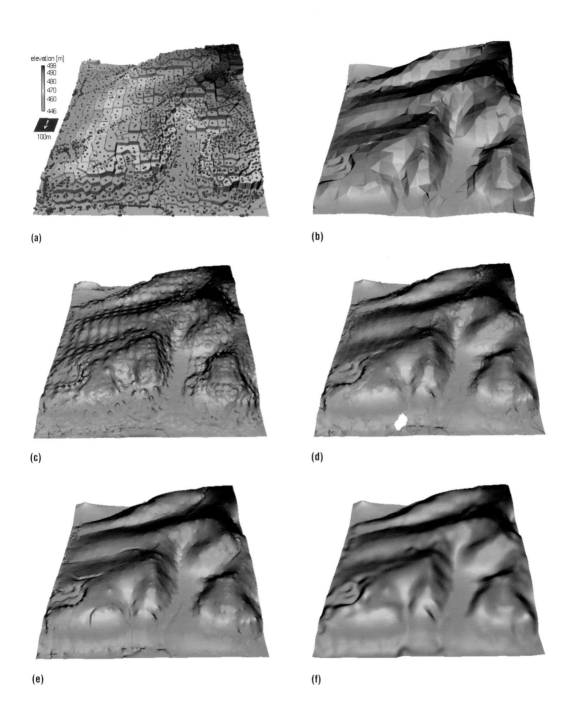

Plate 26
Interpolation of a DEM from scattered point data using methods available in GIS: (**a**) given data and Voronoi polygons; (**b**) TIN-based linear interpolation; (**c**) inverse distance weighting; (**d**) Kriging (spherical variogram); (**e**) spline with tension and stream enforcement; and (**f**) regularised spline with tension and smoothing.

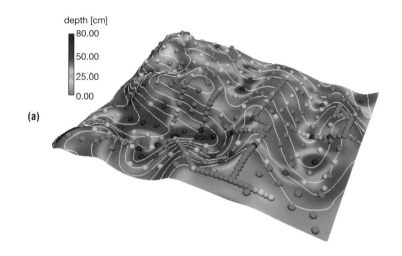

depth [cm]

80.00
50.00
25.00
0.00

(a)

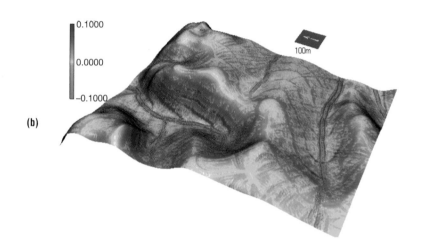

0.1000
0.0000
−0.1000

100m

(b)

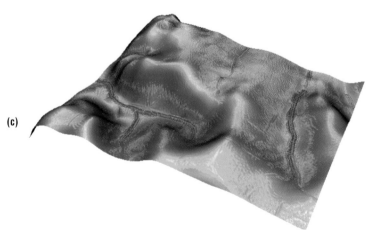

(c)

Plate 27
Influences of interpolation on the results of
an erosion/deposition model: (**a**) observed
depth of colluvial deposits in the sampling
points and as interpolated raster shown in
colour; (**b**) artificial erosion/deposition
pattern predicted from a DEM interpolated
from contours by spline with tension set too
high; and (**c**) realistic erosion/deposition
pattern predicted using a DEM interpolated
from contours by regularised spline with
tension and smoothing.

Plate 28

Interpolation of a 50-m resolution DEM (700 × 300) from a large dataset with complex topography using a regularised spline with tension.
Inset shows the principle of segmented processing based on quadtrees and the detail of given contours.

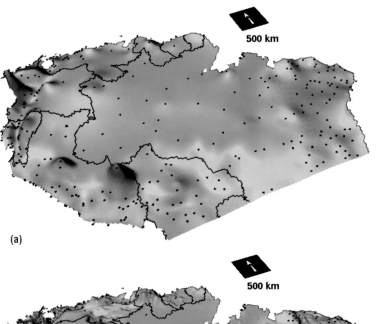

(a)

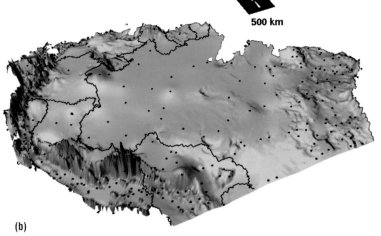

(b)

Plate 29
Interpolation of annual precipitation in tropical
South America using: (**a**) bivariate regularised
spline with tension and smoothing; (**b**) trivariate
interpolation with incorporation of the influence
of topography.

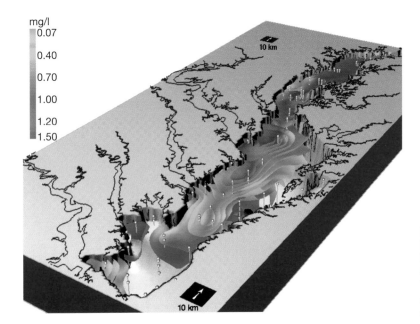

Plate 30
Snapshot from a result of spatio-temporal
interpolation (3-D + time) of nitrogen
concentrations in Chesapeake Bay (January
1991). For the complete animated result see
*http://www.cecer.army.mil/grass/viz/ches.
html.*

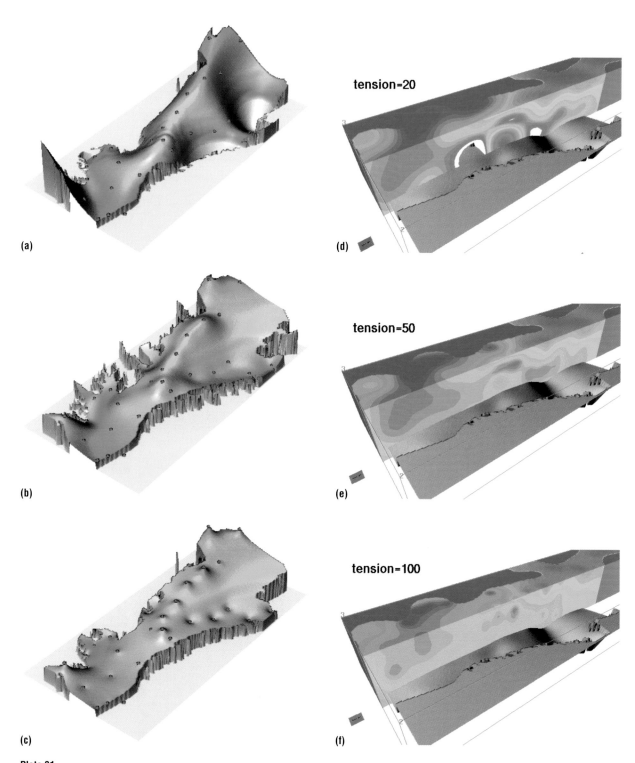

(a)

(b)

(c)

tension=20

(d)

tension=50

(e)

tension=100

(f)

Plate 31
Influence of tension on the surface (**a–c**) and volume (**d–f**) models of nitrogen concentrations in the middle section of Chesapeake Bay.
Low tension (**a, d**) leads to overshoots; high tension (**c, f**) creates extrema in data points. Appropriate tension (**b, e**) was found by
minimising the cross-validation error.

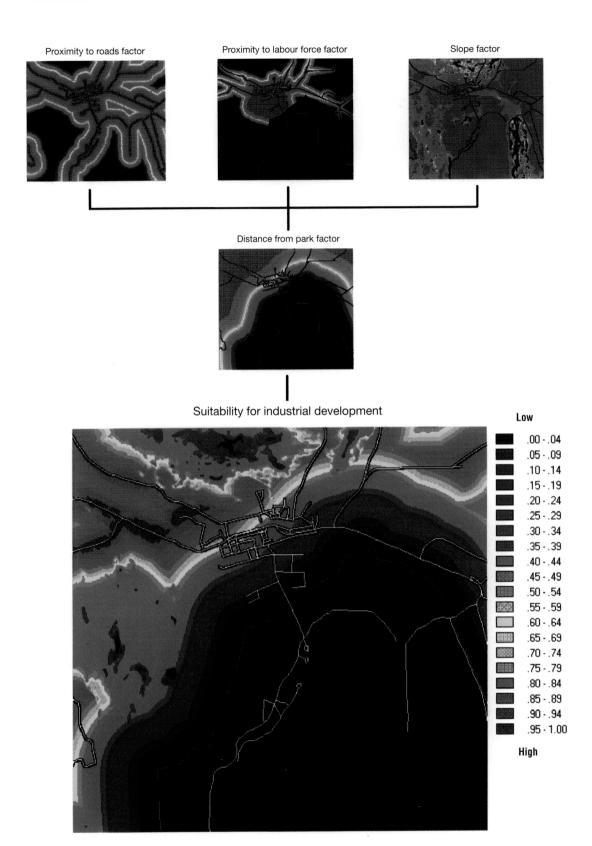

Plate 32 Multi-criteria evaluation using varying degrees of trade-off.

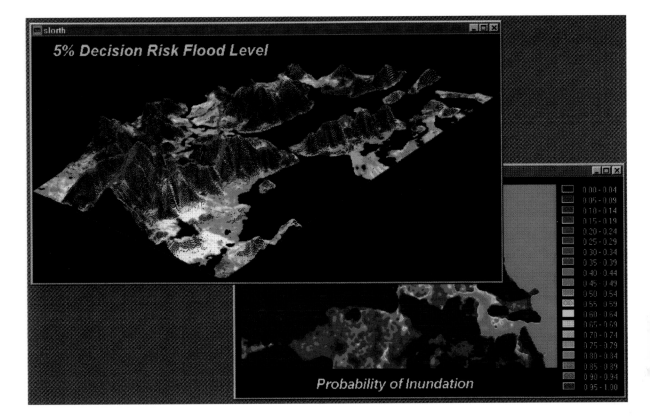

Plate 33
Decision risk: (bottom) the probability (risk)
that an area would be flooded by a rise in
sea level if one were to assume that it would
not, based on errors in measured elevation;
(top) the flood expected at a 5% decision
risk level, draped upon a colour-composite
LANDSAT-TM (Bands 3, 4, 5) image of a
region in north-central Vietnam.

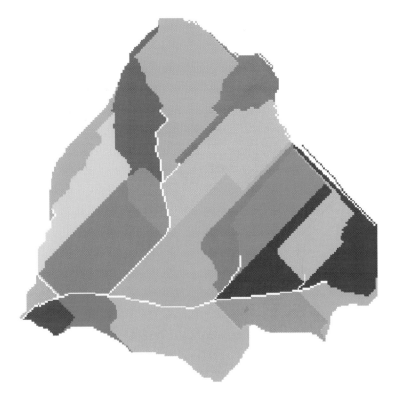

Plate 34
Watershed drainage network and drainage
area partition produced by a graph
optimisation algorithm in the GRASS
algorithm *r.watershed*. Note the bias towards
straight-line drainage connections to
minimise the cumulative cost function.

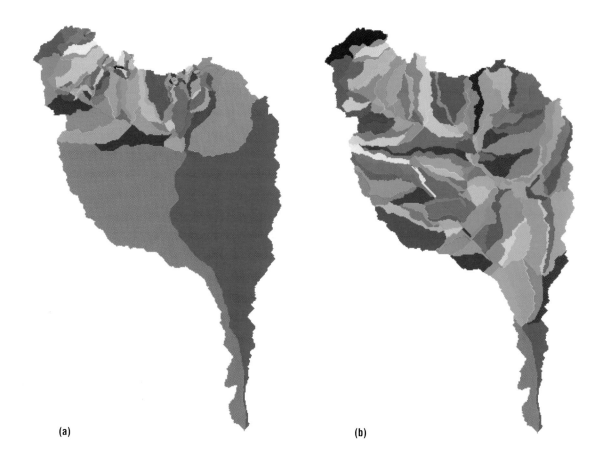

(a)   (b)

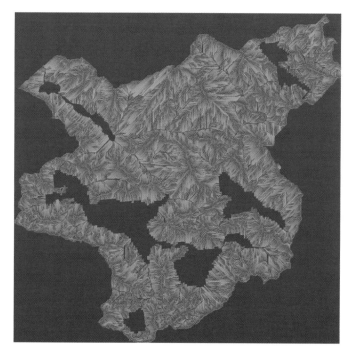

Plate 35 (above)
Partitions of the South Platte River basin produced by
two methods of pruning the full drainage direction
tree: (**a**) simple thresholding of the drainage area
accumulation image; (**b**) adaptively pruning the
drainage tree using the surface spherical variance.
Both images have approximately the same number of
partition units.

Plate 36
Explicit incorporation of lakes and wetlands into a
drainage system involves first identifying and
labelling flat bottomland features, then building the
upland drainage structure surrounding them rather
than attempting to route fictional paths through them.
Site is the Turkey Lakes Experimental Watershed, a
ten square kilometre catchment in northern Ontario.
DEM is 5-metre resolution.

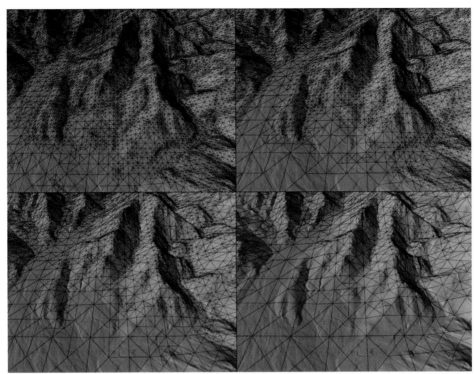

Plate 37

Level of detail management with regular triangulation.

(*Source:* Georgia Tech Virtual GIS project.)

Plate 38

The Georgia Tech Virtual GIS project.

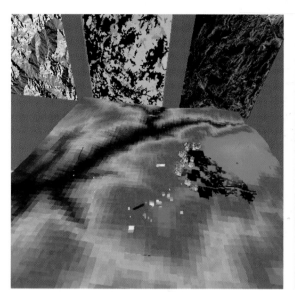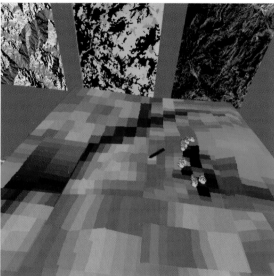

Plate 39
Multi-scale spatial simulation: forest fire model running on top of two terrains with different resolutions.
(*Source*: Gonçalves and Diogo 1994)

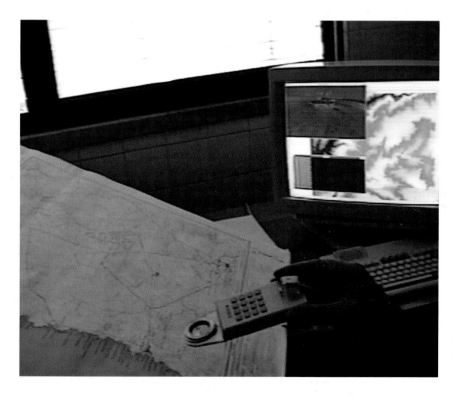

Plate 40
Virtual GIS Room project
interface components.
(*Source*: New University of Lisbon)

Introduction

THE EDITORS

A recurrent theme throughout the first part of this book is that the inherent complexity of the geographical world makes it virtually impossible for any digital representation to be complete, however limited its scope. Although some exceptions exist (we can, for example, create a perfect digital representation of the latitude of the Equator, or a line on the Earth's surface that is by definition straight), there will otherwise be differences between the database contents and the phenomena they represent. Various terms are used to describe these differences, depending on the context. Differences can exist because of errors of measurement, while the term 'uncertainty' seems more appropriate if the digital representation is simply incomplete. More generally, one might simply refer to the 'quality' of the representation.

If data quality is an important property of almost all geographical data, then it must affect the decisions made with those data. In general, the poorer the quality of the data, the poorer the decision. Bad decisions can have severe consequences, as when an ambulance is sent to the wrong location, or a school is inadvertently built over an abandoned storage facility for hazardous waste. Geographical data are often used for regulatory purposes, or to resolve disputes: the custodians of such data are clearly exposed to potential liability if the data are shown to be in error.

Despite what appear to be obvious arguments in favour of explicit treatment of data quality in GIS, and despite substantial research into appropriate methods, much GIS practice continues to proceed as if data were perfect. Results of GIS analysis – whether in the form of tables, maps, or displays – rarely show estimates of confidence, or other indicators of the effects of data quality. In part, such attitudes have been inherited from cartographic practice, since it is often difficult to determine the quality of mapped information. In part, they may also reflect a general tendency to give computers more credit than they deserve – to believe that because numbers or maps have emerged as if by magic from digital black boxes, they must necessarily be reliable.

This section contains four chapters that together represent the state of the geographical data quality art, or, more accurately, science. Howard Veregin presents in Chapter 12 an overview of the components of data quality; their interactions and dependencies; and the efforts that have been made in recent years to embed them in standards. From the perspective of the data producer, quality refers to the difference between the actual characteristics of the product and the relevant specifications that define it, or the claims made about it. Information on quality is immensely useful in managing the production process, particularly if the results of quality analysis point back to suspect sources. On the other hand, details of the production process may be of only marginal interest to a potential user of the data, who is concerned solely with whether the data meet particular requirements. Data quality can thus vary from user to user, depending on respective needs; and the effective measurement and documentation of data quality against needs that are often poorly defined can be an immensely complex and frustrating process.

The problems of determining data quality have been further complicated in recent years by the growth of new communication technologies. These have made it far easier for data to be found, accessed, and shared. The user of a geographical dataset may now be many steps removed from the producer. User and producer may be from entirely different backgrounds, with very little in the way of shared terminology or culture. Even if the data are well documented, the lack of effective systems for documentation, in the form of metadata, may leave the user with an incomplete or incorrect understanding of the meaning of the data. For example, if the units of measurement of a variable are not documented, or if the documentation is not transferred to the user, then from the perspective of the user the data are now subject to a further source of inaccuracy. To the user at the end of a long chain of communication, data quality is most appropriately defined as a measure of the difference between the database's contents and the user's understanding of their true values. The same collection of bits can have different levels of quality, both increasing and decreasing, as it passes from one custodian to another.

In Chapter 13, Peter Fisher discusses alternative models of uncertainty. The traditional scientific concept of measurement error, which accounts for differences between observers or measuring instruments, turns out to be far too simple as a framework for understanding quality in geographical data. Many geographical concepts are incompletely specified – as for example when population density at a point is reported, without specifying the area over which the density was measured – and such incompleteness of specification is an appropriate component of data quality. Many concepts are poorly defined, leading to understandable disagreement between observers. In this context it is useful to distinguish, as Fisher does, between such terms as 'vague', 'fuzzy', and 'probable'. Both fuzzy set theory and probability theory have been found to be useful in modelling uncertainty in geographical data, although their axioms differ in several key respects.

If agreement can be reached on how to measure and express data quality, then such information should be made available to users, preferably by storing it as part of, or in conjunction with, the database. Quality measures that are true of the entire contents are conveniently stored as part of the database metadata, the digital equivalent of documentation. But other quality measures may be true only of parts of the database, such as classes of objects, or individual objects, or even parts of objects, or regions of the study area. In such cases, it is necessary to have 'slots' in the database available to store data quality information in appropriate, meaningful ways. Such slots might take the form of additional data quality attributes of objects, or components of an object class's description; or it might even be necessary to create a complete map of data quality, showing how quality varies across the study area. Thus data quality becomes a significant part of the representation itself.

With adequate information available on data quality, it is possible to determine its effects on the results of GIS analysis, and for decisions made with GIS to reflect the uncertainties present in the base data. This topic of error propagation is the subject of Chapter 14, by Gerard Heuvelink. Several general strategies for error propagation are proposed, at least one of which will be valid in any context. Typically the error propagation is hidden from the user, who sees only a standard GIS function, such as 'compute slope', but is presented with results that include both the requested estimates of slope, and measures of confidence or uncertainty in the results. Software for error propagation is increasingly available in the GIS world, often in the form of specialised 'add-ons' written in a GIS's macro or scripting language.

In the final chapter in this section, Kate Beard and Barbara Buttenfield review techniques that have been developed for visualising uncertainty, issues raised by their use, and problems requiring additional research. Traditional cartographic practice includes remarkably few methods for visualising uncertainty; whether this is because it is difficult to do so within the constraints of map-making, or whether it reflects a human desire to see the world as simpler than it really is, remains a subject of debate. What is beyond doubt, however, is that the continuation of such practices in the world of GIS is both technically and ethically indefensible. The digital world is far more flexible, and Beard and Buttenfield illustrate many of the methods that have been proposed and implemented by the research community. There have been experiments with sound, animation, and use of the third dimension, each with attendant advantages and disadvantages.

Despite such progress on the research front, the issues of dealing with uncertainty remain. GIS has been adopted by individuals and agencies who see its benefits in terms that often include increased accuracy compared to previous methods; yet the data stored in GIS are in most cases no more accurate. Suppose, for example, that the research community were to suggest, on theoretically defensible grounds, that the only effective method for visualising uncertainty would be to present the user with several equally likely versions of how the world might actually look. Uncertainty in soil mapping could be presented by showing ten alternative, equally likely maps of the same area. While this might make perfect sense from the perspective of error theory, it would be almost completely alien to a culture raised on single, apparently exact maps. The problems of coping with uncertainty, and of introducing its effective treatment from a managerial perspective, are the subject of Chapter 45 by Gary Hunter.

12

Data quality parameters

H VEREGIN

This chapter focuses on the definition and measurement of quality components for geospatial databases. A basic distinction is drawn between quality control and truth-in-labelling paradigms. Components of data quality – accuracy, precision, consistency, and completeness – are defined in the context of geographical data. Treatment of quality components in data standards is discussed and the implications of cartographic bias in geospatial data models are briefly addressed. The chapter ends with a discussion of the ways in which institutional values are embedded in geospatial databases and the ways that data quality documentation can help to articulate these values.

1 DEFINING DATA QUALITY

The meaning of 'quality' depends on the context in which it is applied. The term is commonly used to indicate the superiority of a manufactured good or attest to a high degree of craftsmanship or artistry. In manufacturing industries, quality is viewed as a desirable goal to be achieved through management of the production process. Statistical quality control has a relatively long history in manufacturing, where it is used to ensure conformity in products by predicting the performance of manufacturing processes.

Quality is more difficult to define for data. Unlike manufactured products, data do not have physical characteristics that allow quality to be easily assessed. Quality is thus a function of intangible properties such as 'completeness' and 'consistency'. On further reflection, however, these differences are perhaps not as great as they might initially seem. After all, data are the result of a production process, and the manner in which this process is performed clearly affects data reliability. Data consumers can therefore use the same diligence in selecting a database that they might in purchasing an automobile or a pair of shoes.

These comments also apply in the context of geospatial data. Concern for geospatial data quality has burgeoned in recent years for the following reasons:

- Increased data production by the private sector. Historically, mass production of geospatial data was the domain of governmental agencies such as the US Geological Survey (USGS) and the British Ordnance Survey (Rhind, Chapter 56). Unlike these agencies, private companies are not required to conform to known quality standards (Goodchild and Longley, Chapter 40).

- Increased use of GIS as a decision-support tool. This trend has led to realisation of the potential deleterious effects of using poor quality data, including the possibility of litigation if minimum standards of quality are not attained (Onsrud, Chapter 46).

- Increased reliance on secondary data sources. This has been fuelled by a reduction in accessibility and cost constraints resulting from network accessibility and the development of standards for data exchange (Goodchild and Longley, Chapter 40).

These trends have contributed to a reappraisal of the responsibilities of data producers and consumers for data quality. Until quite recently data quality was the responsibility of the producer, and compliance testing strategies were applied in order to sanctify databases meeting official quality thresholds. Compliance testing is a form of quality control that seeks to eliminate error through management of the database production process. However, compliance tests are useful only in a limited range of applications environments. For some applications a particular compliance test may be too lax while for others it may be too restrictive and hence impart unnecessary costs.

Responsibility for assessing whether a database meets the needs of a particular application has therefore shifted to the consumer who is in a position to make such an assessment. This is referred to as determining 'fitness-for-use'. The producer's responsibilities have changed as well. Rather than producing authoritative databases, the producer's role has shifted to data quality documentation or 'truth-in-labelling'. The truth-in-labelling paradigm views error as inevitable and casts the data quality problem in terms of misuse arising from incomplete knowledge of data limitations.

2 DATA QUALITY COMPONENTS

Geographical observations describe phenomena with spatial, temporal, and thematic components (Berry 1964; Sinton 1978). Space, which defines geographical location, is the dominant member of this troika. This dominance is problematic on several levels. First, time is not given sufficient attention. Although poorly accommodated in conventional geospatial data models, time is critical to an understanding of geographical phenomena, not as entities that exist at some location, but as events that appear and disappear in space and time (Peuquet, Chapter 8; Raper, Chapter 5). A second problem is that geographical phenomena are not really about space, but about theme. We can view space (or more precisely space-time) as a framework on which theme is measured. It is true that without space there is nothing geographical about the data, but on the other hand without theme there is only geometry.

These comments set the stage for our discussion of data quality components. Like geographical phenomena, data quality can be differentiated in space, time, and theme. For each of these dimensions, several components of quality (including accuracy, precision, consistency, and completeness) can be identified.

2.1 Accuracy

A useful starting point for discussing accuracy is the entity–attribute–value model, which serves as the conceptual basis for most database implementations

of real-world phenomena. According to this model, 'entities' represent real-world phenomena (such as streets, counties, or hazardous waste sites), 'attributes' specify the relevant properties of these objects (such as width or number of lanes), and 'values' give the specific qualitative or quantitative measurements pertaining to a particular attribute. In this model, error is defined as the discrepancy between the encoded and actual value of a particular attribute for a given entity (see also Fisher, Chapter 13). Accuracy is the inverse of error. This model can be used to define spatial, temporal, and thematic error for a particular entity as, respectively, the discrepancies in the encoded spatial, temporal, and thematic attribute values.

This definition is useful but somewhat limited. What is missing is recognition of the interdependence of space, time, and theme. Geographical phenomena are not just thematic data with space and time attached. They are instead events unfolding over space and time. A change in space or time implies a change in theme, and vice versa. Thus while accuracy can be measured separately for space, time, and theme, these measurements are not necessarily independent. Consider a database dated '1992' that depicts a two-lane road, but assume that in late 1991 the road was converted to a four-lane highway. This is both a thematic error (because in 1992 there were four lanes, not two) and a temporal error (because when the road contained only two lanes the year was at most 1991). Similar types of dependencies exist across space and theme. A classic example is the soil mapping unit delineation problem, in which a mislocated unit boundary is simultaneously a spatial error and a thematic error, since boundary location is defined by variations in thematic attribute value.

The definition of error given above assumes that there is some objective, external reality against which encoded values can be measured (Chrisman 1991). This definition requires not only that 'truth' exists but that it can be observed. Quite apart from any philosophical problems that it raises, this definition is problematic for several reasons. First, the truth may simply be unobservable, as in the case of historical data. Second, observation of the truth may be impractical (because of data cost, for example). Finally, it is possible that multiple truths exist because the entities represented in the database are abstractions rather than real-world phenomena. Indeed many phenomena of interest belong to perceived reality (sometimes referred to as *terrain nominal*: Salgé 1995). Examples include entities that are highly variable (e.g. shorelines) or subjective in nature (e.g. land cover classes interpreted from air photos). In these cases inexactness is a fundamental property of the phenomena under observation (Goodchild 1988b).

Fortunately, objective reality does not need to be articulated in order to perform accuracy assessment. This is because geospatial data are always acquired with the aid of a model that specifies, implicitly or explicitly, the required level of abstraction and generalisation relative to real-world phenomena (Figure 1; Martin, Chapter 6). This conceptual model defines the database 'specification' and it is against this reference that accuracy is assessed (Brassel et al 1995). Accuracy is a relative measure rather than an absolute one, since it depends on the intended form and content of the database. Different specifications can exist for the same general types of geospatial data. To judge the fitness-for-use of the data for some applications, one must not only judge the data relative to the specification, but also consider the limitations of the specification itself (Comité Européen de Normalisation (CEN) 1995).

2.1.1 Spatial accuracy

Spatial accuracy (or 'positional accuracy') refers to the accuracy of the spatial component of a database. Measurement of spatial accuracy depends on dimensionality. Metrics are well defined for point entities, but widely accepted metrics for lines and areas have yet to be developed. For points, error is usually defined as the discrepancy (normally Euclidean distance) between the encoded location and the location as defined in the specification. Error can be measured in any one of, or in combinations of, the three dimensions of space. The most common measures are horizontal error (distance measured in x and y simultaneously) and vertical error (distance measured in z) (Figure 2).

Various metrics have been developed to summarise spatial error for sets of points. One such metric is mean error, which tends to zero when 'bias' is absent. Bias refers to a systematic pattern of error (e.g. error arising from map misregistration). When bias is absent error is said to be random. Another common metric is root mean squared error (RMSE), which is computed as the square root of the mean of the squared errors (see Beard and Buttenfield, Chapter 15). RMSE is commonly used to document vertical accuracy for digital elevation models (DEMs). RMSE is a measure of the magnitude of error but it does not incorporate bias since the squaring eliminates the direction of the error.

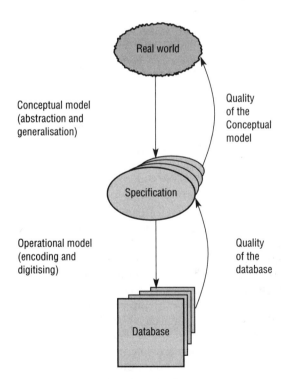

Fig 1. The mediating role of the database specification in assessing data quality.

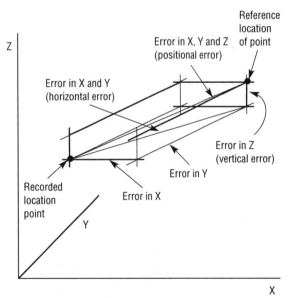

Fig 2. Measuring components of spatial error.

There is a close analogy between classical approaches to error and error in the location of a point. Horizontal error is a 2-dimensional extension of the classical error model in which error in position is defined in terms of a bell-shaped probability surface (Goodchild 1991a). Thus it is possible to perform statistical inference tests and derive confidence limits for point locations (American Society of Civil Engineers 1983; American Society for Photogrammetry 1985). For lines and areas the situation is more complex since there is no simple statistical measure of error that can be adopted from statistics. Errors in lines arise from the errors in the points that define those lines. However, as these points are not randomly selected the errors present at points cannot be regarded as somehow typical of errors present in the line (Goodchild 1991b).

Error is usually defined for lines using some variant of the epsilon band. The epsilon band is defined as a zone of uncertainty around an encoded line within which there is a certain probability of observing the 'actual' line. As yet there is no agreement as to the shape of the zone and the distribution of error within it. Early models assumed that the zone was a uniform 'sausage' within which the distribution of error was uniform (Blakemore 1983; Chrisman 1982). More recent studies show that both the distribution and the band itself might be non-uniform in shape (Caspary and Scheuring 1993; Honeycutt 1986) (Figure 3).

2.1.2 Temporal accuracy

Temporal accuracy has not received much attention in the literature, just as time itself is not dealt with explicitly in conventional geospatial data models. Temporal accuracy is often equated with 'currentness' (Thapa and Bossler 1992). In fact the two concepts are quite distinct. Temporal accuracy refers to the agreement between encoded and 'actual' temporal coordinates. Currentness is an application-specific measure of temporal accuracy. A value is current if it is correct in spite of any possible time-related changes in value. Thus currentness refers to the degree to which a database is up to date (Redman 1992). To equate temporal accuracy with currentness is to state, in effect, that to be temporally accurate a database must be up to date. Clearly this is not the case since a database can achieve a high level of temporal accuracy without being current. Indeed historical studies depend on the availability of such data.

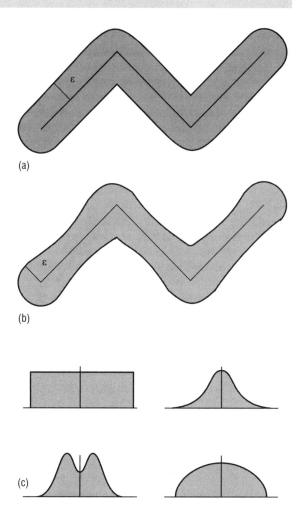

Fig 3. (a) Early models of the epsilon band show a uniform 'sausage' of width epsilon, ε, surrounding the encoded line; (b) more recent studies suggest that the band may be non-uniform in width; and (c) four of the many possible distributions of error around the encoded line.

Assessment of temporal accuracy depends on the ability to measure time objectively using a standard temporal coordinate system. However, standards are not universally accepted (Parkes and Thrift 1980). Another impediment to the measurement of temporal accuracy is that time is often not dealt with explicitly in geospatial databases. Temporal information is often omitted, except in databases designed for explicitly historical purposes. This assumes that observations are somehow 'timeless' or temporally invariant. The implications of this omission are potentially quite significant, especially for features with a high frequency of change over time.

2.1.3 Thematic accuracy

Metrics of thematic accuracy (or 'attribute accuracy') vary with measurement scale. For quantitative attributes, metrics are similar to those used to measure spatial accuracy for point features (e.g. RMSE). Quantitative attributes can be conceived as statistical surfaces for which accuracy can be measured in much the same way as for elevation. For categorical data most of the research into data quality has come from the field of classification accuracy assessment in remote sensing. This work was carried out initially to devise methods to assess the accuracy of classification procedures. Accuracy assessment is based on the selection of a sample of point locations, and a comparison of the land cover classes assigned to these locations by the classification procedure with the classes observed at these locations on a reference source (usually 'ground truth'). A cross tabulation of the results (the 'classification error matrix') permits accuracy assessment (Aronoff 1985; Genderen and Lock 1977).

Various metrics summarising the information in the error matrix have been developed (proportion correctly classified, kappa, user's and producer's accuracies, etc.). These metrics are useful for assessing overall thematic accuracy. The classification error matrix contains additional information on the frequency of various types of misclassification, e.g. which pairs of classes tend most often to be confused. In addition, the matrix permits assessment of errors of omission (omission of a location from its 'actual' class) and errors of commission (assignment of a location to an incorrect class).

2.2 Precision or resolution

Precision refers to the amount of detail that can be discerned. It is also known as granularity or resolution. The latter term is commonly used in GIS and related fields, and is adopted here to avoid confusion with the statistical concept of precision as observational variance. All data are of limited resolution because no measurement system is infinitely precise. Resolution is also limited because geospatial databases are intentionally generalised. Generalisation includes elimination and merging of entities, reduction in detail, smoothing, thinning, and aggregation of classes. Generalisation is inevitable because, at best, geospatial databases can encompass only a fraction of the attributes and their relationships that exist in the real world (Weibel and Dutton, Chapter 10).

Resolution affects the degree to which a database is suitable for a specific application. The resolution of the database must match the level of detail required in the application. Low resolution does not have the same negative connotation as low accuracy. Low resolution may be desirable in certain situations, such as when one wishes to formulate general models or examine spatial patterns at a regional level.

Resolution is also important because it plays a role in interpreting accuracy. For example, two databases may have approximately equal spatial accuracy levels, but if their spatial resolutions are significantly different then the accuracy levels do not denote the same level of quality. One would generally expect accuracy and resolution to be inversely related, such that a higher level of accuracy will be achieved when the specification is less demanding.

2.2.1 Spatial resolution

The concept of spatial resolution is well developed in the field of remote sensing, where it is defined in terms of the ground dimensions of the picture elements, or pixels, making up a digital image (Figure 4). This defines the minimum size of objects on the ground that can be discerned. The concept is applicable without modification to raster databases. For vector data, the

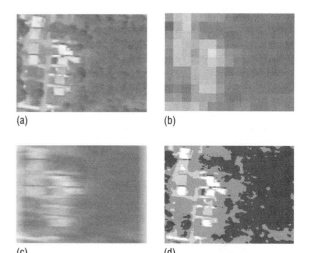

(a) (b)

(c) (d)

Fig 4. (a) A portion of a video image (Akron, Ohio) with spatial resolution of 1 metre, temporal resolution of 1/30 of a second, and thematic resolution of 8 bits (255 unique values); (b) the same image as in (a) but with spatial resolution degraded to 10 metres; (c) the same image as in (a) but with temporal resolution degraded, thus showing the effects of along-track movement of the sensing platform; and (d) the same image as in (a) but with thematic resolution degraded to four unique values.

smallest feature that can be discerned is usually defined in terms of rules for minimum mapping unit size which depend on map scale.

Spatial resolution is related to, but distinct from, the concept of the spatial sampling rate. Resolution refers to the fineness of detail that can be observed while the sampling rate defines the ability to resolve patterns over space. For remotely sensed images, resolution refers to the pixel size (ground area resolved) and sampling rate to the spaces between pixels. Thus in theory one could mix high spatial resolution with low sampling rate (small pixels with large gaps between them) or low spatial resolution with high sampling rate (large pixels that overlap). Normally, resolution and sampling rate are approximately equal.

2.2.2 Temporal resolution

Temporal resolution refers to the minimum duration of an event that is discernible. It is affected by the interaction between the duration of the recording interval and the rate of change in the event. Events with a lifetime less than the sampling interval are generally not resolvable. At best they leave a 'smudge' like pedestrians on nineteenth-century daguerreotypes. This has been referred to as the 'synopticity' problem (Stearns 1968). A shorter recording interval implies higher temporal resolution, just as faster film has given us the ability to photograph quickly moving objects (Figure 4).

For geospatial data, the situation is more complicated because interactions between spatial and thematic resolution must also be considered. In general one cannot resolve any event which, during the time interval required for data collection, changes location in space by an amount greater than the spatial resolution level. Likewise, one cannot resolve any event for which theme changes to a degree that would be discernible given the thematic resolution level (Veregin and Hargitai 1995).

There is a clear distinction between resolution and sampling rate in the temporal domain. Sampling rate refers to the frequency of repeat coverage while resolution refers to the time collection interval for each measurement. For example, motion pictures have a resolution of perhaps a thousandth of a second (one frame) but a sampling rate of 24 frames per second. Geosynchronous satellites are capable of much higher sampling rates than sun-synchronous satellites (repeat coverage several times per minute vs several times per month). Resolution, however, is a function of the time required to obtain spectral reflectance data for one pixel.

2.2.3 Thematic resolution

In the thematic domain, the meaning of resolution depends on measurement scale. For quantitative data, resolution is determined by the precision of the measurement device (Figure 4). For categorical data, resolution is defined in terms of the fineness of category definitions. Land cover classification systems used in remote sensing are useful models to illustrate resolution. These systems define the level of detail in taxonomic definitions in terms of the spatial resolving power of the remote sensing system. This illustrates the interdependence between space and theme when extracting spatial information (land cover class boundaries) from thematic information (spectral reflectance data).

2.3 Consistency

Consistency refers to the absence of apparent contradictions in a database. For geospatial data the term is used primarily to specify conformance with certain topological rules (Kainz 1995). These rules vary with dimensionality; for example, only one point may exist at a given location, lines must intersect at nodes, polygons are bounded by lines, etc. Elimination of topological inconsistencies is usually a prerequisite for GIS processing (Dowman, Chapter 31), such that most databases are topologically 'cleaned' before being released.

Topological consistency is one aspect of consistency in the spatial domain. Spatial inconsistencies can also be identified through redundancies in spatial attributes. For example, an entity might have the value 'Delaware' for the attribute 'state' but the value 'Lincoln' for the attribute 'county'. This is inconsistent since there is no Lincoln county in Delaware. In this case redundancy is partial: the state 'Delaware' eliminates the possibility of the county 'Lincoln', but the county 'Lincoln' does not necessarily imply the state 'Maine' since Maine is only one of 24 states containing a Lincoln County. On the other hand, redundancy may be complete (e.g. state is implied completely by the Federal Information Processing Standard (FIPS) state code), since there is a unique state code for each state. Non-redundancy implies that there is independence between two attributes such that meaningful consistency constraints do not exist (Redman 1992).

Little work has been done on consistency in the temporal domain, although a framework for temporal topology has been developed (Langran 1992). For example, since at a given location only one event can occur at one time, an inconsistency exists if a different entity appears at the same location on two maps of the same date. Since events have a duration, this idea can be extended to identify events that exhibit temporal overlap.

In the thematic domain, the ability to identify inconsistencies requires a level of redundancy in thematic attributes – for example, the three sociodemographic variables 'population', 'mean household size', and 'total number of households'. Of course, the identification of an inconsistency does not necessarily imply that it can be corrected or that it is possible to identify which attribute is in error. Note also that the absence of inconsistencies does not imply that the data are accurate. Thus consistency is appropriately viewed as a measure of internal validity. Despite the potential to exploit redundancies in attributes, tests for thematic consistency are almost never carried out.

2.4 Completeness

Completeness refers to the relationship between the objects in the database and the 'abstract universe' of all such objects. Selection criteria, definitions, and other mapping rules used to create the database are important determinants of completeness. This definition requires a precise description of the abstract universe since the relationship between the database and the abstract universe cannot be ascertained if the objects in the universe cannot be described. The abstract universe can be defined in terms of a desired degree of abstraction and generalisation (i.e. a concrete description or specification for the database). This leads to the realisation that there are in fact two different types of completeness. 'Data completeness' is a measurable error of omission observed between the database and the specification. Data completeness is used to assess data quality, which is application-independent. Even highly generalised databases can be complete if they contain all of the objects described in the specification. 'Model completeness' refers to the agreement between the database specification and the abstract universe that is required for a particular database application (Brassel et al 1995). Model completeness is application-dependent and therefore an aspect of fitness-for-use. It is also a component of 'semantic accuracy' (Salgé 1995).

Additional distinctions are required. The definitions of completeness given above are examples of 'feature or entity completeness'. In addition we can identify 'attribute completeness' as the degree to which all relevant attributes of a feature have been encoded. A final type of completeness is 'value completeness' which refers to the degree to which values are present for all attributes (Brassel et al 1995).

Feature completeness can be defined over space, time, or theme. Consider a database depicting the locations of buildings in the state of Minnesota that were placed on the National Register of Historic Places as of 1995. This database would be incomplete if it included only buildings in Hennepin County (incompleteness in space, since Hennepin County covers only a portion of Minnesota), or only buildings placed on the Register by June 30 (incompleteness in time, since buildings may have been added after June 30), or only residential buildings (incompleteness in theme, due to the omission of non-residential buildings).

As this example shows, completeness is typically defined in terms of errors of omission. However, completeness may also include errors of commission (CEN 1995). Following on the previous example, errors of commission would occur if the database contained buildings in Wisconsin, buildings added to the Register in 1996, or historic districts as well as buildings.

3 DATA QUALITY STANDARDS

A concern for data quality issues is clearly expressed in the development of data transfer and metadata standards. Such standards have been developed at both national and international levels in support of mandates for data acquisition and dissemination. Data quality documentation plays a key role in many standards due to the realisation that an understanding of quality is essential to the effective use of geospatial data (see also Salgé, Chapter 50).

US readers will be most familiar with SDTS (the Spatial Data Transfer Standard) and the Content Standards for Digital Geospatial Metadata developed by the FGDC (Federal Geographic Data Committee). SDTS is a data transfer standard designed to facilitate dissemination and sharing of data. It provides standard definitions of data elements, a standardised format for data transfer, and descriptive metadata about database contents. In 1992 SDTS was adopted by the National Institute of Standards and Technology as a Federal Information Processing Standard (FIPS-173) (Fegeas et al 1992).

The FGDC was established to promote coordinated development and dissemination of geospatial data. Its membership includes numerous US federal government departments and independent agencies. The FGDC has been involved in several activities related to geospatial data quality, including the development of the metadata content standards. Metadata describe the contents of a database. The FGDC standards provide a common set of terminology and a common structure for geospatial metadata (FGDC 1994). The FGDC standards were approved in 1994, and use of these standards is one of the minimum requirements for serving as a node in the National Geospatial Data Clearinghouse of the National Spatial Data Infrastructure (NSDI) (Morain and Budge 1996).

The FGDC standards follow SDTS in terms of recommendations for data quality information to be reported and tests to be performed. The five components of data quality in SDTS are listed in Table 1. Text-based documentation is the norm, although other formats are also permitted including numerical measures and even interactive graphics through online resources.

Many organisations have also created internal standards that contain data quality information. For example, the USGS DEM standard includes descriptors of horizontal and vertical accuracy. Standards have been adopted or are in development at national and international levels as well. Examples include the National Transfer Format (NTF) developed by the Association for Geographic Information (AGI) and adopted as the official British standard (BS7666) in 1992; the Digital Geographic Information Exchange Standard (DIGEST) developed by military service agencies from a number of NATO countries; the International Hydrographic Organisation (IHO) standard for nautical charts; and the draft standard of the CEN. Interested readers should consult Salgé (Chapter 50), Cassettari (1993), and Moellering (1991) for more details.

A major limitation of data quality standards is that they do not necessarily lend themselves to specific software implementations (see Guptill, Chapter 49). Standards provide models for data documentation but not a mechanism whereby users of disparate GIS packages can implement these models for database documentation. A related problem is that standards treat data quality as essentially static. While some accommodation is made for changes in quality as a result of data transformations, there is no mechanism to automatically update quality components as data are

Table 1 Data quality components in SDTS.

Component	Description
Lineage	Refers to source materials, methods of derivation and transformations applied to a database. • Includes temporal information (date that the information refers to on the ground). • Intended to be precise enough to identify the sources of individual objects (i.e. if a database was derived from different source, lineage information is to be assigned as an additional attribute of objects or as a spatial overlay).
Positional accuracy	Refers to the accuracy of the spatial component. • Subdivided into horizontal and vertical accuracy elements. • Assessment methods are based on comparison to source, comparison to a standard of higher accuracy, deductive estimates or internal evidence. • Variations in accuracy can be reported as quality overlays or additional attributes.
Attribute accuracy	Refers to the accuracy of the thematic component. • Specific tests vary as a function of measurement scale. • Assessment methods are based on deductive estimates, sampling or map overlay.
Logical consistency	Refers to the fidelity of the relationships encoded in the database. • Includes tests of valid values for attributes, and identification of topological inconsistencies based on graphical or specific topological tests.
Completeness	Refers to the relationship between database objects and the abstract universe of all such objects. • Includes selection criteria, definitions and other mapping rules used to create the database.

passed through GIS processing steps. While source data may be adequately documented, derived data frequently are not. Finally, because standards such as SDTS provide such a rich collection of information about data quality, users may find it difficult to ascertain fitness-for-use. Likewise the unstructured nature of text-based descriptions means that data quality documentation is difficult to update automatically in a GIS environment.

Data quality standards also fall short of providing the kinds of assurances demanded by agencies that need to limit liability risks (Goodchild 1995). For example, SDTS follows the 'truth-in-labelling' paradigm in which the data quality report makes no *a priori* assumptions about quality requirements. While SDTS documentation might contain statements that the data meet some minimum accuracy standard, SDTS itself does not provide for the definition of data quality objectives necessary in the development of quality assurance/quality control (QA/QC) programs.

Efforts are underway to establish QA/QC programs within agencies that produce geospatial data. Such programs are based on the development of standard operating procedures that allow specific data quality objectives to be realised (Stone et al 1990). To some extent such QA/QC programs mirror the way in which traditional map accuracy standards such as National Map Accuracy Standards (NMAS) are implemented. The NMAS guarantee of a minimal level of positional accuracy is achieved through standard operating procedures that are known to yield the required accuracy levels, coupled with a limited amount of actual compliance testing. Such approaches focus on managing the production process rather than on statistical measurement of quality.

4 METADATA SYSTEMS

Like data quality standards, metadata systems are concerned with documentation of data quality components. The essential difference is that metadata systems emphasise the operational component rather than conceptual issues. Most commercial GIS packages perform a certain amount of metadata documentation. Some metadata is essential in order that data are processed correctly. For example, raster systems need to record the number of rows and columns of cells in each layer, while vector systems need to record the spatial

coordinate system. Often these metadata are propagated forward as new layers are derived (see also Church, Chapter 20).

Only a few commercial GIS packages offer the capability to document data quality. An example is Idrisi version 4.1 which allows users to store information on the five components of data quality defined in SDTS. These data are stored along with other metadata in the documentation file that accompanies each raster layer and are propagated forward to derived layers. The software also performs rudimentary error propagation modelling by transforming metadata for certain data quality components.

For the majority of systems, however, tracking of data quality is the responsibility of the user. This has led to the independent development of software packages that document layers with metadata, update the lineage of layers automatically and perform propagation of data quality components (Veregin 1991). Some systems are quite advanced. Geolineus is an intelligent system that intercepts GIS commands and dynamically builds a graphical representation of the data processing flow and derived layers (Lanter 1991). This allows the user to visualise the flow of data processing steps and the linkages between source and derived data. At the same time Geolineus automatically propagates metadata, including data quality elements. This replaces the traditional approach in which updating of metadata is the sole responsibility of the user, such that it is often not performed at all (Goodchild 1995). Geolineus also stores information about data dependencies to facilitate metadata analysis. Examples of metadata analysis include assessment of processing complexity, analysis of the adequacy of data sources, propagation of error, and the identification of optimal strategies for enhancing derived data quality (Lanter and Surbey 1994; Lanter and Veregin 1992; Veregin and Lanter 1995).

5 CARTOGRAPHIC BIAS

The ability to produce a geospatial database presupposes a model that defines rules for simplifying real-world complexity. Despite their apparent sophistication, geospatial databases reflect many of the same biases as analogue cartographic data. This is true not only because geospatial databases are often produced by digitising paper maps, but because the models embedded in GIS are essentially digital translations of analogue models (Burrough and Frank 1995). Thus in the vector data model geographical phenomena are differentiated according to their dimensionality. Only points, lines, and areas are permitted and these three classes are assumed to be mutually exclusive even though the dimensionality of many entities is known to be scale-dependent (Hutchinson and Gallant, Chapter 9). Dimensionality, originally applied in cartography as a heuristic for representation and symbolisation, has been reified in GIS as a fundamental property of geographical phenomena.

The finite limits to cartographic fidelity imply that maps must represent the real world selectively – that is, they can represent only a subset of all possible entities in the real world and must portray these entities in a generalised way as a function of map scale and purpose. The model is a highly abstract one that assumes that entities exist unambiguously in the real world. In some cases this is true, as with roads, counties, and other anthropogenic phenomena. However, in many cases the phenomena of interest have imprecise geographical expression. These phenomena belong to the perceived world rather than the real world and are inherently inexact and imprecise (Burrough 1986). Such phenomena are accommodated only clumsily in the cartographic model, through the introduction of concepts such as mapping unit 'purity' and 'minimum mapping unit size' which acknowledge that the real world is more complex than cartographic data would allow.

In theory geospatial data are not constrained to the same degree as paper maps. Many authors argue that geospatial technology is liberating as it facilitates new modes of representation and offers freedom from the constraints of scale imposed by pen-and-ink technology. An example is the raster model, which evolved in such fields as television and remote sensing, and represents a significant break from the traditional object-based cartographic model (Goodchild 1988a). It is perhaps not surprising then that many alternate models of geospatial data, such as the field-based model, probabilistic surfaces, and models based on fuzzy set theory, are raster based. These models are able to accommodate imprecision and uncertainty more easily than conventional cartographic models, and are thus more appropriate for many geographical phenomena.

Technology has also loosened the restriction that maps serve the dual purposes of storage and communication. For paper maps, content depends on the communication goal. The desire to communicate a particular message leads to selective enhancement and exaggeration of certain features and elimination or displacement of others. In geospatial databases the storage and communication roles can be more easily separated. This means that data can be collected in as raw a form as possible, and representations can be created to achieve any particular communication objective without altering the contents of the database. An additional advantage is that it is easier to quantify accuracy for raw data than for abstract cartographic representations (Goodchild 1988c).

These problems would not come to the fore if GIS were used only as an electronic map drawer. However, GIS has enormously extended the uses of geospatial data. Once data make their way into GIS they typically begin a process of metamorphosis in which they are transformed and merged with other data in support of queries, analyses, and decision-making models. Unfortunately there is no guarantee that the data are suitable for such applications. This problem is sometimes referred to as 'use error' (Beard 1989). Despite the advances we have made in understanding components of data quality, we have made almost no progress in the development of rules and heuristics to assess fitness-for-use and prevent use error (see Beard and Buttenfield, Chapter 15).

6 GIS, SOCIETY, AND DATA QUALITY

What is the essence of a geospatial database? Is it a faithful image of reality or a rhetorical device designed to convey a particular message? Is it an impartial representation of objective truth or a manifesto for a set of beliefs about the world? This is a central issue in the burgeoning 'GIS and society' debate in which research on data quality has many important implications (Pickles, Chapter 14).

According to some critics, technologies such as GIS have led to the ascendance of a new geospatial science focused on the goal of producing ultimately truthful and objective representations of reality. This goal is seen as a byproduct of the new technological means with its appeals to neo-positivism, reductionism, instrumentalist thinking, and naive empiricism in which 'reality' is uncontested and objectively measurable (e.g. Harley 1991; Wood 1992). According to this view, producers of geospatial databases make no allowance for the possibility that these databases embed specific social and institutional values. As such, GIS promulgates the myth of an objective science which always produces the best delineations of reality (Harley 1989).

While there is some foundation to this critique, it would be unfair to suggest that producers of geospatial data are unaware of the limitations of these data. Like their manually-produced map counterparts, geospatial data are not intended to be miniature replicas of 'reality'. Rather they emphasise some aspects of the environment and suppress others in an effort to convey a particular message (Martin, Chapter 6; Raper, Chapter 5). What is contained in a database is a function not only of the nature of the external environment but also the values of the society and institution within which the database was constructed (Turnbull 1989). Values are embedded at the modelling stage, where they impact on database content, and at the representation stage where they affect database form.

Values are not always embedded deliberately. Broad social values are often taken for granted and may not be consciously recognised. Hence databases often unintentionally reflect and legitimate the social order (Harley 1989). Broad social values form the backdrop for more specific values that reflect institutional characteristics. Perhaps the most significant of these is institutional mandate, which defines institutional mission for data collection and dissemination. For specific databases, mandate is formalised as a set of design guidelines that outline the rules for data collection, encoding, and representation.

Unlike broad social values, values deriving from institutional mandate can be articulated, documented, and communicated to database consumers through the medium of metadata. This communication process is important since it affects the consumer's understanding of the limitations of a database and facilitates its appropriate use. Especially useful in this context is the concept of the 'specification' describing the intended contents of the database. The specification is the reference standard against which the database is compared in order to assess completeness and other data quality components. The specification concept explicitly recognises that each database has a particular set of objectives and that embedded in these objectives is the formal expression of the values associated with institutional factors.

What are the implications for the debate over values? First, geospatial databases are not intended to be accurate mirrors of reality. Rather, they are designed to conform to a database specification which could just as easily be a description of perceived reality. Second, geospatial data producers are generally aware of the significance of values. The database specification is in fact a formal statement of the values that are embedded in a given database. Third, values can be communicated to database consumers who can then use this information to assess the appropriateness of the database for a particular task. Knowledgeable map users have of course always been aware of data limitations.

These are important conclusions since the alternatives are not particularly attractive. For example, some critics have claimed that given the dependence on social values it is not possible to distinguish between competing representations of the same geographical space. Thus it has been argued that the distinction between propaganda and truth is artificial and must be dismantled, as must the arbitrary dualism between art and science (Harley 1989). According to this view, all representations are equally valid since they are all expressions of one's personal values, or the values of one's culture, or the values of one's institution, any one of which has no more claim to legitimacy than any other. This anarchistic epistemology implies that we have no agreed standard of reference and no basis for communicating biases and assumptions. On the other hand, if databases are to be more than just personal artistic diversions and are to convey information rather than simply express the values and viewpoints of their creator, then they must be able to convey their meaning to a broad spectrum of users.

References

American Society for Photogrammetry (Committee for Specifications and Standards, Professional Practice Division) 1985 Accuracy specification for large-scale line maps. *Photogrammetric Engineering and Remote Sensing* 51: 195–9

American Society of Civil Engineers (Committee on Cartographic Surveying, Surveying, and Mapping Division) 1983 *Map uses, scales, and accuracies for engineering and associated purposes*. New York, American Society of Civil Engineers

Aronoff S 1985 The minimum accuracy value as an index of classification accuracy. *Photogrammetric Engineering and Remote Sensing* 51: 99–111

Beard M K 1989b Use error: the neglected error component. *Proceedings, AutoCarto 9*: 808–17

Berry B 1964 Approaches to regional analysis: a synthesis. *Annals of the Association of American Geographers* 54: 2–11

Blakemore M 1983 Generalisation and error in spatial databases. *Cartographica* 21: 131–9

Brassel K, Bucher F, Stephan E-M, Vckovski A 1995 Completeness. In Guptill S C, Morrison J L (eds) *Elements of spatial data quality*. Oxford, Elsevier Science: 81–108

Burrough P A 1986 *Principles of geographical information systems for land resources assessment*. Oxford, Clarendon Press

Burrough P A, Frank A U 1995 Concepts and paradigms in spatial information: Are current geographical information systems truly generic? *International Journal of Geographical Information Systems* 9: 101–16

Caspary W, Scheuring R 1993 Positional accuracy in spatial databases. *Computers, Environment and Urban Systems* 17: 103–10

Cassettari S 1993 *Introduction to integrated geoinformation management*. London, Chapman and Hall

Chrisman N R 1982 A theory of cartographic error and its measurement in digital databases. *Proceedings, AutoCarto 5*: 159–68

Chrisman N R 1991b The error component in spatial data. In Maguire D J, Goodchild M F, Rhind D W (eds) *Geographical information systems: principles and applications*. Harlow, Longman/New York, John Wiley & Sons Inc. Vol. 1: 165–74

CEN (Comité Européen de Normalisation) 1995 *Geographic information – data description – quality* (draft). Brussels, CEN Central Secretariat

FGDC Federal Geographic Data Committee 1994 *Content standards for digital geospatial metadata (June 8)*. Washington DC, Federal Geographic Data Committee

Fegeas R G, Cascio J L, Lazar R A 1992 An overview of FIPS 173, the Spatial Data Transfer Standard. *Cartography and Geographic Information Systems* 19: 278–93

Genderen J L van, Lock B F 1977 Testing land-use map accuracy. *Photogrammetric Engineering and Remote Sensing* 43: 1135–7

Goodchild M F 1988a Stepping over the line: technological constraints and the new cartography. *The American Cartographer* 15: 311–19

Goodchild M F 1988b The issue of accuracy in global databases. In Mounsey H (ed.) *Building databases for global science*. London, Taylor and Francis: 31–48

Goodchild M F 1991a Issues of quality and uncertainty. In Müller J-C (ed.) *Advances in cartography*. Oxford, Elsevier Science: 111–39

Goodchild M F 1991c Keynote address. *Proceedings, Symposium on Spatial Database Accuracy*: 1–16

Goodchild M F 1995b Sharing imperfect data. In Onsrud H J, Rushton G (eds) *Sharing geographic information*. New Brunswick, Center for Urban Policy Research: 413–25

Guptill S C 1993 Describing spatial data quality. *Proceedings, Sixteenth International Cartographic Conference*: 552–60

Harley J B 1989 Deconstructing the map. *Cartographica* 26: 1–20

Harley J B 1991 Can there be a cartographic ethics? *Cartographic Perspectives* 10: 9–16

Honeycutt D M 1986 'Epsilon, generalisation, and probability in spatial databases'. Unpublished manuscript

Kainz W 1995 Logical consistency. In Guptill S C, Morrison J L (eds) *Elements of spatial data quality*. Oxford, Elsevier Science: 109–37

Langran G 1992 *Time in geographic information systems*. London, Taylor and Francis

Lanter D 1991 Design of a lineage-based meta-database for GIS. *Cartography and Geographic Information Systems* 18: 255–61

Lanter D, Surbey C 1994 Metadata analysis of GIS data processing: a case study. In Waugh T C, Healey R G (eds) *Advances in GIS research*. London, Taylor and Francis: 314–24

Lanter D, Veregin H 1992 A research paradigm for propagating error in layer-based GIS. *Photogrammetric Engineering and Remote Sensing* 58: 526–33

Moellering H (ed.) 1991 *Spatial database transfer standards: current international status*. Oxford, Elsevier Science

Morain S, Budge A 1996 The National Spatial Data Infrastructure – why should you care? *GIS World* 9(8): 32–4

Parkes D N, Thrift N J 1980 *Times, spaces, and places: a chrono-geographic perspective*. New York, John Wiley & Sons Inc.

Redman T C 1992 *Data quality*. New York, Bantam

Salgé F 1995 Semantic accuracy. In Guptill S C, Morrison J L (eds) *Elements of spatial data quality*. Oxford, Elsevier Science: 139–51

Sinton D F 1978 The inherent structure of information as a constraint in analysis. In Dutton G (ed.) *Harvard papers on geographic information systems*. Reading (USA), Addison-Wesley

Stearns F 1968 A method for estimating the quantitative reliability of isoline maps. *Annals of the Association of American Geographers* 58: 590–600

Stone H F, Boyle S L, Hewitt M J III 1990 Development of an EPA quality assurance program for geographic information systems and spatial analysis. *GIS/LIS 90*: 814–19

Thapa K, Bossler J 1992 Accuracy of spatial data used in geographic information systems. *Photogrammetric Engineering and Remote Sensing* 58: 835–41

Turnbull D 1989 *Maps are territories*. Chicago, University of Chicago

Veregin H 1991 *GIS data quality evaluation for coverage documentation systems*. Las Vegas, Environmental Monitoring Systems Laboratory, US Environmental Protection Agency

Veregin H, Hargitai P 1995 An evaluation matrix for geographical data quality. In Guptill S C, Morrison J L (eds) *Elements of spatial data quality*. Oxford, Elsevier Science: 167–88

Veregin H, Lanter D 1995 Data-quality enhancement techniques in layer-based geographic information systems. *Computers, Environment and Urban Systems* 19: 23–36

Wood D 1992 *The power of maps*. New York, Guilford Press/ London, Routledge

13

Models of uncertainty in spatial data

P F FISHER

Spatial information is rife with uncertainty for a number of reasons. The correct conceptualisation of that uncertainty is fundamental to the correct use of the information. This chapter attempts to document different types of uncertainty – specifically error, vagueness, and ambiguity. Examples of these three types are used to illustrate the classes of problems which arise, and to identify appropriate strategies for coping with them. The first two categories are well documented and researched within the GIS field, and are now recognised in many varied contexts. The third has not been so widely researched. Cases are also identified where uncertainty is deliberately introduced into geographical information in order to anonymise individuals. Examples are given where both error and vagueness can be applied to the same phenomenon with different understandings and different results. Methods to address the problems are identified and are explored at length.

1 INTRODUCTION

'The universe, they said, depended for its operation on the balance of four forces which they identified as charm, persuasion, uncertainty and bloody-mindedness.'

Terry Pratchett (1986)

acuracy *n.* An absence of erors. 'The computer offers both speed and acuracy, but the greatest of these is acuracy' (*sic*)

Kelly-Bootle (1995)

The handling of large amounts of information about the natural and built environments, as is necessary in any GIS, is prone to uncertainty in a number of forms. Ignoring that uncertainty can, at best, lead to slightly incorrect predictions or advice and at worst can be completely fatal to the use of the GIS and undermine any trust which might have been put in the work of the system or operator. It is therefore of crucial importance to all users of GIS that awareness of uncertainty and error should be as widespread as possible. Fundamental to such understanding is the nature of the uncertainty, in its different guises. This is the subject of this chapter. A minimal response should be that users of the GIS be aware of the possible complications to their analysis caused by uncertainty, and at best present the user of the analysis with a report of the uncertainty in the final results together with a variety of plausible outcomes. A complete response to uncertainty is to present the results of a full modelling exercise which takes into account all types of uncertainty in the different data themes used in the analysis. It seems that neither response is widespread at present, and in any case the tools for doing the latter are currently the preserve only of researchers.

This chapter explores the developing area of the conceptual understanding (modelling) of different types of uncertainty within spatial information. These are illustrated in Figure 1. At the heart of the issue of uncertainty is the problem of defining both the class of object to be examined (e.g. soils) and the individual object (e.g. soil map unit) – the so-called problem of definition (Taylor 1982). Once the conceptual modelling identifies whether the class of objects to be described is well or poorly defined the nature of the uncertainty as follows:

1 If both the class of object and the individual are well defined then the uncertainty is caused by errors and is probabilistic in nature;

2 If the class of object or the individual is poorly defined then additional types of uncertainty may be recognised. Some have been explored by GIS researchers and others have not:

   a If the uncertainty is attributable to poor definition of class of object or individual object, then definition of a class or set within the universe is a matter of *vagueness*, and this can conveniently be treated with fuzzy set theory.

   b Uncertainty may also arise owing to ambiguity (the confusion over the definition of sets within the universe) owing, typically, to differing classification systems. This also takes two forms (Klir and Yuan 1995), namely:

     i Where one object or individual is clearly defined but is shown to be a member of two or more different classes under differing schemes or interpretations of the evidence, then *discord* arises;

     ii Where the process of assigning an object to a class at all is open to interpretation, then the problem is *non-specificity*.

In the context of spatial databases, only vagueness as expressed by fuzzy set theory and error as represented by probability theory have been researched, and these are the primary focus of the discussion below. The list is necessarily not exhaustive: however, the volume of research and the amount of interest in this area continues to increase.

If a chapter had been written in this form for the first edition of this book, it would have focused on only one variety of uncertainty, namely error (Chrisman 1991). A few years later there are two equally important strands to be discussed. Although the strands discussed here seem to explain the majority of the long-recognised causes of uncertainty in spatial information, it is already possible to identify other types of uncertainty that should be addressed in future research.

2 THE PROBLEM OF DEFINITION

The principal issue of geographical uncertainty is the understanding of the collector and user of the data as to the nature of that uncertainty. There are three facets to this, namely uncertainty in measurement of attributes, of space, and of time. In order to define the nature of the uncertainty of an object within the dimensions of space and time, a decision must be made as to whether or not it is

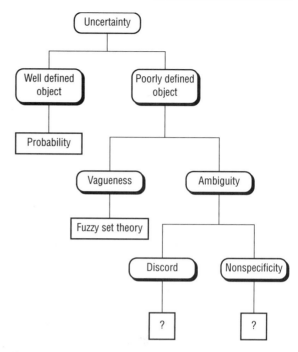

Fig 1. A conceptual model of uncertainty in spatial data (adapted from Klir and Yuan 1995: 268).

clearly and meaningfully separable from other objects in whichever dimension is of interest – ideally it will be separable in both. This is a complex intellectual process, one which draws on the history and the critical appraisal of subject-specific scientists. This conceptual model has been complicated and muddied by conventions which influence the perception of geographical information. Foremost among these is the historical necessity of simplification of information for map production; what Fisher (1996) denotes the paradigm of 'production cartography'. Equally important are the concepts of classification, commonly based on hierarchies, in which objects must fall into one class or another, and of computer database models in which objects are treated as unique individuals and form the basis to analysis.

If a spatial database is to be used, or to be created from scratch, then investigators or users have to ask themselves two apparently simple questions:

1 Is the class of objects to be mapped (e.g. soils, rocks, ownership, etc.) clearly separable from other possible classes?

2 Are the geographical individuals within the class of objects clearly and conceptually separable from other geographical individuals within the same class?

If it is possible to separate unequivocally the phenomenon to be mapped into mappable and spatially distinct objects using the spatial distribution of some individual attribute or collection of attributes, at a given time, then there is no problem of definition. A phenomenon which is well defined should have diagnostic properties for separating individuals into classes based on attributes and into spatially contiguous and homogenous areas.

If it is not possible to define the spatial extent of an object to be mapped or analysed, there is a problem of definition, and it can be said to be 'vague' (Williamson 1994). In this circumstance, while specific properties may be measured and these measurements may be precise, no combination of properties allows the unequivocal allocation of individual objects to a class, or even the definition of the precise spatial extent of the objects. Most spatial phenomena in the natural environment share this problem of definition to some extent. Error analysis on its own does not help with the description of these classes, although any properties which are measured may be subject to errors just as they are in other cases.

2.1 Examples of well-defined geographical objects

In developed countries *census geographies* tend to be well defined; even in less developed countries the geographical concepts are generally well defined, if less clearly implemented. They usually consist of a set of regions each with precise boundaries within which specific attributes are enumerated (Openshaw 1995). The areas at the lowest level of enumeration (city blocks, enumeration districts, etc.) are grouped with specific instances of other areas at the same level to make up higher level areas, which in turn are grouped with other specific areas to form a complete and rigid hierarchy (e.g. see Martin, Chapter 6). The attributes to be counted within the areas are typically based on property units, individuals, and households: although the definitions of 'household' may differ between different surveys (Office of National Statistics 1997) and there is rarely any perfect correspondence between households and property units (e.g. houses in multiple occupation), each definition is nevertheless quite transparent and unambiguous. The data collection process in the western world relies on a certain level of cooperation and literacy amongst those being counted, and while there are frequently legal sanctions for non-cooperation these cannot easily be enforced if people are reluctant to cooperate. The primary errors associated with the US Census of Population arise out of underenumeration of groups such as illegal immigrants and the homeless (Bureau of the Census 1982).

A second example of a well-defined geographical phenomenon in western societies is *land ownership*. The concept of private ownership of land is fundamental to these societies; therefore the spatial and attribute interpretation of that concept is normally quite straightforward in its spatial expression. The boundary between land parcels is commonly marked on the ground, and marks an abrupt and total change in ownership. In point of fact, at least in the UK, the surveyed boundary is only deemed indicative of the actual position of the boundary, and so any property boundary has a defined uncertainty in position, otherwise it would require resurveying every time the boundary marker is rebuilt (Dale and McLaughlin 1988). Even in instances of collective ownership in which two groups may own two adjoining parcels and one person may belong to both groups, the question of ownership and responsibility remains clear in law.

Well-defined geographical objects are essentially created by human beings to order the world they occupy. They exist in well-organised and established political and legal realms. Some other objects in our built and natural environments may seem to be well defined, but they tend to be based on a single measurement, and close examination frequently shows the definition to be obscure. For example, the land surface seems well defined, and it should be possible to determine its height above sea level rigorously and to specified precision. But even the position of the ground under our feet is being brought into question. This is caused by the increasing availability of elevation models derived from photogrammetry to sub-centimetre precision, when the actual definition of the land surface being mapped must come into question, and whether the field was ploughed or the grass was cut, become serious issues in defining the so-called land surface. Most, if not all, other geographical phenomena are similarly poorly defined to some extent.

2.2 Examples of poorly-defined geographical objects

In aboriginal societies the concept of ownership is much less clear than in western society. There are many different native cultures, but many have a conception of the land owning the people, and responsibility for nurturing the land is a matter of common trust within a group (Native North Americans and Australians, for example: Young 1992). Areas of responsibility are less well defined, with certain core areas for which a group or an individual may be responsible (e.g. the sacred sites of the Australian Aborigines: Davis and Prescott 1992), and other regions for which no one is actually responsible but many groups may use (so-called 'frontier zones'). Among both North American and Australian native groups, the spatial extents of these core and peripheral areas have been shown to be well known to the groups concerned, although they may not be marked, precisely located, or fixed over time (Brody 1981; Davis and Prescott 1992). There are therefore acknowledged divisions of space, but the spatial location of the divider may be uncertain. The extent of the zones of uncertainty can be resource dependent, so that when resources are plentiful there may be relatively precise boundaries, and when scarce there may be very diffuse frontiers (Davis and Prescott 1992; Young 1992). Alternatively, ties of kinship between groups may create less specific frontiers, and lack of kinship hard boundaries (Brody 1981). These aboriginal territories have much in common with the documented 'behavioural neighbourhoods' of western individuals. Such neighbourhoods are also poorly defined both spatially and temporally: they may be discontinuous and will inevitably overlap with others, and while possibly unique to an individual or family, may nonetheless make up part of a geographical region that is occupied by a group.

Complexity is also inherent in the mapping of vegetation (Foody 1992). The allocation of a patch of woodland to the class of oak woodland, for example – as opposed to any other candidate woodland type – is not necessarily easy. It may be that in that region a threshold percentage of trees need to be oak for the woodland to be considered 'oak', but what happens if there is one per cent less than that threshold? Does it really mean anything to say that the woodland needs to be classed to a different category? Indeed, the higher level classification to woodland at all has the same problems. Mapping the vegetation is also problematic since in areas of natural vegetation there are rarely sharp transitions from one vegetation type to another, rather an intergrade zone or ecotone occurs where the dominant vegetation type is in transition (Moraczewski 1993). The ecotone may occupy large tracts of ground. The attribute and spatial assignments may follow rules, and may use

indicator species to assist decisions, but strict deterministic rules may trivialise the classification process without generating any deeper meaning.

In discussion of most natural resource information we typically talk about central concepts and transitions or intergrades. Figure 2 shows a scatter plot of some remotely-sensed (LANDSAT) data from Band 3 and Band 5 (which record the amounts of reflected electromagnetic radiation in the wavelength ranges 0.63–0.69 and 1.55–1.75 μm, respectively). This is part of the information used in the assignment of pixels in an image to land covers. The conceptualisation of the land covers is as Boolean objects (discussed below), and yet it is clear from Figure 2 that there are no natural breaks in the distribution of points in the 2-dimensional space shown. This is typical of satellite imagery. Although LANDSAT actually records information in seven spectral bands which can give identification to some natural groups of pixels, the number of identifiable groups very rarely corresponds with the number of land cover types being mapped (Campbell 1987). The classification process involves the identification of prototypical values for land cover types, and the extension of that mapping from the attribute dimensions shown to the spatial context. Conceptually, the same basic process is executed in almost all traditional mapping operations, and the problem of the identification of objects is fundamental. It is apparent from Figure 2 that the intergrades (all possible locations in attribute space which are between the prototype or central concepts) are more commonly and continuously occupied than the prototypical classes.

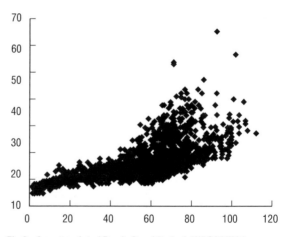

Fig 2. A scatterplot of Bands 3 and 5 of a LANDSAT TM image.

The problem of identification may be extended into locations. Figure 3 shows a soil map of part of the Roujan catchment in France with numbers indicating soil map units (soil types) and the shading indicating the extent of boundary intergrades between types. The width of intergrades is based on the knowledge of soil surveyors who prepared the map (Lagacherie et al 1996).

Within natural resource disciplines the conceptualisation of mappable phenomena and the spaces they occupy is rarely clear cut, and is still more rarely achieved without invoking simplifying assumptions (see also Veregin, Chapter 12). In forestry, for example, tree stands are defined as being clearly separable and mappable; yet trees vary within stands by species density, height, etc., and often the spatial boundary between stands is not well defined (Edwards 1994). Although theorists may recognise the existence of intergrades, the conceptual model of mapping used in this and other natural resource disciplines accepts the simplification and places little

importance on them, although the significance has not been assessed. In other areas, such as soil science and vegetation mapping, some of the most interesting areas are at the intergrade, and these are rightly a focus of study in their own right (Burrough 1989; Burrough et al 1992; Lagacherie et al 1996). The interest in intergrades as boundaries is not a preserve of natural resouce scientists, however, and in discussion of urban and political geography considerable attention is paid to these concepts (Prescott 1987; Batty and Longley 1994).

3 ERROR

If an object is conceptualised as being definable in both attribute and spatial dimensions, then it has a *Boolean* occurrence; any location is either part of the object, or it is not. Yet within GIS, for a number of reasons, the assignment of an object or location to the class may be expressed as a probability. There are

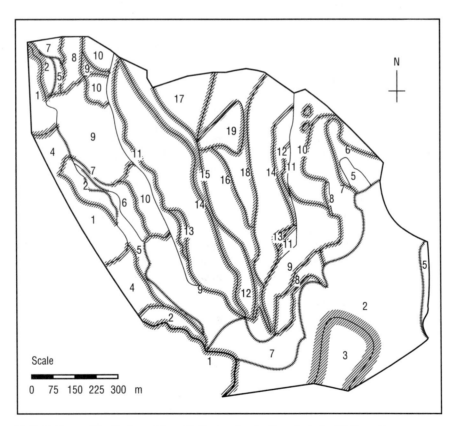

Fig 3. Soil map of the Roujan catchment in France showing the extent of soil intergrades (after Lagacherie et al 1996: 281).

any number of reasons why this might be the case. Three are briefly discussed here:

1 probability owing to error in the measurement;
2 probability because of the frequency of occurrence;
3 probability based on expert opinion.

Errors occur within any database, and for any number of reasons; some reasons are given in Table 1. They are given more complete treatment by Fisher (1991b) and Veregin (Chapter 12). The simplest to handle are those associated with measurement, because well-advanced error analysis procedures have been developed (Heuvelink, Chapter 14; Heuvelink et al 1989; Heuvelink and Burrough 1993; Taylor 1982). If the true value of a property of an object were precisely known, then it would be possible to identify the distribution of 'real world' measurement error by making repeated measurements of the property (which would each differ from the true value by a variable measurement error). It would then be possible to estimate the distribution of the error in its measurement, and thus to develop a full error model of the measurement error. This is, in fact, the basis of the 'root mean square' reporting of error in digital elevation models (see also Beard and Buttenfeld, Chapter 15). Yet there are many instances in which such reductionist measures of error are over-simplistic and aspatial, failing to identify the spatial distribution of the error in GIS-based modelling (Monckton 1994; Walsby 1995).

A further means of describing aspatial error is to create a confusion matrix which shows the cover-type actually present at a location crosstabulated against the cover-type identified in the image classification process. Typically the matrix is generated for a complete image. It reports errors in the allocation of pixels to cover types (Campbell 1987; Congalton and Mead 1983). However, the confusion matrix is of limited use if the precise interpretation of either the classification process or the ground information is not clear cut.

A different view of probability is based on the frequency of the occurrence of a phenomenon. The classic applications of probability in this area include weather and flood forecasting. Floods of a particular height are identified as having a particular return period which translates as a particular probability of a flood of that level occurring.

A third view of probability is as a manifestation of subjective opinion, where an expert states a 'gut feeling' of the likelihood of an event occurring. Much

Table 1 Common reasons for a database being in error.

Type of error	Cause of error
Measurement	Measurement of a property is erroneous
Assignment	The object is assigned to the wrong class because of measurement error by field, or laboratory scientist, or by surveyor
Class generalisation	Following observation in the field and for reasons of simplicity, the object is grouped with objects possessing somewhat dissimilar properties
Spatial generalisation	Generalisation of the cartographic representation of the object before digitising, including displacement, simplification, etc. (see Weibel and Dutton, Chapter 10)
Entry	Data are miscoded during (electronic or manual) entry to a GIS
Temporal	The object changes character between the time of data collection and of database usage
Processing	In the course of data transformations an error arises because of rounding or algorithm error

geological and soil mapping is actually the result of Boolean classification of subjective probability, since it is impracticable to observe directly either of these phenomena across the entire countryside: rather inference is made using sampled points such as outcrops and auger borings. Between those locations it is expert opinion as to what is there; so long as a Boolean model of soil and rock occurrence is applied, the map is implicitly a matter of the expert's maximum probability (Clarke and Beckett 1971).

Probability has been studied in mathematics and statistics for hundreds of years. It is well understood, and the essential methods are well documented. There are many more approaches to probability than the three described here. Probability is a subject that is on the syllabus of almost every scientist qualified at degree level, and so it pervades the understanding of uncertainty through many disciplines. It is not, however, the only way to treat uncertainty.

4 VAGUENESS

In contrast with error and probability which are steeped in the mathematical and statistical literature, vagueness is the realm of philosophy and logic and has been described as one of the fundamental challanges to those disciplines (Williamson 1994; Sainsbury 1995). It is relatively easy to show that a concept is 'vague', and the classic pedagogic exposition uses the case of the 'bald' man. If a person with no hair at all is considered bald, then is a person with one hair bald? Usually, in any working definition of 'bald', the answer to this would be 'yes'. If a person with one hair is bald, then is a person with two hairs bald: again, 'yes'. If you continue the argument, one hair at a time, then the addition of a single hair never turns a bald man into a man with a full head of hair. On the other hand, you would be very uncomfortable admitting that someone with plenty of hair was bald, since this is illogical (Burrough 1992; Burrough 1996; Zadeh 1965). This is known as the *Sorites Paradox* which, little by little, presents the logical argument that someone with plenty of hair is bald! A number of resolutions to the paradox have been suggested, but the most widely accepted is that the logic employed permits only a Boolean response ('yes' or 'no') to the question. A graded response is not acceptable. And yet there is a degree to which a person can be bald. It is also possible that the initial question is false, because 'bald' would normally be qualified if we were examining it in detail, so we might ask whether someone was 'completely bald', and we might define that as someone with no hair at all. Can we ever be certain that individuals have absolutely no hair on their heads? Furthermore, where on their neck and face is the limit of the head such that we can judge whether there is any hair on it? You are eventually forced to admit that by incremental logical argument, it is impossible to specify whether someone is 'completely', 'absolutely', 'partially', or 'not at all' bald, given a count of hairs on their head, even if the count is absolutely correct. So no matter the precision of the measurement, the allocation to the set of people is inherently vague.

The Sorites Paradox is one way which is commonly used to define vague concepts. If a concept is 'Sorites susceptible', it is vague. Many geographical phenomena are 'Sorites susceptible', including concepts and objects from the natural and built environments (e.g. see Band, Chapter 37). When, exactly, is a house a house; a settlement, a settlement; a city, a city; a podsol, a podsol; an oak woodland, an oak woodland? The questions always revolve around the threshold value of some measurable parameter or the opinion of some individual, expert or otherwise.

Fuzzy set theory was introduced by Zadeh (1965) as an alternative to *Cantor* (*Boolean*) *sets*, and built on the earlier work of Kaplan and Schott (1951). Membership of an object to a Cantor set is absolute, that is it either belongs or it does not, and

membership is defined by integer values in the range {0,1}. By contrast, membership of a fuzzy set is defined by a real number in the range [0,1] (the change in type of brackets indicates the real and integer nature of the number range). Definite membership or non-membership of the set is identified by the terminal values, while all intervening values define an intermediate degree of belonging to the set, so that, for example, a membership of 0.25 reflects a smaller degree of belonging to the set than a membership of 0.5. The object described is less like the central concept of the set.

Fuzzy memberships are commonly identified by one of two methods (Robinson 1988):

1 the *Similarity Relation Model* is data driven and involves searching for patterns within a dataset similarly to traditional clustering and classification methods, the most widespread method being the Fuzzy *c* Means algorithm (Bezdek 1981). More recently, fuzzy neural networks have been employed (Foody 1996);
2 the *Semantic Import Model*, in contrast, is derived from a formula or formulae specified by the user or another expert (Altman 1994; Burrough 1989; Wang et al 1990).

Many studies have applied fuzzy set theory to geographical information processing. There are several good introductions to the application of fuzzy sets in geographical data processing, including books by Leung (1988) and Burrough and Frank (1996) – see also Eastman (Chapter 35).

Fuzzy set theory is now only one of an increasing number of soft set theories (Pawlak 1982), in contrast to hard, Cantor sets. However, a number of authorities consider that fuzzy set theory is mistakenly used for problems which more correctly fall within the realm of subjective probability (Laviolette and Seaman 1994). They have, however, primarily addressed fuzzy logic rather than fuzzy sets, and illustrated their arguments with Boolean conditions and decisions. As such, they have failed to address the nature of the underlying set and any inherent vagueness which may be present, as Zadeh (1980) has shown. Moreover, Kosko (1990) has argued that fuzzy sets are a superset of probability.

5 AMBIGUITY

The concepts and consequences of ambiguity (Figure 1) in geographical information are not well researched. Ambiguity occurs when there is doubt as to how a phenomenon should be classified because of differing perceptions of it. Two types of ambiguity have been recognised, namely *discord* and *non-specificity*. In other areas of study some partial solutions have been suggested, but they are not reviewed here because of the lack of specific research with geographical information.

Within geography the most obvious form of discord through ambiguity is in the conflicting territorial claims of nation states over particular pieces of land. History is filled with this type of ambiguity, and the discord which results. Examples in the modern world include intermittent and ongoing border conflicts and disagreements in Kashmir (between India and China) and the neighbouring Himalayan mountains (between China and India). Similarly, the existence or non-existence of a nation of Kurds is another source of discord. All represent mismatches between the political geography of the nation states and the aspirations of people (Horn, Chapter 67; Prescott 1987; Rumley and Minghi 1991).

As has already been noted, many if not most phenomena in the natural environment are also ill-defined. The inherent complexity in defining soil, for example, is revealed by the fact that many countries have slightly different definitions of what a soil actually constitutes (cf. Avery 1980; Soil Survey Staff 1975), and by the complexity and the volume of literature on attempting to define the spatial and attribute boundaries between soil types (Webster and Oliver 1990; Lagacherie et al 1996). Furthermore, no two national classification schemes have either the same names for soils or the same definitions if they happen to share names. This causes many soil profiles to be assigned to different classes in different schemes, as shown in Table 2 (see also Isbell 1996; Soil Classification Working Group 1991; Soil Survey Staff 1975). Within a single country this is not a problem, yet ambiguity arises in the international efforts to produce supra-national or global soil maps. The individual national classifications cause considerable confusion in the process and the classification scheme becomes part of the national identity within the context. There is also rarely a one-to-one correspondence between classification systems (soil type x in this classification corresponds to soil type a in that), but rather a many-to-many classification (soil types a and b correspond broadly to soil type x, but some profiles of soil type a are also soil types y and z). This leads to different

placement of soil boundaries in both attribute and spatial dimensions, and generates considerable problems in mapping soils across international and interstate boundaries (FAO/UNESCO 1990; Campbell et al 1989), as has been exemplified in the creation of the Soil Map of the European Communities (Tavernier and Louis 1984).

Several measures of social deprivation have been suggested which are based upon information from the UK Census of Population (Table 3). Enumeration areas are assigned to one class or another, and the classes have been used in the allocation of resources for a range of social and economic programmes. The fact that there are different bases to the measurement of deprivation means that enumeration areas may be afforded special policy status using one indicator, but not using another, and this is a source of potential discord.

Ambiguity through non-specificity can be illustrated from geographical relations. The relation 'A is north of B' is itself non-specific, because the concept 'north of' can have at least three specific meanings: that A lies on exactly the same line of longitude and towards the north pole from B; that A lies somewhere to the north of a line running east to west through B; or, in common use, that A lies in the sector between perhaps north-east and north-west, but is most likely to lie between north-north-east and north-north-west of B. The first two definitions are precise and specific, but equally valid. The third is the natural language concept which is itself vague. Any lack of definition as to which should be used means that uncertainty arises in interpreting 'north of'.

Arguably, soil classification is a process whereby modern schemes have removed the problem of non-specificity which was inherent in earlier schemes and replaced it by supposedly objective, globally applicable diagnostic criteria. The remaining problems arise out of creating Boolean boundaries in a vague classification environment and the problem of discord.

None of this should be taken to imply that ambiguity is inappropriate or intrinsically 'wrong'. The England and Wales soil classification scheme at the scale of England and Wales is possibly the most relevant classification scheme for the soils in that country. Similarly, the United States Department of Agriculture scheme (Soil Taxonomy) was the best scheme for the US when it was finalised in 1975 (although it does claim a global application). The problem of ambiguity arises when we move to a higher level, and data from the British Soil Survey

Table 2 Alternative soil classification schemes for global and national use.

US Classification (Soil Survey Staff 1975)	Australian Classification (Isbell 1996)	Soil Map of the World (FAO/UNESCO 1990)		British Soil Classification (Avery 1980)
Entisol	Anthroposol	Fluvisol	Kastanozem	Terrestrial raw soil
Inceptisol	Organosol	Gleysol	Chernozem	Hydric raw soil
Spodosol	Podsol	Regosol	Phaeozem	Lithomorphic soil
Mollisol	Hydrosol	Lithosol	Greyzem	Pelosol
Oxisol	Kurosol	Arenosol	Cambisol	Brown soil
Ultisol	Sodosol	Rendzina	Luvisol	Podzolic soil
Alfisol	Chromosol	Ranker	Podzoluvisol	Ground-water gley soil
Aridisol	Calcarosol	Andosol	Podzol	Surface-water gley soil
Histosol	Ferrosol	Vertisol	Planosol	Man-made soil
Vertisol	Dermosol	Solonchak	Acrisol	Peat soil
	Kandosol	Solonetz	Nitosol	
	Rudosol	Yermosol	Ferrasol	
	Tenosol	Xerosol	Histosol	

Table 3 Measures of social deprivation used in the UK, with the associated census variables used in their calculation (Openshaw 1995).

Variable	Jarman	Townsend	Department of the Environment
Unemployment	X	X	X
No car		X	
Unskilled	X		
Overcrowding (more than 1 person per room)	X	X	X
Lacking amenities			X
Not owner occupied		X	
Single-parent household	X		X
Children under 5 years old	X		
Lone pensioners	X		
Ethnic minorities	X		

have to be fused with data from neighbouring countries or countries further afield. In preparing the Soil Map of the European Community, for example, the FAO/UNESCO classification was employed with some amendments.

In a like manner, there is nothing wrong with there being three different methods of defining deprived regions in Britain. Deprivation is a social construct and any quantitative index can only be an approximation which is deemed relevant and acceptable within its own terms of reference. If the constituent attributes of a particular index happen not to be measured in another country, that index simply ceases to have international application. (In fact, with regard to the use of the Jarman Index

within the UK, housing indicators of deprivation replace ethnic indicators in Wales.) Ambiguity nevertheless does come into play in the allocation of social and economic programme resources, and can lead to contention between local, national, and (in the case of EU programmes) supra-national, politicians over the issue of the basis to financial support.

6 CONTROLLED UNCERTAINTY

Many agencies distribute and allow access to spatial information which is degraded deliberately through creating uncertainty. Two examples of this are discussed (see also Heuvelink, Chapter 14, and Hunter, Chapter 45, for a discussion of the management of uncertainty).

If the exact locations of rare or precious objects such as nesting sites of endangered birds or archaeological sites are recorded in a dataset, any more widely distributed versions may introduce a systematic or random error introduced into the locational component. This may be done by only reporting information for large areal aggregations (e.g. 4 km^2 in the county flora of Leicestershire, England: Primavesi and Evans 1988; and 100 km^2 grid in the state flora of Victoria, Australia, distributed on CD-ROM: Viridians 1996). In some cases both systematic and random elements are introduced in order to protect the phenomenon reported, and although the error may be inconvenient, the consequences of not introducing it may be worse.

Uncertainty is also deliberately introduced into census data in order to preserve confidentiality. If only a few people living within any one enumeration area have a particular characteristic – for example high income – and incomes are reported, it may be very easy to identify exactly which person that is. This is not socially acceptable, and so most census organisations withhold or falsify small counts. For example, in the USA, data for areas with small counts are withheld (Bureau of the Census 1982), whereas in the UK small counts have had a random value between +1 and -1 added (Dewdney 1983).

7 DISTINGUISHING BETWEEN VAGUENESS AND ERROR

Appropriate conceptualisation of uncertainty is a prerequisite to its modelling within GIS. In this section two areas of previous study are examined, and the reasons for the use of either vague or error models of uncertainty are discussed.

7.1 Viewshed

The viewshed is a simple operation within many current GIS, which, in its usual implementation, reports those areas in a landscape which are in view and those which are not (coded 1 and 0 respectively), whether in a triangulated grid or dataset (De Floriani and Magillo, Chapter 38; De Floriani et al 1986; Fisher 1993). Fisher (1991a) has shown how, for a variety of reasons, the visible area is very susceptible to error in the measurement of elevations in the Digital Elevation Model (DEM). (While Fisher used a rectangular grid in his 1991 study, the same would be true for a triangulated model.) The database error is propagated into the binary viewshed because of error in the elevation database (Fisher 1991a) and uncertainty in determination of visibility because of variation between different algorithms (Fisher 1993). Fisher (1992, 1993, 1994) has proposed that it is possible to define the error term from the Root Mean Squared Error (RMSE) for the DEM such that the error has a zero mean and standard deviation equal to the RMSE. This is not in fact true and provides insufficient description of the error for a fully justifiable error model since the mean error may be biased (non-zero) and must have spatial structure. Spatial structure of the error may be identified

through spatial autocorrelation measures or full specification of the variogram of the error field (Journel 1996). If the error field is generated using this method then it can be added to the DEM, yielding a revised DEM which includes the known error. If the viewshed is determined over that DEM with error, then a version of the Boolean viewshed is generated. If the process is repeated, then a second version of the Boolean viewshed, a third, a fourth, and so on are generated. If each Boolean viewshed image is coded as 0 and 1 indicating areas which are out-of-sight and in-sight, then using map algebra to find the sum of Boolean viewsheds, a value between 0 and the number of realisations will be found for all locations depending on the number of realisations in which that location is visible. Dividing by the number of realisations will then give an estimate of the probability of that location actually being within the viewshed. The probability of any pixel being visible from the viewing point, or the probability of the land rising above the line of sight somewhere between the viewer and the viewed is given by:

$$p(x_{ij}) = \frac{\sum_{k=1}^{n} x_{ijk}}{n} \tag{1}$$

where
$p(x_{ij})$ is the probability of a cell at row i and column j in the raster image being visible; and x_{ijk} is the value at the cell of the binary-coded viewshed in realisation k such that k takes values 1 to n.

This is illustrated in Plate 9.

In contrast, using a Semantic Import Model it is possible to define a number of different fuzzy viewsheds (Fisher 1994; note that the term is used incorrectly by Fisher 1992) from a family of equations relating the distance from the viewer to the viewed to the fuzzy membership function (Plate 10). Any number of different circumstances can be described, and two are included here: Equation 2 represents normal atmospheric conditions, and Equation 3 describes the visibility through fog.

$$\mu(x_{ij}) = \begin{cases} 1 & for\ d_{vp} \to ij \leq b_1 \\ \dfrac{1}{\left(1+\left(\dfrac{d_{vp} \to ij - b_1}{b_2}\right)^2\right)} & for\ d_{vp} \to ij > b_1 \end{cases} \tag{2}$$

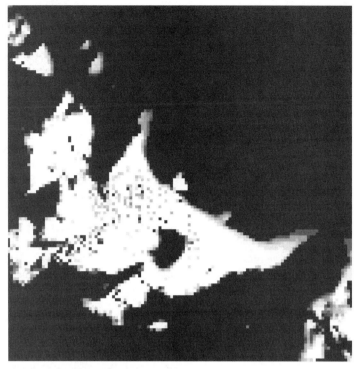

Fig 4. Probable viewshed based on Equation 1.

Fig 5. Fuzzy viewshed based on Equation 2.

$$\mu(x_{ij}) = \begin{cases} 1 & \textit{for } d_{vp} \rightarrow_{ij} \leq b_1 \\ 0 & \textit{for } d_{vp} \rightarrow_{ij} > b_1 + 2 \bullet b_2 \\ \sin\left(\left(\dfrac{d_{vp} \rightarrow_{ij} - b_1}{2 \bullet b_2}\right)\dfrac{\pi}{2}\right) & \textit{for } d_{vp} \rightarrow_{ij} > b_1 \end{cases} \quad (3)$$

where

$\mu(x_{ij})$ is the fuzzy membership at the cell at row i, column j;

$d_{rp} \rightarrow_{ij}$ is the distance from the viewpoint to row i, column j;

b_1 is the radius of the zone around the viewpoint where the clarity is perfect, and the target object can be seen at the defined level of detail;

b_2 is the distance from b_1 to fuzzy membership of 0.5, sometimes called the cross-over point.

The distinction between fuzzy and probable viewsheds is that the first describes the probability of a location being visible, while the second portrays the degree to which objects can be distinguished. Thus there is an objective definition of the first, and only subjective versions of the second which may describe group or even personal circumstances.

7.2 Remote sensing

Classification of remotely-sensed data has been a major source of land cover and land-use information for GIS. The basic methods, based on a number of discriminant functions from numerical taxonomy, are well known and widely documented (Campbell 1987). The assumptions implicit in this approach are threefold:

1 the cover type itself is a well-defined phenomenon with clear breaks reflected by there being more similarity within cover types than between them;

2 the digital numbers recorded in the original satellite image allow the discrimination of land cover/use types, mapping on a one-to-one basis between reflectance and cover type;

3 the area of the pixel on the ground can be identified as having a single cover type (that area can be assigned to one and only one land-cover or land-use).

From these assumptions it is possible to allow the conceptualisation of both the spatial extent of the pixel and the land cover attributes to be determined as Boolean concepts. Therefore uncertainties can be described by probability, and functional methods such as the maximum likelihood classifier are applicable. Unfortunately, all the assumptions are made for the convenience

of the operator, and none matches the actual situation either pragmatically or theoretically.

It is a fact of life that the spatial extent of geographical objects is not coincident with the image pixel, hence the class types on the ground are often hard to define precisely (many are Sorites susceptible), and the digital numbers do not show greater similarity within cover type than between. Therefore, arguably, fuzzy set theory (as an expression of concepts of vagueness) is a more appropriate model for working with satellite imagery and has been the subject of a number of explorations (Foody 1992, 1996; Fisher and Pathirana 1991; Goodchild et al 1994). Both Foody (1992, 1996) and Fisher and Pathirana (1991) have shown that the fuzzy memberships extracted from digital images can be related to the proportion of the cover types within pixels. This can be seen as a step towards a full interpretation of the fuzzy memberships derived from the imagery, since in the work reported the land covers analysed are still well-defined Boolean concepts; the vagueness is introduced by the sensor characteristics (Fisher and Pathirana 1991; Foody 1996). On the other hand, Foody (1992) uses the fuzzy sets to examine a zone of intergrade between vegetation communities, where both the communities and the intergrade are vague concepts.

The confusion between land cover and land use is also problematic (see also Barnsley, Chapter 32). Land use has a socioeconomic dimension to it, which cannot be sensed from satellites. Land cover, on the other hand, pertains to directly observable physical properties of the Earth's surface, and so can be classified directly. Indeed, one reason for the poor results of classification accuracy is the confusion in the conceptualisation of this transformation, and the opacity of the relationship between the surface reflectance of land covers and land use. The most successful attempts at land-use mapping from satellite imagery have adopted rule-based (Wang et al 1991) or graph theoretic approaches to the problem (Barnsley, Chapter 32; Barr and Barnsley 1995), and a combination of fuzzy set theory with these other methods may well further improve the results.

Within remote sensing, it can therefore be seen that the conceptualisation of the problem is the controlling influence. If the assumptions as to the spatial and attribute discrimination of land cover within a pixel noted above are accepted, then there is a Boolean mapping between land cover and digital number which can be extracted by classification, and uncertainty can be expressed probabilistically. If they cannot be accepted, then the Sorites susceptibility of the subjects of mapping indicates their vagueness, and so fuzzy set theory is a more appropriate approach to analysis. A clear conceptualisation of the nature of the phenomenon to be mapped and the approach to be taken is essential to the successful analysis of satellite imagery.

8 CONCLUSION: UNCERTAINTY IN PRACTICE

Through citing a number of different examples, this chapter has argued that within geographical information there are a number of different causes of and approaches to uncertainty. Anyone using uncertain information (i.e. the overwhelming majority of GIS users) needs to think carefully about the possible sources of uncertainty, and how they may be addressed. Uncertainty is a recurrent theme throughout many of the chapters of this book (e.g. Hunter, Chapter 45; Martin, Chapter 6; Raper, Chapter 5); the particular contribution of this chapter is to relate our conceptualisation of the nature of uncertainty to GIS-based data models. Analysis without accommodating data uncertainty (both error and vagueness) can quite severely limit its usefulness. Yet an appropriate conceptualisation of uncertainty and the application of related analytical methods creates a rich analytical environment where decision making based on spatial information is facilitated not only by objective orderings of alternatives but also by giving confidence in those alternatives. New analytical products are beginning to appear as a result of processing, and not ignoring, uncertainty (Burrough 1989; Burrough et al 1992; Davidson et al 1994; Wang et al 1990).

It is crucial to the correct use of geographical information systems that all aspects of uncertainty should be accommodated. This can only be achieved through awareness of the issues and a thorough and correct conceptualisation of uncertainty. The subject of uncertainty in spatial information has developed rapidly, and is still changing, particularly with the increasing use and exploration of alternative, soft set theories (Pawlak 1982).

References

Altman D 1994 Fuzzy set theoretic approaches for handling imprecision in spatial analysis. *International Journal of Geographical Information Systems* 8: 271–89

Avery B W 1980 *Soil classification for England and Wales (higher categories).* Harpenden, Soil Survey Technical Monograph 14

Barr S, Barnsley M 1995 A spatial modelling system to process, analyse, and interpret multi-class thematic maps derived from satellite sensor images. In Fisher P F (ed) *Innovations in GIS 2.* London: Taylor and Francis: 53–65

Batty M, Longley P 1994 *Fractal cities: a geometry of form and function.* London/San Diego, Academic Press

Bezdek J C 1981 *Pattern recognition with fuzzy objective function algorithms.* New York, Plenum Press

Brody H 1981 *Maps and dreams; Indians and the British Columbia frontier.* Harmondsworth, Penguin

Bureau of the Census 1982 *Census of Population and Housing.* Washington DC, US Department of Commerce

Burrough P A 1989 Fuzzy mathematical methods for soil survey and land evaluation. *Journal of Soil Science* 40: 477–92

Burrough P A 1992a Are GIS data structures too simple minded? *Computers & Geosciences* 18: 395–400

Burrough P A 1996 Natural objects with indeterminate boundaries. In Burrough P A, Frank A U (eds) *Geographic objects with indeterminate boundaries.* London, Taylor and Francis: 3–28

Burrough P A, Frank A U (eds) 1996 *Geographic objects with indeterminate boundaries.* London, Taylor and Francis

Burrough P A, MacMillan R A, Deursen W van 1992 Fuzzy classification methods for determining land suitability from soil profile observations and topography. *Journal of Soil Science* 43: 193–210

Campbell J B 1987 *Introduction to remote sensing.* New York, Guilford Press

Campbell W G, Church M R, Bishop G D, Mortenson D C, Pierson S M 1989 The role for a geographical information system in a large environmental project. *International Journal of Geographical Information Systems* 3: 349–62

Chrisman N R 1991b The error component in spatial data. In Maguire D J, Goodchild M F, Rhind D W (eds) *Geographical information systems: principles and applications.* Harlow, Longman/New York, John Wiley & Sons Inc. Vol. 1: 165–74

Clarke G P, Beckett P 1971 *The study of soils in the field,* 5th edition. Oxford, Clarendon Press

Congalton R G, Mead R A 1983 A quantitative method to test for consistency and correctness in photointerpretation. *Photogrammetric Engineering and Remote Sensing* 49: 69–74

Dale P F, McLaughlin J D 1988 *Land information management.* Oxford, Oxford University Press

Davidson D A, Theocharopoulos S P, Bloksma R J 1994 A land evaluation project in Greece using GIS and based on Boolean fuzzy set methodologies. *International Journal of Geographical Information Systems* 8: 369–84

Davis S L, Prescott J R V 1992 *Aboriginal frontiers and boundaries in Australia.* Melbourne, Melbourne University Press

De Floriani L, Falcidieno B, Pienovi C, Allen D, Nagy G 1986 A visibility-based model for terrain features. *Proceedings, Second International Symposium on Spatial Data Handling.* Columbus, International Geographical Union: 235–50

Dewdney J G 1983 Census past and present. In Rhind D W (ed.) *A census user's handbook.* London, Methuen: 1–15

Edwards G 1994 Characteristics and maintaining polygons with fuzzy boundaries in geographic information systems. In Waugh T C, Healey R G (eds) *Advances in GIS research: Proceedings Sixth International Symposium on Spatial Data Handling.* London, Taylor and Francis: 223–39

FAO/UNESCO 1990 *Soil map of the world: revised legend.* FAO, Rome, World Soil Resources Report 60

Fisher P F 1991a First experiments in viewshed uncertainty: the accuracy of the viewable area. *Photogrammetric Engineering and Remote Sensing* 57: 1321–7

Fisher P F 1991b Data sources and data problems. In Maguire D J, Goodchild M F, Rhind D W (eds) *Geographical information systems: principles and applications.* Harlow, Longman/New York, John Wiley & Sons Inc. Vol. 1: 175–89

Fisher P F 1992 First experiments in viewshed uncertainty: simulating the fuzzy viewshed. *Photogrammetric Engineering and Remote Sensing* 58: 345–52

Fisher P F 1993 Algorithm and implementation uncertainty in the viewshed function. *International Journal of Geographical Information Systems* 7: 331–47

Fisher P F 1994a Probable and fuzzy models of the viewshed operation. In Worboys M (ed.) *Innovations in GIS 1.* London, Taylor and Francis: 161–75

Fisher P F 1996a Concepts and paradigms of spatial data. In Craglia M, Couclelis H (eds) *Geographic information research: bridging the Atlantic.* London, Taylor and Francis: 297–307

Fisher P F, Pathirana S 1991 The evaluation of fuzzy membership of land cover classes in the suburban zone. *Remote Sensing of Environment* 34: 121–32

Foody G M 1992 A fuzzy sets approach to the representation of vegetation continua from remotely-sensed data: an example from lowland heath. *Photogrammetric Engineering and Remote Sensing* 58: 221–5

Foody G M 1996 Approaches to the production and evaluation of fuzzy land cover classification from remotely-sensed data. *International Journal of Remote Sensing* 17: 1317–40

Goodchild M F, Chi-Chang L, Leung Y 1994 Visualising fuzzy maps. In Hearnshaw H M, Unwin D J (eds) *Visualisation in geographical information systems.* Chichester, John Wiley & Sons: 158–67

Heuvelink G B M, Burrough P A 1993 Error propagation in cartographic modelling using Boolean logic and continuous classification. *International Journal of Geographical Information Systems* 7: 231–46

Heuvelink G B M, Burrough P A, Stein A 1989 Propagation of errors in spatial modelling with GIS. *International Journal of Geographical Information Systems* 3: 303–22

Isbell R F 1996 *The Australian soil classification.* Australian Soil and Land Survey Handbook 4, CSIRO, Collingwood

Journel A 1996 Modelling uncertainty and spatial dependence: stochastic imaging. *International Journal of Geographical Information Systems* 10: 517–22

Kaplan A, Schott H F 1951 A calculus for empirical classes. *Methodos* 3: 165–88

Kelly-Bootle S 1995 *The computer contradictionary,* 2nd edition. Cambridge (USA), MIT Press

Klir G J, Yuan B 1995 *Fuzzy sets and fuzzy logic: theory and applications.* Englewood Cliffs, Prentice-Hall

Kosko B 1990 Fuzziness vs probability. *International Journal of General Systems* 17: 211–40

Lagacherie P, Andrieux P, Bouzigues R 1996 The soil boundaries: from reality to coding in GIS. In Burrough P A, Frank A U (eds) *Geographic objects with indeterminate boundaries.* London, Taylor and Francis: 275–86

Laviolette M, Seaman J W 1994 The efficacy of fuzzy representations of uncertainty. *IEEE Transactions on Fuzzy Systems* 2: 4–15

Leung Y C 1988 *Spatial analysis and planning under imprecision.* New York, Elsevier Science

Monckton C G 1994 An investigation into the spatial structure of error in digital elevation data. In Worboys M (ed.) *Innovations in GIS 1.* London, Taylor and Francis: 201–11

Moraczewski I R 1993 Fuzzy logic for phytosociology 1: syntaxa as vague concepts. *Vegetatio* 106: 1–11

Office of National Statistics 1997 *Harmonised concepts and questions for government social surveys.* London, Her Majesty's Stationery Office

Openshaw S (ed.) 1995a *Census users' handbook.* Cambridge (UK), GeoInformation International

Pawlak Z 1982 Rough sets. *International Journal of Computer and Information Sciences* 11: 341–56

Pratchett T 1986 *The light fantastic.* Gerrards Cross, Colin Smythe

Prescott J R V 1987 *Political frontiers and boundaries.* London, Allen and Unwin

Primavesi A L, Evans P A 1988 *Flora of Leicestershire.* Leicester, Leicestershire County Museum Service

Robinson V B 1988 Some implications of fuzzy set theory applied to geographic databases. *Computers, Environment, and Urban Systems* 12: 89–98

Rumley D, Minghi J V (eds) 1991 *The geography of border landscapes.* London, Routledge

Sainsbury R M 1995 *Paradoxes,* 2nd edition. Cambridge (UK), Cambridge University Press

Soil Classification Working Group 1991 *Soil classification, a taxonomic system for South Africa.* Memoirs on Agricultural Natural Resources of South Africa 15, Pretoria

Soil Survey Staff 1975 *Soil taxonomy: a basic system of soil classification for making and interpreting soil surveys.* USDA Agricultural Handbook 436. Washington DC, Government Printing Office

Tavernier R, Louis A 1984 *Soil map of the European Communities.* Luxembourg, Office of Offical Publications of the European Communities

Taylor J R 1982 *An introduction to error analysis: the study of uncertainties in physical measurements.* Oxford, Oxford University Press/Mill Valley, University Science Books

Viridians 1996 *Victorian flora database CD-ROM.* Brighton East, Victoria, Viridians Biological Databases

Walsby J C 1995 The causes and effects of manual digitising on error creation in data input to GIS. In Fisher P F (ed.) *Innovations in GIS 2.* London, Taylor and Francis: 113–22

Wang F, Hall G B, Subaryono 1990 Fuzzy information representation and processing in conventional GIS software: database design and application. *International Journal of Geographical Information Systems* 4: 261–83

Wang M, Gong P, Howarth P J 1991 Thematic mapping from imagery: an aspect of automated map generalisation. *Proceedings of AutoCarto 10.* Bethesda, American Congress on Surveying and Mapping: 123–32

Webster R, Oliver M A 1990 *Statistical methods in soil and land resource survey.* Oxford, Oxford University Press

Williamson T 1994 *Vagueness.* London, Routledge

Young E 1992 Hunter-gatherer concepts of land and its ownership in remote Australia and North America. In Anderson K, Gale F (eds) *Inventing places; studies in cultural geography.* Melbourne, Longman: 255–72

Zadeh L A 1965 Fuzzy sets. *Information and Control* 8: 338–53

Zadeh L A 1980 Fuzzy sets versus probability. *Proceedings of the IEEE* 68: 421

14

Propagation of error in spatial modelling with GIS

G B M HEUVELINK

Most GIS users are now well aware that the accuracy of GIS results cannot naively be based on the quality of the graphical output alone. The data stored in a GIS have been collected in the field, have been classified, generalised, interpreted or estimated intuitively, and in all these cases errors are introduced. Errors also derive from measurement errors, from spatial and temporal variation, and from mistakes in data entry. Consequently, errors are propagated or even amplified by GIS operations. But exactly how large are the errors in the results of a spatial modelling operation, given the errors in the input to the operation? This chapter describes the development, application, and implementation of error propagation techniques for quantitative spatial data. Techniques considered are Taylor series approximation and Monte Carlo simulation. The theory is illustrated using a case study.

1 INTRODUCTION

One of the most powerful capabilities of GIS, particularly for the earth and environmental sciences, is that it permits the derivation of new attributes from attributes already held in the GIS database. For example, elevation data in the form of a digital elevation model (DEM) can be used to derive maps of gradient and aspect (Hutchinson and Gallant, Chapter 9); or digital maps of soil type and gradient can be combined with information about soil fertility and moisture supply to yield maps of suitability for growing maize (Burrough 1986). The many basic types of function used for derivations of this kind are often provided as standard functions or *operations* in many GIS, under the name of 'map algebra' (Burrough 1986; Tomlin 1990).

In practice, many GIS operations are used in sequence in order to compute an attribute that is the result of a (computational) model. For instance, the channel flow at the outlet of a watershed can be computed after the relevant hydrological processes have been translated into mathematical equations, thus after reality has been approximated by a suitable computational model. Using GIS for the evaluation of computational models is identified here by the term *spatial modelling* within GIS.

To date, most work on spatial modelling with GIS has been concentrated on the business of deriving computational models that operate on spatial data, on the building of large spatial databases, and on linking computational models with the GIS. However, there is an important additional aspect that has long received too little attention. This concerns the issue of data quality and how errors in spatial attributes propagate through GIS operations.

1.1 The propagation of errors through GIS operations

It can safely be said that no map stored in a GIS is truly error-free. Note that the word 'error' is used here in its widest sense to include not only 'mistakes' or 'blunders', but also to include the statistical concept of error meaning 'variation' (Burrough 1986). An extensive account of important error sources in GIS has been given in a previous chapter (Veregin, Chapter 12).

When maps that are stored in a GIS database are used as input to a GIS operation, then the errors in the input will *propagate* to the output of the operation. Therefore the output may not be sufficiently reliable for correct conclusions to be drawn from it. Moreover, the error propagation continues when the output from one operation is used as input to an ensuing operation. Consequently, when no record is kept of the accuracy of intermediate results, it becomes extremely difficult to evaluate the accuracy of the final result.

Although users may be aware that errors propagate through their analyses, in practice they rarely pay attention to this problem. Perhaps experienced users know that the quality of their data is not reflected by the quality of the graphical output of the GIS, but they cannot truly benefit from this knowledge because the uncertainty of their data still remains unknown. No professional GIS currently in use can present the user with information about the confidence limits that should be associated with the results of an analysis (Burrough 1992; Forier and Canters 1996; Lanter and Veregin 1992).

The purpose of this chapter is to present a methodology for handling error and error propagation in (quantitative) spatial modelling with GIS. Note that this chapter mainly deals with the propagation of *quantitative attribute* errors in GIS, where in addition it is assumed that spatially referenced data are represented as fields, not as objects (Goodchild 1992). However, many of the results presented in this chapter can be generalised and are thus valuable for the general problem of error propagation in GIS. For instance, Wesseling and Heuvelink (1993) have applied the same methodology to spatial objects and the propagation of positional errors can also be studied using a similar approach (Griffith 1989; Keefer et al 1991; Stanislawski et al 1996). The propagation of categorical errors is more difficult because in such circumstances error probability distributions cannot easily be reduced to a few parameters. Some recent work in this area is given by Forier and Canters (1996), Goodchild et al (1992), Lanter and Veregin (1992), and Veregin (1994, 1996). Recent applications of error propagation in spatial modelling are described by Finke et al (1996), Haining and Arbia (1993), Heuvelink and Burrough (1993), Leenhardt (1995), Mowrer (1994), and Woldt et al (1996).

2 DEFINITION AND IDENTIFICATION OF A STOCHASTIC ERROR MODEL FOR QUANTITATIVE SPATIAL ATTRIBUTES

Before considering the propagation of error one must first give a suitable definition of error. An 'error' in a quantitative attribute can be conveniently defined as the difference between reality and our representation of reality (i.e. the map). For instance, if the nitrate concentration of the shallow groundwater at some location equals 68.6 g/m^3, while according to the map it is 62.9 g/m^3, then there will be no disagreement that in this case the error is 68.6 – 62.9=5.7 g/m^3. Generalising this example, let the true value of a spatial attribute at some location x be $a(x)$, and let the representation of it be $b(x)$. Then, according to the definition, the error $v(x)$ at x is simply the arithmetical difference $v(x) = a(x)-b(x)$.

We consider the situation in which the true value $a(x)$ is unknown, because if it were known, then error could simply be eliminated by assigning $a(x)$ to $b(x)$. What is known exactly is the representation $b(x)$, because this is the estimate for $a(x)$ that is available from the map. The error $v(x)$ is also not known exactly, but we should have some idea about the range or distribution of values that it is likely to take. For instance, we may know that the chances are equal that $v(x)$ is positive or negative, or we may be 95 per cent confident that $v(x)$ lies within a given interval.

Knowledge about the error $v(x)$ is thus limited to specifying a range or distribution of possible values. This type of information can best be conveyed by representing the error as a *random variable $V(x)$*. Note that notation using capitals is introduced here, in order to distinguish random variables from deterministic variables. Typically, a random variable is associated with the outcome of a probabilistic experiment, such as the throw of a die or the number drawn in a lottery. But a random variable is equally suited to model the concept of *uncertainty* (Fisher, Chapter 13). For instance, since we do not know the true nitrate concentration of the shallow groundwater, we may think that it is a value drawn from a large set of values that surround the estimated value of 62.9 g/m^3. Although we are aware that the attribute has only one fixed, deterministic value $a(x)$, our uncertainty about $a(x)$ allows us to treat it as the outcome of some random mechanism $A(x)$. We must then proceed by specifying the rules of this random mechanism, by saying how likely each possible outcome is. This will be done more formally in the next section.

2.1 Definition of the stochastic error model

Consider a quantitative spatial attribute

$A(.)=\{A(x) \mid x \in D\}$

that is defined on the spatial domain of interest D. Refer to the value of $A(.)$ at some location $x \in D$ as $A(x)$. The error model introduced in the previous section thus becomes:

$$A(x) = b(x) + V(x) \quad \text{for all } x \in D \quad (1)$$

where $A(x)$ and $V(x)$ are random variables and $b(x)$ is a deterministic variable. Note that $A(.)$ and $V(.)$ are not random variables but random fields, in the geostatistical literature also termed random functions (Cressie 1991; Journel and Huijbregts 1978).

Let us first consider the error at location x only. Denote the mean and variance of $V(x)$ by $E[V(x)]=\mu(x)$ and $Var(V(x))=\sigma^2(x)$. The mean $\mu(x)$ is often referred to as the systematic error or bias, because it says how much $b(x)$ systematically differs from $A(x)$. The standard deviation $\sigma(x)$ of $V(x)$ characterises the non-systematic, random component of the error $V(x)$. In standard error analysis, it is often assumed that errors follow the normal (Gaussian) distribution (Taylor 1982), but this is not always sensible. For instance, in geology, hydrology, and soil science, many attributes are skewed and the errors associated with them may be described more adequately using a lognormal distribution.

Next consider the spatial and multivariate extension of the error model. Although a complete characterisation of the error random field $V(.)$ would require its entire finite-dimensional distribution (Cressie 1991: 52), here we only define its first and second moments, which are assumed to exist. Let x and x' be elements of D. The (spatial auto-)correlation $\rho(x,x')$ of $V(x)$ and $V(x')$ is defined as:

$$\rho(x, x') = \frac{R(x, x')}{\sigma(x)\,\sigma(x')} \quad (2)$$

where $R(x,x')$ is the covariance of $V(x)$ and $V(x')$. Clearly, when $x=x'$ then covariance equals variance, so $R(x,x)=\sigma^2(x)$ and $\rho(x,x)=1$ for all $x \in D$.

When there are multiple attributes $A_i(x)$ and errors $V_i(x)$, $i=1,\ldots,m$, then for each of the attributes an error model $A_i(x)=b_i(x)+V_i(x)$ is defined, where the error $V_i(x)$ follows some distribution with mean $\mu_i(x)$ and variance $\sigma_i^2(x)$. Let $\rho_{ij}(x,x')$ be the (spatial cross-) correlation of $V_i(x)$ and $V_j(x')$, defined as:

$$\rho_{ij}(x,x') = \frac{R_{ij}(x, x')}{\sigma_i(x)\,\sigma_j(x')} \quad (3)$$

where $R_{ij}(x,x')$ is the covariance of $V_i(x)$ and $V_j(x')$. The cross-covariance function $R_{ij}(.\,,.)$ thus defines the covariance of different attribute errors, possibly at different locations.

To illustrate that errors in spatial attributes are often correlated, consider the example of soil pollution by heavy metals, such as is the case in the river Geul valley, in the south of the Netherlands (Leenaers 1991). Consider the concentration of lead and cadmium in the soil, maps of which are obtained from interpolating point observations. In this case the interpolation errors $V_{lead}(x)$ and $V_{cadmium}(x)$ are likely to be positively correlated, because *unexpectedly* high lead concentrations will often be accompanied by *unexpectedly* high cadmium concentrations. Unforeseen low concentrations will also often occur simultaneously. Heuvelink (1993) derives these error correlations mathematically for geostatistical interpolation.

The observation that errors in spatial attributes are often correlated is important because in what follows we will see that presence of non-zero correlation can have a marked influence on the outcome of an error propagation analysis.

2.2 Identification of the error model

To estimate the parameters of the error random field $V(.)$ in practice, certain stationarity assumptions have to be made (Cressie 1991: 53). This can be done in various ways. The most obvious way is to impose the assumptions directly on $V(.)$. This is acceptable when inference on $V(.)$ is based solely on observed errors at test points. For instance, to assess the error standard deviation of an existing DEM it may be sensible to assume that $\sigma(.)$ is spatially invariant, so that it can be estimated by the root mean squared error (RMSE), computed from the differences between the DEM and the true elevation at the test points (Fisher 1992). In addition, it may be sensible to assume that the spatial autocorrelation $\rho(x,x')$ is a (decreasing) function of only the distance $|x-x'|$. If sufficient test points are available (say 60 or more), then $\rho(.)$ can be estimated using geostatistical tools (Cressie 1991; Pannatier 1996).

However, in many situations it is not very sensible to impose the stationarity assumptions directly on the error map $V(.)$. In many situations $V(.)$ is the residual from mapping an attribute from point observations, and where the spatial variability of the attribute has been identified prior to, and has been incorporated in, the mapping. In order to avoid inconsistencies, the error model parameters should then be derived from the spatial variability of the attribute and the mapping procedure used. The spatial variability of the attribute may be characterised using a discrete, continuous, or mixed model of spatial variation, but in all three cases the mapping and error identification will involve some form of Kriging (Heuvelink 1996). The advantage of Kriging is that it not only yields interpolated values but that it also quantifies the interpolation error. For a discussion of Kriging, see Cressie (1991) or Mitas and Mitasova (Chapter 34).

3 THE THEORY OF ERROR PROPAGATION

The error propagation problem can now be formulated mathematically as follows. Let $U(.)$ be the output of a GIS operation $g(.)$ on the m input attributes $A_i(.)$:

$$U(.) = g(A_1(.), ..., A_m(.)) \tag{4}$$

The operation $g(.)$ may be one of various types, such as a standard filter operation to compute gradient and aspect from a gridded DEM (Carter 1992), a pedotransfer function to predict soil hydraulic properties from basic soil properties (Finke et al 1996), or a complex distributed runoff and soil erosion model (De Roo et al 1992). The objective of the error propagation analysis is to determine the error in the output $U(.)$, given the operation $g(.)$ and the errors in the input attributes $A_i(.)$. The output map $U(.)$ also is a random field, with mean $\xi(.)$ and variance $\tau^2(.)$. From an error propagation perspective, the main interest is in the uncertainty of $U(x)$, as contained in its variance $\tau^2(.)$.

It must first be observed that the error propagation problem is relatively easy when $g(.)$ is a linear function. In that case the mean and variance of $U(.)$ can be directly and analytically derived. The theory on functions of random variables also provides several analytical approaches to the problem for non-linear $g(.)$, but few of these can be resolved by simple calculations (Helstrom 1991). In practice, these analytically-driven methods nearly always rely on numerical methods for a complete evaluation. Thus for the general situation analytical methods are not very suitable. In this context, two alternative methods will now be discussed.

For practical purposes the discussion hereafter will be confined to point operations, i.e. GIS operations that operate on each spatial location x separately. This is no real restriction because non-point operations can be handled by minor modification (Heuvelink 1993). For notational convenience, the spatial index x will be dropped. It will also be assumed that the errors V_i have zero mean. This is because unbiasedness conditions are usually included in the mapping of the A_i.

3.1 Taylor series method

The idea of the Taylor series method is to approximate $g(.)$ by a truncated Taylor series centred at $\bar{b} = (b_1, ..., b_m)$. In case of the first order Taylor method, $g(.)$ is linearised by taking the tangent of $g(.)$ in $\bar{b}$. The linearisation greatly simplifies the error analysis, but only at the expense of introducing an approximation error.

The first order Taylor series of $g(.)$ around $\bar{b}$ is given by:

$$U = g(\bar{b}) + \sum_{i=1}^{m} (A_i - b_i) g_i'(\bar{b}) + \text{remainder} \tag{5}$$

where $g_i'(.)$ is the first derivative of $g(.)$ with respect to its i-th argument. By neglecting the remainder in Equation 5 the mean and variance of U are given as (Heuvelink et al 1989):

$$\xi \approx g(\bar{b}) \tag{6}$$

$$\tau^2 \approx \sum_{i=1}^{m} \sum_{j=1}^{m} \rho_{ij} \sigma_i \sigma_j g_i'(\bar{b}) g_j'(\bar{b}) \tag{7}$$

Thus the variance of U is the sum of various terms, which contain the correlations and standard deviations of the A_i and the first derivatives of $g(.)$ at $\bar{b}$. These derivatives reflect the sensitivity of U for changes in each of the A_i. From Equation 7 it also appears that the correlations of the input errors can have a marked effect on the variance of U. Note also that Equation 7 constitutes a well known result from standard error analysis theory (Burrough 1986: 128–31; Taylor 1982).

To decrease the approximation error invoked by the first order Taylor method, one option is to extend the Taylor series of $g(.)$ to include a second order term as well (Heuvelink et al 1989). This is particularly useful when $g(.)$ is a quadratic function, in which case the second order method is free of approximations and the first order method is not. The case study provides an example.

Another method comparable to the first order Taylor method has been proposed by Rosenblueth (1975). This method estimates ξ and τ^2 from 2^m function values of $g(.)$, evaluated at all 2^m corners of a hyperquadrant in m-dimensional space. Unlike the Taylor method, this method does not require that $g(.)$ is continuously differentiable.

3.2 Monte Carlo method

The Monte Carlo method (Hammersley and Handscomb 1979; Lewis and Orav 1989) uses an entirely different approach to analyse the propagation of error through the GIS operation (Equation 4). The idea of the method is to compute the result of $g(a_1,...,a_m)$ repeatedly, with input values a_i that are randomly sampled from their joint distribution. The model results form a random sample from the distribution of U, so that parameters of the distribution, such as the mean ξ and the variance τ^2, can be estimated from the sample.

The method thus consists of the following steps:

1 repeat N times:
   a generate a set of realisations a_i, $i=1,...,m$
   b for this set of realisations a_i, compute and store the output $u=g(a_1,...,a_m)$
2 compute and store sample statistics from the N outputs u.

A random sample from the m inputs A_i can be obtained using an appropriate random number generator (Lewis and Orav 1989; Ross 1990). Note that a conditioning step will have to be included when the A_i are correlated. One attractive method for generating realisations from a multivariate Gaussian distribution uses the Cholesky decomposition of the covariance matrix (Johnson 1987).

Application of the Monte Carlo method to error propagation with non-point operations requires the simultaneous generation of realisations from the random fields $A_i(.)$. This implies that spatial correlation will have to be accounted for. Various techniques can be used for stochastic spatial simulation, an attractive one being the sequential Gaussian simulation algorithm (Deutsch and Journel 1992).

The accuracy of the Monte Carlo method is inversely related to the square root of the number of runs N. This means that to double the accuracy, four times as many runs are needed. The accuracy thus slowly improves as N increases.

3.3 Evaluation and comparison of error propagation techniques

The main problem with the Taylor method is that the results are only approximate. It will not always be easy to determine whether the approximations involved using this method are acceptable. The Monte Carlo method does not suffer from this problem, because it can reach an arbitrary level of accuracy.

The Monte Carlo method brings with it other problems, however. High accuracies are reached only when the number of runs is sufficiently large, which may cause the method to become extremely time consuming. This will remain a problem even when variance reduction techniques such as Latin hypercube sampling are employed. Another disadvantage of the Monte Carlo method is that the results do not come in an analytical form.

As a general rule it seems that the Taylor method may be used to obtain crude preliminary answers. These should provide sufficient detail to be able to obtain an indication of the quality of the output of the GIS operation. When exact values or quantiles and/or percentiles are needed, the Monte Carlo method may be used. The Monte Carlo method will probably also be preferred when error propagation with complex operations is studied, because the method is easily implemented and generally applicable.

3.4 Sources of error contributions: the balance of errors

When the error analysis reveals that the output of $g(.)$ contains too large an error then measures will have to be taken to improve accuracy. When there is a single input to $g(.)$ then there is no doubt where the improvement must be sought, but what if there are multiple inputs to the operation? Also, how much should the error of a particular input be reduced in order to reduce the output error by a given factor? These are important questions that will now be considered.

To obtain answers to the questions above, consider Equation 7 again, which gives the variance of the output U using the first order Taylor method. When the inputs are uncorrelated, this reduces to:

$$\tau^2 \approx \sum_{i=1}^{m} \sigma_i^2 \, (g_i'(\bar{b}))^2 \tag{8}$$

Equation 8 shows that the variance of U is a sum of parts, each to be attributed to one of the inputs A_i. This *partitioning property* allows one to analyse how

211

much each input contributes to the output variance. Thus from Equation 8 it can directly be seen how much τ^2 will reduce from a reduction of σ_i^2. Clearly the output will mainly improve from a reduction in the variance of the input that has the largest contribution to τ^2. Note that this need not necessarily be the input with the largest error variance, because the sensitivity of the operation $g(.)$ for the input is also important. Note also that Equation 8 is derived under rather strong assumptions. When these assumptions are too unrealistic it may be advisable to derive the error source contributions using a modified Monte Carlo approach (Jansen et al 1994).

In the introduction, it was noted that a GIS operation is often in effect a computational model. Consequently, not only will *input error* propagate to the output of a GIS operation, but *model error* will as well. In practice, model error will often be a major source of error and should therefore be included in the error analysis. Ignoring it would severely underestimate the true uncertainty in the model output. Model error can be included by assigning errors to model coefficients or by adding a residual error term to the model equations.

If a reduction of output error is required, it will not necessarily be sensible to improve the input with the highest error contribution. This is because the cost of reducing input error may vary from attribute to attribute. However, in many cases it will be most rewarding to strive for a *balance of errors*. When the error in an attribute has a marginal effect on the output, then there is little to be gained from mapping it more accurately. In that case, extra sampling efforts can much better be directed to an input attribute that has a larger contribution to the output error. For instance, if a pesticide leaching model is sensitive to soil organic carbon and less so to soil bulk density, then it is more important to map the former more accurately (Loague et al 1989).

The example of the pesticide leaching model draws attention to the fact that a balance of errors must also include model error. It is clearly unwise to spend much effort on collecting data if what is gained is immediately thrown away by using a poor model. On the other hand, a simple model may be as good as a complex model if the latter needs lots of data that cannot be accurately obtained. This is why researchers in catchment hydrology have raised the question of whether there is much benefit to be gained from developing ever more complex models when the necessary inputs cannot be evaluated in the required spatial and temporal resolution (Beven 1989; Grayson et al 1992).

4 APPLICATION TO MAPPING SOIL MOISTURE CONTENT WITH LINEAR REGRESSION FOR THE ALLIER FLOODPLAIN SOILS

As part of a research study in quantitative land evaluation, the World Food Studies (WOFOST) crop simulation model (Diepen et al 1989) was used to calculate potential crop yields for floodplain soils of the Allier river in the Limagne rift valley, central France. The moisture content at wilting point (Θ_{wp}) is an important input attribute for the WOFOST model. Because Θ_{wp} varies considerably over the area in a way that is not linked directly with soil type, it was necessary to map its variation separately to see how moisture limitations affect the calculated crop yield.

Unfortunately, because Θ_{wp} must be measured on samples in the laboratory, it is expensive and time-consuming to determine it for a sufficiently large number of data points for Kriging. An alternative and cheaper strategy is to calculate Θ_{wp} from other attributes which are cheaper to measure. Because the moisture content at wilting point is often strongly correlated with the moisture content at field capacity (Θ_{fc}) and the soil porosity (Φ), both of which can be measured more easily, it was decided to investigate how errors in measuring and mapping these would work through to a map of calculated Θ_{wp}. The following procedure was used to obtain a map of the mean and standard deviation of Θ_{wp}.

The properties Θ_{wp}, Θ_{fc} and Φ were determined in the laboratory for 100 cc cylindrical samples taken from the topsoil (0–20 cm) at 12 selected sites shown as the circled points in Figure 1.

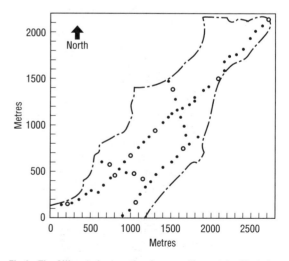

Fig 1. The Allier study area showing sampling points. Circled sites are those used to estimate the regression model.

These results were used to set up a pedotransfer function, relating Θ_{wp} to Θ_{fc} and Φ, which took the form of a multiple linear regression:

$$\Theta_{wp} = \beta_0 + \beta_1\, \Theta_{fc} + \beta_2\, \Phi + \varepsilon \tag{9}$$

The coefficients β_0, β_1, and β_2 were estimated using standard ordinary least squares regression. The estimated values for the regression coefficients and their respective standard deviations were $\hat{\beta}_0 = -0.263 \pm 0.031$, $\hat{\beta}_1 = 0.408 \pm 0.096$, $\hat{\beta}_2 = 0.491 \pm 0.078$. The standard deviation of the residual ε was estimated as 0.0114. The correlation coefficients of the regression coefficients were $\rho_{01} = -0.221$, $\rho_{02} = -0.587$, $\rho_{12} = -0.655$. The regression model explains 94.8 per cent of the variance in the observed Θ_{wp}, indicating that the model is satisfactory. Note that presence of spatial correlation between the observations at the 12 locations was ignored in the regression analysis.

Sixty-two measurements of Θ_{fc} and Φ were made in the field at the sites indicated in Figure 1. From these data experimental variograms were computed. These were then fitted using the linear model of coregionalisation (Journel and Huijbregts 1978). For the purposes of this study the input data for the regression were mapped to a regular 50×50 m grid using block co-Kriging with a block size of 50×50 m. The block co-Kriging yielded raster maps of means and standard deviations for both Θ_{fc} and Φ, as well as a map of the correlation of the block co-Kriging prediction errors. Figure 2 displays these maps. Note that there are clear spatial variations in the correlation between the block co-Kriging errors.

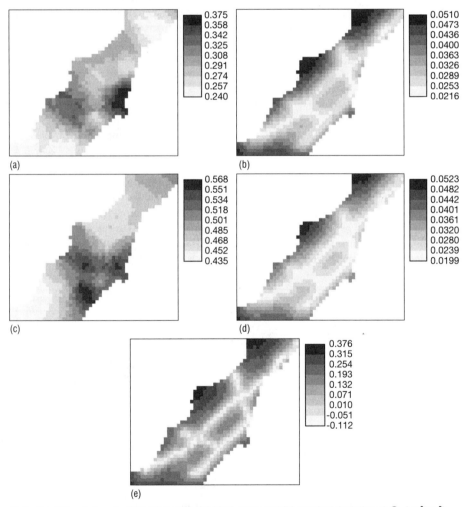

Fig 2. Kriging results for the Allier study area (50 x 50 m grid): (a) block mean and (b) standard deviation of Θ_{fc} (cm^3/cm^3), (c) block mean and (d) standard deviation of Φ (cm^3/cm^3), (e) correlation of co-Kriging prediction errors of Θ_{fc} and Φ.

The maps of Θ_{fc} and Φ were substituted in the regression Equation 9 yielding maps of the attribute Θ_{wp} and the associated error. The operation is a quadratic function and therefore the second order Taylor was considered the most appropriate error propagation technique. Because the model coefficients and the field measurements were determined independently, the correlation between the $\hat{\beta}_i$ and the co-Kriging prediction errors was taken to be zero. The results are given in Figure 3. The accuracy of the map of Θ_{wp} is reasonable: the standard deviation in Θ_{wp} rarely exceeds 25 per cent of the mean. These maps could be used as the basis of a subsequent error propagation analysis in the WOFOST crop yield model.

If an uncertainty analysis with WOFOST would show that the errors in Θ_{wp} cause errors in the output of WOFOST that are unacceptably large, then the accuracy of the map of Θ_{wp} would have to be improved. In order to decide how to proceed in such a situation, the contribution of each individual error source was determined using the partitioning property discussed in the previous section. Figure 4 presents the results and these show that both Θ_{fc} and Φ form the main source of error. Only in the immediate vicinity of the data points is the model a meaningful source of uncertainty, as would be expected because there the co-Kriging variances of Θ_{fc} and Φ are the smallest.

Thus the main source of error in Θ_{wp} is that associated with the Kriging errors of Θ_{fc} and Φ. Improvement of the quality of the map of Θ_{wp} can thus best be done by improving the maps of Θ_{fc} and Φ, by taking more measurements over the study area. The variograms of Θ_{fc} and Φ could be used to assist in optimising sampling (McBratney et al 1981). This technique would allow one to judge *in advance* how much improvement is to be expected from the extra sampling effort.

5 DISCUSSION AND CONCLUSIONS

Error propagation in spatial modelling with GIS is a relevant research topic because rarely if ever are the data stored in a GIS completely error-free. In this chapter several methods were described for analysing the propagation of errors. None of these methods is perfect: some do not apply to all types of operations, others are extremely time consuming or involve large approximation errors. However, in practice there will often be at least one method that is appropriate for a given situation. Thus the methods are in a sense complementary, and as a group in almost all cases they enable one to carry out an error propagation analysis successfully.

Unfortunately, at present the majority of GIS users still has no clear information about the errors associated with the attributes that are stored in the GIS. This is an important problem because an error propagation analysis can only yield sensible results if the input errors have realistic values. Often there will only be crude and incomplete estimates of input error available. This lack of information is perhaps the main reason why error propagation analyses are still the exception rather than the rule in everyday GIS practice. It is essential that map makers become aware that they should routinely convey the accuracy

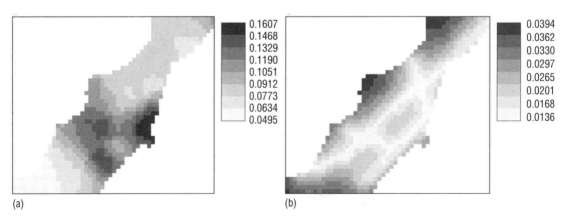

(a)   (b)

Fig 3. Results of the error propagation: (a) block mean and (b) standard deviation of Θ_{wp} (cm³/cm³) as obtained with the regression model.

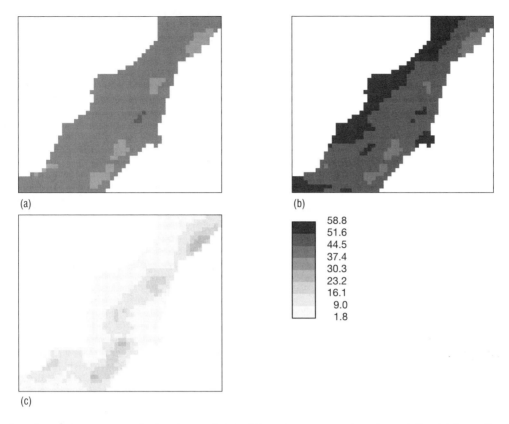

58.8
51.6
44.5
37.4
30.3
23.2
16.1
9.0
1.8

Fig 4. Maps showing the relative contributions (per cent) of the different input errors to the variance of Θ_{wp}: (a) due to Θ_{fc}, (b) due to Φ, and (c) due to the regression model.

of the maps they produce, even when accuracy is less than expected. At the same time, it is important that GIS manufacturers increase their efforts to add error propagation functionality to their products.

It is important to note that an error analysis offers much more than the computation of output error. The partitioning property of an error analysis allows one to determine how much each individual input contributes to the output error. Information of this sort may be extremely useful, because it allows users to explore how much the quality of the output improves, given a reduction of error in a particular input. Thus the improvement foreseen due to intensified sampling can be weighed against the extra sampling costs.

The partitioning property can also be used to compare the contributions of input and model error. With the advent of GIS, and the many computational models that often come freely with it, there is an increased risk of disturbing the balance

between input and model error. When there is no protection against improper use then ignorant users will be tempted to apply models to improper scales, use them for purposes for which they were not developed, or combine them with data that are too uncertain (Heuvelink 1998). These problems can only be tackled when users become more aware of the issue of spatial data quality and when error propagation analysis becomes a routine instrument available to the GIS community.

References

Beven K 1989 Changing ideas in hydrology – the case of physically-based models. *Journal of Hydrology* 105: 157–72

Burrough P A 1986 *Principles of geographical information systems for land resources assessment*. Oxford, Clarendon Press

Burrough P A 1992b Development of intelligent geographical information systems. *International Journal of Geographical Information Systems* 6: 1–11

Carter J R 1992 The effect of data precision on the calculation of slope and aspect using gridded DEMs. *Cartographica* 29: 22–34

Cressie N A C 1991 *Statistics for spatial data*. New York, John Wiley & Sons Inc.

De Roo A P J, Hazelhoff L, Heuvelink G B M 1992 Estimating the effects of spatial variability of infiltration on the output of a distributed runoff and soil erosion model using Monte Carlo methods. *Hydrological Processes* 6: 127–43

Deutsch C V, Journel A G 1992 *GSLIB: geostatistical software library and user's guide*. New York, Oxford University Press

Diepen C A van, Wolf J, Keulen H van, Rappoldt C 1989 WOFOST: a simulation model of crop production. *Soil Use and Management* 5: 16–24

Finke P A, Wösten J H M, Jansen M J W 1996 Effects of uncertainty in major input variables on simulated functional soil behaviour. *Hydrological Processes* 10: 661–9

Fisher P F 1992 First experiments in viewshed uncertainty: simulating fuzzy viewsheds. *Photogrammetric Engineering and Remote Sensing* 58: 345–52

Forier F, Canters F 1996 A user-friendly tool for error modelling and error propagation in a GIS environment. In Mowrer H T, Czaplewski R L, Hamre R H (eds) *Spatial accuracy assessment in natural resources and environmental sciences*. Fort Collins, USDA Forest Service General Technical Report RM-GTR-277: 225–34

Goodchild M F 1992a Geographical data modeling. *Computers and Geosciences* 18: 401–8

Goodchild M F, Sun G, Yang S 1992 Development and test of an error model for categorical data. *International Journal of Geographical Information Systems* 6: 87–104

Grayson R B, Moore I D, McMahon T A 1992 Physically based hydrologic modelling: 2. Is the concept realistic? *Water Resources Research* 28: 2659–66

Griffith D A 1989 Distance calculations and errors in geographic databases. In Goodchild M F, Gopal S (eds) *Accuracy of spatial databases*. London, Taylor and Francis: 81–90

Haining R P, Arbia G 1993 Error propagation through map operations. *Technometrics* 35: 293–305

Hammersley J M, Handscomb D C 1979 *Monte Carlo methods*. London, Chapman and Hall

Helstrom C W 1991 *Probability and stochastic processes for engineers*. New York, Macmillan

Heuvelink G B M 1993 'Error propagation in quantitative spatial modelling: applications in geographical information systems'. PhD thesis. Utrecht, Netherlands Geographical Studies 163

Heuvelink G B M 1996 Identification of field attribute error under different models of spatial variation. *International Journal of Geographical Information Systems* 10: 921–36

Heuvelink G B M 1998 Uncertainty analysis in environmental modelling under a change of spatial scale. *Nutrient Cycling in Agro-Ecosystems* 50: 257–66

Heuvelink G B M, Burrough P A 1993 Error propagation in cartographic modelling using Boolean methods and continuous classification. *International Journal of Geographical Information Systems* 7: 231–46

Heuvelink G B M, Burrough P A, Stein A 1989 Propagation of errors in spatial modelling with GIS. *International Journal of Geographical Information Systems* 3: 303–22

Jansen M J W, Rossing W A H, Daamen R A 1994 Monte Carlo estimation of uncertainty contributions from several independent multivariate sources. In Grasman J, Straten G van (eds) *Predictability and non-linear modelling in natural sciences and economics*. Dordrecht, Kluwer: 334–43

Johnson M E 1987 *Multivariate statistical simulation*. New York, John Wiley & Sons Inc.

Journel A G, Huijbregts C J 1978 *Mining geostatistics*. London, Academic Press

Keefer B J, Smith J L, Gregoire T G 1991 Modeling and evaluating the effects of stream mode digitizing errors on map variables. *Photogrammetric Engineering and Remote Sensing* 57: 957–63

Lanter D P, Veregin H 1992 A research paradigm for propagating error in layer-based GIS. *Photogrammetric Engineering and Remote Sensing* 58: 825–33

Leenaers H 1991 Deposition and storage of solid-bound heavy metals in the floodplains of the river Geul (The Netherlands). *Environmental Monitoring and Assessment* 18: 79–103

Leenhardt D 1995 Errors in the estimation of soil water properties and their propagation through a hydrological model. *Soil Use and Management* 11: 15–21

Lewis P A W, Orav E J 1989 *Simulation methodology for statisticians, operations analysts, and engineers* Vol. 1. Pacific Grove, Wadsworth & Brooks/Cole

Loague K, Yost R S, Green R E, Liang T C 1989 Uncertainty in pesticide leaching assessment in Hawaii. *Journal of Contaminant Hydrology* 4: 139–61

McBratney A B, Webster R, Burgess T M 1981 The design of optimal sampling schemes for local estimation and mapping of regionalized variables: 1. Theory and method. *Computers and Geosciences* 7: 331–4

Mowrer H T 1994 Monte Carlo techniques for propagating uncertainty through simulation models and raster-based GIS. In Congalton R G (ed) *Proceedings of the international symposium on spatial accuracy of natural resource databases*. Washington, American Society for Photogrammetry and Remote Sensing: 179–88

Pannatier Y 1996 *VARIOWIN: software for spatial data analysis in 2D*. New York, Springer

Rosenblueth E 1975 Point estimates for probability moments. *Proceedings of the National Academy of Sciences of the United States of America* 72: 3812–14

Ross S M 1990 *A course in simulation*. New York, MacMillan

Stanislawski L V, Dewitt B A, Shrestha R L 1996 Estimating positional accuracy of data layers within a GIS through error propagation. *Photogrammetric Engineering and Remote Sensing* 62: 429–33

Taylor J R 1982 *An introduction to error analysis: the study of uncertainties in physical measurement*. Oxford, Oxford University Press/Mill Valley, University Science Books

Tomlin C D 1990 *Geographic information systems and cartographic modeling*. Englewood Cliffs, Prentice-Hall

Veregin H 1994 Integration of simulation modelling and error propagation for the buffer operation in GIS. *Photogrammetric Engineering and Remote Sensing* 60: 427–35

Veregin H 1996 Error propagation through the buffer operation for probability surfaces. *Photogrammetric Engineering and Remote Sensing* 62: 419–28

Wesseling C G, Heuvelink G B M 1993 Manipulating quantitative attribute accuracy in vector GIS. In Harts J, Ottens H F L, Scholten H J (eds) *Proceedings EGIS 93*. Utrecht, EGIS Foundation: 675–84

Woldt W, Goderya F, Dahab M, Bogardi I 1996 Consideration of spatial variability in the management of non-point source pollution to groundwater. In Mowrer H T, Czaplewski R L, Hamre R H (eds) *Spatial accuracy assessment in natural resources and environmental sciences*. Fort Collins, USDA Forest Service General Technical Report RM-GTR-277: 49–56

217

15
Detecting and evaluating errors by graphical methods

M K BEARD AND B P BUTTENFIELD

Both uncertainty and errors are inherent in spatial databases. The processes of observing, measuring, interpreting, classifying, and analysing data give rise to systematic and random errors. Some errors may be quite large (blunders) and easily detectable. Other errors and uncertainties in spatial data are more subtle and are not easily detected or evaluated. Casual users of GIS may not be aware of their presence or even the possibility of their existence. These are the most problematic and the ones we must try hardest to illuminate. Graphical methods in conjunction with error analysis provide a means for identifying both gross and subtle errors and evaluating the uncertainty in geographical data. This chapter outlines a rationale for the use of graphical methods, highlights several historical and recent examples, develops a framework linking error analysis and graphical methods, and points to research challenges for the future and the potential for new techniques arising from technical innovations.

1 INTRODUCTION

Several other chapters in this volume discuss error and uncertainty in spatial databases and the importance of making these known to GIS users (Fisher, Chapter 13; Heuvelink, Chapter 14; Hunter, Chapter 45; Veregin, Chapter 12). This chapter focuses on revealing error in geographical data and GIS by graphical means, used in conjunction with error analysis methods.

1.1 Rationale for graphical methods

Almost 200 years ago William Playfair began the serious use of graphs for looking at data. Many of the same issues which motivated Playfair to develop graphical solutions then are present today. Graphical methods for error detection and evaluation are motivated by several factors including physiological, technical, and institutional concerns. First, it is generally accepted that the human information processing system has strong acuity for visualisation and an exceptional ability to recognise structure and relationships. Representing information in a form that matches our perceptual capabilities (mainly visual) makes the process of getting information and digesting it easier and more effective (Gershon and Brown 1996). Second, spatial structure is more easily expressed and grasped through graphic or cartographic representation. Third, graphical methods are a fast communication channel and one capable of carrying high volumes. These characteristics make graphical methods highly suitable for human comprehension of the complex, multidimensional aspects of spatial data quality.

In terms of technical motivations, the growing interest in digital libraries and the National Spatial Data Infrastructure (NSDI) provide new impetus for documenting spatial information reliability. More and more spatial data and geographical information processing resources are rapidly becoming accessible over the Internet. The implications of this for data evaluation are profound. Intelligent use of such resources requires a substantial investment in metadata (data describing the data; Guptill, Chapter 49) and the availability of sense-making tools to digest large volumes of metadata and data. Graphical methods may provide the most efficient means to evaluate the quality of large volumes of geographic data as they become available through digital libraries and NSDI.

Institutional motivations centre around several national and international standards efforts. The US standard (the Spatial Data Transfer Standard or SDTS; Morrison 1992) categorised an initial set of data quality components, and recent completion of a Metadata Content Standard (FGDC 1995) extended and expanded representation guidelines for various categories of spatial data error. Adoption of standards similar to SDTS in other nations demonstrates international recognition of the complexities encountered in detecting and managing errors. In France for example, the effort generated MEGRIN standards (Salgé, Chapter 50; Salgé et al 1992).

1.2 Limitations of graphical methods

Graphical methods are not always either an effective solution or a substitute for conventional numerical analytical tools. Some researchers have suggested that graphic design for data analysis and presentation is still largely unscientific (Cox 1978). Graphical methods like other communication channels are open to misinterpretation. The cartographic literature is full of examples of techniques and their possible

misinterpretations (Monmonier 1991; Robinson et al 1985). As MacEachren (1994) suggests, data exploration tools allow us to identify potentially meaningful patterns that we might otherwise miss, but these tools cannot always determine the probability that the pattern we see is real. Despite limitations in graphic methods we can find several compelling examples of effective use of graphical techniques.

2 EXAMPLES OF GRAPHICAL METHODS

This section reviews several examples of graphic techniques used to detect, evaluate, and display errors. Several disciplines have contributed to these developments including cartography, spatial statistics, statistical graphics, scientific visualisation, and spatial error modelling.

2.1 Graphical methods in statistics

Tukey (1977) was the force behind exploratory data analysis (EDA) and many of the well-known graphical methods for exploring data. These methods highlight unusual values which may in fact

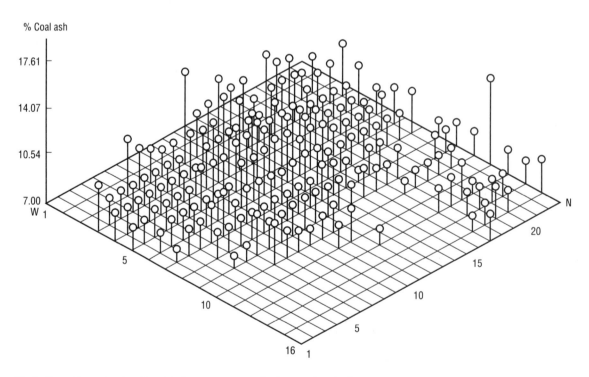

Fig 1. Three-dimensional scatter plot of core measurements of per cent coal ash.
Source: Cressie 1991

be errors. Tukey's work has been carried on and expanded by others (Becker et al 1987; Chambers et al 1983; Cleveland 1993). The shortcoming of most of these aspatial methods is that they do not consider spatial dependencies and as a result do not detect values which may be unusual in a spatial context. Cressie (1991) identifies some EDA methods which overcome this limitation (see Figures 1 and 2). These methods do not provide absolute indication of error but highlight potentially suspect values. Anselin (Chapter 17) describes exploratory methods for spatial data in more detail.

2.2 Graphical methods in cartography

The traditions of map making have included remarkably few methods for displaying uncertainty. Reliability diagrams were an early attempt to display variation in source documents used to compile maps (Wright 1942). More theoretical treatments were applied to projection distortion (Imhof 1964; Maling 1973; Tissot 1881).

2.3 Graphical methods related to GIS

Graphical means for detecting errors are now being explored and implemented in GIS software. Many recent examples illustrate expansion of Bertin's (1983) graphical framework (visual variables). New visual variables have been explored, including defocusing of features (MacEachren 1994; McGranaghan 1993) and development of multivariate symbols (Hancock 1993). Much of the recent change can be attributed to the emergence of new visualisation technologies (voxel-based 'true' 3-dimensional displays, animation, hypermedia).

Specific examples of tools developed for integration with GIS include those implemented by Fisher (1994a, 1994b), Goodchild et al (1994), Hunter and Goodchild (1995), MacEachren et al (1993), Mitasova et al (1995), and Paradis and Beard (1994). MacEachren et al (1993) developed a reliability visualisation tool (RVIS) which supports several options for viewing data and metadata (reliability). Reliability estimates in this system are based on Kriging residuals and cross validation. The display options

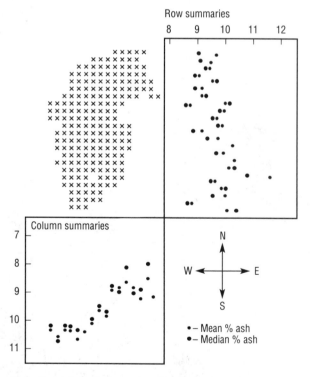

Fig 2. Comparison of mean and median summaries of non-stationarity. Units are in per cent coal ash. Comparison of mean and mean summaries for rows and columns can highlight atypical observations.
Source: Cressie 1991

221

include side-by-side, overlay and merged displays. The merged displays make use of several different bivariate mapping schemes. RVIS also includes a focusing tool which allows a user to specify interactively a subset of the data for further analysis. By interacting with a slider bar, an analyst can select a data range and reliability range to be displayed (Figure 3).

Fisher developed a technique referred to as error animation to view the reliability of classified imagery (Fisher 1994b) and soil maps (Fisher 1994c). For classified imagery the technique uses over-all, producer, and user accuracies as a foundation. The uncertainty inherent in the assignment of a pixel to a class is conveyed by making the value or colour of a pixel proportionate to the strength of it belonging to a particular class. In the case of soil maps the process uses randomisation to display either primary soil type or an inclusion at any particular pixel location on the map. The changing and random location of the inclusions is meant to convey the

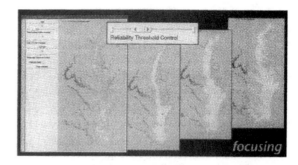

Fig 3. An implementation of the operation of focusing. The set of images illustrates the use of threshold settings for data reliability. *Source:* MacEachren et al 1993

impression that the location of inclusions is not precisely known.

Goodchild et al (1992) use a fuzzy classifier as a foundation for creating multinomial probability fields. The fuzzy class memberships become

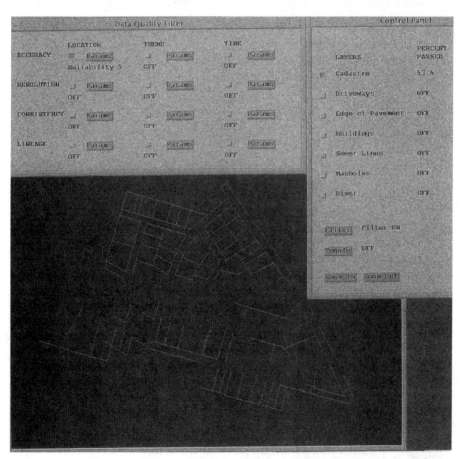

Fig 4. An implementation of the data quality filter showing parcels filtered for positional accuracy. Lines not shown indicate those not meeting the threshold set by the user.

parameters of an error model, and the range of possibilities is defined by realisations of the model. A display of realisations of the error model can inform users of the potential variation.

The data quality filter (Paradis and Beard 1994) allows a user to specify a data quality parameter (e.g. positional accuracy), a quality measure (e.g. root mean square error, RMSE) and a threshold value. The filter is applied to the data and only data meeting this threshold are displayed (Figure 4). Users are informed of how many data did not meet their specified threshold and they can toggle the display to see data which did not meet this cut-off level (Figure 5). Users can interactively adjust the threshold value in order to experiment with data that satisfy higher or lower constraints.

Hunter and Goodchild (1995) describe a probability mapping approach for representing the uncertainty of the horizontal position of a nominated terrain elevation value. They compute the probability of a cell exceeding or being exceeded by a threshold value. Once the probability for a cell has been computed it can be displayed according to different schemes. Plate 9 displays a colour ramp which shows the probability of a cell exceeding the chosen 350-metre elevation.

The visualisation tools developed by Mitasova et al (1995) incorporate multidimensional interpolation, visualisation of the resulting model, and predictive accuracy of model results using cross-validation. The visualisation tools allow cross-validation error to be viewed separately from the data, at a single depth, across different depths, at different times (Plate 10) and in combination with the data (Plate 11). These tools have been incorporated in the GRASS GIS software system. This work includes several novel visualisations but points out the many difficulties in trying to visualise reliability and data simultaneously particularly for 3-dimensional data representations (Plate 12).

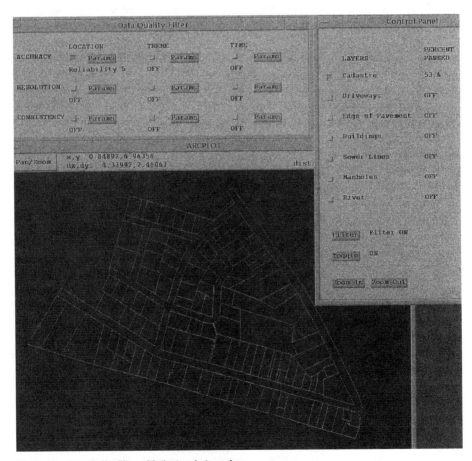

Fig 5. The data quality filter with the toggle turned on.

3 CHALLENGES IN GRAPHIC ERROR DETECTION AND EVALUATION

Visualisation of data is a 'demanding display problem' (Robertson 1991; see also Kraak, Chapter 11) which becomes even more demanding when the display must also include error and uncertainty and address the special characteristics of spatial data. The challenges include: (1) graphic design issues; (2) metadata issues; (3) error analysis issues; and (4) user satisfaction issues.

3.1 Graphic design issues

Graphic detection and evaluation of spatial data error and uncertainty create particular problems for graphic design. Firstly, they require a representation of space or linkage of aspatial displays to a spatial representation (Monmonier 1989) so users can be informed of the spatial distribution of errors or uncertainty. Spatial displays provide users with information on whether errors are regular, random, or clustered in space and may help users to comprehend the cause of errors. Depending on the dimensionality of the data the spatial representation may be 2- or 3-dimensional. Two-dimensional displays restrict views of the full 3-dimensional space but 3-dimensional displays add substantial cognitive and computational costs.

Second, graphic displays need to allow for both implicit and explicit displays of uncertainty. Uncertainty can be conveyed implicitly with visual variables which suggest uncertainty (e.g. fog, unfocused displays, unsaturated colours: McGranaghan 1993). The alternative is to quantify and display the uncertainty explicitly. Quantification of the uncertainty requires a solid understanding arrived at through error analysis. The graphic challenge lies in communicating the quantity visually. Cleveland and McGill (1984) provide a very helpful theoretical and empirical presentation of elementary perceptual tasks and their accuracy in judging quantitative values. For the most part visual variables convey only nominal and ordinal information. However interaction with a graphic can provide access to numerical values. A maxim of Tufte (1983)

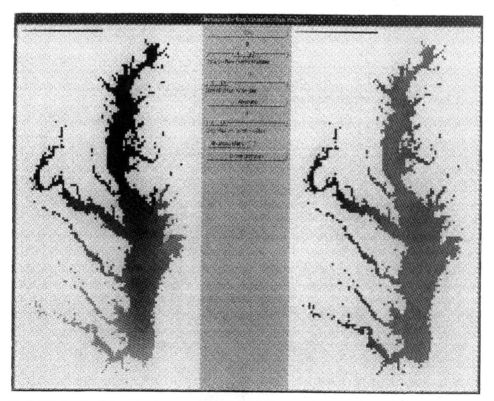

Fig 6. Side-by-side displays showing the same saturation/value range for both data and reliability. The representations for data and reliability only differ in hue and thus can be compared efficiently.

is that text and images can and should be combined. Thus we should not hesitate to incorporate numbers in displays when they may be appropriate.

Third, the graphic display should allow a data distribution and its reliability to be displayed independently or jointly. Complex graphic design issues arise in trying to display data and reliability together so users can observe correlations in the patterns. MacEachren (1994) offers three possibilities for joint display of data and reliability: (1) side-by-side images; (2) composite images; and (3) sequenced images. Each of these options presents its own graphic challenges.

In side-by-side displays the viewer must interpret two images simultaneously. Bertin's (1983) image theory applies in that we need to reduce the number of variables the viewer must process in relating the two images. This suggests that the image representing the data closely matches the image representing the data reliability. The images should be the same size and have the same coordinate scales. In addition, operations in side-by-side images should be linked. A zoom or pan operation in one image should be reflected in the other. MacEachren et al (1993) address some of these issues in their RVIS system. After experimenting with using value to represent the data and saturation to represent reliability they determined that an identical combined value/saturation scale for both data and reliability with dark equal to 'more' was the most effective representation (see Figure 6).

The composite mode of displaying data and reliability together requires overlay of contrasting visual variables, bivariate or multivariate mapping. Bertin (1983) proposes that discrimination of multiple variables in overlaid distributions can be maximised by expressing different data variables with symbols of different dimensions (point, line, area). Mitasova et al (1995) and MacEachren et al (1993) adopt Bertin's strategy by using point symbols to display reliability estimates in combination with coloured surface representations of data values. Brewer (1994) offers several useful strategies for effective use of colour in bivariate maps (see Plate 13).

The third method displays data and reliability images in sequence. When images are sequenced other considerations come into play such as the interval of time between display of the alternating images. The visual frame of reference must also remain constant between images so user attention can be devoted to the changes in the data.

Other options include linked displays and multiple version displays (Anselin, Chapter 17). For disjoint but linked images (Monmonier 1989) there must be common visual cues for the same variable in different contexts. In geographical brushing, a data value in one window is highlighted in the same colour as its representation in another view in order that users may pair up the representations. In multiple version displays we do not have separate visual variables or separate depictions for data and reliability. Instead we display multiple realisations which by their differences indicate a range of uncertainty in the data. They can be displayed as small multiples as described by Tufte (1983), or sequenced using animation (Dibiase et al 1992). Examples of geostatistical simulations are shown by Englund (1993). Uncertainty in this case is expressed implicitly by showing the range of possible variations.

Iteration (Cleveland 1993) is a particularly useful concept for viewing errors or uncertainty and it can serve several purposes. Displaying data in combination with error and uncertainty can quickly exhaust the available graphic variables. Very complex symbols can be designed (for an example see McGranaghan 1993) but these can easily become too complex and detract from interpretation of the intended message. The objective is to create more, but less complex, displays.

Since visual displays themselves can generate misleading information several iterations of a display can help to convey the uncertainty that has arisen out of map design decisions. MacEachren (1994) recommends generating multiple views rather than trying to pick a single best method of representation. In this case the iterations are over changes in visual variables, class breaks, and other map design components rather than the data. This is in contrast to viewing several simulated results in which the underlying data distribution changes, as discussed above. Viewers should be aware of possibilities for both. The challenge here is to create and display several graphic images quickly without the user losing track of his or her goal.

3.2 Metadata issues

Spatial data are frequently poorly documented. Information on how the data were collected, the sampling design, and whether any compilation or processing steps were performed on the data is usually minimal or missing. This is a serious problem for error analysis as without this information there is little basis on which to proceed. Fortunately the standards efforts mentioned above are having some impact and geographic datasets are appearing with more documentation (and see Salgé, Chapter 50). Several issues remain on how such metadata can be effectively stored with the data and maintained as the data evolve (see Goodchild and Longley, Chapter 40; Guptill, Chapter 49).

3.3 Error analysis issues

In most cases errors and uncertainty in spatial data are not simply detected by displaying the raw data (although examples of this are possible). As Tufte (1983) points out, graphics gather their power from content and interpretation beyond the immediate display of numbers. Thus good graphic design and, by association, effective detection and evaluation are highly dependent on effective error analysis.

Simply plotting the data can work as an error detection device because we often have some expectation about the pattern we will see. Deviations from this pattern suggest errors. Thus we can say that all error detection requires some model or reference framework, either implicit or explicit, from which departures can be determined. These may include (1) a known or postulated distribution for a set of observations; (2) a hypothesised or assumed relationship; (3) an expected set or range of values; or (4) an independent (and more accurate) set of observations. These models and frameworks can range from simple and inexpensive to complex and expensive.

Statistics provide one framework for describing and modelling error and uncertainty and thus for visualising it (Fisher, Chapter 13; Goodchild et al 1994). In standard statistics, errors and their significance are characterised by their distance from the central trend in values. This has some limitations as a method for detecting errors since outliers may in fact be unusual values and not errors. However statistical methods make clear that in order for detection to be possible we must first establish some expected distribution for values. In the case of spatial data we can add departures from assumed stationarity of mean or stationarity of dependence as the basis for detection of possible errors. For example, we should be suspicious of observations when they are unusual with respect to their neighbours.

In the case of raw data a range of exploratory techniques can be applied to identify outliers, detect blunders, and perform preliminary identification of data structure and statistical properties. Exploratory techniques are most appropriate where observational data are not obtained by formal means, where measures are not very precise (often made on nominal or ordinal scales), or where real repetition is neither feasible nor practical (Haining 1990). These are common characteristics of many spatial datasets. Cressie (1993) outlines some exploratory techniques for spatial data.

Detection can also rely on establishment of a set of consistency rules – rules indicating ranges of expected values or expected relationships between values. Many GIS apply a set of topological rules deriving from a map model as the basis for automated geometric error detection. Topological rules such as the requirement that all chains begin and end with a node, or that all polygons must close, are applied against the data and any geometric configurations which deviate from these rules are flagged. Most GIS editing packages support graphic highlighting of these inconsistencies for their easy visual detection as well as display of their spatial distribution. Bicking and Beard (1995) describe a formalisation for detection of inconsistencies in attribute data by analysing symbol encoding and symbol relationships.

Other error detection methods require ground truth data or other sources of higher accuracy for their computation (e.g. RMSE: Fisher, Chapter 13; Hutchinson and Gallant, Chapter 9; Veregin, Chapter 12). Root mean square error measures the error between a mapped point and a measured ground position. A limitation of this measure is that the error standard deviation is spatially invariant. There is no information about variation in positional accuracy at individual points. For survey measurements additional information contained as redundancies within the survey network can allow for computation of positional error for individual points within the network. Comprehensive ground checks are expensive, however.

Error and uncertainty in spatial data are not static. New error and uncertainty can occur as data are subjected to geographical information processing operations. Detection is thus an ongoing process which should be continually informed by metadata. Ideally, processes applied to the data should be known to utilise a specific graphic technique. For example, before using Tissot's indicatrix to evaluate projection distortion, we must first know what projection was originally used. If this information is missing, then other techniques must be used.

In the more frequent case where processes are unknown, simulations can be applied to generate information for graphic display. Several researchers have looked at simulating error through geostatistical (Englund 1996), Monte Carlo (Openshaw et al 1991), or similar techniques (Heuvelink, Chapter 14; Goodchild et al 1992) to generate multiple realisations. The set of realisations generated by the simulation provides a distribution from which we can compute a variance and confidence limits. These simulations can be quite computationally demanding.

In the evaluation of errors or uncertainties we assume that having detected them, then users wish to assess their significance. For evaluation a new set of requirements arises. We need to know (1) the context of use, and (2) a model and possibly a hypothesis to determine significance. Cross-validation is one common method used to assess statistical prediction. In cross-validation, observations are iteratively deleted and the remaining data are used to predict deleted observations. Repeating this over many deleted subsets allows an assessment of the variability of prediction error.

Fuzzy classifiers have recently become sources of error descriptions and error models (Burrough 1989; Leung et al 1992). In a raster representation fuzzy classifiers provide a means of describing uncertainty by associating each pixel with a vector of class memberships. Goodchild et al (1992) describe an error model based on the vector of probabilities for a pixel's class membership. These methods can create quite large processing and/or large storage overheads.

In summary it should be evident that substantial costs and processing can be required to generate information that can be graphically displayed to identify error or uncertainty in data. The form and content of graphic displays will be highly dependent on the success of the error analysis. In most cases it cannot be assumed that such analyses will have been carried out on the data. The implications are that GIS or other visualisation software packages must either include error analysis tools or data producers must subject their data to these analyses and store the results alongside the data.

3.4 User satisfaction issues

User satisfaction issues relate to the packaging around the graphic and error analysis tools. The interface to these tools should, as always, be intuitive and easy to use. Users should be able to get the error information they require without losing sight of their original application goals. The ideal graphic displays are those which are simple, relevant, and unambiguous. Uncertainty in the data should not be mapped to an uncertainty in the graphics in a way which requires the user to search hard or spend a long time interpreting the results. For most users the evaluation of uncertainty and error is a step on the path to some further goal rather than an end in itself. Thus error analysis and graphic display should not be a long involved process for the user.

One pertinent issue is how much error information users are really interested in. For example, will users be satisfied with depiction of the existence (location of errors or uncertainty), or will they desire more extensive information such as the rank, magnitude, or significance of errors? Computing the magnitude and significance of errors or uncertainty requires more processing of the data while the display of quantitative values requires some very thoughtful consideration of the visual display.

4 FRAMEWORK FOR GRAPHICAL METHODS

Several frameworks for mapping data to graphic displays have been developed in the past by researchers from different fields (Bertin 1983; Mackinlay 1986; Robertson 1991). While error and uncertainty are inherent in spatial data, indicators of these are not sitting at the surface waiting to be displayed. They must be extracted and revealed through error analysis. Because of this dependency, error analysis and graphical methods are necessarily bundled together. The scope of the proposed framework thus deviates from previous frameworks in that it creates a two-phase mapping. The first mapping is between data, an application context, and a suite of appropriate error analysis methods. The second mapping is between the outcome of the error analysis and graphical display methods.

The framework organises information around three basic components: (1) the data; (2) the context of the analysis; and (3) error analysis/graphical methods. Ideally this framework should support the categorisation of the above examples and also indicate what additional error analysis/graphical methods may be required for a given combination of data characteristics and context. In future it might also provide a foundation for an automated visualiser that, given certain inputs, could respond with an appropriate suite of error analysis and error visualisation methods.

4.1 Data characteristics

Characteristics of the data to be considered in the framework include: (1) status – whether the data are raw or processed and, if processed, what processes and parameters were applied; and (2) observed spatial, thematic, or temporal dimensions of the data.

4.1.1 Data status
Some indication of data status is necessary to assign an appropriate error analysis method. Data status is essentially a lineage report (FGDC 1995). It should describe how the data were collected, how they were compiled and what processes were applied along with any associated parameters, and whether field checks were carried out. For example it would be important to know that ground control information had been gathered if one wished to compute RMSE for positional accuracy.

4.1.2 Data dimensions
The error analysis and graphic display will also depend on what dimensions of the data were observed. The range of possible dimensions includes the three spatial dimensions x, y, and z, several attribute dimensions $A_1 ... A_n$ and time, t. An observation could be a 2- or 3-dimensional spatial observation in which only geometry was observed (a survey measurement), a single or multivalued spatial observation or estimate in which geometry and attributes were observed or estimated (e.g. soil colour and texture at location P), or a single or multivalued space-time observation (e.g. observations on surface temperature and precipitation at the same station at the same time intervals). The presence of these dimensions along with their level of measurement provides essential information for error analysis and display.

4.2 Context characteristics

The context description should indicate the environment in which the error analysis might be carried out, specifying at least (1) the task – error detection or evaluation; (2) the desired dimensions of the error analysis – spatial, thematic, temporal, or combination; and (3) the user types.

The first distinction is between detection and evaluation tasks. The error analysis could involve exploratory methods to detect blunders or unusual values. At the simplest level, detection may be accomplished by plotting the data and relying on the human eye to do the detection. Evaluation methods may also be exploratory but extend toward confirmatory data analysis. Evaluation tasks would include tests for the significance of the errors.

4.2.1 Dimensions of the analysis
The second specification of context is the type of error analysis a user wishes to conduct. The possible dimensions for error analysis are spatial, thematic, temporal, or a combination of two or more of these. For example, the only information that may interest a user may be the error or uncertainty in the location of an observation. With this information the error analysis can be restricted to those methods appropriate for detecting and evaluating positional errors or uncertainty. This information needs to be combined with the available data dimensions. The observed dimensions might restrict a desired positional error analysis to two dimensions rather than three.

As another example, a user may be interested in the accuracy of an estimated attribute in space, in which case another set of error analysis/graphical methods will apply (e.g. cross-validation). Additionally, if users are interested in the reliability of an attribute in space and time, a set of methods to support this analysis can be identified. In this latter situation, animation may be a logical method to detect errors or to evaluate the uncertainty of attributes in space and time.

4.2.2 User types
The user type should also influence the selection of error analysis and graphic methods, but user types are generally too broad to categorise. As an example of why user types can be important, take two prototypical users and their associated tasks for which a set of error analysis/graphic tools might require special consideration. These are the data producer/distributor and the data browser in a digital spatial library.

Data producers which are also data distributors need robust error detection and correction tools that can operate quickly and effectively on large volumes of data. Data producers will be dealing primarily with raw data for which the major objective will generally be blunder detection and correction. The error detection and evaluation will be context-independent so a review of all dimensions of the data (space, theme, time) may apply. A goal for the data producers might be to save the results of the error, analysis and graphical displays as metadata for transfer with the data to end users.

Digital libraries may be the new setting for error analysis. Data browsers will likely search for data, and evaluate them to determine if they will be sufficiently reliable for particular applications. Detection and evaluation tasks may both apply, and the error analysis and graphical methods will need to be fast since users may be paying for connection time. Additional constraints and challenges arise in the context of a client-server environment. Both error analysis and graphics will need to be simple and efficient in order to work over the possible range of client configurations. Other technical issues arise such as: will the error analysis and graphic tools reside with the client or with the server?

4.3 Error analysis/graphic display options

To organise the error analysis techniques within the framework we assign them several attributes which relate them to the data and their context. These attributes include: (1) the data status level for which the techniques are applicable; (2) the data dimensions for which they are applicable; (3) the analysis tasks (detection, evaluation) for which they are applicable; (4) their computational complexity; (5) their cost; and (6) the applicable user groups.

The examples that follow distinguish only two data status levels: raw and processed. As more specific metadata becomes available and as error analysis methods evolve it should be possible to make this relationship more specific (e.g. error models linked to specific spatial operations). Some error analysis methods are general and can apply to the analysis of any dimension (e.g. plotting). Others may apply to a single dimension (e.g. adjustment computation for the spatial dimension). A collection of error analysis methods could address purely the accuracy or uncertainty of the thematic dimension of observations. More pertinent are methods that indicate errors in thematic variables as they are distributed in space. Users may be interested in errors in the observed value at these locations or in

estimating values and determining the reliability of the estimates. Cross-validation, Kriging prediction error, and simulation would apply in this latter case (see Anselin, Chapter 17).

Other error analysis methods may also apply specifically to the temporal dimension. Temporal errors are difficult to detect because they can be easily confounded with other errors. They can be detected by an observation being out of sequence or in a future time, through inconsistencies among relationships to other known events, or through aberrations in a measured attribute value. For example a recorded January temperature of 80°F in Maine could be indicative of an error in the recorded time. Similar to thematic variables, space-time errors or uncertainties are of particular interest.

An example set of error analysis methods along with their attributes and the tasks they are suited for is shown in Table 1. The curly brackets used under 'applicable dimensions' indicate that the analysis method applies to combined dimensions rather than to dimensions individually. The underline indicates the dimension of primary interest. Computational complexity could be measured more precisely but it is classified here simply by category. Reading through the table, it can be seen for example that plotting as an error analysis technique applies to raw data, can be applied to the analysis of all dimensions, serves the detection task, and has low computational complexity.

Each error analysis method produces an output which can be characterised. Given that the goal is to display the output graphically, the characteristics of interest are: (1) the level of measurement of the result; and (2) the spatial object to which the result attaches (point, line, pixel, surface, etc.). The graphic problem here can be expressed as one of representing k variables in an n-dimensional field using a fixed set of spatial object representations (points, lines, pixels, surfaces). The range of possible variables which need to be displayed either separately or jointly includes: (1) the observed data values; (2) the errors in or reliability of the observed values; (3) estimated data values; and (4) the reliability of estimated values. Depending on the interests of a user any one of these four may be displayed independently or in some combination. If displays of data and reliability are combined it is necessary to know the characteristics of both. The dimension of the field in which these variables are displayed can be two or three (simulated). The choice will be influenced by the observed data dimensions and characteristics of the error analysis result.

Table 1 Error analysis methods and their corresponding dataset and context characteristics.

Error analysis method	Data status	Applicable dimensions	Tasks	Computational complexity
Plots	raw	x,y,z,a,t	detection	low
Consistency checks	raw	x,y,z,a,t	detection	low
Ground truth checks	processed	x,y,z,a	detection, evaluation	low
Adjustment computation	raw	x,y,z	detection, evaluation	low–moderate
Cross validation	processed	{x,y,z,a}	evaluation	moderate
Fuzzy classification	processed	{x,y,z,a}	evaluation	moderate
Simulation	processed	x,y,z,a,t	detection, evaluation	high

Table 2 provides the information to link the outcome of the error analysis to a graphic display. It creates the second mapping between characteristics of the error analysis results and graphic display options. The table identifies the level of measurement of the output and the spatial object representation to which the output may attach. These two attributes give sufficient information to display the error information independently of the data. Several previously developed frameworks (Bertin 1983; Robertson 1991) provide the structure for working from this level to assign appropriate visual variables. The more specific objective of Table 2 is to guide the choice of graphic display mode if the data and their reliability are to be displayed together. In a joint display some spatial representation of the error analysis output must be shown in combination with a spatial representation of the data. The key therefore lies in combining the two spatial representations. The graphic modes in the table refer specifically to the graphic techniques for combining data and the reliability representations discussed earlier in the chapter. These include the side-by-side, composite, and sequenced images as described by MacEachren (1994) and the small multiples discussed by Tufte (1983). A composite map is the first choice since it is visually most efficient. The user can focus on one image rather than flick back and forth between two or more as in side-by-side or sequenced images. However the efficiency of the composite image breaks down as the number of variables or the complexity of the spatial representation increases. For example when both the data and the error analysis result are surfaces, a composite image of the two becomes difficult to encode graphically. When this occurs two simple side-by-side images are preferable. Table 2 includes the same error analysis methods which appear in Table 1, except for the plot which is excluded since it already exists as a graphic form.

Table 2 Basis for associating error analysis output with graphic display modes.

Error analysis method	Level of measurement	Applicable spatial object	Spatial object evaluated	Graphic model
Consistency check	nominal	point, line, area	point, line, area	composite
Ground control check	real	point, pixel, set of pixels	point, pixel, set of pixels	composite
Adjustment computation	real	point	point	composite
Cross validation	real	point surface	point surface	composite side by side
Fuzzy classification	real	pixel	pixel	small multiple animation
Simulation	nominal real	surface	surface	small multiple animation

To help make sense of this table a few explanations of the table entries may be offered. The row entry for consistency checks indicates that the output of this analysis takes the form of a nominal value (consistent or inconsistent). It is known from Table 1 that this analysis method can apply to any dimension. Column three indicates the spatial representation to which the error analysis result applies. If checking the consistency of the spatial dimension the result applies to the point, line, or area under analysis. If checking the consistency of attributes or times, it is assumed that the outcome can be associated with a point, line, or area. The fourth column refers back to the dataset being evaluated. It indicates the spatial representation (point, line, area) associated with the dimension being evaluated. Column five indicates the graphic mode. In this case a logical choice of graphic model is a composite. An inconsistency detected as the result of the consistency check can be displayed using a visual variable appropriate for nominal valued data such as colour hue or shape.

In the case of adjustment computations the outcomes are real-valued deviations around a point. The spatial object being evaluated is a point and the result of the computation refers to a point. The logical graphical choice is a composite in which for example the data points are shown in combination with their error ellipses. In the case of the fuzzy classifier the spatial object being evaluated is a cell, therefore, the outcome of the analysis is a vector of real numbers (probabilities) which can be associated with a cell. In the case of a simulation the entire dataset, referred to here as a surface, is being evaluated. The results will share the same level of measurement as the input data and take the form of multiple new surfaces, is being evaluated. To illustrate the error or uncertainty it is necessary to display several surfaces, so small multiples or sequenced images are appropriate.

5 FUTURE RESEARCH IN GRAPHICAL METHODS

The constraints on advances in this area are not technological. Technological possibilities have outstripped the ability to understand and model uncertainty and error in spatial data. The most pressing needs still lie in advancing error models for spatial data, the development of error propagation techniques, and enforcement or encouragement of

better documentation of datasets (see Heuvelink, Chapter 14). Maps of mean estimates and estimation variance (common in Kriging) present a limited view of error and uncertainty (Hunter, Chapter 45; Cressie 1991; Hunter and Goodchild 1995). In general, graphical presentation of such error descriptors is inappropriate since maps of error descriptions are not possible realisations of stochastic error models. Instead, presentation of a sample of realisations, by animation or simultaneous display, may create the only sound understanding of uncertainty and its implications. Software functions to provide these must be implemented in existing GIS, concurrent with integration of spatial statistical models. Their use and comprehension must be informed by continued empirical testing (not simply of visual variables, but of cognitive use patterns and establishment of user comprehension).

Advantages of feature-oriented approaches to data quality representation have been determined, but more needs to be learned about the computational and data-volume overhead that their inclusion in a database may generate. Layer-based GIS functions must continue to proliferate in part because error modelling remains based upon layer data models, and in part because the layer model is most efficient for raster-based terrain and imagery. It is evident that errors that accrue differentially with specific GIS operations (buffering, overlay, coordinate conversion, etc.) may also depend on the data theme, resolution, and timeframe. Understanding of these differences needs to be formalised. Continued research must reinforce the development of data models and error models that spatially refine information for input to visualisation techniques.

Several error detection and evaluation methods for spatial data are computationally complex. To create these in timeframes acceptable to users will be challenging. The alternative of computing such information in advance and storing it has additional problems. The storage overhead may be quite substantial if the goal is to store several realisations of an error model. This approach also assumes knowledge of what information the users will want. There is, of course, the possibility of being less responsive to users.

Development of error models and more appropriate error analysis will only improve with better data documentation. While efforts are underway to improve metadata documentation, there is still have a long way to go. Most of the metadata for spatial data archives are being created after the fact. This is an arduous and error-prone process. Metadata collection needs to start prior to data collection and continue parallel with the data lifespan (Beard 1996).

As databases become distributed and shared by multiple users, the need for users to detect and anticipate error information becomes critical. In this context the search for information and assessment of the quality of spatial data is outside a traditional GIS. The implications of these developments are that error analysis and graphical display software must be able to function independently of GIS. Future research should be directed toward interoperable components which could be easily recombined. Users could then select tools that would apply to a specific analysis context rather than having to support a large package which tried to incorporate all error analysis and display functions.

References

Beard M K 1996 A structure for organising metadata collection. *Proceedings, Third International Conference/Workshop on Integrating GIS and Environmental Modeling, Sante Fe.* Santa Barbara, NCGIA. CD and *http://www.ncgia.ucsb.edu*

Becker R A, Cleveland W S, Wilkes A R 1987 Dynamic graphics for data analysis. *Statistical Science*: 355–95

Bertin J 1983 *Semiology of graphics: diagrams, networks, maps.* Madison, University of Wisconsin Press

Bicking B, Beard M K 1995 Toward implementing a formal approach to automate thematic accuracy checking for digital cartographic datasets. *Proceedings AutoCarto* 12: 355–62

Brewer C A 1994 Colour use guidelines for mapping and visualisation. In MacEachren A, Taylor D R F (eds) 1994 *Visualisation in modern cartography*. Oxford, Elsevier Science: 123–48

Burrough P A 1989 Fuzzy mathematical methods for soil survey and land evaluation. *Journal of Soils Science* 40: 477–92

Chambers J M, Cleveland W S, Kleiner B, Tukey P 1983 *Graphical methods for data analysis.* Boston, Duxbury Press

Cleveland W S 1993 *Visualising data.* Murray Hill, AT&T Bell Laboratories

Cleveland W S, McGill R 1984 Graphical perception: theory, experimentation and application to the development of graphical methods. *Journal of the American Statistical Association* 79: 531–53

Cox D R 1978 Some remarks on the role in statistics of graphical methods. *Applied Statistics* 27: 9

Cressie N A C 1993 *Statistics for spatial data,* revised edition. New York, John Wiley & Sons Inc.

Dibiase D, MacEachren A M, Krygier J, Reeves C 1992 Animation and the role of map design in scientific visualisation. *Cartography and Geographic Information Systems* 19: 201–14, 265–6

Englund E 1996 Spatial simulation: environmental applications. In Goodchild M F, Parks B O, Steyart L T (eds) *Environmental Modeling with GIS.* New York, Oxford University Press: 432–7

FGDC (Federal Geographic Data Committee) 1995 *Content standards for digital geospatial metadata (June 8).* Washington DC, Department of the Interior and *http://www.fgdc.gov*

Fisher P 1994b Visualisation of the reliability in classified remotely sensed images. *Photogrammetric Engineering and Remote Sensing* 60: 905–10

Fisher P 1994c Visualising the uncertainty of soil maps by animation. *Cartographica* 30: 20–7

Gershon N, Brown J R 1996 The role of computer graphics and visualisation in the global information infrastructure. *IEEE Computer Graphics and Applications* 16: 60

Goodchild M F, Buttenfield B, Wood J 1994 Introduction to visualising data validity. In Hearnshaw H, Unwin D (eds) *Visualisation in geographic information systems.* Chichester, John Wiley & Sons: 141–9

Goodchild M F, Sun G, Yang S 1992 Development and test of an error model for categorical data. *International Journal of Geographical Information Systems* 6: 87–104

Haining R P 1990 *Spatial data analysis in the social and environmental sciences.* Cambridge (UK), Cambridge University Press

Hancock J R 1993 Multivariate regionalisation: an approach using interactive statistical visualisation. *Proceedings AutoCarto 11 Minneapolis*: 218–27

Hunter G J, Goodchild M F 1995 Dealing with error in spatial databases: a simple case study. *Photogrammetric Engineering and Remote Sensing* 61: 529–37

Imhof E 1964 Beiträge zur Geschichte der topographischen Kartographie. *International Year Book of Cartography* 4: 129–54

Leung Y, Goodchild M F, Lin C C 1992 Visualisation of fuzzy scenes and probability fields. *Proceedings, Fifth International Symposium on Spatial Data Handling, Charleston*: 480–90

MacEachren A M 1994a *Some truth with maps: a primer on symbolization and design.* Washington DC, Association of American Geographers

MacEachren A M, Howard D, Wyss M von, Askov D, Taormino T 1993 Visualising the health of Chesapeake Bay: an uncertain endeavor. *Proceedings GIS/LIS 93 Minneapolis*: 449–58

Mackinlay J 1986 Automating the design of graphical presentations of relational information. *ACM Transactions on Graphics* 5: 110–41

Maling D H 1973 *Coordinate systems and map projections.* London, George Philip

McGranaghan M 1993 A cartographic view of data quality. *Cartographica* 30: 8–19

Mitasova H, Mitas L, Brown W, Gerdes D P, Kosinovsky I, Baker T 1995 Modeling spatially and temporally distributed phenomena: new methods and tools for GRASS GIS. *International Journal of Geographical Information Systems* 9: 433–46

Monmonier M 1989 Geographic brushing: enhancing exploratory analysis of the scatterplot matrix. *Geographical Analysis* 21: 81–4

Monmonier M 1991a *How to lie with maps.* Chicago, University of Chicago Press

Morrison J L 1992 Implementing the Spatial Data Transfer Standard – introduction. *Cartography and Geographic Information Systems* 19: 277

Openshaw S, Charlton M, Carver S 1991 Error propagation: a Monte Carlo simulation. In Masser I, Blakemore M (eds) *Handling geographical information.* Harlow, Longman/New York, John Wiley & Sons Inc.: 78–101

Paradis J, Beard M K 1994 Visualisation of data quality for the decision-maker: a data quality filter. *Journal of the Urban and Regional Information Systems Association* 6: 25–34

Robertson P K 1991 A methodology for choosing data representations. *IEEE Computer Graphics and Applications* 11: 56–67

Robinson A H, Sale R D, Morrison J, Muehrcke P 1985 *Elements of cartography,* 5th edition. New York, John Wiley & Sons Inc.

Salgé F, Smith N, Ahonen P 1992 Towards harmonized geographical data for Europe: MEGRIN and the needs for research. *Proceedings, Fifth International Symposium on Spatial Data Handling, Charleston*: 294–302

Tissot A 1881 *Mémoire sur la représentation des surfaces et les projections des cartes geographiques.* Paris, Gauthier Villars

Tufte E R 1983 *The visual display of quantitative information.* Cheshire (USA), Graphics Press

Tukey J W 1977 *Exploratory data analysis.* Reading (USA), Addison-Wesley

Wright J K 1942 Map-makers are human: comments on the subjective in maps. *Geographical Review* 32: 527–54

Introduction

THE EDITORS

It is clear from the discussion of GIS principles thus far that we now live in a data-rich world in which a vast and increasing array of geographical phenomena are represented in digital form. GIS-based data models are by definition selective abstractions and the data used to build them are error prone, yet they can lay the foundation for legitimate context-sensitive inputs to generalisable analysis and forecasting. The contributions to this section set out to identify how GIS allows us to summarise the properties of spatial distributions, inductively solve spatial problems, and contribute towards spatial decision-making.

The established paradigm for quantitative description and generalisation about geographical phenomena has been to use spatial statistics. What makes spatial statistics distinct from its parent discipline is its concern with observations which are located near to one another in space and which, as a consequence, tend to share similar attribute values – in Anselin's words (Chapter 17): 'the phenomenon where locational similarity (observations in spatial proximity) is matched by value similarity (correlation)'. Spatial and geostatisticians have developed a range of specialised methods and techniques for dealing with such cases. The emergence and chronological development of spatial statistics in the pre-GIS era is the first theme considered in the contribution by Art Getis (Chapter 16).

An important emergent debate within GIS has been the continuing relevance of spatial statistical approaches. Briefly, the key arguments may be summarised as: first, spatial statistics developed in what all of the contributors to this section would recognise as a 'data-poor' era, in which statistics were based upon few (by present-day standards) observations; second, this paucity of data made computation a fairly straightforward procedure; and, third, the geography of areal units was fixed (and usually coarse), and not itself subjectable to the

range of transformations and sensitivity analyses that have been outlined in previous contributions to this Principles part of the book. There is evident consensus among the contributors to this section that spatial analysis has received far too little attention in the development of GIS, yet they have some different views as to whether and how spatial statisticians can continue to contribute practical spatial analysis skills to GIS. On the one hand, Getis (Chapter 16) and Luc Anselin (Chapter 17) set out some of the enduring contributions that spatial statistics is making to GIS, especially in the area of exploratory spatial data analysis (ESDA). On the other hand, the views of Stan Openshaw and Seraphim Alvanides (Chapter 18) and Manfred Fischer (Chapter 19) lean towards the view that the changes associated with the development of GIS require a more root-and-branch reappraisal of the practice of spatial analysis in GIS. Richard Church (Chapter 20) presents a review of the ways in which GIS is being applied to locational analysis problems in GIS.

The review by Getis (Chapter 16) charts the considerable progress that has been made in developing ESDA spatial statistics through the media of GIS. However, he laments the relative lack of progress in developing spatial hypothesis testing within GIS. By implication, he seems to sound a warning that the media of GIS are in danger of overwhelming the message of spatial statistical analysis as conventionally understood. Indeed, developments in scientific visualisation and ESDA appear to have contributed little to our incomplete theoretical and statistical understandings of the ways that observations should be differentially weighted across space (i.e. the effects of distance) and the effects of boundaries and edges on the results of spatial statistical analysis.

To dwell upon such (possibly unresolvable) statistical issues might be seen as admonishing failure, when the media of GIS have been demonstrably effective in exploring locational scenarios and visualising spatial outcomes. Data exploration within GIS is the domain of ESDA techniques – defined by Anselin (1994) as being used 'to describe and visualise spatial distributions, identify atypical locations (spatial outliers), discover patterns of spatial association (spatial clusters) and suggest different spatial regimes and other forms of spatial instability or spatial non-stationarity'. Anselin's contribution to this volume develops the views that GIS has become data rich but theory poor, and that ESDA statistics can be used to structure, visualise, and explain a wide array of geographical data. His comprehensive review builds upon Getis' conceptions of spatial autocorrelation (i.e. geostatistical and spatial weights formulations), and in this context goes on to identify important domains of ESDA as pertaining to identification of local patterns of spatial association within global patterns.

For Openshaw and Alvanides, by contrast, the dominant impression is that we are seemingly unable to structure and analyse the vast quantities of spatial data that are now available, and that this failure reflects our continued adherence to the 'pre-GIS' spatial statistical analysis paradigm. Given that the development of GIS and of modern databases was substantially technology-led, it is at least intuitively plausible that analysis might be developed through the same guiding force. Thus for Openshaw and Alvanides (Chapter 18) the way forward lies through broad-based techniques of 'geocomputation' – that is, 'the adoption of a large-scale computationally-intensive approach to the problems of physical and human geography in particular, and the geosciences in general'. This kind of approach has undoubtedly had a profound impact through demonstration and exploration of modifiable areal unit effects, and the geocomputational paradigm has clear application in allowing spatial analysts to create zone designs that satisfy particular constellations of constraints. Building upon this, Openshaw and Alvanides see generic solutions emerging from the use of 'intelligent' pattern-seeking techniques which might become integral to GIS. The implication is that new computational techniques may be used to search for new theories and to generate new knowledge using applied, problem-solving approaches.

Fischer (Chapter 19) takes a wider perspective on the emergence of what he terms 'computational intelligence' (CI) technologies in relation to classical spatial statistics. He shares many of the expressed doubts of Openshaw and Alvanides that conventional spatial statistics can be adapted to accommodate the richness of the GIS environment, and instead advocates a CI paradigm (involving artificial life, evolutionary computation, and neural networks) based essentially upon geocomputation. He develops an extended case study involving the use of a neural net model for satellite image classification, and shows how spatial analysis proceeds through model specification, estimation, and testing phases. His exposition is lucid and non-technical, and he is at pains to dispel suspicion about the 'mystique and metaphorical jargon' hitherto associated with CI techniques. Whether the widest GIS audience will share his confidence that use of the 'universal language of mathematics' alone will dispel such scepticism is a moot point: indeed, this raises important issues about the gulf between 'machine-intelligent' spatial analysis of digital abstractions and scientific theory and reasoning as conventionally understood.

We have seen in the first section to this part of the book how ontology ('meta-theory') prescribes particular detailed approaches to analysis, and that no stage in scientific reasoning can be considered in isolation. The emphasis in much of the second part was on demonstrating how digital data provide only imperfect, incomplete, and error-prone representations of reality. Together this would seem to require that choice of spatial analytical method is rational, informed, and sensitive to context, rather than being data led in a naive empiricist way. Just as Goodchild and Longley (Chapter 40) argue later in the Technical issues section, that volume of data is not a substitute for scientific rigour, so the paradigm of geocomputation needs to demonstrate why and in what circumstances spatial analysts should have 'confidence' (both broadly and narrowly defined) in its substantive findings. Openshaw and Alvanides cite a number of high-profile and celebrated case studies which have adopted what has come to be described as a geocomputational approach (notably in the identification of clusters of diseases), yet none appears wholly to have withstood scientific scrutiny: as such, the jury must still be out regarding the advisability of wholesale reliance upon geocomputational approaches. It is beyond doubt that the world has never been as data rich but, in the realm of spatial analysis, there have been a number of false dawns before – as our comments in the Introduction on the coverage of artificial intelligence in the first edition of this book testifies. Thus while Openshaw and Alvanides conjecture that 'it may be possible to compute our way out of the data swamp', it remains a moot point as to how and why we may have become lured into it in the first place. It remains to be seen whether and how far the protagonists of computational approaches are able to rebuff concerns that they are indulging in uninformed pattern-seeking empiricism in the absence of clear theoretical guidance as conventionally understood. Elsewhere, Openshaw and Openshaw (1997) have begun to demystify new ways of viewing our digital world although, at the other end of the spectrum, Curry (1995: 82) has concluded that 'to develop an understanding of the data adequate to a resolution of the problems which arise in the production of a GIS would very likely render those systems irrelevant'. There is still some way to go, both with regard to demystifying technique and to understanding data.

An oft-rehearsed but nevertheless resonant theme running through all of these contributions concerns the integration of spatial analytic functionality into proprietary GIS. Neither spatial statistical models, nor geocomputational methods, nor refined ESDA techniques form part of proprietary GIS. This is in part because of user ignorance about the range of simplifying assumptions that routine usage brings, and in part because (as a consequence) vendors are unlikely to prioritise functionality for which there are no strong user demands. Given the very small likelihood of fully integrated spatial analytical GIS in the foreseeable future, Getis, Openshaw and Alvanides, and Anselin each explore a number of options for the close and loose coupling of GIS to specialist spatial analysis packages, as well as the potential role of the Internet as a platform for integrating GIS and spatial analysis.

Some of the earliest applications of spatial analysis involved the calculation of statistical moments and distributions for the classic locational models of geography, models which were used in the pre-GIS era to identify the best location for industrial and service facilities. The enduring relevance of locational modelling to GIS is the focus of the contribution by Church (Chapter 20). A conventional facet to this problem has been the use of GIS to locate single facilities with respect to spatial patterns of demand and, through use of the overlay model, to identify corridors linking the different sites involved in activities in the most cost-efficient way. In recent years, progress has been made towards solving multiple-site location problems, most notably instances in which sites may need to be relocated in response to very short-term changes in demand (as in the relocation of 'on call' ambulances to cover for vehicles which are already attending emergencies). This echoes Getis' sentiment of continuity of approach in spatial analysis, but what has changed here with the innovation of GIS is the richness of the data which can be brought to bear on site location problems, the computational support for new and complex location–allocation algorithms, and the visual quality of the computer environment for data exploration, investigation of scenarios and decision support. Location modelling also provides a good exemplar of the wider problems that remain on the spatial analysis agenda, namely: the compatibility of data structures/data quality issues; the representation of spatial patterns of demand, and the screening process used to identify sites; the size and scale of the elemental units used to specify location–allocation problems; and the ways in which errors are created and propagated in formulation of problems. GIS is clearly having far-reaching impacts upon the specification, estimation, and testing of spatial relationships, and the Applications part of this book (in Volume 2) provides evidence of the practical relevance of these techniques (e.g. Cova, Chapter 60; Gatrell and Senior, Chapter 66).

References

Anselin L 1994a Exploratory spatial data analysis and geographic information systems. In Painho M (ed.) *New tools for spatial analysis*. Luxembourg, Eurostat: 45–54

Curry M R 1995a GIS and the inevitability of ethical inconsistency. In Pickles J (ed.) *Ground truth: the social implications of geographic information systems*. New York, Guilford Press

Openshaw S, Openshaw C A 1997 *Artificial intelligence in geography*. Chichester, John Wiley & Sons

16

Spatial statistics

A GETIS

The field of spatial statistics is based on the assumption that nearby georeferenced units are associated in some way. More and more, the GIS community needs to draw upon the work of the spatial statistician to help find meaning in spatial data. The precursors of current spatial statistical researchers include those who sought to describe areal distributions, the nature of spatial interactions, and the complexities of spatial correlation. The spatial statistical methods in current use, and upon which research is continuing, include: spatial association, pattern analysis, scale and zoning, geostatistics, classification, spatial sampling, and spatial econometrics. In a time-space setting, scale, spatial weights, and spatial boundaries are especially difficult problem areas for further research. Those working in GIS welcome comprehensive packages of spatial statistical methods integrated into their software.

1 INTRODUCTION

It is common for statisticians to confine their attention to *data description*, including exploratory analysis, and *induction*, the development of generalisations about a defined population on the basis of a sample drawn from that population. Map-oriented researchers have long been interested in data description and induction, usually searching the statistics literature for ideas on how to extract as much information as possible from georeferenced data. The search is often directed toward biometry, psychometry, geology, econometrics, and statistics (see also Fischer, Chapter 19). A relatively small area of original research that cuts across these fields can be attributed to the work of spatial statisticians, who can be distinguished by their attention to georeferenced or spatial data. In recent years, some of this work has been spurred by the development of GIS. In this chapter, spatial data analysis with particular emphasis on the uses and applications of spatial statistics in a GIS environment is discussed.

Spatial statistics can be considered a distinct area of research. Traditional statistical theory bases its

models on assumed independent observations. Although common sense tells us that in most real-world situations independence among observations on a single variable is more the exception than the rule, independence is still a suitable benchmark from which to identify statistically significant non-independent phenomena. The field of spatial statistics is based on the non-independence of observations; that is, the research is based on the assumption that nearby units are in some way associated (Tobler 1979). Sometimes this association is because of a spatial spillover effect, such as the obvious economic relationship between city and suburb. Sometimes the association is a distance decline effect; that is, as distance increases from a particular observation, the degree of association between observations lessens. An example is the influence of an earthquake; its effect declines with distance from the epicentre.

Statistics, in general, and spatial statistics with its emphasis on location, are the glue that holds much of our research efforts together. In the search for a high degree of certainty, we look to statistics. As the GIS community matures, it will draw more and more upon the work of the statistician to help to find meaning in spatial data and in the development of GIS functionality.

The approach in this chapter is to trace briefly and selectively, in section 2, the pre-GIS contributions of map-oriented researchers to spatial statistics. In section 3, brief synopses of statistical analytical devices that spatial analysts use are provided. The distinction is made between the work of the spatial statisticians (those statisticians, biologists, econometricians, atmospheric/oceanic scientists, and geographers who seek to identify the statistical parameters of spatially distributed variables) and the geo-statisticians (those statisticians, geologists, and mining engineers who analyse their data at a number of scales in terms of spatial trend and correlation). Each subsection contains examples of, and key references to, current research. Finally, in section 4, some of the problems and challenges that face spatial researchers are outlined, with a reflection on the nature of statistical work in a GIS environment.

2 PRE-GIS USES OF SPATIAL STATISTICS

Geography has a long history of development of clever cartographic devices that allow for particularly insightful views of spatial data. From Lalanne's (1863) hexagonal railway patterns to the map transformations of Tobler (1963), the pre-GIS literature is filled with interesting ideas designed to enable spatial data to 'speak for themselves'. The desire to make maps a useful part of analysis led pre-computer geographical writers to try to find ways to depict spatial distributions of data in imaginative ways. It was just a short step from interesting depictions on maps to statistical tests on their significance relative to some supposition about the meaning of the maps. Now we have powerful computers and extensive software that guide us toward the production of new and unusual maps. Concomitantly, new statistical devices have been developed, albeit at a slower rate of growth, to answer questions about mapped patterns. Knowing that the spatial perspective is an important aspect of knowledge, analysts seek better ways to depict data on maps and to test hypotheses based on some expected pattern form or structure. Four themes can be considered antecedents of what has become the modern statistical analysis of spatial data.

2.1 Statistical analysis of areal distributions

Although the roots of his work go back to the nineteenth century, Neft (1966), working under the direction of Warntz, was the first to produce a comprehensive, mathematically consistent system for describing areal distributions. Drawing on the work of Carey (1858), Mendeleev (1906), the location theorists – Thünen (1826), Weber (1909), Christaller (1935), and Lösch (1954) – and the ideas of the social physicists – particularly those of Stewart (1950) and Warntz and Neft (1960) – Neft described the statistical moments of areal distributions. For point distributions, he produced statistical measures of skewness and kurtosis of average position (various centroids), spatial dispersion, and surfaces. In addition, he addressed one of the challenging tasks still very much on the agenda of current spatial statistical researchers: producing valid measures of statistical association of spatial variables.

2.2 Spatial interaction

There is no more important topic for the spatial analyst who deals with human issues than the study of the interaction of activities in one place with those in another. Research in this area has a long and distinguished history, dating back to Carey (1858), Ravenstein (1885), Reilly (1929), Zipf (1949), and Stewart and Warntz (1958). The famous Newtonian formula, $m_1 m_2 / d^2$, where m_1 and m_2 are measures of mass at sites 1 and 2, and d is the distance separating the masses at those sites, was the foundation stone. Modified by spatial theorists, this physical law has been used to great advantage to study and to predict a wide variety of human spatial interactions, such as transportation movements, the spread of information, and the potential for economic growth. Modern expositions of the theory and statistical estimation procedures make up a significant portion of modern transportation and marketing literature (Birkin et al, Chapter 51; Gatrell and Senior, Chapter 66). The work of Wilson (1967) must be singled out as a relatively recent attempt to derive practical spatial interaction theory. Rather than depend on physical science analogies, Wilson devised probabilistic laws that described possible human movement. Nowadays an important use of GIS is to allow for the manipulation of data so that parameters that describe movement can be calibrated and evaluated.

2.3 Spatial correlation

Before the 1960s, only a modest literature had developed in geography on perhaps the most challenging spatial question: in an unbiased way, how is one to account for the correlation in spatially distributed variables? The fundamental ideas concerning the measurement of, and testing for, spatial autocorrelation were spawned in geography by Robinson (1956), and Thomas (1960) saw the difficulties in dealing with dependent unequally sized units. Through their work and that of others, the modifiable areal unit problem was addressed and spatial residuals from regression were evaluated. It was during this period that the statisticians Moran (1948) and Geary (1954) developed their measures of spatial autocorrelation. Building on the work of Moran (1948) and Krishna Iyer (1949), Dacey (1965) addressed the issue of the possible association among contiguous spatial units. These join count statistics led to the work of Cliff and Ord, whose monograph 'Spatial Autocorrelation' (1973) opened the door to a new era in spatial analysis. In section 3, we outline the link between the Cliff–Ord work and modern approaches to spatial statistical analysis.

2.4 Hypotheses about settlement patterns

Much of the excitement in the University of Washington's Department of Geography during the late 1950s and early 1960s centred on understanding and testing the theories of the economic geographer, Walter Christaller, and the economist, August Lösch. From the standpoint of spatial statistics, of note is the work of Dacey (1963), who by taking the lead from the plant ecologists such as Clark and Evans (1954), tested various statistical distributional theories using sets of georeferenced data that represented the location of towns in a settlement system. From this work, a point pattern 'industry' developed that featured the work of King (1962), Getis (1964), Harvey (1966), Clark (1969), and Rogers (1969).

3 SPATIAL STATISTICS IN CURRENT USE

The types of statistical methods popular today are a function of both the nature of the problems being studied and the availability of computers. Seven areas of research are listed that are particularly favoured. Each is described in terms of the kinds of problems being solved, their general formulation (if not discussed in detail elsewhere in this book), and their usefulness to the GIS community of analysts. In addition, current research themes are noted together with key references. Such areas of inquiry as spatial neural nets, spatial fuzzy sets, and simulated annealing are just now being developed and are discussed by Fischer (Chapter 19).

3.1 Spatial association

The Cliff–Ord monograph enabled researchers to assess statistically the degree of spatial dependence in their data, and, in so doing, to search for additional or more appropriate variables, and to avoid many of the pitfalls that arise from autocorrelated data. Many GIS contain the Cliff–Ord routines that allow for the calculation of spatial autocorrelation. Much of present-day interest in spatial analysis derives directly from the 1973 Cliff-Ord monograph and the authors' subsequent (1981) more complete discussion. These shed light on the problem of model mis-specification owing to autocorrelation and demonstrated statistically how one can test residuals of a regression analysis for spatial randomness. They explicated the nature of the spatial weight matrix and provided step-by-step procedures for applying statistical tests on Moran's I and Geary's c, the two major autocorrelation statistics.

Finding the degree of spatial association (autocorrelation) among data representing related locations is fundamental to the statistical analysis of dependence and heterogeneity in spatial patterns. Like Pearson's product–moment correlation coefficient, Moran's statistic is based on the covariance among designated associated locations, while Geary's takes into account numerical differences between associated locations. The tests are particularly useful on the mapped residuals of an ordinary least squares regression analysis. Statistically significant spatial autocorrelation implies that the regression model is not properly specified and that one or more new variables should be entered into the regression model.

Mantel (1967), Hubert (1979), and Getis (1991) have shown that statistics of this nature are special cases of a general formulation, gamma, that is defined by a matrix representing possible locational associations (the spatial weights matrix) among all points, multiplied by a matrix representing some specified non-spatial association among the points. The non-spatial association may be an economic, social, or other relationship. When the elements of these matrices are similar, high positive autocorrelation ensues. Gamma describes spatial association based on covariances (Moran's statistic, I), or subtraction (Geary's statistic, c), or addition (the G statistic of Getis and Ord 1992). These statistics are global insofar as all measurements between locations are taken into account simultaneously. Aspinall (Chapter 69) provides examples in the realm of landscape conservation.

When the spatial weights matrix is a column vector, gamma becomes local; that is, association is sought between a single point and all other points (I_i, c_i, G_i). Research on local statistics has been especially active recently because they lend themselves to kernel-type analyses in a GIS where datasets are large (Anselin 1995; Getis and Ord 1992; Ord and Getis 1995). Local statistics have been used to classify remotely-sensed data (Getis 1994), and to show associations between neighbourhoods' crime rates (Anselin 1993) and countries' conflict propensities (O'Loughlin and Anselin 1991).

Some current research themes in this area are:

- the identification of spatial spillover or nuisance autocorrelation (Anselin and Griffith 1988; Anselin 1990a; Anselin and Rey 1991);
- characteristics of the structure of spatial weight matrices (Griffith 1988; Anselin 1986; Boots and Kanaroglou 1988);
- heterogeneity issues in local measurements of spatial association (Bao and Henry 1996);
- determining the exact distribution of spatial autocorrelation statistics (Tiefelsdorf and Boots 1994);
- alternatives to the Cliff–Ord approach (Kelejian and Robinson 1995);
- multivariate spatial association (Wartenberg 1985).

3.2 Pattern analysis

Popular in the 1960s was point pattern analysis based on the spatial homogeneous Poisson process (see also Fischer, Chapter 19). It was common to find a researcher working at a light table making measurements from numbered points to the first nearest neighbour of each point. Now, with the use of digitised georeferenced data, we are easily able to take measurements from all points to all other points. In addition, measurements of line segments, distances between line intersections, areas, and characteristics of areas such as perimeter length, neighbouring areas, and so on, are basic within most GIS.

Pattern analysis in the spatial sciences grew out of an hypothesis-testing tradition, not out of the extensive pattern recognition literature. Nearest neighbour work continues today, but the work of Clark and Evans (1954) has now been modified for the sake of unbiasedness to take into account the length of the perimeters of study areas (Donnelly 1978) and the distance to study area boundaries (refined nearest neighbour analysis: Diggle 1979; Boots and Getis 1988).

In recent years, point pattern analysis has regained its vigour as an area of study as a result of the ability of computers to handle large numbers of objects. Statistical approaches are usually based on hypotheses of complete spatial randomness (CSR), that is, the theoretical pattern is assumed to be representative of: (a) objects that are located independently of each other; and (b) a study area where each location has an equal chance of receiving an object. The pattern analyst tests hypotheses about the spatial characteristics of point, line, or area patterns. These geometric forms represent everything from the location of individuals suffering from an infectious disease to the shape of hardened basalt flows (Boots and Getis 1988).

Related to pattern analysis is the continuing interest that ecologists have in studying plant and animal distributions. Surprisingly, only in recent years have plant ecologists become aware that because of dependence among nearby observations, a particular pattern of plants may not represent a suitable sample for model testing (Franklin 1995). A set of key references in this area may be found in Potvin and Travis (1993).

Perhaps the most important developments in recent years are the applications of K-function analysis to the study of point patterns, and the use of Voronoi polygons to study spatial tessellations (Boots, Chapter 36; Okabe et al 1992). In addition, fractals study is a promising area for pattern analysis (Batty and Longley 1994).

3.2.1 The idea behind the use of K-function analysis

The K-function is the ratio of the sum of all pairs of points within a pre-specified distance, d, of all points to the sum of all pairs of points regardless of distance. The function is adjusted to take into account distances that are closer to the boundary of the study area than to d. The original K-function by

Ripley (1977) was modified by Besag (1977) to take into account the need to stabilise variance, and Getis (1984) generalised the formula to include the weighting of points, such that the sum of pairs of points became the sum of the multiples of the weights associated with each member of a pair of points. Diggle (1983) has done much to exploit this formulation to show many new features of patterns. For example, not only can one easily show the difference between an existing pattern and a random pattern but one can also develop theoretical expectations for other than random patterns. In addition, patterns divided into different point types (marked patterns) can be studied easily. For testing purposes, an envelope of possible outcomes under the hypothesis of say, randomness, is usually constructed by means of a Monte Carlo simulation. Studies of the spatial distribution of vegetation dominate the empirical literature of K-function analysis (Diggle 1983), but the method has been used for the study of human population distribution (Getis 1983) and disease distribution (Morrison et al 1996). Recently, Gatrell et al (1996) showed that the K-function can be used as an indicator of time-space clustering; that is, one simultaneously finds pairs of points separated by designated units of time and distances in space. This approach is particularly useful for identifying disease clustering over time.

3.2.2 Successful applications to spatial phenomena
Some themes of current interest in pattern analysis are:

- the development and testing of time-space pattern models (Griffith 1996; Gatrell et al 1996; Jacquez 1995);
- search for pockets of extreme values in large data-sets (Ord and Getis 1995; Haslett et al 1991);
- development of pattern models based on differences, absolute differences, and similarities between nearby observations (Getis and Ord 1996).

3.3 Scale and zoning (the modifiable areal unit problem)

The problem of scale effects was made particularly clear by the results of Openshaw and Taylor's (1979) study of voting behaviour in Iowa. They showed that the level of spatial aggregation and arrangement of spatial units (zoning) has a marked effect on the correlation of variables. Fotheringham and Wong (1991) identified the extent of the spatial bias in a multivariate regression analysis and Fotheringham et al (1995) carried out similar

research in a p-median problem context. The most comprehensive treatment to date is that of Arbia (1989), who identified the relationship between levels of autocorrelation and spatial unit aggregation. A recent study by Holt et al (1996) shows the scale problem to be an area selection problem. Some themes being pursued include:

- spatial aggregation biases (Okabe and Tagashira 1996; Tobler 1989);
- the relationship of spatial autocorrelation to scale differences (Arbia et al 1996);
- the effect of different zoning on results of various types of analyses (Openshaw 1996; Green and Flowerdew 1996);
- identifying scale effects by use of principal axis factor analysis (Hunt and Boots 1996);
- scale effects on parameters of spatial models (Amrhein and Reynolds 1996; Wrigley et al 1996).

This theme is developed by Openshaw and Alvanides (Chapter 18).

3.4 Geostatistics

The *variogram* (or semivariogram) (Cressie 1991) plays a useful role as the function that describes spatial dependence for a regional (georeferenced) variable. The term 'intrinsic stationarity' is used to indicate the natural increase in variance between observations of a regional variable as distance increases from each observation. The semivariance – a measure of the variance as distance increases from all points or areas (blocks) – eventually reaches a value equal to the variance for the entire array of data locations, regardless of distance. Clearly, at zero distance from a point, the semivariance is also zero, but the semivariance increases until, at a distance called the range and a semivariance value called the sill, it is equal to the variance. The function describing the semivariance is usually spherical, exponential, or Gaussian.

The variogram is essential for *Kriging*, which is a technique for estimating the value of a regional variable from adjacent values while considering the dependence expressed in the variogram. There are many kinds of kriging, each designed to give the highest possible confidence to the estimation of a variable at non-data locations. If there is no bias in the variogram, and all required assumptions are met, the kriged values, as opposed to trend surface, triangulated irregular network (TIN), or other estimation devices, will be optimal.

A large amount of literature has developed in geostatistics. The definitive text by Cressie (1991) details many instances where the geostatistical approach has proved helpful. These include studies of soil-water tension, wheat yields, acid deposition, and sudden infant death syndrome. Aspinall (Chapter 69) and Wilson (Chapter 70) discuss applications in landscape conservation and agriculture, respectively. The variogram has now been introduced into several GIS, and programs that can be interfaced with GIS are available to help construct variograms and to apply the kriging process (GS+ 1995; GEO-EAS 1988; S+SpatialStats 1996). The geographical literature on practical applications is building rapidly. Of particular interest is the work of Oliver and Webster (1990).

3.5 Classification

Interest in this problem rises or falls depending on the challenges presented by the subject matter and the type of data used. As part of any image analysis of remotely-sensed data, grouping algorithms are needed. Supervised and unsupervised classification schemes have been developed that allow for pixel values to be identified with a particular category of, say, land cover. Spectral, regression tree, autocorrelation, neural network, and fuzzy logic schemes have been adapted to deal with the problems of aggregation. Themes being pursued include:

- evaluation of neural pattern classifiers (Fischer, Chapter 19; Fischer et al 1997);
- the degree of supervision needed in finding statistically significant groupings (Gong and Howarth 1990);
- effects of resolution and sensitivity on various classification schemes (Marceau et al 1994);
- incorporation of non-remotely-sensed data in decision tree algorithms (Michaelsen et al 1996);
- application of classification routines to the results of spectral-unmixing (Mertes et al 1995);
- classification routines applied to hyperspectral and high spatial resolution data (Barnsley, Chapter 32; Barnsley and Barr 1996).

3.6 Sampling issues

Just as the jury selection process affects the outcome of a trial, so does the sampling scheme influence research results. Spatial sampling is a particularly difficult problem to deal with, since the idea (unlike many jury selection processes) is to select an unbiased sample, but finding independent observations is impossible. Spatial sampling requires that the researcher recognise the degree of dependence in the data. Very often, the surfaces from which samples are taken are complex and oddly shaped, presenting difficult problems to overcome in the statistical analysis. For many years, considerable effort was given to making sense from small samples. The challenge now is to make sense of large datasets (Fischer, Chapter 19; Openshaw and Alvanides, Chapter 18), and one means of so doing is to sample from them. Research in this area includes:

- line transects and variable circular plots, including kernel sampling (Quang 1992);
- network sampling (Faulkenberry and Garoui 1991);
- cluster and systematic sampling (Thompson 1992);
- spatial sample size (Ripley 1981; Goodchild and Gopal 1989; Haining 1990);
- strip and stratified adaptive cluster sampling (Thompson 1992);
- heterogeneous data sampling (Griffith et al 1994).

3.7 Spatial econometrics

The fundamental work in this area can be traced to Paelinck (1967; see also Paelinck and Klaassen 1979). Anselin has made spatial econometrics accessible to a wide audience with his text (1988) and software (Chapter 17; 1992). In addition, texts by Haining (1990), Griffith (1988), and Upton and Fingleton (1985) have helped to widen the appeal of these methods in geography. As Anselin says, the approach is 'model driven'; that is, the focus is on regression parameter estimation, model specification, and testing when spatial effects are present. Regression models constitute the leading approach for the study of economic and social phenomena. The assumptions required for the basic linear regression model, however, do not satisfy the needs of spatial regression models, which must take into account spatial dependence and/or spatial heterogeneity. Spatial dependence occurs when there is a relationship between observations of one or more variables at one point in space with those at another point in space, while spatial heterogeneity results from data that are not homogeneous – for example, population by areas which vary considerably by size and shape.

A number of spatial autoregressive models have been developed that include one or more spatial weight matrices that describe the many spatial associations in the data. The models include either a single general stochastic autocorrelation parameter, a series of autocorrelation parameters, one for each independent variable conditioned by spatial effects (dependency or heterogeneity), an error term autocorrelation parameter, or some combination of these. Parameter estimation procedures can be complex. The usual approach is to use diagnostic statistics to test for dependence and/or heteroscedasticity among the spatially weighted variables or the error term. Fortunately, SpaceStat, designed for the exploration and testing of spatial autoregressive models, is sufficiently user friendly to allow for the development of final autoregressive models (see Anselin, Chapter 17, for a general discussion). In addition, the package has been linked explicitly to several GIS, including ArcView (1995) and Idrisi (Eastman 1993).

Several other approaches have been taken to specify the influence of spatial effects in a regression model environment. Casetti's (1972) expansion method is designed to increase the number of variables in a regression model to take into account secondary, but influential, spatial variables, such as the x, y coordinates of georeferenced variables. This approach uses the parameters of the expansion variables as the indicators of the spatial effects.

In another development, Getis (1990, 1995) suggests transforming the spatially autocorrelated model into one without spatial autocorrelation embedded within it. By filtering out the spatial autocorrelation, the ordinary least squares model can be estimated and evaluated using R^2. By use of the Getis–Ord statistics mentioned earlier, variables are transformed to become relatively free of dependency effects. The filtered spatial components are re-entered into the regression equation as separate spatial variables.

The list of recent research themes, many of which can be found in the volume edited by Anselin and Florax (1995), can be divided into two parts: spatial modelling and estimation. The spatial modelling themes are:

- robust approaches to testing spatial models (Anselin 1990b);
- mis-specification effects in spatial models (Florax and Rey 1995; Hepple 1996);
- data problems in spatial econometric modelling (Haining 1995);
- the general linear model and spatial autoregressive models (Griffith 1995);
- multiprocess mixture (space-time) models (LeSage 1995);

- adaptive filtering and dependence filtering for spatial models (Foster and Gorr 1986; Getis 1995).

Parameter estimation is a subject central to model development. For models having spatial parameters or variables, a number of issues have arisen. The robustness, consistency, and reliability of parameters is a function of underlying theoretical distributions. The assumption of asymptotic normality has been called into question in some cases, and in others, sample sizes must be large before normality assumptions can be invoked. Bayesian approaches have been introduced in order to bring more information to bear on parameter estimation. Maximum likelihood procedures are fundamental to spatial model estimation, but data screening and filtering have been suggested as ways to simplify estimation. Current research includes:

- estimation of regression parameters in spatially constructed regression equations (Florax and Folmer 1992; Kelejian and Robinson 1993);
- estimating space-time probit models (McMillen 1992);
- estimating logit models with spatial dependence (Dubin 1995);
- spatial parametric instability (Casetti and Poon 1995);
- small sample properties of tests for spatial dependence (Anselin and Florax 1995a).

4 PROBLEMS, CHALLENGES, AND FUTURE DIRECTIONS

At the heart of spatial science are the statistical and mathematical techniques that allow for confirmatory statements to be made about the relationship between variables in a spatial setting. The thrust in recent years has been to develop more and better ways to describe data (see Anselin, Chapter 17). The exploratory data analysis movement has given researchers a bevy of fast ways to view data. Much of this work has been created as a response to the large and detailed datasets that are becoming available. At the same time, relatively few new methods have been offered to allow for the confirmation of hypotheses, which is well behind in the race for new understanding of spatial phenomena (Anselin and Getis 1992).

A number of barriers hamper the modellers and others seeking verification of their suppositions.

Many of these obstacles derive from flawed data. Problems include poor data quality, inadequate data coverage, incompatible datasets, inappropriate data, and the inability to handle large datasets (see the contributions to Section 1(b) of this book). While the main goal of spatial statistical analysis is to assist in data interpretion, it cannot improve on flawed data (Goodchild 1992). In addition, at least three further obstacles stand in the way of confirmatory analysis; these are described below.

4.1 Scale

Although much has been written about the nature of the scale problem, there are few useful suggestions for dealing with it. Perhaps the most often suggested solution is to attempt to solve the scale problem at a number of scales in the hope that a certain robustness to the process will allow results to be generalised to a number of spatial scales. In order to understand this issue, however, more research is needed on the nature of distributional parameters when data are aggregated. A wider review of scale issues is provided by Weibel and Dutton (Chapter 10).

4.2 Spatial weights effects

Identifying and describing spatial association is the goal of much research. The intrinsic stationarity of the variogram represents an empirically derived theory of spatial effects. In essence, it is the spatial weights matrix of the autoregressive models and the spatial association statistics. Thus, the spatial weights matrix is the manifestation of our understanding of spatial association. Too often, the contiguity spatial weights matrix is chosen simply because no further understanding of distance, or interaction, or association is assumed. The spatial association statistics, such as I, c, and G, could be put to good use as indicators of the appropriateness of particular spatial weights matrices. To understand better issues such as dependence and spillover, generalisation between global and local scales, and data heterogeneity and homogeneity, appropriate mathematical constructs – such as eigenvectors – must be related to the form and structure of our data. In addition, types of variable – economic, social, physical – must be related to the geometry of their spatial representation.

4.3 Boundary effects

Related to the above two problem areas is the issue of boundary effects. In spatial studies, the delineation of

boundaries bear heavily upon results. Although many truncated probability distributions have been derived, they have not been used to good effect to account for spatial boundary conditions. A number of statistical procedures, such as refined nearest neighbour analysis and K-function analysis, take into consideration the effect of boundaries, but spatial scientists have yet to consider boundary effects systematically. Stochastic approaches to modelling take into account impervious, reflecting, and other types of boundary conditions, but this work has not yet entered the mainstream of spatial science.

The problems discussed above describe at most half of the challenge. Increasingly, the temporal dimension is becoming a part of formerly static models of spatial human and physical processes. Deeper understanding usually comes from the study of differences in space as well as time. Bringing these two fundamental dimensions into a modelling framework where parameters can be estimated is a considerable challenge.

5 GIS AND SPATIAL STATISTICS

With regard to GIS, this volume makes clear how well suited these systems are for the exploration and manipulation of spatial data. Initially, the contributions to GIS were in the form of commands that allowed for the rectification of inconsistencies between a number of coverages (spatial variables) of the same geographical region. Much was made of the fundamental data model, that is, raster or vector. The main purpose was to link georeferenced datasets that are either in pixel or polygonal spatial form so that various combinations of variables could be mapped. As sophistication increased, functions were developed that allowed for new data to be derived from the various coverages, and for back and forth movement between data models.

For the most part, however, testing of hypotheses using statistical methodology was left for non-GIS statistical packages. It was quite enough to develop the technology and the functions that allow for data manipulation. Naturally, exploratory analytical functions were developed. A great deal of progress has been made in this regard, mainly from the standpoint of graphical summaries of data distributions together with simple summary measures like means and standard deviations. The need for more sophisticated analyses, voiced by many academics, is now getting a hearing in GIS literature

(Longley and Batty 1996a). Analysts are now beginning to take advantage of the data processing and data manipulation qualities of GIS to help create and test models using statistical methodology. A number of packages have been developed that enable researchers to interface with GIS-formatted datasets. Some of these are: S+Gislink links S+SpatialStats with ARC/INFO (1996), SpaceStat links with ArcView (Anselin 1997), Bailey and Gatrell's *Interactive Spatial Data Analysis* (1995), and Regard (Haslett et al 1990). Other packages, such as GS+, can be adapted to GIS requirements.

The effect of the new technology on spatial statistical analysis has led to a broadening of the process of hypothesis testing (Getis 1993). Heretofore, the hypothesis-testing process was straightforward, with little opportunity to recast hypotheses while in the testing process. Now, the approach is much more flexible. Note that in Figure 1 a step has been added to the traditional approach of hypothesis guided inquiry, and most steps have been expanded to include more opportunities to assess data from different vantage points. The added step, *data manipulation*, presents researchers with opportunities to use larger samples, view data over a series of map scales, and generally be in a stronger position to carry out statistical tests by means of simulations, sensitivity analyses, and bootstrap methods. Note, however, that each of these aproaches broadly adheres to what Goodchild and Longley (Chapter 40) term the 'linear project design'.

The flurry of activity in recent years has led to the publication of a number of edited volumes and special journal issues that provide examples of the various themes that are designed to wed spatial statistical analysis with GIS. Included among these are books edited by Fotheringham and Rogerson (1994), Frank and Campari (1993), Fischer and Nijkamp (1993), Fischer et al (1996), Longley and Batty (1996b), and Fischer and Getis (1997). Given the attention paid to this subject, in the next years we might expect a full-fledged statistical package, in the SPSS sense, integrated with the most comprehensive GIS.

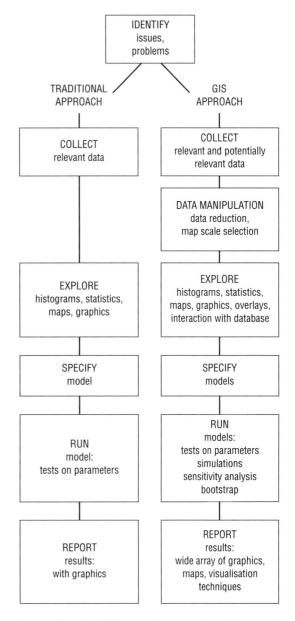

Fig 1. Traditional and GIS approaches to spatial statistics analysis.

References

Amrhein C G, Reynolds H 1996 Using spatial statistics to assess aggregation effects. *Geographical Systems* 3: 143–58

Anselin L 1986 Non-nested tests on the weight structure in spatial autoregressive models: some Monte Carlo results. *Journal of Regional Science* 26: 267–84

Anselin L 1988 *Spatial econometrics: methods and models.* Dordrecht, Kluwer

Anselin L 1990a What is special about spatial data? Alternative perspectives on spatial data analysis. In Griffith D A (ed.) *Statistics, past, present and future.* Ann Arbor, Institute of Mathematical Geography: 63–77.

Anselin L 1990b Some robust approaches to testing and estimation in spatial econometrics. *Regional Science and Urban Economics* 20: 141–63

Anselin L 1992 *SpaceStat: a program for the analysis of spatial data.* NCGIA, Santa Barbara, University of California

Anselin L 1993 *Exploratory spatial data analysis and geographic information sysytems.* West Virginia University, Regional Research Institute, Research Paper 9329

Anselin L 1995 Local indicators of spatial association – LISA. *Geographical Analysis* 27: 93–115

Anselin L, Bao S 1997 Exploratory spatial data analysis linking SpaceStat and ArcView. In Fischer M M, Getis A (eds) *Recent developments in spatial analysis: spatial statistics, behavioural modelling, and computational intelligence.* Berlin, Springer

Anselin L, Florax R J G M (eds) 1995 *New directions in spatial econometrics.* Berlin, Springer

Anselin L, Getis A 1992 Spatial statistical analysis and geographic information systems. *Annals of Regional Science* 26: 19–33

Anselin L, Griffith D A 1988 Do spatial effects really matter in regression analysis? *Papers of the Regional Science Association* 65: 11–34

Anselin L, Rey S 1991 Properties of tests for spatial dependence in linear regression models. *Geographical Analysis* 23: 112–31

Arbia G 1989 *Spatial data configuration in statistical analysis of regional economic and related problems.* Dordrecht, Kluwer

Arbia G, Benedetti R, Espa G 1996 Effects of the MAUP on image classification. *Geographical Systems* 3: 159–80

ArcView v 2.1 1996 *The geographic information system for everyone.* Redlands, ESRI

Bailey T C, Gatrell A C 1995 *Interactive Spatial Data Analysis.* Harlow, Longman/New York, John Wiley & Sons Inc.

Bao S, Henry M 1996 Heterogeneity issues in local measurements of spatial association. *Geographical Systems* 3: 1–13

Barnsley M J, Barr S L 1996 Inferring urban land-use from satellite sensor images using kernel-based spatial reclassification. *Photogrammetric Engineering and Remote Sensing* 62: 949–58

Batty M, Longley P 1994 *Fractal cities: a geometry of form and function.* London/San Diego, Academic Press

Besag J 1977 Discussion following Ripley. *Journal of the Royal Statistical Society* B 39: 193–5

Boots B N, Getis A 1988 *Point pattern analysis.* Newbury Park, Sage

Boots B N, Kanaroglou P S 1988 Incorporating the effect of spatial structure in discrete choice models of migration. *Journal of Regional Science* 28: 495–507

Carey H C 1858 *Principles of social science.* Philadelphia, Lippincott

Casetti E 1972 Generating models by the expansion method: applications to geographic research. *Geographical Analysis* 4: 81–91

Casetti E, Poon J 1995 Econometric models and spatial parametric instability: relevant concepts and an instability index. In Anselin L, Florax R J G M (eds) *New directions in spatial econometrics.* Berlin, Springer: 301–21

Christaller W 1935 *Die zentralen Orte in Süddeutschland.* Jena, G Fischer

Clark P J, Evans F C 1954 Distances to nearest neighbor as a measure of spatial relationships in populations. *Science* 121: 397–8

Clark W A V 1969 Applications of spacing models in intra-city studies. *Geographical Analysis* 1:391–9

Cliff A D, Ord J K 1973 *Spatial autocorrelation.* London, Pion

Cliff A D, Ord J K 1981b *Spatial process: models and applications.* London, Pion

Cressie N 1991 *Statistics for spatial data.* Chichester, John Wiley & Sons

Dacey M F 1963 Order neighbor statistics for a class of random patterns in multidimensional space. *Annals of the Association of American Geographers* 53: 505–15

Dacey M F 1965 A review of measures of contiguity for two and *k*-color maps. In Berry B J L, Marble D F (eds) *Spatial analyses: a reader in statistical geography.* Englewood Cliffs, Prentice-Hall: 479–95

Diggle P J 1979 Statistical methods for spatial point patterns in ecology. In Cormack R M, Ord J K *Spatial and temporal analysis in ecology.* Fairland, International Cooperative Publishing House: 95–150

Diggle P J 1983 *Statistical analysis of spatial point patterns.* London, Academic Press

Donnelly K P 1978 Simulations to determine the variance and edge effect of total nearest neighbour distance. In Hodder I (ed.) *Simulation methods in archaeology.* Cambridge (UK), Cambridge University Press: 91–5

Dubin R 1995 Estimating logit models with spatial dependence. In Anselin L, Florax R J G M (eds) *New directions in spatial econometrics.* Berlin, Springer: 229–42

Eastman J R 1993 *Idrisi: a geographical information system.* Worcester (USA), Clark University

Faulkenberry G D, Garoui A 1991 Estimating a population total using an area frame. *Journal of the American Statistical Association* 86: 445–9

Fischer M M, Getis A (eds) 1997 *Recent developments in spatial analysis: spatial statistics, behavioural modelling, and computational intelligence.* Berlin, Springer.

Fischer M M, Gopal S, Staufer P, Steinnocher K 1997 Evaluation of neural pattern classifiers for a remote sensing application. *Geographical Systems* 4: 195–223

Fischer M M, Nijkamp P (eds) 1993 *Geographic information systems, spatial modelling, and policy evaluation.* Berlin, Springer

Fischer M M, Scholten H, Unwin D (eds) 1996 *Spatial analytical perspectives on GIS in environmental and socio-economic sciences.* London, Taylor and Francis

Florax R J G M, Folmer H 1992 Specification and estimation of spatial linear regression models: Monte Carlo evaluation of pre-test estimators. *Regional Science and Urban Economics* 22: 405–32

Florax R J G M, Rey S 1995 The impacts of mis-specified spatial interaction in linear regression models. In Anselin L, Florax R J G M (eds) *New directions in spatial econometrics.* Berlin, Springer: 111–35

Foster S A, Gorr W L 1986 An adaptive filter for estimating spatially varying parameters: application to modeling police hours in response to calls for service. *Management Science* 32: 878–89

Fotheringham A S, Densham P J, Curtis A 1995 The zone definition problem in location–allocation modelling. *Geographical Analysis* 27: 60–77

Fotheringham A S, Rogerson P (eds) 1994 *Spatial analysis and GIS.* London, Taylor and Francis

Fotheringham A S, Wong D W S 1991 The modifiable areal unit problem in multivariate statistical analysis. *Environment and Planning A* 23: 1025–44

Frank A U, Campari I (eds) 1993 *Spatial information theory: a theoretical basis for GIS.* Berlin, Springer

Franklin J 1995 Predictive vegetation mapping: geographic modelling of biospatial patterns in relation to environmental gradients. *Progress in Physical Geography* 19: 474–99

Gatrell A C, Bailey T C, Diggle P J, Rowlinson B S 1996 Spatial point pattern analysis and its application in geographical epidemiology. *Transactions, Institute of British Geographers* 21: 256–74

Geary R 1954 The contiguity ratio and statistical mapping. *The Incorporated Statistician* 5: 115–45

GEO-EAS 1988, Las Vegas, United States Environmental Protection Agency, Environmental Monitoring Systems Laboratory

Getis A 1964 Temporal land-use pattern analysis with the use of nearest neighbor and quadrat methods. *Annals of the Association of American Geographers* 54: 391–9

Getis A 1983 Second-order analysis of point patterns: the case of Chicago as a multi-center urban region. *Professional Geographer* 35: 73–80

Getis A 1984 Interaction modelling using second-order analysis. *Environment and Planning A* 16: 173–83

Getis A 1990 Screening for spatial dependence in regression analysis. *Papers of the Regional Science Association* 69: 69–81

Getis A 1991 Spatial interaction and spatial autocorrelation: a cross-product approach. *Environment and Planning A* 23: 1269–77

Getis A 1993 GIS and modelling prerequisites. In Frank A U, Campari I (eds) *Spatial information theory: a theoretical basis for GIS.* Berlin, Springer: 322–40

Getis A 1994 Spatial dependence and heterogeneity and proximal databases. In Fotheringham A S, Rogerson P (eds) *Spatial analysis and GIS.* London, Taylor and Francis: 105–20

Getis A 1995 Spatial filtering in a regression framework: experiments on regional inequality, government expenditures, and urban crime. In Anselin L, Florax R J G M (eds) *New directions in spatial econometrics.* Berlin, Springer: 172–88

Getis A, Ord J K 1992 The analysis of spatial association by use of distance statistics. *Geographical Analysis* 24: 189–206

Getis A, Ord J K 1996 Local spatial statistics: an overview. In Longley P, Batty M (eds) *Spatial analysis: modelling in a GIS environment.* Cambridge (UK), GeoInformation International: 269–85

Gong P, Howarth P J 1990 An assessment of some factors influencing multispectral land cover classification. *Photogrammetric Engineering and Remote Sensing* 56: 597–603

Goodchild M F 1992 Geographical information science. *International Journal of Geographical information Systems* 6: 31–45

Goodchild M F, Gopal S 1989 *Accuracy of spatial databases.* London, Taylor and Francis

Green M, Flowerdew R 1996 New evidence on the modifiable areal unit problem. In Longley P, Batty M (eds) *Spatial analysis: modelling in a GIS environment.* Cambridge (UK), GeoInformation International

Griffith D A 1988 *Advanced spatial statistics: special topics in the exploration of quantitative spatial data series.* Dordrecht, Kluwer

Griffith D A 1995 The general linear model and spatial autoregressive models. In Anselin L, Florax R J G M (eds) *New directions in spatial econometrics.* Berlin, Springer: 273–300

Griffith D A 1996a Computational simplifications for space-time forecasting within GIS: the neighbourhood spatial forecasting model. In Longley P, Batty M (eds) *Spatial analysis: modelling in a GIS environment.* Cambridge (UK), GeoInformation International

Griffith D A, Haining R, Arbia G 1994 Heterogeneity of attribute sampling error in spatial datasets. *Geographical Analysis* 26: 300–20

249

GS+ Geostatistics for the environmental sciences 1995 Plainwell, Gamma Design Software

Haining R P 1990 *Spatial data analysis in the social and environmental sciences*. Cambridge (UK), Cambridge University Press

Haining R P 1995 Data problems in spatial econometric modelling. In Anselin L, Florax R J G M (eds) *New directions in spatial econometrics*. Berlin, Springer: 156–71

Harvey D 1966 Geographic processes and the analysis of point patterns. *Transactions, Institute of British Geographers* 40: 81–95

Haslett J, Bradley R, Craig P, Unwin A, Wills G 1991 Dynamic graphics for exploring spatial data with application to locating global and local anomalies. *The American Statistician* 45: 234–42

Haslett J, Wills G, Unwin A 1990 SPIDER [Regard] – an interactive statistical tool for the analysis of spatially distributed data. *International Journal of Geographical Information Systems* 4: 285–96

Hepple L W 1996 Directions and opportunities in spatial econometrics. In Longley P, Batty M (eds) *Spatial analysis: modelling in a GIS environment*. Cambridge (UK), GeoInformation International

Holt D, Steel D G, Tranmer M 1996 Area homogeneity and the modifiable areal unit problem. *Geographical Systems* 3: 181–200

Hubert L J 1979 Matching models in the analysis of cross-classifications. *Psychometrika* 44: 21–41

Hunt L, Boots B 1996 MAUP effects in the principal axis factoring technique. *Geographical Systems* 3: 101–22

Jacquez G M 1996 Disease cluster statistics for imprecise space-time locations. *Statistics in Medicine* 15: 873–85

Kelejian H H, Robinson D P 1993 A suggested method of estimation for spatial interdependent models with autocorrelated errors, and an application to a county expenditure model. *Papers in Regional Science* 72: 297–312

Kelejian H H, Robinson D P 1995 Spatial correlation: a suggested alternative to the autoregressive model. In Anselin L, Florax R J G M (eds) *New directions in spatial econometrics*. Berlin, Springer: 75–95

King L J 1962 A quantitative expression of the pattern of urban settlement in selected areas of the United States. *Tijdschrift voor Economische en Sociale Geografie* 53: 1–7

Krishna Iyer P V A 1949 The first and second moments of some probability distributions arising from points on a lattice, and their applications. *Biometrika* 36: 135–41

Lalanne L 1863 Untitled. *Comptes Rendus des Séances de l'Academie des Sciences*, 57(July–Dec): 206–10

LeSage J P 1995 A multiprocess mixture model to estimate space-time dimensions of weekly pricing of certificates of deposit. In Anselin L, Florax R J G M (eds) *New directions in spatial econometrics*. Berlin, Springer: 359–97

Longley P, Batty M 1996a Analysis, modelling, forecasting, and GIS technology. In Longley P, Batty M (eds) *Spatial analysis: modelling in a GIS environment*. Cambridge (UK), GeoInformation International: 1–15

Longley P, Batty M (eds) 1996b *Spatial analysis: modelling in a GIS environment*. Cambridge (UK), GeoInformation International

Lösch A 1954 *The economics of location*. New Haven, Yale University Press

Mantel N 1967 The detection of disease clustering and a generalised regression approach. *Cancer Research* 27: 209–20

Marceau D J, Gratton D J, Fournier R A, Fortin J P 1994 Remote sensing and the measurement of geographical entities in a forest environment: 2. The optimal spatial resolution. *Remote Sensing of the Environment* 49: 105–17

McMillen D P 1992 Probit with spatial autocorrelation. *Journal of Regional Science* 32: 335–48

Mendeleev D I 1906 *K poznaniyu rossii* (Russian information). St Petersburg, A S Suvorina

Mertes L K, Daniel D L, Melack J M, Nelson B, Martinelli L A, Forsberg B R 1995 Spatial patterns of hydrology, geomorphology, and vegetation on the floodplain of the Amazon River in Brazil from a remote sensing perspective. *Geomorphology* 13: 215–32

Michaelsen J, Schimel D S, Friedl M A, Davis F W, Dubayah R C 1996 Regression tree analysis of satellite and terrain data to guide vegetation sampling and surveys. *Journal of Vegetation Science* 5: 673–86

Moran P A P 1948 The interpretation of statistical maps. *Journal of the Royal Statistical Society B* 10: 243–51

Morrison A C, Getis A, Santiago M, Rigau-Peres J G, Reiter P 1996 Exploratory space-time analysis of reported dengue cases during an outbreak in Florida, Puerto Rico, 1991–92. *American Journal of Tropical Medicine*

Neft D S 1966 *Statistical analysis for areal distributions*, Monograph Series, 2, Philadelphia, Regional Science Research Institute

Okabe A, Boots B, Sugihara K 1992 *Spatial tessellations: concepts and applications of Voronoi diagrams*. New York, John Wiley & Sons Inc.

Okabe A, Tagashira N 1996 Spatial aggregation bias in a regression model containing a distance variable. *Geographical Systems* 3: 77–100

Oliver M A, Webster R 1990 Kriging: a method of interpolation for geographical information systems. *International Journal of Geographic Information Systems* 4: 313–32

O'Loughlin J, Anselin L 1991 Bringing geography back to the study of international relations: dependence and regional context in Africa, 1966–78. International Interactions 17: 29–61

Openshaw S 1996 Developing GIS-relevant zone-based spatial analysis methods. In Longley P, Batty M (eds) *Spatial analysis: modelling in a GIS environment*. Cambridge (UK), GeoInformation International

Openshaw S, Taylor P 1979 A million or so correlation coefficients: three experiments on the modifiable a real unit problem. In Bennett R J, Thrift N J, Wrigley N (eds) *Statistical applications in the spatial sciences*. London, Pion

Ord J K, Getis A 1995 Local spatial autocorrelation statistics: distributional issues and an application. *Geographical Analysis* 27: 286–306

Paelinck J 1967 *L'efficacité de la politique économique regionale*. Namur, Faculté des Sciences Economiques: 58

Paelinck J, Klaassen L 1979 *Spatial econometrics*. Farnborough, Saxon House

Potvin C, Travis J 1993 Concluding remarks: a drop in the ocean... . *Ecology* 74: 1674–6

Quang P X 1993 Nonparametric estimators for variable circular plot surveys. *Biometrics* 49: 837–52

Ravenstein E G 1885 The laws of migration. *Journal of the Royal Statistical Society* 48: 52

Reilly W J 1929 Methods for the study of retail relationships. *University of Texas, Bulletin*, 2944

Ripley B D 1977 Modelling spatial patterns. *Journal of the Royal Statistical Society B* 39: 172–94

Ripley B D 1981 *Spatial statistics*. New York, John Wiley & Sons Inc.

Robinson A H 1956 The necessity of weighting values in correlation of areal data. *Annals of the Association of American Geographers* 46: 233–6

Rogers A 1969 Quadrat analysis of urban dispersion: 1. Theoretical techniques. *Environment and Planning* 1: 47–80

S+SpatialStats 1996 Seattle, MathSoft, Inc.

Stewart J Q 1950 The development of social physics. *American Journal of Physics* 18: 239–53

Stewart J Q, Warntz W 1958 Macrogeography and social science. *Geographical Review* 48: 167–84

Thomas E N 1960 Maps of residuals from regression: their characteristics and uses in geographic research. *State University of Iowa, Department of Geography, Report*, 2

Thompson S K 1992 *Sampling*. New York, John Wiley & Sons Inc.

Thünen J H von 1826 *Der isolierte Staat in beziehung auf Landwirtschaft und Nationalokonomie*. Jena, G Fischer

Tiefelsdorf M, Boots B 1994 The exact distribution of Moran's *I*. *Environment and Planning A* 27: 985–99

Tobler W R 1963 Geographic area and map projections. *Geographical Review* 53: 59–78

Tobler W R 1979 Cellular geography. In Gale S, Olsson G (eds) *Philosophy in geography*. Dordrecht, Reidel: 379–86

Tobler W R 1989 Frame independent spatial analysis. In Goodchild M G, Gopal S (eds) *The accuracy of spatial databases*. London, Taylor and Francis: 115–22

Upton G J, Fingleton B 1985 *Spatial statistics by example*, Vol. 1. Chichester, John Wiley & Sons

Warntz W, Neft D S 1960 Contributions to a statistical methodology for areal distributions. *Journal of Regional Science* 2: 47–66

Wartenberg D 1985 Multivariate spatial correlation: a method for exploratory geographical analysis. *Geographical Analysis* 17: 263–83

Weber A 1909 *Über den Standort der Industrien*. Tubingen

Wilson A G 1967 A statistical theory of spatial distribution models. *Transportation Research* 1: 253–69

Wrigley N, Holt D, Steel D G, Tranmer M 1996 Analysing, modelling, and resolving the ecological fallacy. In Longley P, Batty M (eds) *Spatial analysis: modelling in a GIS environment*. Cambridge (UK), GeoInformation International: 25–40

Zipf G K 1949 *Human behavior and the principle of least effort*. Reading (USA), Addison-Wesley.

17
Interactive techniques and exploratory spatial data analysis

L ANSELIN

This chapter reviews the ideas behind interactive and exploratory spatial data analysis and their relation to GIS. Three important aspects are considered. First, an overview is presented of the principles behind interactive spatial data analysis, based on insights from the use of dynamic graphics in statistics and their extension to spatial data. This is followed by a review of spatialised exploratory data analysis (EDA) techniques, that is, ways in which a spatial representation can be given to standard EDA tools by associating them with particular locations or spatial subsets of the data. The third aspect covers the main ideas behind true exploratory spatial data analysis, emphasising the concern with visualising spatial distributions and local patterns of spatial autocorrelation. The geostatistical perspective is considered, typically taken in the physical sciences, as well as the lattice perspective, more familiar in the social sciences. The chapter closes with a brief discussion of implementation issues and future directions.

1 INTRODUCTION

Recent developments in computing hardware and GIS software have made it possible to interact directly with large spatial databases and to obtain almost instantaneous results for a wide range of GIS operations. The sophistication in storage, retrieval, and display provided by the rapidly evolving GIS technology has created a demand for new tools to carry out spatial analysis in general and spatial statistical analysis in particular (see, among others, Anselin and Getis 1992; Bailey 1994; Goodchild 1987; Goodchild et al 1992; Openshaw 1991). This demand grew out of an early awareness that the implementation of 'traditional' spatial analysis techniques was insufficient to address the challenges faced in a GIS environment (Goodchild and Longley, Chapter 40). The latter is often characterised by vast numbers of observations (hundreds to several thousands) and 'dirty' data, and some go so far as to completely reject 'traditional' spatial analysis that is based on statistical inference

(Openshaw and Alvanides, Chapter 18; Fischer, Chapter 19; Openshaw 1990, 1991). While this rather extreme viewpoint is not shared by many, it is widely recognised that many of the geographical analysis techniques of the 1960s fail to take advantage of the visualisation and data manipulation capabilities embodied in modern GIS. Specifically, most spatial statistical techniques, such as tests for spatial autocorrelation and spatial regression models, are primarily *static* in nature, allowing only limited interaction between the data, the models, and the analyst. In contrast, *dynamic* or *interactive* approaches to data analysis stress the user interaction with the data in a graphical environment, allowing direct manipulation in the form of instantaneous selection, deletion, rotation, and other transformations of data points to aid in the exploration of structure and the discovery of patterns (Buja et al 1996; Cleveland 1993; Cleveland and McGill 1988).

The importance of EDA to enhance the spatial analytical capabilities of GIS has become widely recognised (Anselin 1994; Anselin and Getis 1992; Bailey and Gatrell 1995; Fotheringham and Charlton 1994). The EDA paradigm for statistical analysis is based on a desire to let the data speak for themselves and to impose as little prior structure upon them as possible. Instead, the emphasis is on creative data displays and the use of simple indicators to elicit patterns and suggest hypotheses in an inductive manner, while avoiding potentially misleading impressions given by 'outliers' or 'atypical' observations (Good 1983; Tukey 1977). Since spatial data analysis is often characterised as being 'data rich but theory poor' (Openshaw 1991), it would seem to form an ideal area for the application of EDA. However, this is not a straightforward exercise, since the special nature of spatial data, such as the prevalence of spatial autocorrelation, may invalidate the interpretation of methods that are based on an assumption of independence, which is the rule in mainstream EDA (Anselin 1990; Anselin and Getis 1992). Hence, the need has arisen to develop specialised methods of exploratory spatial data analysis (ESDA) that take the special nature of spatial data explicitly into account (for recent reviews, see Anselin 1994; Anselin and Bao 1997; Bailey and Gatrell 1995; Cook et al 1996; Cressie 1993; Majure and Cressie 1997).

This chapter reviews the ideas behind interactive and ESDA and their relation to GIS. Many of the ESDA techniques have been developed quite recently and this remains an area of very active research. Therefore, the emphasis will be on general principles, rather than on specific techniques. The latter will only be used to illustrate the overall framework and no attempt is made to cover a comprehensive set of methods. The bulk of the chapter considers three important aspects of the integration of ESDA and interactive methods with GIS. First, an overview is presented of the principles behind interactive spatial data analysis, based on insights from the use of dynamic graphics in statistics and their extension to spatial data. This is followed by a review of spatialised EDA techniques, that is, ways in which a spatial representation can be given to standard EDA tools by associating them with particular locations or spatial subsets of the data. The third aspect covers the main ideas behind true exploratory spatial data analysis, emphasising the concern with visualising spatial distributions and local patterns of spatial autocorrelation (Getis, Chapter 16). The chapter closes with a brief discussion of implementation issues and future directions.

2 PRINCIPLES OF INTERACTIVE SPATIAL DATA ANALYSIS

The principles behind interactive spatial data analysis can be traced back to the work on dynamic graphics for data analysis in general, originated by the statistician John Tukey and a number of research groups at AT&T Bell Laboratories. An excellent review of the origins of these ideas is given in the collection of papers edited by Cleveland and McGill (1988), and early discussions of specific methods are contained in the papers by, among others, Becker et al (1987), Becker and Cleveland (1987), and Stuetzle (1987). More recent reviews of methods for the dynamic analysis of high-dimensional multivariate data and other aspects of interactive statistical graphics can be found in papers by, among others, Becker et al (1996), Buja et al (1991, 1996), Cleveland (1993), and Cook et al (1995).

Dynamic graphical methods started as enhancements to the familiar static displays of data (e.g. histograms, bar charts, pie charts, scatterplots), by allowing direct manipulation by the user that results in 'immediate' change in a graph (see Elshaw Thrall and Thrall, Chapter 23, for some examples). This had become possible by the availability of workstations with sufficient computational power to generate the statistical graphs without delays and to allow interaction with the data by means of an input device (light pen or mouse). The overall motivation was to involve the human factor more directly in the exploration of data (i.e. exploiting the inherent capabilities of the brain to detect patterns and structure), and thereby gain richer insights than possible with the traditional rigid and static display. This was achieved by allowing the user to delete data points, highlight (brush) subsections of the data, establish links between the same data points in different graphs, and rotate, cut through, and project higher-dimensional data. Furthermore, the user and not a preset statistical procedure determined which actions to perform. Interactive statistical procedures become particularly effective when datasets are large (many observations) and high-dimensional (many variables), situations where characterisation of the data by a few numbers becomes increasingly unrealistic (for an early assessment see, for example, Andrews et al 1988: 75). While dynamic graphics for statistics were originally mostly experimental and confined to research environments, they have quickly become pervasive features of the EDA capability in modern commercial statistical software packages.

An important aspect of dynamic graphics is the representation of data by means of multiple and simultaneously available 'views', such as a table, a list of labels, a bar chart, pie chart, histogram, stem and leaf plot, box plot, or scatterplot. These views are shown in different windows on a computer screen. They are linked in the sense that when a location in any one of the windows (e.g. a bar on a bar chart or a set of points on a scatterplot) is selected by means of a pointing device (brushing), the corresponding locations in the other windows are highlighted as well (see Becker et al 1987). While geographical locations have always played an important role in dynamic graphics (see the many examples of Cleveland and McGill 1988), it is only recently that the 'map' was introduced explicitly as an additional view of the data, for example by Haslett et al (1990, 1991), MacDougall (1991), and Monmonier (1989).

The most comprehensive set of tools to date that implement dynamic graphics for exploring spatial data is contained in the Regard (formerly Spider) software of Haslett, Unwin and associates, which runs on a Macintosh platform (see also Bradley and Haslett 1992; Haslett and Power 1995; Unwin 1994). Regard, and its successor Manet (Unwin et al 1996) allow for the visualisation of the distribution and associations between data for any subset of locations selected on a map display. Similarly, for any subset of data highlighted in a non-spatial view, such as a category in a histogram, the corresponding locations are highlighted on the map. This is illustrated in Figure 1, where attention focuses on suggesting promising multivariate relations pertaining to electoral change in the new German Bundesländer (formerly East Germany). Six types of dynamically linked views of the data are included, consisting of a map with highlighted constituencies, a bar chart,

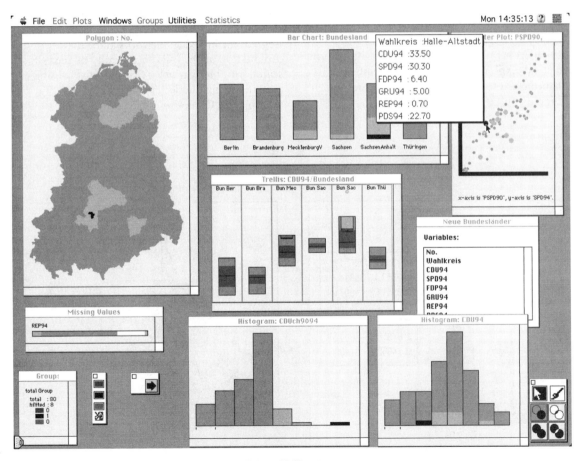

Fig 1. Interactive dynamic graphics for exploring spatial data with Manet.

conditional (trellis) plot, histogram, scatterplot and missing value chart, as well as lists with variable names and values observed at a specific location. (For details on the Manet approach, see Unwin et al 1996 and *http://www1.Math.Uni-Augsburg.de/~theus/Manet/ManetEx.html.*) While highly dynamic in its statistical graphics, the Spider–Regard–Manet approach is still somewhat limited in terms of the spatial aspects of the data, in the sense that it is based on a fixed map and does not take advantage of GIS functionality, such as specialised data models to facilitate spatial queries and overlays (see also Hazelhoff and Gunnink 1992).

Several ideas from the methodology of dynamic statistical graphics are reflected in the design of current GIS and mapping software. For example, the ArcView GIS (ESRI 1995b) is organised around several linked 'views' of the data (a map, a table, and several types of charts). These allow a limited degree of dynamic interaction in the sense that a selection made in any of the views (spatial selection of features on a map, records in a table) is immediately reflected in all other views. While Version 2.1 is rather limited in terms of its built-in statistical (exploratory) analysis capabilities, enhancements to make ArcView into a tool for interactive ESDA have been developed by

linking it to other software modules. For example, at the Statistics Laboratory of Iowa State University, a 2-directional link was established between the XGobi dynamic graphics software of Buja et al (1991, 1996) and ArcView (Cook et al 1996; Majure et al 1996a, 1996b; Symanzik et al 1994, 1995, 1996; *http://www.gis.iastate.edu/XGobi-AV2/XGobi-AV2.html*). Similarly, the SpaceStat software for spatial data analysis of Anselin (1992, 1995a) was linked with ArcView in a Microsoft Windows environment (Anselin and Bao 1996, 1997; *http://www.rri.wvu.edu/utilities.htm*). In many respects, these and similar efforts achieve a functionality close to that of Regard, although not as seamless and considerably slower in execution. For example, in Figure 2, ArcView scripts were used to construct a histogram for the median values of housing in West Virginia counties, linked to a map (a view in ArcView). Using a selection tool to click on a given bar (interval) in the histogram, the relevant counties in the map are highlighted (for further details on the dataset and the procedures, see Anselin and Bao 1996, 1997). In contrast to Regard, the linked frameworks allow the exploitation of the full functionality of the GIS to search for other variables that may display similar patterns, using queries and spatial overlays (for example, see Cook et al 1996).

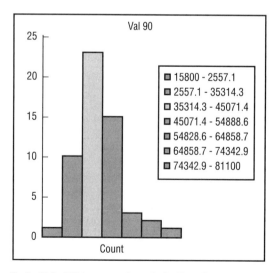

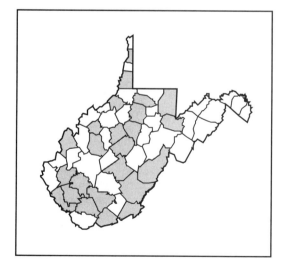

Fig 2. Linked histogram and map in ArcView–SpaceStat.

3 SPATIALISED EXPLORATORY DATA ANALYSIS

Whilst a widely available commercial implementation of interactive and dynamic spatial data analysis integrated with a GIS does not exist at the time of writing, the use of EDA with GIS has become fairly common. For example, in the 'archaeologist's workbench' of Farley et al (1990) and Williams et al (1990), standard EDA tools such as box plots and scatterplots were applied to geographical data, by exporting information from a GIS to a statistical package (a 1-directional link). However, the latter is not ESDA in the sense used by Cressie (1993) and Anselin (1994), but rather non-spatial EDA applied to spatial data (see also Anselin and Getis 1992).

Spatialised EDA (Anselin 1994) is one step closer to true ESDA in the sense that location is combined with a graphic description of the data in the form of a bar chart, pie chart, or various icons. The most familiar example of this may be the positioning of Chernoff faces at geographical locations on a map, such as coordinates of cities or centroids of states, as illustrated by Fotheringham and Charlton (1994) and Haining (1990: 226) (but for a critical assessment see Haslett 1992). The facility to add bar charts and pie charts to areal units on a map is by now a familiar feature in many commercial GIS and mapping packages.

A more meaningful combination of location and data description is obtained when summaries of spatial distributions are visualised for different subsets in the data, providing initial insight into spatial heterogeneity (i.e. different for spatial subsets in the data, such as a north–south differential) or suggesting a spatial trend (a systematic variation of a variable with location, such as an east–west trend). For example, Haining (1990: 224) organises box plots for standardised mortality rates by distance band away from the centre of the city, revealing a clear spatial trend. Similarly, spatialised EDA techniques may be used to carry out a form of exploratory spatial analysis of variance, in which the interest centres on differences in central tendency (mean, median) of the distribution of a variable between spatial subsets (or spatial regimes) in the data. In Figure 3 this is illustrated for the West Virginia data. Two box plots refer respectively to counties at the outer rim and inner counties (generated by applying a spatial selection operation in a GIS). A comparison of the two graphs suggests a systematically higher value for counties at the rim, although a few counties in either group do not fit the pattern. In an interactive data analysis, this could easily be addressed by sequentially removing or adding counties to one or the other subset, providing the groundwork for a spatial analysis of variance (for other examples see Anselin et al 1993). However, it is well recognised that potential spatial autocorrelation among these observations could invalidate the interpretation of any analysis of variance or regression analysis. Therefore, techniques only qualify as true ESDA when this is addressed explicitly.

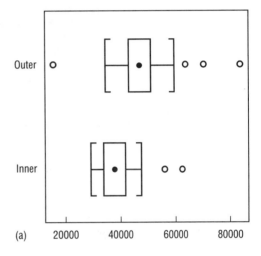

(a)

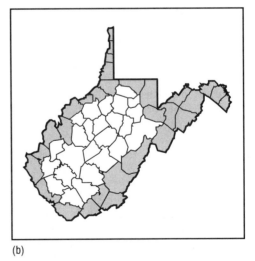

(b)

Fig 3. Exploratory spatial analysis of variance.

4 EXPLORATORY SPATIAL DATA ANALYSIS

ESDA can be broadly defined as the collection of techniques to describe and visualise spatial distributions, identify atypical locations (spatial outliers), discover patterns of spatial association (spatial clusters), and suggest different spatial regimes and other forms of spatial instability or spatial non-stationarity (Anselin 1994; see also Beard and Buttenfield, Chapter 15). Central to ESDA is the concept of spatial autocorrelation, that is, the phenomenon where locational similarity (observations in spatial proximity) is matched by value similarity (correlation).

Spatial autocorrelation has been conceptualised from two main perspectives, one prevalent in the physical sciences, the other in the social sciences. Following Cressie's (1993) classification, the so-called *geostatistical* perspective considers spatial observations to be a sample of points from an underlying continuous spatial distribution (surface). This is modelled by means of a variogram, which expresses the strength of association between pairs of locations as a continuous function of the distance separating them (for comprehensive reviews see Cressie 1993 and Isaaks and Srivastava 1989). By contrast, in the so-called lattice perspective, spatial locations are discrete points or areal units, and spatial data are conceptualised as a single realisation of a spatial stochastic process, similar to the approach taken in the analysis of time series. Essential in the analysis of lattice data is the concept of a *spatial weights matrix*, which expresses the spatial arrangement (topology, contiguity) of the data and which forms the starting point for any statistical test or model (for extensive reviews see Cliff and Ord 1981; Cressie 1993; Haining 1990; Upton and Fingleton 1985).

Juxtaposed on the distinction between the geostatistical and lattice perspective is that between global and local indicators of spatial association. Global indicators, such as the familiar Moran's I and Geary's c spatial autocorrelation statistics, summarise the overall pattern of dependence in the data into a single indicator (see Getis, Chapter 16). A major practical drawback for GIS analysis is that these global indicators are based on a strong assumption of spatial stationarity, which, among others, requires a constant mean (no spatial drift) and constant variance (no outliers) across space. This is not very meaningful or may even be highly misleading in analyses of spatial association for hundreds or thousands of spatial units that characterise current GIS applications. The main contribution of ESDA with respect to GIS lies therefore in visualising local patterns of spatial association, indicating local non-stationarity and discovering islands of spatial heterogeneity (Anselin 1994; Cressie 1993). In the remainder of this section, first some techniques are considered to visualise spatial distributions, with a particular focus on identifying outliers and atypical observations. These techniques are more specialised than the methods for visualisation for GIS discussed by Kraak (Chapter 11). This is followed by a short review of ESDA techniques to visualise and assess spatial autocorrelation, for both geostatistical and lattice perspectives.

4.1 Visualising spatial distributions

Many of the spatialised EDA techniques described above can be successfully applied to gain insight into the distribution of data across locations in a GIS. These methods can also be integrated in a dynamic interactive framework in a fairly straightforward way, for example as in the Manet software. A more explicit focus on identifying spatial outliers is offered by the so-called *box map*, the extension of a familiar quantile choropleth map (a standard feature in most GIS and mapping software) with highlighted upper and lower outliers, defined as observations outside the 'fences' in a box plot (Cleveland 1993). A box map can easily be implemented in many current GIS and mapping packages (e.g. Anselin and Bao 1997). By comparing box maps for different variables using overlay operations in a GIS, an initial look at potential multivariate associations can be obtained (e.g. see Talen 1997). Other approaches to identify outliers in spatial data can be envisaged as well, for example by constructing spatial queries for those locations whose values exceed some criterion of 'extremeness'. Such devices can be readily implemented in most currently available commercial GIS.

A more rigorous approach, geared towards the geostatistical perspective, consists of the estimation of a spatial cumulative distribution function (SCDF), that is, a continuous density function for all observations in a given region. This is implemented in the ArcView–XGobi linked framework mentioned earlier. The linkage allows users to highlight regions of the data on a map in ArcView and to find an SCDF plot in XGobi, to brush areas on the map to find the corresponding subset in the SCDF, and to brush quantiles of the estimated SCDF and find the matching locations on the map. For example, in Figure 4 (from Majure et al 1996a), the two SCDF functions for forest health indicators in the graph on the left-hand side correspond to the two large sub-regions of New England states in the map on the

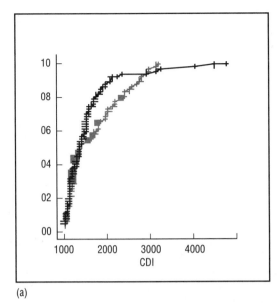

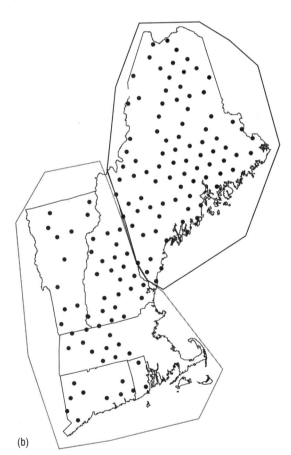

(a)

(b)

Fig 4. Spatial cumulative distribution function (SCDF) in ArcView–Xgobi.
Source: Majure et al 1996a

right. An advantage of this form of linkage is that the GIS can be used to overlay other data onto the sample points, in order to suggest potential multivariate associations. Clearly, an approach such as SCDF could be integrated into a more comprehensive Manet-type dynamic interactive ESDA framework, although this has not been implemented to date.

4.2 Visualising spatial autocorrelation: the geostatistical perspective

The main focus of ESDA in geostatistics is on identifying 'unusual' and highly influential (pairs of) locations in order to obtain more robust estimates of the variogram. Such locations are referred to as spatial outliers, or pockets of local non-stationarity, and they require closer scrutiny before proceeding with geostatistical modelling or spatial prediction (Kriging). The basic tools are outlined by Cressie (1993) and include the *variogram cloud*, the *variogram box plot* and the *spatial lag scatterplot*. A variogram cloud is a scatterplot of squared differences (or of square root absolute differences: see Cressie 1993) between all pairs of observations, sorted by distance band. An implementation of this device in an interactive dynamic graphics framework consisting of ArcView and XGobi is illustrated in Figure 5 (from Majure et al 1996). By brushing points in the cloud plot, lines are drawn between pairs of observations on the map, suggesting potential regions that are spatial outliers. A similar, but more encompassing approach is implemented in the Regard software, where the variogram cloud is included as one of the linked views of the data to facilitate a search for local pockets of spatial

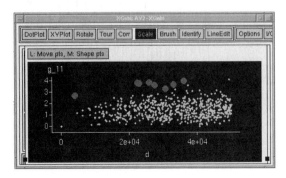

Fig 5. Brushed variogram cloud plot in ArcView–Xgobi.
Source: Majure et al 1996a

259

non-stationarity (Bradley and Haslett 1992; Haslett 1992; Haslett et al 1991; Haslett and Power 1995). The spatial lag scatterplot (also referred to as a lagged scatterplot) and the variogram box plot provide two different summary views of the information in the cloud plot. The spatial lag scatterplot focuses on the observation pairs that belong to a given distance class, that is, a subsection of the variogram cloud between two distances. The value observed at each point is plotted against the value observed at the 'lagged' point (a point separated from it by a distance belonging to the given distance band). The spatial lag scatterplot identifies potential influential locations as points that are far-removed from the 45 degree line (Majure and Cressie 1997). The variogram box plot consists of a box plot for each distance band in the variogram cloud, as in the left-hand side of Figure 6, illustrating the spatial dependence in the West Virginia housing values. For several distance bands, outliers may be identified as points outside the fences of the box plot. These outliers can be associated with the pairs of locations to which they correspond, as in the right-hand side of Figure 6, typically obtained in an interactive manner (and in a way similar to the procedure illustrated in Figure 5).

Extensions of both types of plots are possible in many ways, for example by using robustified measures of squared difference, by focusing on different directions (anisotropy), or by including multiple variables (for extensive examples see Majure and Cressie 1997). ESDA techniques based on the geostatistical perspective can be found in many academic as well as a number of commercial geostatistics software packages (e.g. S+SpatialStats, MathSoft 1996a), although the linkage to GIS is still limited or non-existent at the time of writing.

4.3 Visualising spatial autocorrelation: the lattice perspective

Central in the lattice perspective to spatial autocorrelation is the concept of a *spatial weights matrix* and associated spatially lagged variable or *spatial lag*. The non-zero elements of the spatial weights matrix indicate for each location which other locations potentially interact with it (the so-called spatial neighbours). Furthermore, the value of the non-zero elements is related to the relative strength of this interaction (for technical details see Cliff and Ord 1981; Haining 1990; Upton and Fingleton 1985). A spatial lag is constructed as a weighted average (using the weights in the spatial weights matrix) of the values observed for the neighbours of a given location (see Anselin 1988).

The matching of the value observed at a location with its spatial lag for a given spatial weights matrix provides useful insight into the local pattern of spatial association in the data. More precisely, when a high degree of positive spatial autocorrelation is present, the observed value at a location and its spatial lag will tend to be similar. Spatial outliers will tend to be characterised by very different values for the location and its spatial lag, either much higher or much lower in the location compared to the average for its neighbours. The association between a variable and its spatial lag can be visualised by means of so-called *spatial lag pies* and *spatial lag bar charts* (Anselin 1994;

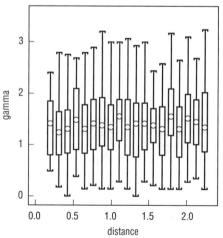

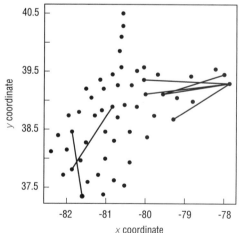

Fig 6. Variogram box plot with outlier pairs identified by location.

Anselin et al 1993; Anselin and Bao 1997). Both of these are made up of visual devices (size of the pie or length of the bar) that indicate the relative value of the spatial lag compared to the value at a location, as illustrated in Figure 7. Other visualisation schemes are possible as well, for example based on the difference, absolute difference, squared difference, or ratio between the value observed at a location and its spatial lag. These devices can be implemented in most GIS and mapping software in a straightforward way. In addition to the usual zooming and querying facilities available in an interactive GIS, the use of spatial lag pies or spatial lag bar charts could be made dynamic by allowing an interactive definition of the spatial weights matrix. It is envisaged that systems implementing these ideas will be available in the near future.

A more formal approach towards visualising spatial association can be based on the concept of a *Moran scatterplot* and associated *scatter map* (Anselin 1994, 1995b, 1997). It follows from the interpretation of the Moran's *I* statistic for spatial autocorrelation as a regression coefficient in a bivariate spatial lag scatterplot. More precisely, in a scatterplot with the spatial lag on the vertical axis and the value at each location on the horizontal axis, Moran's *I* corresponds to the slope of the regression line through the points. When the variables are expressed in standardised form (i.e. with mean zero and standard deviation equal to one), this allows for an assessment of both global spatial association (the

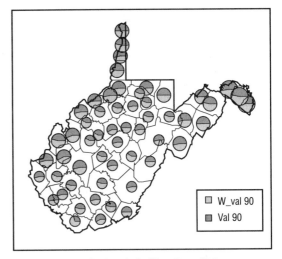

Fig 7. Spatial lag pie chart in ArcView–SpaceStat.

slope of the line) as well as local spatial association (local trends in the scatterplot). The latter is obtained by the decomposition of the scatterplot into four quadrants, each corresponding to a different type of spatial association: positive association between high values in the upper right and between low values in the lower left quadrants; negative association between high values surrounded by low values in the lower right and the reverse in the upper left quadrant. An illustration of this decomposition for the West Virginia data is given in Figure 8. The spatial locations that correspond to

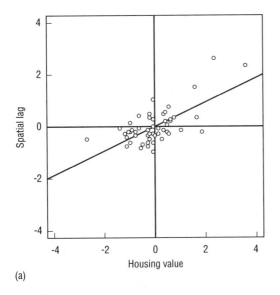

(a)

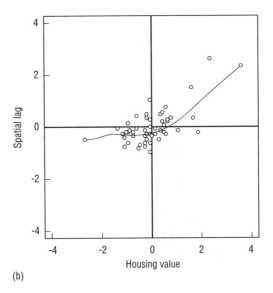

(b)

Fig 8. Moran scatterplot with linear and loess smoother.

the points in the scatterplot can be found in a linked map, where each quadrant is represented by a different shade or colour, as in Figure 9. By interactively identifying particular points in the graph (e.g. extreme values), the corresponding location can be shown on the map. This is a straightforward extension of the notion of brushing scatterplots to assess local spatial association.

Two additional interpretations of the Moran scatterplot are useful in an interactive ESDA setting. One is to identify outliers or high leverage points that unduly influence the slope of the regression line (i.e. the measure of global spatial association). Such outliers can be found by means of standard regression diagnostics and are easily identified on a map in a linked framework. They can also be related to the significance of local indicators of spatial association (LISA) statistics (Getis, Chapter 16; Anselin 1995b; Getis and Ord 1992; Ord and Getis 1995). In conjunction with a map of significant LISA statistics, the Moran scatterplot provides the basis for a substantive interpretation of spatial clusters or spatial outliers (further details are given by Anselin 1995b, 1996). A second interpretation is to consider the extent to which a non-linear smoother (such as a loess smoother; Cleveland 1979) approximates the linear fit in the scatterplot. Strong non-linear patterns may indicate different spatial regimes or other forms of local spatial non-stationarity. For example, on the right-hand side of Figure 8, the loess function suggests two distinct slopes in the graph, one considerably steeper than the other. The Moran scatterplot and

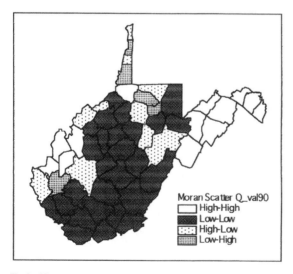

Fig 9. Moran scatter map.

associated map (Figure 9) can easily be implemented in a dynamic graphics setting, for example using the ArcView–SpaceStat linked framework.

5 IMPLEMENTATION AND FUTURE DIRECTIONS

To date a fully interactive ESDA functionality is not yet part of commercial GIS. However, several partial implementations exist, where a spatial statistical 'module' is added to an existing GIS (a point also made by Aspinall, Chapter 69; Boots, Chapter 36; Fischer, Chapter 19; and Getis, Chapter 16). Early discussions of these approaches were primarily conceptual, and a number of different taxonomies for integration have been advanced, primarily focusing on the nature of the linkage – closely coupled versus loosely coupled – and the types of statistical function that should be included (e.g. Anselin and Getis 1992; Goodchild et al 1992). Building on the general framework outlined by Anselin and Getis (1992), a schematic overview of the interaction between different analytical functions of a GIS is given in Figure 10 (based on Anselin 1998; see Getis, Chapter 16; and Goodchild and Longley, Chapter 40, for related conceptual schema). Following the usual classification of GIS functionality into four broad groups (input, storage, analysis, and output), the analysis function can be further subdivided into selection, manipulation, exploration and confirmation. Anselin et al (1993) considered the first two of these to form a 'GIS module' while the latter two formed a 'data analysis module' to emphasise the practical division of labour between typical commercial GIS software and the specialised (add-on) software needed to carry out spatial data analysis. However, this distinction is becoming increasingly irrelevant, since many statistical software packages now have some form of mapping (or even GIS) functionality, and a growing number of (spatial) statistical functions are included in GIS software. More important than classifying these functions as belonging to one or other module is to stress their interaction and the types of information that must be exchanged between them, as illustrated by the linkages in Figure 10. While many other taxonomies are possible, the main point of the classification in Figure 10 is that selection and manipulation (shown on the left) are present in virtually all advanced systems and have become known as 'spatial analysis' in the commercial world (e.g. ESRI 1995c: Lesson 8). By contrast, the spatial data analysis functions (shown on the right) are essentially absent in commercial systems.

The essence of any integration as in Figure 10 is that spatial information (such as location, topology, and distance) must be transferred from the GIS to the statistical module and location-specific results of the statistical analysis must be moved back to the GIS for mapping. Apart from the self-contained approach taken in Spider-Regard-Manet, most implementations to date of ESDA functionality in a GIS are extensions of existing systems by means of macro-language scripts. This typically hides the linked nature of the analysis routines from the user. Recent examples are extensions of ARC/INFO with non-spatial EDA tools, such as scatterplots (e.g. Batty and Xie 1994), and routines for the computation of global and local indicators of

spatial association (e.g. Ding and Fotheringham 1992; Bao et al 1995). An alternative is a closely-coupled linkage between two software packages that allow remote procedure calls (in Unix) or dynamic data exchange (in a Microsoft Windows environment). This approach is taken in the only commercial implementation that exists to date of an integrated data analysis and GIS environment, the S+Gislink between the S-Plus statistical software and the ARC/INFO GIS (MathSoft 1996b). On Unix workstations a bi-directional link is established that allows data to be passed back and forth in their native format. In addition, the linkage allows users to call S-Plus statistical functions from within ARC/INFO. A similar approach is taken in the ArcView–XGobi integration at the Statistics Laboratory of Iowa State University. A much looser

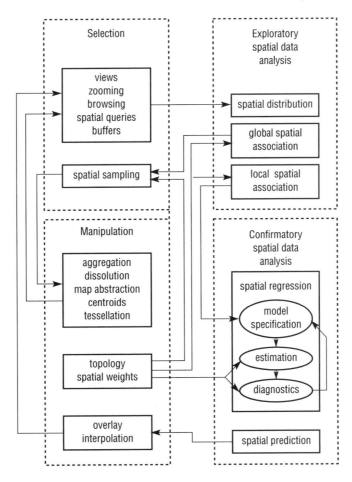

Fig 10. Spatial analysis in GIS.

coupling is implemented in the SpaceStat–ArcView linkage. Both of these efforts focus explicitly on ESDA, while the S-Plus–ARC/INFO linkage pertains primarily to traditional non-spatial EDA.

Several promising research directions are being pursued in the quest to develop more powerful tools for spatial analysis in GIS in general, and interactive spatial data analysis in particular. Highly relevant ongoing efforts include the use of the Internet to facilitate interactive mapping and visual data exploration (e.g. the Iris framework of Andrienko and Andrienko 1996; and see Batty, Chapter 21), the extension of data mining techniques to spatial data (e.g. Ng and Han 1994), and the use of massive parallel computing for the estimation of local indicators of spatial association (e.g. Armstrong and Marciano 1995). The extent of commercial and academic research activity devoted to methodological and computational facets will likely lead to a much-enhanced ESDA functionality in the GIS of the near future. This is an area of rapid change, and it is hoped that the general principles outlined in this chapter may provide a basis for the interpretation and assessment of future developments.

References

Andrews D F, Fowlkes E B, Tukey P A 1988 Some approaches to interactive statistical graphics. In Cleveland W S, McGill M E (eds) *Dynamic graphics for statistics*. Pacific Grove, Wadsworth: 73–90

Andrienko N, Andrienko G 1996 *IRIS, a knowledge-based system for visual data exploration*. See also *http://allanon.gmd.de/and/java/iris/Iris.html*

Anselin L 1988 *Spatial econometrics: methods and models*. Dordrecht, Kluwer

Anselin L 1990a What is special about spatial data? Alternative perspectives on spatial data analysis. In Griffith D A (ed.) *Spatial statistics, past, present, and future*. Ann Arbor, Institute of Mathematical Geography: 63–77

Anselin L 1992 *SpaceStat: a program for the analysis of spatial data*. Santa Barbara, NCGIA, University of California

Anselin L 1994a Exploratory spatial data analysis and geographic information systems. In Painho M (ed.) *New tools for spatial analysis*. Luxembourg, Eurostat: 45–54

Anselin L 1995a *SpaceStat version 1.80 user's guide*. Morgantown, Regional Research Institute, West Virginia University

Anselin L 1995b Local indicators of spatial association – LISA. *Geographical Analysis* 27: 93–115

Anselin L 1997 The Moran scatterplot as an ESDA tool to assess local instability in spatial association. In Fischer M, Scholten H, Unwin D (eds) *Spatial analytical perspectives on GIS in environmental and socio-economic sciences*. London, Taylor and Francis: 111–25

Anselin L 1998 GIS research infrastructure for spatial analysis of real estate markets. *Journal of Housing Research* 8

Anselin L, Bao S 1996 *SpaceStat.apr user's guide*. Morgantown, Regional Research Institute, West Virginia University

Anselin L, Bao S 1997 Exploratory spatial data analysis: linking SpaceStat and ArcView. In Fischer M, Getis A (eds) *Recent developments in spatial analysis – spatial statistics, behavioural modelling and neurocomputing*. Berlin, Springer

Anselin L, Dodson R, Hudak S 1993 Linking GIS and spatial data analysis in practice. *Geographical Systems* 1: 3–23

Anselin L, Getis A 1992 Spatial statistical analysis and geographic information systems. *Annals of Regional Science* 26: 19–33

Armstrong M P, Marciano R 1995 Massively parallel processing of spatial statistics. *International Journal of Geographical Information Systems* 9: 169–89

Bailey T C 1994 A review of statistical spatial analysis in geographical information systems. In Fotheringham A S, Rogerson P (eds) *Spatial analysis and GIS*. London, Taylor and Francis: 13–44

Bailey T C, Gatrell A C 1995 *Interactive spatial data analysis*. Harlow, Longman/New York, John Wiley & Sons Inc.

Bao S, Henry M, Barkley D, Brooks K 1995 RAS: a regional analysis system integrated with ARC/INFO. *Computers, Environment, and Urban Systems* 18: 37–56

Batty M, Xie Y 1994a Modelling inside GIS: part I. Model structures, exploratory spatial data analysis and aggregation. *International Journal of Geographical Information Systems* 8: 291–307

Becker R A, Cleveland W S 1987 Brushing scatterplots. *Technometrics* 29: 127–42

Becker R A, Cleveland W S, Shyu M-J 1996 The visual design and control of Trellis display. *Journal of Computational and Graphical Statistics* 5: 123–55

Becker R A, Cleveland W S, Wilks A R 1987 Dynamic graphics for data analysis. *Statistical Science* 2: 355–95

Bradley R, Haslett J 1992 High interaction diagnostics for geostatistical models of spatially referenced data. *The Statistician* 41: 371–80

Buja A, Cook D, Swayne D F 1996 Interactive high-dimensional data visualization. *Journal of Computational and Graphical Statistics* 5: 78–99

Buja A, McDonald J A, Michalak J, Stuetzle W 1991 Interactive data visualisation using focusing and linking. In Nielson G M, Rosenblum L (eds) *Proceedings of Visualisation 91*. Los Alamitos, IEEE Computer Society Press: 155–62

Cleveland W S 1979 Robust locally weighted regression and smoothing scatter plots. *Journal of the American Statistical Association* 74: 829–36

Cleveland W S 1993 *Visualizing data*. Summit, Hobart Press

Cleveland W S, McGill M E (eds) 1988 *Dynamic graphics for statistics*. Pacific Grove, Wadsworth

Cliff A, Ord J K 1981b *Spatial processes: models and applications*. London, Pion

Cook D, Buja A, Cabrera J, Hurley C 1995 Grand tour and projection pursuit. *Journal of Computational and Graphical Statistics* 4: 155–72

Cook D, Majure J, Symanzik J, Cressie N 1996 Dynamic graphics in a GIS: a platform for analysing and exploring multivariate spatial data. *Computational Statistics* 11: 467–80

Cressie N A C 1993 *Statistics for spatial data*, revised edition. New York, John Wiley & Sons Inc.

Ding Y, Fotheringham A S 1992 The integration of spatial analysis and GIS. *Computers, Environment, and Urban Systems* 16: 3–19

ESRI 1995b *ArcView 2.1, the geographic information system for everyone*. Redlands, ESRI

ESRI 1995c *Understanding GIS, the ARC/INFO method*. Redlands, ESRI Inc.

Farley J A, Limp W F, Lockhart J 1990 The archaeologist's workbench: integrating GIS, remote sensing, EDA and database management. In Allen K, Green F, Zubrow E (eds) *Interpreting space: GIS and archaeology*. London, Taylor and Francis: 141–64

Fotheringham A S, Charlton M 1994 GIS and exploratory spatial data analysis: an overview of some research issues. *Geographical Systems* 1: 315–27

Getis A, Ord J K 1992 The analysis of spatial association by use of distance statistics. *Geographical Analysis* 24: 189–206

Good I J 1983 The philosophy of exploratory data analysis. *Philosophy of Science* 50: 283–95

Goodchild M F 1987 A spatial analytical perspective on geographical information systems. *International Journal of Geographical Information Systems* 1: 327–34

Goodchild M F, Haining R P, Wise S et al 1992 Integrating GIS and spatial analysis – problems and possibilities. *International Journal of Geographical Information Systems* 6: 407–23

Haining R P 1990 *Spatial data analysis in the social and environmental sciences*. Cambridge (UK), Cambridge University Press

Haslett J 1992 Spatial data analysis – challenges. *The Statistician* 41: 271–84

Haslett J, Bradley R, Craig P, Unwin A, Wills G 1991 Dynamic graphics for exploring spatial data with applications to locating global and local anomalies. *The American Statistician* 45: 234–42

Haslett J, Power G M 1995 Interactive computer graphics for a more open exploration of stream sediment geochemical data. *Computers and Geosciences* 21: 77–87

Haslett J, Wills G, Unwin A 1990 SPIDER – an interactive statistical tool for the analysis of spatially distributed data. *International Journal of Geographical Information Systems* 4: 285–96

Hazelhoff L, Gunnink J L 1992 Linking tools for exploratory analysis of spatial data with GIS. *EGIS 92, Proceedings Third European Conference on Geographical Information Systems*. Utrecht, EGIS Foundation: 204–13

Isaaks E H, Srivastava R M 1989 *An introduction to applied geostatistics*. Oxford, Oxford University Press

MacDougall E B 1991 A prototype interface for exploratory analysis of geographic data. *Proceedings, Eleventh Annual ESRI User Conference* Vol. 2. Redlands, ESRI Inc.: 547–53

Majure J, Cook D, Cressie N, Kaiser M, Lahiri S, Symanzik J 1996a Spatial CDF estimation and visualisation with applications to forest health monitoring. *Computing Science and Statistics* 27: 93–101

Majure J, Cressie N 1997 Dynamic graphics for exploring spatial dependence in multivariate spatial data. *Geographical Systems*

Majure J, Cressie N, Cook D, Symanzik J 1996b GIS, spatial statistical graphics, and forest health. *Proceedings, Third International Conference/Workshop on Integrating GIS and Environmental Modeling, Santa Fe, 21–26 January*. Santa Barbara, NCGIA.

MathSoft 1996a *S+SpatialStats user's manual, version 1.0*. Seattle, MathSoft, Inc.

MathSoft 1996b *S+Gislink*. Seattle, MathSoft, Inc.

Monmonier M 1989 Geographic brushing: enhancing exploratory analysis of the scatterplot matrix. *Geographical Analysis* 21: 81–4

Ng R, Han J 1994 *Efficient and effective clustering methods for spatial data mining*. Technical Report 94–13. Vancouver, University of British Columbia, Department of Computer Science

Openshaw S 1990 Spatial analysis and geographical information systems: a review of progress and possibilities. In Scholten H, Stillwell J (eds) *Geographical information systems for urban and regional planning*. Dordrecht, Kluwer: 153–63

Openshaw S 1991c Developing appropriate spatial analysis methods for GIS. In Maguire D, Goodchild M F, Rhind D (eds) *Geographical information systems: principles and applications*. Harlow, Longman/New York, John Wiley & Sons Inc. Vol. 1: 389–402

Ord J K, Getis A 1995 Local spatial autocorrelation statistics: distributional issues and applications. *Geographical Analysis* 27: 286–306

Stuetzle W 1987 Plot windows. *Journal of the American Statistical Association* 82: 466–75

Symanzik J, Majure J, Cook D 1996 Dynamic graphics in a GIS; a bidirectional link between ArcView 2.0 and XGobi. *Computing Science and Statistics* 27: 299–303

Symanzik J, Majure J, Cook D, Cressie N 1994 Dynamic graphics in a GIS: a link between ARC/INFO and XGobi. *Computing Science and Statistics* 26: 431–35

265

Symanzik J, Megretskaia I, Majure J, Cook, D 1997
Implementation issues of variogram cloud plots and spatially
lagged scatterplots in the linked ArcView 2.1 and XGobi
environment. *Computing Science and Statistics* 28

Talen E 1997 Visualizing fairness: equity maps for planners.
Journal of the American Planning Association

Tukey J W 1977 *Exploratory data analysis*. Reading (USA),
Addison-Wesley

Unwin A 1994 REGARDing geographic data. In Dirschedl P,
Osterman R (eds) *Computational statistics*. Heidelberg,
Physica: 345–54

Unwin A, Hawkins G, Hofman H, Siegl B 1996
Interactive graphics for data sets with missing values –
MANET. *Journal of Computational and Graphical
Statistics* 5: 113–22

Upton G J, Fingleton B 1985 *Spatial data analysis by example*.
New York, John Wiley & Sons Inc.

Williams I, Limp W, Briuer F 1990 Using geographic
information systems and exploratory data analysis for
archeological site classification and analysis. In Allen K,
Green F, Zubrow E (eds) *Interpreting space: GIS and
archaeology*. London, Taylor and Francis: 239–73

18

Applying geocomputation to the analysis of spatial distributions

S OPENSHAW AND S ALVANIDES

Developments in IT and GIS have combined to create an enormously data-rich world. The need now is to develop GIS-relevant spatial analysis tools that will assist endusers in making good use of their spatial information. The problems are: (a) the absence of many appropriate tools; and (b) the new types of analysis required as a consequence of the innovation of GIS. The hope is that a combination of GIS databases, high performance computers, artificial intelligence and a geocomputation paradigm will together provide a generic workable solutions strategy.

1 INTRODUCTION

The GIS revolution has created an immense wealth of spatial information in a large number of different application areas. The emphasis in GIS upon database creation and systems building will soon have to be replaced by a new concern for applications using spatial analysis and modelling. The development of GIS can be largely regarded as the computerisation of pre-existing manual procedures and established technologies that were already fairly mature in research terms. Thus no great innovation was required in order to underpin the GIS revolution, although doubtless much did occur as the software systems developed. However, when the focus of attention switches to spatial analysis and modelling then it is a very different story. Frequently there have been no existing useful manual procedures to computerise, and over the last 20 years there has been very little relevant new research focused specifically on the special needs of GIS. Moreover, an increasing number of the emerging analysis tasks are novel and have not previously attracted much or any attention – for example, the exploration of very large spatial datasets for completely unknown patterns and relationships, or the real-time analysis of live spatial databases for emerging patterns, 'hot spots' (see Getis, Chapter 16), and anomalies of interest.

The quest for improved analysis of spatial distributions is predicated upon three considerations:

1 a general desire to make use of the information resources created by GIS;
2 attempts to gain competitive advantages or other benefits from investments in information technology (of which GIS is a component);
3 hardware developments that are trivialising the costs of computation and are hence creating new ways of analysing spatial distributions.

One might argue that it is scandalous that so many key databases are not currently being properly analysed – be they pertaining to morbidity, mortality, cancer, or crime incidence – while commercial concerns and government agencies probably waste many millions (if not billions) of dollars or ECUs by poor spatial data management and inefficient locational decisions (Openshaw 1994d). Yet perhaps the users cannot be blamed for not using tools that are unavailable! The problem is essentially a longstanding failure to evolve distinctly geographical-data-appropriate tools and styles of analysis and modelling – albeit with a small number of exceptions. The legacy of statistical methods (Getis, Chapter 16) may not be helpful, at least partly because its inherent limitations need to be properly understood. In essence, no amount of apparent statistical sophistication should be allowed to hide the fact that much of spatial statistics is very limited in what it can do, and is even more limiting in its view of spatial information and the handling of their special properties. It also needs to be appreciated that the 'post GIS-revolution' world of the late 1990s is quite different from the primeval (by contemporary standards) computing and data-poor environments in which many of the existing spatial analysis and modelling technologies were developed. Are these old legacy technologies and their latter-day offspring still appropriate, or is a new period of basic research and development needed to create the spatial analysis tools likely to be required in the late 1990s and beyond? There is no denial that advances have been made in improving some spatial statistical methods (Getis, Chapter 16), but these advances – such as developing local versions of global statistics – are of limited usefulness in practice.

This chapter attempts to address some of these concerns by focusing on the changes in the computational environment that has occurred during the 1990s. It argues that large-scale computation can now be used as a paradigm for solving some of the major spatial analysis problems that are relevant to GIS. However, if computational power is to be useful, then there also has to be a clear understanding of what the requirements and the user needs are. This leads us to a brief typology of alternative approaches and a brief illustrative case study based on one method in which high performance computing is combined with GIS data and artificial intelligence (AI) tools to develop better ways of engineering zoning systems as a decision support, analysis, modelling, and data management tool.

2 ADVANCES IN 'HIGH PERFORMANCE COMPUTING'

There is now considerable excitement in many traditional sciences about developments in supercomputing or high performance computing (HPC). Computation is now regarded as a scientific tool of equal importance to theory and experimentation, since fast computers have stimulated new ways of doing science via large-scale computer-based experimentation, simulation, and numerical approximation. There is equally a case for thinking that a supercomputing-based paradigm is

also relevant to many areas of GIS, but it should be appreciated that this involves much more than attempts to revamp basic GIS functions using parallel computing. HPC is defined as computer hardware based on vector or parallel processors (or some mixture) that offers at least one order of magnitude increase in computing power over that available from a mid 1990s workstation. In fact, as highly parallel processors take over from the earlier vector machines, the performance gain from using leading edge HPC hardware is more usually at least two orders of magnitude. This whole area is now developing at a rapid rate with most HPCs having a two- to three-year life cycle. For example, in 1996 it was possible to buy for £500000 (US $800000) a parallel machine with equivalent performance to one costing about 10 times as much only a few years earlier (see also Longley et al, Chapter 1). The computing world is in the throes of a major technological change; that of highly parallel supercomputing (Hillis 1992). A highly or massively parallel processor (MPP) is a computing system with multiple central processing units (CPUs) that can work concurrently on a single task. This idea is not new but it was only in the mid 1990s that the technology matured sufficiently for multiple CPUs to become the dominant future HPC machine architecture (see also Batty, Chapter 21). A nice feature of MPPs is that both processing capacity and memory are scaleable – if you want more computer power, then simply add more processors. If spatial analysis tools and models are also scaleable then running them on more processors reduces computer wall-clock times in a linear way (Turton and Openshaw 1996).

Openshaw (1994b) has suggested that by 1999 it is quite likely that available HPC hardware will be 10^9 times faster (and bigger in memory) than what was common during the 'Geography Quantitative Revolution' years of the 1960s (when many of the current so-called spatial statistical methods were developed: see Getis, Chapter 16); 10^8 times faster than hardware available during the mathematical modelling revolution of the early 1970s (on which virtually all of the so-called 'intelligent' model based spatial decision support systems employed in today's GISs were based: Birkin et al 1996); 10^6 times since the GIS revolution of the mid 1980s (a time of considerable neglect of quantitative geography), and at least a further 10^2 times faster than what in 1994 was Europe's fastest civilian supercomputer – the Edinburgh Cray T3D.

A widespread problem is that many potential users appear to have failed to appreciate what these developments in HPC mean. For instance, the Edinburgh Cray T3D has 512 processors and has a peak theoretical performance of 76.8 gigaflops – but what does that mean? One way of answering this question is to create a benchmark code that can be run on the widest possible range of computer hardware, ranging from a PC, UNIX workstations, vector supercomputers, and massively parallel machines. The widely used scientific benchmark codes measure machine performance in terms of simple matrix algebra problems, but it is not clear whether this is relevant in a GIS context. Openshaw and Schmidt (1997) have developed a social science benchmark code based on a scaleable spatial interaction model which can be run on virtually any serial and parallel processor. Table 1 provides a preliminary assessment of the performance of some current HPC hardware in terms of processing speed relative to the performance of a 486 66 MHz PC. At the time of writing, the best performance for small problem sizes was the SGI Onyx (a workstation with multiple CPUs), followed by the vector processor (the Fujitsu VPX240) which was about an order of magnitude more expensive. However, once problem sizes increase then soon there is no alternative to the massively parallel Cray T3D with speed gains of about 1335 times for a 10 000 by 10 000 zone matrix (equivalent to the UK ward level journey-to-work or migration data from the 1991 Census). This run took 2.4 seconds of wall-clock time (compared with 18 hours on a workstation), while the even larger 25 000 by 25 000 benchmark required 13 seconds. Note that HPC is not just about speed but also memory. The larger memory sizes required in these latter two runs reflect problems that have previously simply been uncomputable. It is interesting that during the 1990s machine speeds have been doubling every 1.5 to 2 years, and that this is expected to continue for at least another 10 years. One way of explaining what these changes in HPC hardware mean is to ask: how would you approach the spatial analysis challenges presented by GIS if that workstation on your desk were about 5000 times faster and bigger?

The criticism that few GIS end-users will ever be able to afford HPC hardware is irrelevant for two reasons. Firstly, what is possible using, for example, a mid 1990s national HPC research centre machine will within five years be affordable and 'do-able' using many workstations – even earlier if 'workstation farms' are used. Note that five years is probably also the lead time for the research and development cycle of new spatial analysis tools. Second, it is possible that the need for highly specialised but generic analytical functions will be met through the development of *embedded systems*. Embedded systems are unifunctional hardware that typically employ multiple CPUs. They are currently mainly used in signal processing. However, there is no reason why they cannot be programmed to perform specialised spatial analysis functions which may need large amounts of processing power. Such a system could appear to the GIS user as a call to a subroutine, except that the subroutine is in fact located somewhere else on the Internet and is not software but an integrated hardware and software

Table 1 Relative performance of a selection of available HPC hardware on a social science benchmark code in relation to 486 PC.

Hardware	Problem size: numbers of origin and destinations					
	Number of processors	100 by 100	500 by 500	1000 by 1000	10 000 by 10 000	25 000 by 25 000
Massively parallel						
Cray T3D	64	88				
	128		241	258		
	256			545	665	
	512				1335	1598
Parallel						
SGI Onyx	4	218	221	192	np	np
SGI Power Challenge	4	51	66	63	np	np
Vector supercomputer						
VPX240	1	162	195	196	np	np
Cray J90	8	8	35	39	np	np
Workstations						
SGI Indy	1	10	10	9	np	np
HP9000	1	14	12	10	np	np
Sun Ultra 2	1	18	17	16	np	np
Personal computers						
133MHz Pentium	1	3	4	4	np	np

Note: Benchmark problem sizes greater than 1000 by 1000 are not possible on a 486 PC. The times are estimated using linear interpolation which provides a good statistical fit to a range of smaller-sized problems.

system (see Maguire, Chapter 25). This form of heterogeneous distributed GIS is possible now and is one way of handling highly specialist but generic needs for spatial analysis functionality. The problem at present is that there is not yet a single example of such a system in operation, and not many ideas about the nature of the spatial analysis technology that should be run on them. On the other hand, the good news is that the languages and software tools needed to develop portable and future-proofed parallel applications are now quite well developed. Very significant here is the recent international standardisation of both a Highly Parallel FORTRAN (HPF) compiler and of the message passing interface (MPI).

3 A GEOCOMPUTATION PARADIGM FOR GIS

The rise of scaleable parallel hardware dramatically increases the opportunities within GIS for large-scale spatial analysis, using new approaches that seek to solve some of the traditional problems by switching to a more computationally intensive paradigm (see Openshaw 1997 for a review). There are now new ways of approaching spatial analysis using what has been termed a geocomputational paradigm. Geocomputation is itself a relatively new term, defined as the adoption of a large-scale computationally intensive approach to the problems of physical and human geography in particular, and the geosciences in general. Geocomputation is a paradigm that is clearly relevant to GIS, but also goes far beyond it. Spatial data manipulation on parallel supercomputing may well involve a return to flat data held in massive memory spaces, rather than recursive relational and hierarchically structured databases held on disk (see Worboys, Chapter 26). It also involves the development and application of new computational techniques and algorithms that are dependent upon, and can take particular advantage of, supercomputing. The motivating factors are threefold:

1 developments in HPC stimulating the adoption of a computational paradigm to problem solving, analysis, and modelling;
2 the need to create new ways of handling and analysing the increasingly large amounts of spatial information about the world stored in GIS;
3 the increased availability of AI tools and Computational Intelligent methods (Bezdek

1994) that already exist and are readily applicable to many areas of GIS (Openshaw and Openshaw 1997). Geocomputation also involves a fundamental change of style with the replacement of computational minimising technologies that reflect an era of hand calculation by a highly computationally intensive one. It also brings with it some grand ambitions about the potential usefulness that may well result from the fusion of virtually unlimited computing power with smart AI-based technologies that have the potential to open up entirely new perspectives on the ways in which we do geography and perform GIS applications (see Openshaw 1994a, 1995). This new emphasis on geocomputation is an unashamedly applied, problem-solving approach. The challenge is to create new tools which are able to suggest or discover new knowledge and new theories from the increasingly spatial data-rich world in which we live.

4 WHAT SORT OF HPC-POWERED GIS-RELEVANT SPATIAL ANALYSIS TOOLS ARE NEEDED?

A longstanding difficulty with GIS-Relevant Spatial Analysis (GRSA) is the lack of any consensus as to what it means, what its requirements are, what its users want now, and what its users would want if only they knew it were possible (or if the methods existed to stimulate demand). The situation has not improved much over the last five years (see Openshaw 1991). Far too often GRSA is equated with whatever old or new statistical technology a researcher happens to be familiar with, or with what a largely unskilled enduser thinks is required based on knowledge of what a proprietary GIS vendor provides. Yet GIS offers far more than a source of data that can be run with pre-GIS methods! 'Yes', that can be done, and 'yes' it can be useful; but GIS presents a much deeper challenge to spatial analysts. The question 'what kind of spatial analysis do researchers and academics want in GIS?' has to be tempered by the feasibility constraint of 'what kinds of spatial analysis can be implemented in or with GIS?' and the sensibility constraint 'what types of spatial analysis is it sensible to provide for GIS and its user community?'. Another set of general design constraints reflect other very important but hitherto neglected considerations, such as who are the likely users, what it is (in generic terms) that they want, and what sort of analysis technology they can handle given fairly low levels of statistical knowledge and training in the spatial sciences. Table 2 summarises many of the principal design questions. It is noted that the abilities of users are very important and the future viability of whatever technologies are proposed will ultimately depend on the extent to which the methods can be safely packaged for use by non-experts. There is a considerable mismatch between the criteria identified in Table 2 and the capabilities of existing spatial statistical and spatial modelling tools: for example, statistical packages are really of potentially very limited use outside of a research organisation and, in any case, they lack the power to cope with most of the analysis problems created by GIS.

Table 2 Basic design questions.

What kinds of spatial analysis are:

- relevant to GIS data environments?
- sensible given the nature of GIS data?
- reflect likely enduser needs?
- compatible with the GIS style?
- capable of being used by endusers?
- add value to GIS investment?
- can be regarded as an integral part of GIS?
- offer tangible and significant benefits?

A most important challenge is to identify and develop generic spatial analysis tools which are appropriate for use with spatial data in GIS environments. Table 3 shows the ten basic 'GISability' criteria that spatial analysis methods should ideally attempt to meet (Openshaw 1994e). It is important to recognise that GIS creates its own spatial analysis needs and that these needs make it a special and different sub-field of spatial statistics. It is within this context that spatial analysis tools need to be regenerated, rediscovered, or created anew. The present is a good time to tackle these problems.

The debate as to whether these methods should be accessed from within or without a GIS package is irrelevant. There is no reason to insist on only one

Table 3 Openshaw's 10 basic 'GISability' criteria.

1. Can handle large N values
2. Study region invariant
3. Sensitive to the nature of spatial data
4. Mappable results
5. Generic analysis
6. Useful, and valuable
7. Interfacing problems invariant
8. Ease of use and understandable
9. Safe technology
10. Applied rather than research-only technology

Source: Openshaw 1994f

form of integration or interfacing, leaving aside the obvious point that to be a GRSA tool the spatial analysis operation has to be called from, and ultimately end up within, a GIS environment. In an era of heterogeneous distributed computing there is no longer any need for all of the systems to be on the same machine (see Coleman, Chapter 22). Equally, the extent to which methods are perceived as having to be run within a GIS environment is often overplayed, since the special properties that a GIS can offer spatial analysis amount to little more than spatial data and consistently defined contiguity lists. Thus much system complexity can be avoided by the simple expedient of separating out the different components needed by the analysis process and developing a high-level system to call a GIS here, a model or analysis tool there, or a map drawer when one is needed.

These GRSA criteria can be converted into a series of researchable topics that would appear to have considerable relevance in the late 1990s. Participants at a workshop on 'New Tools for Spatial Analysis', held in Lisbon in 1993, were asked to think about the research themes that might be the most useful in the spatial analysis area. A summary of the suggestions that emerged after several hours of discussion and debate spread over a three-day period is given in Table 4. At present there is no funding to develop any of these themes, and this is one of the reasons why spatial analysis relevant to GIS is so backward.

271

Table 4 A research agenda for spatial analysis in the mid 1990s.

Theme 1: Toolkits for spatial analysis in GIS

Theme 2: Methods and tools for handling uncertainty in spatial data

Theme 3: Methods of automated and exploratory spatial analysis

Theme 4: Data driven modelling and data mining of GIS databases

Theme 5: Statistical aspects of model evaluation and choice

Theme 6: Confidentiality of spatial data and zone design

Theme 7: Impediments to the development and use of spatial analysis methods

Source: Adapted from Openshaw and Fischer 1995

5 A TYPOLOGY OF SPATIAL ANALYSIS TECHNOLOGIES

The developments in HPC environments and the increasing interest in geocomputation as a paradigm relevant to GIS offer a useful perspective on the current state of GRSA methods. A threefold typology is suggested:

1 Type One methods are based on computationally limited technology. Most conventional statistical methods are of this type. It is true that some require supercomputers to invert rank N matrices, where N is the number of spatial observations which can be very large. However, this is still computationally limited technology if, for example, there are only a small number of spatial origins, or if only a few variables can be handled.

2 Type Two methods are computationally intensive, but in a dumb manner. The early uses of supercomputers in spatial analysis resulted in the development of 'brute force'-based exploratory pattern and relationship detectors. For example, they include the Geographical Analysis Machines (GAM) of Openshaw et al (1987), and the GAM/K version of Openshaw and Craft (1991) who used Cray 1, 2, and Cray X-MP and Y-MP vector supercomputers to power a large-scale exploratory search. These methods were originally criticised by some spatial statisticians (Besag and Newell 1991) who suggested simpler variations. Subsequent testing appears to have demonstrated the superiority of the GAM/K

variant, as documented in Openshaw 1997b, although the results were five years late in being published! The Geographical Correlates Exploration Machine (GCEM) of Openshaw et al (1990) is another type of 'brute force' search for localised patterns and geographical associations. Since 1990, both methods can be run on workstations. A few variations on the GAM theme in particular have been suggested by other researchers – for example, Fotheringham and Zhan (1996) describe a procedure almost identical to part of the GAM. Note though that the quality of the results does depend on the purpose of the experiment or investigation. For example, if the objective is to test hypotheses then typically several million Monte Carlo significance tests with repeat replication may be required in order to handle multiple testing problems. The process is highly parallel but probably still needs the next generation of machine. However, if the objective is to use a GAM style of approach as a descriptive tool, indicating areas where to look or perform more detailed work, then this can be avoided. The advantages of this style of approach are essentially those of automating an exploratory spatial analysis search function as well as obviating the need to have prior knowledge about where to look for localised spatial patterning. Type Two methods were used in the first spatial analyses to recognise the importance of searching for localised patterning rather than global patterns.

3 Type Three methods are computationally intensive but also computationally intelligent. The difficulty with the Type Two methods is that as problem sizes increase (e.g. as a consequence of improved data resolution) and as the dimensionality of the data increases from two spatial dimensions to multiple data domains, then this technology breaks down. The answer is to switch to a smart search strategy. Openshaw and Cross (1991) described the use of a genetic algorithm to move hypercircles around a multidimensional map in the search for crime clustering. This technology has been developed further into suggestions for exploratory analysis that can operate in space, in time, and in multivariate data domains. The resulting database exploring creatures (termed Space Time Attribute Creatures – STACs) are described in Openshaw (1994e, 1995). Interpretation of the outcome of analysis can be aided by using computer animation to follow the search behaviour of the STACs as they go on a data pattern hunting safari. This has been developed into a prototype system called MAPEX (Map Explorer: Openshaw and Perrée, 1996). The hope is that the inexperienced or unskilled endusers can visualise and discover what is happening in their databases by viewing a library of computer movies illustrating different amounts and types of geographical patterning. The resulting user-centred spatial analysis system uses computationally intensive methods in order to ensure that the results that the users see have been processed to remove or at least reduce the effects of multiple testing and other potential complications. The users are assumed not to be interested in *p*-values and Type One or Type Two errors, but merely need to know only where the patterns are strongest and whether they can be trusted. The computer animation provides a useful communication tool. The expansion of this technology into the multiple data domains occupied by STACs and its integration into a standard GIS environment is currently underway. The zone design problem of Openshaw (1978) is another example of a Type Three problem which is of practical significance. Its wider application has been delayed until both digital boundaries were routinely available and HPC hardware speeds had increased sufficiently to make zone design a practical proposition.

The challenge for the geocomputational and HPC future is how to evolve more Type Three methods, which can handle rather than ignore the challenges of performing more intelligent spatial analysis, using computational and AI technologies. Spatial analysis needs to become more intelligent and less reliant on the skills of the operator, and this can only be achieved in the long term by a movement towards Type Three methods. The ultimate aim is to develop an intelligent partnership between user and machine, a relationship which currently lacks balance. Many of the statistical and computational components needed to create these systems exist. What has been lacking is a sufficient intensity of understanding of the geography of the problems and of the opportunities provided by GIS in an HPC era. A start has been made but much work still needs to be done.

6 SPATIAL ANALYSIS INVOLVING COMPUTATIONAL ZONE DESIGN

Geographers have been slow to appreciate the importance of spatial representation in their attempts to describe, analyse and visualise patterns in socioeconomic data. The effects of scale and/or aggregation of zones upon the nature of the mappings that are produced are well known (Openshaw 1984; Fotheringham and Wong 1991). Yet this very significant source of variation is usually ignored because of the absence of methods in existing GIS packages that are able to handle it. Once this mattered much less, inasmuch as users had no real choice since they were constrained to use a small number of fixed zone based aggregations. GIS has removed this restriction and as the provision of digital map data improves, so users are increasingly exposed to the full range of possible zoning systems. Openshaw (1996: 66) explains the dilemma as follows: 'Unfortunately, allowing users to choose their own zonal representations, a task that GIS trivialises, merely emphasises the importance of the MAUP. The user modifiable areal unit problem (UMAUP) has many more degrees of freedom than the classical MAUP and thus an even greater propensity to generate an even wider range of results than before.' The challenge is to discover how to turn this seemingly impossible problem into a useful tool for geographical analysis. What it means is that the same microdata can be given a very large number of broadly equivalent but different spatial representations. As a consequence, it is no longer possible to 'trust' any display or analysis of zone-based spatially aggregated data that just happens to have been generated for a zoning system. Users have to start seriously worrying about the nature of the spatial representations contained in the zoning systems they use. It is argued that this problem mainly affects socioeconomic information where, because of confidentiality constraints and lingering data restrictions, attention is often limited to the display and analysis of data that has been spatially aggregated one or more times.

The only alternative to an 'as is' spatial representation is to develop zone design as a spatial engineering tool. The zone design problem can be formulated as a non-linear constrained integer combinatorial optimisation problem. Openshaw (1977) defines this so-called automated zone design problem (AZP) as optimise $F(X)$ where X is a zoning system containing an aggregation of N original zones into M regions ($M<N$) subject to the members of each region being internally connected and all N zones being assigned to a region. The $F(X)$ can be virtually any function that can be computed from the M region data: for instance, it could be a simple

statistic or a mathematical model that is fitted to the data. Typically, F(X) is non-linear, it could possess multiple suboptima, it need not be globally convex, and it is probably discontinuous (because of the contiguity and coverage constraints on X). Additionally, the user may wish to impose extra constraints either on the regions created by X (e.g. shape or size) and/or on the global data generated by the set of M regions (e.g. normality or spatial autocorrelation properties). This is achieved by converting F(X) into a penalty function. The Powell-Fletcher penalty function method has been found to be very effective (Fletcher 1987). This involves optimising the new function:

$$\Phi(X,\sigma,\theta) = F(X) + 0.5\sum_{i}^{N}\sigma_i(C_i(X) - \theta_i)^2$$

where

$\Phi(X,\sigma,\theta)$ is a penalty function that is dependent on X, σ_i and θ_i;
σ_i and θ_i are a series of parameters that are estimated to ensure gradual satisfaction of the constraints;
$C_i(X)$ is a constraint violation which is some function of the zoning system;
$F(X)$ is defined as previously.

This AZP was solved for small problems in the mid 1970s (Openshaw 1977, 1978). The technology was revived in the early 1990s when the increased availability of digital map data and the GIS revolution highlighted the importance of the problem. Larger datasets required better algorithms for optimising these functions. Openshaw and Rao (1995) compared three different methods and suggested that simulated annealing was the best choice. Unfortunately, they also found that simulated annealing took about 100 times longer to run than the other methods, while the use of additional constraints would have added another factor of 30 or so. Attempts were made to speed-up the simulated annealing approach by switching to parallel supercomputers. The immediate difficulty was the need for a fully parallel simulated annealing algorithm relevant to AZP types of problems. After considerable effort a hybrid simulated annealer with multiple adaptive temperatures controlled by a genetic algorithm was developed and was shown to work very effectively (Openshaw and Schmidt 1996).

It is now possible to use zone design to re-engineer all types of zoning systems with five principal areas of application:

1 to demonstrate the MAUP by seeking minimum and maximum function value zoning systems – this also helps prove the lack of simple minded objectivity in GIS;
2 to design zoning systems with particular properties that are believed to be beneficial for certain applications – for example, electoral re-districting or to minimise data confidentiality risks;
3 as a spatial analysis tool – for example, the zoning system acts as a pattern detector tuned to spot particular patterns;
4 as a visualisation tool – for example, to make visible the interaction between a model and the data it represents;
5 as a planning aid – for example, to define areas of maximum but equal accessibility or regions which are comparable because they share common properties.

These five application areas have been illustrated using the ARC/INFO-based Zone Design System (ZDES: Openshaw and Rao 1995; Alvanides 1995). The system attempts to routinise zone design using a number of generic zone design functions, and has been designed as a portable add-on to ARC/INFO.

7 ZONE DESIGN ANALYSIS OF SPATIAL DISTRIBUTIONS

This section provides an example of a geocomputation approach to zone design that demonstrates some of the potential capabilities of ZDES as a spatial analytical tool. Consider a problem that involves the analysis of non-white population data for England and Wales. This is currently of interest to some telecoms companies seeking to establish retail networks offering cut-price long-distance telephone calls. The 1991 Census data for persons born outside the UK may provide one surrogate indicator of this potential market demand for cheap long-distance calls. Figure 1(a) shows the 54 county zones that cover England and Wales while Figure 1(b) displays a choropleth map of the ethnic population. The key underlying geographical question is whether Figure 1(b) provides a 'meaningful' spatial representation of the Census data. Certainly, the visual patterns appear to identify some areas of apparent concentrations. Yet distortions introduced by differences in county sizes may blur some of the patterns and diminish local concentrations by averaging them out. The map is the outcome of highly complex distortion of the data, caused by its aggregation into counties. Additionally, the use of counties as the object of study introduces an arbitrary geography, since there is no reason to suppose that it has any relevance whatsoever to the factors governing the distribution of ethnic communities in the UK. Indeed the principal attraction of the county zone is the convenience and ease of using a standard geography! Different zoning systems may be expected to offer different levels of data distortion, and some may tell a different story as a consequence. It is interesting, therefore, to explore some of the alternative patterns of the same 1991 Census data aggregated from the underlying 9522 census wards as an exercise in spatial analysis by zone design.

First consider what happens if an attempt is made to create a new set of 54 different 'county' regions that have approximately equal ethnic population counts, and then remap the ethnicity rates (see Figures 2(a) and 2(b)). The constant shading of Figure 2(b) shows the effects of the equal size function – the three small light zones are formed by contiguity islands in the data. One problem is the very intricate zonal boundary patterns formed by the ZDES optimiser as it attempted to create regions of equal ethnic population size in Figure 2(a). In fact, the more efficient the zoning system optimiser becomes so the more intricate are the resulting boundaries because of the imposed constraint of equal population sizes. The simulated annealer used here was very successful in optimising the function, but produced extremely crenulated and irregularly-shaped regions. Perhaps this also says something about the fine-scale spatial distribution of the ethnic population at the ward level – that is, the pattern is concentrated but discontinuous hence the need to link widely separated areas together in order to meet the population size restriction. It is a matter for further research as to whether there is an optimal spatial scale (or level of aggregation) at which the zones suddenly become more regular. This would be an interesting question to try to answer. Another problem here is that there are potentially many different zoning systems that will yield areas of nearly, or approximately, equal population size. It may also be necessary to introduce shape constraints.

A more useful spatial analytic function would be some measure of the spatial dispersion of the population around a set of region centroids. This is broadly equivalent to a type of large location–allocation problem. The objective is to minimise the global sum of the population weighted distances to each of the ward centroids. This function is expressed as follows:

$$\text{minimise} \qquad \sum_{j}^{M} \sum_{i \in j}^{N} P_i D_{ij}$$

where

P_i is the population value for ward i;
D_{ij} is the distance from ward i to the population centroid of region j of which i is a member. (Note that this centroid depends on the current membership of region j.)

The restriction on the second summation indicates that the summation only occurs for wards that belong to region j. This is a way of partitioning the zoning system of N Census wards into M regions.

Figure 3 shows the results of minimising this local spatial dispersion function. This map is the outcome of a contest between different parts of the UK, as the ZDES algorithm seeks to trade-off population gains in some parts of the country against losses in others. The resultant zoning system is thus a visualisation of the tension (or interaction) between the objective function, the Census ward zonation and aggregation effects. Those parts of the UK with the largest ethnic populations have relatively small regions. The map in Figure 3(b) seems to offer a more sensitive view of the distribution of ethnic populations compared with Figure 1(b), and it highlights some of the areas where aggregation effects have seemingly removed some of the patterns. This approach is developed further in Figure 4(a), which shows the results for the same objective function but subject to constraints that the total population of all the regions should be at least 75 per cent of the average. This involves solving a penalty function version of ZDES. On this application the simulated annealer needed a Cray T3D to produce the results within a convenient time – that is, one hour. The patterns in Figure 4(b) are considerably more disjointed than in Figure 3(b) with more intricate boundary resolution in the principal areas of ethnic concentration.

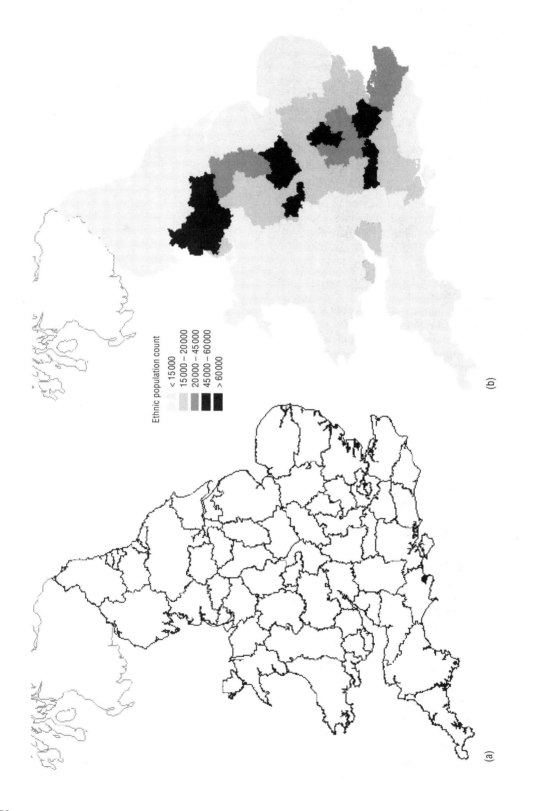

Fig 1. (a) The 54 counties in England and Wales; (b) distribution of ethnic population for counties in England and Wales.

Ethnic population count

< 15 000
15 000 – 20 000
20 000 – 45 000
45 000 – 60 000
> 60 000

(a)

(b)

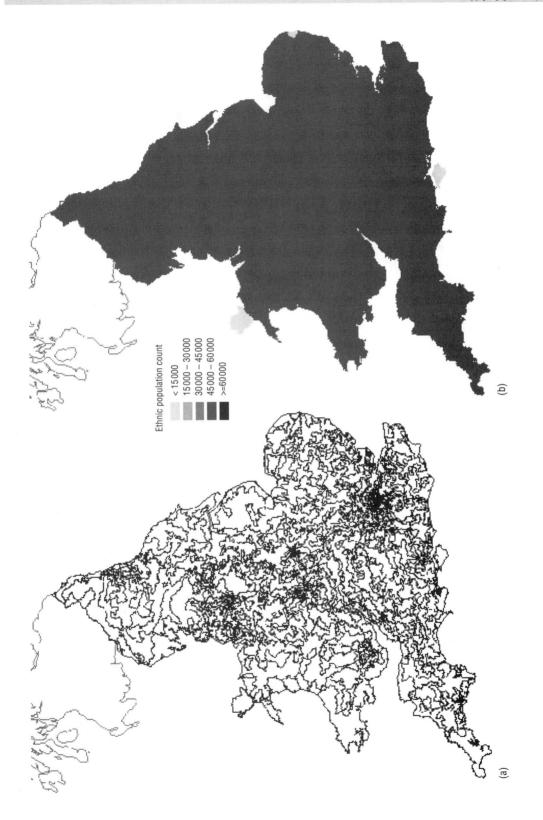

Fig 2. (a) Equal ethnic population regions in England and Wales; (b) distribution of ethnic population in England and Wales for equal population regions.

Ethnic population count

< 15000
15000 – 30000
30000 – 45000
45000 – 60000
>=60000

(a)

(b)

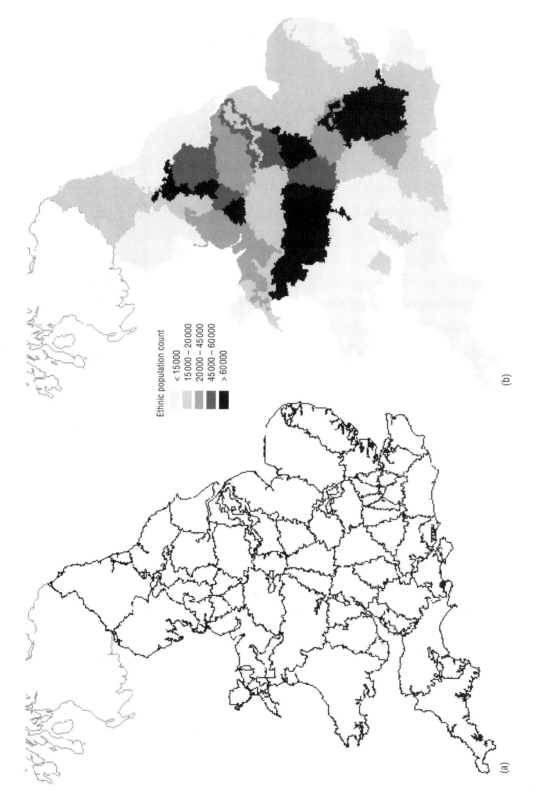

Ethnic population count

<15000
15000 – 20000
20000 – 45000
45000 – 60000
>60000

(b)

(a)

Fig 3. (a) Regions in England and Wales that minimise population weighted distances; (b) distribution of ethnic population for accessibility regions.

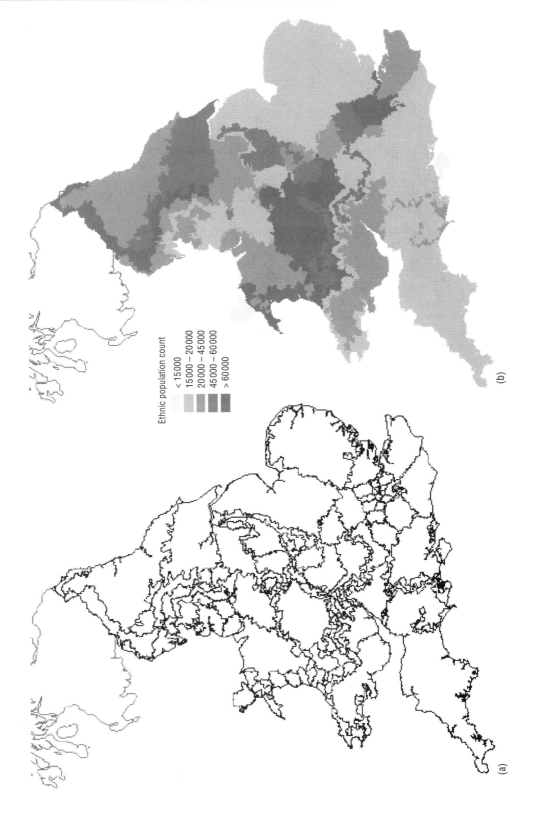

Ethnic population count

< 15 000
15 000 – 20 000
20 000 – 45 000
45 000 – 60 000
> 60 000

(b)

(a)

Fig 4. (a) Regions in England and Wales that minimise constrained population weighted distances; (b) distribution of ethnic population for constrained accessibility regions.

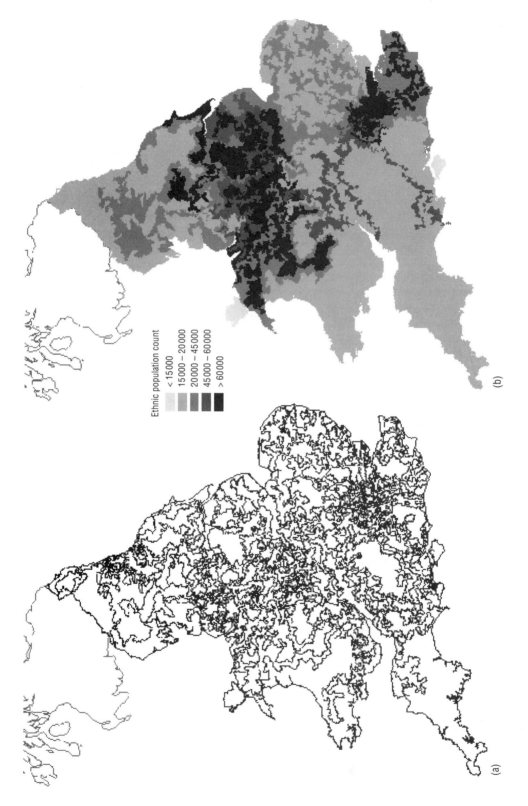

Ethnic population count

<15000
15000 – 20000
20000 – 45000
45000 – 60000
> 60000

(b)

(a)

Fig 5. (a) Regions in England and Wales that minimise a constrained spatial interaction model; (b) distribution of ethnic population for constrained spatial interaction regions.

A final demonstration involves converting the previous objective function into a measure of population potential as a form of crude spatial interaction model. The function is as follows:

$$\text{minimise} \quad \sum_{j}^{M} \sum_{i \in j}^{N} P_i D_{ij}^{-2}$$

subject to minimum size constraints. The resulting zonation is shown in Figure 5(a). The region boundaries are once again very crenulated, especially in areas of high ethnic populations. The resulting population distribution is mapped in Figure 5(b). Once again the underlying impression is that of local area concentrations surrounded by gaps. This may be a reflection of areas of localised negative spatial autocorrelation. The deficiency of the administrative scheme, represented here by the county geography, is demonstrated by a series of descriptive statistics. Table 5 shows some statistics concerning the ethnic population per zone for each of the different models illustrated earlier. The administrative county geography scores worse than any other scheme with zones extending beyond 700 000 ethnic residents and an extreme standard deviation figure (113 750). The equal ethnic population geography demonstrates the best statistical properties for a representative administration scheme, with a maximum ethnic population very close to the mean ethnic population value (54 650 residents). However, the boundaries of the zones are rather obscure for any policy making exercise and the same problem occurs with a constrained interaction geography, as we saw in Figures 3(a) and 5(a) respectively.

A trade-off between a comprehensible geography and satisfactory statistical properties is the constrained accessibility model illustrated in Figure 4(a). In this case the weighted distance minimisation function works as a shape constrained thus producing relatively compact zones, while the population constraint function restricts the maximum population of zones to

176 200 and retains the standard deviation low enough for the zones to be comparable. Possibly the 'best' zoning systems that would represent this type of socioeconomic data would be very different from the conventional administrative zoning scheme and this forms the subject of ongoing research (Openshaw and Alvanides 1997b).

8 CONCLUSIONS

The case for developing a wide range of GRSA tools is very compelling. GIS is creating an immensely data-rich environment. The technology for spatial data capture, management, and handling has far outstripped the available tools for its analysis. The hope now is that it may be possible to compute our way out of the data swamp by developing new generations of intelligent spatial analysis tools which are better able to cope with the conflicting requirements of large volumes of data, geographical reasonableness, and the endusers. The need is undeniable and there are various ways forward involving the use of zoning systems as data, pattern, and model visualisers; and the development of smart GIS database explorers. The technological basis for these exists: HPC, AI toolkits, computational statistics, large spatial databases, and well-developed GIS. Powerful zone design algorithms have been developed. Artificial life-based 'creatures' or 'agents' can be created that are able to move around space-time-attribute GIS databases under their own control in an endless search for patterns and relationships of possible interest. Computer animation provides the basis for users 'watching' what is happening in highly complex geo-cyberspaces projected on to a 2-dimensional map (Openshaw and Perrée 1996). The basic methods will probably run on a workstation, but as the complexity of the data domain increases or as greater use is made of computational statistics in order to improve performance, so they will need HPC. There is a suggestion that the evolution of new types of spatial analysis technology is about to start.

The challenge then is to solve the principal outstanding spatial analysis problems by developing various geocomputational approaches, to demonstrate they work on a range of generic problems, and then to make them available either within or without current GIS software. This task is becoming increasingly urgent and requires geographers (in particular) to use their understanding of the geography

Table 5 Statistical properties of different geographies for ethnic population.

Geography	Maximum	Mean	Standard deviation
Administrative counties	703 700	46 850	113 750
Equal ethnic population	58 650	54 650	13 200
Population accessibility	320 900	54 650	74 700
Constrained accessibility	176 200	55 700	39 850
Constrained interaction	273 200	55 700	51 500

of the problems to create a new approach to developing geographical analysis methods. A start is being made (here as well as elsewhere in this book) but much still remains to be done.

References

Alvanides S 1995 'The investigation of a Zone Design System for reconstructing census geographies'. Unpublished dissertation 3031 available from School of Geography, University of Leeds

Besag J, Newell J 1991 The detection of clusters in rare diseases. *Journal of the Royal Statistical Society* A 154: 143–55

Bezdek J C 1994 What is computational intelligence? In Zurada J M, Marks R J, Robinson C J (eds) *Computational Intelligence: Imitating Life*. New York, IEEE: 1–12

Birkin M, Clarke G P, Clarke M, Wilson A G (eds) 1996 *Intelligent GIS: location decisions and strategic planning*. Cambridge (UK), GeoInformation International

Fletcher R 1987 *Practical Methods of Optimisation*. Chichester, John Wiley & Sons

Fotheringham A S, Wong D W S 1991 The modifiable areal unit problem in multivariate statistical analysis. *Environment and Planning A 23*: 1025–44

Fotheringham A S, Zhan F B 1996 A comparison of three exploratory methods for cluster detection in spatial point patterns. *Geographical Analysis* 28: 200–18

Hillis W D 1992 What is Massively Parallel Computing and Why is it important? In Metropolis N, Carlo Rota G (eds) *A new era in computation*. Cambridge (USA), MIT Press: 1–15

Openshaw S 1977 A geographical solution to scale and aggregation problems in region-building, partitioning, and spatial modelling. *Transactions of the Institute of British Geographers* 2: 459–72

Openshaw S 1978 An optimal zoning approach to the study of spatially aggregated data. In Masser I, Brown P J B (eds) *Spatial representation and spatial interaction*. Boston (USA), Martinus Nijhoff: 95–113

Openshaw S 1984 *The modifiable areal unit problem*. Concepts and Techniques in Modern Geography 38. Norwich, Geo-Books

Openshaw S 1991c Developing appropriate spatial analysis methods for GIS. In Maguire D, Goodchild M F, Rhind D (eds) *Geographical information systems: principles and applications*. Harlow, Longman/New York, John Wiley & Sons Inc. Vol. 1: 389–402

Openshaw S 1994a A concepts-rich approach to spatial analysis, theory generation, and scientific discovery in GIS using massively parallel computing. In Worboys M F (ed) *Innovations in GIS 1*. London, Taylor and Francis: 123–38

Openshaw S 1994b Computational human geography: towards a research agenda. *Environment and Planning A* 26: 499–505

Openshaw S 1994d GIS crime and spatial analysis. *Proceedings of GIS and Public Policy Conference*. Ulster, Ulster Business School: 22–35

Openshaw S 1994e Two exploratory space-time-attribute pattern analysers relevant to GIS. In Fotheringham A S,

Rogerson P (eds) *Spatial Analysis and GIS*. London, Taylor and Francis: 83–104

Openshaw S 1994f What is GISable spatial analysis? In *New Tools for Spatial Analysis*. Luxembourg, Eurostat: 157–62

Openshaw S 1995 Developing automated and smart spatial pattern exploration tools for geographical information systems applications. *The Statistician* 44: 3–16

Openshaw S 1996a Developing GIS relevant zone based spatial analysis methods. In Longley P, Batty M (eds) *Spatial analysis: modelling in a GIS environment*. Cambridge (UK), GeoInformation International: 55–73

Openshaw S 1997 Supercomputing in Geographical Research. *Proceedings of Conference on IT in the Humanities*. London, British Association

Openshaw S, Alvanides S 1997 Designing zoning systems for representation of socioeconomic data. In Frank I, Raper J, Cheylan J (eds) *Time and motion of socioeconomic units*. GISDATA series. London, Taylor and Francis

Openshaw S, Charlton M, Wymer C, Craft A 1987 A mark I Geographical Analysis Machine for the automated analysis of point datasets. *International Journal of Geographical Information Systems* 1: 335–58

Openshaw S, Craft A 1991 Using Geographical Analysis Machines to search for evidence of clusters and clustering in childhood leukaemia and non-Hodgkin Lymphomas in Britain. In Draper G (ed) *The geographical epidemiology of childhood leukaemia and non-Hodgkin lymphomas in Great Britain, 1966–83*. London, Her Majesty's Stationery Office: 89–103

Openshaw S, Cross A 1991 Crime pattern analysis: the development of Arc/Crime. *Proceedings of AGI Annual Conference*. Birmingham (UK)

Openshaw S, Cross A, Charlton M 1990 Building a prototype geographical correlates exploration machine. *International Journal of Geographical Information Systems* 3: 297–312

Openshaw S, Fischer M M 1995 A framework for research on spatial analysis relevant to geostatistical information systems in Europe. *Geographical Systems* 2: 325–37

Openshaw S, Openshaw C A 1997 *Artificial intelligence in geography*. Chichester, John Wiley & Sons

Openshaw S, Perrée T 1996 User-centred intelligent spatial analysis of point data. In Parker D (ed.) *Innovations in GIS 3*. London, Taylor and Francis: 119–34

Openshaw S, Rao L 1995 Algorithms for re-engineering 1991 Census geography. *Environment and Planning A* 27: 425–46

Openshaw S, Schmidt J 1996 Parallel simulated annealing and genetic algorithms for re-engineering zoning systems. *Geographical Systems* 3: 201–20

Openshaw S, Schmidt J 1997 A Social Science Benchmark (SSB/1) Code for serial, vector, and parallel supercomputers. *International Journal of Geographical and Environmental Modelling*

Turton I, Openshaw S 1996 Modelling and optimising flows using parallel spatial interaction models. In Bougé L, Fraigniaud P, Mignotte A, Roberts Y (eds) *Euro-Par 96 Parallel Processing Vol. 2* Lecture Notes in Computer Science (1124). Berlin, Springer: 270–5

19

Spatial analysis: retrospect and prospect

M M FISCHER

This chapter briefly reviews spatial analysis as a technology for analysing spatially referenced data. Spatial data analysis techniques are important and are becoming even more so as the supply of spatial data increases. Novel new modes of computation, known collectively as 'computational intelligence technologies' will meet some of the new analysis needs that have been stimulated by GIS. Computational intelligence technologies in general and neural networks in particular provide novel, sophisticated, and interesting models and methods which are potentially applicable to a wide range of applications. They are thus seen as the way forward to analyse the data-rich environments of contemporary GIS.

1 INTRODUCTION

Spatial analysis is a technology which typically requires two types of information about spatial objects: attribute and locational information. The scope of discussion here will be restricted to methods and techniques for spatial data analysis (SDA), often referred to as spatial analysis in the strict sense. Smoothing techniques such as areal interpolation (Flowerdew and Green 1991), Kriging (Isaaks and Srivastava 1989), and kernel smoothing methods (Silverman 1986) as well as (locational and attribute) error assessment (Goodchild and Gopal 1989) and the modifiable areal unit problem (Openshaw and Alvanides, Chapter 18; Fotheringham and Wong 1991; Openshaw 1984) are precluded from the discussion, even though they are often fundamental steps and problems in spatial analysis. These issues are addressed in part by other authors in this volume (e.g. Martin, Chapter 6; Openshaw and Alvanides, Chapter 18; Goodchild and Longley, Chapter 40).

The chapter is organised into six sections. It is useful to begin by trying to understand the salient features which make spatial analysis special and different from other forms of data analysis (section 2). Section 3 briefly deals with the development of SDA and describes some significant achievements to date. This is followed by a discussion of Openshaw's

(1994a) basic rules for identifying future 'GISable' spatial analysis technology (section 4). Leading from this, section 5 argues for a very different non-conventional style of approach based upon novel modes of computation – which are collectively known as 'computational intelligence' (CI) technologies – as laying the foundations for a new generation of useful and more powerful SDA tools relevant to data-rich spatial data environments. In particular, neural networks, the single most important component of CI-driven spatial analysis, are seen to offer spatial analysts rich and interesting classes of novel data driven non-linear models, and are deemed to be applicable to a wide range of application domains. The potential of this approach is exemplified in two classical spatial analytic tasks: spatial interaction modelling and pattern classification (section 6). In the concluding section some major aspects of this new paradigm are summarised and directions for further research are outlined.

2 WHAT MAKES SPATIAL DATA ANALYSIS SPECIAL?

Given the diversity of analytical perspectives within GIS it is difficult to define SDA as anything more specific than a body of methods and techniques for analysing 'events' at a variety of spatial scales, the results of which depend upon the spatial arrangement of the 'events' (Goodchild et al 1992; Haining 1994). 'Events' may be represented as point, line, or area 'objects' or 'spatial primitives' which are located in geographical space and possess a set of (one or more) other attributes. Location, topology, spatial arrangement, distance, and spatial interaction become the focus of attention in SDA activities. The outcomes of analysis are: detection of patterns in spatial data; exploration and modelling of relationships between such patterns; enhanced understanding of the processes that might be responsible for the observed patterns; and improved ability to predict and control events arising in geographical space. It is the explicitly spatial focus of spatial analysis that distinguishes SDA from other forms of data analysis (Goodchild et al 1992).

It follows that two different types of information are integral to SDA:

- *Locational* (geometric/topological) information about the spatial objects of concern which are generally described by means of their position on a map or using geographical coordinate systems. The spatial objects utilised in most spatial analyses are statistical areas such as census tracts, or points which are sampled from continuous geographical space (Martin, Chapter 6). For some types of spatial analysis it is common practice to represent areas by points (2-dimensional discrete representation of space).
- *Attribute* information about the spatial objects of interest. Two types of attributes may be distinguished: primary attributes (e.g. socioeconomic characteristics, physical properties); and secondary attributes of, or relations between, spatial objects (e.g. flows of information, capital, goods, or people).

SDA employs a wide range of tools ranging from spatial autocorrelation measures, through nearest neighbour methods, *K*-functions, spatial classification and regionalisation methods, to spatial extensions of conventional statistical techniques such as regression models. In principle, we may distinguish between those SDA techniques that use locational information alone and those that use both locational and attribute information. The first class of techniques is essentially concerned with the analysis of spatial distributions (Longley and Batty 1996) and includes techniques such as point pattern analysis

(Goodchild et al 1992). The second class includes techniques such as spatial regression models, and utilises both locational and attribute data in order to assess the spatial variation in attribute measurements. This class may be further disaggregated into techniques and methods that deal with primary attributes (interval and/or categorical scale) and those that deal with secondary attributes (relations). The latter includes spatial interaction models, interregional input–output accounting systems of various kinds, and log-linear models: all of these generally rely on 2-dimensional discrete, rather than continuous, geographical spaces.

The crucial role of geographical location of objects, both in an absolute and a relative sense (spatial arrangement), has profound implications for the way in which they can be analysed (Anselin and Getis 1993). In fact, location leads to two different types of spatial effects: *spatial dependence* (often referred to as spatial autocorrelation) and *spatial heterogeneity*. The first directly results from Tobler's (1979) 'First Law of Geography' where 'everything is related to everything else, but near things are more related than distant things'. Thus, spatial dependence implies that the data for particular spatial units are related and similar to data for other nearby spatial units (Getis 1992). Spatial dependence caused by a variety of measurement problems (e.g. the arbitrary delineation of spatial units of observation, the problem of spatial aggregation, the presence of spatial externalities, and spillover effects) poses particular challenges for conventional statistical analysis since this assumes that units of observation are statistically independent of one another (Anselin and Getis 1993; Griffith 1993). The second and equally important spatial effect – spatial heterogeneity or non-stationarity – is related to spatial differentiation which follows from the intrinsic uniqueness of each location, as is evident in spatial regimes for variables, functional forms, or model coefficients (Anselin 1994a). These special features of spatial data render classical statistical methods unreliable unless they have been modified to accommodate the spatial problems at hand. The complications are similar to those found in time series analysis but are exacerbated by the multi-directional, 2-dimensional nature of dependence in space rather than uni-directional nature in time (Griffith 1993).

3 SPATIAL DATA ANALYSIS: ORIGINS AND PROGRESS

The origins of SDA lie in the development of quantitative geography and regional science, and date back to the early 1960s. The use of quantitative (mainly statistical) methods and techniques to analyse the pattern and form of geographical objects (points, lines, areas, and surfaces: Martin, Chapter 6) depicted on maps or defined by coordinates in 2- or 3-dimensional space characterised this early research. Later on, more emphasis was placed on the inherent properties of geographical space, on spatial choice processes, and the spatial-temporal evolution of complex spatial systems.

Many of the SDA techniques were developed in the 1960s and 1970s, in an era of limited computing power, small datasets, and rudimentary computer graphics. Today, as a consequence, current implementations take only limited advantage of the data storage, retrieval, and visualisation capabilities of GIS. Early attempts to implement SDA techniques in a computational environment relied on source code programming, especially FORTRAN. The 1970s saw the advent of statistical software packages such as BMDP, SPSS, and SAS, which soon became the primary applications medium, even although these packages were, and are still, based on statistical techniques which are fundamentally non-spatial in nature. Even today, much SDA activity remains embedded in the aspatial environment of software packages such as SAS, Minitab, Systat, SPSS, S-Plus, and GLIM (Goodchild et al 1992).

In the early days of SDA there was strong momentum behind the *spatial geometric view* with its strong emphasis on point pattern analysis, quadrat analysis, and nearest neighbour methods (Dacey 1960; Getis 1964; Haggett et al 1977; Rogers 1965). This approach used locational data only and attempted to describe the spatial patterning of point objects (events) by comparing observed patterns with those that might theoretically be expected from various normative models, especially those based upon spatial randomness. More recently, geographers (e.g. Getis, Chapter 16; Getis 1983) have realised that better descriptions of point patterns may be obtained by using *second-order* methods, that is methods which describe the relative positioning of pairs of points (Diggle 1983). One such method is to compute a (multivariate) K-function (Ripley 1977) which examines all inter-point distances rather than just those separating nearest neighbours. A comparison of the observed K-function with those derived from possible explanatory models (e.g. those based upon the Poisson model of spatial randomness) over the study area permits assessment of whether the observed occurrences are likely to have arisen from the processes underlying such models (Getis and Boots 1978). By extension, such SDA tools are now available to 'explain' one pattern of particular interest in terms of others (in the multivariate case), as well as to deal with situations involving space-time patterns.

The interest in point pattern analysis was complemented by the *application of standard statistical techniques* to spatial data (Berry and Marble 1968; Haggett 1965; King 1969). In particular, spatial analysts have used the general linear model (e.g. multiple regression analysis), factor and principal components analysis, regional taxonomic methods (spatial classification and regionalisation), multidimensional scaling, discriminant analysis, and trend surface analysis. Only a small number of these SDA tools (notably regionalisation methods with spatial contiguity constraints) were actually developed from first principles within the spatial sciences rather than being based on methods and techniques adapted from other disciplines. One consequence has been that very few SDA techniques have taken into account the special characteristics of spatial data when invoking statistical assumptions, particularly when modelling using statistical packages. Increasing awareness of the problems caused by spatial heterogeneity and dependence, and their effects upon the validity of conventional statistical tools, has led to the development of a large body of methods and techniques (e.g. Anselin and Griffith 1988).

Despite the very large number of rather diverse contributions three major areas can be identified where significant progress has been made in the last decades (see also Getis, Chapter 16):

- *Spatial dependence and heterogeneity descriptors.* The problem of spatial heterogeneity and dependence has received substantial attention in recent times (Cliff and Ord 1981). Spatial analysts concerned with spatial dependence now have a number of tools available. Important measures include Moran's I and Geary's c (Cliff and Ord 1981), semi-variogram parameters, and generalised measures of spatial autocorrelation.

Such measures are of use in a general exploratory sense to summarise the overall existence of a stable pattern of spatial dependence in attribute data, to establish the validity of various stationarity assumptions prior to modelling, and to identify possible forms of a spatial model for the data. They are extremely useful for small datasets, but only of very limited use in the context of large and very large GIS datasets where several regimes of spatial association might be present (Anselin 1997a). It is only very recently that a focus on detecting local rather than global *patterns of association* has been developed to provide a more appropriate perspective. Examples of these descriptors are the distance-based *G*-statistics of Getis and Ord (1992) which can easily be implemented into a GIS-framework (Anselin et al 1993; Ding and Fotheringham 1992). The idea behind these descriptors has been extended to a general class of 'local indicators of spatial analysis' (termed LISA: Anselin, Chapter 17; Anselin 1995).

- *Spatial regression modelling.* In essence, spatial regression models may be viewed as spatial extensions to the familiar family of standard elementary linear regression models for non-spatially related cross-sectional data. This extension is typically achieved by means of a ($N \times N$) matrix of spatial weights (typically a first-order binary contiguity matrix) and a spatial autoregressive structure for the error terms, where N denotes the number of observations (spatial units). N is usually quite small and thus represents only a coarse level of spatial resolution. This has significant implications for the correct specification, estimation, and testing of spatial linear regression models. Following the pioneering work of Getis and Boots (1978), Paelinck and Klaassen (1979), Cliff and Ord (1981) among others, considerable progress has been made in various directions: the refinement of the original framework of spatial linear process models, with a special focus on estimation and testing (the development of new and alternative tests and estimators for various types of spatial linear regression models), the development of more complex models that incorporate different contributions of spatial dependence and heterogeneity, and extensions from a purely cross-sectional to a space–time context. This progress is manifest in a series of recent monographs and edited volumes on spatial statistics and spatial

econometrics by Upton and Fingleton (1985), Anselin (1988), Griffith (1988), Arbia (1989), and Cressie (1993) among others, but its dissemination into practice has been hampered by the lack of readily available software (Anselin and Hudak 1992).

- *Discrete spatial data analysis.* The mainstream tradition in SDA has been focused on aggregated spatial data. One area where scholarly interest has been growing in the last two decades is the area of discrete or categorical data analysis (Wrigley 1985). Logistic/logit regression models and (quasi) log-linear models for spatial contingency tables (Aufhauser and Fischer 1985) are the primary workhorses of discrete SDA. The family of statistical models used for discrete SDA is a part of Nelder and Wedderburn's (1972) unified family of generalised linear models, in which a response variable is assumed to come from the exponential family of probability distributions (with the normal, Poisson, binomial, and multinomial distributions).

In spite of these various technical advances, the flurry of results on methods and techniques in SDA has had only limited impact outside the research community. To a large extent this state of affairs is attributable to the lack of readily available software that incorporates explicitly spatial tests and estimators. Currently, none of the popular statistical or econometric packages includes any tools for spatial data analysis, and the only generally available program that performs a range of spatial statistical techniques is Anselin's SpaceStat (1992; Anselin, Chapter 17). The same holds true to a large extent for commercial GIS. Consequently, the actual application of appropriate spatial data analytic techniques has been very limited, even within the academic community of geographers and regional scientists. In contrast (and as documented throughout these volumes) there has been recent and very rapid growth in the availability and richness of spatial data as a consequence of the GIS data revolution, making the somewhat esoteric area of SDA of considerable potential interest. The momentum behind developments in GIS, however, is not the academic arena with its theoretical and methodological interests in knowledge acquisition, but rather the concern to analyse spatial databases for a variety of applied purposes.

4 THE NEW ANALYSIS NEEDS – OPENSHAW'S CRITERIA FOR IDENTIFYING FUTURE SDA TECHNOLOGY

The next few years seem set to provide a unique opportunity for spatial analysts to enter a new era in the development of novel SDA styles. New analysis needs are being created and stimulated as a by-product of developments in GIS technology. GIS is creating extremely data rich and multi-domain, but theory poor and hypothesis-free, environments which are different from those within which computational SDA techniques have hitherto been applied.

While there is a general consensus that the lack of SDA functionalities in current GIS seriously limit the usefulness of GIS as a research tool to analyse spatial data and relationships (Anselin and Getis 1993; Fischer and Nijkamp 1992; Goodchild 1987; Openshaw 1991), there is little agreement about the kinds of SDA techniques and methods that are most relevant to GIS environments. Openshaw (1991, 1994f) and Openshaw and Alvanides (Chapter 18) suggest several criteria that aim to distinguish between 'GISable' and 'GIS-irrelevant' technology. These relevancy criteria provide a useful guide to the new analysis needs, without specifying in detail how such SDA methods might be developed. The most important criteria for 'relevance' are:

- A GISable SDA tool should be able to handle *large* and *very large numbers* (from a few tens to millions) of *spatial objects* without difficult and thus meet the large-scale data processing needs in GIS.
- GIS relevant SDA techniques should be sensitive to the *special nature* of spatial information.
- The most useful GISable SDA techniques and models will be *frame independent* (i.e. invariant under different spatial partitionings of a study region).
- GIS relevant SDA should be a *safe technology* (i.e. the results should be reliable, robust, resilient, error and noise resistant, and not based in any important way on standard distributions).
- GISable SDA techniques should be *useful in an applied sense* (i.e. they should focus upon spatial analysis tasks that are relevant to GIS environments).
- The results of SDA operations should be *mappable* in order to afford understanding and insight, since GIS is a highly visual and graphics-oriented technology.

These criteria make it apparent that future GISable spatial analysis technology will be *data driven* rather than theory driven in nature, and essentially *exploratory* rather than inferential in a conventional spatial hypothesis-testing sense. There is a clear need for a quantitative exploratory style of spatial analysis which can complement the map-oriented nature of GIS. Exploratory spatial data analysis (ESDA; Anselin, Chapter 17), a spatial extension of mainstream exploratory data analysis, provides a useful means of generating insights into (global and local) patterns and associations within spatial datasets. The search process is controlled by the user in a highly interactive graphical environment as, for example, in Regard (Unwin 1993). The use of ESDA techniques, however, is generally restricted to expert users interacting with the data displays and statistical diagnostics to explore spatial information, and to fairly simple low dimensional datasets.

In view of these limitations, it becomes evident that we urgently need novel exploration tools which are sufficiently automated and powerful to cope with the data richness-related complexity of exploratory spatial analysis of large (multiple gigabyte) datasets (Openshaw 1995). The need is for tools that intelligently allow the user to sift through large quantities of spatial data, to simplify multivariate data, and efficiently and comprehensively to explore for patterns and relationships against a background of data uncertainty and noise.

From this perspective the question of how to link SDA technology and GIS (Anselin and Getis 1993; Fischer et al 1996; Goodchild et al 1992) becomes less important than the need to rethink spatial analysis technology fundamentally, to adopt the most useful and relevant technologies for solving problems in new data-rich environments, and to demonstrate the utility of novel approaches to spatial analysis (Openshaw and Fischer 1995).

5 COMPUTATIONAL INTELLIGENCE – A NEW PARADIGM FOR SPATIAL ANALYSIS

Novel modes of computation which are collectively known as CI-technologies hold some promise to meet the needs of SDA in data-rich environments. Following Bezdek (1994) we use the term 'computational intelligence' in the sense that the lowest-level forms of intelligence stem from the capacity to process numerical (low-level) data, without explicitly using knowledge in an artificial intelligence sense. CI tolerates imprecision and uncertainty in large-scale real-world problems in order to achieve tractability, robustness, computational adaptivity, low cost, real-time speed approaching human-like turnaround and error rates which approximate to human performance.

Artificial life, evolutionary computation, and neural networks are the major representative components in this arena. Artificial life is a methodological approach incorporating evolutionary principles: it is based on population rather than individual simulation, simple rather than complex specifications, bottom-up rather than top-down modelling, and local rather than global control (Langton 1989). It has great potential to develop novel exploratory approaches capable of efficiently and comprehensively exploring large spatial databases for patterns and relationships, as illustrated in Openshaw (1994e). Evolutionary computation (genetic algorithms, evolutionary programming, and evolutionary strategy) derives from biology and has proved its merit in treating hard optimisation problems where classical optimisation algorithms (e.g. hill climbers and simplex) and less classical ones (e.g. simulated annealing) tend to be inappropriate. Evolutionary computation might be adopted in SDA, for example to improve the quality of results of spatial optimisation problems (e.g. optimal sizing: Birkin et al 1995), route choice, and zone design problems.

No doubt, CI is currently best suited to systems which can efficiently process information in a massively parallel way and which can 'learn' by adjusting certain parameters. This neural network view is extremely attractive in a world where information abounds, as in the case of large spatial databases. Neural networks are likely to become the single most important component of a CI-driven SDA program (Fischer 1997). The recent re-emergence of neural-network-based approaches has been accomplished by a massive expansion of research, spanning a range of scientific disciplines – perhaps wider than any other contemporary intellectual endeavour. Much of the recent interest of computational geographers in neural network modelling (e.g. Leung 1997; Openshaw 1993) stems from the growing realisation of the limitations of conventional tools as vehicles for exploring patterns and relationships in GIS and remote-sensing

environments and from the consequent hope that these limitations may be overcome by judicious use of neural net approaches.

Neural networks (connectionist models) are parallel distributed information processing structures consisting of simple, but generally non-linear processing elements (which can possess a local memory and can carry out localised information processing operations with adaptive capabilities), massively interconnected via unidirectional signal conduction paths called connections. Each connection has a weight associated with it that specifies the strength of this link. Each processing element (PE) can receive any number of incoming connections and has a single output connection which can branch into copies to form multiple output connections, where each carries the same signal. The information processing active within each PE can be defined arbitrarily with the restriction that it has to be completely local – that is, it has to depend only on the current values of the input signals arriving at the PE and on values stored in the PE's local memory (Hecht-Nielsen 1990). Characteristically, two mathematical functions are active at each PE. The first integrates the connection weights with the inputs arriving via the incoming connections which impinge upon the PE. Each PE then typically applies a transfer (activation) function to the value of the integrator function and produces its output signal. A common choice is the logistic function in the case of continuous network inputs (see Fischer 1995, 1997).

Although a vast variety of neural network models exist, and more continue to appear as research continues, many of them have common topological characteristics, PE properties, and training (learning) heuristics. Three basic entities characterise a neural network (Fischer and Gopal 1993):

- the network topology or interconnection of its PEs (called the neural networks architecture);
- the characteristics of its PEs;
- the method of determining the weights at the connections (called the training or learning strategy).

Different interconnection strategies lead to different types of neural net architectures (e.g. feedforward versus recurrent) which require different learning (training) strategies. At the most fundamental level two categories of training may be distinguished, namely supervised and unsupervised. In supervised learning the network is trained on a training set consisting of a sequence of input and target output data. Training is accomplished by adjusting the network weights so as to minimise the difference between the desired and actual network outputs. Weight adjustment is based on the definition of a suitable error function, which is then minimised with respect to the weights and biases in the network using a suitable algorithm (e.g. gradient descent or global optimisation). Alternatively, unsupervised learning (also called self-organisation) requires only input data in order to train the network. During the training process the network weights are adjusted so that similar inputs produce similar outputs. This is accomplished by a training algorithm that extracts statistical regularities from the training set, representing them as the values of network weights (Fischer and Gopal 1994b; Fischer 1995). Prior knowledge may be used to specify the properties of the network learning methods. Bootstrap techniques, for example, may be used for estimating the bias of network parameters.

Multilayer feedforward networks (perceptrons and radial basis function networks) have emerged as the most attractive neural network architecture for various spatial analysis tasks (Fischer and Gopal 1994a; Gopal and Fischer 1996, 1997; Leung 1997). Analytical results show that two-layer (one hidden layer) feedforward networks are very capable of approximating arbitrary mappings in the presence of noise. However, they do not provide more than very general guidance on how this can be achieved, and what guidance they do offer suggests that network training will be difficult. Consequently, there is an urgent need to develop application domain-specific methodologies which provide more specific guidelines for judicious use of neural network approaches in SDA.

One critical issue for a successful application of neural-network-based spatial analysis is the complex relationship between learning (training) and generalisation. It is important to stress that the ultimate goal of network training is not to create an exact representation of the training data itself, but rather to build a model of the process which generates the data in order to achieve a good generalisation (out-of-sample) performance of the model. One method of optimising the generalisation performance of a model is to control its effective complexity where complexity is measured in terms of network parameters.

The attraction of neural-network-based SDA essentially stems from the following features:

- representational flexibility and freedom from linear model design constraints;
- inbuilt ability (via net representation and training) to incorporate rather than ignore the special nature of spatial data;
- robustness and fault tolerance to deal with noisy data and missing or fuzzy information;
- efficiency of large spatial datasets analysis, raising the prospect of being able to process finer resolution data or to carry out real-time analysis;
- inbuilt capability to adapt the connection weights to changes in the surrounding environment (learning);
- improved generalisation (out-of-sample performance) capabilities;
- potential to improve the quality of results by reducing the number of rigid assumptions and shortcuts introduced by conventional methodologies.

6 APPLICATION DOMAINS AND EXAMPLES OF NEURAL-NETWORK-BASED SPATIAL ANALYSIS

Neural network models in general, and feedforward neural network models in particular, can provide novel, elegant, and extremely valuable classes of mathematical tools for SDA, based on sound theoretical concepts. They may be viewed as non-linear extensions of conventional spatial statistical models such as regression models, spatial interaction models, linear discriminant functions, and pattern recognition techniques (Fischer and Gopal 1994a; Fischer et al 1997). They are particularly appropriate to two major domains (Fischer 1994):

- as *universal function approximators* in spatial regression, spatial interaction modelling, spatial choice, and space–time series analysis;
- as *pattern recognisers and classifiers* of large datasets (e.g. census small area statistics, high-resolution remote sensing data).

Feedforward neural network model building may be considered as a three-stage process, as outlined in Fischer and Gopal (1994a) and applied to telecom traffic modelling by Gopal and Fischer (1996):

- identification of a specific model from a family of two-layer feedforward networks which are characterised by specific types of non-linear processing elements;

- estimation of the network parameters of the selected neural network model and the model optimisation (using regularisation theory, network pruning, or cross-validation) for the given training set;
- testing and evaluating the out-of-sample (generalisation) performance of the model.

There is little doubt that neural pattern classifiers have an important role to play in high dimensional problems of pattern recognition and classification of massive quantities of data, for example associated with national classifications based on census small area statistics or with spectral pattern classification problems using RS satellite imagery. For example, Fischer and Gopal (1996) illustrate the virtues of NN classification *vis à vis* its conventional ML counterpart in a pixel-by-pixel supervised spectral pattern classification of a Landsat-5 Thematic Mapper image of Vienna. The task of discriminating between *a priori* defined urban land cover categories is challenging because urban areas comprise a complex spatial assemblage of disparate land-cover types – including built structures, numerous vegetation types, bare soil and water bodies – each of which has different spectral reflectance characteristics. However, the results suggest that neural network classifiers in general and a fuzzy ARTMAP classifier in particular are very powerful tools for classifying remotely-sensed imagery if non-linearity is encountered in the dataset. Indeed in the Vienna application it has an outstanding out-of-sample classification accuracy of 99.26 per cent on the pixels testing dataset. This error rate is less than 1/15 that of the two-layer perception, 1/20 that of the Gaussian maximum likelihood classifier and 1/30 that of the radial basis function network. Inspection of the classification error matrices reveals that the fuzzy ARTMAP classifier accommodates more easily a heterogeneous class label such as 'densely built-up residential areas' to produce a visually and numerically correct urban land cover map, even given smaller numbers of training pixels. In particular the normal maximum likelihood classifier tends to be sensitive to the purity of land cover signatures and performs poorly if they are not pure. Another serious problem with the normal classifier is its long processing time if RS data of a large area are to be analysed – which is a

common feature in GIS environments. This problem will be exacerbated given anticipated increases in data volumes from planned multichannel satellites (Barnsley, Chapter 32; Dowman, Chapter 31).

7 CONCLUSIONS AND PROSPECTS

GIS technology has already greatly increased the remit of SDA. Conventional SDA tools are generally not sufficiently powerful to cope with the new analysis needs. SDA is entering a new era of data-driven exploratory searches for patterns and relationships. CI technologies in general and neural networks in particular provide an interesting and powerful paradigm to meet the new challenges, yet one that is likely to evolve slowly rather than instate radical change within a short timeframe. The driving forces behind this change are the large amounts of GIS-based spatial data that are now available, the availability of attractive and novel CI tools, the rapid growth in computational power (especially that delivered through massively parallel computers), and the new emphasis on exploratory data analysis and modelling.

Neural networks provide not only novel and extremely valuable classes of data-driven mathematical tools for a series of spatial analysis tasks, but also an appropriate framework for re-engineering our well-established SDA techniques to meet the new large-scale data processing needs in GIS. Application of neural network models to spatial datasets holds the potential for fundamental advances in empirical understanding across a broad spectrum of application fields in spatial analysis. To realise these advances, it is important to adopt a principled rather than an ad hoc approach where spatial statistics and neural network modelling have to work together. The most important challenges in the next years will be twofold: first, to develop specific methodologies for particular application domains; second, to gain deeper theoretical insights into the complex relationship between learning and generalisation. These are of critical importance for the success of real-world applications.

The mystique and metaphorical jargon promulgated by the field may have the effect of lessening the amount of serious attention given to the new neural networks paradigm. Nevertheless many aspects of the study of neural networks lend themselves to rigorous mathematical analysis, and this provides a sound foundation on which to base a study of the capabilities and limitations of neural network systems and applications. Casting the analysis in the universal language of mathematics makes it possible to dispel much of the mystique (White 1992). A start has been made for a neural-network-based SDA, but much remains to be done.

References

Anselin L 1988 *Spatial econometrics: methods and models*. Dordrecht, Kluwer

Anselin L 1992 *SpaceStat, a program for the analysis of spatial data*. Santa Barbara, NCGIA, University of California

Anselin L 1994a Exploratory spatial data analysis and geographic informations systems. In Painho M (ed.) *New tools for spatial analysis*. Luxembourg, Eurostat: 45–54

Anselin L 1995b Local indicators of spatial association – LISA. *Geographical Analysis* 27: 93–115

Anselin L 1997a The Moran scatterplot as an ESDA tool to assess local instability in spatial association. In Fischer M M, Scholten H, Unwin D (eds) *Spatial analytical perspectives on GIS*. London, Taylor and Francis: 111–25

Anselin L, Dodson R, Hudak S 1993 Linking GIS and spatial data analysis in practice. *Geographical Systems* 1: 3–23

Anselin L, Getis A 1993 Spatial statistical analysis and geographic informations systems. In Fischer M M, Nijkamp P (eds) *Geographic information systems, spatial modelling, and policy evaluation*. Berlin, Springer: 35–49

Anselin L, Griffith D 1988 Do spatial effects really matter in regression analysis? *Papers of the Regional Science Association* 65: 11–34

Anselin L, Hudak S 1992 Spatial econometrics in practice: a review of software options. *Regional Science and Urban Economics* 22: 509–36

Arbia G 1989 *Spatial data configuration in statistical analysis of regional economic and related problems*. Dordrecht, Kluwer

Aufhauser E, Fischer M M 1985 Log-linear modelling and spatial analysis. *Environment and Planning A* 17: 931–51

Berry B J L, Marble D F (eds) 1968 *Spatial analysis: a reader in statistical geography*. Englewood Cliffs, Prentice-Hall

Bezdek J C 1994 What is computational intelligence? In Zurada J M, Marks R J, Robinson C J (eds) *Computational intelligence imitating life*. New York, IEEE: 1–12

Birkin M, Clarke M, George F 1995 The use of parallel computers to solve non-linear spatial optimisation problems: an application to network planning. *Environment and Planning A* 27: 1049–68

Cliff A D, Ord J K 1981a Spatial and temporal analysis: autocorrelation in space and time. In Wrigley N, Bennett R J (eds) *Quantitative geography: a British view*. London, Routledge: 104–10

Cressie N A C 1993 *Statistics for spatial data*, revised edition. New York, John Wiley & Sons

Dacey M F 1960 A note on the derivation of nearest neighbour distances. *Journal of Regional Science* 2: 81–7

Diggle P J 1983 *Statistical analysis of spatial point patterns*. London, Academic Press

Ding Y, Fotheringham A S 1992 The integration of spatial analysis and GIS. *Computers, Environment, and Urban Systems* 16: 3–19

Fischer M M 1994 Expert systems and artificial neural networks for spatial analysis and modelling: essential components for knowledge-based geographical information systems. *Geographical Systems* 1: 221–35

Fischer M M 1995 Fundamentals in neurocomputing. In Fischer M M, Sikos T T, Bassa L (eds) *Recent developments in spatial information, modelling and processing*. Budapest, Geomarket Co.: 31–41

Fischer M M 1997 Computational neural networks – a new paradigm for spatial analysis. *Environment and Planning A* 29

Fischer M M, Gopal S 1993 Neurocomputing – a new paradigm for geographic information processing. *Environment and Planning A* 25: 757–60

Fischer M M, Gopal S 1994a Artificial neural networks. A new approach to modelling interregional telecommunication flows. *Journal of Regional Science* 34: 503–27

Fischer M M, Gopal S 1994b Neurocomputing and spatial information processing. From general considerations to a low dimensional real-world application. *New tools for spatial analysis*, Luxembourg, Eurostat: 55–68

Fischer M M, Gopal S 1996 Spectral pattern recognition and fuzzy ARTMAP classification. In Zimmermann H-J (ed.) *Proceedings, Fourth European Congress on Intelligent Techniques and Soft Computing*, Vol. 3: Aachen, Verlag Mainz: 1664–8

Fischer M M, Gopal S, Staufer, P, Steinnocher K 1997 Evaluation of neural pattern classifiers for a remote sensing application. *Geographical Systems* 4: 195–223, 231–2.

Fischer M M, Nijkamp P 1992 Geographic information systems and spatial analysis. *The Annals of Regional Science* 26: 3–12

Fischer M M, Scholten H J, Unwin D 1996 Geographic information systems, spatial analysis and spatial modelling. In Fischer M M, Scholten H J, Unwin D (eds) *Spatial analytical perspectives on GIS*. London, Taylor and Francis

Flowerdew R, Green M 1991 Data integration: Statistical methods for transferring data between zonal systems. In Masser I, Blakemore M (eds) *Handling geographical information*. Harlow, Longman/New York, John Wiley & Sons Inc.: 18–37

Fotheringham A S, Wong D W S 1991 The modifiable areal unit problem in multivariate statistical analysis. *Environment and Planning A* 23: 1025–44

Getis A 1964 Temporal land-use pattern analyses with the use of nearest neighbour and quadrat methods. *Annals of the Association of the American Geographers* 54: 391–8

Getis A 1983 Second-order analysis of point patterns: the case of Chicago as a multi-center urban region. *The Professional Geographer* 35: 73–80

Getis A 1992 *Spatial dependence and proximal databases. Paper presented at 39th North American Meeting of the RSAI, Chicago, 14 November 1991*

Getis A, Boots B 1978 *Models of spatial processes.* Cambridge (UK), Cambridge University Press

Getis A, Ord K 1992 The analysis of spatial association by use of distance statistics. *Geographical Analysis* 24: 189–206

Goodchild M F 1987 A spatial analytical perspective as geographical information systems. *International Journal of Geographical Information Systems* 1: 327–34

Goodchild M F, Gopal S (eds) 1989 *Accuracy of spatial databases.* London, Taylor and Francis

Goodchild M F, Haining R P, Wise S 1992 Integrating GIS and spatial analysis: problems and possibilities. *International Journal of Geographical Information Systems* 6 : 407–23

Gopal S, Fischer M M 1996 Learning in single hidden-layer feedforward network models. *Geographical Analysis* 28: 38–55

Gopal S, Fischer M M 1997 Fuzzy ARTMAP – a neural classifier for multispectral image classification. In Fischer M M, Getis A (eds) *Recent developments in spatial analysis – spatial statistics, behavioural modelling, and neurocomputing.* Berlin, Springer

Griffith D A 1988 *Advanced spatial statistics: special topics in the exploration of quantitative spatial data series.* Dordrecht, Kluwer

Griffith D A 1993 Which spatial statistics techniques should be converted to GIS functions? In Fischer M M, Nijkamp P (eds) *Geographic information systems, spatial modelling, and policy evaluation.* Berlin, Springer: 101–14

Haggett P 1965 *Locational analysis in human geography.* London, Edward Arnold

Haggett P, Cliff A D, Frey A E 1977 *Locational methods in human geography,* 2nd edition. London, Edward Arnold

Haining R P 1994 Designing spatial data analysis modules for geographical information systems. In Fotheringham A S, Rogerson P (eds) *Spatial analysis and GIS.* London, Taylor and Francis: 45–63

Hecht-Nielsen R 1990 *Neurocomputing.* Reading (USA), Addison-Wesley

Isaaks E H, Srivastava R M 1989 *An introduction to applied geostatistics.* Oxford, Oxford University Press

King L J 1969 *Statistical analysis in geography.* Englewood Cliffs, Prentice-Hall

Langton C G (ed.) 1989 *Artificial life. The proceedings of an interdisciplinary workshop on the synthesis and simulation of living systems.* Reading (USA), Addison-Wesley

Leung Y 1997 Feedforward neural network models for spatial pattern classification. In Fischer M M, Getis A (eds) *Recent developments in spatial analysis – spatial statistics, behavioural modelling, and neurocomputing.* Berlin, Springer

Longley P, Batty M 1996a Analysis, modelling, forecasting, and GIS technology. In Longley P, Batty M (eds) *Spatial analysis: modelling in a GIS environment.* Cambridge (UK), GeoInformation International: 1–15

Nelder J A, Wedderburn R W M 1972 Generalised linear models. *Journal of the Royal Statistical Society A* 135: 370–84

Openshaw S 1984 The modifiable areal unit problem. *Concepts and Techniques in Modern Geography* 38. Norwich, Geo-Books

Openshaw S 1991a A spatial analysis research agenda. In Masser I, Blakemore M (eds) *Handling geographical information: methodology and potential applications.* Harlow, Longman: 18–37

Openshaw S 1993 Modelling spatial interaction using a neural net. In Fischer M M, Nijkamp P (eds) *Geographic information systems, spatial modelling, and policy evaluation.* Berlin, Springer: 147–64

Openshaw S 1994e Two exploratory space–time-attribute pattern analysers relevant to GIS. In Fotheringham A S, Rogerson P (eds) *Spatial analysis and GIS.* London, Taylor and Francis: 83–104

Openshaw S, 1994f What is GISable spatial analysis? In *New tools for spatial analysis.* Luxembourg Eurostat: 36–44

Openshaw S 1995b Developing automated and smart spatial pattern exploration tools for geographical systems applications. *The Statistician* 44: 3–16

Openshaw S, Fischer M M 1995 A framework for research on spatial analysis relevant to geo-statistical information systems in Europe. *Geographical Systems* 2: 325–37

Paelinck J, Klaassen L 1979 *Spatial econometrics.* Farnborough, Saxon House

Ripley B D 1977 Modelling spatial patterns. *Journal of the Royal Statistical Society B* 39: 172–212

Rogers A 1965 A stochastic analysis of the spatial clustering of retail establishments. *Journals of the American Statistical Association* 60: 1094–103

Silverman B W 1986 *Density estimation for statistics and data analysis.* London, Chapman and Hall

Tobler W R 1979a Cellular geography. In Gale S, Olsson G (eds) *Philosophy in geography.* Dordrecht, Reidel: 379–86

Unwin A 1993 *Interactive statistical graphics and GIS – current status and future potential.* Position Paper, Workshop on Exploratory Spatial Data Analysis and GIS, NCGIA, Santa Barbara, 25–27 February

Upton G J, Fingleton B 1985 *Spatial statistics by example.* New York, John Wiley & Sons Inc.

White H (ed.) 1992 *Artificial neural networks. Approximation and learning theory.* Oxford, Blackwell

Wrigley N 1985 *Categorical data analysis for geographers and environmental scientists.* Harlow, Longman

20

Location modelling and GIS

R L CHURCH

Location modelling involves the search for the best location of one or more facilities to support some desired function. Examples range from retail site location to the location of multiple ambulance dispatch points. GIS has played a large role in the siting of single facilities, including rights-of-way for roads and transmission lines. Its use in multi-facility location problems is a relatively recent development. This chapter describes some of the history of location search as supported by GIS. It also discusses some of the current impediments to the application of location models, issues associated with the integration of location models into GIS, and future needs in GIS functionality to support location models.

1 WHAT IS A LOCATION PROBLEM?

Webster's Dictionary defines location as (1) position in space, and (2) an area marked off for a specific purpose. Hence, a location problem would involve identifying a specific position or place for a specific function or activity. There are several common types of location problem associated with GIS. The most common involves the measurement of where something exists. Given adequate time, it is possible to measure the location of virtually anything on the Earth. This type of problem can be called a location measurement problem. The second type involves the search for an appropriate location for an activity. This problem can be called the locational search problem. It is common to refer to the locational search problem as a facility location problem. This type of problem can involve the placement of one activity (e.g. a retail store) or the placement of a set of interrelated facilities (e.g. fire stations to serve an urban area). Such problems are called single and multi-facility location problems. Other chapters in this book provide examples in the contexts of business and service planning (Birkin et al, Chapter 51), transportation (Waters, Chapter 59), and electoral districting (Horn, Chapter 67).

2 EARLY HISTORY AND DEVELOPMENT

Both locational measurement and locational search problems are important. The focus of this chapter will be on the problem of locational search. One of the early applications which helped to start the development of GIS involved locational search problems over large regions. In the 1970s many states began the development of geographical databases (in raster format) for the storage and retrieval of environmental and planning data. Examples of such systems include LUNR, a land-use and natural resource inventory of the State of New York, and the MAGI (Maryland Automated Geographic Information) database of Maryland. As an historical side note, Dangermond of ESRI (Environmental Systems Research Institute) was a principal developer of the MAGI database. During the mid 1970s there was considerable debate as to whether the electrical utilities could keep up with the increasing demand for electricity. Further, many questioned whether enough suitable sites for power plants existed. Out of this concern, the State of Maryland funded the Maryland Power Plant Siting Project and the development of a simple grid-based GIS. The State was divided into approximately

293

30 000 cells, each measuring 91.8 acres. In each cell 52 variables were measured, including land-use, land cover, proximity to a water source, seismicity etc. Each of these variables was converted to a subscore function and added together to form a site suitability score (Dobson 1979). From this, a map was produced which identified those cells which scored higher than one standard deviation above the mean. This process is depicted in Figure 1. The Planning Office of the State of Maryland still manages and updates this database, although it is now stored in both raster and vector format.

Another location problem which has evolved with GIS is the location of rights-of-way for roads and transmission power lines. It is interesting to note that much of the development of computerised modelling for such linear facilities has been coordinated with some type of GIS, principally raster. McHarg (1969) in *Design with Nature* developed a process which represents a precursor to the classic overlay process in GIS. McHarg's process involved developing a colour acetate map sheet for each basic theme that is used in the location of a corridor. The shading of a given map was from light to dark (high suitability to low suitability). When all of the acetate sheets were laid on top of each other, and placed over a light-table, the areas that had little colour were considered to be the best in terms of suitability. McHarg's process called for the tracing of the most direct route connecting the desired terminus points and which travelled through as much lightly coloured area as possible. Most

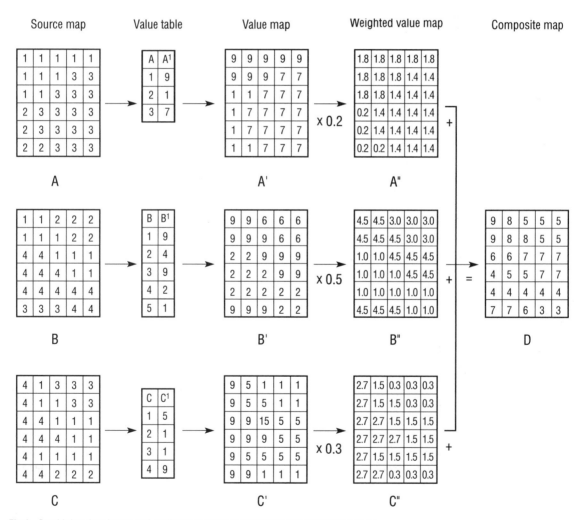

Fig 1. Combining data layers and attributes in order to determine site suitability scores.

subsequent corridor routing applications have involved raster-based systems. Information associated with various attributes is typically converted to subscores and combined into suitability scores for compatibility with use for a corridor. Next, a network is defined where the nodes represent the centroids of the cells. This is depicted in Figure 2. The arcs depict possible directions which can be taken by the corridor through a given cell. Goodchild (1977) and Huber and Church (1985) have shown that considerable error may exist in the unneeded elongation of a route based upon the limits made in the compass directions an arc may take from a given centroid. Arc costs or weights are usually defined as some weighted combination of the suitability scores through which the arc passes. In Figure 3, a case is depicted where the arc has a width of influence, and the weights are associated with the footprint of the corridor in a given cell. Given a network, complete with arcs and associated traverse costs, a shortest path algorithm can be applied in order to identify the least cost (most suitable) route across the landscape. The beauty of this process is that it can be solved optimally for large problems. An example is given in Figure 4, which involves a landscape of more than 16 000 cells and 256 000 arcs and where the solution time is but a few minutes on even modest workstations. Because such a process is fast, it can be repeated for a wide variety of parameter weights associated with different components of the suitability score, allowing the user to test the sensitivity of the routing to various levels of importance weights. The more efficient a solution

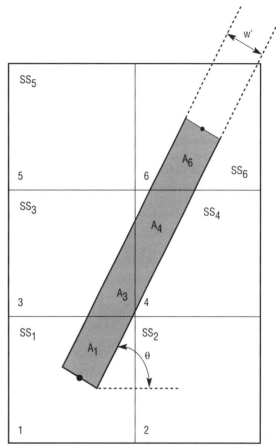

Arc value = $A_1{}^*SS_1+A_3{}^*SS_3+A_4{}^*SS_4+A_6{}^*SS_6$

Fig 3. Determining the weighted cost for an arc.

algorithm can be made for a given application, the more flexibility there is in designing a user interface to allow for sensitivity analysis and general spatial exploration (Church et al 1992).

Single-site location analysis can often be approached with some form of ranking or scoring process, which is based upon a set of attributes. From a landscape of potential site scores, areas are screened or filtered out if their scores are too low to be acceptable (as in the power plant siting example). This process is generally well supported in a number of GIS products. Documented applications abound in GIS conference proceedings associated with the use of such a site search process. In fact a number of companies provide data and services to assist in such general site searches. Further, special scoring processes like the Analytic Hierarchy Process have been linked to GIS in systems like Idrisi (1996).

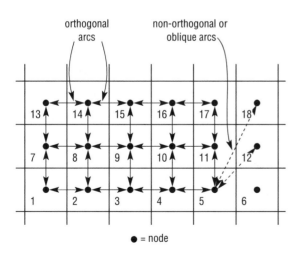

Fig 2. Depicting arc directions on a raster system.

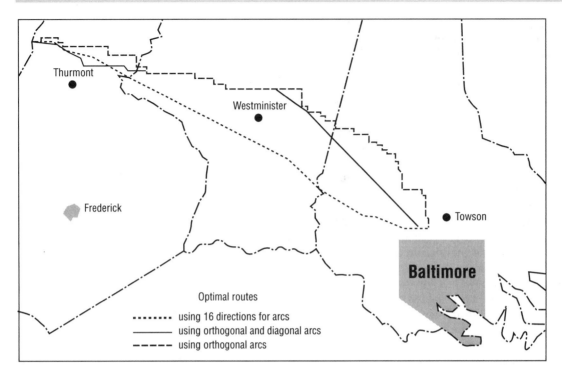

Optimal routes

∙∙∙∙∙∙∙ using 16 directions for arcs
————— using orthogonal and diagonal arcs
– – – – using orthogonal arcs

Fig 4. An efficient route using State of Maryland data.

Because such site searches can be performed with a modest set of GIS functions, it is both well supported in software and widely applied.

Beyond the search for an appropriate site or the search for a corridor across a landscape, a third major location search problem exists. This third major category represents simultaneous multiple-site selection. To explain this, consider the placement of ambulances for the purpose of emergency response. One common approach to developing a deployment plan is called System Status Management (SSM; Stout 1989). A deployment plan for SSM is established for each hour of the week. For example, the City of Denver has an SSM plan for the time interval of Tuesday 2–3 am. This plan is based on providing a specific number of ambulances in order to meet the expected demand with a very high probability (for example to provide 95 per cent response within 8 minutes). It also details how ambulances should be positioned during that hour, when all are available, when all but one are available, when all but two are available, etc. As ambulances are called for service during that hour, ambulances are repositioned to provide the best coverage possible according to the SSM plan. When

ambulances are finished with a call, some or all are repositioned in order to deploy better what is currently available. To develop an SSM for this one hour alone, it is necessary to identify a location plan for p ambulances, p–1 ambulances, p–2 ambulances, etc. To do this requires the solution of a series of location problems, each involving the location of a multiple number of units (facilities). The multi-facility location problem will be discussed at greater length in the next several sections of this chapter.

3 DEVELOPING A FRAMEWORK FOR LOCATION MODEL CLASSIFICATION AND GIS

Most location models are variants of four general classes: median, covering, capacitated, and competitive. A median model involves locating a fixed number of facilities in such a manner that the average distance from any user to their closest facility is minimised. Covering models involve locating facilities in order to cover all or most demand within some desired service distance (often called the maximum service distance). The idea is that the more users who are served relatively close to a facility, the better the service. For example, in ambulance deployment a common goal is to serve at least 90 per cent of the population within 8 minutes. Classic median models are based upon the assumption that there are enough resources at each facility to handle whatever demand needs to be served. Thus, everyone is assumed to be served by their closest facility. Capacitated facility models place some limit on what can be accomplished at each facility (e.g. the number of units that can be manufactured, the amount of demand that can be served or assigned, the volume of garbage that can be handled per day etc.). Finally, competition models involve the case where a competitor has the capability to readjust to any location decisions other competitors make over predefined time frames. If you locate a new branch which exploits a poorly served area of your competitor, your competitor may over a period of time relocate a branch or locate a new one to recoup some of that lost market. Thus, in making a location decision, it becomes necessary to analyse what response your competitor may make to keep from a loss of business. Median, covering, and capacitated models are usually addressed as classical optimisation models, whereas competition models are often addressed by game theory and simulation. Good reviews of location research can be found in the literature (Beaumont 1981; Brandeau and Chiu 1989; Eiselt 1992; ReVelle 1987; Schilling et al 1993).

How a location problem is defined involves the spatial relationships between what is defined as a demand and what is defined as a facility. Even though demand is often spread across space (e.g. a neighbourhood, a census tract, or an apartment building), it is often represented as a single point. In fact, most models are based upon the assumption that demands are represented as points, although there are some notable exceptions to this general rule in the location science literature (see Wesolowsky and Love 1971, where demand is represented by continuous rectangular areas). Facility sites can be defined as points, lines, or areas. For example, in a corridor location problem, the facility usually represents a curvilinear facility like a roadway that connects two prespecified terminus points.

For continuous surface problems, facility locations are often allowed to be anywhere. In terms of network models, facility sites are most often described as nodes although considerable theory has been developed concerning the location along arcs

or links as well. More often than not, both demand and facility sites are represented as discrete points since most solution algorithms have been developed for such cases. Miller (1996) has presented a classification of location models based upon the geometric representation of demand (or clients) and the geometric representation of facilities. For example, demands represented as polygons and facilities represented by polygons can be used to define a polygon–polygon location problem involving the location of a set of polygons to serve a set of weighted polygons. An example of such a problem can be found in very large-scale integration chip design and in production layout. Miller argues that GIS allows for better opportunities to represent location model features, such as facility size and shape, and that such improvements could have the potential to increase the relevance and flexibility of facility location models. Should this prove to be true, then GIS will become an integral part of many location model approaches in the future.

4 INTRODUCTION TO BASIC PROBLEMS IN THE IMPLEMENTATION OF LOCATION MODELS WITH GIS

Most single-site location search problems are solved by enumerating all possibilities. Even though there may be many possible sites, scoring and selecting the top sites is not a very computationally burdensome task (see Dobson 1979 for a classic application of site search in GIS). In contrast, the task of simultaneously selecting a configuration of sites to accomplish some objective is, indeed, significant. Recall from the above discussion that the City of Denver might have up to ten ambulances in operation at a given hour. Suppose 200 good deployment locations for ambulances have been identified in the city. This means that there are:

$$\binom{200}{10} = \frac{200!}{(10!)(200-10!)} = 2.245 \times 10^{16}$$

possible configurational plans (i.e. the number of distinct ways 200 sites can be selected ten at a time) that involve siting exactly ten ambulances. This is obviously a large number. In fact, this number is so large it is impossible to enumerate and evaluate each configuration within the lifetime of a given computer. Thus, enumerating all possible configurations and selecting the best solution is not a feasible approach. Consequently, the problem needs to be solved through the development of some type of algorithm or heuristic. To approach this type of problem, it is common first to develop a mathematical formulation for the problem. Then, solution techniques are designed to solve the problem and take advantage of its structure. This is described in the next section at greater length for two very popular location model constructs.

Many of the multi-facility location models are considered to be computationally complex (Garey and Johnson 1979). Beyond the computational task involved in solving a location model, there are several important issues related specifically to model integration and GIS:

1 *Compatibility of data structures.* There can be differences between the data structure that has been designed to best support a location model algorithm and the principal data model used in a GIS (Densham and Rushton 1991). As an example, to support a location heuristic in ARC/INFO, designers have an intermediate process create a data string that is read and used by the solution process. That is, the GIS data structure is not used directly by the solution process. Hence, the solution process and the GIS are loosely coupled by the exporting and importing of data and results.

2 *Demand representation and site identification.* In order to search for a solution to most location models, there must be a defined set of demand areas (represented by points, lines, or areas) as well as a set of predefined sites. GIS can serve an important role in the definition of demand areas and facility sites. A number of constructs are possible (as described in the previous section), although demand areas and facility sites are often represented as points. GIS can add significant value to a given application when data gathered for another purpose are available to characterise demand or the feasibility of specific sites or regions.

3 *Aggregation.* Before GIS datasets were available for location study, problem size was relatively small. Since information is characterised in a rather exact way in GIS, this detail has begun to be represented in site and demand definition. For example, in the past it might be common to represent a city by census tracts alone. Now, even enumeration districts or possibly block faces can be used to characterise and represent demand. In one recent application by a public utility, individual houses represent the basic

unit of demand. Thus, the demand surface is often characterised to a relatively fine level of detail by hundreds if not thousands of demand points. The same can be true for the representation of potential facility sites (whether they be areas, points, or line segments). Since many solution approaches cannot easily handle thousands of demand points and sites, some type of data aggregation is necessary. Goodchild (1979) was the first to demonstrate that data aggregation can have a great effect in the absolute location of specific facilities. Issues and properties of aggregation have been addressed by numerous researchers in geographical analysis (e.g. Current and Schilling 1987, 1990; Hillsman and Rhoda 1978). The current study of aggregation techniques can be aided by use of GIS.

4 *Error propagation.* Even though error in data is a fact of life, it is important to understand how data error may propagate in any procedure that is used, e.g. aggregation. Further, a specific aggregation scheme may introduce error in problem representation. Veregin (Chapter 12; 1995) has analysed the propagation of error in GIS analysis for classical GIS functions like overlay (and see Heuvelink, Chapter 14; Openshaw and Alvanides, Chapter 18). How error propagates and the extent that it may alter results in location models is in need of further research. It is necessary to analyse such occurrences, as there can be diminished reliability in the final results.

5 *Visualisation.* Exploring and comparing results from a location model can be enhanced by visual aids (Densham 1994). This much is known. Just what types of visualisation scheme are the best is still subject to research (Arentze et al 1996). One popular visualisation is the presentation of a map with demand allocated to located facilities by directed lines. This 'spider-like' plot provides an easy-to-understand view of where the facilities are, which demand is served by which facility (i.e. service regions), as well as the potential area differences in service regions (Armstrong et al 1992). There is a major need to identify which possible views aid an analyst in generating and searching for the best solution. This is also an element in which GIS can be of significant value in location modelling, as many attributes can be presented simultaneously with model results. For example, assume a set of cellular telephone communication sites have been located. With elevation data, the location pattern can be draped on a surface depicting elevation (see Fry, Chapter 58).

5 INTEGRATING TWO CLASSIC LOCATION MODELS INTO GIS: P-MEDIAN AND MAXIMAL COVERING

The p-median location model involves the location of a fixed number p of facilities. The objective is to locate the p facilities in such a manner that the total weighted distance of serving all demand is minimised. Weighted distance for a demand point represents the amount of demand multiplied by the distance to the closest facility. For example, if demand is measured in terms of the number of trips that need to be made by users of the facility, then weighted distance represents the total mileage involved in going to the facility. For a fixed level of demand, minimising total weighted distance is equivalent to minimising average distance. This model form can address many different types of application, from locating schools and health clinics to locating road maintenance garages and emergency response vehicles. Because this model captures the essence of locating a set of facilities to serve an area by maximising accessibility, it has become a popular model for application. This model can be formulated mathematically using the following notation (ReVelle and Swain 1970):

i = index of demand areas or nodes,
p = the number of facilities to be located,
j = index of potential facility sites,
a_i = the amount of demand at area/node i
d_{ij} = the shortest distance between demand i and facility site j,

$$x_{ij} = \begin{cases} 1, \text{ if demand at } i \text{ assigns to a facility at } j \\ 0, \text{ otherwise} \end{cases}$$

$$x_{jj} = \begin{cases} 1, \text{ if demand node } j \text{ self-assigns, meaning a} \\ \quad \text{facility is allocated to site } j \\ 0, \text{ otherwise} \end{cases}$$

The p-median model formulation is thus:

Minimise $Z = \sum_i \sum_j a_i d_{ij} x_{ij}$

Subject to
each demand must assign to a facility:

$\sum_j x_{ij} = 1 \quad \forall i$

each demand assignment is restricted to what has been located:

$x_{ij} \le x_{jj} \quad \forall i,j, i \ne j$

exactly p-facilities are located:

$\sum_j x_{jj} = p$

and integer restrictions on decision variables:

$x_{ij} = 0, 1 \quad \forall i,j.$

The above formulation can be classified as a binary-integer programming problem, and represents the case where each node is both a point of demand and a potential facility site. The formulation can easily be tailored to restrict facilities or demand to specific subsets of nodes. The model determines how each demand node is to be served. The sense of the objective will ensure that each node is served by its closest facility. It is important to recognise that it is not necessary to require that the assignment variables x_{ij}, where $i \ne j$, be integer. This keeps the number of integer variables equal to the number of nodes that are listed as potential facility sites. The above model can be solved directly using a general purpose integer linear programming solution procedure. Because the formulation represents a rather tight structure, an optimal relaxed linear programming (LP) solution often meets the integer constraints. Consequently, a branch and bound process is frequently not required. Unfortunately, the model is quite large in terms of the number of variables and constraints (n^2 variables and $n^2 - n + 1$ constraints). Using a general-purpose maths programming package is, therefore, limited to problems of less than several hundred nodes. This means that it is unrealistic to integrate GIS and general linear programming/integer programming (LP/IP) software to solve p-median location models.

Minimising average distance can still leave individual demand points long distances from their closest facility. In the context of emergency facilities, such long distances may be viewed as too far for adequate service. For example, a suburban neighbourhood should be within a mile and a half of a fire station to be considered adequately covered by fire protection services. In planning emergency services, average distance does not capture the urgency of the service, and is commonly replaced by the use of a maximal distance or time standard. A demand is defined as *covered* if a facility has been located within some predefined service standard (e.g. maximal service distance or time). Even though all demand points are to be served, the focus is to provide as many as possible with a level of service which meets some minimally acceptable standard

299

(called 'service coverage'). The development of location models utilising the coverage concept have taken two principal directions: location set covering (Toregas and ReVelle 1972) and maximal covering (Church and ReVelle 1974). Location set covering models minimise the number of facilities needed to cover all demand. The maximal covering location model locates a fixed number of facilities in a manner that coverage is maximised. The maximal covering location model and the location set covering model form the basis of a large class of location models.

Minimising weighted distance and maximising coverage are perhaps two of the most popular objectives that have been developed in location science. It is important to note that a covering model, like maximal covering, can be represented in a completely different model structure from that of a median problem. Consider:

$$y_i^2 = \begin{cases} 1, \text{ if demand } _i \text{ is not covered} \\ 0, \text{ otherwise} \end{cases}$$

$$x_i = \begin{cases} 1, \text{ if a facility is located at site } j \\ 0, \text{ otherwise} \end{cases}$$

$N_i = \{ j \mid d_{ij} \le S \}$, the set of sites which can provide coverage to demand i.

The maximal covering location model formulation:

Minimise $Z = \sum_i a_i y_i$

Subject to
defining if demand i is covered:

$\sum_{j \in N_i} x_j + y_i \ge 1 \ \forall \ i,$

locating exactly p facilities:

$\sum_j x_j = p;$

and integer restrictions:

$x_j = 0, 1 \ \forall j$ and

$y_i = 0, 1 \ \forall i$

The above formulation is very compact and contains only $n+1$ constraints and $2n$ variables. It is structured to minimise the amount of demand not covered. This is mathematically equivalent to maximising the number that can be provided coverage. The maximal covering model is easy to solve optimally for problems involving thousands of nodes using general purpose integer programming software.

There are two important theoretical issues associated with the median and covering models. First, most covering models can be cast as a special form of some type of a median problem (Church and Weaver 1986). This can be done for the maximal covering model as follows:

$$d_{ij} = \begin{cases} 0, \text{ if } d_{ij} \le \text{S} \\ 1, \text{ otherwise} \end{cases}$$

By using the distance values d_{ij} in the p-median problem, the sense of the objective is transformed from minimising weighted distance to minimising the number that are served beyond the distance S. Therefore, any maximal covering problem can be solved using a p-median solution technique by first making a distance transformation (Church and ReVelle 1976; Hillsman 1984). It is, therefore, not surprising to see that the design of the location–allocation module in ARC/INFO, for example, utilises a p-median solution technique and provides the capability to transform a covering model into a median problem format.

The second major issue is that the p-median and maximal covering problems are non-deterministic polynomial (NP)-hard. Practically, this means that specific instances of the p-median and maximal covering problems will be difficult if not impossible to solve optimally in some reasonable amount of computer time. The fact that many vehicle routing, location, and districting problems fall into the class of NP means that applications involving such spatial optimisation problems integrated into GIS must be designed to involve potentially difficult problem instances. Since general purpose solution procedures are only capable of solving median problems that are less than several hundred nodes and since median and covering problems can be represented by thousands of nodes in a GIS, it is virtually impossible to consider solving such problems optimally. There are two approaches that have gained favour in the modelling literature: heuristics and Lagrangian relaxation with limited branch and bound. Lagrangian relaxation with branch and bound has been employed to solve the p-median problem to an exact optimum or to identify a solution within a known percentage of optimality. This technique requires a special purpose code and can be quite sensitive to parameter settings in a specific application. Lagrangian relaxation has been used to solve problems of sizes up to a thousand nodes (Beasley 1993). Heuristics have been developed to solve the *p*-median problem for three principal reasons: (1) to solve problems whose size falls outside the range of an algorithm; (2) to solve problems considerably faster than an algorithm; and (3) to reduce the costs of implementation and application.

Several types of heuristic have been designed to solve the *p*-median problem, and some of them can be easily implemented into GIS. Such heuristics include: vertex substitution, simulated annealing, greedy adding and dropping, semi-greedy, GRASP, genetic, TABU, and hybrids like GRIA (global regional interchange algorithm: Densham and Rushton 1992). Details of these procedures are provided by Church and Sorensen (1996). The procedure that has received the widest recognition is the simple vertex substitution process of Teitz and Bart (1968). The Teitz and Bart procedure starts with a pattern of *p* facilities. At each step of the heuristic, a candidate site is selected and tested to see if it can be used as a substitute for one of the current facility sites. If any such substitution can be made which yields an improvement in the objective, then a switch (or substitution) is made. The switch that is made is the one which yields the best improvement in weighted distance. In one test, Rosing et al (1979) found that the Teitz and Bart heuristic performed flawlessly, repeatedly identifying optimal solutions with different starting solutions for a 49-node problem. This led the researchers to believe that the process was virtually independent of the starting solution. It is now recognised that even though such a process is relatively robust at finding good, if not optimal, solutions, its performance is controlled by the potential existence of many local optima (Church and Sorensen 1994).

It makes sense to pick a heuristic like Teitz and Bart for solving median and covering models and integrate it into a GIS, based on the premises that it is easy to explain, relatively easy to code, produces very good results and is relatively fast at converging to a final answer. The only problem is that it will not always generate the same final solution unless it is initiated with the same starting solution. Such behaviour is not a common characteristic in GIS, as users expect GIS functions to produce the same answer, unless changes in the data are made. One design approach to this problem is to develop a procedure to determine a starting solution which will always be the same for the same dataset. In this case, the heuristic will always terminate with the same result on the same dataset. This is, in fact, the

approach taken by ESRI in developing a location-allocation module for ARC/INFO. A second approach is to acknowledge the limitations of the heuristic and allow the capability of restarting the heuristic with multiple starting configurations, until a relatively simple stopping rule is reached (Church and Sorensen 1994). Doing this can increase the probability of identifying the optimal solution to a given *p*-median problem. Unfortunately, such a process may not always produce the same result.

As stated above, median and covering models have been used for many types of planning problem. An example is given in Plate 14, which depicts the results of a reserve site selection problem for biodiversity protection in southern California. The problem involves the selection of a set of sites which contains as many species as possible. In essence, this is a form of the maximal covering location problem as defined for reserve site selection (Church et al 1996). The result was produced by the location–allocation module of ARC/INFO which was applied to a special logical network of sites and species presence (Gerrard et al 1996). The complete application involves a user interface programmed in the ARC macro language. By customising specific applications, it is possible to aid analysts in data exploration, testing model sensitivity, and visualising results.

6 WHAT'S OPTIMAL WHEN SOLVING A LOCATION MODEL?

Twenty years ago, location problem sizes in geographical analysis were relatively small (i.e. less than 100 nodes, see Hillsman 1984; Swain 1974). One of the principal reasons for this was the difficulty in collecting, storing, and retrieving relevant data. Much of this can now be easily provided by GIS. Instead of having a list of feasible sites, it is now possible to rate them based upon multiple attributes. Many of the classic location–allocation models do not explicitly take advantage of this type of information. Brill (1979) has stated that when a model is solved optimally, that optimal solution may be inferior (i.e. not Pareto optimal) when another objective or criterion is introduced. This means that optimality can be defined for only what has been explicitly included in a model. To support the search for solutions which meet the needs of the decision-maker, it is important to produce close-to-optimal solutions (i.e. solutions which meet high levels of explicitly stated objectives as well as good performance in terms of those issues that are still a part of the decision-making process but not included in the model. The bottom line is that GIS provides a richer data fabric, upon which more complicated location models need to be defined and solved. Integrated solution techniques need to be capable of searching for high-performing alternatives, instead of just one solution (Lombard and Church 1993). This area of analysis represents a new frontier for GIS and decision support.

7 LOOKING TO THE FUTURE

GIS has had a substantial impact in the field of location model application, in that it has presented a valuable way to organise spatial data for locational search processes. Such searches include retail site location, emergency services location, and factory location. GIS has also allowed many applications to share data easily within a 'corporate' organisation, sometimes creating unanticipated value for location model applications. The integration of multi-facility location models, like p-median and maximal covering, has just begun and is generally oriented towards the location of public facilities. Although few systems currently support a multi-facility model (like ARC/INFO), this will undoubtedly change. Features like algorithms to solve capacitated facility location problems are needed to meet the needs of many industries.

The biggest issue facing the GIS community in facility location planning is that most location models and their solution counterparts have been developed and tested for green-field planning problems. That is, the models are typically defined and the algorithms designed to solve the case when all of the facilities are new. Even though all current GIS implementations have the capability to fix existing sites into the solution, they do not have the capability to move existing sites when to do so is relatively superfluous to the solution. That is, sites can only be fixed into a solution or out of a solution. This capability only supports one side of what can be called the brown-field planning problem. In brown-field planning (i.e. adding to, taking away, or transforming an existing configuration) there must be the capability to solve for a new configuration which maintains much of what currently exists and which adds or moves

specific facilities to better locations. This can be defined formally for the p-median problem using the following notation:

$E = \{\, j \mid \text{site } j \text{ currently houses a facility}\,\}$

$R = $ the number of existing facilities that are to be closed

$P = $ the number of existing facilities

$P^i = $ the number of facilities to be added

We can then define a version of the brown-field median location problem by replacing the p-facility constraint $\left(\sum_j x_j = P\right)$ with the following two constraints:

(a) Close R of the current facilities:

$$\sum_{j \epsilon E} x_{jj} = P - R$$

(b) Open P new facilities in addition to the R being closed:

$$\sum_{j \epsilon E} x_{jj} = P$$

The above type of model form is relatively flexible, allowing for some existing sites to be closed and/or new ones to be added. At this time, few if any solution algorithms have been designed or tested to consider such important nuances. The fact that planners usually need to incorporate existing facilities into a new pattern, either whole or in part, can only be approached with this added model construct. Expanding GIS location routines to be capable of supporting this type of functionality should be a high priority.

References

Arentze T A, Borgers W J, Timmermans H J P 1996 Design of a view-based DSS for location planning. *International Journal of Geographical Information Systems* 10: 219–36

Armstrong M P, Densham P J, Lolonis P, Rushton G 1992 Cartographic display to support locational decision making. *Cartography and Geographic Information Systems* 19: 154–64

Beasley J 1993 Lagrangian heuristics for location problems. *European Journal of Operations Research* 65: 383–99

Beaumont J R 1981 Location–allocation problems in the plane: a review of some models. *Socioeconomic Planning Sciences* 15: 217–29

Brandeau M L, Chiu S S 1989 An overview of representative problems in location research. *Management Science* 35: 645–74

Brill E D Jr 1979 The use of optimisation in public sector planning. *Management Science* 25: 413–21

Church R L, Loban S R, Lombard K 1992 An interface for exploring spatial alternatives for a corridor location problem. *Computers and Geosciences* 18: 1095–1105

Church R L, ReVelle C S 1974 The maximal covering location problem. *Papers of the Regional Science Association* 32: 101–18

Church R L, ReVelle C S 1976 Theoretical and computation links between the *p*-median, location set covering, and the maximal covering location problems. *Geographical Analysis* 8: 406–15

Church R L, Sorensen P 1996a *Integrating normative location models into GIS: problems and prospects with the* p-*median model*. Technical Report 94–5. Santa Barbara, NCGIA

Church R L, Stoms D M, Davis F W 1996 Reserve selection as a maximal covering location problem. *Biological Conservation* 76: 105–12

Church R L, Weaver J R 1986 Theoretical links between median and coverage location problems. *Annals of Operations Research* 6: 1–19

Current J R, Schilling D A 1987 Elimination of source A and B errors in *p*-median location problems. *Geographical Analysis* 19: 95–110

Current J R, Schilling D A 1990 Analysis of errors due to demand aggregation in the set covering and maximal covering location problems. *Geographical Analysis* 22: 116–26

Densham P J 1994 Integrating GIS and spatial modelling: visual interactive modelling and location selection. *Geographical Systems* 1: 203–19

Densham P J, Rushton G 1991 *Designing and implementing strategies for solving large location–allocation problems with heuristic methods*. Technical Report 91–10. Santa Barbara, NCGIA

Densham P J, Rushton G 1992 A more efficient heuristic for solving large *p*-median problems. *Papers in Regional Science* 71: 307–29

Dobson J E 1979 A regional screening procedure for land use suitability analysis. *Geographical Review* 69: 224–34

Eiselt H A 1992 Location modelling in practice. *American Journal of Mathematical and Management Sciences* 12: 3–18

Garey M R, Johnson D S 1979 *Computers and intractability: a guide to the theory of NP-completeness*. New York, W H Freeman and Co.

Gerrard R A, Stoms D A, Church R L, Davis F W 1996 Using GIS models for reserve site selection. *Transactions in GIS*

Goodchild M F 1977 An evaluation of lattice solutions to the problem of corridor location. *Environment and Planning A* 9: 727–38

Goodchild M F 1979 The aggregation problem in location–allocation. *Geographical Analysis* 11: 240–55

Hillsman E L 1984 The *p*-median structure as a unified linear model for location–allocation analysis. *Environment and Planning A* 16: 305–18

Hillsman E L, Rhoda R 1978 Errors in measuring distances from populations to service centers. *Annals of Regional Science* 7: 74–88

Huber D L, Church R L 1985 Transmission corridor location modelling. *Journal of Transportation Engineering* 111: 114–30

Idrisi 1996 WEIGHT command. *http://www.idrisi.clarku.edul PRODUCTS/specindx.htm#DECISION SUPPORT*.

Lombard K, Church R L 1993 The gateway shortest path problem: generating alternative routes for a corridor location problem. *Geographical Systems* 1: 25–45

McHarg I L 1969 *Design with nature*. New York, The Natural History Press

Miller H J 1996 GIS and geometric representation in facility location problems. *International Journal of Geographical Information Systems* 10: 791–816

ReVelle C S 1987 Urban public facility location. In Mills E S (ed) *Handbook of regional and urban economics*, Vol 11. Amsterdam, Elsevier Science: 1053–96

ReVelle C S, Swain R 1970 Central facilities location. *Geographical Analysis* 2: 30–4

Rosing K E, Hillsman E L, Rosing-Vogelaar H 1979 A note comparing optimal and heuristic solutions to the *p*-median problem. *Geographical Analysis* 11: 86–9

Schilling D A, Jayaraman V, Barkhi R 1993 A review of covering problems in facility location. *Location Science* 1: 25–55

Stout J 1989 Peak-load staffing: what's fair for personnel and patients? *Journal of Emergency Medical Systems* 14: 73–4

Swain R 1974 A parametric decomposition approach for the solution of uncapacitated location problems. *Management Science* 21: 189–98

Teitz M B, Bart P 1968 Heuristic methods for estimating the generalised vertex median of a weighted graph. *Operations Research* 16: 953–61

Toregas C, ReVelle C S 1972 Optimal location under time or distance constraints. *Papers of the Regional Science Association* 28: 133–43

Veregin H 1995 Developing and testing of an error propagation model for GIS overlay operations. *International Journal of Geographical Information Systems* 9: 595–619

Wesolowsky G O, Love R F 1971 Location of facilities with rectangular distances among point and area destinations. *Naval Research Logistics Quarterly* 18: 83–90

PART 2

TECHNICAL ISSUES

Contents

(a) GIS architecture issues

Introduction   307–308
The Editors

21. New technology and GIS   309–316
    M Batty

22. GIS in networked environments   317–329
    D J Coleman

23. Desktop GIS software   331–345
    S Elshaw Thrall and G I Thrall

24. GIS interoperability   347–358
    M Sondheim, K Gardels, and K Buehler

25. GIS customisation   359–369
    D J Maguire

(b) Spatial databases

Introduction   371–372
The Editors

26. Relational databases and beyond   373–384
    M F Worboys

27. Spatial access methods   385–400
    P van Oosterom

28. Interacting with GIS   401–412
    M J Egenhofer and W Kuhn

29. Principles of spatial database analysis and design   413–424
    Y Bédard

(c) Technical aspects of GIS data collection

Introduction   425–426
The Editors

30. Spatial referencing and coordinate systems   427–436
    H Seeger

31. Encoding and validating data from maps and images   437–450
    I J Dowman

32. Digital remotely-sensed data and their characteristics   451–466
    M Barnsley

33. Using GPS for GIS data capture   467–476
    A F Lange and C Gilbert

(d) Data transformation and linkage

Introduction   477–479
The Editors

34. Spatial interpolation   481–492
    L Mitas and H Mitasova

35. Multi-criteria evaluation and GIS   493–502
    J R Eastman

36. Spatial tessellations   503–526
    B Boots

37. Spatial hydrography and landforms   527–542
    L Band

38. Intervisibility on terrains   543–556
    L De Floriani and P Magillo

39. Virtual environments and GIS   557–565
    J N Neves and A Câmara

40. The future of GIS and spatial analysis   567–580
    M F Goodchild and P A Longley

Introduction

THE EDITORS

We have already described how the real and absolute costs of computing have continued their precipitous fall of recent years, and how the attendant developments in computer graphics and visualisation have provided necessary (although not in themselves sufficient) conditions for putting GIS principles into practice. Yet this is only part of the story, and here in this Technical Issues part of the book we will explore changes in GIS architecture, issues of data collection and database management, and developments in the ways in which data are transformed and linked together. We begin with an extended discussion of the many ways in which GIS are configured, and the wide range of interactions with computers in general and GIS in particular.

In the opening chapter, Michael Batty provides a succinct yet wide-ranging review of the development of computer technology, in which he emphasises the unpredictability of change and our inability to anticipate it – a point we return to below with respect to the technical predictions made in the first edition of this book. He sees technology as also providing a stimulus to the proliferation of digital data sources, and the consequent increased richness of digital analysis. It seems inevitable that the pace of technical change will continue to accelerate, and that computation will continue to be a fast-changing and diversifying medium. Batty's own general predictions are that data-rich computing will soon be distributed across global networks, that decentralisation of software will bring changes in the way in which it is both licensed and used, that there will be rapid technical advance in voice input and output, and that there will be knock-on consequences for the development of new kinds of virtual realities.

The prospects for GIS are equally dramatic. The vendors of proprietary systems have constantly to adapt to change by packaging software and data into niche market solutions, by facilitating data exchange (particularly across networked environments), and by developing novel approaches to business solutions and consulting.

Software and information exchange is crucial to these technical developments. One of the most far-reaching (yet unanticipated – even in the first edition of this book!) changes in recent years has been the development of data and software transfer across the Internet. David Coleman describes in Chapter 22 how sophisticated network environments and browser systems have developed from the early distributed computing systems of local- and wide-area networks.

An accompanying development has been that as software converges across platforms, so geographical components are becoming more pervasive within general-purpose software – for example, spreadsheets now have GIS capability and vice versa. This is seen most clearly in the development of so-called 'desktop' GIS (Elshaw Thrall and Thrall, Chapter 23) which has come to embrace a range of general-purpose application and consumer mapping products, including digital atlases, digital gazetteers, geographically-enabled spreadsheets, and thematic mapping products. Such 'shrink-wrapped' products lack most, or even all, of the analytical capabilities of 'true GIS', yet are clearly geographical software products. In these circumstances, any distinct identity of GIS inevitably begins to blur. Taken together with the developments in networking, this suggests that GIS is becoming both more specialised in its range of possible applications, and more pervasive in its wider usage as a background technology. The broad picture is that software is both breaking up – fragmenting on the desktop so that users can construct all kinds of tailor-made applications from individual elements – and coming together – in that vendors are providing non-traditional functions within their traditional software, 'hooks', and linkages to other related software and openness to networked environments.

The implication of all of this is that software will emerge which is extremely basic to the computational environment and that programming languages will develop which enable this software to be connected in diverse ways. This is almost a full circle back to the early days of computer cartography and GIS when researchers wrote their own FORTRAN programs, although the elemental building blocks are very different. Now, there is the prospect of GIS users assembling reusable software 'modules' using the highest of high-level languages, in order to develop highly customised solutions to their problems. Already this is possible at a somewhat lower level using the various macro languages which are available within GIS. In the future, graphics, data elements, routing algorithms, modelling methods, and so on are likely to be packaged in whatever manner the user requires, using software which simply exists within appropriate environments – possibly based on networks rather than on desktops. The vendor response to this has been recognition of the need for vendor-neutral computing standards, and so-called 'open systems'. This is the theme of 'GIS interoperability' which is explored by Mark Sondheim, Kenn Gardels, and Kurt Buehler in Chapter 24.

There have been other sea changes in the organisation of the GIS industry. For example, the longest-established GIS companies, such as Intergraph, were initially primarily hardware developers, but are now more concerned with software and consultancy provision; while more recent entrants, such as ESRI, are increasingly involved with the development of desktop systems, network platforms, consulting, and data provision. Overlain onto this is the drive across the industry to develop new niche markets around reusable software modules. This is leading GIS vendors to work closely with their clients to develop customised system specifications, designs, and implementations, with or without the assistance of in-house consultancy services. The means and methods of GIS customisation in a wide range of applications are discussed here by David Maguire (Chapter 25).

This is the second edition of *Geographical Information Systems*, and it is perhaps appropriate to contrast the contemporary technological setting with the first edition's projections for 'GIS 2000' (Rhind et al 1991: 320–2). The first edition did, of course, draw attention to the changes in computer architectures consequent upon the growth of computer power – specifically, the shift away from mainframe and mini computers to PCs and workstations – but did not anticipate the development of the Internet. A second prediction was that the diffusion and wide uptake of GIS would likely lead to dramatic falls in real software

prices. While we might quibble whether the prediction of a 'fully functional GIS for about \$500 in the mid 1990s' was strictly speaking realised, each of the contributions to this section demonstrates that GIS has become a more affordable and routine technology. Indeed if our definition of GIS is drawn more loosely than Susan Elshaw Thrall and Grant Thrall's (Chapter 23) 'true GIS' (e.g. to include Microsoft and MapInfo GI products), the number of systems actually in use by the year 2000 now seems set to exceed the first edition prediction of 580 000. The range of application areas has continued to diversify, to the point at which it is perhaps more instructive to comment on areas in which GIS has not been used than to enumerate those in which it has: this is a theme that is picked up in the Applications part to this edition.

Which new technologies were not anticipated in the first edition? As suggested above, it failed entirely to predict the emergence of the Internet and the World Wide Web from its early civilian use in inter-university electronic mail. Second, there was at best only a hazy conception of the many ways in which GIS functionality would be packaged into specific 'shrink-wrapped' applications for the current generation of low-cost, application-specific desktop systems. Third, and related to this, there was still a sense in the first edition of even the most rudimentary GIS applications being the preserve of 'GIS specialists'. The book failed to anticipate the degree that this would be overcome by the development of vastly improved graphical user interfaces. Fourth, there was perhaps an over-emphasis upon the anticipated differentiation of proprietary products at the expense of the drive towards GIS interoperability, a development which has gained impetus with network-based data transfer and the use of the Internet as a GIS platform. Fifth and finally, these technological changes have led to organisational changes in the ways in which GIS is customised to specific applications, and the development of new application tools. These themes are all developed here in this section of the second edition.

Reference

Rhind D W, Goodchild M F, Maguire D J 1991 Epilogue. In Maguire D J, Goodchild, M F, Rhind D W (eds) *Geographical information systems: principles and applications.* Harlow, Longman/New York, John Wiley & Sons Inc. Vol. 2: 313–27

21

New technology and GIS

M BATTY

GIS is in the vanguard of a sea change in the meaning of technology. GIS technologies mainly comprise software rather than hardware but data and organisational design increasingly form part of this wider perspective. In this chapter, we trace the evolution of this technology, specifically focusing upon the emergence of desktop GIS, the development of GIS across networks, and the move to interfaces based on virtual reality systems which link 2-dimensional mapping to 3-dimensional visualisations. We conclude by noting that GIS itself, like many contemporary software technologies, is diffusing into the very infrastructure of the systems to which it is applied, thus posing radical opportunities for this technology to be central to the real-time management and organisation of postindustrial society.

1 ABOUT TECHNOLOGY

Our common conception of technology is still based on machines developed a century or more ago in a bygone age. Hofstadter (1985: 492) illustrates the point quite cogently when he says: '. . .we all have a holdover image from the Industrial Revolution that sees machines as clunky iron contraptions, gawkily moving under the power of some loudly chugging engine'. GIS technology however is very different in that increasingly it no longer refers to the material world – the computational hardware – but to software and its engineering, and more recently to its application in diverse contexts. GIS technology clearly involves software but it also includes the data which comprise applications and the kinds of organisation in which GIS is embedded. Hardware, software, 'dataware,' and 'orgware' thus comprise this technology but there is another spin to this technology in sight. As digital information comes to dominate more and more activities of everyday life, GIS is fast becoming part of the infrastructure of society itself. In this chapter, we will extend our image of this technology into the very infrastructure which is often the subject of GIS applications themselves, thus posing a conundrum of usage which is the hallmark of late twentieth-century technology. No longer hard but soft, no longer material but

ethereal: in this sense, GIS is at the vanguard of what technology will come to portray in the twenty-first century which is a very different conception from that on which modern science was founded.

Before we discuss the characteristic technology, it is also worth noting the longstanding distinction between science and technology and how this relates to GIS. Commentators on the history of science and technology have frequently made the point that technologies often develop quite independently from the science which they imply and they point to the age of steam which developed in parallel to the physics and mechanics which provides its theoretical justification. In one sense, the development of GIS mirrors the same kinds of disjunction in that although the software technologies depend upon computer hardwares which would never have been invented but for breakthroughs in the most esoteric of science, GIS software owes something to developments in computational science, particularly to database theory, but not much, so far, to geographic theory per se. In fact, as we run headlong towards postindustrial society, the traditional distinctions between science and technology seem increasingly irrelevant and in themselves may be only appropriate to past times. As Metropolis (1993: 121), one of the founders of atomic physics, has recently said it is in computer science '. . . that real scientific progress is nowadays to be found. Computer science . . . has changed and continues to change the face of the world more thoroughly and more drastically than did any of the great discoveries in theoretical physics'.

What has never been understood in any broad sense throughout the entire history of computing is that digital computation is universal. The inventors of the field such as Turing and von Neumann only barely perceived the point that when everything can be reduced to bits, everything can be subject to computation and therefore the technology will ultimately manifest itself in every form of media (Negroponte 1995). Over the last half century this point has been driven home by successive waves of hardware developments: miniaturisation enabled the earliest forms of numerical computation to be swamped first by text-based, then graphics-based applications, and then by multimedia systems and entire computer environments such as those based on virtual reality which now represent the cutting edge (Neves and Câmara, Chapter 39). Computing is characterised by many layers of activity, each developing and evolving different manifestations of digital technology. Mainframes gave way to minicomputers, thence to micros and workstations at one level while super and parallel computers have evolved at another (see Openshaw and Alvanides, Chapter 18). Handheld digital devices from cameras to palmtop organisers are emerging at the grass roots embodying much of the software such as spreadsheets and even simple GIS which were evolved at higher levels. In another dimension, computing and telecommunications have rapidly converged, first from the need to interact directly with machines over distance, but more recently, with computation itself becoming embodied within networks themselves.

For GIS, which essentially began as part of the 1980s wave of graphical computing, the key technological issue has until recently revolved around miniaturisation. Ever faster graphics processing has enabled GIS software to move from remote systems to the desktop. Only a decade ago, this technology was essentially remote in that users, although networked, largely used such technology in non-interactive or non-immediate form. Miniaturisation has enabled GIS computing to move to the more personal domain, from collectively- to individually-based computation (Batty 1995). Current changes in these technologies are

emphasising applications across the Internet with dramatic consequences for all aspects of GIS computing. And last but not least, GIS technology is beginning to leak into the very infrastructure which GIS seeks to understand and describe, as the technology begins to find uses in much more routine applications, such as in-vehicle navigation, the delivery of local services, gazetteers, real estate use, and so on (Graham and Marvin 1996).

The rest of this chapter will review four key aspects of this technology. We will sketch the evolution of GIS through the hardware and software revolutions, and then describe the move to the desktop. This represents the state-of-the-art, but the cutting edge is very different. The move of these technologies to the Internet, and the idea of urban infrastructure embodying GIS represent dramatic changes in the nature of these technologies and we have only just begun to see the prospects ahead. In the previous edition of this book (published in 1991), there was hardly any mention of networks (see Coleman, Chapter 22), and certainly no sense that GIS was part of the very infrastructure it is designed to understand and monitor (see also Waters, Chapters 59). Desktop GIS (see Elshaw Thrall and Thrall, Chapter 23) had not come of age, for the state-of-the-art was based on workstation applications and the technology was much less diffuse. Needless to say, we can barely anticipate what is in store in the next decade although real-time applications with real-time data inputs are likely to dominate the future, as the gap between actual systems and the software used to understand them narrows. Before we speculate on the future, however, we must trace the evolution of GIS through its hardware and software revolutions.

2 GIS THROUGH THE HARDWARE AND SOFTWARE REVOLUTIONS

Inevitably, any general discussion of GIS requires some definition. Here we see the technology as embodying the representation of geographical phenomena in digital form where the data can be visualised in at least two different ways or 'views'. The traditional distinction is between cartographic visualisation, and the manipulation of spatial data in a form that exploits its spatial nature. Cartographic visualisation by itself defines computer cartography while spatial data analysis per se is rarely considered as GIS. Normally it is the intersection of these different forms of analysis which defines GIS as software with the ability to look at spatial data through two or more views (see also Longley et al, Chapter 1). In its early history, computer cartography was somewhat separate from spatial data analysis but towards the end of the mainframe era of computing, various efforts were made (particularly at the Harvard Laboratory for Computer Graphics, and at the Royal College of Art) to integrate the visual with the mathematical analysis of spatial data (Chrisman 1988). The logic of this was based on the notion that spatial data have their own unique form which is essential to the analysis of deep geographical structure but is also essential to its 'correct' visualisation and mapping.

It took a long time to put these ideas together, but during the 1970s work on spatial data analysis slowly proceeded, drawing on developments in spatial statistics and emergent database theory. Developments in spatial analysis in this area were orientated towards geometric operations involving points, lines and polygons, interpolation, contouring, and point-in-polygon algorithms such as those found elsewhere in this book (e.g. Dowman, Chapter 31; Martin, Chapter 6). It was not until computer hardware really got to grips with graphics in the 1980s that GIS came to be defined as it is conceived in this book. Data analysis of the geometric attributes of space was independent of its visualisation, which in the 1960s and 1970s was through computer cartography and computer draughting. Almost as soon as computers were developed in the 1940s, graphics were invoked, first as a by-product of monitoring the working of the machine using oscilloscopes, but mainly in the 1950s and 1960s using line plotters where the computer drawing was carried out off-line from the computation itself. What marked the late 1960s in the emergent field of GIS was the use of the normal output device attached to mainframe and minicomputers – the line printer – for mapping in which the resolution of the device was the size of the characters themselves. Some remarkably good thematic maps were displayed in this form using ingenious forms of overprinting but true GIS had to await the development of the more personalised technologies of the 1980s.

The graphics technology which was cutting edge in the 1970s was a development of the oscilloscope – the vector graphics device – in which each line, and

each point in any filled area of the drawing was swept out on the light sensitive surface of the scope – the phosphor – by an electron beam. This technology was not used very much for computer cartography, where the emphasis was upon the kind of precision more akin to line plotting, but it was widely used in scientific visualisation and it was thus a surprise when the technology changed so dramatically after the invention of the microprocessor. Miniaturisation began almost as soon as computers were conceived with the invention of the transistor at Bell Laboratories in 1948, the integrated circuit in 1957, and the microprocessor or 'computer on a chip' in 1971 at Intel. Many anticipated that computers would get ever smaller but few realised that graphic images could be directly associated with computer memory or that computers would pervade society in the way they have done. In fact, each great wave of computing has been unforeseen by the establishment. The remote computing industry represented by IBM and DEC did not foresee personal computing (indeed they positively resisted it), while none foresaw the amazing rise of the Internet, network computing and the drift of computation into the ether. In the 1980s, it was cheap computer memory that made better graphics achievable and this led directly to better graphics interfaces as well as improved image processing, of which GIS was a direct beneficiary. The desktop environment has followed as personal computers have become ever more powerful and as workstations have moved to the desktop.

The 1980s was the time when computer cartography finally merged with spatial data analysis. Early versions of proprietary software such as ARC/INFO did make use of plotter devices for visual outputs, but it was the advent of cheap raster graphics devices in the early 1980s that made interactive use a possibility. GIS then began to take off. In terms of data input, digitisers gradually came into use, although many cartographic data were input through the keyboard. With the advent of raster graphics, scanners have become widely used, as have remotely-sensed images. However, the current distinction between vector- and raster-based GIS does not reflect the earlier distinction between vector and graphics devices. Most GIS now deal with data which can be represented in either raster or vector form, but whether raster or vector, the usual way that data are visualised is on a raster device. This distinction is beginning to blur as new algorithms are being developed to translate raster into vector and vice versa, although the distinction remains, largely because of the type of data and the way they are input rather than the way they are output (see also Martin, Chapter 6).

GIS software however remained quite rudimentary during this period. In essence, operations on cartographic data represented most of the functionality of the software. The most significant functions were those which pertained to geometry with the concept of layers and overlay central to the operations known as spatial modelling. Models pertaining to the attributes of the geographical system being represented remained separate from GIS, for its logic is related to spatial geometry rather than to any system model based on the functioning of those attributes. This is a tension within the field. GIS is a generic technology and thus its functions must be applicable to different types of system. Modelling, analysis, and design usually pertain to specific systems to which GIS must be applied and insofar as this technology can embrace the relevant functionality pertaining to different systems of interest, it is through connecting the technology to other software (Batty and Xie 1994). Programming of GIS software systems has reflected the development of programming per se with early versions being in FORTRAN or more structured languages such as Pascal, later ones in C and its variants. Of late, new systems such as Smallworld make use of object-orientation (OO: see Worboys, Chapter 26) while recent desktop packages such as ArcView embody OO scripting languages which enable their interfaces to be customised and some new functionality to be programmed in.

3 THE MOVE TO THE DESKTOP

The state-of-the-art in GIS technology when the first edition of this book was produced was workstation GIS, usually networked in client-server or file-server mode, where spatial data was usually viewed using raster graphics devices but with minimal customisation in terms of graphical interfaces. Windows systems were first introduced on workstations in the mid 1980s based on MIT's X standard although the Apple Macintosh broke the mould in personal computing around the same time in its adoption of the interface developed at Xerox-Parc. However, GIS technologies still tended to output graphics either in a single window – virtually

offline from the main processing – or for final print on line plotters. There was some ability to customise the interface (see Maguire, Chapter 25), and a typical plot frame using the ArcPlot window from ARC/INFO (circa 1991) is illustrated in Plate 15 (Batty 1994). Some customisation of the output was possible using rudimentary drawing functions in the package through the Arc Macro Language (AML).

Desktop GIS (see Elshaw Thrall and Thrall, Chapter 23) emerged rapidly following the introduction of Windows 3.1 as the replacement to DOS in the late 1980s. Prior to that, desktop GIS such as PC-ARC/INFO was usually based on stripped-down versions of workstation or even mainframe/mini packages or rather crude graphics interfaces built on top of DOS such as early versions of the educational raster-based GIS Idrisi (see Eastman, Chapter 35 for more recent use of this software). The main feature of desktop GIS is in fact integrated functionality based on the WYSIWYG principle (What You See Is What You Get). This will be illustrated with the package MapInfo which began life as a desktop mapping package and has rapidly moved 'upscale' in acquiring ever more functionality, thus now approaching the capability of workstation GIS but with everything on the desktop. Figure 1 shows a thematic map output from a simple GIS based on the 32 boroughs comprising Greater London. On the screen are the three main windows or views which MapInfo offers – the map window, the browser which is the table containing the attribute data for each borough, and a graph of data which in this case are 1991 populations. Also shown are two buttonpads which provide the user with short cuts to the various functions within the package, and to basic drawing tools which can be used for editing the map geometry and for presentation. Like most desktop GIS, MapInfo has a structured query function (see Egenhofer and Kuhn, Chapter 28) which enables the user to search for various combinations of attributes and to highlight these as spatial objects in any of the open windows (through hotlinking) and as a short cut to information in the browser stored as an information function on the main buttonpad and activated using point and click. In MapInfo, apart from the usual notion of structuring the data in layers and seeing the data as map layers, there are the usual thematic mapping capabilities which can be used to combine and derive new layers. Unlike Idrisi or ArcView, there is no elaborate overlay capability as such, thus betraying the program's origins in thematic mapping. But like all desktop GIS, newer releases are incorporating ever more sophisticated functions only found hitherto in workstation GIS.

Other developments involve workstation GIS acquiring desktop features with new user-friendly front-ends being developed (see also Schiffer, Chapter 52). ArcView was originally intended to be such a front-end to ARC/INFO although the package now exists in its own right. Desktop GIS is now open to other kinds of software through dynamic data and other exchange mechanisms, and through the addition of programming languages within the GIS which opens it to other software on the same machine or network. Furthermore, what were once very different software packages are beginning to converge. Spreadsheets are adding map capability and it is only a matter of time before wordprocessing packages add GIS-like functions –

drawing and animation functions are already a part of such software. But perhaps the most interesting development is the further move down the hardware hierarchy to putting GIS on handheld devices such as personal digital assistants like the Apple Newton. The package Local Expert, essentially GIS software for navigating or streetfinding in different cities for use by tourists or local residents, is an example. Local Expert, typical output from which is illustrated in Figure 2 for London, enables the user to find information on restaurants, places of historic interest, and so on using a particularly bulletproof user interface. Much of its functionality is similar to that found in more professional desktop GIS, although the emphasis at this level is upon structured query language (SQL)-type interrogation, route finding and the measurement of shortest paths. Nevertheless, it is clear that by the time the next edition of this book is written, this kind of application will be widespread.

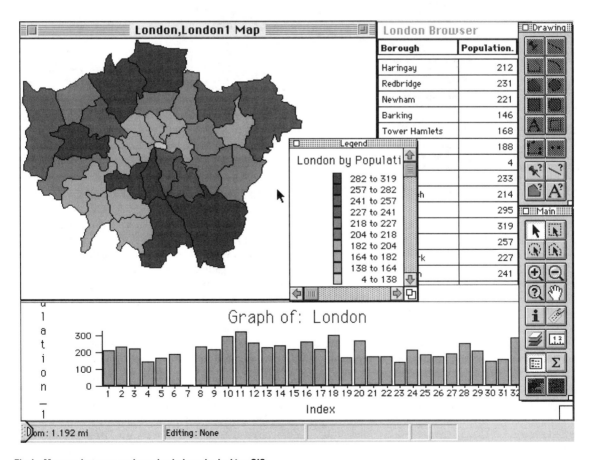

Fig 1. Mapper, browser, and graph windows in desktop GIS.

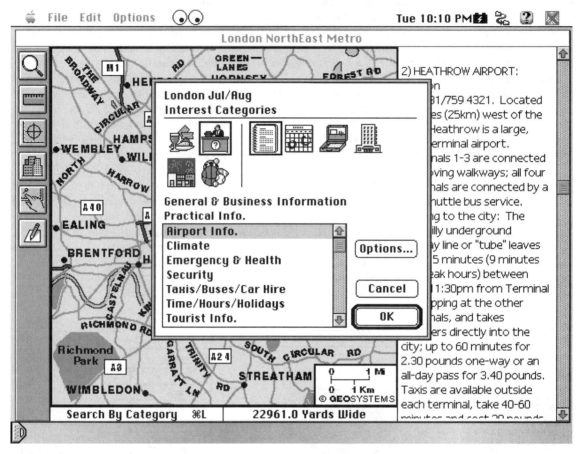

Fig 2. Information and street finders based on low cost GIS.

4 THE MOVE TO THE INTERNET

The convergence of computers and telecommunications was anticipated as a means of linking computers and their users remotely, but what was never anticipated was the extent to which computation itself would fuse with communications (see also Coleman, Chapter 22). The movement of software to the World Wide Web (WWW), for example, is a startling and unexpected development which was deemed unlikely even five years ago. At one level, what is currently happening is simply an extension of the opening of software on the desktop to other software across the Internet. With the costs of memory still dropping fast, and new memory still soaking up ever more graphics, this trend has migrated to networks. The fact that the Internet has been effectively free to users in educational environments, and low cost to others, has fostered a

massive increase in shareware of a very high quality. What is amazing is the high quality 3-dimensional and multimedia software which now exists in the public domain and this is threatening quite well-established vendors of proprietary software. Three-dimensional graphics is being particularly taxed by these developments with the advent of virtual reality software over the Internet, but GIS shareware is on the horizon as a variety of developers is beginning to develop graphical interfaces to locational data in the form of virtual reality interfaces such as animated maps.

These developments are so fundamental that hardware itself is beginning to change. Several hardware-software companies have announced the 'NC' or network computer which will function with no more software than a web browser and some communications protocols (e.g. see Dale and McLaren, Chapter 61; Sugarbaker, Chapter 43). The implication is that all the software ever required will be available on the Internet to be downloaded to the user or leased in some form. What goes for software also goes for data. Archives of data in the public domain are appearing such as the US Population Census archive at the Lawrence Berkeley Laboratory (*http://cedr.lbl.gov/cdrom/lookup/*) and already there are applications which take such public domain data and explore it using appropriate GIS shareware. Animated maps are appearing. For example, the BigBook 3D browser is an experimental animated map query system to the Yellow Pages which will display 3-dimensional visualisations of city blocks from queries made in terms of the 2-dimensional map (*http://vrml.bigbook.com/bb3d/*).

These developments in network computing when coupled with developments in 3-dimensional computing and multimedia, specifically in virtual reality systems, provide dramatic possibilities for GIS technologies. Powerful 3-dimensional and animation software coupled with digital video provides the raw material to creating 3-dimensional interfaces to the kinds of 2-dimensional databases which are central to GIS. At University College London, a group has created a virtual interface to an information system of buildings in the college. This system was created for programming and making inventories of room use. An animated interface was constructed as an interactive Quick Time movie based on sequences of digital video of college buildings. This was made operational within Hypercard. The essence of the Hypercard stack is the use of hotlinks to access other software such as various types of GIS used to portray the data. Once the system had been constructed, the various links within the software were exploited, first to open the system to other software on the desktop. For example, the local software was linked to electronic atlases as well as more local map software. An example of the interface is illustrated in Plate 16, where the AUTOCAD model of the college gives the entry points to the video which accesses the building information system from which an electronic atlas and other GIS can be accessed.

However, what can exist on the desktop can also exist on the net, and the system has now been extended to the web where the Virtual Reality Markup/Modelling Language (VRML) has been used to provide an interactive model of the college based on the AUTOCAD model. If you have the correct version of the Netscape browser and plugins such as 3D Live, then you can use the model to access data

and other WWW pages associated with departments in the college from anywhere in the world. If you log onto *http://www.ge.ucl.ac.uk/* then there are several options to run the interface under QT, VRML, and other graphics interfaces. With the most basic of browsers, some animation is possible. An illustration of the VRML model is given in Plate 17 where the façades of the college buildings have been texture mapped. The intention is to extend this kind of interface to a much bigger model of Central London and the City to provide a virtual interface to GIS data.

Finally, the Internet introduces yet another dimension to GIS. Most software is written for individual users. Applications may involve some team work or group problem-solving, but the notion of using software collectively is a very recent idea in GIS at least. Networks provide the means for such interactive use. In the design sciences and the arts, the idea of the virtual design studio is well advanced where users engage in group design using common software but also use common repositories and archives for various solutions or designs (Day 1994; Wojtowicz 1995). The idea of the digital pinboard is central to such developments. The same kinds of use might be envisaged for GIS at the level of the construction of a GIS, its use in analysis, and where appropriate its use in problem solving, design, and decision making. The use of GIS in this context is encapsulated in spatial decision support systems and their extension in a collaborative mode is under way (Batty and Densham 1996; Densham 1991). The immediate future is likely to see substantial developments of these ideas especially where the hardware is an NC, and the software and data are available over the Internet.

5 GIS AS URBAN INFRASTRUCTURE

The brief history of GIS has involved a rapid broadening of the technology, from hardware and software to data and organisational design, and from individual to collective use across networks. The diffusion of the technology has also changed its nature, and this diffusion has now reached the point where many routine functions which make society work on a day-to-day basis are being endowed with some GIS functionality. This is what we mean by GIS leaking out into the social infrastructure in parallel to many other software technologies. It is the passage of GIS from professional use to lower level routine use that changes this balance. In one sense, all science is part of the infrastructure although it forms such a small proportion of everyday activity and is concentrated in such a small elite that the notion of science being affected by the very activity of science itself still remains an alien one. However, the idea that GIS is being affected by the wider all-pervasive activity of GIS is clearer. Networking is central to this in that the kinds of inputs and outputs to GIS are beginning to be available in real-time.

In Figure 3, we show how GIS is becoming part of the infrastructure through the kinds of links which are being formed between hitherto disparate activities. The real-time delivery of data makes real-time GIS a feasible proposition for appropriate uses such as traffic management, emergency services, even just-in-time geodemographics for marketing. Very basic uses such as navigation and information finding within cities by individuals is also on the horizon, as the data used by handheld devices can be updated through wireless technology. In fact, the entire continuum of GIS usage from the most immediate and the most routine to the most remote and long term is being integrated though network and wireless technologies. All kind of devices and networks – cable, TV, telephone, computer – are merging and this is making software technologies in any one of these accessible and relevant to the others (Graham and Marvin 1996).

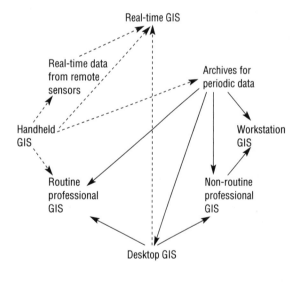

Fig 3. The evolution of GIS technologies.

A major impact of new computer technologies on cities is the fact that patterns of use, location, and movement are beginning to change. In a global economy, patterns of accessibility in cities are becoming very different from those on which much geographical theory and some GIS applications are based. GIS itself is fuelling these changes as the software is being used for all kinds of navigation, for the delivery of local services, for welfare provision, for real estate acquisition, and so on. Electronic commerce and geodemographics in business is a direct user of GIS, which in turn is changing the data for GIS and even the functions which are embodied within GIS. Such immediacy from applications has for long characterised those technologies of the industrial age, but a new mindset is required to understand similar kinds of change associated with software technologies. This kind of complexity marks out the postindustrial world from any hitherto, and in extending technology in this fashion to embrace the city, society and economy, it heralds a redefinition of the meaning of technology itself.

References

Batty M 1994 Using GIS for visual simulation modeling. *GIS World* 7: 46–8

Batty M 1995 Planning support systems and the new logic of computation. *Regional Development Dialogue* 16: 1–17

Batty M, Densham P 1996 Decision support. GIS and urban planning. *Sistema Terra* 5: 72–6

Batty M, Xie Y 1994c Urban analysis in a GIS environment: population density modelling using ARC/INFO. In Fotheringham A S, Rogerson P A (eds) *Spatial analysis and GIS*. London, Taylor and Francis: 189–219

Chrisman N R 1988 The risk of software innovation: a case study of the Harvard lab. *American Cartographer* 15: 291–300

Day A 1994 From map to model. *Design Studies* 15: 366–84

Densham P J 1991 Spatial decision support systems. In Maguire D J, Goodchild M F, Rhind D W (eds) *Geographical information systems: principles and applications*. Harlow, Longman/New York, John Wiley & Sons Inc. Vol. 1: 403–12

Graham S, Marvin S 1996 *Telecommunications and the city: electronic spaces, urban places*. London, Routledge

Hofstadter D R 1985 *Metamagical themas: quests for the essence of mind and pattern*. New York, Basic Books

Metropolis N 1993 The age of computing. In Metropolis N, Rota G C (eds) *A new era of computation*. Cambridge (USA), MIT Press: 119–30

Negroponte N 1995 *Being digital*. New York, Basic Books

Wojtowicz J (ed) 1995 *Virtual design studio*. Hong Kong, Hong Kong University Press

22

Geographical information systems in networked environments

D J COLEMAN

This chapter examines basic approaches to distributed architectures in GIS, with special emphasis on exploitation of local- and wide-area networks and, more recently, the Internet and wireless communications. The first section reviews the design models inherent in host-based systems, distributed networks, and field-based systems. The second section describes selected applications of these design models, with particular emphasis placed on Internet-based geographical information retrieval and computer-supported cooperative work. The final section deals with system performance considerations in networked environments and describes new developments affecting the performance of future systems.

1 INTRODUCTION

Technical design models supporting mainstream GIS technology have evolved from early host-based efforts on mainframe and mini-computers, through stand-alone systems operating on personal computers and workstations, and on to today's distributed computing environments across local- and wide-area networks. Each advance has extended overall flexibility in terms of the relative location of users, processing capabilities, and data storage units (see also Batty, Chapter 21). Recent advances in broadband and wireless communications technologies – as well as the dramatic increase in Internet usage and extension of Internet browsing technology – promise to extend further the reach and range of GIS users working in offices or laboratories, in the field or at home.

This chapter examines the basic approaches to distributed architectures in GIS, with special emphasis on the exploitation of local- and wide-area networks and, more recently, the Internet and wireless communications. The first section of the chapter reviews the design models inherent in host-based systems, distributed networks, and field-based systems. The second section describes specific applications of the latest design models, with particular emphasis placed on emerging issues

associated with Internet-based geographical information retrieval and computer-supported cooperative work. The final section of the chapter deals with the issue of system performance in networked environments under a variety of conditions. After a brief summary of considerations involved in objectively assessing system performance, the author describes new developments affecting the performance of future systems.

2 HOST-BASED SYSTEMS AND EARLY PERSONAL COMPUTING ENVIRONMENTS

2.1 Host-based systems

Early centralised computing environments were characterised by small numbers of large-scale mini- or mainframe computers, with shared storage devices attached via hardware input/output channels and multiple users connected via terminals possessing varying levels of on-board 'intelligence' or processing power (Katz 1991). Such environments defined the architecture for most major data processing and information systems applications until the mid 1980s, including the GIS installations found in major forestry organisations, utilities, municipalities, and land records management programmes.

This architecture implied greater control over data integrity and system security; the database was managed centrally, with responsibility for system and data administration entrusted to experienced data processing specialists. However, performance of such systems would often degrade in unpredictable ways when growing numbers of users demanded computing resources and database access. Conflicts with system administrators over development and maintenance priorities also often resulted in dissatisfaction among endusers in many large organisations.

2.2 Stand-alone, PC-based systems

By 1986, PC-based GIS software packages had begun moving geoprocessing out of the hands of information system managers (Miller 1990). Besides their low cost, these systems offered more predictable response times since the user was the only one on the system. However, the proliferation of stand-alone PC-based systems meant it was much more difficult to share data among several different people in the organisation. Also, the enduser often had to become his or her own system and database administrator. While PC-based systems undoubtedly accounted for the dramatic growth in GIS usage through the late 1980s, they also put greater onus on managers in large organisations to keep effective track of the data being collected and processed by an increasing number of endusers with little experience in routine data management procedures.

3 COMPUTER NETWORKING AND DISTRIBUTED SYSTEMS

Beginning in the mid 1980s, higher-performance workstations connected through local and wide area networks (LANs and WANs) became a viable alternative to host-based and stand-alone configurations. Connectionless LAN and 'LAN-interconnect' services began displacing earlier connection-oriented services which required a dedicated link between the enduser and the host computer (Figure 1).

Based on packet-switching transmission protocols which did not require dedicated connections between user and host, networking permitted users to share access to scarce equipment resources (e.g. printers, plotters, databases etc.), made possible inter-site communication applications like electronic

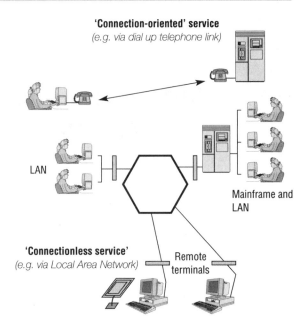

'Connection-oriented' service
(e.g. via dial up telephone link)

LAN

Mainframe and LAN

'Connectionless service'
(e.g. via Local Area Network)

Remote terminals

Fig 1. Examples of alternative wide-area networking services: 'connection-oriented' vs 'connectionless' models.

mail and file transfer, and enabled distributed processing. It also allowed users to expand their facilities with a degree of vendor independence and to incorporate special purpose processors, storage units, or input/output devices as required. The term *distributed computing* was coined to describe a situation where processing tasks and data are distributed among separate hardware components connected by a network, with all these various components capable of being accessed in a relatively transparent manner (Champine et al 1980).

Distributed systems today generally offer a combination of: (1) greater access to GIS data stored and managed on a central server; (2) faster response time due to local computing; (3) tighter system security; (4) less complexity; and (5) in many cases, lower-cost computing solutions than more traditional mainframe or minicomputer solutions. At a higher level, the notion of distributed computing often provides a better fit to the complex structures and often multidisciplinary nature of modern organisations and offers greater user involvement in information management activities.

Early investigators of formal network-based approaches to distributed GIS database management systems (DBMS: e.g. Ezigbalike et al 1987; Webster 1988) laid out excellent foundations for future research, but were limited by both the distributed DBMS technology then commercially available at reasonable cost, and the relatively limited use of LANs or WANs by the GIS community at that time. The subject has been revisited more recently by Ingoldsby (1991) and Laurini (1994), among others.

3.1 The client–server model

Most distributed computing today is based on a client–server architecture (Katz 1991). In this model, a collection of workstations (or *clients*) relies on one or more *servers* residing elsewhere on the network for access to data files, application software and, in certain cases, more powerful computing resources. Such servers are really high-volume storage devices with processors which have been optimised to provide high-speed retrieval of large disk-based data or database files. In such an environment, the data retrieval aspects of a database query can be carried out largely independently of the data processing and display tasks.

The time required to execute such a data retrieval operation depends on the individual performance of the storage, processing, and communications components involved in the client–server system. While improvements in all three of these technologies have made key contributions, it has been the introduction of remote file management services which finally provided the transparency required for distributed computing. Remote file system implementations like Sun Microsystems' Network File System (or NFS) on UNIX workstations, PC-NFS and Novell's Netware on PC-compatibles, and Appletalk on Macintoshes are now present in many organisations.

3.2 Network architecture

A computer network architecture can be considered as a set of functions, interfaces, and protocols which enables devices to communicate with one another on-line. The architecture is composed of a layered collection of communication, networking, and application functions implemented such that – while each layer is designed to operate independently – higher-level operations are built on functions provided by the lower layers (Chorafas 1980).

Network protocols are the formal sets of rules or specifications for coding messages exchanged between two communication processes on a network (Voelcker 1986). Protocols govern data control and format across a network, and a variety of protocols exist to ensure that these communications are conducted effectively.

Two different stacks or suites of layered approaches are in common use today:

- OSI: the seven-layer Open System Interconnect suite of protocols developed by the International Standards Organisation;
- TCP/IP: the four-layer Transmission Control Protocol/Internet Protocol suite originally developed for the ARPAnet research network in the USA.

Originally developed by computer user groups and European telephone companies, the OSI model helped unify world telephony and provided a clear framework and explanation of the functions required for computer communications. However, it was regarded by some as being too cumbersome for high-speed networks (Wittie 1991).

By comparison, the TCP/IP suite of protocols became a *de facto* standard by the early 1980s as a result of its early use in the implementation of the US Defense Department-funded Internet. Its longer-term popularity was secured through subsequent bundling with the 1983 release of the Berkeley UNIX 4.2 operating system. While TCP/IP protocols do not precisely fit into the more general OSI model, the functions performed by each OSI layer correspond to the functions of each part of the TCP/IP protocol suite and provide a good framework for visualising the respective relationships between the various protocols.

3.3 Local area networks

LANs physically or logically connect together multiple workstations, terminals, and peripheral devices via a single cable or shared medium (Pretty 1992). Through the 1980s alone, over 100 000 LANs were set up in offices and laboratories around the world to link workstations to printers, share files and send electronic mail (Wittie 1991). LAN usage in the general computing community grew at a rate of 80 per cent per year between 1985 and 1991 (Pretty 1992), with networks extending into schools, libraries, laboratories, and offices around the developed world using telephone lines, optical fibres, and satellite links.

319

Several accepted and standardised types of LAN technology now share the market, including IEEE 802.3 (CSMA/CD or Ethernet), IEEE 802.4 (Token Bus), IEEE 802.5 (Token Ring), and ANSI FDDI (fibre distributed data interface). To date, the Ethernet (Figure 2) and Token Ring technologies have dominated the market.

3.4 Metropolitan and wide area networks

Until the early 1990s, most wide-area networking and LAN-interconnect services did not deliver the performance required either to move large GIS data files quickly or to offer real-time access to remote users over very long distances (Craig et al 1991). Where data delivery was an issue, magnetic tapes and diskettes were viewed as the media of choice for the distribution or exchange of data among GIS data producers and users (Newton et al 1992).

Modern metropolitan area networks (MANs) and WANs employ different protocols and technology in order to offer speeds comparable to LANs while operating over greater distances. By 1992, fibre-based packet switching services like SMDS in North America and FASTPAC in Australia began providing high-speed (34–45 Mbit/sec) links between LANs across and between major cities. Today,

dedicated FDDI networks connect users within limited areas at rates up to 100 Mbit/sec, and higher-speed asynchronous transfer mode (ATM) services promise to support a wide range of real-time multi-media applications across long distances.

Results of the performance testing experiments described by Coleman (1994) indicated that future users of these broadband services would be able to access remote disks, processors, and output devices at near-LAN levels of performance. Today, such services have already influenced the nature of data delivery and software support practices to a small but growing group of users in the GIS community (e.g. Annitto and Patterson 1995; Streb 1995).

3.5 The Internet

Especially since 1993, it has been the connectionless services built atop the Internet which have come to define the current paradigms for mass market network usage. The Internet is a collection of interconnected campus, local, provincial, national, and corporate networks in more than 150 countries. Originally designed to connect researchers in university, government, and industrial defence establishments, the Internet is now estimated to

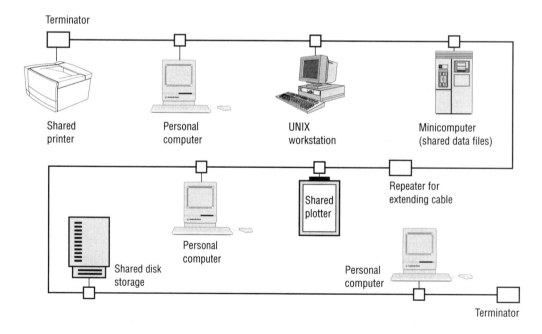

Fig 2. Simplified Ethernet local area network.

service more than 20 million users through a loose collection of more than 5000 registered networks (Reinhardt 1994; Thoen 1995). While electronic mail, file transfer, and remote login services originally accounted for the majority of Internet traffic, it has been ease-of-use and the multimedia capabilities of the World Wide Web (WWW) that have attracted most of the mass-market attention and resultant increases in usage. Over 145 000 Web sites have appeared since its introduction in 1993, and this number is continuing to rise monthly (Webcrawler Survey 1996).

3.6 Wireless, field-based systems

Recent advances in wireless communications, notebook computing, and the integration of GIS and global positioning system (GPS) technology have extended the application of traditional GIS databases into field operations (Lange and Gilbert, Chapter 33). Using normal telephone lines and, more recently, cellular telephone modems, field users tying into enterprise-wide networks may now access information resources previously available only from within the branch office or headquarters. New implementations are extending functionality beyond traditional records management and modelling roles into such areas as property inspection, field updating, facilities maintenance, customer service, and emergency response (Elliot 1994).

At least one industry source predicts the mobile computer/mobile field automation market will grow from $30 billion in 1995 to $80 billion by the year 2001 (FieldWorks Inc. 1996). FieldWorks estimates the 1995 total market for 'ruggedised' portable computers at $500 million with an annual growth rate of 35 per cent.

This predicted growth in the ruggedised portable market is based on current trends within large service firms and government organisations to automate the data collection and communication capabilities of field personnel through greater use of reliable mobile computers. As growth in notebook over desktop sales suggests, purchasers want systems which combine the functionality of desktops with the ability to fulfil specific applications beyond word processing and spreadsheet manipulation. The integration and use of GIS software and GPS technology within mobile computers for field data collection and database updating is one example of such extended functionality.

4 APPLYING THESE NEW MODELS

4.1 Enterprise computing implementations

Large utilities and municipalities were among the first to identify the operational requirements to integrate smaller GIS- and facilities management-related databases on stand-alone systems with larger corporate databases residing on mainframes (Popko 1988). As LANs became more widespread, enterprise computer systems evolved from host-based (or 'single-tier') to two-tier environments with networked PCs and workstations in a client-server architecture replacing connected terminals. This modular approach is attractive to many organisations since it allows them to take quicker advantage of price/performance improvements on modular hardware components and the other advantages of network computing described earlier (Mimno 1996).

While advantageous in many respects, users have found that these two-tier environments may not necessarily retain some of the advantages found in centralised host-based implementations, notably availability, expandability, and reliability of service (Strand 1995). As a result, newer three-tier environments have been developed to facilitate the placement of applications and data in locations in order to optimise these three factors (see Figure 3).

Existing mainframes at Tier 3 – possibly connected over long distances via MANs and WANs – are used to support legacy applications and provide access to large databases. To implement a client/server computing strategy, additional layers of hardware and software are added as a front end to the host computer. These layers consist of shared servers at Tier 2 (interconnected by high-speed LANs), and LAN-based PCs and laptop computers at Tier 1 (Mimno 1996). GIS and desktop mapping applications are included at this first tier, although the data may reside on either the Tier 2 servers or even the Tier 3 hosts.

This enterprise information architecture is highly flexible and can be modified easily to accommodate changes in business requirements. As more large customers adopt this approach, commercial GIS software firms and mainstream DBMS vendors alike are modifying their offerings in response to customer demands. In particular, spatial data management models and processes are becoming a more intrinsic component of an organisation's larger information management architecture. ESRI's Spatial Database Engine and MapInfo's 1996

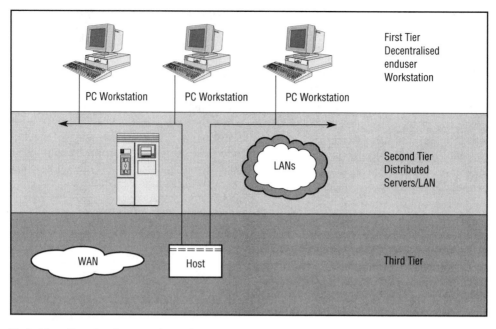

Fig 3. Three-tier enterprise computing environments.
Source: Strand 1995

acquisition of the Spatialware server technology from Unisys Corporation are two examples of how GIS vendors are providing even tighter high-performance links into mainstream relational DBMS packages like ORACLE and others, while based on *de facto* industry standards for object linking and embedding (OLE), structured query language (SQL), third- and fourth-generation languages, and ODBC. Table 1 contains a summary of recent industry offerings in this regard.

At the same time, the international database standards community (through ISO/IEC/JTC1/SC21 – WG3) is extending the proposed SQL MM ('MultiMedia') to accommodate spatial relationships, indexing arrangements, and operators typically

found in GIS packages. Mainstream commercial database vendors are already incorporating some such enhancements into their own RDBMS offerings, with ORACLE's SDO and Informix/Illustra's Spatial DataBlades offerings being two such examples. These developments, combined with the recent formal attention being paid to interoperability between software packages (Sondheim et al, Chapter 24), suggest that future enterprise implementations may rely less on 'full-solution' packages maintained by a single vendor or consortium of vendors. Rather, large institutional users may opt for a series of smaller, lower-cost software components or applications possessing limited functionality on their own, but which seamlessly interact with other database, word processing or advanced modelling applications which may be present in those organisations (see Elshaw Thrall and Thrall, Chapter 23).

To make optimum use of this flexibility, application developers must have the ability to allocate data and processes anywhere in the multi-tier enterprise information architecture – and also reallocate them as operational requirements change. At the time of writing, concerns have been expressed that first-generation client/server application development tools like Visual Basic,

Table 1 Tighter linkages between GIS and RDBMS: 'Spatial Middleware' (from Costello 1996).

Vendor	Client Software product	Middleware Product
Vision Intergraph Corporation	AutoCAD MicroStation	Vision Server Jupiter
ESRI	ArcView MapObjects	Spatial Database Engine (SDE)
MapInfo Corporation	MapInfo	SpatialWare

PowerBuilder, and others may not have the capabilities and performance required to support more complex, enterprise-wide applications adequately (Mimno 1996). While it is likely that newer versions of these and similar tools will address many of these concerns, they must still evolve as the needs and practices of different user groups in these large organisations change over time.

4.2 Data access and delivery

Dozens of major land information and resource inventory programs around the world have demonstrated on-line organisation and electronic distribution of government imagery, mapping, and related land information across proprietary networks. Early examples included Land Information Alberta (McKay 1994) and the Manitoba Land Information Utility (Oswald 1994) projects in Canada, the early commercial ImageNet service in the United States, and the Land Ownership and Tenure System (LOTS) in South Australia (Sedunary 1988). Numerous other early examples may also be found elsewhere in these countries and in Europe.

Originally, such projects offered connection-oriented access to secure databases utilising either direct LAN-interconnect services or high-speed modems. Especially since late 1993, however, the Internet-based WWW has emerged as an alternative means of accessing, viewing, and distributing spatial information. Used in combination with modern Web 'browsing' software packages like Netscape and Microsoft's Internet Explorer, the WWW is emerging as a mainstream tool for

- the distribution of public-domain spatial information;
- online ordering of commercial datasets;
- indexing and cataloguing of related spatial datasets available off-line (Dawe 1996).

An analysis of more than 25 home pages classified WWW usage in the GIS community into four overlapping categories (Coleman and McLaughlin 1997). Setting aside those sites focusing on product or program advertising, the remaining three categories included:

1 *Data distribution*. At such sites, users may search for specific spatial information features or datasets based on either keyword searches or Boolean queries of existing database fields (e.g. Crossley 1994; Nebert 1994; Pleuwe 1994). In some cases, the user may obtain only pointers to

datasets stored off-line (e.g. Marmie 1995). In more sophisticated implementations, the user may retrieve image, map graphics, or attribute information covering a given area of interest (e.g. Conquest and Speer 1996).

2 *Custom map creation and display*. These sites represent the results of special-purpose development projects aimed at the composition, display, and downloading of user-defined custom map products. They may or may not involve the use of a GIS package running in the background. Currently limited in terms of the extent and variety of data coverage, they nevertheless represent an exciting development in providing 'just-in-time' mapping to the general public (Pleuwe 1997).

3 *GIS/WWW integration*. These sites represent the results of integrating front-end query capabilities (supported through standard Internet and WWW interfaces and protocols) with the capabilities of commercial DBMS and GIS software packages residing in the background. User-defined queries are translated into corresponding SQL commands and passed to the 'back-end' GIS database for handling. The resulting response is passed back through the gateway to the user and, in the case of map-based responses, the resulting map is translated into a graphics or bit-map format suitable for fast transmission and viewing across the Internet. User-developed examples of such GIS/WWW integration are documented by Conquest and Speer (1996), Nebert (1996) and Liederkerke et al (1995), among others. Software vendors offering WWW interfaces to spatial data servers or map-based spatial data browsers include Autodesk, ESRI, Genasys, MapInfo, and Universal Systems.

The lines distinguishing these categories are fading quickly. Various electronic telephone and business directories offered in Australia (*http://www.whitepages.com.au*), Chile (*http://www.chilnet.cl/index.htm*), and the USA (*http://www.bigbook.com*) now allow users both to obtain contact information and to find the location of selected individuals or businesses on custom-generated electronic street maps (see Elshaw Thrall and Thrall, Chapter 23). Other creative early examples of how these systems may appeal to general users include:

1 The 'Tripquest' service (Figure 4) (*http://www.mapquest.com*), which enables users to specify a starting point and destination of a

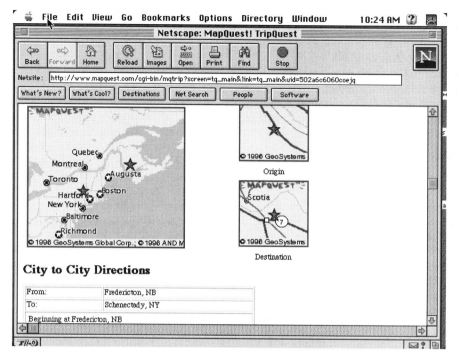

Fig 4. 'Tripquest' service.

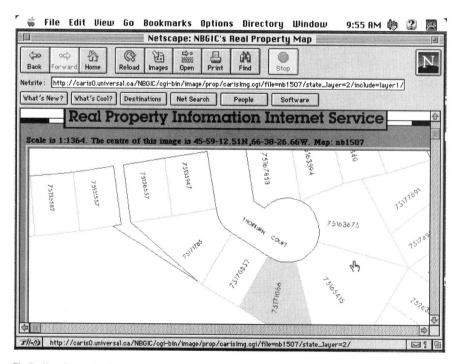

Fig 5. New Brunswick Geographic Information Corporation Property Information Service.

road trip, then receive detailed driving instructions based on a road network database contained at the server site.

2 The LandData BC site *(http://www.landdata. gov.bc.ca)*, which offers on-line access to a wide variety of mapping and land information products from different provincial government departments in British Columbia, Canada.

3 The New Brunswick Geographic Information Corporation Property Information Server (Figure 5), which allows NBGIC subscribers to access and download property ownership, registration, assessment, and parcel-centred mapping information covering any property in New Brunswick, Canada via keyword search, database query or map-based index (Arseneau et al 1996).

The advent of Java – a portable, object-oriented Internet language developed by Sun Microsystems Inc. – promises to remove many of the constraints inherent in early WWW protocols and further extend the capabilities of Web-based data browsing systems (Strand 1995). By moving much of the requisite display, processing, and analysis functionality to the client end of the Internet connection, performance delays due to server load and Internet bandwidth limitations may be greatly reduced (Figures 6 and 7).

Early examples of Java-based applications in GIS have tended to focus on limited tasks, including improved client-based spatial data browsing (Mapguide 1996) and network analysis engines (Fetterer et al 1996). More recent research efforts (Choo 1996) are aimed at using Java to provide an even wider suite of GIS display and modelling capabilities across the Internet. With a number of

vendors and standards organisations (e.g. the Open GIS Consortium) now including it as an important component of an overall network-based GIS architecture, Java-based technology may eventually be one of the keys to developing more open systems of distributed processing of geographical data.

At the time of writing, more extensive and well-maintained lists of WWW sites falling into all these categories may be found at the 'Metadata and WWW Mapping' home page maintained by Katz *(http://www.blm.gov/gis/nsdi.html)* and the 'GeoWeb' site *(http://wings.buffalo.edu/geoweb)* maintained by Pleuwe.

4.3 Collaborative production and group decision-making

Hardware, software, and procedures to support computer-supported cooperative work (CSCW) have been discussed and compared for more than 30 years (Englebart 1963). In the corporate world, shared access to corporate resources and innovative new approaches to collaborative production using groupware tools like Lotus Notes (Coleman 1995) and, more recently, the WWW (Pilon et al 1996) are now being investigated.

When considered in combination with emerging groupware products, the developments in GIS and broadband communications mentioned earlier may also offer a new approach to both collaborative group decision-making and digital map production – enabling some processes to be conducted concurrently rather than sequentially (for examples see Churcher and Churcher 1996; Coleman and Brooks 1995; Faber et al 1995; Karacapilidis et al 1995).

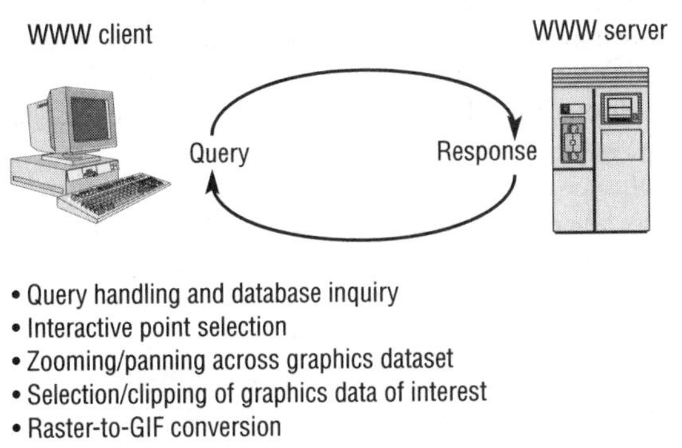

WWW client   WWW server

Query   Response

- Query handling and database inquiry
- Interactive point selection
- Zooming/panning across graphics dataset
- Selection/clipping of graphics data of interest
- Raster-to-GIF conversion

Fig 6. Internet spatial data request handling using HTTP.

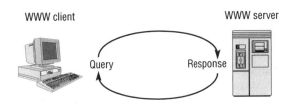

WWW client   WWW server

Query   Response

- Downloading of selected datasets
- Interactive point selection
- Zooming and panning
- Selected analytical functions

Java moves processing load off server and increases functionality of client

Fig 7. Internet spatial data request handling using Java language.

325

Network-based collaborative applications depend on more than just electronic mail. Specifically, they are predicated on the presence of a shared 'database' or collection of files, and rules which permit the definition of group members' roles, task status reporting and tracking, and gateways to electronic mail and other sources of data. Such systems should permit the organisation of correspondence, comments, reports etc. associated with a project or product and should support the management of multiple versions of objects (e.g. images, vector-based charts, video, and sound). Finally, a limited number of applications may require two or more remote users to be able to view the same file simultaneously, modify or add comments to specific entities in the same file where necessary, and communicate via voice, video, or e-mail while making these changes.

A number of organisations are already prototyping the WWW as the medium for collaborative decision-making (Gordon et al 1996), enhanced design, and concurrent engineering applications. Moreover, recent extensions to WWW viewers like Netscape are now introducing much of the communication and threaded-discussion, file transfer, data management and simultaneous data viewing functionality required in these applications (Forrester Research Inc. 1996; Ziegler 1995).

However, while some major corporations have made significant productivity gains through adoption of early groupware packages. The costs and cultural changes involved in such adoptions are likely to blunt the near-term impact of CSCW in the marketplace. Technically, vigorous adoption of collaborative technology will depend on:

- acceptance of the Internet or corporate intranets as the preferred medium for communications and information delivery;
- acceptance of common standards for threaded discussions and calendar management;
- subsequent development of a wider variety of inexpensive and compatible components or 'plug-in' tools capable of operating atop the standards extended from or similar to Netscape's ONE and Microsoft's COM (Forrester Research Inc. 1996).

Culturally, successful groupware implementations are likely to be shaped more by prevailing organisational cultures and constraints than by the technology itself. Since organisations work in different ways, it is best to let them customise their own groupware applications using modular toolkits rather than expecting any single shrink-wrapped package or technology to serve their needs adequately (Schrage 1995).

5 SYSTEM PERFORMANCE IN NETWORKED ENVIRONMENTS

Forthcoming generations of system architecture and broadband telecommunication technology promise to change fundamentally the way many organisations manage, transfer, and utilise their spatial data. However, before adopting such technology, both suppliers and potential customers require a clear understanding of potential network usage and the performance, capacity, and cost trade-offs involved in existing and emerging applications. A study carried out for the Government of Canada (IDON 1990) suggested that designing and building cost-effective GIS networks within and between federal government organisations would require defensible answers to six important groups of questions. These questions concerned the kind of information to be moved, the data volumes involved, response-time requirements of the users, the distance the data were being transferred, the frequency or regularity of data transmissions and, perhaps most important, the available funds which could be devoted to such operations.

Given the ability of today's WWW servers to track the volume and nature of user transactions at and between given sites, it is now easier to address such questions than it was seven years ago. Further, given the ubiquity of the Internet today, the justification of such a network may appear a moot exercise. However, the continuing requirement for 'hard numbers' reinforces the need to determine defensibly the performance of a particular application or group of applications across a network under known conditions, in order to predict the performance of the same application(s) under other conditions.

Numerous authors have suggested the fundamental importance of performance determination within the overall framework of the system life cycle and structured system and database design processes (e.g. Ferrari et al 1983; Jain 1991). However, while performance analysis has formed an important component of many GIS selection processes since the early 1980s, test procedures and results have largely remained unavailable because of the constraints of commercial confidentiality and competitive pressures.

Researchers have been developing more systematic and rigorous approaches to the determination of GIS performance on stand-alone configurations since the mid 1980s (e.g. Goodchild and Rizzo 1986; Hawke 1991; Marble and Sen 1986; Wagner 1991). Investigations at the University of Edinburgh (Gittings et al 1993) and the University of Tasmania (Coleman 1994) have extended these studies into client/server environments across LANs and WANs. The latter two references in particular provide extensive reviews of the literature related to GIS performance testing.

A key limitation of GIS performance testing research to date has been its lack of extendibility. Absolute performance figures are tied closely to the hardware configurations employed. Further, even under controlled network conditions, it is extremely problematic to model end-to-end GIS performance and to predict how response times will be affected by changes in server load and network data traffic.

As an increasing number of applications take advantage of the Internet – which, in itself, can send data packets across randomly changing combinations of high- and low-speed network connections with ever growing traffic loads – it will be increasingly difficult to make any claims (apart from criticisms) concerning application performance in a given setting. Further, if events over the past 15 years are any indication, the storage, memory, and processing requirements of application software and operating systems have a tendency to expand to surpass eventually the capabilities of a given generation of computers and networks alike.

That being said, the following advances in technology and services will – along with many others – help improve GIS application performance over the next five years:

- the increasing optimisation of high-performance disk arrays and parallel processors for use in networked GIS environments (Sloan 1996);
- the proliferation of fibre-based broadband communication services in both the workplace and the home;
- the introduction of higher-speed satellite and cellular data communication links into areas not served by normal services;
- the introduction of higher-speed Internet services (e.g. the proposed 'Internet II') and value-added service offerings which improve performance by 'detouring' subscriber traffic around high-volume links and nodes;

- the accelerating introduction of special plug-ins, Java 'applets' and other modular software components which balance client and server processor loads and further optimise data traffic between workstations on a network;
- (perhaps most importantly) improvements in the collections of underlying algorithms and procedures defining the application software packages themselves, removing many existing bottlenecks and optimising older code to run on newer systems.

References

Annitto R, Patterson B 1994 A new paradigm for GIS data communications. *Journal of the Urban and Regional Information Systems Association* 7: 64–7

Arseneau B, Kearney A, Quek S, Coleman D 1997 Internet access to real property information. *Proceedings GIS 97 Conference, Vancouver.* GIS World Inc.

Champine G A, Coop R D, Heinselman R C 1980 *Distributed computer systems.* Amsterdam, Elsevier Science

Choo Y K 1996 'Interactive distributed geographical information systems (IDGIS)'. Unpublished Master's Research Project, Department of Rangeland Ecology and Management, Texas A&M University. *http://starr-www.tamu.edu/choo/idgis/intro.html*

Chorafas D 1980 *Computer networks for distributed information systems.* New York, Petrocelli Books

Churcher M, Churcher C 1996 GROUPARC: a collaborative approach to GIS. In Pascoe R T (ed.) *Proceedings, Eighth Annual Colloquium of the Spatial Information Research Centre.* Dunedin, University of Otago: 156–63

Coleman D J 1994 'Geographic information systems performance in a broadband communications environment'. Unpublished PhD dissertation, Department of Surveying and Spatial Information Science, University of Tasmania

Coleman D J 1995 An overview of groupware. In Coleman D J, Khanna R (eds) *Groupware technology and applications.* Englewood Cliffs, Prentice-Hall

Coleman D J, Brooks R 1995 Applying collaborative production approaches to GIS data collection and electronic chart production. Santa Barbara, NCGIA and see *http://www.ncgia.ucsb. edu/research/i17/i17papers.html*

Coleman D J, McLaughlin J D 1997a Information access and usage in a spatial information marketplace. *Journal of Urban and Regional Information System*s 9(1)

Conquest J, Speer E 1996 Disseminating ARC/INFO dataset documentation in a distributed computing environment. *Proceedings 1996 ESRI User Conference.* Redlands, ESRI. *http://www.esri.com/resources/userconf/proc96/TO200/PAP16 5/P165.HTM*

Costello B 1996 Trends in client-server architecture in GIS environments. Slide from invited presentation at 1996 Geomatics Atlantic Conference, Fredericton, New Brunswick, Canada

Craig W J, Tessar P, Ali Khan N 1991 Sharing graphic data files in an open system environment. *Journal of the Urban and Regional Information Systems Association* 3: 20–32

Crossley D 1994 WAIS through the Web: discovering environmental information. Paper presented Second International WWW Conference Mosaic and the Web, Chicago, 17–20 October

Dawe P 1996 'An investigation of the Internet for spatial data distribution'. Unpublished Master of Engineering report, Department of Geodesy and Geomatics Engineering, University of New Brunswick, Fredericton, Canada

Elliot W 1994 Moving AM/FM to the field: technology, trends and success stories. *Proceedings of 1994 AM/FM International Conference*: 845–57

Englebart D 1963 A conceptual framework for the augmentation of man's intellect. In Howerton P (ed.) *Vistas in information handling*. Washington DC, Spartan Books 1: 1–29

Ezigbalike I, Cooper R, McLaughlin J 1987 A query management strategy for land information network implementation. *Proceedings of 1987 Annual Meeting of the Urban and Regional Information Systems Association (URISA)* 3: 105–12

Faber B G, Wallace W, Cuthbertson J 1995 Advances in collaborative GIS for land-resource negotiation. *Proceedings of GIS 95 Ninth Annual Symposium on Geographic Information Systems*: 183–9

Ferrari D, Serazzi G, Zeigner A 1983 *Measurement and tuning of computer systems*. Englewood Cliffs, Prentice-Hall

Fetterer A, Goyal B, Agarwal N 1996 Background: Java interactive routing for Minneapolis: explanation of WebRoute project. Spatial Database Laboratory, Department of Computer Science, University of Minnesota. *http://www.ggrweb.com/geojava/*

FieldWorks Inc. 1996 Eden Prairie. *http://www.canaska.com/fw.htm*

Forrester Research Inc 1996 Teams on the Internet. *The Forrester report* 7(6): 2–8. Cambridge (USA), Forrester Research Inc. *http://www.forrester.com*

Gittings B M, Sloan T M, Healey R G, Dowers S, Waugh T C 1993 Meeting expectations: a view of GIS performance issues. In Mather P M (ed) *Geographical information handling*. Chichester and New York, John Wiley & Sons

Goodchild M F, Rizzo B 1986 Performance evaluation and workload estimation for geographic information systems. *Proceedings, Second International Symposium on Spatial Data Handling*: 497–509

Gordon T, Karacapilidis N, Voss H 1996 Zeno: a mediation system for spatial planning. *Proceedings of ERCIM workshop on CSCW and the Web, Sankt Augustin, Germany, 7–9 February. http://orgwis.gmd.de/W4G/proceedings/zeno.html*

Hawke D 1991 Characterising the performance of geographic information systems. *Proceedings of Third Annual Colloquium of the Spatial Information Research Centre, Dunedin*. University of Otago

IDON Corporation 1990 *Final report: a federal geographic information systems (GIS) network requirements study*. Contract report prepared for the Data Communications and Networking Subcommittee, Interagency Committee on Geomatics, Government of Canada

Ingoldsby T R 1991 Transparent access to geographically related data from heterogeneous networked systems. *Proceedings of 1991 Annual Conference of the Urban and Regional Information Systems Association* 3: 14–24

Jain R 1991 *The art of computer systems performance analysis: techniques for experimental design, measurement, simulation, and modeling*. New York, John Wiley & Sons Inc.

Karacapilidis N, Papadias D, Egenhofer M 1995 Collaborative spatial decision making with qualitative constraints. *Proceedings of Third International ACM Workshop on Advances in Geographic Information Systems*: 53–9

Katz R 1991 *High performance network and channel-based storage*. Sequoia 2000 Technical Report 91/2. Computer Science Division, Department of Electrical Engineering and Computer Sciences, University of California, Berkeley

Laurini R 1994 Sharing geographic information in distributed databases. *Proceedings of 1994 Annual Conference of the Urban and Regional Information Systems Association (URISA)*: 441–54

Liederkerke M van, Jones A, Graziani G 1996 The European Tracer Experiment system: where GIS and WWW meet. *Proceedings of 1996 ESRI User Conference*. Redlands, ESRI

MapGuide (1996) Mill Valley, Autodesk Inc. *http://www.mapguide.com*

Marble D F, Sen L 1986 The development of standardized benchmarks for spatial database systems. *Proceedings, Second International Symposium on Spatial Data Handling*: 488–96

Marmie A 1995 Promoting and providing GIS data via the Internet. *Proceedings, 1995 ESRI User Conference*. Redlands, ESRI. *http://www.esri.com/resources/userconf/proc95/to250/p2101.html*

McKay L 1994 Data brokering in land information – Land Information Alberta. *Proceedings, 22nd Annual Conference of AURISA* 1: 171–82

Miller A 1990 GIS in a networking environment. *Proceedings of GIS 90, International Symposium on Geographic Information Systems*: 255–62

Mimno P 1996 *Building enterprise-class client/server applications*. Lisle, Dynasty Technologies, Inc. *http://www.dynasty.com/product/mimno_wp.htm*

Nebert D D 1994 Serving digital information through the WWW and Wide Area Information Server technology. Paper presented at Second International WWW Conference Mosaic and the Web, Chicago

Nebert D D 1996 Supporting search for spatial data on the Internet: what it means to be a clearinghouse node. *Proceedings of 1996 ESRI User Conference, Palm Springs* Redlands, ESRI. *http://www.esri.com/resources/ userconf/proc96/TO100/PAP096/P96.htm*

Newton P W, Zwart P R, Cavill M 1992 Inhibitors and facilitators in high-speed networking of spatial information systems. In Newton P W, Zwart P R, Cavill M E (eds) *Networking spatial information systems*. London, Belhaven

Oswald R 1994 Manitoba Land Related Information System land information utility. *Proceedings, 1994 Conference of the Urban and Regional Information Systems Association (URISA)*: 179–92

Pilon D, Whalen T, Palmer C 1996 *Communications toolkit*. Technical Paper. Network Services and Interface Design Laboratory, Communication Research Centre, Industry Canada. *http://debra.dgbt.doc.ca/~daniel/cscw.text.html*

Pleuwe B 1994 The GeoWeb project: using WAIS and the World Wide Web to aid location of distributed datasets. Paper presented at WWW 94 Conference, Chicago, 17–20 October. *http://wings.buffaloedu/~plewe/paperwww.html*

Pleuwe B 1997 *GIS online: information retrieval, mapping, and the Internet*. Santa Fe, Onward Press

Popko E 1988 *An enterprise implementation of AM/FM*. Internal paper. Geo-Facilities Information Systems Application Center, IBM Corporation, Houston, Texas

Pretty R W 1992 LANs, MANs, and WANs: introductory tutorial on computer networking. In Newton P W, Zwart P R, Cavill M E (eds) *Networking spatial information systems*. London, Belhaven Press

Reinhardt A 1994 Building the data highway. *Byte Magazine* 19: 46–74

Schrage M 1995 Groupware requires much more than bandwidth. *Business Communications Review*: 35–8

Sedunary M E 1984 LOTS and the nodal approach to a total land information system. In Hamilton A C, McLaughlin J D (eds) *Proceedings of 'The Decision Maker and Land Information Systems' – FIG Commission III International Symposium*

Sloan T M 1996 'The impact of parallel computing on the performance of geographical information systems'. Unpublished MPhil dissertation, Department of Geography, University of Edinburgh

Strand E J 1995 GIS thrives in three-tier enterprise environments. *GIS World* 8: 38–40

Streb D 1995 Using cable television to distribute GIS access. *Proceedings, 1995 ESRI User Conference*. Redlands, ESRI

Thoen B 1995 Web GIS: toy or tool? *GIS World* (October) *http://www.csn.net/~bthoen/webgis.html*

Voelcker J 1986 Helping computers communicate. *IEEE Spectrum* (March): 61–79

Wagner D F 1991 'Development and proof-of-concept of a comprehensive performance evaluation methodology for geographic information systems'. Unpublished PhD dissertation, Department of Geography, Ohio State University, Columbus

Webcrawler Survey 1996 *World Wide Web size and growth*. Global Network Navigator, Inc. *http://webcrawler.com/WebCrawler/Facts/size.html*

Webster C J 1988 Disaggregated GIS architecture. Lessons from recent developments in multi-site database management systems. *International Journal of Geographical Information Systems* 2: 67–80

Wittie L D 1991 Computer networks and distributed systems. *IEEE Computer*: 67–76

Ziegler B 1995 Internet software poses threat to Notes. *The Wall Street Journal* (7 November)

23

Desktop GIS software

S ELSHAW THRALL AND G I THRALL

This chapter reviews the emergence of desktop GIS, which is defined as products which may have begun as features of workstation GIS but which have been spun off to create new and useful niche market products. As such, desktop GIS is seen to include: digital atlases; interactive street displays and route finding software; mapping on the Internet; spreadsheet and database mapping; clip art and readymade maps; thematic maps; and so-called 'true' desktop GIS. Geographically-enabled programming languages are also considered from the standpoint of add-ons to desktop GIS.

1 INTRODUCTION

GIS keeps getting bigger, encompassing more as technology develops and our knowledge base increases. As GIS grows, so pieces of what had previously been viewed as exclusively the territory of GIS are being sliced away to become market niches in their own right. Those who use new market niche products may even be unaware of the background GIS technology from which it developed and may have no need ever to be directly involved with GIS per se (see also Batty, Chapter 21).

What, then, is 'desktop GIS'? There are a number of ways in which to answer this question. From one viewpoint, desktop GIS is simply that part of the newly emergent desktop computer market that is not developing for any specific market niche. From this perspective, desktop GIS is no more than a software and technology marketing term. And yet in order to understand desktop GIS trends a second, more analytical perspective is required. In this chapter, we choose to define desktop GIS as inclusive of those products that may have begun as features within workstation or desktop GIS software programs and technology, but which have been spun off to create more useful products within new market niches.

How is the market for desktop GIS developing? Thrall, interviewed for the industry magazine *Geo Info Systems* in 1996, made the following five year forecast:

'I expect mainstream [desktop] GIS will be mainstream application software. GIS will lose its distinguishable identity in the market-place. If a map makes sense in an application, then a map will be in that application. If GIS functionality makes sense in an application, then that application will contain GIS functionality. GIS will become application niche software and application niche software will seamlessly include GIS. Customers will expect it. In five years, only us old-timers will remember when GIS was a technology with a culture separate and distinct from mainstream software.' (Thrall and Trudeau 1996: 12)

From the vantage point of the general software user, GIS will be all-pervasive and unrecognisable as a unique technology with a unique history. From the vantage point of the academic, the software developer, and the more sophisticated user, understanding of the development of desktop GIS technology will be one means of distinguishing between casual users and experts.

In this chapter we present an overview of desktop GIS. Our broad definition of desktop GIS will enable us to discern the market trends more readily. Our purpose is to present a categorisation of desktop GIS, and to provide examples of each. This will contribute towards clarification of what desktop GIS is today and how it is distinguished from previous GIS technology. Table 1 presents eight categories of desktop GIS, in increasing order of complexity. As new markets emerge, so new categories will be added: as technology and the markets change, so the ordering of the categories will change.

Table 1 Categories of desktop GIS technology.

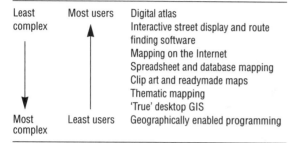

Least complex	Most users	Digital atlas
		Interactive street display and route finding software
		Mapping on the Internet
		Spreadsheet and database mapping
		Clip art and readymade maps
		Thematic mapping
		'True' desktop GIS
Most complex	Least users	Geographically enabled programming

Novice users of desktop GIS might consider any software that fits within any of the categories of Table 1 to be 'GIS'. Conversely, novices may believe that their requirements can only be met by what we refer to as 'true' GIS software; or, because they are producing thematic maps they may believe that they are working within a 'true GIS environment'. Such differences in part reflect the rapidity of development of the field: what may once have been a central component of desktop GIS may have evolved into a niche which is peripheral to 'true' GIS, but which is nevertheless an important emerging new market segment. For example, GIS functionality and geographical data have been key to increasing the productivity of those who need customised maps: however, the new category of readymade maps (which includes no GIS functionality) may now better suit the needs of this user group. The core components of each of the classes of geography software identified in Table 1 will now be discussed in turn.

2 THE DIGITAL ATLAS

The digital atlas has developed from its traditional analogue counterpart using GIS technology (see also Salgé, Chapter 50). Today the digital atlas is typically distributed on CD-ROM. The illustrative products included in Table 2 serve the same purpose as printed atlases. In terms of production costs, the first (master) copy of the digital version may cost as much or more than the first master copy of its printed version counterpart. However, the marginal

cost of producing each additional digital atlas, especially when very large mass-market volumes are considered, becomes insignificant. The digital atlas has further advantages over its printed counterpart by virtue of being compact and lightweight – so that, for example, information can even be accessed with a CD-ROM-equipped laptop computer. The digital atlas may also include a computer-automated 'find' feature: printed atlases require one to go to an index, read cryptic references to the page and row and column that a map feature may be found on, and then – armed with such information – the atlas reader must locate the feature on the map. The digital atlas will find the feature for the user and show the user where the feature is on the appropriate map. Digital street atlases like those discussed below may even allow the user to enter an address and the digital version will then zoom into the likely location of the address. It is traditional for the printed atlas to include some thematic maps and charts. The digital version may allow the user to select from more themes, and to vary the manner in which the theme is displayed.

Table 2 Simple mapping products and digital atlases.

Small Blue Planet	Now What Software
Maps 'n' Facts	Broderbund
Picture Atlas of the World	National Geographic
Expert Maps Gold	Expert Software
World Atlas	Software Toolworks

To sum up, the digital geographical technology offers advantages over traditional printed technology in terms of convenience of use, reduced size and weight, and lower cost. The lower marginal production cost is likely to mean that the product is more frequently updated: by contrast, the traditional atlas in a library may previously only have been replaced when its pages became worn from years of use.

3 INTERACTIVE STREET DISPLAY AND ROUTE FINDING

Interactive street and route finding software is a variation on the digital atlas that is a separate and new emerging technology. Several of the products that are available in the USA, for example, are listed in Table 3. DeLorme has become dominant in the

US market with this type of product because it entered the market early, its street maps are attractively designed and easy to read, it is inexpensive, and (very important for a product of this type) it uses data compression technology to include the software and the geographical data on a single CD-ROM.

Table 3 Interactive street display and route finding.

Global Explorer	DeLorme	multimedia with photographic images
Map 'n' Go	DeLorme	includes route finding
Street Atlas USA	DeLorme	based on US Census TIGER/Line

Beyond the basic functions of street display and route finding, there are also products (e.g. those of Sony Inc.) that combine features from Table 3 with hardware. The hardware is ergonomically designed for ease of use in an automobile. These products may include Global Positioning Systems (GPS) technology (see Lange and Gilbert, Chapter 33) with interactive voice, whereby the device will select the best route based upon current traffic conditions and will audibly inform the driver where to turn (see also Waters, Chapter 59).

4 MAPPING ON THE INTERNET

Simple descriptive maps have also become a feature used to measure user access, also known as 'hits', to World Wide Web (WWW) sites on the Internet. Users may access these 'Where is it?' maps in either of two ways. First, the user utilises a 'find' feature to locate an Internet address, using a similar procedure to that used to interrogate a CD-ROM street atlas. Advertisers will often pay to have their logo and universal reference locator (URL) jump to their site prominently displayed alongside a 'Where is it?' map. Second, the 'Where is it?' map feature may be part of a WWW page that provides the user with a street map so that some desired destination can be found (Figure 1).

Fig 1. 'Where is it?' map feature from a WWW page.

'Where is it?' maps can be downloaded or printed thereby making available to the user custom atlas type features. An example of two kinds of 'Where is it?' map is provided in Table 4. The Vicinity site uses Etak Corporation geographical data and software to create the maps; the maps are used as a draw to their site for the display of advertising. Union Street Links uses on-line maps to inform the user of business locations in order to encourage patronage.

The ability to locate point addresses on a map is considered a standard feature of GIS software; the 'Where is it?' map separates out this feature from the general purpose GIS, allowing the user to display the location of an address or object. As discussed below, to be 'true GIS' the software must be capable of transforming and manipulating the data thereby effectively creating new data. 'Where is it?' maps are highly valuable but are limited to the display of stored geographical information. At the same time, if the program user's needs require only 'Where is it?' capability, then there is no need to use any more complex GIS software. Table 5 includes a list of software vendors that provide the capability to program WWW sites to include GIS features such as the 'Where is it?' capability.

A second type of mapping on the Internet is analogous to placing the geographical and attribute data and GIS software in a client/server configuration. The actual software and the data reside on a server which the user accesses via the Internet. The users (clients) request maps and the information is returned to them. Many of the GIS user tasks, such as opening tables and formatting the

Table 4 Examples of 'Where is it?' maps on the Internet.

Name of	Owner	Description	URL	WWW site
Vector	Vicinity Inc.	Address locator on demand		http://www.vicinity.com
Union St Links	G&S Thrall	Displays location of business		http://www.afn.org/~links/

Table 5 Vendors of Internet enabled mapping software.

Caliper	http://www.caliper.com
ESRI	http://www.esri.com
MapInfo	http://www.mapinfo.com
Sylvan Ascent	http://www.sylvanmaps.com

Note: see authors' Web page for sample list of URLs using Internet enabled mapping software: http://www.afn.org/~thrall/gitwebs.htm

layers, are performed by the server. Proper software design is imperative, because many of those that access such Internet sites will be novices to geographical technology.

5 GEOGRAPHICALLY-ENABLED SPREADSHEETS

The thematic mapping capability has become mainstream with its introduction as an adjunct to mass consumer market spreadsheets (Thrall 1996d). For example, in the mid 1990s the alliance between Microsoft Corporation and MapInfo Corporation provided users of the Microsoft Excel spreadsheet software with the capability to generate simple thematic maps (Table 6). Similar products are available from Autodesk and Intergraph. While highly limited in terms of numbers of mapping features and options, the Excel/MapInfo product is a good example of how a mapping software program can be designed to be highly user friendly (see also Birkin et al, Chapter 51). The product also includes a limited amount of coarse resolution geographical boundaries, including those of the major land forms, many nations (e.g. the USA, Canada, Mexico, Western Europe, Australia), states and provinces of North America, and US counties. Excel does not include smaller boundary files like census tracts or zip codes.

With Microsoft's Excel, the user selects a range of cells to map. One of the columns of selected cells in the range enables the data to be linked to a geographical object such as a US county. For the attribute data to be geographically enabled, the field with the geographical identifier must follow a rigid standardised convention. The other columns in the

Table 6 Geographically-enabled spreadsheets.

Borland	*Quattro Pro*	Strategic Mapping Inc.
Lotus 1-2-3	*Lotus 1-2-3*	Strategic Mapping Inc.
Microsoft	*Excel*	MapInfo Inc.

Note: In 1996, Strategic Mapping Inc. was purchased by ESRI (Environmental Systems Research Institute)

selected range contain the attribute data to be used to provide values to the mapped themes. All mapping is done by linking attribute data to polygon map objects. Excel cannot map with user-specified latitude and longitude coordinates. Excel does not allow the user to customise polygon boundaries, and polygons may not be joined or buffered. As such, the polygon areas that can be mapped with Excel are highly limited. Yet these very limitations, which make it simple for the novice to use, in combination with its widespread distribution by Microsoft, have created a lucrative market for GIS data vendors. A variety of boundary files and attribute data can be purchased separately and then used within Excel.

Wessex Inc. has released an 'after market' product designed to extend the range of attribute data easily accessible by a mapping program to include the entire US Census of Population and US Census of Housing. Geographical boundary files for census tracts are provided with the Wessex product. Wessex's products include ProFiler (the US Census of Population data), a set of TIGER/Line files (the US Census boundary files in MapInfo format), First Map (for desktop mapping with Excel), and First Street (ArcView Software from ESRI (Environmental Systems Research Institute), Wessex's ProFiler, and Wessex's boundary files in ArcView format).

Lotus Corporation has also linked with a GIS software vendor to provide mapping capability similar to that described above for the Excel/MapInfo product. The choice of Lotus to have aligned with Strategic Mapping Incorporated (SMI) may have been unfortunate since SMI has subsequently ceased to exist and has sold its geographical software division to ESRI – best known for the GIS program ARC/INFO. However, what may appear to have been a mistake for Lotus – to align itself with a troubled GIS software vendor (SMI) – may in the final analysis appear to be fortuitous given the quality and strength of product offered by ESRI. For more discussion on ESRI's takeover of SMI see Thrall (1996a) and see Longley et al (Chapter 1; Chapter 72) for a discussion of rationalisation and change in the GIS software industry.

The mapping capabilities of spreadsheet programs are highly user friendly, but primitive in mapping capability. Prior to the introduction of mapping in spreadsheet software there were perhaps 100 000 users of geographical mapping software in the world, with varying skills and interests. Now with geographical technology included in mass market spreadsheet products the numbers of persons becoming familiar with geographic technology may exceed 20 million. As those high numbers of users proceed through the learning curve they will demand a greater variety of mapping operations, together with the ability to manipulate geographically enabled data. The scene is thus set for GIS to become part of the mainstream mass market computer software industry, in contrast to its past which appears increasingly esoteric and peripheral in comparison. At the time of writing, several database vendors including ORACLE, Informix, Computer Associates, Sybase, and IBM have already announced planned products that will allow the user to work with spatial data (Sonnen 1996).

6 CLIP ART AND READYMADE MAPS

In this section we will discuss clip art maps and a newly emerging product that we refer to here as 'readymade maps'. In Table 1 they could have been listed as two separate product categories. They share a number of common characteristics and are both based upon a newly emergent technology in a highly dynamic market. Indeed, the vendors of what we call readymade maps advertise their products as clip art maps. However, the authors of this chapter believe that once the consumers understand the capabilities of the mapping products, readymade maps will be identified as a new product category (see Thrall and McLean 1997).

Readymade maps combine the technology of clip art imagery with GIS. Clip art has long been a common feature of desktop publishing where the digital artist may begin to construct a computer image by importing an image from a catalogue of images. Clip art usually allows simple manipulation of the image such as changing the colour or deleting a part of the image. Clip art may also stand alone without modification and is used extensively in the design of brochures and other forms of advertising. Clip art maps allow simple editing of a map by exporting it to a graphics program which is used to manipulate it.

The readymade map is a more-powerful variation of the standard clip art map. Readymade maps differ from clip art maps in the magnitude of the file size and in the complexity of the data file. Clip art is typically one layer of information while readymade maps typically have many layers. Clip art images are typically small while readymade maps can require as many as 64 megabytes of RAM memory just for their editing. Clip art is designed to be brought into a text document – for example as an image in a Microsoft Word text file. Readymade maps, by contrast, generally must be imported into a graphics program to be edited. Most readymade maps are distributed as an ensemble of geographical data and a software program to display them. Readymade map software can also often allow the geographical data to be exported in a variety of formats for subsequent editing.

With readymade maps, the user can do much more than change the colour or insert a map into text. With readymade maps, the user can use advanced features such as masks which allow the addition of boundaries, physical features such as rivers and mountains, graticules, scale bars, shadowing, and so on. Readymade maps can also be edited at a 'sub-image' level using a product like Macromedia's Freehand which allows the map to be edited in layers. Some readymade maps allow the importation of multiple files that automatically register to one another at the same scale and in the correct location without manual manipulation. For example, if a file of a country is opened, then other imported countries will align correctly and at the same scale as the original opened country. Table 7 gives examples of some readymade map vendors.

Usually the quality of readymade maps requires the highest quality printer and a desktop computer specifically configured for high resolution graphics production. These are truly maps ready to publish and bear comparison with the finest maps published anywhere today.

Table 7 Some readymade map vendors and their products.

Mountain High Maps	Digital Wisdom	High resolution relief maps of countries of the world
Cool Maps	Digital Wisdom	Relief maps for desktop publishing produced using visually exciting colours and projections
Globe Shots	Digital Wisdom	Views of the earth from space: many of the earth images are animated for easy inclusion in multimedia

Note: Based upon Thrall and McLean (1997)

Clip art maps and readymade maps provide basic descriptive geography for the publisher and graphics artist/cartographer. The fundamental characteristic that differentiates clip art maps and their more powerful cousin, readymade maps, from, say, a digital atlas is that the digital atlas is targeted towards the end or final consumer whereas clip art maps and readymade maps are intended to be modified using the map program itself or by using a program such as Adobe Photoshop or Macromedia Freelance to prepare the map for the enduser. The finished map may remain in digital form. Macromedia includes the ability to distribute electronic images produced within its software program on diskette. With Macromedia Director comes the ability to produce multimedia productions and to distribute the results over the Internet. By acquiring the desktop GIS capability to edit maps, clipart maps and readymade maps have established a market niche for themselves.

7 THEMATIC MAPPING

Thematic mapping begins one small step beyond the categories of the digital atlas, 'Where is it?' maps, and readymade maps. Thematic mapping software can be quite complex and sophisticated in its data manipulation functionality. Many 'true' desktop GIS software programs include thematic mapping capabilities; however, these capabilities are generally highly limited. 'True' desktop GIS software has greater focus upon the manipulation of spatial data while thematic mapping software has greater focus toward the display of spatial data. Software specialising in thematic mapping generally offers a much greater range of thematic mapping features than is available from desktop GIS software programs. Thus users of GIS who desire more complex thematic mapping capability may process and manipulate their data within the GIS, and then transfer the results of analysis to a thematic mapping program for presentation.

Thematic mapping capability includes the ability to produce shaded choropleth maps where the data ranges are represented by different colours, shading, hatching patterns or dot densities, or by graduated symbols. In dot density maps, dots are randomly scattered within polygons, although the total number of dots represents the number of observations. In graduated symbol maps, the symbol size is proportional to the value of the observation (see Martin 1996; and Kraak, Chapter 11, for a

general discussion of digital symbolisation). Some thematic mapping software also includes the ability to construct isoline and 3-dimensional shaded relief maps, or prism maps in which a polygon can be extended 3-dimensionally above the surface – the height above the surface being proportional to some attribute data value. Thematic mapping software may also allow the mapping of two variables as bivariate maps. However, even though some software allows a virtually unlimited number of themes to be simultaneously displayed, it should be recognised that a map generally loses interpretability when more than two themes are displayed. Each of these types of maps is discussed in fuller detail below in the GIS software section.

Thematic mapping software often requires data input in a very rigid data format and field layout. Thematic mapping software programs also generally have very limited capability for modification of the data. Thus while many GIS software programs include a variety of thematic mapping functionalities, most thematic software programs cannot be classified as true GIS software. Table 8 gives some examples of thematic mapping products.

Table 8 Some thematic mapping products.

Product	Manufacturer	Description
Surfer	Golden Software	Processes the user's data into isolines or contour lines. Provides a robust variety of spatial interpolation features as well as a reasonable set of default options for the novice. For a review and further discussion of Surfer see Thrall (1995a).
Map Viewer	Golden Software	Allows the user's data to be associated with US county and US state boundary files for the display of 3-dimensional prism maps as well as other forms of thematic map that are popular among endusers. Program users may provide their own boundary files. (See Elshaw Thrall 1997.)

8 'TRUE' DESKTOP GIS

'True' desktop GIS software programs include features which allow the user to achieve similar results to all of the above desktop technologies, but also include additional capabilities. True desktop

GIS software programs allow the user to access information using spatial logic, to modify geographically-enabled data, and to visualise the results as a map. This category of desktop software allows the user to query data and map objects using enhanced structured query language (SQL: see Egenhofer and Kuhn, Chapter 28) database operations. The user can also query individual map objects.

It is routine for desktop GIS software to include the capabilities for the input of textual spatial queries, including SQL queries, polygon joins, point-in-polygon operations, and buffering. Each of these uses is discussed in detail below. Table 9 identifies some of the leading desktop GIS software programs.

Who are the users of feature-rich desktop geographical software technology? The user of desktop GIS today is quite a different kind of individual to the user who is reliant upon mainframe and workstation GIS. Large GIS platforms like ARC/INFO or Intergraph are now used primarily by governmental organisations and large research universities. Governmental organisations have not traditionally been highly price sensitive or high productivity-motivated, and therefore they have been able to afford the high expense of large platform software and hardware, as well as the high learning curve required of highly technically trained specialised workers. ESRI, which produces ARC/INFO, has suggested that the learning curve to become fully proficient in ARC/INFO is roughly three years (Thrall 1995b; Thrall et al 1995). Thus GIS such as the mainframe systems of ARC/INFO and Intergraph have traditionally been used by the public sector.

Private businesses are generally unwilling to allocate the high level of expenditure necessary to establish mainframe and workstation GIS. Businesses are also generally unwilling to commit themselves to adding employees who require years of training in a software program which can only indirectly enhance their decision making. In short, businesses are unwilling to become hostage to an exotic technology operated by irreplaceable workers. The negatives of the mainframe and workstation environments make private businesses the primary audience for desktop GIS since, other than the ubiquitous desktop computer, there are no special hardware requirements. Moreover, powerful GIS software of unprecedented user-friendliness can be obtained at prices similar to those of bundled business office software (wordprocessing, spreadsheet, database management, presentation designers, etc.) from major manufacturers. Inexpensive GIS software now comes as standard with the basic geographically-enabled data required for most business decisions (see also Coleman, Chapter 22; Sugarbaker, Chapter 43).

Private businesses will not readily adopt research frontier technology, yet until quite recently GIS has resided within this realm. However, with mapping now included with conventional spreadsheet software, businesses no longer perceive mapping and geographical technology as being the research frontier. Most businesses which do adopt GIS perceive it as an aid in what Thrall has referred to as the first stage of GIS reasoning; namely, the employment of GIS technology to represent spatial phenomena for descriptive purposes (Thrall 1995c). Businesses may understand the need for spatial data visualisation. They may also use GIS for fast manipulation of large volumes of spatial data. For them GIS saves the many weeks or years that might be needed to organise and visualise their own data in association with data that describe their market. Prior to the innovation of desktop GIS, the cost and complexity of such descriptive representations made all but a few businesses stay away from the technology. Desktop GIS has changed all that, so now having this information is considered to be a necessary input for proper business decisions.

Some businesses which better understand geographical concepts may have proceeded to Thrall's second and third stages of GIS reasoning,

Table 9 Leading desktop GIS software programs.

ArcView 3	ESRI (Environmental Systems Research Incorporated)	Redlands, USA
Atlas GIS*	ESRI (Environmental Systems Research Incorporated)	Redlands, USA
Autodesk World	Autodesk Corporation	San Rafael, USA
GeoMedia	Intergraph Corporation	Huntsville, USA
MapInfo	MapInfo Corporation	Troy New York
Maptitude	Capliper Corporation	Newton, USA

* Note that Atlas GIS orginally published by Strategic Mapping Incorporated is now under the ownership of ESRI so the future of Atlas GIS is doubtful.

namely using GIS for explanation and prediction. The ability to visualise large amounts of geographical data has made it easier for businesses to understand correlation and causality. In order to enhance explanation and prediction, mathematical constructs such as gravity models (Haynes and Fotheringham 1984) have been programmed to use data that have been spatially summarised or modified using desktop GIS software (see Birkin et al, Chapter 51; Waters, Chapter 59). Geographical prediction with desktop GIS software is akin to 'what-if' scenarios performed with conventional spreadsheet software. Forecasts of spatial trends and processes, such as the changing market for a business firm, can be created with the desktop GIS software. It is in this context that Thrall (1995c) has written that 'relevance and marketability are key to understanding market-driven GIS today . . .'.

Business utilisation of desktop GIS software has not substantially gone beyond descriptive applications. There has been very little documented use of GIS to enhance business judgement and decision strategy. Business remains largely unaware of the contemporary capabilities of geographical technology and methodology. Today it is essentially the sizzle and pizzazz of a descriptive map that sells GIS. Given that the use to which GIS is put by business has been highly limited, the cost of implementation of GIS in a business environment must be low, in order to overcome the perceived limitations of the technology. As awareness of capabilities increases, business will demand more geographical capability from their GIS software and employees. As a consequence, we anticipate bright prospects for appropriately trained economic geographers. However, knowing the commands of the GIS software does not geographically enable the personnel using the software. By analogy, knowing the software features of a spreadsheet program does not by itself make one an expert in finance or accounting. The need for personnel who are trained in geographical reasoning is a prerequisite for the continued expansion of the desktop GIS industry.

Business requirements for high productivity that accompany inexpensive and user-friendly turnkey GIS solutions translates into a demand for prepackaged GIS data. And the market can respond by offering commercial business data because the volume is high and in the USA at least the production costs are very low (but see Rhind, Chapter 56). Free availability of high quality geographical and geographically enabled attribute data (Smith and Rhind, Chapter 47) has given the USA a competitive edge over those countries where access to similar data is highly restrictive or very costly.

Businesses have not adopted GIS in the same way that governments and related large public institutions have. Still, the chronology of the adoption by business of geographical technology mirrors the earlier process of adoption by government. Outside of military applications (see Swann, Chapter 63), those government agencies and divisions that were the first to adopt GIS were in the fields of natural resources or physical geography (e.g. Hutchinson and Gallant, Chapter 9). Only later did government agencies and divisions concerned with human geography applications adopt GIS. Similarly, those businesses in the natural and earth resources industries, such as forestry, were the first to adopt geographical technology, including desktop GIS. It has been a recent phenomenon that businesses have adopted GIS for use within the fields related to human geography; examples include real estate appraisal and investment, market analysis for retail outlets, and banking (see Birkin et al, Chapter 51; Longley and Clarke 1995).

8.1 True desktop GIS features and functions

Desktop GIS can be used to perform a variety of operations, though many of these operations are not unique to GIS. For example, adding new street line segments to a base map is a feature shared between GIS (computer assisted drawing), and software designed for and limited to digitising. In this section we discuss those features and operations that differentiate desktop GIS software from other software. These features include generating thematic maps, performing spatial queries or selections, joining polygons, performing point-in-polygon operations, and buffering (see Thrall and Marks 1994).

A fundamental application of desktop GIS software is the creation of thematic maps. Although thematic mapping is but one facet to the visualisation capabilities of GIS (see Kraak, Chapter 11), the thematic mapping feature is one that has been carved out from the GIS technology and sold on the market place as a separate product (as described in section 7 above). The advantage of separate smaller products is simplicity of software design, cost of software production and support maintenance, and a quicker learning curve for the program user; the disadvantage is that the product is much more limited in scope.

The GIS thematic map presents an underlying motif of data in spatial form. Data that can be mapped can be numeric or nominal. Nominal data are data with no numeric value that are described by name (Star and Estes 1990: 28): for example, various types of crimes (murder, assault, rape, robbery), or various types of underground cables (electric lines, cable television lines, fibre optic cables). The thematic maps that can be generated on desktop computers fall into several general categories: maps of individual values, choropleth maps, dot density maps, and graduated symbol maps. Another type combines two data themes in the same map and is referred to as a bivariate map. Some thematic mapping programs specialise in isoline, wire frame, and prism maps (see also Beard and Buttenfield, Chapter 15; Kraak, Chapter 11).

Individual value maps show different colours for each unique value of the data. Individual value maps are used primarily for the display of nominal data. A different colour or pattern is used to represent each data value. The attribute data can be associated with any form of geographic object, namely points, lines, or polygons. For example, a crime occurrence has a location and a classification; the colour of a point being red may indicate the crime as murder; likewise, assault could be represented by a brown dot, or a fibre optic cable line can be represented by a blue line and electrical lines represented by black lines. Individual value maps can also present numeric data, though often a map of this type suffers from loss of interpretability. Consider a map of Europe where the colours represent unique population density figures of the countries; each country would then have a different colour since while countries may have similar population density, no two countries will have identical population density values.

Choropleth maps (see Figure 2) are used to present ranges of values by colour or shading or hatching. Appropriate data ranges can usually be

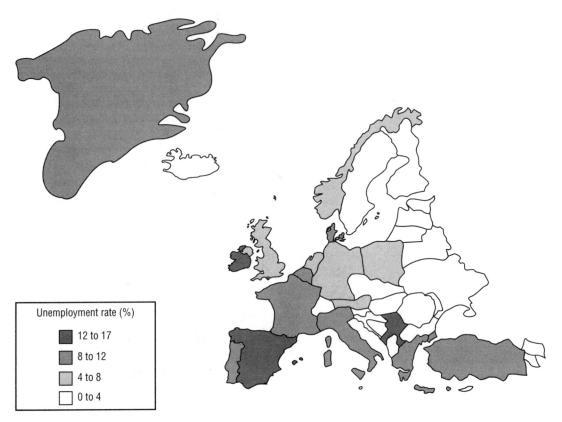

Unemployment rate (%)

- 12 to 17
- 8 to 12
- 4 to 8
- 0 to 4

Fig 2. An illustrative choropleth range map.

identified by the user where: there are the same number of observations in each range; the ranges themselves are of the same size; or the user can customise the ranges to any convenient values. Each value range then is represented by a particular colour, and successive ranges are denoted using lighter or darker shading of the same colour, or by a different hatching pattern.

Dot density maps (see Figure 3) represent raw data values that reside within a polygon. With a dot density map the software assigns one point for each incremental range of data values, such as one point for every 1000 people in a state or province or county. Desktop GIS software generally displays the dots as being randomly scattered as opposed to clustered where the density of observations might be the greatest.

Finally, the graduated symbol map (see Figure 4) displays a symbol of varying size where size represents the magnitude of the attribute data value. For example, a large aeroplane symbol could represent a major regional airport while a small aeroplane symbol could represent a minor local private airport. The size of symbol could be graded by thousands of passengers served.

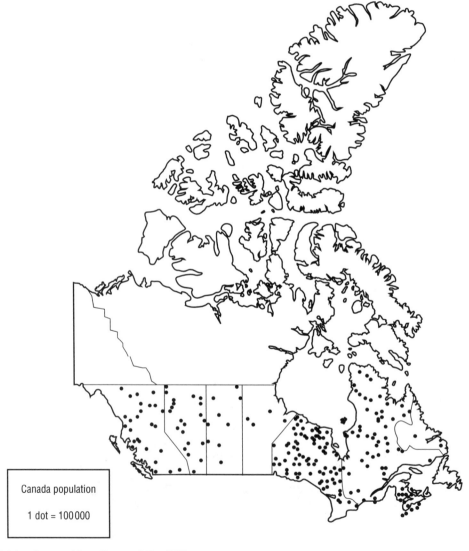

Canada population

1 dot = 100 000

Fig 3. Dot density map of Canadian population 1990.

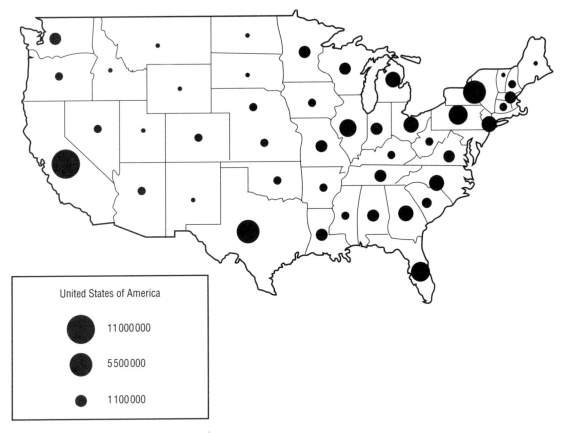

Fig 4. Graduated symbol map of number of housing units per US State 1990.

A bivariate map displays two themes, or two data variables. For example a map may have different symbols representing the type of crop grown in an area (like a corn stalk for corn or wheat shaft for wheat) and then the size of the symbol can represent the amount of yield for the particular crop. Bivariate maps typically use a combination of different symbols with different colours or sizes or the same symbol showing different sizes and colours to represent the two themes. Bar chart maps and pie chart maps are variations on the theme of bivariate maps. With bar chart or pie chart maps the user can have different bars or pie slices represent different fields for each record.

Isoline maps show varying data ranges by means of lines. For example, an isotherm map may show temperature ranges. The lines are labelled with the data value they represent. From these isoline maps, 3-dimensional wire frames can be mathematically interpolated and drawn. Some mapping programs can generate wire frame maps (see Figure 5) also offer features of sun inclination showing on the wire

frame, perhaps with a solid or shaded drape over the frame, to give a more realistic 3-dimensional effect (Thrall 1995a; Dowman, Chapter 31).

Another form of GIS produced thematic map, known as a prism map (see Figure 6), is a variation of a bar chart map. The height of the bar (prism) is proportional to the attribute value being mapped. But the prism itself is more than a plain rectangular 2-dimensional or 3-dimensional bar; it is in the form of the polygon with which the attribute is associated. The effect is to generate a map of varying plateaux or elevations in the shape of the polygons comprising the map.

8.2 True desktop GIS spatial selections

All true desktop GIS software programs include the capability to perform spatial operations. A spatial operation is a database query that is performed on spatial data using spatial criteria (see also Martin, Chapter 6). For example, a regular aspatial query

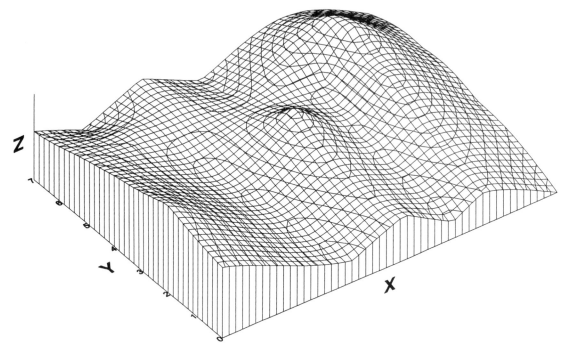

Fig 5. A wire frame map.

might be to select those home sales records for which the value of the sale is above a certain amount. A spatial query might be to select those homes for sale which are within a five kilometre radius of a given school. Again, by itself the capability to perform this operation does not distinguish true GIS from some GIS niche products.

Spatial selections in GIS are usually accomplished using tools selected using a toolbox icon, or via typed queries entered using a dialogue box. Tools for spatial queries using desktop GIS fall into three general categories: selection or pointer-like tools; circular or 'radius select' tools; and polygon (non-circular area) tools. The selection or

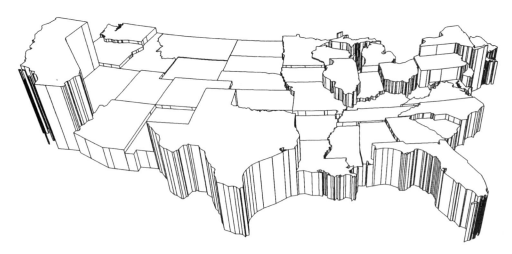

Fig 6. A prism map depicting the relative population sizes of the contiguous United States.

pointer tool is typically in the shape of an arrow head (and is a normal tool found, for example, in windows-based wordprocessors). It is used to point at map objects, and by clicking the main mouse button the user may select items. In a true desktop GIS, the items selected are spatial objects, namely points, lines and polygons, and the associated labels. Generally, the layer that includes the item being selected must be active. Many true desktop GIS software programs allow only one map layer to be active at a time.

The radius selection tool allows the user to draw circular areas around a point such that all records falling within a circle are selected. For example, in order to find all major cities in a database that fall within a 100-km radius of New York City, the software user would click on the central point of the City and drag a radius of 100 km from it. All records falling within the circle defined by the radial distance would be selected (but see Birkin et al, Chapter 51, for a discussion of the limitations of such tools).

The polygon selection tool is similar to the radius selection tool in that it allows the user to draw polygons around areas and to select those records that fall within the polygon. The polygon tool does not limit the user to a circular area, but rather allows the user to construct any regular or irregular closed shape.

Spatial queries can be performed using logic and words as well as pointing and selecting with a tool. Spatial querying using words is known as 'word query'. All regular database queries in true desktop GIS can be performed using word queries, including simple queries using the relational operators (less than, less than or equal to, greater than, etc.) and complex compound queries using the logical operators (and, or, not). In addition to these standard database operators, true desktop GIS software programs include geographic operators such as 'contains', 'falls within', and 'intersects'. 'Contains' is an operator where one object contains another, such as a circle inside of a square. 'Falls within' is an operator where one object falls spatially within another object such as a point residing within a circle. 'Intersects' is an operator where part of one object overlaps or intersects part of another object such as a line that passes through a circle.

Structured query language (SQL) word queries allow the user to perform selection operations which are more complex than simply pointing and clicking

with a mouse (see also Egenhofer and Kuhn, Chapter 28). SQL is a standard database query language. When using a regular word query with the geographical operators, the operators work on fields that are actually in the database. SQL queries can work on fields that are not explicitly in the database, but which can be implicitly calculated from fields and spatial information in the database. For example, a regular word query can be performed on the population of the countries of the world (e.g. 'select countries whose population exceeds 50 million'). A SQL query can be performed on any field so long as that field can be created using a combination of information from other fields. Thus, for example, queries can be made about population density, even though population density is not one of the fields in the database, since population is known and a desktop GIS can calculate the area of a county polygon. The desktop GIS software will create a new data field – population density – which can subsequently be saved to the database. For extended discussions of this and related topics, see Oosterom (Chapter 27) and Worboys (Chapter 26).

8.3 True desktop GIS polygon joins

Another common use of desktop GIS is to join polygons. New polygons in the same or in a new layer are created by joining existing polygons – for example, county polygons might be joined to create sales territories for a sales force, and the newly created layer saved as the 'sales territories' layer. Each map object normally has attribute data assigned to it, such as the population of a county. When joining polygons, the user can choose to discard the attribute data or to save them after modification using a mathematical operator – for example, the user may require the software to add the population of each county that is joined to form a new attribute, the 'population of the sales territory'. For further details see Thrall (1992) and Martin (Chapter 6).

8.4 True desktop GIS point-in-polygon operations

The term 'point-in-polygon search' was originally coined to describe the geometrical operations used to locate points relative to vectorised line boundaries. Point-in-polygon operations also involve performing mathematical operations on attribute data fields, for points that fall within the same polygon. For example, if crimes are identified as points on a map, then the number of crimes within a census tract (polygon) can be measured by commanding the software to count the number of points by polygon. If the crime is burglary and the value loss from the burglary is included as an attribute field, then the software can be used to calculate the total loss from burglary within a census tract. Standard mathematical forms of aggregation include performing counts, averaging, multiplying, calculating standard deviations or variance, and so on.

8.5 True desktop GIS buffering

Buffering is the creation of polygons that surround other points, lines, or polygons. Individual buffers can be created around individual objects, or multiple objects can be buffered to act as one buffer area. The user may wish to create buffers to exclude a certain amount of area around a point, line, or polygon, or to include only the buffer area in a study. For example, in using GIS to help determine the possible sites for a new water well, the location of chemical factories may have a 10-km buffer drawn around them so that these buffered areas are excluded in the list of possible sites. The user of desktop GIS can indicate how many line segments are to be used to make up the boundary of the buffer, thereby controlling the accuracy of the buffer boundary. Buffering of points is performed in a way analogous to the 'radius select' or 'polygon select' tools described above. Buffering of lines and polygons can be accomplished by using the buffer function in the desktop GIS program.

8.6 True desktop GIS programming

GIS is foremost a *spatial* database management tool. One of the uses of database management is application development through the use of the programming language feature of the database management program. General purpose GIS programs are seen by many business managers as being too difficult to learn. They want push-button GIS capability in the software where no special knowledge on how to use a GIS is required. Special purpose GIS or niche GIS is being demanded by those without the knowledge or time to proceed up a steep GIS learning curve. Niche GIS programs are programmed using either

true desktop GIS, or written using a geographically-enabled software language.

True desktop GIS software has language capabilities that allow the development of specific or niche applications. As a variation on this theme, the many true desktop GIS software vendors offer the capability to add GIS functionality to programs written in languages such as Visual Basic and C++, or written in database management languages such as PowerBuilder, Visual Fox Pro or Access.

9 GEOGRAPHICALLY-ENABLED PROGRAMMING LANGUAGES

Some desktop GIS have add-on products that allow the user to program with a language embedded within the GIS software. This capability generally requires a fee in addition to the price of the true desktop GIS software. Also, if copies of the resulting program are distributed to other users, additional fees are generally payable.

MapInfo Corporation sells a programming language known as MapBASIC. MapBASIC is similar to the generic BASIC with the addition of geographical operators such as 'is located within' that are used for the distinct geographical operations. Caliper Corporation also offers a stand alone geographically enabled language known as the 'geographic information systems developers kit' (GISDK).

Other true desktop GIS software programs allow the programming of modules using standard programming languages such as Visual C++ or Visual Basic. They permit the use of the generic languages to write modules that can be used in conjunction with the GIS. The modules are called from the GIS as standard dynamic link libraries (DLLs) or using object linking and embedding (OLE). Often, however, communicating with DLLs or using OLE requires knowledge of advanced programming procedures.

If the applications programmer is not starting with a pre-existing GIS program, but is writing a program that will have GIS capabilities, a language such as Visual Basic or C++ will probably be used for the primary program. A commercial off-the-shelf add-on module can then be used to write the GIS component of the program. Standard components called VBXs (16 bit components) or OCXs (32 bit components) exist that allow functions to be added to the Visual Basic or C++ program. Examples of functionality added to software programs include the capabilities of drawing and charting, spreadsheets, wordprocessing, database management, telecommunications, and so on. These add-on modules, called custom controls, save the programmer the time needed to program them from scratch, and are usually moderately priced and very easy to use. The VBX or OCX feature is thus an added tool in the programmer's toolbox. The programmer can add the feature to the program with a click of a mouse on the tool. For further discussion on the role of GIS programming tools and the impact on the GIS industry see Thrall (1996b, 1996c).

10 CONCLUSION

Desktop GIS has changed immeasurably since its inception as a province of desktop computing. It has grown to include many new features and capabilities. The start-up costs of learning 'traditional' GIS software have never been small and with added features and functionality comes a steeper learning curve. Yet users of spatial software have increasingly demanded more user-friendliness. The response to these demands has been that parts of what has always been considered GIS are being spun off into new market niche software. Thus while 'traditional' GIS has been growing in capabilities, its stature is at the same time being eroded by niche software and other generic spreadsheets and databases – as vendors discover that these are symbiotic with certain GIS features. GIS software vendors are finding opportunities to form cooperative ventures with generic software vendors, and those that fail to form such cooperative ventures may in time find their own market lost to niche software vendors.

References

Elshaw Thrall S 1997 MapViewer: first impressions software review. *Geo Info Systems* 7 (November)

Haynes K, Fotheringham A S 1984 *Gravity and spatial interaction models.* Newbury Park, Sage Publications

Longley P, Clarke G 1995 *GIS for business and service planning.* Cambridge (UK), GeoInformation International

Martin D 1996 *Geographic information systems: socioeconomic applications,* 2nd edition. London, Routledge

Sonnen D 1996 Spatial information management: an emerging market. *Map World* 1: 24

Star J L, Estes J E 1990 *Geographic information systems.* Englewood Cliffs, Prentice-Hall

Thrall G 1992 Using the JOIN function to compare census tract populations between census years. *Geo Info Systems* 2: 78–81

Thrall G 1995a Surfer: review of 3-dimensional surface modelling software. *Journal of Real Estate Literature* 4: 73–5

Thrall G 1995b New generation of mass-market GIS software: a commentary. *Geo Info Systems* 5: 58–60

Thrall G 1995c The stages of GIS reasoning. *Geo Info Systems* 5: 46–51

Thrall G 1996a Battle builds for business GIS market. *Geo Info Systems* 6: 46–7

Thrall G 1996b SylvanMaps/OCX: first impressions software review. *Geo Info Systems* 6: 47–8

Thrall G 1996c Modular component programming: the foundations of GIS applications. *Geo Info Systems* 6: 45–6

Thrall G 1996d Maps, data and mapplets: first impressions software review. *Geo Info Systems* 6: 48–9

Thrall G, McLean M 1997 Blurring the lines between GIS and desktop publishing: first impressions software review. *Geo Info Systems* 7: 49–53

Thrall G, Marks M 1994 Functional requirements of a geographic information system for performing real estate research and analysis. *Journal of Real Estate Literature* 1: 49–61

Thrall G, Trudeau M 1996 Java and Applets: breakfast of champions. *Geo Info Systems Showcase* 6: 10–15

Thrall G, Valle J del, Elshaw Thrall S 1995 Review of GUI-based GIS software products. *Geo Info Systems* 5: 60–5

24

GIS interoperability

M SONDHEIM, K GARDELS, AND K BUEHLER

The term 'interoperability' has a range of meanings, but all focus on the ability to move easily from one system to another. This chapter reviews various strategies that may be used to achieve degrees of interoperability, including common exchange formats, conversions, common interfaces, and common database models. Geographical data modelling lies at the core of the interoperability issue, since agreement is needed on the representational framework. The chapter also covers issues of geographical metadata, common catalogues, and other tools to support search and retrieval across systems.

1 INTRODUCTION

Interoperability has been a goal of the computer industry for years and is essentially one view of the push for open systems. The term 'open systems' usually implies the intention to adhere to vendor-neutral computing standards, with the added benefit of producing a more level playing field among software and hardware companies. As a means of achieving open systems, interoperability has centred on common communications infrastructures, application programming interfaces in the public domain, and a common architecture for defining objects and transporting them across networks. Open computing and interoperability are as vital to geographical information systems as to any other area of information technology.

In the GIS arena interoperability has been a serious concern since the late 1970s. Problems with incompatible computing environments have been compounded by the inherent complexity of geographical information and the many ways in which it can be modelled. Nevertheless, some interoperability schemes are viable today, and more powerful ones are under development. This chapter reviews the general notion of interoperability, strategies for GIS interoperability, geospatial data models, and high-level concepts about information sharing.

2 GENERAL NOTION OF INTEROPERABILITY

Where applications require more than the operations available on a single desktop and dataset, data users and providers must achieve a new set of objectives regarding heterogeneous data and distributed computing resources.

- Data producers must ensure that their data are readily accessible and understandable to potential data consumers.
- Users must be able to identify and locate relevant information, and know whether a given set of data is germane to their work.
- Queries to dispersed sites must be formulated and processed in a manner meaningful to both data server and application client.
- Geodata from one source must be capable of being integrated with data from another, in terms of both structure and semantics.
- Display and analytical functions must be associated with particular data models and made available to the requester.

In some situations, meeting these requirements is relatively straightforward: a distributed application may be designed in a top-down fashion across an enterprise, based on common technology and a common semantic framework. The enterprise standardises on a given technical infrastructure, typically provided by a single vendor or systems integrator. Individuals in the enterprise use a distributed database and associated applications, built on a limited set of data models and targeting particular business needs. This approach of common semantics and technology may be scaled up to include a group or community of organisations as long as all of the participants are willing and able to adopt the new system. Although such scenarios meet the primary goals of interoperability, they are usually described as integrated systems, rather than interoperability solutions.

Interoperability usually refers to bottom-up efforts (Litwin et al 1990), neither imposed by a central authority nor driven by a single application. The systems and data models found among the users are heterogeneous, having been developed independently of one another; consequently, systems re-engineering may be required to meet basic interoperability requirements. Two significant challenges must be met:

1 the autonomous systems must be able to exchange data and to handle queries and other processing requests;
2 they must be able to make use of a common understanding of the data and requests.

The first requirement implies that a common set of services must be universally available and accessible through network communications. The second dictates that a common formal language and a common model representation be used (UCGIS 1996). The semantics of the data, as conveyed through the language and model, are either used directly by each system, or are transformed to local semantic constructs which can be interpreted directly by the users' applications.

3 GIS INTEROPERABILITY STRATEGIES

3.1 Direct translation

Over the last two decades, much practical interoperability work has boiled down to exchanging geographical and computer-aided-design data by using bi-directional translators operating on vendor-specific data formats. The architecture of most translators amounts to a data reader, a correlation table defining the correspondence between input and output data types and a writer. The correlation table, which may be hard-wired into the translator, defines

how given data types and values in the input stream should relate to data types and values in the output stream (Figure 1). Consequently, where the input and output data models are dissimilar, a good match may not be possible, resulting in loss of information. Often such translators are most successful when tailored for specific datasets.

A more powerful approach to translation is shown in Figure 2. In this case the software maintains an internal data model which is semantically more rich than either the input or output models. The input data types and values are mapped to types and values of this internal model. The transfer data types play essentially the same role as a common format; however, in this case, the intermediary form is a transient in-memory representation. Once the data are in the transfer data types, they may be redefined if desired through a series of transformations and geoprocessing steps available in the translation software. The software may even support feature redefinition in which a number of input features define a single output feature, or vice versa. Practically, it becomes possible to consider very different input and output models and to infer matches impossible with direct correlation. This kind of processing may be termed smart translation or semantic translation.

Consider forest stand data in a given system with the attributes stored partly in labels and partly in column-aligned ASCII tables, and with the area for each stand represented by a point and a set of bounding arcs which are also shared by the adjacent areas. Now assume that the data are desired in another system where each area is represented as an independent polygon with no inside point, and with all attributes in a dbf file (a common standard file format for databases). Additionally, users may want to simplify the line work and amalgamate small polygons with their most similar neighbours. It may also be desired to take a series of forest stand files, turn them into a seamless coverage, and then clip the result to a given watershed boundary. To carry out such translations is not possible with the first architecture, but is with the second.

Fig 1. Simple translation through correlation.

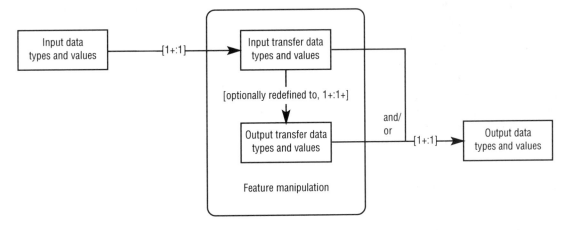

Fig 2. Semantic translation through feature manipulation and correlation.

Regardless of architecture, direct translation depends on the ability to read and write commercial formats. Some vendors have made their format specifications publicly available, but others have not. In the latter case, reverse engineering may be possible, but long-term maintenance of the translator may be costly because of undisclosed format changes by the vendors as their products evolve. Another option for vendors is to provide software interfaces (application programming interfaces, APIs) to their data management applications, as discussed in section 3.3.

3.2 Common transfer format

A variation of simple translation is the use of a basic format such as DXF or DLG (for a more detailed discussion of exchange standards see Salgé, Chapter 50) as part of a two-step process. Data in System A format are first translated to the intermediate format and from there a second translation converts them to the System B format. As used, these basic formats represent a lowest common denominator between the two end systems, possibly resulting in information loss. Additional limitations are that there may be no means of including the definitions of the types of data within the file, of handling metadata, of addressing updates, or of translating some kinds of data at all.

A number of efforts have centred on developing loss-less file exchange standards, based on non-proprietary file formats. Some of these standards have been aimed at resolving specific problems, such

as a government agency receiving topographic data from private contractors in a neutral format (IEF in Israel or MOEP in British Columbia). Some have been designed to work within given disciplines or areas of endeavour. The International Association of Geodesy and other groups working with the Global Positioning System (GPS: see Lange and Gilbert, Chapter 33) recommend use of the receiver independent exchange format (RINEX) for the exchange of GPS data for post-processing. The International Hydrographic Organisation has developed S–57 as a transfer standard for digital hydrographic data, with the long-term intention of replacing traditional nautical charts around the world with electronic navigation charts supplied in S–57 format.

Other loss-less file-based efforts have been much more general in intent and have included sophisticated data models for geographical data. A clean separation is made between the logical model and the data encoding, the physical representation in a file. The Spatial Data Transfer Standard (SDTS) is a standard from the US government (US Geological Survey 1996) that provides a logical data model and a physical encoding (ISO 8211). Sponsored by the Digital Geographic Information Working Group, associated with NATO, the Digital Geographic Information Exchange Standard (DIGEST) is a collection of file exchange standards for different types of data (Digital Geographic Information Working Group 1995a). It has many similarities to SDTS and also refers to ISO 8211 for encoding. The Spatial Archive and Interchange Format (SAIF) is a Canadian standard

349

based on an extensible, object-oriented paradigm and a formal modelling language (Geographic Data BC 1995). SAIF makes use of a zipped ASCII encoding which also allows for the direct inclusion of binary data. SDTS, DIGEST, and SAIF all provide support to developers through APIs to data files in their respective data encodings (see section 3.3).

Using the approach of a common format, each type of data holding is mapped to a common model, as manifest in a common format. Figure 3 shows some typical situations where the geometry and traditional attributes may be stored together or separately. As well, different data models may exist in each of these cases; for example, Model 2 and Model B may be quite dissimilar, even though they have similar implementations. The figure also shows the common model as implemented in a single file of a given format. Although this may be the case, an alternative is a series of files in given formats associated with particular types of data.

Use of a common format typically involves two operations: translating a file in a given format to an intermediary file in a common format, and then translating that intermediary file to a third file in the desired format. Only one step is required, of course, if the data are available in the common format, or if the common format is the desired format. Carrying this argument one step further, if the start and end formats are identical then translation is never required. In fact, there are many situations where Format X to Common Format to Format X is worthwhile, if the translation software provides quality assurance benefits or simply redefinition within the same format.

Work has also been underway to extend the data-centric view of files to a more processing-centric view. A technical committee of the Comité Européen de Normalisation (CEN) is examining geographical information (CEN/TC 287 1996) with the intent of developing a transfer method based on a data model and encoding, as with the standards described above. However, it also makes use of EXPRESS (ISO 10303-11), a relationally-oriented language appropriate for handling traditional attributes. Through EXPRESS, query and update operations may be supported. Another committee (ISO/TC 211) under the auspices of the International Organisation for Standardisation (ISO) is working on an approach similar to that of CEN (ISO/TC 211 Secretariat 1996), but with the intention to include operators and services to enable logical model to logical model transformations. As a separate endeavour, the FMEBC (Geographic Data BC 1996) is a sophisticated translation software package associated with the SAIF standard and specifically SAIFLite (a pared-down version of SAIF for operational use). Through a scripting language, the FMEBC supports model-to-model transformations, geometric restructuring, geometric and semantic filtering, datum and projection transformations, various quality control operations, clipping, overlay etc.

3.3 Published interfaces

Moving away from files, formats and encoding entirely, attention now turns to interoperability through interfaces. With such an approach, the

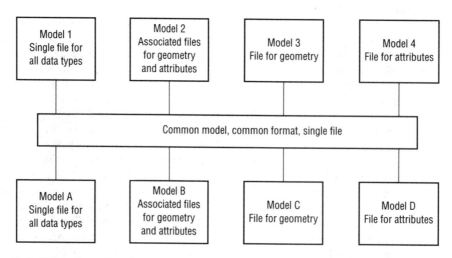

Fig 3. Data exchange through a common data model and format.

internal structure of the data is irrelevant; instead, the emphasis is on the behaviour presented by the interface. In other words, the interface is able to provide or accept data in response to a request. How it does so does not matter. There is no assumption that the data behind the interface must match the data provided to it or by it. Instead of making the specification of a proprietary format publicly available, vendors may choose to develop an API and to allow other developers to read and write their data through the API. An API can hide complexity, provide a number of capabilities supported by the application, and lessen the chances of inadvertently degrading database integrity. Moreover, it can be designed to respond to queries as part of a data access server. Translation software, as well as other applications, may interface to data holdings through an API (see also Worboys, Chapter 26).

Although APIs offer obvious advantages, they also have several disadvantages. They may be very dependent on particular technological environments, thus limiting their use in general interoperability scenarios. Because of marketing and engineering considerations, most APIs cannot operate independently of the main GIS application. Thus, if proprietary data files can be detached or exported from a GIS and shipped to another system, the receiver will not have access to the source vendor's API to help read the data. Only some GIS products have associated APIs, and typically, those that are available have not been created to any standard specification. Data received through different vendor APIs may conform to very different models, making integration of the data problematic.

An alternative approach to achieving interoperability is to develop an industry-wide common interface, based on distributed computing technologies (see also Worboys, Chapter 26). This is the primary goal of the Open GIS Consortium (OGC). Having coined the term OpenGIS, the OGC has as its mission 'the full integration of geospatial data and geoprocessing resources into mainstream computing and the widespread use of interoperable geoprocessing software and geospatial data products throughout the information infrastructure' (OGC 1996a). The OGC is developing an interface definition referred to as the OpenGIS Specification (formerly known as the Open Geodata Interoperability Specification). Interfaces compliant with this specification can be incorporated directly into new systems and built onto legacy systems.

Two significant components comprise the OpenGIS Specification: the Open Geodata Model (OGM) and the Services Architecture. The OGM is a collection of data types and methods, organised into a hierarchical class library. It is comprehensive, in that it embraces fundamental geospatial (and ultimately spatio-temporal) data types, including their geometric representation, spatial reference, and semantic content. The Services Architecture provides the mechanism by which individual objects and their associated interfaces may be assembled into complex queries, transformations, analytical functions, and presentation directives. It also enables the construction of catalogues that allow users to identify, evaluate, and interpret complex geographical information dispersed throughout a network.

Figure 4 shows a common interface for each of a number of data stores and applications. When a request is made by an application, it travels to the interface of a data store or another application. The interface returns the requested information as containers of information (encapsulated objects or components), in accordance with a distributed computing platform (DCP) specification. Such DCPs have been defined in Unix, Microsoft, and operating system independent environments. Examples include the Common Object Request Broker Architecture (CORBA) from the Object Management Group and the associated OpenDoc component model from the CI Labs consortium; Object Linking and Embedding (OLE), Common Object Model, Distributed Common Object Model and Active X from Microsoft; and Java and Java Beans from SunSoft. Consequently, the OpenGIS Specification must be sufficiently abstract to be able to be implemented in very different technical environments (OGC 1996b).

The common interface approach will allow disparate applications and technologies to interoperate as a single system. Ideally, applications can be built with both data and processing components coming from anywhere on a network. However, to define the common interface, to implement it effectively for day-to-day operational needs, and to gain widespread acceptance of it will take a long-term, concerted effort. Another potential disadvantage is that the common interface may not be capable of delivering objects required for a given application; for example, specific data types or subsets of the data may be of interest, but the interface specification may not have addressed these

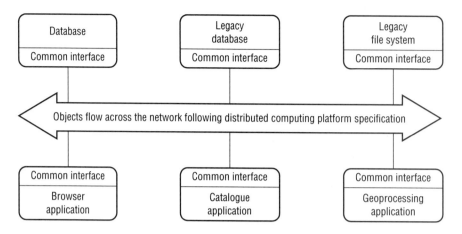

Fig 4. Interoperability through a common interface.

requirements adequately. We may see vendors supporting a common interface and extending it in line with their particular product capabilities. As has happened with SQL (section 3.4 below), the common interface once established may evolve over time to meet increasingly varied and complex needs.

3.4 Database developments

There has been a great deal of interest within the database world in managing atypical, non-tabular data, which together are considered as multimedia. The SQL language, long the backbone of the relational database market, was originally designed to handle only tabular data (Egenhofer and Kuhn, Chapter 28; Worboys, Chapter 26). However, the latest version, SQL3, is comparable to a full programming language and includes the ability to define abstract data types (ADT). An ADT may include functions; a polygon may have one function to return its boundary and another to return the number of holes within the polygon. An ADT is roughly equivalent to an object type in object-oriented technology. Instead of thinking about data in a database as consisting of rows of attributes with primitive domains (integers, real numbers, character strings, and the like), we can now include higher-level data types such as lines and polygons, each of which may have various functions as an inherent part of its definition.

The creation of a standardised set of data types, based on the ADT capability in SQL3, is the basis for the multimedia extensions known as SQL/MM and defined in the SQL3 language. One particular application area included in SQL/MM is Spatial (ISO/IEC JTC 1/SC 21 1996). These developments are significant because they enable databases to store and process a wide variety of data types, including spatial data types, all in the same environment. The enhancement of extended-relational and object-relational databases to geospatial applications parallels the migration of conventional APIs to more object-oriented designs encapsulating both data and methods. SQL/MM Spatial and the OpenGIS Specification are compatible standardisation efforts which will make it relatively easy for database vendors intending to comply with the former to also comply with the latter. Consequently, connectivity between different databases housing geospatial data and between such databases and other software products will become practical through the common interface defined by the OpenGIS Specification.

The different interoperability strategies described above have been developed to meet different technical objectives. Even so, there is a high degree of commonality in the underlying geospatial models. STDS, the first of these developments, strongly influenced aspects of DIGEST, and both influenced SAIF. SQL/MM Spatial and aspects of the OpenGIS Specification are related to SAIF, which in turn was modified to mesh with new ideas brought forward by these initiatives. Even with the diverse requirements driving all of these efforts, the underlying models have much in common. This is good news for organisations wishing to use various interoperability strategies (see Bédard, Chapter 29; Salgé, Chapter 50).

4 GEOGRAPHICAL DATA MODELLING

Data management and analysis demand that basic notions about data types be established through data models. With geographical data this implies that general-purpose concepts of geographical phenomena be available, as well as definitions of space and time. On the one hand, such models must be practical to understand and to implement; on the other, they must be sufficiently sophisticated and robust to describe data types applicable to a wide range of geographical applications. The modelling techniques used will influence the ease with which such concepts can be portrayed and implemented in different systems.

Geographical data modelling may make use of both text-based and graphical techniques. Examples of the former are INTERLIS, for data exchange of cadastral data (GEOHUB 1996), Class Syntax Notation, the data modelling language of SAIF (Geographic Data BC 1995), and SQL/MM Spatial (ISO 1996), designed specifically for database applications and under the auspices of the ISO. Graphical methods in common use include OMT (Rumbaugh et al 1991), the Unified Modelling Language (Booch et al 1996), Syntropy (Cook and Daniels 1994), and extended entity-relationship methods such as Designer/2000 (Oracle 1996). Bédard (Chapter 29) has also developed an approach with icon extensions to represent spatial and temporal constructs. Most of these techniques are object-oriented and provide a great deal of flexibility.

4.1 Abstraction of geographical phenomena

Fundamental to the interoperability problem is that different groups of users have different Earth models, which in turn manifest themselves in the representation of geographical information (Mark, Chapter 7; Martin, Chapter 6; Peuquet, Chapter 8; Raper, Chapter 5). Of significance is how perceived elements of the landscape are described and modelled, and how queries and analytical services on such data are defined and implemented. The most basic questions concern the conceptualisation of geographical entities and the space and time in which they reside. Such entities, existing in the real world, are represented in computer systems by units of data referred to as features or geographical objects, which in turn may be categorised into a general type of object termed Feature (using the OGC terminology, and equivalent to GeographicObject in SAIF).

4.1.1 Features

A *feature* is a representation of a phenomenon in object space, typically modelled through a series of attributes, including position. More specifically, it has a set of properties which includes its spatial representation, manifested through a geometric object and an associated spatial or spatio-temporal reference system. The set of properties may also include other properties, of a non-spatial nature. A number of these properties, in conjunction with the feature type, establish the semantics of the feature. A road segment may be defined as having various attributes including the number of lanes, the surface material, average traffic flows, update history, and a spatial representation as a geometric object such as an arc or curve. The coordinates of the points defining the geometry, and the implied geometric shape between the points, are meaningful in the context of a spatial reference system, including horizontal and vertical data and a planar or curvilinear projection.

A feature may also be composed recursively of other features, such as with a road network composed of roads or a farm consisting of agricultural fields and farm buildings. As described in the next section, some types of feature pertain to phenomena which vary continuously over a spatial extent, as with reflectance values in a satellite image, forest stand classification across a forest, or barometric pressure across a landscape.

Figure 5 provides a more formal definition of a feature (OGC 1966c), based on the OpenGIS Specification and using Syntropy. Every feature has an object identity which must be unique over space and time. It also has a property set, equivalent to a row in a relational database. The property set contains zero to many named properties (i.e. attributes). Each property may be simple (e.g. age represented as an integer) or complex (e.g. chemistry represented by a series of chemical attributes), and optionally may have constraints restricting its possible value. Zero to many properties are geometric; associated with a reference system, they specify the spatial or spatio-temporal extent of the feature. Geometric properties, which are types of properties, include coordinate geometry and may include other properties such as positional accuracy. A feature may be associated with other features, through various relationships, including spatial and temporal relationships. The association may also indicate containment, as previously noted. A user may define feature subtypes specific to his or her dataset or application.

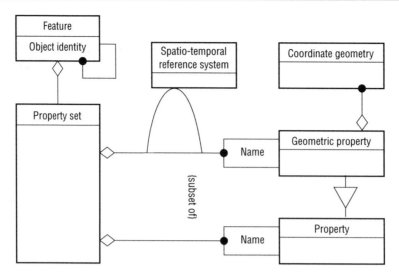

Fig 5. Feature, after the OpenGIS Specification.

4.1.2 Coverages

A point of ongoing discussion in the GIS literature has been the distinction between feature concepts based on discrete objects, and coverage concepts pertaining to continuous phenomena (Peuquet 1984). In the former case, the discrete object is generally treated as homogeneous with respect to its attribute values and usually the spatial boundaries of the object (e.g. the surface of a 3-dimensional object) are directly observable. Spatial location is often conceived as simply another attribute which distinguishes a geographical object from any other kind of object (Worboys, Chapter 26; 1995). Sometimes traditional attributes such as average elevation and slope are derived from the geometry and treated as additional dimensions in the multidimensional attribute space characterising typical analyses.

By contrast, a coverage (cf. the concept of a field discussed by Mark, Chapter 7) is a phenomenon in object space modelled as an attribute range (possibly complex) over some spatio-temporal domain. Coverage concepts imply a continuous surface in 2- or 3-dimensional space, with a given attribute value characterising any point on the surface. The concept of coverage may also be applied to solids (i.e. volumes), in 3-dimensional space. Mathematically, a coverage implies a computable function. Input to the function is any spatial or spatio-temporal position within a given spatial or spatio-temporal extent, and output is a value for a simple or complex attribute. Such a function may exist explicitly as a mathematical expression or it may simply be implied by the data type and structure.

Housing prices across a city could be represented as a trend surface through a polynomial or thin plate spline function. Breaks may exist on such surfaces, as with a cliff on an elevation map. With either prices or elevation, values may be displayed as contour lines or as a field of values at specific locations. A typical partitioned coverage of categorical thematic data shows bounded areas treated as internally homogeneous with discontinuities (i.e. boundaries) separating one area from another. Thus, a forest stand map or a map of countries can considered as defined by a discrete-valued function.

Error or uncertainty may be an intrinsic part of a phenomenon (Fisher, Chapter 13), as with the concentration of a given mineral calculated by a Kriging function, which produces both a predicted surface of the concentration and an associated error surface. These surfaces may be represented as interrelated coverages, or as a single coverage with a pair of predicted and error values as output. Geometrically, a simple surface may be expressed as contours, as a triangulated irregular network, as a set of points at locations of interest, etc. Fuzzy boundaries on categorical maps may imply fuzzy membership; land-use categories across the landscape may be represented by a set of membership functions, each giving the likelihood of belonging to a given category (Fisher, Chapter 13). In this case, the series of individual functions may be treated as a single function, with geographical position as input and a set of likelihood estimates as output.

From a modelling perspective the question is: what is the relationship between feature and coverage? Instead of treating them as two entirely separate types of objects, coverage may be modelled as a specialised type of feature which either explicitly or implicitly stores a function able to provide an attribute value for any point across a given extent. Figure 6 presents a simplified view derived from the OpenGIS Specification (OGC 1996c). For reasons of clarity, the figure shows Geometry instead of Geometric Property, Coordinate Geometry, and Spatial Temporal Reference System (which are inherited from Feature).

Practically, a coverage and its associated function can take on one of several forms:

- a partitioned coverage with each area or patch defined by a discrete value;
- a set of values at specific point locations and with or without breaks or surface discontinuities;
- rasters of images and other similarly structured data;
- networked patches (e.g. triangulated irregular networks) with associated values;
- purely analytical functions;
- combinations of these such as a series of patches with an analytical routine pertaining to each or all patches.

From an interoperability perspective, the first four options should be reasonably easy to handle because they relate to well-known data structures. The fifth

and sixth options pose the problem of representing the analytical functions and their programming interfaces in a universally acceptable way. Technically this requires that an interface to a dataset or application must not only be able to retrieve specific values, but also be able to specify parameters for an analytical method used at the server to generate the values. Alternatively, the methods may be returned to the requesting client, encapsulated with the raw data (see Goodchild and Longley, Chapter 40).

4.2 Space, time, and object identity

Concepts of space and time require (1) spatial and temporal reference systems and (2) geometric and temporal constructs which define position in the context of these systems (Couclelis, Chapter 2). The reference systems define the coordinate space, which typically has either two or three spatial dimensions and optionally a temporal dimension (Peuquet, Chapter 8).

A variety of types of spatial coordinate systems are in use around the world for different applications and in different areas. They include simple rectangular systems used in CAD environments to Earth-related systems defined through combinations of the reference ellipsoid, horizontal adjustment system, vertical adjustment system or surface, and projection. Because of extensive standardisation efforts in the geodetic realm, models of spatial reference systems are quite similar in most geoprocessing applications. A major exception has to do with linear referencing systems. Used extensively in road networks, they are based on driven distance from arbitrary reference points.

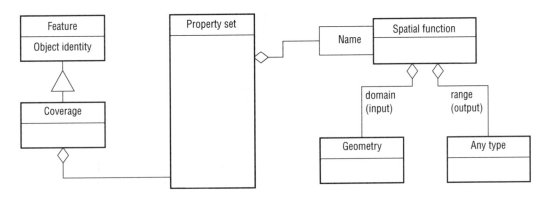

Fig 6. Coverage definition after OpenGIS Specification.

355

Interoperability with transportation and hydrological applications is complicated by the proliferation of linear referencing systems and discrepancies between positions defined in such systems and their correct geographical positions.

Time may be incorporated in two very different ways, either as part of a coordinate system, or as a time, a date, a time stamp (time and date together), or as an interval. The time may be specified with respect to universal time coordinates (UTC) or a local time zone. Less obvious issues of time are to distinguish among, and have access to, the actual time of an event, the time of observation, the time an object first entered a data store, the time the object was modified or replaced in a data store, and the time an object was retired or deleted from a data store (Peuquet, Chapter 8).

Closely tied to those notions of time is the issue of object identity. If a river changes course, we may consider that the geometry of a river segment has changed, or alternatively, we may state that the original segment has retired and been replaced by a new segment (Veregin, Chapter 12). Such concerns are of fundamental interest to interoperability if queries about changes in the landscape are to be answered. The problem is compounded by the requirements of different applications; it may be that a database must be able to present a number of views, depending upon the context of the query.

5 SHARING INFORMATION

5.1 Information communities

Information communities (Lake 1996) are groups of users, both data providers and consumers, who share digital data. To create a Geospatial Information Community or GIC (OGC 1996c) typically requires acceptance of particular feature (and coverage) models, reference systems, and geometries. Where models from user to user are divergent, the users may still form a community, so long as they have the means to exchange information through an interoperability strategy.

A number of organisations have created information communities around particular needs. Three of these are briefly reviewed. The Petrotechnical Open Software Corporation (POSC 1995) is a not-for-profit organisation funded by the oil industry with the objective of establishing and promoting standards for information sharing specifically within the exploration and production sectors of that industry. Part 4 of DIGEST (see section 3.2) defines the Feature and Attribute Coding Catalogue (FACC; Digital Geographic Information Working Group 1995b). The FACC provides standardised definitions and codes for features in ten categories (Culture, Hydrography, Hypsography. Physiography, Vegetation, Demarcation, Aeronautical Information, Cadastral, Special use, and General). It is the intention of the Digital Geographic Information Working Group that all NATO defence forces adopt these definitions and use them with the DIGEST transfer format. Originally developed by private industry and now under the auspices of the Comité Européen de Normalisation, TC 278, the Geographic Data Files standard (GDF: European Commission DGXIII 1996) includes models of road network features (Salgé, Chapter 50). GDF is playing a prominent role, especially in Europe, in the development of intelligent transportation systems.

Efforts are under way to create general-purpose geospatial information communities. Under the Federal Geographic Data Committee, the National Spatial Data Infrastructure (FGDC 1996) is a long-term effort to share geospatial data throughout the USA. It has been structured around two major activities: the Clearinghouse, which is a metadata cataloguing and searching model for enabling access to distributed geodata resources, and the Framework, which is focused specifically on characterising a collection of datasets that are of national interest and whose feature attributes may be standardised (Guptill, Chapter 49).

5.2 Schemas, schema fusion, and schema transformation

A set of data type definitions, i.e. data models, constitutes a *schema*. It provides a systematic and coherent description of the content and organisation of a dataset or collection. It can include the semantic representation of features, the details of attribute names and types, the dictionary defining the structure of geographical objects and their spatial and non-spatial attributes, metadata providing a synoptic view of geographical information, and a thesaurus of related terms. A government agency may define schemas pertaining to the types of data they collect and distribute, and similarly, members of an information community may define a schema meaningful to them.

Key interoperability issues include how schemas are represented, how a given user-defined schema relates to standardised geospatial concepts (e.g. feature, reference system, geometry), and how different user-defined schemas relate to one another. Geospatial standards provide a high degree of semantic consistency insofar as fundamental geospatial concepts are concerned. Outside these concepts, however, semantic consistency remains an issue. *Schema fusion* involves creation of a common schema capable of describing all or most information of interest. At times a more practical alternative is to establish a series of model-to-model transformations, such that a given schema is redefined in a way most appropriate to the recipient. This latter approach is more flexible but only possible with a detailed knowledge of the respective schemas and *schema transformation* software.

5.3 Metadata and catalogues

Metadata, or data about data, may be used to give a high-level description of a data collection, including such items as the feature types, ownership, quality, lineage, spatial referencing, currency, and version (Guptill, Chapter 49). Metadata are intended to serve many objectives, ranging from dataset documentation, to provision of concise definitions of the dataset's structure and organisation, to a basis for browsing and searching (Goodchild and Longley, Chapter 40). One of the more influential developments is the Content Standards for Digital Geospatial Metadata from the US Federal Geographic Data Committee (FGDC 1994). The standard specifies the metadata content only; it does not state how such data should be stored or accessed. It also does not require that schemas or detailed data dictionaries be available.

A *catalogue* is a collection of metadata descriptions which may apply to various levels of aggregation or granularity. For example, it may pertain to collections of datasets (such as all USGS 7.5' quadrangle maps), to a single dataset (all data on a given map), or to given types of data only (all roads on the entire series or on some subset of it). Acting as a gazetteer or index, a catalogue serves as an authoritative reference on one or more data collections and may include a data dictionary and formal data models. Potential data users may browse the catalogue to determine the data's relevance, extent, cost, etc., as well as valid ways to query the

data. The catalogue may even control access through some sort of authentication process. A significant concern with catalogues is to ensure that the metadata they contain is current. If the catalogue includes descriptions of a large number of active datasets, maintenance requires ongoing cooperation from the data providers. Such cooperation is particularly practical if the data providers are all part of an information community.

6 CONCLUSIONS

The GIS interoperability strategies described in this chapter are not mutually exclusive. Data managers and GIS practitioners may use them in complementary ways, as appropriate for specific environments or data holdings. Initially vendors are likely to develop interoperability among their own products. As pressure from specific communities grows and as the underlying technology matures, both vendors and third parties can be expected to offer true, multi-vendor interoperability solutions. It is easy to imagine purchasing criteria which include meeting specific interoperability capabilities.

GIS interoperability developments are significant for another reason as well: they are contributing to the migration away from the monolithic systems which have dominated the GIS market for so long. Databases, browsers, smart translators, and geoprocessing tools are now being coupled through adherence to open standards and specifications. The backbone of much GIS activity is likely to be network-accessible databases, with connectivity requirements met by common interfaces. The big winners will be information communities gearing up to take advantage of these new opportunities.

References

Booch G, Rumbaugh J, Jacobson I 1996 *The unified modelling language for object-oriented development. Documentation set Version 0.9 addendum.* Santa Clara. *http://www.rational.com/ot/uml.html*

CEN/TC 287 Secretariat 1996 *CEN/TC 287 Geographic Information. http://www.statkart.no/sk/standard/cen/*

Cook S, Daniels J 1994 *Designing object systems: object-oriented modelling with syntropy.* Englewood Cliffs, Prentice-Hall

Digital Geographic Information Working Group 1995a *Digital geographic information exchange standard.* Version 1.2a. *http://132.156.33.161/Engineer/DIGEST_1.2a/cover.htm*

Digital Geographic Information Working Group 1995b *Feature attribute and coding catalogue.* Version 1.2a. *http://132.156.33.161/Engineer/DIGEST_1.2a/covers/part4.htm*

European Commission DGXIII 1996 GDF home page. *http://205.139.128.5/ehq/gdf*

FGDC 1994 *Content standards for digital geospatial metadata,* Washington DC, Federal Geographic Data Committee. *http://www.fgdc.gov/metadata*

FGDC 1996 *National Spatial Data Infrastructure,* Washington DC, Federal Geographic Data Committee. *http://www.fgdc.gov/nsdi2.html*

Geographic Data BC 1995 *Spatial archive and interchange format: formal definition.* Release 3.2. *http://www.env.gov.bc.ca/gdbc/saif32*

Geographic Data BC 1996 *Feature manipulation engine BC.* *http://www.env.gov.bc.ca/gdbc/fmebc*

GEOHUB 1996 *INTERLIS.* *http://www.geo.unizh.ch/~keller/geohub.html#tagOO2*

ISO/IEC JTC 1/SC 21 1996 *SQL multimedia and application packages (SQL/MM) – Part 3: spatial.* ISO/IEC JTC 1/SC 21 N 10441, ISO/IEC CD 13249-3:199x (E) SQL/MM:MAD005. *ftp://speckle.ncsl.nist.gov/isowg3/sqlmm/MADdocs/*

ISO/TC 211 Secretariat 1996 *ISO/TC 211 Geographic Information/Geomatic. http://www.statkart.no/isotc211/*

Lake R W 1996 Information communities support information sharing. *GIS World* 9(2): 72–3

Litwin W, Mark L, Roussopoulos N 1990 Interoperability of multiple autonomous databases. *ACM Computing Surveys* 22: 265–93

OGC (Open GIS Consortium) 1996a *Vision and mission statements. http://www.opengis.org/vision.html*

OGC (Open GIS Consortium) 1996b *The OpenGIS guide, introduction to interoperable geoprocessing, Part 1. http://ogis.org/guide/guide1.html*

OGC (Open GIS Consortium) 1996c *The OpenGIS abstract specification. http://www.opengis.org/public/abstract.html*

Oracle Corporation 1996 *Oracle Designer/2000.* Release 1.3

Petrotechnical Open Software Corporation (POSC) 1995 *POSC technical information.* *http://www.posc.org/technical_toc.html*

Peuquet D J 1984 A conceptual framework and comparison of spatial data models. *Cartographica.* 21: 66–133

Rumbaugh J, Blaha M, Premerlani W, Eddy F, Lorensen W 1991 *Object-oriented modelling and design.* Englewood Cliffs, Prentice-Hall

US Geological Survey 1996 *Spatial data transfer standard. http://mcmcweb.er.usgs.gov/sdts/standard.html*

UCGIS (University Consortium for Geographic Information Science) 1996 *The UCGIS research agenda: interoperability. http://www.ucgis.org*

Worboys M F 1995 *GIS, a computing perspective.* London, Taylor and Francis

25

GIS customisation

D J MAGUIRE

Customisation is the process of adapting a generic system to an individual specification. It is generally considered one of the most expensive non-personnel components of a GIS implementation. Because of the limited size and diversity of the GIS market, many GIS software developers have adopted the approach of developing a generic suite of multi-purpose software routines, together with some type of customisation programming capability. This has allowed core GIS software developers to concentrate effort on engineering robust and reliable generic routines. The task of creating specific-purpose, end-user (or vertical application) customisations is usually seen as the domain of application developers.

In the case of desktop and professional level GIS the process of customisation typically involves modification of a standard graphical user interface and extension of the 'out of the box' tools by writing application programs. More sophisticated users may be allowed access to the underlying core GIS capabilities and database. They may be able to extend the core class libraries or reuse objects within their own programs.

Traditionally, GIS software developers have had to develop their own programming languages. However, with the wider incorporation within GIS of industry standard programming environments – such as Visual Basic, Visual C++, and Java – this is changing.

1 INTRODUCTION

GIS software has been used in an extremely wide range of applications: from archaeological site mapping, to managing land assets, to storm runoff prediction, and global zoological analysis. One of the main reasons why it has been possible to employ GIS software in such a diverse range of applications is because of the customisation capabilities that software developers incorporate into their products. These allow application developers to create specific customisations of generic software systems. This inherent flexibility has been one of the major factors in the success of GIS.

This chapter begins with a short history of GIS customisation and the various approaches adopted to customising GIS software systems. The process of GIS customisation is then described in detail, including some consideration of costs. This is followed by a look at the role of software engineering in GIS customisation. Next, examples of the main GIS customisation tools are described. Finally, the conclusion draws the main points together and briefly looks towards the future.

In the early days of GIS-relevant software development (the 1960s and 1970s) all of the software systems produced were specific-purpose and highly tailored to each application. These monolithic (sometimes called 'stovepipe') systems were unique islands of information processing functionality incapable of exchanging data with other systems (see also Batty, Chapter 21; Sondheim et al, Chapter 24). This situation arose for several reasons.

- Each system had to be developed from scratch because there were no other systems or common pools of routines to adapt or extend.
- The market was very small and there was no incentive to develop generic systems which could be extended or adapted, and then sold to other users.

- There was limited expertise within the developer community about what constituted a generic application.
- The hardware and software development tool limitations of the day necessitated that each application be highly optimised to give the best possible performance.

As technology developed and the expertise and market size grew, the effects of each of these limitations was ameliorated. In the 1980s GIS software developers began to create generic GIS software systems capable of customisation and then deployment in multiple application areas. The first example of a successful generic GIS software system was ARC/INFO, from Environmental Systems Research Institute Inc. (ESRI), released in 1981. In the first few releases, however, the capabilities for user customisation were limited. It was not until the release of the ARC Macro Language (AML) as part of ARC/INFO 4.0 in 1987 that the software was really capable of being customised by end-users. In the late 1980s and early 1990s virtually all major GIS software vendors adopted this approach of developing a generic suite of multi-purpose software routines, together with some type of customisation programming capability. This has allowed core GIS software developers to concentrate effort on engineering robust and reliable generic routines. The task of creating specific-purpose, end-user customisations is usually seen as the domain of application developers. These application developers may belong to the core software developer's organisation, a user organisation, or some independent third-party organisation. This approach to software and product development has significant implications for the cost of implementing GIS in organisations and also the levels of technical expertise required by users, as the following discussion will demonstrate.

More recently, with the late 1990s developments in technology (Batty, Chapter 21) and the cumulative increase in expertise and market size, the trend has been more towards systems designed for specific endusers with only limited capabilities for applications development (Elshaw Thrall and Thrall, Chapter 23). Currently, this is more evident at the lower end of the market. The typology in Figure 1 shows how mass market end-user products – such as ESRI's *Business*MAP, MapLinx's MapLinx, and AutoRoute from Microsoft – have extremely limited

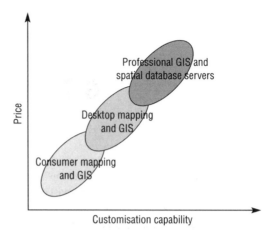

Fig 1. Typology of GIS software based on customisation capability and price. Note that price is approximately inversely proportional to market size.

customisation capabilities. Desktop mapping and GIS software systems, such as ESRI's ArcView and MapInfo's MapInfo, tend to have enduser-orientated customisation capabilities. These typically allow users to change the user interface and add their own macros and programs. Professional, or high end GIS software systems, such as ESRI's ARC/INFO and Spatial Database Engine, Intergraph's MGE and Smallworld's Smallworld GIS, allow customisation to a greater or lesser extent.

Customisation offers both advantages and disadvantages for GIS users and developers. The main advantages for users are that they get systems which incorporate their process-specific business rules and closely match their requirements. As far as developers are concerned, developing a combination of a generic system and a customisation capability is a cost-effective solution for small- to medium-sized markets. It is only in large market sectors that there is a business case to produce a specific-purpose ready-to-run application which does not require customisation capabilities. On the down side, there are some disadvantages of this approach to delivering GIS software solutions. For users, customisation is an expensive and time-consuming exercise (Newell 1993) requiring a considerable degree of input and expertise (e.g. to specify user requirements, perform acceptance testing, and sign off completed customisations). A further problem is that customisations created with high level, user-orientated development languages are inevitably slower than core programs created with a low level, system-orientated language such as C++. Until a given sector (or vertical market) increases to a size sufficient to justify a low level specific-purpose application this will remain the case.

The alternative is bespoke applications development which offers the advantages of an optimised, very specific system with few compromises. Set against this is the fact that these bespoke systems tend to be very expensive (both to build and, more importantly, to maintain), they are very risky projects (because they often start from a low level), and the completed application often cannot be easily adapted, either as the project requirements change or as new projects arise. Furthermore, users with bespoke systems cannot benefit from the continuing general development of commercial-off-the-shelf (COTS) systems for multiple users. These arguments lie behind the decision by most major government and military agencies around the world to move away from proprietary or bespoke systems towards increasing use of COTS solutions: that is, solutions which incorporate as much commercially produced software as possible (see also Bernhardsen, Chapter 41; Meyers, Chapter 57).

2 THE PROCESS OF CUSTOMISATION

Customisation is the process of adapting a system to an individual specification. GIS can be customised in several different ways. In order to explain this it is necessary to describe in general terms the architecture of modern GIS software systems. Figure 2 shows in schematic form a generalised architecture for a desktop or professional GIS software system (see also Elshaw Thrall and Thrall, Chapter 23). Users normally interact with the GIS software via a typically graphical, menu driven, icon-based graphical user interface (GUI). Selections from the GUI make calls to geoprocessing tools (i.e. tools for proximity analysis, overlay processing, or data display). The tools in turn make calls to the data management functions responsible for organising and managing data stored in a database. This three-tier architecture has been widely used (at least conceptually) by many software developers in order to facilitate organisation and management of software development.

It is possible to configure GIS software systems at all three levels. At the GUI level this typically

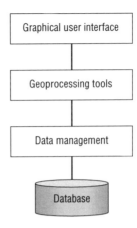

Fig 2. Architecture of a typical desktop or professional GIS software system.

involves configuring the form and appearance of the interface (e.g. adding/removing menu choices and buttons, changing the pattern of icons, and personalising the colour scheme and other characteristics of windows). This is normally carried out using an interactive graphical customisation environment (for an example see Figure 5). At the Tools level customisation involves creating macros to automate frequently required processes and adding new functionality (such as new spatial analysis functions or data translators). This type of customisation can more properly be referred to as programming. More advanced users and software developers are also interested in customising the data management routines within a GIS software system, perhaps to add new datasets to create a new spatial database schema or connect to an external tabular database. These are only a few of the ways in which general-purpose GIS software systems can be customised to create specific-purpose, user-orientated applications.

The above description of the architecture of GIS software systems and the customisation capabilities is useful for explaining the various options available. In practice, however, most of the larger GIS software systems allow users and developers to customise the software at two levels: the application and the core or object code level. In some cases this may be carried out using two programming environments, in others only a single environment is preferred. For example, ESRI's ARC/INFO can be customised at the application (also called 'user' level) by working with the integral fourth generation programming language, AML. Additionally, GIS can be customised at the much lower software development library level using a third generation programming language such as C. In fact the ARC/INFO product is assembled internally by ESRI software development staff who embed the software development library objects in C programs. In contrast, the Smallworld GIS has a single customisation environment, Magik, which is used both internally by Smallworld staff and externally for application customisation by users and developers.

The two level approach adopted by ESRI has the advantages that endusers can use a high level, user-oriented development environment and that they are not exposed to low level programming concepts. The single level approach selected by Smallworld has the advantages that users can gain access to all of the GIS functions and that there is only one development environment for the company to maintain.

3 COSTS OF CUSTOMISATION

It is widely recognised that, along with data capture, customisation is usually the most expensive element of an operational GIS (Antennucci et al 1991; Korte 1996; Smith and Tomlinson 1992).

Table 1 is a generalisation of the approximate breakdown of costs between the various elements of typical GIS implementations based on the author's experience of implementing GIS in over 150 organisations. The 'Low' figures are for a system comprising two server seats (UNIX or Windows NT workstations) and eight clients (desktop PCs). This type of configuration might be found in a small commercial or local government organisation. The 'High' figures are for a corporate implementation comprising 35 UNIX or NT workstations and a UNIX GIS server. This type of configuration might be found in a medium–large local government or utility site. Both configurations run a mixture of desktop and professional GIS software.

Table 1 Breakdown of the percentage costs of typical small desktop (low) and large professional or enterprise (high) operational GIS.

	Low	High
Hardware	22	7
Software	13	12
Data	6	23
Customisation	4	30
Personnel	55	28

Many new users mistakenly believe that hardware and software are the major costs of establishing a GIS. In fact, staff costs are the most expensive. One explanation for this apparent discrepancy is that in many public (and even some private) agencies personnel costs are not included in assessment of GIS implementation costs. It is also often stated by GIS commentators that data are the most expensive component of a GIS (Rhind, Chapter 56). In general data do often comprise a significant proportion of costs. In Table 1 it is assumed that in the Low case the data are purchased and in the High case they are part purchased, part specially captured. If all the data had to be captured for the projects then the figures for data would be several percentage points higher.

4 THE SOFTWARE ENGINEERING APPROACH TO GIS CUSTOMISATION

4.1 GIS software engineering

All GIS implementations, including those involving customisation, have in common the fact that they must meet user requirements and be delivered on time, in budget, and in accordance with quality standards. These goals will be greatly facilitated if a rigorous software engineering approach is adopted and if the process of software development is split up into a series of independent steps which are carried out in sequence. In this so-called software development lifecycle, also referred to as the 'waterfall model', each step is well defined and leads to the creation of a definite product (often a piece of paper), thus allowing the correctness of each step to be checked. More recently, software engineers have questioned the waterfall methodology and have proposed the use of prototyping as an alternative or extension to this approach. These two approaches will be discussed in turn below.

Excellent introductions to software engineering are provided by Bell et al (1992), Flaatten et al (1992) and Gilb (1996). For more critical discussion see Maguire (1994) and Brooks (1995).

4.2 The waterfall model

In the waterfall model (Figure 3), the first stage in the software development lifecycle is to establish user requirements. Essentially, this involves a dialogue between one or more representatives from the user and application developer groups. Initially ideas will be loose and vague in part, but over time they will become clearly defined. This is arguably the most important and sometimes difficult stage of application development.

User requirements need to be formally specified if they are to be of use as a description of the application to be developed. They also form the basis of an acceptance test which will determine if the system meets its requirements. Formal specification is all about describing *what* an application will do rather than *how* it will do it; the latter should be left to the discretion of the programmer.

The design stage involves the application developer creating conceptual, logical, and physical designs of the system. These stages progressively refine the application design from being implementation-independent to being system-specific. In GIS application development the design stage will typically address the user interface, the geoprocessing tools required, and the data management capabilities employed. There are various tools available to assist in this process. These include data flow modelling and various database diagramming techniques such as entity-relation modelling and the object modelling technique (OMT: Date 1995; Rumbaugh et al 1991).

There is now almost universal agreement amongst software developers that applications (indeed all software systems) should be designed and implemented using structured programming techniques (Worboys, Chapter 26). There is also a belief that an application should be created as a series of independent modules. Modular software

development is preferred because it supports software reuse and cooperative development, maintenance, testing, and debugging.

Implementation is all about coding, testing, and debugging the design. In the past this is what people have thought of as 'programming'. The implementation stage usually concludes with a period of acceptance testing and bug fixing before final sign-off by users.

The last stage is the operation and maintenance of an application. This will typically involve a period of user training, system enhancement, bug fixing, and system use. During system use, new user requirements will inevitably arise. This initiates the sequence again (through a 'change order' to the original contract) and so the cycle continues.

In recent years several commentators have questioned the waterfall approach (Flaatten et al 1992). In particular, they have pointed out the length of time it can take to go through the lifecycle using this top-down methodology. Also at issue is the fact that many users do not really understand large specification documents. GIS applications are very visual and until potential users see an application interface they often do not really appreciate what it is and how it will work. They may 'sign off' the document but still be surprised when they see the first release. A further issue is that over long periods user requirements and technology can change, particularly in a fast moving area like GIS. This problem is compounded by the fact that GIS is still relatively new for many users. It is also the case that as new users understand the technology better their ideas and aspirations change, often leading to new and enhanced requirements. Changing the requirements and incorporating new elements into the design has proven to be very expensive for systems based on the waterfall approach (typically the later they are incorporated, the more expensive they become).

4.3 The prototyping approach

Prototyping involves creating working designs of a system rather than designs on paper. Functional requirements documents are still required, however, since they form the basis of a contract document and a series of milestones defining acceptance criteria and a payment schedule. These prototypes are demonstrated and evaluated, and form the basis of future prototypes. The prototyping approach allows closer user involvement in system creation, focusing effort on producing user-oriented systems

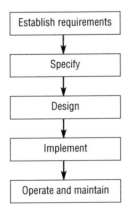

Fig 3. The waterfall approach to software development.

and catering for users' evolving knowledge. As Figure 4 shows, prototyping is an incremental process. Some suggest that it can be represented in the form of a spiral with major issues and decisions taken in the early stages and progressively more detail covered in subsequent iterations.

An important issue associated with prototyping is whether prototypes should be discarded or used as the starting point for the next iteration. Unless evolving systems are built on firm foundations which do not compromise system integrity, there is a danger that the final system will be weak. The increasing use of object-oriented approaches to software development, with support for extensibility and modular replacement/development, has helped to minimise this problem.

Many of the modern application development tools are particularly suited to the prototyping approach. Development environments based on the object-oriented paradigm and those which support interactive graphical development are useful for rapid application development based on prototyping (e.g. Visual Basic, described below, and Delphi). Other advantages of prototyping include the availability of early results and deliverables, as well as reduced costs because of the greater likelihood of developing the application that users actually want.

On the downside, prototyping is less helpful for providing fixed prices for development, nor is it appropriate for some small jobs (although some organisations still prefer this approach). In situations where the application is well understood

and relatively simple, prototyping is generally regarded as inefficient. Experience suggests that it is very difficult to persuade users to discard prototypes even though their underlying architectures may not be suitable for supporting the large mission-critical developments which must be accommodated over the long term.

4.4 Discussion

On balance most technical GIS people would agree that all GIS application development projects require a clear implementation plan which identifies a sequence of well-defined tasks. Whether the classical waterfall approach is adopted, or whether prototyping is used, is open to discussion by the user and developer. Prototyping seems most useful in the areas of user interface design, performance estimation, and functional requirements analysis.

5 GIS APPLICATION DEVELOPMENT TOOLS

The earlier sections of this chapter have examined various approaches to GIS application development. This section looks at the practicalities of implementing an application and the tools and techniques which are available for GIS customisation. The application customisation capabilities of three software systems are discussed. Unfortunately, given the author's background these are all ESRI software systems. It must be emphasised, however, that the principles are common to virtually all GIS software products. First, the customisation capabilities of ARC/INFO, a high end professional GIS, will be discussed. Next, ArcView, a general purpose desktop GIS software product, will be described. Finally, the customisation capabilities of MapObjects, a highly customisable object-based developer product, will be addressed.

5.1 ARC/INFO

ARC/INFO is an example of a high end or professional level, general-purpose GIS software system. Figure 5 shows the interactive graphical application development environment of ARC/INFO. This example shows the software running in a Microsoft Windows environment on a Windows NT workstation. The functionality is similar but the look and feel of the development environment is slightly

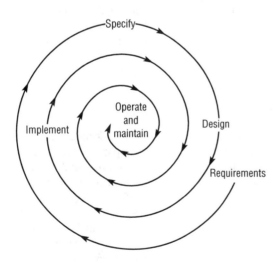

Fig 4. The prototyping approach to software development.

different on UNIX systems (reflecting the different windowing standards of the two operating systems). The display shows a series of windows depicting some of the productivity tools available to help application developers create custom applications. The 'LINE Theme Properties' menu is a complete menu which will be part of the final application (it is a generic menu which will be used to set the symbolisation characteristics of themes or data layers). The 'Untitled – FormEdit' menu and the 'Slider Properties' menu are parts of ARC/INFO's interactive graphical menu-builder called FormEdit. The 'Untitled – FormEdit' window is a widget palette showing the types of widgets which users can place (drag and drop) on menus. The right-hand window of the two shows the property sheet for a slider bar; the second widget down on the left in the FormEdit window is a slider bar (it contains the numbers 750, 0 and 25 000). Changes made using FormEdit can be run immediately in the application to determine their impact and to assist debugging. This greatly facilitates the application development

process. In the bottom left-hand corner of Figure 5 is a text dialogue window in which messages and text dialogue will appear (in this instance the person doing the customisation has opted for white text on a black background). In the bottom right of the screen there is a Text Editor window which is used to edit programs and menu descriptions in ASCII format. These programs are typically attached to menu buttons and executed when a user presses the button in the interface. In this case Microsoft Word is being employed as an editor but the developer could choose any text editor. In the bottom centre there is a file browser window (labelled 'Coverage Manager'). This is used to navigate through the file system in order to locate workspaces and specific files.

Figure 5 has been designed to show the main types of tools available to developers. The figure is not a direct example of how an application developer would actually customise a GIS. Usually only a selection of the windows are open at any one time. Most application developers prefer instead to close or iconise windows until they are needed.

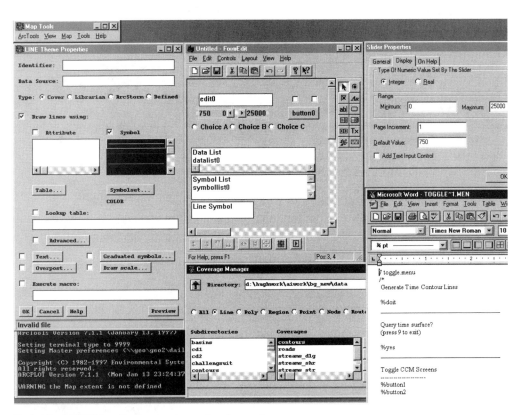

Fig 5. The ARC/INFO application development environment.

Normally menus would be developed using FormEdit and programs would be created with the Text Editor. Both of these are really part of the same run time environment in which the final applications will operate. This allows menus and programs to be tested and debugged interactively before they are released as part of a complete application (perhaps comprising 20–100 menus and 50–500 programs, along with additional elements like 'help' files, documentation, and an install script).

Application developers use this type of environment to create a menu-driven interface for an application. Business rules and process tasks are included as AML programs which are called from the interface. Once an application has been created the menus and code are saved in persistent computer files. Typically, a master program will be called at application start-up. This will then call other programs and menus depending on user selections from the interface and answers to prompts issued by the application. AML is an interpreted language and programs do not need to be compiled into machine object code prior to execution.

5.2 ArcView

A second example of a GIS development environment is provided by ESRI's ArcView desktop GIS. ArcView GIS has been designed and developed to provide stand alone and corporate-wide (using client-server network connectivity) integration of spatial data. ArcView is an object-oriented GIS with an object-oriented customisation language (Avenue – itself based on C++).

Within ArcView's application customisation environment there are various programmer resources which support application development and customisation (Figure 6). This ArcView screen shows six PC windows: the ArcView application itself (ArcView GIS Version 3.0); the Project window (Untitled); a View (View1) containing a World Country Theme (World94.shp) and a Degree Lines Theme (Deg30.shp); the Document Designer (Customize: Untitled); the Script Editor (Script 1); and the Script Manager. These have been designed to be easy to use so that endusers can develop their own applications. Scripts can be created using the integral editor or they can be entered from text files.

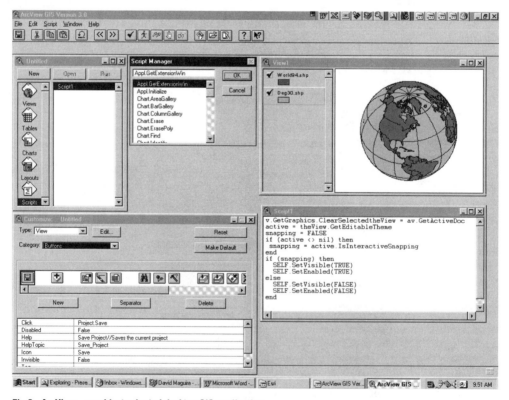

Fig 6. ArcView – an object-oriented desktop GIS application.

They are organised using the Script Manager. Users can attach the scripts to controls in the GUI of each document using the Document Designer (the window labelled 'Customize: Untitled'). This is also used to control the appearance and functionality available within the GUI. Scripts contain requests (messages) to other objects which return an object. Typically scripts are made accessible to users by being placed behind a menu choice or button in the GUI. In event-driven systems such as ArcView, scripts can be run whenever an event occurs (such as resizing of a window, arrival of new data from an external source, or when a user clicks on a button). ArcView additionally has a form interface builder (not shown) used to create multi-widget screen forms.

Like ARC/INFO, Avenue is an interpreted development language and any applications created can be immediately run by the user. Stand-alone applications can be developed and delivered to other users by wrapping the scripts and user interface amendments together as an ArcView Extension (an application which operates within and extends the standard COTS ArcView application).

5.3 MapObjects

ESRI's MapObjects is a collection of GIS objects which conform to Microsoft's Object Linking and Embedding (OLE) / Component Object Model (COM) or ActiveX software specification. Software developers can create applications which embed MapObjects (and other non-GIS objects which conform to the COM specification) within programs written with any industry standard OLE/COM compliant software development environment (e.g. Visual Basic, Visual C++, Visual J++ or Delphi).

Figure 7 shows the Visual Basic application development environment. All elements of the display are part of Visual Basic with the exception of the additional map icon on the bottom right of the left-hand tool palette (this is used to create a MapObjects custom control [an instance of a map window] on the form – the large window in the centre of the screen called Form1). The white area in the centre of Form1 is the map custom control window. When the application is executed this would normally display some geographical data. The form can be thought of as the application user interface. Visual Basic developers create applications by placing controls (maps, buttons, scrolling lists, etc.)

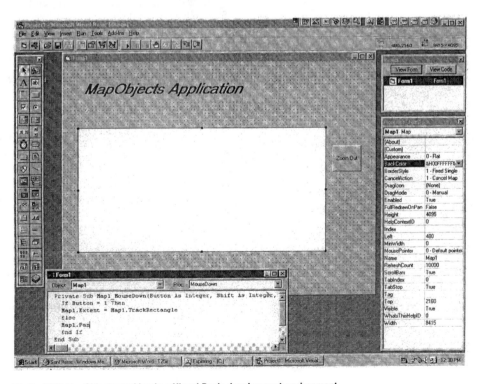

Fig 7. ESRI MapObjects working in a Visual Basic development environment.

on a form. Code is then attached to the controls. The lower, Form1, code window shows the Visual Basic code attached to the map control. This particular program will execute a zoom in or pan operation when a user clicks on the map. The right-hand window, called Properties Form1, is used to set the properties of controls (e.g. font size, type, and colour) on buttons such as the 'Zoom Out' button placed on the form. The top right-hand window, called Project1, is used to organise and register all the windows, controls, and code for the application.

Once a Visual Basic programmer has created an application embedding MapObjects the application can be compiled into a stand-alone executable program which users may run from a menu, icon, or other user interface object.

6 PRACTICAL ISSUES ASSOCIATED WITH GIS CUSTOMISATION

There is clearly more to application development than just obtaining some software and writing code. Any GIS project which adopts a pure technology focus is doomed to failure and customisation projects are no exception.

Because customisation can be a time consuming and expensive activity it is very important to involve senior management and sponsorship at the earliest opportunity. Educating management will help to get their support for the process, as well as further assistance should there be any problems with the implementation.

It is generally recommended that all large customisation projects employ the basic concepts of software engineering as described earlier. While this will not ensure success it will certainly ease the process and provide a framework for assessing progress. A key part of software engineering is specifying the precise scope and content of the system as early as possible. A timetable for deliverables should then be established and agreed upon by the developer(s) and the user(s). If the project is large then staged delivery with 'sign off' should be considered. If either party is inexperienced in using GIS then it is a good idea to adopt a prototyping approach. This will ensure that users learn what is possible with GIS and developers learn what users want to do with the system.

A further important practical recommendation is that the GIS implementation accounting model should budget for requirements analysis, specification, training, documentation, and acceptance testing. It is also as well to remember that coding will typically take only about 15 per cent of the time and that testing and documentation can take up to 30 per cent of the time. Even after a customisation project has been completed there may well be a period of 'institutionalisation' during which the application becomes incorporated into the business practices of an organisation.

Finally, experience suggests that it usually takes about twice as long as inexperienced users first estimate to do customisation. However, if the basic rules and suggestions outlined above are borne in mind then it should be possible to produce reasonably reliable cost and timescale estimates.

7 CONCLUSIONS

This chapter has defined the process of customisation and has shown why GIS customisation is necessary. The formal, well-established steps in the software development lifecycle have been described in outline terms. This included a comparison of the waterfall and prototyping methods. Information has been presented about the main development tools available to application developers. The integrated GIS-specific development environments of ARC/INFO and ArcView have been contrasted with the contemporary industry standard development environment offered by Microsoft's Visual Basic. In essence it seems that users and developers want a GIS customisation environment that is standard across GIS and non-GIS applications, easy to use and highly productive, yet offers access to powerful and sophisticated tools.

As the GIS market grows it is to be expected that more end-user products will be created by core software vendors and third party developers. When this happens, the need for end-user customisation will decrease. There will, however, always be a need for GIS customisation for the simple reasons that GIS is a very diverse field and all users and projects are different in some way.

The utilisation of object technology, such as Microsoft's OLE/COM, has already been cited as a key development in GIS customisation. As the Open Geodata Consortium releases implementation details of their Open Geodata Interoperability Specification (OGIS) and software vendors releases products which conform to it, it is expected that many more developers will be able to develop specific purpose GIS applications and customisations (Sondheim et al, Chapter 24). End-users will benefit enormously from this activity because they will be able to work with ready to run domain specific applications rather than generic GIS products.

References

Antenucci J, Brown K, Croswell, Kevany M 1991 *Geographic information systems – a guide to the technology*. New York, Van Nostrand Reinhold

Bell D, Morrey I, Pugh J 1992 *Software engineering: a programming approach*, 2nd edition. New York, Prentice-Hall

Brooks F 1995 *The mythical man month. Essays on software engineering*. Reading (USA), Addison-Wesley

Date C J 1995 *Introduction to database systems*, 6th edition. Reading, Addison-Wesley

Flaatten P O, McCubbrey D J, O'Riordan P D, Burgess K 1992 *Foundations of business systems*, 2nd edition. Orlando, The Dryden Press

Gilb T 1996 *Principles of software engineering management*, 2nd edition. Reading, Addison-Wesley

Korte G B 1996 Weighing GIS benefits with financial analysis. *GIS World* 9(7): 48–52

Maguire S A 1994 *Debugging the development process*. Redmond, Microsoft Press

Newell R G 1993 Customising a GIS. *GIS Europe* 2(1): 20–1

Rumbaugh J, Blaha M, Premerlani W, Eddy F, Lorensen W 1991 *Object-oriented modelling and design*. Englewood Cliffs, Prentice-Hall

Smith D A, Tomlinson R F 1992 Assessing costs and benefits of geographical information systems: methodological and implementation issues. *International Journal of Geographical Information Systems* 6: 247–56

Introduction

THE EDITORS

Like many other technologies, digital computers have evolved to provide ever more sophisticated environments for their users. Early programmers worked with very simple languages that in principle could do anything, but in practice were limited by the complexity of the necessary programming. Today's programming languages, and application programmer interfaces, allow far more to be achieved with much less effort. While it may have taken a million lines of code to write an early GIS in the 1960s or 1970s, the same could probably have been achieved with at least two orders of magnitude less, had it been possible to take advantage of the sophisticated programming environments available today. Languages like Tcl/Tk, for example, allow easy-to-use graphic interfaces to be constructed quickly that would have taken vastly more programmer effort 20 years ago.

Such progress relies on a simple principle: that if enough commonality can be identified between the needs of a sufficiently large number of users, then it makes sense to embed those common needs in the computing environment. Like the human mind, the digital computer is capable of supporting ever more complex concepts provided they can be constructed from simpler ones (and ultimately a 'hard-wired' base) in well-defined ways. Besides obvious gains in efficiency and productivity, such approaches provide consistency and rigour, offer simplicity by hiding the complex workings of operations from the programmer or user, and allow for uniform approaches to such issues as integrity.

By the mid 1960s, the computer industry had begun to see how this principle might be applied to the datasets processed by digital computers. Computer applications had been growing rapidly in various areas of industry and commerce, and were requiring and producing increasingly complex masses of data. Rather than treat each application as unique, and program its operations from scratch, there appeared to be sufficient commonality in the ways these applications interacted with data to justify the development of generic structures and approaches. Thus the database industry was born, in the form of special software applications to manage the interactions between programs and data. By assigning standard data management operations to generic systems, these so-called database management systems (DBMS) relieved the programmer of much inherently repetitive programming. They also encouraged a more disciplined approach to data management, which was perceived to have its own benefits in terms of increased efficiency and control.

While the database industry is by definition generic, and the characteristics of geographical data and GIS widely acknowledged to be special in many respects, nevertheless by the late 1970s significant efforts were under way to take advantage of database technology in GIS applications. Instead of a monolithic, stand-alone software application, GIS was increasingly perceived as layered, with specialised software working in conjunction with, or conceptually on top of, a standard DBMS. ESRI's ARC/INFO was one of the first of these, released in 1981 and incorporating an existing DBMS into a specialised GIS environment. Today, more and more of the functionality of GIS is assigned to increasingly sophisticated but still generic database products, many of which now include the capability to store and process explicitly spatial data.

These moves towards reliance on underlying DBMS reflect several important priorities and concerns in the GIS industry. First, if GIS and underlying DBMS are at least partially independent, then one DBMS can be easily replaced with another. This is attractive to many GIS customers, who may be able to share the DBMS among many computing applications within the organisation, and value the freedom to update the DBMS independently of the GIS. Second, the DBMS may be perceived as more reliable than less generic approaches to data management, because of the relative size of the DBMS industry – an industry more sophisticated in its approach to data management, with better ways of ensuring data integrity; offering greater interoperability between software environments; and with greater adherence to general standards.

Michael Worboys begins this section with a discussion of database models (Chapter 26). The first generations of database systems, appearing in the 1960s, were regarded as too general for effective use in GIS, and it was not until the emergence of the relational model, with its greater sophistication, that GIS began to adopt database solutions in earnest. The term 'georelational' is often used to describe the particular implementation of the relational model for geographical data, in which geographical relationships between entities become the basis for many of the common keys or linkages between relational tables. Nevertheless, this idea took some time to emerge, and early uses of relational databases in GIS were driven largely by the more general advantages of database systems listed earlier.

Worboys takes the reader beyond the relational model into more recent research and thinking in database systems for GIS, notably the concepts broadly known as 'object-orientation'. Just as the relational model gave GIS users a natural way to represent geographical relationships, object-oriented models provide a natural way to manipulate the various entities found on the geographical landscape, and to describe their behaviours. As Worboys notes, object-oriented databases are in their infancy, and although several successful object-oriented GIS have appeared in recent years, there is still much work to be done in identifying the exact limits of the application of object-oriented thinking in GIS.

The designer of a generic solution to management of data must make decisions based on expectations about usage that will inevitably reflect the needs of the largest segment of users. As a specialised application and a relatively small part of the DBMS market, GIS has its own particular needs that are often difficult to promote in the wider arena of DBMS design. GIS databases tend to be large (a single remotely-sensed image or topographic map can easily require 100 million bytes of storage); and searching for geographical objects based on their locations is inherently multidimensional. DBMS solutions for GIS have often encountered disastrously poor performance, even though it is often possible to 'tune' a modern DBMS for the particular characteristics of a given application. Early users of relational DBMS for GIS found it necessary to develop complex implementation guidelines to ensure minimally

acceptable performance. Unfortunately, it is almost always true that the benefits of generic solutions must be balanced against the inability to optimise a generic design for the specific needs of a complex application.

Spatial indexing offers one of the most powerful tools to affect and improve performance in a GIS application, just as indexing in publishing or library cataloguing affects the usefulness of those fields. In the second chapter of this section, Peter van Oosterom reviews the state of the indexing art in spatial databases. Many indexing schemes have been devised, and it seems unlikely that any one is optimal over any significant domain of GIS applications. Many different schemes have been implemented, but although spatial indexing is often invisible to the user, it seems likely that in those applications where performance is critical, some degree of involvement of the user in the implementation of indexing will always be necessary.

Early DBMS followed one or other of the standard models for databases, but used proprietary languages for interaction with the user. Even though the underlying structure was essentially the same, a user wanting to move from one DBMS product to another often had to learn an entirely new language. The introduction of standard query languages, notably SQL, across entire sections of the DBMS industry led to much greater interoperability between systems, and greatly reduced the complications of training users. Recent efforts to extend SQL to the needs of GIS are reviewed by Worboys, while Max Egenhofer and Werner Kuhn in Chapter 28 give an overview of user interaction in general, comparing the query language approach to other, newer, and more powerful methods of user interface design. As an inherently visual technology, GIS stands to benefit enormously from graphic user interfaces, which offer the potential to make GIS much easier to use, and much easier to learn. Egenhofer and Kuhn review the various metaphors that are guiding contemporary user interface design for GIS, and that make use of an increasing number of distinct media.

In the final chapter in this section, Yvan Bédard adds a distinctly practical flavour to the topics discussed in the previous three. While databases provide the broad framework for describing the geographical world, the specific details of implementation can be critical, in determining performance, and essential to the success of any given application. Generic tools have been developed for database design, and much effort has gone into adapting these to the special needs of GIS.

26

Relational databases and beyond

M F WORBOYS

This chapter introduces the database perspective on geospatial information handling. It begins by summarising the major challenges for database technology. In particular, it notes the need for data models of sufficient complexity, appropriate and flexible human-database interfaces, and satisfactory response times. The most prevalent current database paradigm, the relational model, is introduced and its ability to handle spatial data is considered. The object-oriented approach is described, along with the fusion of relational and object-oriented ideas. The applications of object-oriented constructs to GIS are considered. The chapter concludes with two recent challenges for database technology in this field: uncertainty and spatio-temporal data handling.

1 INTRODUCTION TO DATABASE SYSTEMS

1.1 The database approach

Database systems provide the engines for GIS. In the database approach, the computer acts as a facilitator of data storage and sharing. It also allows the data to be modified and analysed while in the store. For a computer system to be an effective data store, it must have the confidence of its users. Data owners and depositors must have confidence that the data will not be used in unauthorised ways, and that the system has fail-safe mechanisms to cope with unforeseen events. Both data depositors and data users must be assured that, as far as possible, the data are correct. There should be sufficient flexibility to give different classes of users different types of access to the store. Most users will not be concerned with how the database works and should not be exposed to low-level database mechanisms. Data retrievers need flexible methods for establishing what is in the store and for retrieving data according to their requirements and skills. Users may have different conceptions of the organisation of the data in the store. The database interface should be sufficiently flexible to respond equally well to both single-time users with unpredictable and varied requirements, and to regular users with little

variation in their requirements. Data should be retrieved as effectively as possible. It should be possible for users to link pieces of information together in the database to get the benefit of the added value from making the connections. Many users may wish to use the store, maybe even the same data, at the same time and this needs to be controlled. Data stores may need to be linked to other stores for access to pieces of information not in their local holdings.

1.2 Database history

Database management systems have grown out of file management systems that perform basic file handling operations such as sorting, merging, and report generation. During the 1950s, as files grew to have increasingly complex structures, an assortment of data definition products came into use. These became standardised by the Conference on Data Systems and Languages (CODASYL) in 1960 into the Common Business-Oriented Language, that is the COBOL programming language, which separates the definition on file structure from file manipulation. In 1969, the DataBase Task Group (DBTG) of CODASYL gave definitions for data description and definition languages, thus paving the way for hierarchal and network database management systems. The underlying model for these systems is navigational, that is connections between records are made by navigating explicit relationships between them. These relationships were 'hard-wired' into the database, thus limiting the degree to which such databases could be extended or distributed to other groups of users.

The acknowledged founder of relational database technology is Ted Codd, who in a pioneering paper (Codd 1970) set out the framework of the relational model. The 1970s saw the advent of relatively easy-to-use relational database languages such as the Structured Query Language, SQL, originally called the Structured English Query Language, SEQUEL (Chamberlin and Boyce 1974) and Query Language, QUEL (Held et al 1975), as well as prototype relational systems such as IBM's System R (Astrahan et al 1976) and University of California at Berkeley's Interactive Graphics and Retrieval Systems, INGRES (Stonebraker et al 1976).

From the latter part of the 1970s, shortcomings of the relational model began to become apparent for particular applications, including GIS. Codd himself provided extensions to incorporate more semantics (Codd 1979). Object-oriented notions were introduced from programming languages into databases, culminating in prototype object-oriented database systems, such as O_2 (Deux 1990) and ORION (Kim et al 1990). Today, object-oriented systems are well established in the marketplace, as are object-oriented extensions of relational systems, which may be where the future really is. Early developments in object-relational systems are described in Haas et al (1990) and Stonebraker (1986). SQL has developed into the international standard SQL-92, and SQL3 is being developed.

1.3 Data models

The data model provides a collection of constructs for describing and structuring applications in the database. Its purpose is to provide a common computationally meaningful medium for use by system developers and users. For developers, the data model provides a means to represent the application domain in terms that may be translated into a design and implementation of the system. For the users, it provides a description of the structure of the system, independent of specific items of data or details of the particular implementation.

A clear distinction should be made between data models upon which database systems are built, for

example the relational model, and data models whose primary roles are to represent the meaning of the application domains as closely as possible (so-called *semantic data models*, of which entity-relationship modelling is an example: see also Martin, Chapter 6; Raper, Chapter 5). It might be that a semantic data model is used to develop applications for a database system designed around another model: the prototypical example of this is the use of the entity-relationship model to develop relational database applications. The three currently most important data modelling approaches are *record-based*, *object-based* and *object-relational*.

1.4 Human database interaction

Humans need to interact with database systems to perform the following broad types of task:

1 *Data definition*: description of the conceptual and logical organisation of the database, the database schema;
2 *Storage definition*: description of the physical structure of the database, for example file location and indexing methods;
3 *Database administration*: daily operation of the database;
4 *Data manipulation*: insertion, modification, retrieval, and deletion of data from the database.

The first three of these tasks are most likely to be performed by the database professional, while the fourth will be required by a variety of user types possessing a range of skills and experience as well as variable needs requirements in terms of frequency and flexibility of access.

User interfaces are designed to be flexible enough to handle this variety of usage. Standard methods for making interfaces more natural to users include menus, forms, and graphics (windows, icons, mice: see Egenhofer and Kuhn, Chapter 28; Martin 1996). Natural language would be an appropriate means of communication between human and database, but successful interfaces based on natural language have not yet been achieved. For spatial data, the graphical user interface (GUI) is of course highly appropriate. Specialised query languages for database interaction have been devised.

1.5 Database management

The software system driving a database is called the *database management system* (DBMS). Figure 1 shows schematically the place of some of these components in the processing of an interactive query, or an application program that contains within the host general-purpose programming language some database access commands. The DBMS has a query compiler that will parse and analyse a query and, if all is correct, generate execution code that is passed to the runtime database processor. Along the way, the compiler may call the query optimiser to optimise the code so that performance on the retrieval is improved. If the database language expression had been embedded in a general-purpose computer language such as C++, then an earlier precompiler stage would be needed. To retrieve the required data from the database, mappings must be made between the high-level objects in the query language statement and the physical location of the data on the storage device. These mappings are made using the system catalogue. Access to DBMS data is handled by the stored data manager, which calls the operating system for control of physical access to storage devices.

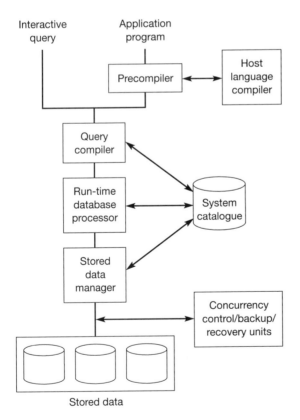

Fig 1. DBMS components used to process user queries.

The logical atom of interaction with a database is the *transaction*, broadly classified as *create*, *modify* (*update*), and *delete*. Transactions are either executed in their entirety (committed) or not at all (rollback to previous commit). The sequence of operations contained in transactions is maintained in a system log or journal, hence the ability of the DBMS to roll back. When a 'commit' is reached, all changes since the last commit point are then made permanent in the database. Thus, a transaction may be thought of as a unit of recovery. The DBMS seeks to maintain the so-called ACID properties of transactions: Atomicity (all-or-nothing), Consistency (of the database), Isolation (having no side-effects and unforeseen effects on other concurrent transactions), and Durability (ability to survive even after system crash).

2 RECORD-BASED DATA MODELS: RELATIONAL DATABASES

2.1 Introduction to the relational model

A record-based model structures the database as a collection of files of fixed-format records. The records in a file are all of the same record type, containing a fixed set of fields (attributes). The early network and hierarchical database systems, mentioned earlier, conform to the record-based data model. However, they proved to be too closely linked to physical implementation details, and they have been largely superseded by the relational model.

A relational database is a collection of tabular relations, each having a set of attributes. The data in a relation are structured as a set of rows. A row, or *tuple*, consists of a list of values, one for each attribute. An attribute has associated with it a domain, from which its values are drawn. Most current systems require that values are atomic – for example they cannot be decomposed as lists of further values – so a single cell in a relation cannot contain a set, list or array of values. This limits the possibilities of the pure relational model for GIS.

A distinction is made between a *relation schema*, which does not include the data but gives the structure of the relation (its attributes, their corresponding domains, and any constraints on the data) and a *relation*, which includes the data. The relation schema is usually declared when the database is set up and then remains relatively

Table 1 Tuples from the Country relation.

Name	Population (millions)	Land area (thousand sq. miles)	Capital
Austria	8	32	Vienna
Germany	81	138	Berlin
Italy	58	116	Rome
France	58	210	Paris
Switzerland	7	16	Bern

Table 2 Tuples from the City relation.

Name	Country	Population (thousands)
Vienna	Austria	1500
Berlin	Germany	3400
Hamburg	Germany	1600
Rome	Italy	2800
Milan	Italy	1400
Paris	France	2100
Zurich	Switzerland	300
Bern	Switzerland	100

Table 3 Tuples from the Country relation after a project operation.

Name	Population (millions)
Austria	8
Germany	81
Italy	58
France	58
Switzerland	7

Table 4 Tuples from the City relation after a restrict operation.

Name	Country	Population (thousands)
Berlin	Germany	3400
Rome	Italy	2800
Paris	France	2100

Table 5 Tuples from the joined Country and City relations.

Name	Country population (millions)	Land area (thousand sq. miles)	Capital	Country city	City population (thousands)
Austria	8	32	Vienna	Austria	1500
Germany	81	138	Berlin	Germany	3400
Italy	58	116	Rome	Italy	2800
France	58	210	Paris	France	2100
Switzerland	7	16	Bern	Switzerland	100

unaltered during the lifespan of the system. A relation, however, will typically be changing frequently as data are inserted, modified and deleted. A *database schema* is a set of relation schemata and a *relational database* is a set of relations, possibly with some constraints. An example of a database schema, used throughout this chapter, comprises two relations Country and City, along with their attributes, as shown:

> Country (Name, Population, Land Area, Capital)
> City (Name, Country, Population).

Tables 1 and 2 show part of an example database according to this schema. Each row of a relation in a relational database is sometimes called a tuple.

The primitive operations that can be supported by a relational database are the traditional set operations of union, intersection, and difference, along with the characteristically relational operations of project, restrict, join, and divide. The structure of these operations and the way that they can be combined is provided by relational algebra, essentially as defined by Codd (1970). The set operations union, intersection, and difference work on the relations as sets of tuples. The project operation applies to a single relation and returns a new relation that has a subset of attributes of the original. For example, Table 3 shows the Country

relation projected onto its Name and Population attributes. The restrict operation acts on a relation to return only those tuples that satisfy a given condition. For example, Table 4 shows a restriction of the City relation, retrieving from the City relation those tuples containing cities with populations greater than two million. The join operation makes connections between relations, taking two relations as operands, and returns a single relation. The relation shown in Table 5 is a join of the Country and City relations, matching tuples when they have the same city names.

2.2 Relational database interaction and SQL

From the outset, there has been a collection of specialised query languages for database interaction. For relational databases, the Structured or Standard Query Language (SQL) is a *de facto* and *de jure* standard. SQL may either be used on its own as a means of direct interaction with the database, or may be embedded in a general-purpose programming language. The most recent SQL standard is SQL-92 (also called SQL2: ISO 1992). There is a large effort to move forward to SQL3.

2.2.1 Schema definition using SQL

The data definition language component of SQL allows the creation, alteration, and deletion of relation schemata. It is usual that a relation schema is altered only rarely once the database is operational. A relation schema provides a set of attributes, each with its associated data domain. SQL allows the definition of a domain by means of a CREATE DOMAIN expression.

A relation schema is created by a CREATE TABLE command as a set of attributes, each associated with a domain, with additional properties relating to keys and integrity constraints. For example, the relation schema City may be created by the command:

```
CREATE TABLE    City
(Name           PlaceName,
Country         PlaceName,
Population      Population,
PRIMARY KEY     (Name)
```

This statement begins by naming the relation schema (called a table in SQL) as City. The attributes are then defined by giving each its name and associated domain (assuming that we have already created domains PlaceName and Population). The primary key, which serves to identify a tuple uniquely, is next given as the attribute Name. There are also SQL commands to alter a relation schema by changing attributes or integrity constraints and to delete a relation schema.

2.2.2 Data manipulation using SQL

Having defined the schemata and inserted data into the relations, the next step is to retrieve data. A simple example of SQL data retrieval resulting in the relation in Table 4 is:

```
SELECT *
FROM City
WHERE Population > 2000000
```

The SELECT clause indicates the attribute to be retrieved from the City relation (* indicates all attributes), while the WHERE clause provides the restrict condition. Relational joins are effected by allowing more than one relation (or even the same relation called twice with different names) in the FROM clause. For example, to find names of countries whose capitals have a population less than two million people, use the expression:

```
SELECT Country.Name
FROM Country, City
WHERE Country.Capital = City. Name
AND City. Population < 2000000
```

In this case, the first part of the WHERE clause provides the join condition by specifying that tuples from the two tables are to be combined only when the values of the attributes Capital in Country and Name in City are equal. Attributes are qualified by prefixing the relation name in case of any ambiguity.

Most of the features of SQL have been omitted from this very brief summary. The documentation on the SQL2 standard is about 600 pages in length. The reader is referred to Date (1995) for a good survey of the relational model and SQL2.

2.3 Relational technology for geographical information

There are essentially two ways of managing spatial data with relational technology: putting all the data (spatial and non-spatial) in the relational database (integrated approach), or separating the spatial from the non-spatial data (hybrid approach). The benefits of using an integrated architecture are considerable, allowing a uniform treatment of all data by the DBMS, and thus not consigning the spatial data to a less sheltered existence outside the database, where integrity, concurrency, and security may not be so rigorously enforced. In theory, the integrated approach is perfectly possible: for example, Roessel (1987) provides a relational model of configurations of nodes, arcs, and polygons. However, in practice the pure relational geospatial model has not up to now been widely adopted because of unacceptable performance (Healey 1991). Essentially, problems arise because of:

1 slow retrieval due to multiple joins required of spatial data in relations;
2 inappropriate indexes and access methods, which are provided primarily for 1-dimensional data types by general-purpose relational systems;

3 lack of expressive power of SQL for spatial queries.

The first problem arises because spatial data are fundamentally complex – polygons being sequences of chains, which are themselves sequences of points. The object-oriented and extended relational models are much better able to handle such data types. With regard to the second problem, extended relational models allow much more flexibility in declaring indexes for different types of data. For the third problem, the limitations of SQL have been apparent for some time in a number of fields (for example, CAD/CAM, GIS, multimedia databases, office information systems, and text databases). SQL3, currently being developed as a standard, promises much in this respect.

3 OBJECT-BASED DATA MODELS

3.1 Introduction and the entity–relationship–attribute approach

The primary components of an object-based model are its *objects* or *entities*. The entity–relationship–attribute (ERA) model and the object-oriented (OO) models are the two main object-based modelling approaches. The ERA approach is attributed to Chen (1976) and has been a major modelling tool for relational database systems for about 20 years. In the ERA approach, an entity is a semantic data modelling construct and is something (such as a country) that has an independent and uniquely identifiable existence in the application domain. Entities are describable by means of their attributes (for example, the name, boundary, and population of a country). Entities have explicit relationships with other entities. Entities are grouped into entity types, where entities of the same type have the same attribute and relationship structure. The structure of data in a database may be represented visually using an ERA diagram. Figure 2 shows an ERA diagram representing the structure of these data in the example database schema in Tables 1 and 2. Entity types are represented by rectangles with offshoot attributes and connecting edges showing relationships. The ERA approach is fully discussed by Bédard (Chapter 29), and so is not considered

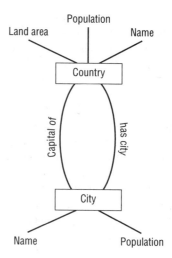

Fig 2. Example of an ERA diagram.

further here.

3.2 The object-oriented approach

3.2.1 Objects, classes, encapsulation, and identity

For many application domains, including GIS, ERA modelling has proved too limited and is being superseded by the OO approach. The OO approach is in use both as a method of semantic data modelling and as a model of data handled by object-oriented programming and database management systems. From the database systems viewpoint, the OO model adapts some of the constructs of object-oriented programming languages to database systems. The fundamental idea is that of encapsulation which places a wrapper around an identifiable collection of data and the code that operates upon it to produce an object. The state of an object at any time is determined by the value of the data items within its wrapper. These data items are referred to as *instance variables*, and the values held within them are themselves objects. This is an important distinction between objects (in the OO sense) and entities (in the ERA sense) which have a two-tier structure of entity and attribute.

The American National Standards Institute (ANSI 1991) Object-Oriented Database Task Group Final Technical Report describes an object as something 'which plays a role with respect to a

378

request for an operation. The request invokes the operation that defines some service to be performed.' The code associated with a collection of data in an object provides a set of methods that can be performed upon it. As well as executing methods on its own data, an object may as part of one of its methods send a message to another object, causing that object to execute a method in response. This highly active environment is another feature that distinguishes between OO and ERA, which is essentially a collection of passive data. An object has both state, being the values of the instance variables within it, and behaviour, being the potential for acting upon objects (including itself). Objects with the same types of instance variables and methods are said to be in the same object class. Figure 3 shows some instance variables and methods associated with classes Country and Polygon and the manner in which the class Polygon is referenced as an instance variable by the class Country. Figure 4 shows in schematic form an object encapsulating state and methods, receiving a message from another object

Country

    Population: Integer
    Name: String
    Capital city: City
    Extent: Polygon

    Update City: City → City
    Insert City: → City
    Delete City: City →

Polygon

    Boundary: Set (Segment)

    Area: Polygon → Real

Fig 3. Part of the class descriptions for *Country* and *Polygon*.

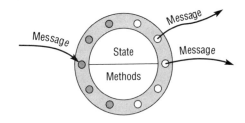

Fig 4. State, methods, and messages of an object.

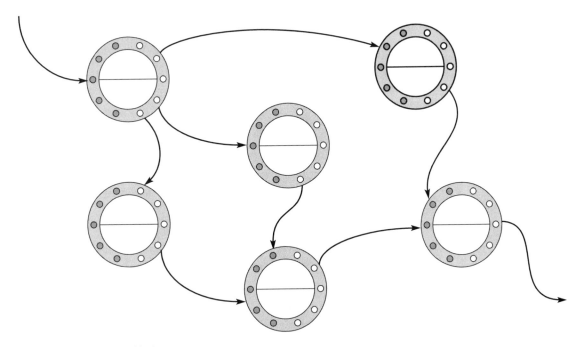

Fig 5. Messages between objects.

and executing methods which result in two messages output. Figure 5 shows the interaction of several objects in response to a message to one of them.

With encapsulation, the internal workings of an object are transparent to users and other objects, which can communicate with it only through a set of predefined message types that the object can understand and handle. To take an example from the real world, I usually do not care about the state of my car under the bonnet (internal state of object class Car) provided that when I put my foot on the accelerator (send message) the car's speed increases (change in the internal state leading to a change in the observable properties of the object). From the viewpoint external to the object, it is only its observable properties that are usually of interest.

3.2.2 Inheritance and composition of objects

Inheritance is an important system and semantic modelling construct, and involves the creation of a new object class by modifying an existing class. Inheritance in an object-oriented setting allows inheritance of methods. Thus, Triangle and Rectangle are subclasses of Polygon. The subclasses inherit all the instance variables and methods from the superclass as well as adding their own. In this example, Triangle and Rectangle may have specialised methods, for example the algorithm implementing the operation Area may be different for Rectangle and Triangle, and each will be different from an Area algorithm for Polygon. This phenomenon, where an operator with the same name has different implementations in different classes, is called *operator polymorphism*. An example of an inheritance hierarchy of spatial object classes is given below (see Figure 8).

Object *composition* allows the modelling of objects with complex internal structures. There are several ways in which a collection of objects might be composed into a new object. *Aggregation* composes a collection of object classes into an aggregate class. For example, an object class Property might be an aggregate of object classes Land Parcel and Dwelling. To quote Rumbaugh et al (1991), 'an aggregate object is semantically an extended object that is treated as a unit in many operations, although physically is made up of several lesser objects'. *Association* groups objects all from the same class into an associated class. For example, an object class Districts might be an association of individual district object classes.

As an illustration of some of these constructs, Figure 6 shows the object class Country (as an abstract object class, represented as a triangle) with three of its instance variables Name, Population, and Area. Variables Name and Population reference printable object class Character String (represented as an oval) and Area references abstract class Polygon. The class Polygon has instance variable Boundary referencing an association of class Segment (the association class shown in the figure as a star and circle). Each segment has a Begin and End Point, and each Point has a Position which is an aggregation (shown as a cross and circle) of printable classes X-coordinate and Y-coordinate.

3.3 Object-oriented database management systems

3.3.1 Making OO persistent

Object-Oriented Programming Languages (OOPLs) such as C++ and Smalltalk provide the capabilities to support the OO approach described above, including the creation, maintenance, and deletion of objects, object classes, and inheritance hierarchies. Object-Oriented Database Management Systems (OODBMS) supplement these capabilities with database functionality, including the ability to support:

1 persistent objects, object classes, and inheritance hierarchies;
2 non-procedural query languages for object class definition, object manipulation, and retrieval;
3 efficient query handling, including query optimisation and access methods;
4 appropriate transaction processing (ACID properties), concurrency support, recovery, integrity, and security.

There are essentially two choices for the developer of an OODBM system: extend a relational system to handle OO, or build a database system around an OO programming language. Both choices have been tried, and section 3.4 on object-extensions to relational technology explores the former. With regard to the latter, object-oriented features will already be supported by the OOPL, so there is the need to add persistency, query handling, and transaction processing. An approach to persistency is to add a new class Persistent Object and allow all database classes to inherit from this class. The class Persistent Object will include methods to:

1 create a new persistent object;
2 delete a persistent object;

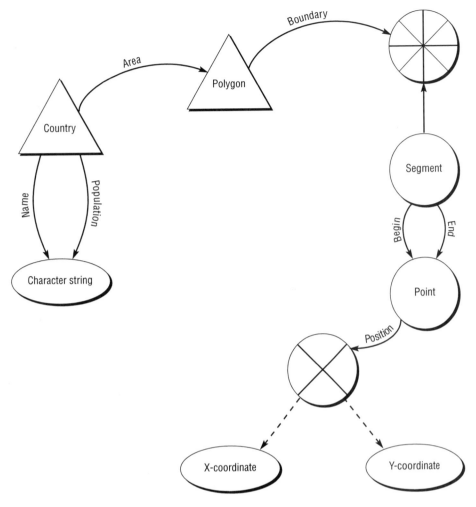

Fig 6. Complex objects.

3 retrieve the state of a persistent object;
4 provide concurrency control;
5 modify a persistent object.

A key benefit of an OODBMS is the support it provides for a unified programming and database environment. However, most current OODBMS treat persistent and non-persistent data differently. A fundamental distinction between RDBs and OODBs is that between call-by-value and call-by-reference. In an RDB, relationships are established by value matching. In our example, to retrieve the population of the capital of Germany, a join between Country and City is made using the value of the Capital field, Berlin. In an OODB (see Figure 7), the connection is made by navigation using the object identifiers (OIDs). The Capital instance variable of Country points to the appropriate City object.

3.3.2 Standardisation of OO systems

The OO approach is more complex than the relational model and has not yet crystallised into a set of universally agreed constructs; even basic constructs like inheritance have been given several different interpretations. Nevertheless, there has been considerable work to arrive at some common definitions.

The Object Management Group (OMG) is a consortium of hardware and software vendors, founded in 1990 with the aim of fostering standards for interoperability of applications within the OO approach. To this end, it has defined the OMG Object Model (see, for example, Kim 1995) many of the concepts of which have been discussed above. An important component of the OMG work is the

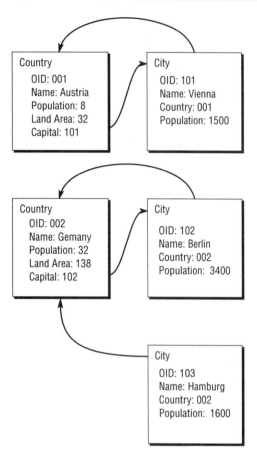

Fig 7. Navigation using object identifiers.

Common Object Request Broker Architecture (CORBA) standard, which specifies an Interface Definition Language for distributed access to objects (see Sondheim et al, Chapter 24). The Object Database Management Group (ODMG) is a consortium of OODBMS vendors, founded in 1990 with the aim of arriving at a commonly agreed OO database interface. The ODMG has defined object definition, manipulation, and query languages, corresponding to data definition and manipulation languages in relational systems.

3.4 Object extensions to relational technology

Object-relational models combine features of object-based and record-based models. They enhance the standard relational model with some object-oriented features, as opposed to OODBMS, that build

database functionality around an OO programming language. Enhancements include complex, possibly user-defined data types, inheritance, aggregation, and object identity. Early work at the University of California at Berkeley on the inclusion of new data types in relational database systems (Stonebraker 1986) led to the POSTGRES DBMS (Stonebraker and Rowe 1986). Parallel developments at the IBM Research Laboratories at San Jose, California resulted in the STARBURST project (reported in Haas et al 1990). These developments have led to contemporary proprietary object-relational systems as well as to the addition of object features to new releases of widely used proprietary relational systems. With regard to query languages that support OO extensions to the relational model, SQL3 is currently under development as an international standard. SQL3 is upwardly compatible with SQL-92, and adds support for objects, including multiple inheritance and operators. The goal for object-relational systems is to provide the wide range of object-oriented functionality that has proved so useful for semantic data modelling and programming systems, while at the same time giving the efficient performance associated with the relational model.

Objects are structured by the relational model as tuples of atomic values such as integers, floats, Booleans, or character strings. This provides only a limited means to define complex data types. Object-relational systems allow non-atomic types. A common extension of the relational model to provide for complex data types is to allow *nested relations*. In a nested relation, values of attributes need not be atomic but may themselves be relations.

Relational database systems provide hashing and B-tree indexes for access to standard, system-provided data types. A major extension that an object-relational system allows is the provision of more appropriate indexes for user-defined types. Object-relational systems provide for the definition of a range of indexes appropriate to a heterogeneous collection of object classes.

3.5 OOGIS

A basic requirement for any OO approach to GIS is a collection of spatial object classes. Figure 8 uses the notation of Rumbaugh et al (1991) to represent an inheritance hierarchy of some basic classes. Class Spatial is the most general class, which is specialised into classes Point and Extent (sets of points). Spatial extents may be classified according to dimension, and examples of classes in one dimension (Polyline) and two dimensions (Polygon) are given. Class Polyline is further specialised into classes Open Polyline and Closed Polyline, the former having two distinct end-points while the latter is joined and has no end-points. Of course, this is just an example of some basic spatial object classes. Table 6 shows some sample methods that will act upon these classes. The name of the method is given, along with the classes upon which it acts and the class to which the result belongs.

The OO approach to geospatial data management is now well established. For some time there have been innovatory proprietary GIS that provide OO programming language support, including spatial object classes, overlaying flat-file, or relational databases. There now exist proprietary GIS that incorporate a full OODBMS. Papers that survey the

Table 6 Inheritance hierarchy of spatial object classes.

Method	Operand	Operand	Result
Equals?	Spatial	Spatial	Boolean
Belongs?	Point	Extent	Boolean
Subset?	Extent	Extent	Boolean
Intersection	Extent	Extent	Extent
Union	Extent	Extent	Extent
Difference	Extent	Extent	Extent
Boundary	Polygon		ClosedPolyline
Connected?	Extent		Boolean
Extremes	OpenPolyline		Set(Point)
Within?	Point	ClosedPolyline	Boolean
Distance	Point	Point	Real
Bearing	Point	Point	Real
Length	Polyline		Real
Area	Polygon		Real
Centroid	Polygon		Point

application of OO to GIS include Egenhofer and Frank (1992); Worboys (1994); Worboys et al (1990).

4 CONCLUSIONS AND CHALLENGES

The purpose of a database is to serve a user community as a data store for a particular range of applications. Regarding geospatial applications, relational databases have fallen short of effectively achieving that purpose for two main reasons:

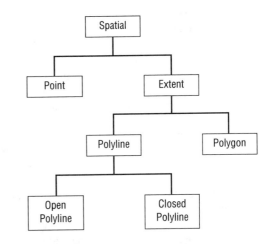

Fig 8. Spatial object class inheritance hierarchy.

1 the relational model has not provided a sufficiently rich set of semantic constructs to allow users to model naturally geospatial application domains;
2 relational technology has not delivered the necessary performance levels for geospatial data management.

This chapter has argued that the OO approach provides part of the solution to these difficulties. It might be that the rapprochement between OO and relational technologies offers the best possible way forward.

There are still significant challenges for the database community in this area, and the chapter concludes by mentioning two of them. First, handling uncertain information has always played a major part in GIS, because many phenomena in the geographical world cannot be represented with total precision and accuracy (Fisher, Chapter 13). Reasoning with uncertain information and managing the associated data in a database remains an important research topic. Deductive databases incorporate logical formalisms, usually subsets of first-order logic, into databases, thereby increasing the expressive power of the query languages and allowing richer semantics for the data models (see, for example, Ceri et al 1990). There has also been work on the fusion of deductive and object technologies for GIS (Paton et al 1996).

Second, the world is in a continual state of change. Classical database technology provides only the capability to manage a single, static snapshot of the application domain. There are two ways in which this can be extended: *temporal* databases manage multiple snapshots (history) of the application domain as it evolves; and *dynamic* databases where a single snapshot changes in step with a rapidly and continuously changing application domain. There are many geospatial applications for both types of extension. Temporal GIS are required to handle such diverse applications as spatio-temporal patterns of land ownership and use, navigation, and global environmental modelling (see Peuquet, Chapter 8). Dynamic systems are required to model such rapidly changing contexts as transportation networks. Problems with development of such systems include the enormous volumes of data required for temporal databases and real-time transaction processing requirements in dynamic systems. These matters are covered in more detail by Bédard (Chapter 29).

References

ANSI 1991 *Object-Oriented Database Task Group final report.* X3/SPARC/DBSSG OODBTG. American National Standards Institute

Astrahan M, Blasgen M, Chamberlin D, Eswaran K, Gray J, Griffiths P, King W, Lorie R, McJones P, Mehl J, Putzolu G, Traiger I, Wade B, Watson V 1976 System R: a relational approach to database management. *ACM Transactions on Database Systems* 1: 97–137

Ceri S, Gottlob G, Tanca L 1990 *Logic programming and databases.* Berlin, Springer

Chamberlin D, Boyce R 1974 SEQUEL: a structured English query language. *Proceedings ACM SIGFIDET Workshop Conference, New York.* ACM Press: 249–64

Chen P P-S 1976 The entity–relationship model – toward a unified view of data. *ACM Transactions on Database Systems* 1: 9–36

Codd E 1970 A relational model for large shared data banks. *Communications of the ACM* 13: 377–87

Codd E 1979 Extending the relational database model to capture more meaning. *ACM Transactions on Database Systems* 4: 397–434

Date C J 1995 *An introduction to database systems*, 6th edition. Reading (USA), Addison-Wesley

Deux O 1990 The story of O₂. *Institute of Electrical and Electronics Engineers Transactions on Knowledge and Data Engineering* 2: 91–108

Egenhofer M J, Frank A 1992 Object-oriented modeling for GIS. *Journal of the Urban and Regional Information Systems Association* 4: 3–19

Haas L M, Chang W, Lohman G M, McPherson J, Wilms P F, Lapis G, Lindsay B, Pirahesh H, Carey M J, Shekita E 1990 Starburst mid-flight: as the dust clears. *IEEE Transactions on Knowledge and Data Engineering* 2: 143–60

Healey R G 1991 Database management systems. In Maguire D J, Goodchild M F, Rhind D W (eds) *Geographical information systems: principles and applications.* Harlow, Longman/New York, John Wiley & Sons Inc. Vol. 1: 251–67

Held G D, Stonebraker M R, Wong E 1975 INGRES: a relational database system. *Proceedings AFIPS 44, Montvale.* AFIPS Press: 409–16

ISO 1992 *Database language SQL.* Document ISO/IEC 9075. International Organisation for Standardisation

Kim W (ed) 1995 *Modern database systems: the object model, interoperability, and beyond.* New York, ACM Press

Kim W, Garza J, Ballou N, Woelk D 1990 Architecture of the ORION next-generation database system. *IEEE Transactions on Knowledge and Data Engineering* 2: 109–25

Martin D J 1996 *Geographic information systems: socioeconomic applications*, 2nd edition. London, Routledge

Paton N, Abdelmoty A, Williams M 1996 Programming spatial databases: A deductive object-oriented approach. In Parker D (ed.) *Innovations in GIS 3.* London, Taylor and Francis: 69–78

Roessel J W van 1987 Design of a spatial data structure using the relational normal forms. *International Journal of Geographic Information Systems* 1: 33–50

Rumbaugh J, Blaha M, Premerlani W, Eddy F, Lorensen W 1991 *Object-oriented modeling and design.* Englewood Cliffs, Prentice-Hall

Stonebraker M 1986 Inclusion of abstract data types and abstract indexes in a database system. *Proceedings 1986 IEEE Data Engineering Conference, Los Alamitos.* IEEE Computer Society: 262–9

Stonebraker M, Rowe L 1986 The design of POSTGRES. *ACM SIGMOD International Conference on Management of Data, New York.* ACM Press: 340–55

Stonebraker M, Wong E, Kreps P 1976 The design and implementation of INGRES. *ACM Transactions on Database Systems* 1: 189–222

Worboys M F 1994b Object-oriented approaches to georeferenced information. *International Journal of*

27

Spatial access methods

P VAN OOSTEROM

This chapter first summarises why spatial access methods are needed. It is important to note that spatial access methods are not only useful for spatial data. Some early main memory spatial access methods are described (section 2), followed by an overview of space-filling curves (section 3). As it is impossible to present all of the spatial access methods described in the recent literature, only the following characteristic families are presented: quadtree, grid-based methods, and R-tree (sections 4–6). Special attention is paid to spatial access methods taking multiple scales into account (section 7). Besides the theory of spatial access methods, the issue of using them in a database in practice is treated in the conclusion of this chapter (section 8).

1 WHY ARE SPATIAL ACCESS METHODS NEEDED?

The main purpose of spatial access methods is to support efficient selection of objects based on spatial properties. For example, a range query selects objects lying within specified ranges of coordinates; a nearest neighbour query finds the object lying closest to a specified object (see Worboys, Chapter 26). Further, spatial access methods are also used to implement efficiently such spatial analyses as map overlay, and other types of spatial joins. Two characteristics of spatial datasets are that they are frequently large and that the data are quite often distributed in an irregular manner. A spatial access method needs to take into account both spatial indexing and clustering techniques. Without a spatial index, every object in the database has to be checked to see whether it meets the spatial selection criterion; a 'full table scan' in a relational database. As spatial datasets are usually very large, such checking is unacceptable in practice for interactive use and most other applications. Therefore, a spatial index is required, in order to find the required objects efficiently without looking at every object. In cases when the whole spatial dataset resides in main memory it is sufficient to know the addresses of the requested objects, as main memory storage allows

random access and does not introduce significant delays. However, most spatial datasets are so large that they cannot reside in the main memory of the computer and must be stored in secondary memory, such as its hard disk. Clustering is needed to group those objects which are often requested together. Otherwise, many different disk pages will have to be fetched, resulting in slow response. In a spatial context, clustering implies that objects which are close together in reality are also stored close together in memory. Many strategies for clustering objects in spatial databases adopt some form of 'space-filling curve' by ordering objects according to their sequence along a path that traverses all parts of the space.

In traditional database systems, sorting (or ordering) of the data forms the basis for efficient searching, as in the B-tree approach (Bayer and Creight 1973). Although there are obvious bases for sorting text strings, numbers, or dates (1-dimensional data), there are no such simple solutions for sorting higher-dimensional spatial data. Computer memory is 1-dimensional but spatial data is 2-dimensional or 3-dimensional (or even higher dimensioned), and must be organised somehow in memory. An intuitive solution is to use a regular grid just as on a paper map. Each grid cell has a unique name, e.g. 'A3', 'C6', or 'D5'. The cells are stored in some order in memory and can each contain a (fixed) number of object references. In a grid cell, a reference is stored to an object whenever the object (partially) overlaps the cell. However, this will not be very efficient because of the irregular data distribution of spatial data: many cells will be empty (e.g. in the ocean), while many other cells will be over-full (e.g. in the city centre). Therefore, more advanced techniques have been developed.

2 MAIN MEMORY ACCESS METHODS

Though originally not designed for handling very large datasets, main memory data structures show several interesting techniques with respect to handling spatial data. In this section the KD-tree (adaptive, bintree) and the BSP-tree will be illustrated.

2.1 The KD-tree

The basic form of the KD-tree stores K-dimensional points (Bentley 1975). This section concentrates on the 2-dimensional case. Each internal node of the KD-tree contains one point and also corresponds to a rectangular region. The root of the tree corresponds to the whole region of interest. The rectangular region is divided into two parts by the x-coordinate of the stored point on the odd levels and by the y-coordinate on the even levels in the tree; see Figure 1. A new point is inserted by descending the tree until a leaf node is reached. At each internal node the value of the proper coordinate of the stored point is compared with the corresponding coordinate of the new point and the proper path is chosen. This continues until a leaf node is reached. This leaf also represents a rectangular region, which in turn will be divided into two parts by the new point. The insertion of a new point results in one new internal node. Range searching in the KD-tree starts at the root, checks

whether the stored node (split point) is included in the search range and whether there is overlap with the left or right subtree. For each subtree which overlaps the search region, the procedure is repeated until the leaf level is reached.

A disadvantage of the KD-tree is that the shape of the tree depends on the order in which the points are inserted. In the worst case, a KD-tree of n points has n levels. The adaptive KD-tree (Bentley and Friedman 1979) solves this problem by choosing a splitting point (which is not an element of the input set of data points), which divides the set of points into two sets of (nearly) equal size. This process is repeated until each set contains one point at most; see Figure 2. The adaptive KD-tree is not dynamic: it is hard to insert or delete points while keeping the tree balanced. The adaptive KD-tree for n points can be built in $O(n \log n)$ time and takes $O(n)$ space for $K=2$. A range query takes $O(\mathrm{sqrt}(n)+t)$ time in two dimensions where t is the number of points found. Another variant of the KD-tree is the bintree (Tamminen 1984). Here the space is divided into two equal-sized rectangles instead of two rectangles with equal numbers of points. This is repeated until each leaf contains one point at the most.

A modification that makes the KD-tree suitable for secondary memory is described by Robinson (1981) and is called the KDB-tree. For practical use, it is more convenient to use leaf nodes containing more than one data point. The maximum number of points that a leaf may contain is called the 'bucket size'. The bucket size is chosen in such a way that it fits within one disk page. Moreover, internal nodes are grouped and each group is stored on one page in order to minimise the number of disk accesses. Robinson describes algorithms for deletions and insertions under which the KDB-tree remains balanced. Unfortunately, no reasonable upper bound for memory usage can be guaranteed.

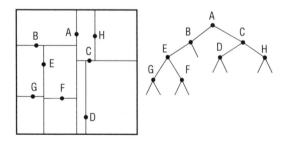

Fig 1. The KD-tree.

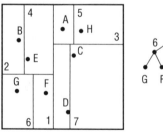

Fig 2. The adaptive KD-tree.

Matsuyama et al (1984) show how the geometric primitives polyline and polygon may be incorporated using the centroids of bounding boxes in the 2-D-tree. Rosenberg (1985) uses a 4-D-tree to store a bounding box by putting the minimum and maximum points together in one 4-dimensional point. This technique can be used to generalise other geometric data structures that are originally suited only for storing and retrieving points. The technique works well for exact-match queries, but is often more complicated in the case of range queries. In general, geometrically close 2-dimensional rectangles do not map into geometrically close 4-dimensional points (Hutflesz et al 1990). The ranges are transformed into complex search regions in the higher-dimensional space, which in turn results in slow query responses.

2.2 The BSP-tree

We begin here by describing the binary space partitioning (BSP)-tree, before presenting a variant suitable for GIS applications: the multi-object BSP-tree for storing polylines and polygons. The original use of the BSP-tree was in 3-dimensional computer graphics (Fuchs et al 1980; Fuchs et al 1983). The BSP-tree was used by Fuchs to produce a hidden surface image of a static 3-dimensional scene. After a preprocessing phase it is possible to produce an image from any view angle in O(n) time, with n the number of polygons in the BSP-tree.

In this chapter the 2-dimensional BSP-tree is used for the structured storage of geometric data. It is a data structure that is not based on a rectangular division of space. Rather, it uses the line segments of the polylines and the edges of the polygons to divide the space in a recursive manner. The BSP-tree reflects this recursive division of space. Each time a (sub)space is divided into two subspaces by a so-called splitting primitive, a corresponding node is added to the tree. The BSP-tree represents an organisation of space by a set of convex subspaces in a binary tree. This tree is useful during spatial search and other spatial operations. Figure 3(a) shows a scene with some directed line segments. The 'left' side of the line segment is marked with an arrow. From this scene, line segment A is selected and space is split into two parts by the supporting line of A. This process is repeated for each of the two sub-spaces with the other line segments. The splitting of space continues until there are no line segments left. Note that sometimes the splitting of a

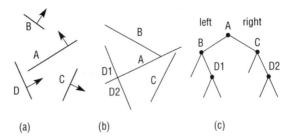

Fig 3. The building of a BSP-tree: (a) 2-dimensional scene; (b) convex sub-spaces; and (c) BSP-tree.

space implies that a line segment (which has not yet been used for splitting itself) is split into two parts. Line D, for example, is split into D1 and D2. Figure 3(b) shows the resulting organisation of the space, as a set of (possibly open) convex subspaces. The corresponding BSP-tree is drawn in Figure 3(c).

The BSP-tree, as discussed so far, is suitable only for storing a collection of (unrelated) line segments. In GIS it must be possible to represent objects, such as polygons. The multi-object BSP-tree (Oosterom 1990) is an extension of the BSP-tree which caters for object representation. It stores the line segments that together make up the boundary of the polygon. The multi-object BSP-tree has explicit leaf nodes which correspond to the convex subspaces created by the BSP-tree. Figure 4(a) presents a 2-dimensional scene with two objects, triangle T with sides abc, and rectangle R with sides defg. The method divides the space in the convex subspaces of Figure 4(b). The BSP-tree of Figure 4(c) is extended with explicit leaf nodes, each representing a convex part of the space. If a convex subspace corresponds to the 'outside' region, no label is drawn in the figure. If no more than one identification tag per leaf is allowed, only mutually exclusive objects can be stored in the multi-object BSP-tree, otherwise it would be possible also to deal with objects that overlap. A disadvantage of this

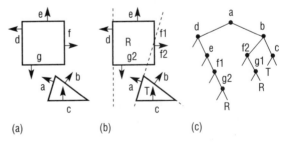

Fig 4. The building of a multi-object BSP-tree: (a) object scenes; (b) convex sub-spaces; and (c) multi-object BSP-tree.

BSP-tree is that the representation of one object is scattered over several leaves, as illustrated by rectangle R in Figure 4. The (multi-object) BSP-tree allows efficient implementation of spatial operations, such as pick and rectangle search.

The choice of which line segment to use for dividing the space very much influences the building of the tree. It is preferable to have a balanced BSP-tree with as few nodes as possible. This is a very difficult requirement to fulfil, because balancing the tree requires that line segments from the middle of the dataset be used to split the space. These line segments will probably split other line segments. Each split of a line segment introduces an extra node in the BSP-tree. However, Paterson and Yao (1989) prove that, if the original line segments are disjoint, then it is possible to build a BSP-tree with $O(n \log n)$ nodes and depth $O(\log n)$ using an algorithm requiring only $O(n \log n)$ time.

3 SPACE-FILLING CURVES

This section presents an overview and some properties of space-filling curves. Space-filling curves order the points in a discrete 2-dimensional space. This technique is also called tile indexing. It transforms a 2-dimensional problem into a 1-dimensional one, so it can be used in combination with a well known data structure for 1-dimensional storage and retrieval, such as the B-tree (Bayer and McCreight 1973). This presentation is based on several papers (Abel and Mark 1990; Goodchild and Grandfield 1983; Jagadish 1990; Nulty and Barholdi 1994) and the book by Samet (1989).

Row ordering simply numbers the cells row by row, and within each row the points are numbered from left to right; see Figure 5(a). Row-prime (or snake like, or boustrophedon) ordering is a variant in which alternate rows are traversed in opposite directions; see Figure 5(b). Obvious variations are column and column-prime orderings in which the roles of row and column are transposed. Bitwise interleaving of the two coordinates results in a 1-dimensional key, called the Morton key (Orenstein and Manola 1988). The Morton key is also known as the Peano key, or N-order, or Z-order. For example, row 2 = 10_{bin} column 3 = 11_{bin} has Morton key 14 = 1110_{bin}; see Figure 5(c). Hilbert ordering is based on the classic Hilbert-Peano curve, as drawn in Figure 5(d). Gray ordering is obtained by bitwise interleaving the Gray codes of the x and y

coordinates. As Gray codes have the property that successive codes differ in exactly one bit position, a 4-neighbour cell differs only in one bit; see Figure 5(e) (Faloutsos 1988). In Figure 5(f) the Cantor-diagonal ordering is shown. Note that the numbering of the points is adapted to the fact that we are dealing with a space that is bounded in all directions; for example, row 3 column 1 has order number 10 instead of 11 and row 3 column 3 has order number 15 instead of 24. Spiral ordering is depicted in Figure 5(g). Finally, Figure 5(h) shows the Sierpinski curve, which is based on a recursive triangle subdivision.

Figure 6 shows the geometric construction of the Peano curve: at each step of the refinement each vertex of the basic curve is replaced by the previous order curve. A similar method, but now also including reflection, is used to construct the reflected binary Gray curve; see Figure 7. The Hilbert curve is constructed by rotating the previous order curves at vertex 0 by –90 degrees and at vertex 3 by 90 degrees; see Figure 8. The Sierpinski curve starts with two triangles; each triangle is split into two new triangles and this is repeated until the required resolution is obtained. Note that the orientation and ordering of the triangles is important; see Figure 9.

Abel and Mark (1990) have identified the following desirable qualitative properties of spatial orderings:

- An ordering is continuous if, and only if, the cells in every pair with consecutive keys are 4-neighbours.
- An ordering is quadrant-recursive if the cells in any valid sub-quadrant of the matrix are assigned a set of consecutive integers as keys.
- An ordering is monotonic if, and only if, for every fixed x, the keys vary monotonically with y in some particular way, and vice versa.
- An ordering is stable if the relative order of points is maintained when the resolution is doubled.

Ordering techniques are very efficient for exact-match queries for points, but there is quite a difference in their efficiency for other types of geometric queries, for example range queries. Abel and Mark (1990) conclude from their practical comparative analysis of five orderings (they do not consider the Cantor-diagonal, the spiral and the Sierpinski orderings) that the Morton ordering and the Hilbert ordering are, in general, the best options. Some quantitative properties of curves are: the total length of the curve, the variability in unit lengths (path between two cells next in order), the average of

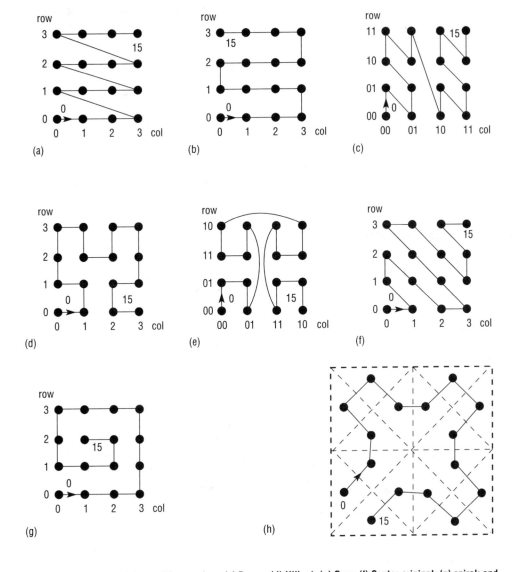

Fig 5. Eight different orderings: (a) row; (b) row prime; (c) Peano; (d) Hilbert; (e) Gray; (f) Cantor-original; (g) spiral; and (h) Sierpinski (triangle).

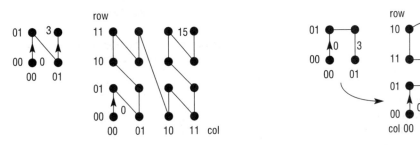

Fig 6. Geometric construction of the Peano curve.

Fig 7. Geometric construction of the Gray curve.

389

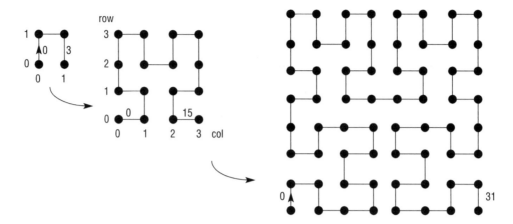

Fig 8. Geometric construction of the Hilbert curve.

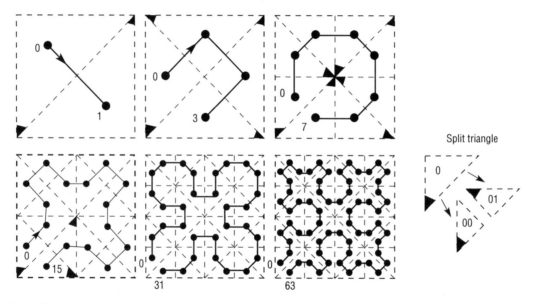

Fig 9. Geometric construction of the Sierpinski curve.

the average distance between 4-neighbours, and the average of the maximum distance between 4-neighbours. Goodchild (1989) proved that the expected difference of 4-neighbour keys of an n by n matrix is $(n+1)/2$ for Peano, Hilbert, row and row-prime orderings, indicating that this is not a very discriminating property. Therefore, Faloutsos and Roseman (1989) suggest a better measure for spatial clustering: the average (Manhattan) maximum distance of all cells within $N/2$ key value of a given cell on a N by N grid; see Table 1. Another measure for clustering is the average

number of clusters for all possible range queries. Note that a cluster is defined as a group of cells with consecutive key value; see Table 2 and Figure 10.

Table 1 Average Manhatten maximum distance of cell within N/2 key value (after Faloutsos and Roseman 1989).

grid	Hilbert	Gray	Peano
2*2	1.00	1.00	1.50
4*4	2.00	2.75	2.75
8*8	3.28	5.00	4.84
16*16	4.89	8.52	7.91

Spatial query

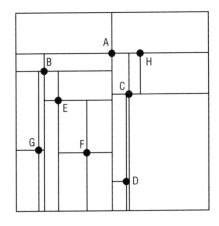

Peano requires 3 ranges

Spatial query

Hilbert requires 2 ranges

Fig 10. Number of clusters for a given range query.

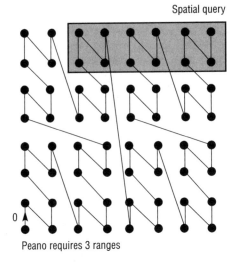

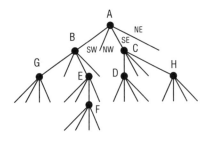

Fig 11. The point quadtree.

Table 2 Average number of clusters for all possible range queries. (After Faloutsos and Roseman 1989.)

grid	Hilbert	Gray	Peano
2*2	1.11	1.11	1.22
4*4	1.64	1.92	2.16
8*8	2.93	4.02	4.41
16*16	5.60	8.71	9.29

4 THE QUADTREE FAMILY

The quadtree is a generic name for all kinds of trees that are built by recursive division of space into four quadrants. Several different variants have been described in the literature, of which the following will be presented here: point quadtree, PR (point region) quadtree, region quadtree, and PM (polygonal map) quadtree. Samet (1984, 1989) gives an excellent overview.

4.1 Point quadtree and PR quadtree

The point quadtree resembles the KD-tree described in section 2. The difference is that the space is divided into four rectangles instead of two; see Figure 11. The input points are stored in the internal nodes of the tree. The four different rectangles are typically referred to as SW (southwest), NW (northwest), SE (southeast), and NE (northeast).

391

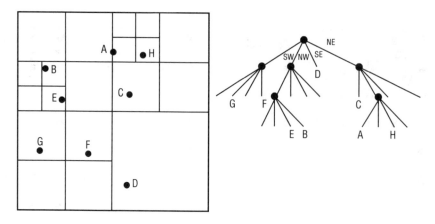

Fig 12. The PR quadtree.

Searching in the quadtree is very similar to the KD-tree: whenever a point is included in the search range it is reported and whenever a subtree overlaps with the search range it is traversed.

A minor variant of the point quadtree is the PR quadtree, which does not use the points of the data set to divide the space. Every time it divides the space, a square, into four equal subsquares, until each contains no more than the given bucket size (e.g. one object). Note that dense data regions require more partitions and therefore the quadtree will not be balanced in this situation; see Figure 12.

4.2 Region quadtree

A very well known quadtree is the region quadtree, which is used to store a rasterised approximation of a polygon. First, the area of interest is enclosed by a square. A square is repeatedly divided into four squares of equal size until it is completely inside (a black leaf) or outside (a white leaf) the polygon or until the maximum depth of the tree is reached (dominant colour is assigned to the leaf); see Figure 13. The main drawback is that it does not contain an exact representation of the polygon. The same applies if the

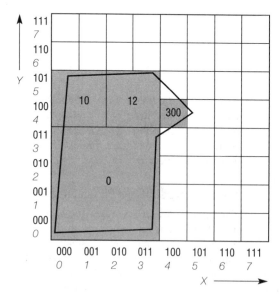

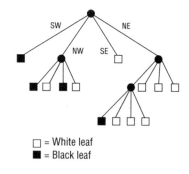

□ = White leaf
■ = Black leaf

(note that SW=0, NW=1, SE=2, NE=3)

Quadcode 0 has Morton range: 0–15
Quadcode 10 has Morton range: 16–19
Quadcode 12 has Morton range: 24–27
Quadcode 300 has Morton range: 48–48

Fig 13. The region quadtree.

region quadtree is used to store points and polylines. This kind of quadtree is useful for storing raster data.

4.3 PM quadtree

A polygonal map, a collection of polygons, can be represented by the PM quadtree. The vertices are stored in the tree in the same way as in the PR quadtree. The edges are segmented into q-edges which completely fall within the squares of the leaves. There are seven classes of q-edges. The first are those that intersect one boundary of the square and meet at a vertex within that square. The other six classes intersect two boundaries and are named after the boundaries they intersect: NW, NS, NE, EW, SW, and SE. For each non-empty class, the q-edges are stored in a balanced binary tree. The first class is ordered by an angular measure and the other six classes are ordered by their intercepts along the perimeter. Figure 14 shows a polygonal map and the corresponding PM quadtree. The PM quadtree provides a reasonably efficient data structure for performing various operations: inserting an edge, point-in-polygon testing, overlaying two maps, range searching and windowing.

The extensive research efforts on quadtrees in the last decade resulted in more variants and algorithms to manipulate these quadtrees efficiently. For example, the CIF quadtree is particularly suited for rectangles. Another interesting example is the linear quadtree: in this representation there is no explicit quadtree, but only an enumeration of the quadcodes belonging to the object; e.g. 0, 10, 12, and 300 in Figure 13. Rosenberg (1985) and Samet (1984) describe these variants.

5 GRID-BASED METHODS

The intuitively attractive approach of organising space by imposing a regular grid has been refined in several different ways, in order to avoid the problems of dealing with irregular distributed data. In this section two different approaches are described: the grid file and the field-tree.

5.1 The grid file

The principle of a grid file is the division of a space into rectangles (regular tiles, grids, squares, cells) that can be identified by two indices, one for the x-direction and the other for the y-direction. The grid file is a non-hierarchical structure. The geometric primitives are stored in the grids, which are not necessarily of equal size. There are several variants of this technique. In this subsection the file structure as defined by Nievergelt et al (1984) is described.

The advantage of the grid file as defined by Nievergelt et al is that, unlike most other grid files, it adjusts itself to the density of the data: however, it is also more complicated. The cell division lines need not be equidistant; for each of x and y there is a 1-dimensional array (in main memory) giving the actual sizes of the cells. Neighbouring cells may be joined into one bucket if the resulting area is a rectangle. The buckets have a fixed size and are stored on a disk page. The grid directory is a 2-dimensional array, with a pointer for each cell to the correct bucket. Figure 15 shows a grid file with linear scales and grid directory. The grid file has good dynamic properties. If a bucket is too full to store a new primitive and it is used for more than one cell, then the bucket may be divided into two buckets. This is a

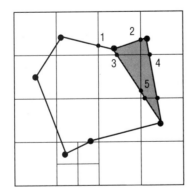

Fig 14. The PM quadtree.

Tree: same as PR Quadtree
2 nodes with their balanced binary trees

node with q-edges

1,2,3	4,5	
vertex tree	NS-tree	SW-tree
1 2 3 other trees empty	4	5 other trees empty

393

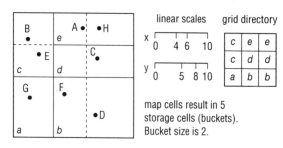

Fig 15. The grid file.

linear scales grid directory

map cells result in 5
storage cells (buckets).
Bucket size is 2.

minor operation. If the bucket is used for only one
cell, then a division line is added to one of the linear
scales. This is a little more complex but is still a minor
operation. In the case of a deletion of primitives, the
merging process is performed in a manner analogous
to the splitting process for insertion.

5.2 The field-tree

The field-tree is suited to store points, polylines, and
polygons in a non-fragmented manner. Several variants
of the field-tree have been published (Frank 1983;
Frank and Barrera 1989; Kleiner and Brassel 1986).
In this subsection attention will be focused on the
partition field-tree. Conceptually, the field-tree consists
of several levels of grids, each with a different
resolution and a different displacement/origin; see
Figure 16. A grid cell is called a field. The field-tree is
not, in fact, a hierarchical tree, but a directed acyclic
graph, as each field can have one, two, or four
ancestors. At one level the fields form a partition and

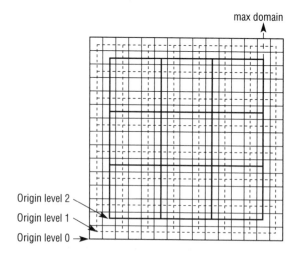

max domain

Origin level 2
Origin level 1
Origin level 0

Fig 16. The positioning of geometric objects in the field-tree.

therefore never overlap. In another variant, the cover
field-tree (Frank and Barrera 1989), the fields may
overlap. It is not necessary at each level that the entire
grid be explicitly present as fields.

A newly inserted object is stored in the smallest
field in which it completely fits (unless its importance
requires it to be stored at a higher level). As a result
of the different displacements and grid resolutions,
an object never has to be stored more than three
levels above the field size that corresponds to the
object size. Note that this is not the case in a
quad-tree-like structure, because here the edges at
different levels are collinear. The insertion of a new
object may cause a field to become too full. In this
case an attempt is made to create one or more new
descendants and to reorganise the field by moving
objects down. This is not always possible. A
drawback of the field-tree is that an overflow page
is sometimes required, as it is not possible to move
relatively large or important objects from an
over-full field to a lower level field.

6 THE R-TREE FAMILY

Instead of dividing space in some manner, it is also
possible to group the objects in some hierarchical
organisation based on (a rectangular approximation
of) their location. This is the approach of the R-tree
and in this section several variants are also described:
the R+-tree, R*-tree, Hilbert R-tree and sphere-tree.

6.1 The R-tree

The R-tree is an index structure that was defined by
Guttman in 1984. The leaf nodes of this multiway
tree contain entries of the form (I, object-id), where
object-id is a pointer to a data object and I is a
bounding box (or an axes-parallel minimal
bounding rectangle, MBR). The data object can be
of any type: point, polyline, or polygon. The internal
nodes contain entries of the form (I, child-pointer),
where child-pointer is a pointer to a child and I is
the MBR of that child. The maximum number of
entries in each node is called the branching factor M
and is chosen to suit paging and disk I/O buffering.
The Insert and Delete algorithms assure that the
number of entries in each node remains between m
and M, where $m \leq \lceil M/2 \rceil$ is the minimum number of
entries per node. An advantage of the R-tree is that
pointers to complete objects (e.g. polygons) are
stored, so the objects are never fragmented.

When inserting a new object, the tree is traversed from the root to a leaf choosing each time the child which needs the least enlargement to enclose the object. If there is still space, then the object is stored in that leaf. Otherwise, the leaf is split into two leaves. The entries are distributed among the two leaves in order to try to minimise the total area of the two leaves. A new leaf may cause the parent to become over-full, so it has to be split also. This process may be repeated up to the root. During the reverse operation, delete, a node may become under-full. In this situation the node is removed (all other entries are saved and reinserted at the proper level later on). Again, this may cause the parent to become under-full, and the same technique is applied at the next level. This process may have to be repeated up to the root. Of course, the MBRs of all affected nodes have to be updated during an insert or a delete operation.

Figure 17 shows an R-tree with two levels and $M = 4$. The lowest level contains three leaf nodes and the highest level contains one node with pointers and MBRs of the leaf nodes. Coverage is defined as the total area of all the MBRs of all leaf R-tree nodes, and overlap is the total area contained within two or more leaf MBRs (Faloutsos et al 1987). In Figure 17 the coverage is A ∪ B ∪ C and the overlap is A ∩ B. It is clear that efficient searching demands both low coverage and low overlap.

6.2 Some R-tree variants

Roussopoulos and Leifker (1985) describe the Pack algorithm which creates an initial R-tree that is more efficient than the R-tree created by the Insert algorithm. The Pack algorithm requires all data to be known *a priori*. The R+-tree (Faloutsos et al 1987),

a modification of the R-tree, avoids overlap at the expense of more nodes and multiple references to some objects; see Figure 18. Therefore, point queries always correspond to a single-path tree traversal. A drawback of the R+-tree is that no minimum space utilisation per node can be given. Analytical results indicate that R+-trees allow more efficient searching, in the case of relatively large objects.

The R*-tree (Beckmann et al 1990) is based on the same structure as the R-tree, but it applies a different Insert algorithm. When a node overflows, it is not split right away, but first an attempt is made to remove p entries and reinsert these in the tree. The parameter p can vary. In the original paper it is suggested that p be set to be 30 per cent of the maximum number of entries per node. In some cases this will solve the node overflow problem without splitting the node. In general this will result in a fuller tree. However, this reinsert technique will not always solve the problem and sometimes a real node split is required. Instead of only minimising the total area, an attempt is also made to minimise overlap between the nodes, and to make the nodes as square as possible.

The Hilbert R-tree uses the centre point Hilbert value of the MBR to organise the objects (Kamel and Faloutsos 1994). When grouping objects (based on their Hilbert value), they form an entry in their parent node which contains both the union of all MBRs of the objects and the largest Hilbert value of the objects. Again, this is repeated on the higher levels until a single root is obtained. Inserting and deleting is basically done using the (largest) Hilbert value and applying B-tree (Bayer and McCreight 1973) techniques. Searching is done using the MBR and applying R-tree techniques. The B-tree technique makes it possible to get fuller nodes, because '2-to-3' or '3-to-4' (or higher) split policies can be used. This results in a more compact tree,

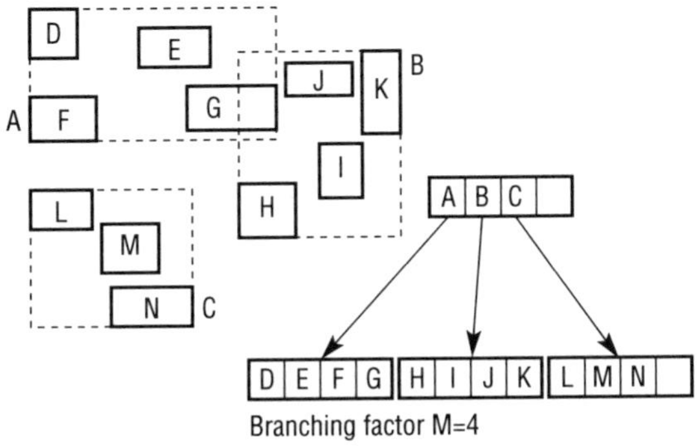

Fig 17. The R-tree.

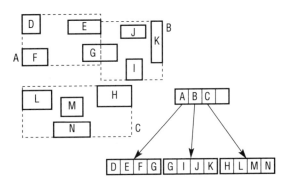

Fig 18. The R+-tree.

395

which is again beneficial for performance. The drawback of the Hilbert R-tree is that the real spatial aspects of the objects are not used to organise them, but instead their Hilbert values. It is possible that two objects which are very close in reality have Hilbert values which are very different. Therefore, these two objects will not end up in the same node in spite of the fact that they are very close; instead each of them may be grouped with other objects (further away in reality, but with closer Hilbert value). This will result in larger MBRs and therefore reduced performance.

The sphere-tree (Oosterom and Claassen 1990) is very similar to the R-tree (Guttman 1984), with the exception that it uses minimal bounding circles (or spheres in higher dimensions, MBSs) instead of MBRs; see Figure 19. Besides being orientation-insensitive, the sphere-tree has the advantage over the R-tree in that it requires less storage space. The operations on the sphere-tree are very similar to those on the R-tree, with the exception of the computation of the minimal bounding circle, which is more difficult (Elzinga and Hearn 1972; Megiddo 1983; Sylvester 1857).

7 MULTISCALE SPATIAL ACCESS METHODS

Interactive GIS applications can be supported even better if importance (resolution, scale) is taken into account in addition to spatial location when designing access methods (see also Weibel and Dutton, Chapter 10). Think of a user who is panning and zooming in a certain dataset. Just enlarging the objects when the user zooms in will result in a poor map. Not only must the objects be enlarged, but they must be displayed with more

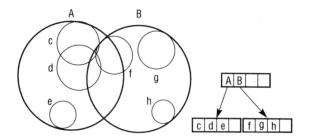

Fig 19. The sphere-tree.

detail (because of the higher resolution), and less significant objects must also be displayed. A simple solution is to store the map at different scales (or levels of detail). This would introduce redundancy with all the related drawbacks of possible inconsistency and increased memory usage. Therefore, geographical data should be stored in an integrated manner without redundancy and, if required, be supported by a special data structure. Detail levels are closely related to cartographic map generalisation techniques. Besides being suited for map generalisation, these multiscale data structures must also provide spatial properties; e.g. it must be possible to find all objects within a specified region efficiently. The name of these types of data structures is reactive data structures (Oosterom 1989, 1991, 1994).

The simplification part of the generalisation process is supported by the binary line generalisation tree (Oosterom and Bos 1989) based on the Douglas–Peucker algorithm (Douglas and Peucker 1973); see Figure 20. The reactive-tree (Oosterom 1991) is a spatial index structure that also takes care of the selection part of generalisation. The reactive-

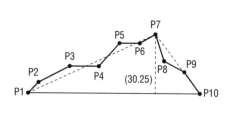

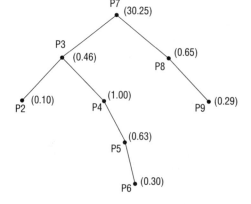

Fig 20. The binary line generalisation tree.

tree is based on the R-tree (Guttman 1984) with the difference that important objects are not stored at leaf level, but are stored at higher levels according to their importance; see Figures 21 and 22. The further one zooms in, the more tree levels must be addressed. Roughly stated, during map generation based on a selection from the reactive-tree, one should try to choose the required importance value such that a constant number of objects will be selected. This means that if the required region is large only the more important objects should be selected, and if the required region is small then the less important objects must also be selected. When using the reactive-tree and the binary line generalisation (BLG)-tree for the generalisation of an area partitioning, some problems are

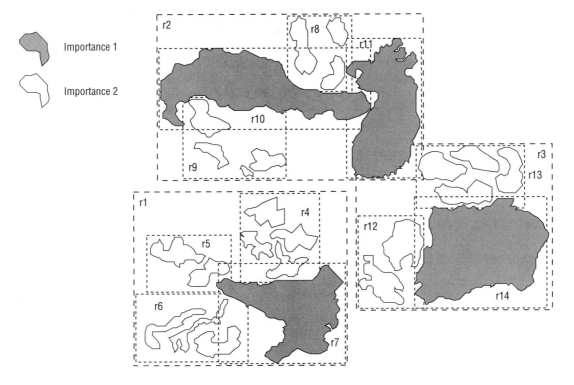

Fig 21. An example of reactive-tree rectangles.

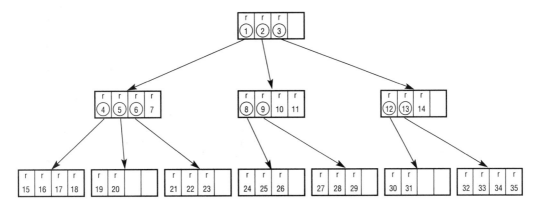

Fig 22. The reactive-tree.

397

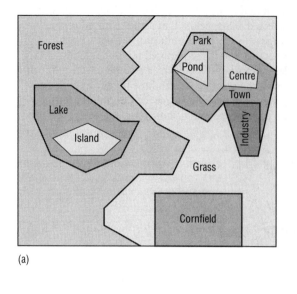

(a)

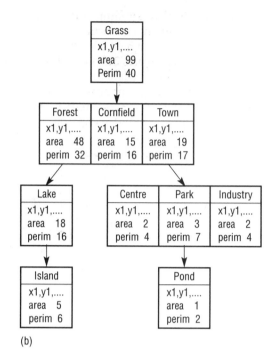

(b)

Fig 23. (a) The scene and (b) the associated GAP-tree.

encountered: gaps may be introduced by omitting small area features and mismatches may occur as a result of independent simplification of common boundaries. These problems can be solved by additionally using the generalised area partitioning (GAP)-tree. Using the reactive-tree, BLG-tree, and the GAP-tree, it is possible to browse interactively through large geographical data sets at very different scales (Oosterom and Schenkelaars 1995).

8 CONCLUSION

Though many spatial access methods have been described in the literature, the sad situation is that only a few have been implemented within the kernel of any (commercial) database and are ready to be used: exceptions are the use of the R-tree in Illustra (Informix) and the use of the Hilbert R-tree in CA-OpenIngres. At the moment several layered 'middleware' solutions are provided; for example, the spatial data engine (SDE) of ESRI (Environmental

Systems Research Corporation). The drawback of the layered approaches is that the database query optimiser does not know anything about the spatial data, so it cannot generate an optimal query plan. Further, all access should be through the layer, yet already many database applications do exist which have direct access to the database. In this situation consistency may become a serious problem. Thus, in practice, possibilities are quite limited and users have to develop their own solutions. An approach for this is the Spatial Location Code (SLC: Oosterom and Vijlbrief 1996), which has been designed to enable efficient storage and retrieval of spatial data in a standard (relational) DBMS. It is used for indexing and clustering geographical objects in a database and it combines the strong aspects of several known spatial access methods (quadtree, field-tree, and Morton code) into one SLC value per object. The unique aspect of the SLC is that both the location and the extent of possibly non-zero-sized objects are approximated by this single value. The SLC is quite general and can be applied in higher dimensions.

References

Abel D J, Mark D M 1990 A comparative analysis of some 2-dimensional orderings. *International Journal of Geographical Information Systems* 4: 21–31

Bayer R, McCreight E 1973 Organization and maintenance of large ordered indexes. *Acta Informatica* 1: 173–89

Beckmann N, Kriegel H-P, Schneider R, Seeger B 1990 The R-tree: an efficient and robust access method for points and rectangles. *Proceedings ACM/SIGMOD, Atlantic City*. New York, ACM: 322-31

Bentley J L 1975 Multi-dimensional binary search trees used for associative searching. *Communications of the ACM* 18: 509–17

Bentley J L, Friedman J H 1979 Data structures for range searching. *Computing Surveys* 11: 397–409

Douglas D H, Peucker T K 1973 Algorithms for the reduction of points required to represent a digitized line or its caricature. *Canadian Cartographer* 10: 112–22

Elzinga J, Hearn D W 1972 Geometrical solutions for some minimax location problems. *Transportation Science* 6: 379–94

Faloutsos C 1988 Gray codes for partial match and range queries. *IEEE Transactions on Software Engineering* SE-14: 1381–93

Faloutsos C, Roseman S 1989 Fractals for secondary key retrieval. *Eighth ACM SIGACT-SIGMOD-SIGART Symposium on Principles of Database Systems (PODS)*: 247–52

Faloutsos C, Sellis T, Roussopoulos N 1987 Analysis of object oriented spatial access methods. *ACM SIGMOD* 16: 426–39

Frank A U 1983 Storage methods for space-related data: the field-tree. Tech. rep. Bericht no. 71. Zurich, Eidgenössische Technische Hochschule

Frank A U, Barrera R 1989 The Field-tree: a data structure for geographic information systems. *Symposium on the Design and Implementation of Large Spatial Databases, Santa Barbara, California*. Berlin, Springer: 29–44

Fuchs H, Abram G D, Grant E D 1983 Near real-time shaded display of rigid objects. *ACM Computer Graphics* 17: 65–72

Fuchs H, Kedem Z M, Naylor B F 1980 On visible surface generation by a priori tree structures. *ACM Computer Graphics* 14: 124–33

Goodchild M F 1989 Tiling large geographical databases. *Symposium on the Design and Implementation of Large Spatial Databases, Santa Barbara, California*. Berlin, Springer: 137–46

Goodchild M F, Grandfield A W 1983 Optimizing raster storage: an examination of four alternatives. *Proceedings AutoCarto 6*: 400–7

Guttman A 1984 R-trees: a dynamic index structure for spatial searching. *ACM SIGMOD* 13: 47–57

Hutflesz A, Six H-W, Widmayer P 1990 The R-file: an efficient access structure for proximity queries. *Proceedings IEEE Sixth International Conference on Data Engineering, Los Angeles, California*. Los Alamitos, IEEE Computer Society Press: 372–9

Jagadish H V 1990 Linear clustering of objects with multiple attributes. *ACM/SIGMOD, Atlantic City*. New York, ACM: 332–42

Kamel I, Faloutsos C 1994 Hilbert R-tree: an improved R-tree using fractals. *VLDB Conference*

Kleiner A, Brassel K E 1986 Hierarchical grid structures for static geographic databases. In Blakemore M (ed.) *AutoCarto London*. London, The Royal Institution of Chartered Surveyors: 485–96

Matsuyama T, Hao L V, Nagao M 1984 A file organization for geographic information systems based on spatial proximity. *Computer Vision, Graphics, and Image Processing* 26: 303–18

Megiddo N 1983 Linear-time algorithms for linear programming in R3 and related problems. *SIAM Journal on Computing* 12: 759–76

Nievergelt J, Hinterberger H, Sevcik K C 1984 The grid file: an adaptable, symmetric multikey file structure. *ACM Transactions on Database System*s 9: 38–71

Nulty W G, Barholdi J J III 1994 Robust multi-dimensional searching with space-filling curves. In Waugh T C, Healey R G (eds) *Proceedings, Sixth International Symposium on Spatial Data Handling, Edinburgh, Scotland*. London, Taylor and Francis: 805–18

Oosterom P van 1989 A reactive data structure for geographic information systems. *Auto-Carto 9, April 1989*. American Society for Photogrammetry and Remote Sensing: 665–74

Oosterom P van 1990 A modified binary space partitioning tree for geographic information systems. *International Journal of Geographical Information Systems* 4: 133–46

Oosterom P van 1991 The Reactive-tree: a storage structure for a seamless, scaleless geographic database. In Mark D M, White D (eds) *AutoCarto 10*. American Congress on Surveying and Mapping: 393–407

Oosterom P van 1994 *Reactive data structures for geographic information systems*. Oxford, Oxford University Press

Oosterom P van, Claassen E 1990 Orientation insensitive indexing methods for geometric objects. *Fourth International Symposium on Spatial Data Handling, Zürich, Switzerland*. Colombus, International Geographic Union: 1016-29

Oosterom P van, Schenkelaars V 1995 The development of an interactive multi-scale GIS. *International Journal of Geographical Information Systems* 9: 489–507

Oosterom P van, Bos J van den 1989 An object-oriented approach to the design of geographic information systems. *Computers & Graphics* 13: 409–18

Oosterom P van, Vijlbrief T 1996 The spatial location code. In Kraak M-J, Molenaar M (eds) *Proceedings, Seventh International Symposium on Spatial Data Handling, Delft, The Netherlands*. London, Taylor and Francis

Orenstein J A, Manola F A 1988 PROBE spatial data modeling and query processing in an image database application. *IEEE Transactions on Software Engineering* 14: 611–29

Paterson M S, Yao F F 1989 Binary partitions with applications to hidden-surface removal and solid modeling. *Proceedings, Fifth ACM Symposium on Computational Geometry*. New York, ACM: 23–32

Robinson J T 1981 The K-D-B-tree: a search structure for large multidimensional dynamic indexes. *ACM SIGMOD* 10: 10–18

Rosenberg J B 1985 Geographical data structures compared: a study of data structures supporting region queries. *IEEE Transactions on Computer Aided Design* CAD-4: 53–67

Roussopoulos N, Leifker D 1985 Direct spatial search on pictorial databases using packed R-trees. *ACM SIGMOD* 14: 17–31

Samet H 1984 The quadtree and related hierarchical data structures. *Computing Surveys* 16: 187–260

Samet H 1989 *The design and analysis of spatial data structures*. Reading (USA), Addison-Wesley

Sylvester J J 1857 A question in the geometry of situation. *Quarterly Journal of Mathematics* 1: 79

Tamminen M 1984 Comment on quad- and octrees. *Communications of the ACM* 27: 248–9

28

Interacting with GIS

M J EGENHOFER AND W KUHN

The user interface of a GIS determines to a large extent how usable and useful that system is for a given task for a user. The usability of GIS has significantly improved in recent years, through changes in our understanding of what system use should achieve, which communication channels it should exploit, and in what form the interaction should occur. This chapter analyses these characteristics of human interaction with GIS, looking back at what has been accomplished in practice, and forward to what can be expected from current research and development efforts. The emphasis lies on an overview of interaction paradigms, modalities, and styles in GIS, rather than on an exemplary discussion of particular systems. While other chapters discuss cognitive, social, institutional, and economic factors affecting GIS usability, the emphasis here is upon the technical aspects, as they are investigated in the field of human–computer interaction.

1 INTRODUCTION

Over the past decade, human interaction with GIS has received increasing attention among researchers, developers, and users. In the 1970s and early 1980s, GIS–user interfaces were dominated by command-style query languages accessible only to expert users. In the second half of the 1980s, graphical user interfaces (GUIs) used the same syntactic structures, but hid them behind icons and menus. This development has primarily improved the familiarity of systems by reducing the need to memorise commands and by providing visual feedback to users.

In the late 1990s, the state-of-the-art in commercial GIS–user interfaces is characterised by the use of windows, icons, menus, and pointing devices (WIMP-style interfaces). These group related operations into understandable chunks and represent them in menus or visually through icons with elementary direct-manipulation operations, like selection by pointing and clicking, or moving by dragging. Their prevailing paradigm of interaction is that of querying a geographical database and presenting the results in maps and tables (Egenhofer and Herring 1993). The map is used as a

presentation medium for query results in their spatial context and as a referencing mechanism to indicate location by pointing (Frank 1993). This paradigm has recently been carried over from GIS to digital map libraries where users browse through datasets in a similar style, providing an abstraction from the physical location of geographical datasets when querying or searching them (Smith 1996). The World Wide Web (WWW) offers easy and wide access to these kinds of geographical data (Coleman, Chapter 22).

The functionality offered by current GIS–user interfaces primarily includes the selection of data layers; the identification of objects by location, name, and elementary spatial relations; and the modification of graphical output parameters such as colours and patterns. Most GIS designers have attempted to provide users with a wide range of functions, allowing them to ask as many as possible of the following generic spatial and temporal questions (adapted from Rhind and Openshaw 1988):

> where is … ?
> what is at location …?
> what is the spatial relation between …?
> what is in a particular spatial relation to …?

what is similar to ...?

where has ... occurred?

what has changed since ...?

what will change if ...?

what spatial pattern(s) exist(s) and where are anomalies?

While current interaction styles represent a significant improvement over the state-of-the-art a decade ago (Frank and Mark 1991), they generally fall short of achieving the usability necessary to solve spatial problems without being a GIS specialist. The user interface itself too often remains an impediment to effective system use in problem-solving or decision-making (Medyckyj-Scott and Hearnshaw 1993).

The proliferation of so-called GIS viewers with reduced functionality, tailored to the inspection of datasets, has reduced the complexity of system use at the inevitable expense of losing more powerful operations (see Elshaw Thrall and Thrall, Chapter 23). Recent developments toward more flexible and adaptable interfaces have eased customisation, but have not resolved this dilemma. In the absence of GIS-specific interface design guidelines, a growing variety of layouts and arrangements have appeared, limiting the possibility for knowledge transfer from one system to the other and improving usability at best marginally. At the same time, modern application programming tools for interface and system design have become widely available and are being used to implement and customise GIS applications. These range from simple macro languages to sophisticated programming languages including such mechanisms as inheritance and polymorphism (Maguire, Chapter 25).

Several ongoing developments are now pointing the way towards interfaces that offer substantial improvements in the usability of GIS. Among them are the focus on specific tasks and on the operations needed to accomplish them (Davies and Medyckyj-Scott 1995). Examples of such tasks include map digitising, where productivity is dramatically influenced by usability (Haunold and Kuhn 1994), and the interaction with car navigation systems where drivers have specific and very limited needs for spatial information, such as distance to and direction of the next turn (Waters, Chapter 59). Virtual reality systems have demonstrated that the concurrent use of multiple modes of interaction dramatically increases the engagement of the user by coming closer to natural

ways of interacting with the world itself rather than with maps or other static models of it (Neves and Câmara, Chapter 39; Jacobson 1992). The primary interaction modes beyond keyboard input and visual presentation are sketching and gesturing in the visual channel; speech and other sound input and output; tactile input and feedback. Novel interaction devices come with displays in a wide variety of sizes and resolutions, enabling interaction that is tailored to special user needs like field portability or group work (Florence et al 1996). The current emphasis on multimedia technologies, in particular on video, CD-ROMs and the WWW, increases both the need and the possibilities for spatial representations and interaction. Scientific visualisation is creating and manipulating worlds of its own, in which motion, perspectives, and multiple representations convey information otherwise hidden in datasets (see Anselin, Chapter 17). The traditional textual query languages assume a new role as 'intergalactic data speak' (Stonebraker et al 1990), supporting interoperability by allowing heterogeneous systems to exchange data with each other (Sondheim et al, Chapter 24). Most of these developments are happening outside of the GIS field and come with their own theories of interaction and collaboration. They are, however, rapidly becoming the determining factors shaping human interaction with GIS.

This chapter reviews the state-of-the-art in GIS query and manipulation languages, measures it against the current understanding of usability requirements, and proceeds to an outlook of how current scientific and technological developments will shape future GIS–user interfaces. A survey of existing query languages or of tools for data capture, data manipulation, and application programming lies beyond its scope. The chapter compiles and integrates the various ways of comparing GIS–user interfaces from a user perspective, offering an introduction to the commonly-used terminology and criteria. It is structured in the following way for classifying interaction languages:

- the interaction paradigms (e.g. querying, browsing, interviewing, analysing, updating, experiencing);
- the interaction modalities (e.g. text, speech and sound, graphics, animation and video, sketching);
- the interaction styles (e.g. command-line, direct manipulation, filtering, delegation).

2 INTERACTION PARADIGMS

People want to do different things with a GIS: some use it as a substitute for a collection of paper maps; others consider it a repository of geographical data that they want to feed into a simulation model; and others think of a GIS as a model of reality in which they want to find interesting places, configurations, and relationships. These differences in the use of a GIS reflect different understandings or paradigms of use (Kuhn 1992). Some of these have traditionally been dominant (querying, browsing, analysing) and are, therefore, better supported by current commercial GIS software than others (interviewing, updating, experiencing).

2.1 Querying

Querying refers to the retrieval of information from a database using a language with well-defined syntax and semantics. The concept is adopted from database systems, where the query language provides a uniform way to access stored data. The idea of a database query is that the user specifies properties of the desired result. For example, if one wanted to know the population of all capitals in the European Union (EU), the query should ask for the names and populations of those cities in Europe that are capitals of a country that is an EU member.

Query languages allow users to retrieve data not only as they were stored, but also in combinations through which information can be obtained that is not directly stored. The combinations are based on logical operators such as conjunction, disjunction, and negation, plus some simple arithmetics. Though different types of query languages exist, the most prominent kind of languages is based on the Structured Query Language (SQL: Worboys, Chapter 26; Melton 1996). Written in SQL, the above example could read as follows:

SELECT name, pop FROM cities, countries
WHERE name = capital AND eumember = 'true'.

The advantage of using a query language rather than low-level programming is that a query language works independently of the contents of a database and, consequently, is useful for different database schemata. This data independence guarantees access without having to worry about storage locations and formats. Furthermore, logical and numerical

computations need not be programmed. The result is a very powerful data retrieval mechanism based on a simple syntax and clean semantics (Frank 1982).

However, these benefits have their limitations. First of all, the above example shows that the user needs to know how the data are structured in the relational tables (what attributes exist in what table and what they are called). Second, most users find it difficult to phrase their requests in expressions involving logical connectors (e.g. AND, OR) whose semantics clash with those of their natural language counterparts (Mark, Chapter 7; Reisner 1981), and systems perform poorly in constructing and comparing different nested queries (Luk and Kloster 1986). Third, different data models may expose different properties that cannot be addressed by one and the same query language. For example, the use of a relational query language for a system based on an object-oriented data model would not support particular object-oriented characteristics, such as identity, encapsulation and inheritance (Atkinson et al 1989). In a similar vein, spatial characteristics have required extensions to query languages, primarily for the inclusion of spatial relations to allow users to make selections based on high-level spatial properties (Egenhofer 1994). Such spatially-extended query languages are based on an extension of the relational algebra to include spatial data types and operators (Güting 1988).

For human interaction with a GIS as a whole, database query languages are obviously biased toward the retrieval of data (Egenhofer and Kuhn 1991). Although the term 'query language' has taken on a broader meaning in the context of GIS (Egenhofer and Herring 1993), most GIS query languages treat geographical data as sets of attributes that can be logically combined in multiple ways. A query is usually considered a separate, standalone interaction and the next query does not build on the context established by the answers to previous queries (Egenhofer 1992). The narrow focus of a query language makes it difficult to support other tasks, such as asserting whether a certain set of facts is true or exploring whether an interesting configuration can be found in a dataset.

2.2 Browsing

With the proliferation of spatial datasets, finding those datasets that are of interest to a user has become an essential task. Consider digital libraries or the whole WWW acting as a huge but unstructured repository for spatial datasets. In such contexts, users commonly have little knowledge about the contents of the available datasets and cannot specify precisely what they want to retrieve. The paradigm of querying does not help here, as it would require exact knowledge of the configurations sought in order to specify appropriate constraints in a query. Rather, users want to be able to browse the data collections, enabling them to recognise rather than having to describe the desired data. The issue is not to find a needle in the haystack, but to look in a haystack for a straw of a shape and texture that the user is unable to specify prior to seeing it. Even if a user were to apply a query language to start a browsing activity, the result set of such a query would usually be very large and require a tedious case-by-case examination.

Browsing, like querying, is content-oriented, as users are looking for datasets that contain specific configurations; however, browsing generally requires additional information at a higher level of abstraction. Traditionally, browsing in GIS has been limited to searching by file names, providing very little and arbitrary information about contents. The current approach is to supply aggregate information and descriptions of dataset properties as metadata. It attempts to ease the user's task of selecting a dataset without having to examine all choices in detail. Emerging metadata standards, however, focus more on what the data producers have to say than on what the data users need to know (Timpf and Raubal 1996). Important descriptors are often unstructured text fields whose content is subject to chance and interpretation. While users can query and read such metadata prior to purchasing, downloading, or examining a dataset, they cannot yet browse them the same way as they would skim newspaper headlines and perceive relevant articles at a glance.

Browsable meta-information tailored to user needs will require search engines incorporating intelligent agents. These can learn about a user's task, mine through geographical datasets, and extract representative subsets, effectively bridging the gap between user needs and data sources.

2.3 Interviewing

The interaction with traditional information systems can be seen as based on the metaphor of an interview with a single person. The user asks questions and the information system responds.

If the interviewed system contradicts itself, the user starts to lose trust in its statements. As long as users perceive each information system as the integrated coherent collection of a single person's knowledge, this paradigm is not challenged.

With a new generation of information systems, however, the understanding of what it means to use an information system may change. These information systems exploit multiple distributed datasets, using them to generate query answers or suggestions supporting the user's decisions. Much like the text-based search engines available today on the WWW, such distributed GIS will face a huge number of dispersed and often isolated data collections. The paradigm for using such information systems changes from the interview with a single person to the interview with multiple and diverse informants or agents, some of them digital, some of them human. These interviewees all hear the same question and they answer when they believe they have something to contribute; however, they are located at different sites such that they generally do not hear each others' answers. Some use this opportunity sparsely and provide input only when they believe that they have something significant to contribute. Others may answer each and every question asked. Some in the audience may be experts in certain areas, whereas others may have a less sophisticated background, but still voice their views. Some may have had recent insights that may affect the answer to a question, while others may contribute their outdated views. Since interviewers hear the views of a group of people, they are expected to make better-informed decisions.

2.4 Analysing

The mere retrieval of stored data is often insufficient since users want to relate and combine data to 'see' patterns and connections among different data elements. Such connections may be visible in reality, but different data collection methods may have hidden them. Or they may be truly discovered through the process of spatial analysis (see Getis, Chapter 16 and Openshaw and Alvanides, Chapter 18 for overviews of competing spatial analysis discovery paradigms). Traditional database query languages offer only modest support for combining data in the form of joining tables over common attribute values. Finding other relationships requires further mathematical analysis, often using statistical or computational methods.

The most common analysis tool for combining data in GIS is the overlay operation. It integrates two or more thematic layers through various analytical operations in order to generate a new layer. Map algebra (Tomlin 1990) has been the common framework for such operations, allowing users to specify (1) what layers to combine, (2) with what operations, and (3) what to do with the result. For example, in order to find the probability of pieces of land near a river being flooded, a layer representing flood risk zones can be overlain with a cadastral layer.

While the source domain of the map overlay metaphor in GIS – stacking transparent sheets on top of each other on a light table (Steinitz et al 1976) – implies an arithmetic multiplication of the layers, the digital environment enables a much larger set of operations. Tomlin's semi-formal model is strictly raster-oriented, but corresponding overlay operations have also been implemented for vector data. Map algebra remains a core functionality in GIS–user interfaces, served by a variety of user interface styles from command line to iconic, direct-manipulation languages (Bruns and Egenhofer 1997; Kirby and Pazner 1990).

More advanced analysis operations rely on sophisticated spatial analysis techniques such as those explored in the spatial analysis chapters of the Principles part of this volume and by Eastman (Chapter 35). Their integration into spatial query languages (Svensson and Zhexue 1991) occurs in highly dispersed ways that have so far shown no common thread toward generic GIS analysis (Yuan and Albrecht 1995). The state-of-the-art solution is to adopt a 'toolbox' metaphor, allowing for an open collection of analysis tools, though often at the expense of limited usability.

2.5 Updating

Few geographical datasets are static and keep their currency and validity over extended periods of time. Updates, corrections, and the generation of 'what if?' scenarios are important operations through which users want to change geographical data. The changes include adding new data and modifying or removing existing data. Adding new data to a dataset means to embed the new data within the setting of existing data (see Peuquet, Chapter 8, for the case of temporal updating; and Goodchild and Longley, Chapter 40, for a discussion of the 'life of a dataset').

Most query languages offer some constructs for elementary updates, though not always in ways that readily support user tasks or are easy to learn and use. Special languages for data manipulation, on the other hand, are much less developed and standardised than query languages. The reason for this is their tight coupling to specific application requirements. For example, a language to design and modify the geometry of a cadastral database requires entirely different operations from one designed to maintain the accuracy of a statistical database (Kuhn 1990).

Operations to update geographical data can be classified according to their scope. At the elementary level, all manipulation involves changing a particular value or adding (deleting) an object. Both of these kinds of operations can apply to data themselves or to the schema of a database, although the latter is not always supported at the end-user interface of a GIS. At a higher level, these elementary operations can be aggregated to changes of value or object collections. Various constraints on these collective changes propagate down from the user's task (e.g. to split a parcel) to the elementary value and object level. In order to maintain the consistency of a database, the commitment of these changes has to be coordinated by transactions (Bédard, Chapter 29). Their granularity can vary from single value changes to modifications sweeping through entire databases. A key problem with current query and manipulation languages is that they operate primarily at lower levels, leaving the management of transactions to the discretion of organisational, rather than purely technical, considerations.

2.6 Experiencing

An entirely different paradigm of interaction with GIS is that of operating within the modelled world itself rather than asking questions about it or writing instructions for changes. Virtual reality, as an immersive variant of direct manipulation, offers this possibility (Neves and Câmara, Chapter 39). In the context of GIS, the idea of 'living' in the model rather than just looking at it has an immediate appeal. Since the phenomena modelled by GIS exist at human or larger scales, it appears natural to provide digital equivalents for human ways of interacting with the world, such as turning one's head, walking, driving, flying, gesturing, and manipulating objects. Systems providing these have been in use for years in applications like flight simulation or games. Their greatest potential for GIS lies in the possibility of overlaying sensory input from the real world with that from one or more models. These ideas have, however, not yet significantly influenced the traditional, map-oriented architectures of GIS interfaces (Kuhn 1991). For some years to come, the interface between our bodies and GIS models is likely to remain the tiny area of our fingertips (on a keyboard or mouse) and overlays will be limited to digital map layers.

3 INTERACTION MODALITIES

Communication between people exploits multiple modalities (spoken, written, gestured, graphical) which map onto different communication channels (visual, auditory, tactile). Each of the modalities has its own strengths and the unavailability of one or the other may seriously impede how people are able to interact with each other and with their environment. A key to successful communication is the appropriate combination and redundant use of different modalities: people gesture while talking, or annotate drawings with spoken or written text. Similar to communication among people, the choice of modalities plays a key role in the success or failure of the interaction between one or more users and a GIS. GIS have often been referred to as early examples of multimedia systems, integrating alphanumerical with map-like and other graphical representations. Ample opportunities for the use of modalities beyond static text and graphics exist and GIS query languages are increasingly making use of them. Almost no theory exists, however, on how to combine multiple modalities appropriately (Egenhofer 1996a).

3.1 Text

Text has been the principal mode for providing instructions to a GIS – be they typed or selected from menus or forms. Unlike in conventional information systems, however, the presentation of query results to a user in textual form has always been secondary to graphical presentations. This is partially so because most GIS replaced map use or production systems in instances where there was little need to convey text to users. Only with the advent of the requirement for data exchange among GIS did the emphasis on

textual information strengthen, and ASCII text files have become a standard way to move data from one system to another. Text is then used to transmit attribute data as well as encoded geometry, but it is another system, not an interactive user, that requests the data.

Text input or output can come in more-or-less-structured form, ranging from free-form natural language texts, to structured tables, to expressions coded in a formal language. Independently of the form, however, textual descriptions of spatial situations are frequently ambiguous and may lead to misinterpretations (cf. the difficulties when using textual directions in way-finding). The use of traditional textual spatial query languages has serious limitations, because geographical concepts are often vague, imprecise, little understood, or not standardised (Fisher, Chapter 13). The dilemma is most apparent in the semantics of spatial terms (Mark, Chapter 7). What does it take to refer to something as a 'mountain' and when would it be more appropriate to call it a 'hill'? Likewise, what paths would qualify to 'cross' the Rocky Mountains and when is a restaurant really 'in town'? These difficulties make most current textual spatial query languages error-prone and difficult to use. They leave the interpretation of terms up to the designers, while users have to comply with their (often hidden) judgements.

On the other hand, the ambiguity inherent in textual descriptions of spatial situations can be exploited to express just that: situations that are not fully determined. Linguists and designers of visual languages have found that icons or sketches (see section 3.5) often over-specify spatial relations (Haarslev and Wessel 1997). For example, one can ask verbally for an object located 'outside' a region, while a corresponding icon always implies a certain distance and especially a direction from the region. In such cases, text is actually more appropriate than graphics or sketches to represent spatial configurations.

3.2 Speech and sound

Some situations of GIS use make it impractical to enter text by typing or selecting; for instance, when working with a mobile GIS in the field. Similarly, it can be preferable to receive output in spoken or other auditory rather than visual form, for instance when driving, assisted by a vehicle navigation system (Waters, Chapter 59). Speech recognition and synthesis techniques are today sufficiently advanced to allow effective communication in situations with a limited vocabulary and few users. An even more limited interaction language allows for a broader range of users (e.g. call-in information systems for public transportation schedules), and vice versa (e.g. text entry systems on a PC).

There are, so far, few examples of using sound in human interaction with GIS (see Shiffer, Chapter 52; Shiffer 1995; Weber 1997). Apart from the intrusiveness of sound in professional as well as private environments, lack of understanding of the role sound plays in cognition (e.g. to identify and locate objects) seems to hinder a broader use of this medium. Considering that sound plays a crucial, though often unconscious, role in our interaction with the world, it can be expected to become more important in GIS–user interfaces. An important reason for further development of auditory interfaces is that they represent the only practical modality supporting visually handicapped users in highly interactive settings.

3.3 Graphics

The strong traditional link between geographical information and graphical communication has led to a higher emphasis on graphics in GIS–user interfaces than in those of most other information systems. These graphics are predominantly maps, which have evolved to become the most sophisticated means of communication about geography during the past 3500 years. Most GIS offer at least some map output to present query results or support data entry. These screen maps are slowly becoming more versatile than simple digital versions of their paper ancestors, offering ways of interacting to refine a query or ask further queries. Users can often point to features on maps to obtain or enter information about them, select map features as input for operations, outline a zoom window to get a map with different contents at a larger scale, or select different layers of information.

In addition to maps, business graphics supply a graphic modality to represent attribute data by graphs and diagrams (e.g. see Elshaw Thrall and Thrall, Chapter 23). Where they occur in isolation, there is nothing that distinguishes them in a GIS from other software that visualises tabular data. Where they are combined with base maps, however, they achieve one of the most refined modes of representing information visually: thematic maps.

The key interaction issue posed by them is how users can choose among symbolisation options and how system designers can provide reasonable default symbolisation (see Kraak, Chapter 11).

3.4 Animation and video

Moving pictures have yet to find their role in communicating geographical information in a GIS (but see Raper, Chapter 5). Attempts at using them have so far concentrated on video clips, offering more intuitive perspectives on buildings and landscapes than maps or textual descriptions can provide (Shiffer 1995; but see Batty, Chapter 21). While such animated views can be very useful (for instance, in a system assisting home buyers), they also consume considerable system and user resources and are difficult to integrate with other parts of a GIS–user interface. Nevertheless, video sequences represent highly valuable information sources within a GIS, and the real issue is to develop appropriate indexing and retrieval systems that can support users in monitoring changes. Time has been noted as a domain of considerable research and development efforts regarding GIS languages. These efforts focus on modelling aspects, but query languages will look very different depending on what kinds of models they support.

3.5 Sketching

Sketching has been used primarily for design tasks as they occur in computer-aided design (CAD) systems. SketchPad (Sutherland 1963) and ThingLab (Borning 1986) were initial approaches to formulate spatial constraints graphically, a principle that was later introduced to GIS interfaces (Egenhofer 1996b; Kuhn 1990). Sketching has also been explored to describe consistency constraints in spatial databases through the construction of situations that would establish unacceptable database states (Pizano et al 1989) or the definition of spatial relations by examples (Petersen and Kuhn 1991). These approaches have confirmed that sketches, like all graphic representations, are good at describing single configurations, but fail in scenarios that require multiple geometric specifications ('this or that', 'this without that') or topological information only ('across', 'outside'). Despite these shortcomings, sketching offers great potential for interaction with GIS when it is given its appropriate role (describing unique, but not exactly determined situations) and combined with other modalities (primarily speech; Egenhofer 1996a, 1996b). It also shows the way to a much more prominent role for gestures in interaction. The very limited understanding of how people assign meaning between speech, gestures, and other means of communication, however, still hinders the development of broadly usable interaction techniques along these lines.

4 INTERACTION STYLES

Independently of the chosen interaction paradigms and modalities, query and manipulation languages can be classified according to their interaction styles. The style of an interaction captures how users express queries or updates and how they receive results: by written commands, direct manipulation, dynamic queries, or delegation.

4.1 Command-line input

Command-line systems represented the state-of-the-art before GUIs became available in the early 1980s. Their syntax can be formally defined, making command interpreters and compilers relatively easy to create. For all their learnability and other usability problems, command-line systems have some definite advantages, particularly for experienced users. Macro commands are easy to write by grouping commands in a text file. Adding programming constructs, such as variables, functions, branching, and looping, can make the power and flexibility of a complete programming language available to the user. However, command-line interfaces have poor cognitive characteristics. The users are interacting through text only and a screen full of text has too high a density of information. Textual objects are more difficult to identify and locate on a screen than graphical objects. Also, typing is physically tiring and error-prone.

The major problems with command-line interfaces are: (1) determining the appropriate command for a task; (2) remembering its name and the names of variables; and (3) entering commands in the correct syntax. The last two problems are significantly reduced in form-based interfaces, which often come as parts of direct manipulation interfaces, but do not constitute a dominating GIS interaction style by themselves.

4.2 GUIs, WIMPs, and direct manipulation

Contemporary GUIs are essentially extensions of window-based operating systems. Such window environments depend on high-quality bit-mapped raster displays and some kind of pointing device (mouse, track ball, joystick etc.). This dependency explains why the terms WIMP (windows, icons, menus, and pointing devices) and GUI are often used interchangeably. GUIs have a much stronger visual component than command-line interfaces and have therefore also been called visual interfaces. The windows, icons, and menus determine the visual characteristics of the user interface.

Icons are symbolised pictorial representations for objects or operations in a user interface. The small space allocated to their pictorial part necessitates careful design and testing. Many icons found in current GIS–user interfaces are more important as place holders for commands (exploiting the user's spatial memory) than as symbols whose meaning can be understood on first sight. Icons can be enhanced through the use of animation or sound.

While most menus are textual, menus with graphics or icons, for instance in the form of tool bars, have become more popular. Dynamic or contextual menus constrain a user to allowable actions at any given time, significantly reducing the chance for errors. Windows allow users to switch rapidly and coherently between multiple tasks or multiple parts of a single task. Various strategies exist for managing the organisation of several screen windows.

Pointing and typing characterise the physical aspects of GUIs. Pointing is used to select objects and operations, typically within a window. For example, a map algebra or SQL expression may be created in a form by pointing to various icons and objects on a map. The users are still composing a command line, but instead of typing, they recognise and select tokens and the system does the typing.

The interaction style supported by GUIs or WIMPs has been introduced in practice by the Xerox Star development (Smith at al 1982) and in theory by Ben Shneiderman (Shneiderman 1983). Although the principles of direct manipulation are by now commonplace, they are far from always being satisfied in GIS interfaces and warrant careful consideration in every design process of the need for:

- visual presentation of objects and operations;
- visual presentation of results;
- rapid, incremental, and reversible operations;

- selection by pointing rather than typing;
- immediate and continuous feedback.

All variations of direct manipulation share a few key qualities: the appearance of and interaction with a system are based on metaphors and multiple metaphors need to be combined (Kuhn 1995; Mark 1992). When moving text in a wordprocessor or dragging a document icon to a folder, users are engaged in a multi-modal activity. What they see and what they do are closely coupled, both physically and conceptually.

Direct manipulation is an appropriate interaction style for the primarily visual operations in a GIS, like zooming, panning or map overlays. Historically, the process of map overlay has been a visual and tactile operation, presenting a rich source domain for direct manipulation metaphors. By enforcing a visual representation of data and operations, direct manipulation also fosters exploratory data analysis (Anselin, Chapter 17). Exploration requires a dynamic, absorbing, and engaging task environment. Users need to become less aware of the existence of the user interface and more immersed in their analytical tasks. Such genuinely empowering environments are still rare. The use of metaphors and direct manipulation alone does not automatically lead to them. The metaphors must draw on the visual and physical characteristics inherent in the user's understanding of a task.

4.3 Filtering

The consequent application of direct manipulation principles to querying led to a variety of interaction techniques that can best be characterised as interactive filters for spatial data. While traditional GIS query languages apply direct manipulation to the composition of queries which are then sent as commands to the database, interactive filters are directly evaluated while the user sets some parameters, with update rates in the order of 100 milliseconds. The effect is that users have a much greater sense of control over the database and an opportunity to 'play' with the data. This supports to a large extent the paradigm of exploration, and supports browsing as well as exploratory data analysis. Examples for such filtering techniques are dynamic queries, dynamic filters, and magic lenses.

Dynamic queries give users interactive control over the setting of query parameters, usually in the form of sliders or buttons (Shneiderman 1994).

They were invented with GIS as one of the key applications in mind. An example is a dynamic 'home finder' with sliders for distances to two places, number of bedrooms, and cost, plus buttons for home type and home features. While the user manipulates these, a map displays the location of homes satisfying the criteria. Clicking at these points reveals detailed descriptions of homes.

Dynamic filters continuously control the density of information shown on screen and provide panning and zooming techniques to focus on portions of the displayed contents (Ahlberg and Shneiderman 1994). Magic lenses (Stone et al 1994) are yet another filtering tool that allows users to change the presentation of objects over which the lens is laid. For example, a portion of a topographic map can change into a weather map or a population map when a special lens is applied to it.

The exploratory nature of the filtering interaction style allows for discovering patterns in the data, forming and testing hypotheses about correlations, and identifying outliers. As such, filtering is a practical approach to interactive data mining (Fayyad et al 1996). Technical problems are posed by the bottleneck of accessing databases and displaying the data rapidly. A more fundamental issue is the application-specific nature of each interface: a home finder looks quite different from, say, a cancer rate visualisation tool. This shows again the persistent trade-off between generality and usability of interaction languages.

4.4 Delegation

Delegation is a style of interaction founded on a special, terse form of communication, where the system takes on the role of an agent or assistant to the user. It has the advantage of establishing a restricted, fairly simple, and familiar communication protocol. Delegation is often seen as an antithesis to direct manipulation (Negroponte 1989) and gets naturally associated with speech-based interaction. It has the potential to compensate for the negligence of the auditory channel in visual interfaces. By its terseness, delegation suits current speech recognition technology quite well. On the other hand, the difficulties of knowing what to ask for and of being able to express it, as well as the decoupling of input and output in the interaction process, have so far prevented successful delegation interfaces in GIS. A more fundamental obstacle is the lack of formal theories for talking about space (Mark and Gould 1991).

5 LOOKING AHEAD

The overview of various kinds of GIS interaction languages in this chapter has been written from a human–computer interaction perspective. It focused on paradigms, modalities, and styles of interaction, as developed in theory and practice over the past two decades. Today, user interface design and interaction research are increasingly driven by broader cognitive, social and economic concerns (Mark and Frank 1991; Nyerges et al 1995). Cognitive sciences investigate how people think and communicate about their applications or about space and how this affects the usability of technology (Mark, Chapter 7). Social studies focus on the role of technology in society, which is strongly influenced by system usability. Economic approaches such as business re-engineering study the core processes of organisations and how they can better be supported by interactive systems and shared databases (Campbell, Chapter 44). The impact of these developments goes beyond traditional GIS applications to areas where geographical information plays a role in decision-making, but GIS cannot yet be applied due to their complexity.

A conclusion emerging from our analysis of interaction issues is that GIS will probably never offer a common query language, let alone common ways of manipulating data. Such generic approaches attempt too much for too many different settings. Indeed, today's market shows a departure altogether from the idea of a GIS platform common to the large and rapidly expanding spectrum of applications. A wider perspective on geoprocessing is emerging, considering it an integral part of enterprise computing. This idea has forged the Open GIS Consortium (OGC, see *http://www.opengis.org*), a cooperative effort between researchers, developers, and users to standardise object-oriented software interfaces for open, interoperable GIS (Sondheim et al, Chapter 24). It is based on the consensus that standardising software interfaces (in the sense of an 'intergalactic data speak', i.e. relatively low-level data retrieval languages along the lines of extended SQL) while diversifying user interfaces will significantly improve overall usability.

In this scenario, integration will occur with other tools at the user sites (databases, spreadsheets, groupware, workflow management, etc.) rather than with GIS in other application domains. Application programming interfaces will support specific classes of GIS tasks and users much better than today. This

change in thinking about GIS architectures is almost certain to affect GIS user interfaces at least as much as the introduction of visual interfaces.

With the integration of geoprocessing into mainstream computing comes also a chance to turn things around and use spatial forms of interaction in non-spatial applications (Kuhn 1996). Using spatial metaphors to structure interaction with non-spatial information is at least as old as the desktop metaphor (Smith et al 1982). More recent examples include the use of a landscape metaphor to visualise, explore, and query non-spatial data (Wise and Thomas 1995). With novel interaction paradigms like that of experiencing, and modalities like speech and gestures further developing, entirely new interaction styles along the lines of filtering will develop within and outside the GIS field. Space and spatial interaction, as fundamental categories of human cognition, will be one of their key characteristics.

References

Ahlberg C, Shneiderman B 1994 Visual information seeking: tight coupling of dynamic query filters with starfield displays. In Adelson B, Dumais S, Olson J (eds) *Human factors in computing systems, CHI 94 conference proceedings*. Boston, ACM Press: 313–17

Atkinson M, Bancilhon F, DeWitt D, Dittnick K, Maier D, Zdonik S 1989 The object-oriented database system manifesto. *Proceedings, First International Conference on Deductive and Object-Oriented Databases, Kyoto, Japan*. Amsterdam, Elsevier Science

Borning A 1986 Defining constraints graphically. In Mantei M, Orbeton P (eds) *Human factors in computing systems, CHI 86 conference proceedings*. Boston, ACM Press: 137–43

Bruns T, Egenhofer M 1997 User interfaces for map algebra. *Journal of the Urban and Regional Information Systems Association* 9: 44–54

Davies C, Medyckyj-Scott D 1995 Feet on the ground: studying user-GIS interaction in the workplace. In Nyerges T L, Mark D M, Laurini R, Egenhofer M J (eds) *Cognitive aspects of human–computer interaction for GIS*. Dordrecht, Kluwer: 123–41

Egenhofer M J 1992 Why not SQL! *International Journal of Geographical Information Systems* 6: 71–85

Egenhofer M J 1994 Spatial SQL: a query and presentation language. *IEEE Transactions on Knowledge and Data Engineering* 6: 86–95

Egenhofer M J 1996a Multi-modal spatial querying. In Kraak M-J, Molenaar M (eds) *Advances in GIS research II (Proceedings, Seventh International Symposium on Spatial Data Handling)*. London, Taylor and Francis: 785–99

Egenhofer M J 1996b Spatial query-by-sketch. In *VL 96: IEEE Symposium on Visual Languages*. IEEE Computer Society: 60–7

Egenhofer M J, Herring J 1993 Querying a geographical information system. In Medyckyj-Scott D, Hearnshaw H (eds) *Human factors in geographical information systems*. London, Belhaven Press: 124–36

Egenhofer M J, Kuhn W 1991 Visualising spatial query results: the limitations of SQL. In Knuth E, Wegner L (eds) *IFIP Transactions A-7: Visual Database Systems, II, Proceedings of IFIP TC2/WG2.6 Second Working Conference on Visual Database Systems, Budapest, Hungary*. Amsterdam, North-Holland: 5–18

Fayyad U, Haussler D, Stolorz P 1996 Mining scientific data. *Communications of the ACM* 39: 51–7

Florence J, Hornsby K, Egenhofer M 1996 The GIS WallBoard: interactions with spatial information on large-scale displays. In Kraak M-J, Molenaar M (eds) *Advances in GIS research II (Proceedings of Seventh International Symposium on Spatial Data Handling)*. London, Taylor and Francis: 449–63

Frank A U 1982 MAPQUERY: database query language for retrieval of geometric data and its graphical representation. *ACM SIGGRAPH* 16: 199–207

Frank A U 1993 The user interface is the GIS. In Medyckyj-Scott D, Hearnshaw H (eds) *Human computer interaction and geographic information systems*. London, Belhaven Press: 3–14

Frank A U, Mark D M 1991 Language issues for GIS. In Maguire D J, Goodchild M F, Rhind D W (eds) *Geographic information systems: principles and applications*. Harlow, Longman/New York, John Wiley & Sons Inc. Vol. 1: 147–63

Güting R 1988 Geo-relational algebra: a model and query language for geometric database systems. In Schmidt J, Ceri S, Missikoff M (eds) *Advances in database technology – EDBT 88 International Conference on Extending Database Technology, Venice, Italy*. New York, Springer 303: 506–27

Haarslev V, Wessel M (1997) Querying GIS with animated spatial sketches. In Tortona J (ed.) *Thirteenth IEEE Symposium on Visual Languages, Capri, Italy*. IEEE Society

Haunold P, Kuhn W 1994 A keystroke level analysis of a graphics application: manual map digitising. In Adelson B, Dumais S, Olson J (eds) *Human factors in computing systems, CHI 94 conference proceedings*. Boston, ACM Press: 337–43

Jacobson B 1992 The ultimate user interface. *Byte* 17: 175–82

Kirby K C, Pazner M 1990 Graphic map algebra. In Kishimato H, Brassel K (eds) *Proceedings, Fourth International Symposium on Spatial Data Handling, Zurich, Switzerland*. Columbus, International Geographical Union 1: 413–22

Kuhn W 1990 Editing spatial relations. In Kishimoto H, Brassel K (eds) *Proceedings, Fourth International Symposium on Spatial Data Handling, Zurich, Switzerland*. Columbus, International Geographical Union 1: 423–32

Kuhn W 1991 Are displays maps or views? In Mark D, White D (eds) *Proceedings, Tenth International Symposium on Computer-Assisted Cartography (AutoCarto 10)*. American Congress of Surveying and Mapping: 261–74

Kuhn W 1992 Paradigms of GIS use. *Proceedings, Fifth International Symposium on Spatial Data Handling*. Columbus, International Geographical Union 1: 91–103

Kuhn W 1995 7±2 questions and answers on metaphors for GIS user interfaces. In Nyerges T L, Mark D M, Laurini R, Egenhofer M J (eds) *Cognitive aspects of human–computer interaction for geographic information systems*. Dordrecht, Kluwer: 113–22

Kuhn W 1996 Handling data spatially: spatialising user interfaces. In Kraak M-J, Molenaar M (eds) *Advances in GIS research II (Proceedings Seventh International Symposium on Spatial Data Handling)*. London, Taylor and Francis: 877–93

Luk W S, Kloster S 1986 ELFS: English language from SQL. *ACM Transactions on Database Systems* 11: 447–72

Mark D M 1992 Spatial metaphors for human–computer interaction. In Cowen D (ed.) *Proceedings, Fifth International Symposium on Spatial Data Handling*. International Geographical Union 1: 104–12

Mark D M, Frank A U (eds) 1991 *Cognitive and linguistic aspects of geographic space*. NATO ASI Series D: Behavioural and Social Sciences, Vol. 63. Dordrecht, Kluwer

Mark D M, Gould M 1991 Interaction with geographic information: a commentary. *Photogrammetric Engineering and Remote Sensing* 57: 1427–30

Medyckyj-Scott D, Hearnshaw H M (eds) 1993 *Human factors in geographical information systems*. London, Belhaven Press

Melton J 1996 SQL language summary. *ACM Computing Surveys* 28: 141–3

Negroponte N 1989 An iconoclastic view beyond the desktop metaphor. *International Journal of Human–Computer Interaction* 1: 109–13

Nyerges T L, Mark D M, Laurini R, Egenhofer M J (eds) 1995 *Cognitive aspects of human-computer interaction for geographic information systems*. NATO ASI Series – Series D: Behavioural and Social Sciences. Dordrecht, Kluwer

Petersen J K, Kuhn W 1991 Defining GIS data structures by sketching examples. *ACSM/ASPRS Annual Convention*. American Congress on Surveying and Mapping 2: 261–9

Pizano A, Klinger A, Cardenas A 1989 Specification of spatial integrity constraints in pictorial databases. *IEEE Computer* 22: 59–71

Reisner P 1981 Human factors studies of database query languages: a survey and assessment. *ACM Computing Surveys* 13: 13–31

Rhind D W, Openshaw S 1988 The analysis of geographical data: data rich, technology adequate, theory poor. *Proceedings, Fourth International Working Conference on Statistical and Scientific Database Management*. Berlin, Springer: 427–54

411

Shiffer M J 1995b Geographic interaction in the city planning context: beyond the multimedia prototype. In Nyerges T L, Mark D M, Laurini R, Egenhofer M (eds) *Cognitive aspects of human–computer interaction for geographic information systems*. Dordrecht, Kluwer: 295–310

Shneiderman B 1983 Direct manipulation: a step beyond programming languages. *IEEE Computer* 16: 57–69

Shneiderman B 1994 Dynamic queries for visual information seeking. *IEEE Software* 11: 70–7

Smith D C, Irby C, Kimball R, Verplank B, Harslem E 1982 Designing the Star user interface. *Byte* 7: 242–82

Smith T 1996 Alexandria Digital Library. *Communications of the ACM* 4: 61–2

Steinitz C, Parker P, Jordan L 1976 Hand-drawn overlays: their history and prospective uses. *Landscape Architecture* 66: 444–55

Stone M, Fishkin K, Bier E 1994 The movable filter as a user interface tool. In Adelson B, Dumais S, Olson J (eds) *Human factors in computing systems, CHI 94 conference proceedings*. Boston, ACM Press: 306–12

Stonebraker M, Rowe L A, Lindsay B, Gray J, Carey M, Beech D 1990 Third-generation database system manifesto. *SIGMOD Record* 19: 31–44

Sutherland I 1963 SketchPad: a man-machine graphical communication system. *Proceedings of AFIPS Spring Joint Computer Conference*: 329–46

Svensson P, Zhexue H 1991 Geo-SAL: a query language for spatial data analysis. In Günther O, Schek H-J (eds) *Advances in spatial databases – second symposium, SSD 91*. New York, Springer 525: 119–40

Timpf S, Raubal M 1996 Experiences with metadata. In Kraak M-J, Molenaar M (eds) *Advances in GIS research II (Proceedings Seventh International Symposium on Spatial Data Handling)*. London, Taylor and Francis: 815–27

Tomlin C D 1990 *Geographic information systems and cartographic modeling*. Englewood Cliffs, Prentice-Hall

Weber C 1997 The representation of spatio-temporal variation in GIS and cartographic displays: the case for sonification and auditory data representation. In Egenhofer M, Golledge R (eds) *Spatial and temporal reasoning in* GIS. New York, Oxford University Press: 74–85

Wise J A, Thomas J 1995 Visualising the non-visual: spatial analysis and interaction with information from text documents. *Proceedings of IEEE Visualisation 95:* 51–8

Yuan M, Albrecht J 1995 Structural analysis of geographic information and GIS operations: from a user's perspective. In Frank A U, Kuhn W (eds) *Spatial information theory – a theoretical basis for GIS*. Lecture notes in computer science 988. Berlin, Springer: 107–22

29

Principles of spatial database analysis and design

Y BÉDARD

This chapter covers the fundamentals of spatial database analysis and design. It begins by defining the most important concepts: 'spatial database', 'analysis', 'design', and 'model'; and continues with a presentation of the rationale supporting the use of formal methods for analysis and design. The basic elements and approaches of such methods are described, in addition to the processes used. Emphasis is placed on the particularities of spatial databases and the improvements needed for non-spatial methods and tools in order to enhance their efficiency. Finally, the chapter presents a set of tools, called CASE (computer-assisted software engineering), which are built to support the formal analysis and design methods.

1 INTRODUCTION

This chapter examines the principles of spatial database analysis and design. These two critical phases of system development remain informal in most small GIS projects. However, when one needs to build a large system, to facilitate its maintenance or to work with a team of information technology specialists, formal engineering-like methods must be adopted.

Formal methods for database analysis and design have been developed to master complex problems. They provide fundamental principles and well-defined steps aimed at improving the efficiency of the database development process and the quality of the result. These methods have existed since the 1970s and rely heavily on models and dictionaries. They have been supported for more than ten years by computer assisted software engineering (CASE) tools which facilitate the building of these models and dictionaries. Although a growing number of GIS specialists master formal methods, they rarely use CASE tools to create their spatial database schema and dictionary or to generate GIS code. This is partly because existing methods and tools must be extended to become truly effective with spatial databases.

The following sections cover the fundamentals of spatial database analysis and design. After defining basic concepts, the rationale of formal analysis and design is presented. This is followed by a discussion of the methods used, and by a description of the process to follow for spatial databases. Characteristics of spatial databases are discussed along with examples. Finally, CASE tools are presented and discussed in a spatial database context.

2 BASIC DEFINITIONS AND CONCEPTS

For the purpose of this chapter, 'spatial database' refers to any set of data describing the semantic and spatial properties of real world phenomena (temporal properties are also possible). Such spatial databases can be implemented in a GIS, in a computer-assisted design (CAD) system coupled with a database management system (DBMS), in a spatial engine accessed through an application programming interface (API), and sitting on top of a DBMS, in a universal (object-relational) server with spatial extension, in a web server with spatial viewer, etc. These spatial databases can use flat file, hierarchical, network, relational, object-oriented, multidimensional or hybrid structures. In addition, they can be organised in very diverse architectures such as stand-alone GIS, client-server solutions, intranets or spatial data warehouses. Gone are the days of a spatial database implemented solely on a stand-alone GIS.

413

Many definitions exist with regards to 'analysis' and 'design', sometimes in contradiction with each other, sometimes very specific to a full life-cycle system development method (see Carmichael 1995 and Olle 1991 for a comparison of methods). This chapter uses the most common and intuitive definitions of analysis and design, but also recognises the fuzziness of the distinction between them (Jacobson and Christerson 1995). In a spatial database context, analysis is the action of understanding and describing what the users need for their spatial database. Thus, it results in a formal and detailed database requirements specification. Similarly, design is the action of defining and describing how the analysis result will be implemented in the selected technology. It is where we consider practical issues such as the limitations of the technology used to manage the spatial database, the desired performance and flexibility, the implementation of security requirements, etc. Design, therefore, results in a formal and detailed programming specification.

Another fundamental definition is that of 'models' since formal models are the thinking tools as well as the final results of database analysis and design. In this context, models are formal representations of something that needs to be understood, remembered, communicated, and tested; they are purposeful surrogates built at a given level of abstraction to include only what is relevant to the system being developed. When a model represents how users' reality is organised in terms of objects, properties, relationships, and processes, then it is described as an 'analysis model', and represents what users want to be implemented in their spatial databases. The analysis model is also called the business model, conceptual model, user's model, and sometimes specification model. When a model represents the database internal structure and related processes as implemented, then it is described as a 'design model'. This latter is also called the implementation model, internal model, or physical model, although the level of detail may vary.

3 THE RATIONALE FOR FORMAL ANALYSIS AND DESIGN METHODS

For any database, analysis and design models determine what can be done easily, with difficulty, or not at all, once the system has been implemented. However, the impact of bad models appears to be higher for spatial databases than it is for non-spatial databases. Consequently, and especially when considering the high cost of spatial data and the long delays in acquiring them, an organisation's return on investment is very sensitive to good analysis and design.

Using a formal method to complete such tasks provides guidance, supports the thinking process and encourages consistent communication and documentation. There are several such methods. The most recent are based on the object-oriented paradigm (Worboys, Chapter 26; Booch 1994; Coad and Yourdon 1991a, 1991b; Cook and Daniels 1994a; Jacobson et al 1993; Martin and Odell 1993; Rumbaugh et al 1991; Shlaer and Mellor 1991). Although these methods have been created to support any type of software development and built to support object-oriented (OO) programming, they can be used efficiently for database development. They all rely on solid theoretical concepts and have all been tested over and over again so they have acquired formal rigour and proved their utility. Once a formal method is mastered, it is faster and better results are delivered, especially when supported by CASE tools (see section 5). Mastering a method also helps developers to solve the most important problems before computerisation; the sooner the problems are solved, the less expensive they are to solve. Finally, good documentation facilitates maintenance (e.g. adding new data types and new processes, migrating to new equipment) as well as software reuse; it also frees the system from its dependency upon individuals. It has been recognised for several years that the higher cost of such a formal approach is lower than the continuous hidden costs of chaotic data and processes. Figure 1 illustrates the impact of good analysis and design methods on the efforts to build and maintain a spatial database.

Use of formal analysis and design methods is also being pushed by the increasing complexity of the spatial database development process. The recent evolution of the software industry indicates that spatial databases are becoming mainstream solutions seamlessly integrated with non-spatial corporate data. This is happening more and more, with new categories of tools outside the traditional GIS packages. In comparison to non-spatial databases, these solutions offer a higher level of diversity both within and across categories. The complexity of spatial objects is also inherently higher; issues such as geometry, spatial reference systems, movement, spatial precision, spatial

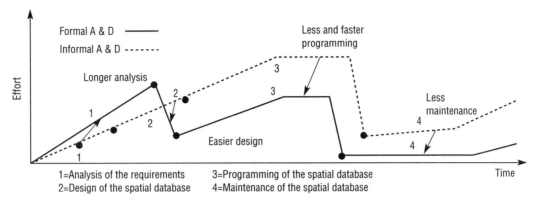

Fig 1. Impact of formal analysis and design methods on efforts to build and maintain a spatial database.

integration, metadata management, database versioning, data quality analysis, and so on may have a tremendous impact when designing a database for spatial querying, spatial analysis, spatial data exchange, and system interoperability (Oosterom, Chapter 27). A look at the ISO SQL3/MM spatial standard (ISO 1996a), the ISO TC211 document *Geographic Information – Rules for Application Schema* (ISO 1996b), and the *Open GIS Virtual Geodata Model* (OGC 1996) rapidly convinces the reader of this inherent complexity of spatial data (see Sondheim et al, Chapter 24). Such complexity adds to the omnipresent need for very high performance, a need which encourages database structures which are denormalised and difficult to understand. Thus, spatial database analysis and design must rely more than ever on formal methods to cope efficiently with such levels of complexity, especially in large projects. Consequently, spatial database developers have a higher need to split the problem between analysis and design and to deliver separate models. According to previous definitions, analysis focuses on real-world issues independent of the technology, while design focuses and depends completely on the technology selected. Such a separation helps to better understand users' needs, to structure the database, to facilitate maintenance, and to encourage metadata management and software reuse. This split is essential for multi-platform environments and systems interoperability – one of today's most challenging issues for spatial databases (Sondheim et al, Chapter 24). Such a 'divide and conquer' strategy has been used for over 20 years in database design (ANSI/SPARC 1975). It is clearly defined in most formal methods, including OO methods, in spite of a different claim in their

first years. (The initial claim stated that there is a perfect one-to-one mapping between the objects of the analysis model and the objects implemented with OO programming tools. However, with real and large projects, it became obvious that the 'analysis-to-design' translation was not straightforward, even with the best OO systems. In addition, most of today's commercial database technologies still rely on the relational approach, with or without OO extensions, thus offering only limited and indirect capability to support OO concepts; this is especially true for spatial databases.) As clearly stated by Cook and Daniels (1994b), 'an important question is the extent to which the activities of analysis and design can be merged. The simplistic approach is to say that object-oriented development is a process requiring no transformations, beginning with the construction of an object model and progressing seamlessly into object-oriented code . . . While superficially appealing, this approach is seriously flawed. It should be clear to anyone that models of the world are completely different from models of software.'

Today's practice is to use at least two levels of models, separating the 'what' from the 'how' and leading to more robust and reusable results. Depending on the formal method being used, these levels are based either on different modelling techniques (e.g. entity/relationship schema for the conceptual level, relational schema for the logical level, structured query language, SQL, code for the physical level) or on the same technique (e.g. the additive approach where more details are introduced while going from the analysis level to the implementation level). Batini et al (1992) offer an excellent description of the common three-levels approach used with the entity/relationship

415

paradigm. Cook and Daniels (1994b) explain their three levels in an OO paradigm. Rumbaugh (1996) clearly presents the layered additive approach common in the OO community (Worboys, Chapter 26). When applied correctly, the multi-level approach is quite powerful since it allows developers to work at different levels of abstraction for different purposes. Integrating all the above mentioned issues into only one model makes the work much less efficient and the result less reusable. Consequently the GIS community is also embracing the multi-level approach in major efforts such as the OGIS Object Model (OGC 1996) and the ISO-TC211 rules for application schema (ISO 1996b).

While doing analysis and design, spatial database developers benefit from a method with a higher power of expression than traditional methods. Research in recent years has made high-level

modelling more efficient and semantically complete, including for spatial databases (e.g. Bédard et al 1996; Caron and Bédard 1993; Golay 1992; ISO 1996b; Pantazis and Donnay 1996). However, spatial database analysts still need to improve their modelling techniques to depict better the spatial and temporal properties of geographic features (e.g. Figure 2) and to include geometric considerations better. This is needed to help the developer to convince the user that he or she understands how every piece of information is semantically and geometrically defined and related to the others, how it is used, what are the allowed values, where it comes from, how reliable it is, etc. Similarly, spatial database designers must convince users that they have created a solution offering the best way to map every element of the analysis model into a piece of code supported by the client's technology. To do so

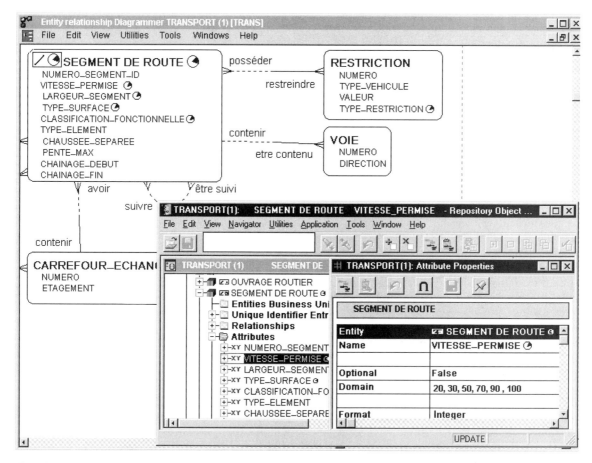

Fig 2. Extract of an analysis model made with Oracle Designer 2000 (Oracle Corporation 1996) where spatial and temporal pictograms (see Figures 4 and 5) are added to indicate the geometry and temporality of geographical features (model in development for Québec Ministry of Transportation at the time of writing the present chapter).

efficiently and to facilitate communication, especially in the multipurpose context of many spatial databases, one must extend the traditional methods, as explained in the next section.

Finally, database developers rely more and more on CASE packages. The arrival of such tools has been a big incentive for databases developers to embrace formal methods, especially when the tools automatically generate code from the models and vice versa (cf. reverse engineering). In the GIS arena, a similar movement has started. For example, there already exist graphical schema builders with an integrated data dictionary for the design of the spatial database (e.g. Intergraph MGE). However, we need to go further and to automate this process right from the analysis model. Although no complete tool exists for spatial databases outside university laboratories (e.g. Orion, developed in 1992 by the author), we can expect that the present international standardisation efforts and the strong demand for GIS interoperability will push the development of such extended methods and tools.

4 FORMAL METHODS FOR SPATIAL DATABASE ANALYSIS AND DESIGN

A formal analysis and design method is a set of guidelines and rules to capture the semantics of users' reality and to build a spatial database supporting it. It is used for thinking, documenting, and communicating in a consistent and coherent manner via models. Thus, modelling is the foundation of analysis and design. Any model is built out of a deliberately limited but sufficiently powerful and crisply defined set of constructs. These constructs, along with a simple notation and a small set of rules, constitute a 'formal language' (also called a formalism). Such a formalism can have a textual notation, a graphical notation, or a mix of the two. Human cognitive research and psychology have shown that graphical languages are more efficient for synthetic views and textual languages for detailed descriptions (see Figure 2 which shows both graphical and textual notations). Cognitive sciences have also shown that combining both graphical and textual languages is necessary to achieve clear understanding. This fact is recognised by formal methods, since they offer graphical notations to create, present, validate, and manipulate models and use textual details in dictionaries and programming code. However, the graphical notations are the most

visible part of a method and may mistakenly be thought of as *the* method. This is misleading since graphical notations only reflect a part of the underlying constructs proposed by a method.

The basic constructs are very similar in nature across formal methods of the same type but differ among types (relational, entity relationship, and object oriented). The relational approach relies on one basic construct called a relation which is a table of columns (attributes) and rows (occurrences of a phenomenon) manipulated with a relational algebra. The elegance of the relational approach lies with its simplicity, while its popularity relies on the fact that most commercial DBMS have implemented the relational structure. However, it is widely known that 'the relational model is limited with respect to semantic content (i.e. expressive power) and there are many design problems which are not naturally expressible in terms of relations. Spatial systems are a case where the limitations become clear.' (Worboys et al 1990)

The entity/relationship approach utilises more constructs, such as 'entity', 'attribute', and 'relationship'. This provides a better expressive power; however, few DBMS support the entity/relationship structure. Also, an entity/relationship schema must be translated into a relational schema to be implemented in a relational DBMS. Worboys et al (1990) mention that 'many systems may be modelled using entities, attributes and relationships, including systems with a dominating spatial component … However, experience has shown that for many systems the initial set of modelling constructs (entity, attribute, and relationship) is inadequate'. In fact, the last few years have witnessed the addition of several extensions to entity/relationship constructs, including aggregation, generalisation, geometry, and temporality (e.g. Modul-R: Bédard et al 1996; Caron and Bédard 1993).

Finally, the OO approach relies on: (1) 'objects' encompassing 'properties' (or attributes) with the 'operations' modifying data (also called methods and procedures); (2) on 'relationships' between objects; (3) on aggregation of objects into more complex objects; and (4) on generalisation or specialisation of the types of objects to more general or more specific types, respectively. The OO approach also uses 'states', 'events', and 'messages' to show the behaviour of objects. Such an integrated description leads to richer database analysis and design (see the Unified Modelling Language for the

417

most recent and robust constructs: Booch et al 1996; Worboys 1995). It is undoubtedly the most powerful modelling paradigm nowadays. Part of the success of the OO approach lies in the fact that 'the object-oriented paradigm, which originated from programming languages, has been successfully applied to the analysis and design and even earlier phases of system development' (Magrogan et al 1996).

Like entity/relationship, it is possible to extend OO methods with spatial and temporal 'plug-ins' to increase efficiency (Figure 3). The new constructs must include abstraction mechanisms powerful enough to facilitate the analysis phase. At the design phase they must also map to the built-in proprietary components of GIS, CAD, universal servers, and similar tools on the market. It must be remembered that these generic structures and operators deal with geometric primitives, graphic properties, spatial reference systems, topological relationships, spatial operators, and so on which do not exist in traditional DBMS.

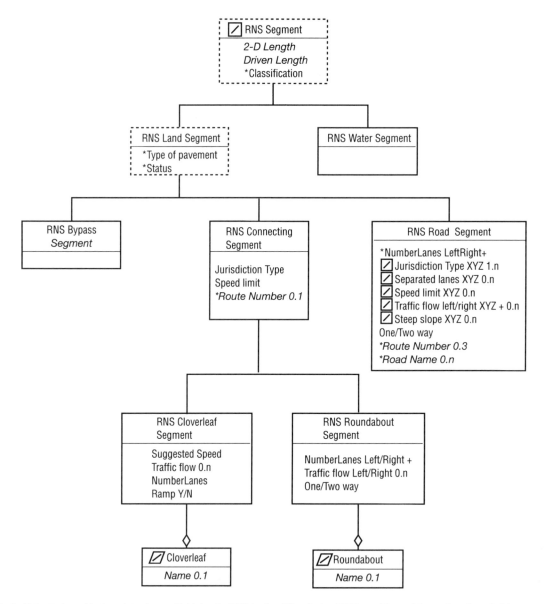

Fig 3. Extract of an object model using spatial 'plug-ins' (Bédard and Sondheim 1996); see Figure 4 for explanation of pictograms.

While at the design level the offered primitives must be considered, at the analysis level these complexities are hidden from the user; only the general geometric information is relevant (e.g. does the user want to have this type of object represented on the map? using which type of shape?). In fact, during the analysis phase, users do not have to deal with the intricacies of points vs nodes vs vertices, lines vs arcs vs polylines, metric vs topology, and so on; they should only deal with houses, lots, streets, and similar concepts of interest. Accordingly, methods should be extended with the proper geometric constructs and coding rules. Their availability in commercial CASE tools will improve the efficiency of analysis and design for spatial databases.

All of the constructs mentioned above result from the fundamental abstraction concepts used by humans to understand the world where they live: classification, association, aggregation, and generalisation. These are used by the system analyst who tries to understand the users' perceptions of their reality (e.g. the types of objects they deal with, their properties, how they relate to each other, how they are geometrically represented, how they behave, how they are used to provide new information, how they are spatially related). These are also the abstraction mechanisms used by the system designer to build efficient programming code on the selected spatial database technology. The process used when applying these abstraction mechanisms is a subtle one. It calls for creativity as well as observation and rigour. When a formal method is used it adds an engineering-like rationality to a work of art. The next paragraphs explain some technical elements of the analysis and design processes for spatial databases.

To start the analysis phase, the different types of features which are of interest to the users (e.g. house, street, owner, contract) must be identified; these features may or may not exist in their present systems. Then the semantic properties and identifiers of these features (e.g. the value, style, and address of the feature 'house') should be selected. In the case of geographical features, the types of geometry must be identified as well as a spatial reference system. Bédard et al (1996) present such geometries. They combine a dimensional pattern (0-dimensional, 1-dimensional, 2-dimensional, 3-dimensional) with a composition pattern (simple, complex, alternate, multiple). More specifically, the 'simple' pattern is used when a geographic feature is geometrically represented by only one occurrence of a given

dimension (e.g. a park represented by a single polygon, i.e. a 2-dimensional primitive); this is the most frequent situation. The 'complex' pattern indicates geometric aggregations (e.g. hydrographic networks made of 1-dimensional rivers and 2-dimensional lakes). The 'alternative' pattern indicates mutually exclusive geometries (e.g. parks digitised as points if smaller than 500 square metres or as polygons if larger). Finally, the 'multiple' pattern is the rarest of all. It indicates that more than one shape must be digitised for each occurrence of a geographical feature; this happens when the desired shapes cannot be deduced from each other. Consider, for example, the feature 'city' which is represented by a polygon on certain maps and by a point located downtown on other maps; the point cannot be derived from the polygon and thus requires its own digitising. Figure 4 shows the graphical notation used for these patterns in the Modul-R method while Figure 2 illustrates their use in a model.

For spatio-temporal databases (e.g. temporal GIS), the types of temporality needed for each type of feature must be added. These temporalities follow the same logic as the spatial patterns, i.e. dimensional pattern (instantaneous 0-dimensional, durable 1-dimensional) and composition pattern (simple, complex, alternate, multiple). These patterns apply to the existence of the feature, its presence, its functionality, and its evolution (semantic and geometric). (See Figures 2 and 5.)

Once the desired features are defined with their semantic, geometric and temporal properties, the analyst must identify the relationships of interest between these features. These include semantic relationships (e.g. house 'is owned by' owner) as well as semantically significant spatial and temporal relationships (e.g. house 'is on' lot). Integrity

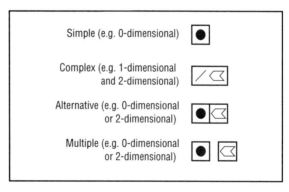

Fig 4. Graphical notation used for spatial patterns.

419

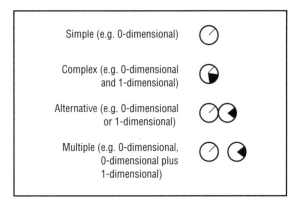

Simple (e.g. 0-dimensional)

Complex (e.g. 0-dimensional
and 1-dimensional)

Alternative (e.g. 0-dimensional
or 1-dimensional)

Multiple (e.g. 0-dimensional,
0-dimensional plus
1-dimensional)

Fig 5. Graphical notation used for temporal patterns; an example of a spatio-temporal pattern is also presented, that is one of a 0-dimensional object (e.g. an emergency vehicle) with a position which sometimes varies continuously and sometimes remains stable.

constraints must also be specified as they restrict the content of the spatial database. There are attribute rules (e.g. lists of values for nominal attributes, ranges for numeric attributes), inter-attribute rules (e.g. building.area smaller than lot.area), and inter-object rules (e.g. lot existed before building). The latter include the cardinalities of relationships (e.g. houses built on only one lot, but a lot may have 0 to N houses).

All along this process the analyst must build a schema and dictionary which are sufficiently complete for the programmers while keeping the result understandable to the users. Such a compromise between two contrasting objectives can be accomplished through the building of views, that is, exact or modified-but-compatible subsets of the global model and dictionary. Views also help to divide the problem into smaller and more manageable parts. The analyst must also decide the level of detail for the model and dictionary:

'it is not always possible or desirable to capture every nuance and restriction in a model: similarly, there is always a danger in producing a convoluted model that captures every small detail at the expense of general understandability. As a basic principle, therefore, we follow the maxim: "If you must choose between undermodelling and overmodelling, choose the undermodel and add textual commentary".' (Booch et al 1996)

The analyst verifies the logic, coherence, and completeness of his or her model, dictionary, and views. At this point the model is normalised if the approach is relational or entity/relational. Normalisation is used to eliminate data redundancies and dependencies, to facilitate the maintenance of the integrity of the database, and to build application-independent structures. If the analyst works with an extended entity/relationship or an OO approach, then objects can be generalised with common properties and relationships to create supertypes. In the case of OO methods, operations may be added and the dynamic models may be built. The last verifications are made with additional interviews and site visits, with formal walk-throughs with the users, or with testing scenarios.

With regard to the design phase, one needs new basic constructs. These constructs vary widely among spatial database management systems, since they depend on their primitives. For example, relational systems require foreign keys to materialise relationships, object-oriented systems require messages to activate operations, and commercial GIS require specific links between geometric primitives and semantic objects. As stated by Günther and Lamberts (1994), 'for the geometric representation there exists a large variety of spatial data structures that each support a certain class of spatial operators'. It may also be required to optimise the database structure with careful denormalisations of relational tables or with proper fusion and separation of objects. In particular, denormalisation is a very popular technique for spatial databases because of the large amount of spatial data handled for every single request, coupled with the need for very high performance (Egenhofer and Frank 1992). This is especially true when one wants to accelerate spatial analysis. However, alternative ways should be investigated to improve the performance of the system, either through better programming of queries and procedures or through better indexing, clustering, and buffer management.

To facilitate the preceding steps of analysis and design, different techniques may be used, like building throw-away and evolved prototypes to validate user requirements. One may also include 'use cases' or 'scenarios' (Jacobson and Christerson 1995) and 'CRC Cards' (Wilkinson 1995) which detail how users interact with a database. Finally, database 'patterns' offer well-documented and tested solutions to common problems (Coplien 1996; Fowler 1995; Gamma et al 1995).

In spite of such aids, certain steps of analysis and design lead to revisiting the models, redefining objects, adding attributes, etc. Such trips back happen naturally because the more we advance into the details, the more we understand the subtleties. Thus, analysis and design are not clear-cut sequential processes, but rather are incremental and iterative processes. In the past, most methods and CASE tools forced the developer to create very definite breaks between phases, and going back was difficult. Nowadays, with CASE tools offering reverse engineering, and OO methods suggesting additive models, it has become natural to iterate. In spite of this iterative nature and the resulting fuzzy boundary between the analysis and design processes, there remain two clearly distinct results: the technology-independent analysis model describing the users' spatial reality as understood once the project is complete, and the design model describing the spatial database as developed on the selected technology.

5 SOFTWARE SUPPORTING SPATIAL DATABASE ANALYSIS AND DESIGN

In the first years of formal methods, all schemas and dictionaries were made by hand. As a result, the drawing and editing process took more time than the thinking process, and manually keeping the coherence among evolving models, views, and dictionary proved to be an almost impossible task. This problem was solved by the new category of software created in the mid-1980s, namely CASE. These packages offered drawing, dictionary, checking, and reporting functions which accelerated the creation and modification of the schemas, views, and reports suggested by formal methods. Using

such tools greatly increased productivity, especially when one CASE could import and export results to another CASE used in the preceding or subsequent step (Bédard and Larrivée 1992). According to Roux (1991), 'the objectives of CASE tools are, in order of importance: to reduce maintenance considerably, to provide real quality, to speed up delivery, to develop at less cost' (author's translation). Figure 6 illustrates the impact of CASE tools on formal methods of database development.

The first CASE tools copied what was done by hand and specialised in one type of schema or in one task of the development process (e.g. an entity/relationship tool, a data flow diagram drawer, a screen painter). Most of them were not compatible with other CASE tools automating the preceding and following tasks, resulting in limited productivity gains. Nowadays, most CASE tools support all the schemas, views, and reports suggested by a formal method, and the result of a task situated early in the development process can be used by another task situated later in the process. This is done via a common dictionary (also called encyclopaedia, repository, or information resources dictionary system) which maintains the coherence among models.

As opposed to diagramming, drawing, and CAD packages such as Visio, Corel Draw and AutoCAD respectively, CASE tools store the meaning of each construct of a method and of each piece of a model (e.g. objects, properties, relations). They use this intelligence to enforce the rules of a method and to control the behaviour of the constructs. For example, a CASE tool may refuse an object in the dictionary which is not depicted in a schema, or may automatically fill parts of the dictionary from a schema. When an object is moved in a diagram, the software keeps intact its relationships with the other objects. When one tries to draw a relationship

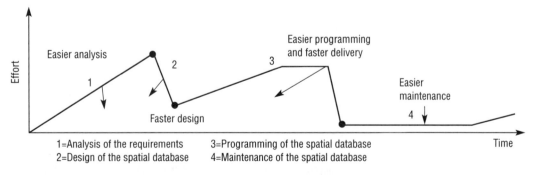

Fig 6. Impact of CASE tools on formal methods of spatial database analysis and design.

between two other relationships, the software refuses this operation if it is not allowed by the method. When one uses the same name for different objects, the software requires a change, and so on. Products such as ERWin, S-Designor, Sylverrun, IEF, Designer2000, OMTool, and Rational Rose are a few examples of intelligent CASE tools covering the entire development and maintenance cycle. Like GIS, some CASE tools cost thousands of dollars while others cost a few hundreds, and the latest trend is to embed limited CASE engines within programming environments (e.g. Delphi, Visual Basic).

In the case of spatial databases, present commercial CASE tools can be used. However, extensions for spatial constructs, rules, and code generation are needed. Therefore, spatial database developers must turn towards a more advanced category of CASE tools: metaCASE tools. These are specialised development packages working at the metamodel level; they can be programmed to accept the new constructs and rules of a new method and afterwards to run like a CASE tool for that method. Several large organisations rely on such a solution to have a CASE tool adapted to their proprietary method. This was used to build our spatial CASE tool called Orion (Figure 7). ObjectMaker and Paradigm Plus are such products. In comparison to traditional CASE tools, they offer a large number of methods. Some recent CASE tools offer similar but limited extension capabilities. MetaCASE and extensible CASE products represent the best choice for spatial databases.

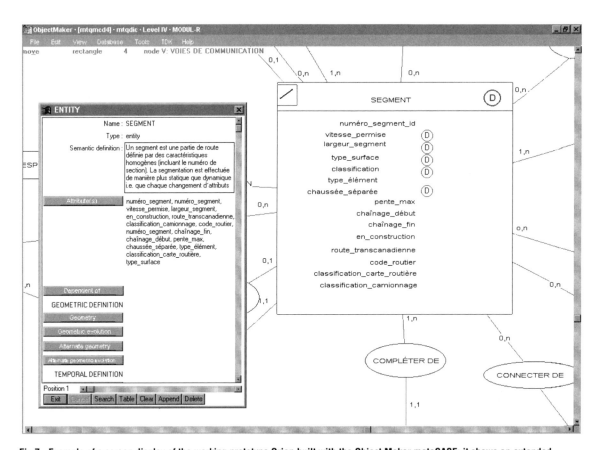

Fig 7. Example of a screen display of the working prototype Orion built with the Object Maker metaCASE; it shows an extended entity/relationship analysis model with spatial and temporal pictograms, as well as an extract of the extended data dictionary which is partly filled automatically from the schema; it is used for automatic code generation for Intergraph GIS.

6 CONCLUSION

This chapter has presented the fundamental elements of spatial database analysis and design. Good analysis and design rely on formal methods. Such methods help us to work in a more rigorous manner directed by principles, techniques, and guidelines. This is particularly important for the analysis phase: 'although it represents only a small proportion of the total development effort, its impact on the final system is probably greater than any other phase . . . Analysts estimate that a change that costs $1 to fix in the requirements stage will cost $10 in design, $100 in construction, and $1000 in implementation!' (Moody 1996).

Formal methods rely on specific constructs, models, and processes like the ones presented in this chapter. The most recent methods are based on the object-oriented paradigm, and they are the most powerful. Like other methods, they are supported by CASE tools to facilitate the building of models, dictionaries, documentation, and maintenance. Present methods and tools may already be used for the analysis and design of spatial databases. However, recent research and standardisation efforts will help to adapt these methods and tools to spatial information technology. This should further encourage the development of spatial databases which are robust and flexible, faster to build, easier to maintain, and closer to interoperability. Finally, this will help GIS developers to enter the mainstream of information technologies which is a natural evolution.

References

ANSI/SPARC 1975 ANSI/X3/SPARC Study Group on data base management systems, interim report. *ACM SIGFIDET* 7: 3–139

Batini C, Ceri S, Navathe S B 1992 *Conceptual database design, an entity–relationship approach*. Redwood City, Benjamin Cummings

Bédard Y, Caron C, Maamar Z, Moulin B, Vallière D 1996 Adapting data models for the design of spatio-temporal databases. *Computers, Environment, and Urban Systems* 20: 19–41

Bédard Y, Larrivée S 1992 Développement des systèmes d'information à référence spatiale: vers l'utilisation d'ateliers de génie logiciel. *Journal of the Canadian Institute of Surveying and Mapping* 46: 423–33

Bédard Y, Sondheim M 1996 *Road Network System data model. Technical report*. Geographic Data. *http://www.env.gov.bc.ca/gdbc/rns*

Booch G 1994 *Object-oriented analysis and design with applications*, 2nd edition. Redwood City, Benjamin Cummings

Booch G, Rumbaugh J, Jacobson I 1996 *The Unified Modelling Language for object-oriented development, documentation set version 0.9 addendum*. Santa-Clara, Rational Software Corporation. *hhtp://www.rational.com/ot./uml.html*

Carmichael A (ed.) 1995 *Object development methods*. New York, SIGS Books

Caron C, Bédard Y 1993 Extending the individual formalism for a more complete modelling of urban spatially referenced data. *Computers, Environment, and Urban Systems* 17: 337–46

Coad P, Yourdon E 1991a *Object-oriented analysis*, 2nd edition. Englewood Cliffs, Prentice-Hall

Coad P, Yourdon E 1991a *Object-oriented design*. Englewood Cliffs, Prentice-Hall

Cook S, Daniels J 1994a *Designing object systems: object-oriented modelling with Syntropy*. Englewood Cliffs, Prentice-Hall

Cook S, Daniels J 1994b Software isn't the real world. *Journal of Object-Oriented Programming* (May): 2–28

Coplien J O 1996 *Software patterns*. New York, SIGS Books

Egenhofer M J, Frank A U 1992 Object-oriented modelling for GIS. *Journal of the Urban and Regional Information Systems Association*: 3–19

Fowler M 1995 *Analysis patterns: reusable object models*. Reading (USA), Addison-Wesley

Gamma E, Helm R, Johnson R, Vlissides J 1995 *Design patterns: elements of reusable object-oriented software*. Reading (USA), Addison-Wesley

Golay F 1992 'Modélisation des systèmes d'information à référence spatiale et de leurs domaines d'utilisation spécialisés, aspects méthodologiques, organisationnels et technologiques'. PhD thesis, École Polytechnique Fédérale de Lausanne, Switzerland

Günther O, Lamberts J 1994 Object-oriented techniques for the management of geographic and environmental data. *The Computer Journal* 37: 16–25

ISO 1996a *SQL multimedia and application packages (SQL/MM) part 3: spatial*. SQL/MM: MAD-005. International Organisation for Standardisation

ISO 1996b *Geographic information – rules for application schema*. ISO 15046-9. ISO TC211: WG2 N030 Geographic information-Geomatics, Working group 2. International Organisation for Standardisation

Jacobson I, Christerson M 1995 Modelling with use cases: a confused world of OOA and OOD. *Journal of Object-Oriented Programming* (September): 15–20

Jacobson I, Christerson M, Jonsson P, Overgaard G 1993 *Object-oriented software engineering*. Reading (USA), Addison-Wesley

Magrogan P J, Schardt J A, Chronoles M J 1996 Object-oriented conceptualisation. Report on object analysis and design. *Journal of Object-Oriented Programming* (September): 54–63

Martin J, Odell J J 1993 *Principles of object-oriented analysis and design*. Englewood Cliffs, Prentice-Hall

Moody D 1996 The seven habits of highly effective data modellers. *Database Programming and Design* (October): 57–64

OGC (Open GIS Consortium) 1996c *The Open GIS abstract specification. http://www.openings.org/public/abstract.html*

Olle T W 1991 *Information systems methodologies*, 2nd edition. Reading (USA), Addison-Wesley

Oracle Corporation 1996 *Oracle Designer/2000, Release 1.3*

Pantazis D, Donnay J P 1996 *La conception de SIG, méthode et formalisme*. Paris, Hermès

Roux F G 1991 Comment le génie vient au logiciel. *L'informatique professionnelle* 93(April): 19–26

Rumbaugh J 1996 Layered additive models: design as a process of recording decisions. *Journal of Object-Oriented Programming* (March–April): 21–48

Rumbaugh J, Blaha M, Premerlani W, Eddy F, Lorensen W 1991 *Object-oriented modelling and design*. Englewood Cliffs, Prentice-Hall

Shlaer S, Mellor S 1991 *Object lifecycles: modelling the world in states*. Englewood Cliffs, Prentice-Hall

Wilkinson N 1995 *Using CRC cards: an informal approach to object-oriented development*. New York, SIGS Books

Worboys M F 1995 *GIS: a computing perspective*. London, Taylor and Francis

Worboys M F, Hearnshaw H, Maguire D J 1990 Object-oriented data modelling for spatial databases. *International Journal of Geographical Information Systems* 4: 369–8

Introduction

THE EDITORS

Prior to the widespread development of digital data capture, data were captured by manual recording of attribute measurements and then reconciling them to some kind of georeferenced frame. This section begins with a 'back-to-basics' rendition of the impact of the science of geodesy upon the way in which we position ourselves on the Earth's surface. In Chapter 30 Hermann Seeger's clear message is that in an age in which surveying techniques have been apparently de-skilled and reference frameworks globalised, GIS users will nevertheless ignore the scientific basis to global positioning at their peril. After a description of the principles used to locate points on the Earth's surface, Seeger reviews the techniques available for conversion of coordinate data between different positioning systems. With respect to GIS usage, he goes on to suggest that a range of potential complexities and gross errors may arise in geographical positioning and that, in particular, conflation of conventional map information with that derived from global positioning systems (GPS: see Lange and Gilbert, Chapter 33) may create considerable practical problems. The message is thus a need for a good understanding of the science of geodesy coupled with adequate knowledge of the genealogy of the maps from which other GIS-based data have been derived. Goodchild and Longley (Chapter 40) return to this issue in the context of the quest for 'perfect positioning' at the end of the Technical Issues Part.

The early years of GIS were characterised more by rapid developments in computer hardware and software, and as such GIS was technology – rather than data – led. Many of the data for early GIS applications were captured from existing analogue sources which had already been reconciled with spatial referencing systems. Even though many of the chapters elsewhere in this book (especially those discussing spatial analysis in Section 1c) have made much of the subsequent creation of huge digital datasets and the emergence of a digital information economy, it is important to remember that the costs

of data are still very often the most significant component of a GIS (see also Bernhardsen, Chapter 41; Rhind, Chapter 56). Moreover, a recurring theme through this book has been the interdependencies between the conception, measurement, representation, and analysis of geographical phenomena, with the implication that involvement (or at least acquaintance) with each of these successive stages is likely to lead to more sensitive and intelligent geographical information handling (e.g. Goodchild and Longley, Chapter 40). Founding a GIS upon dubious or poor quality data is simply not good enough – and such data will have obvious repercussions of the 'garbage in – garbage out' variety when used in analysis or decision-making. If source data were originally captured in a different medium, and were intended for use in different ways, then we need to understand the characteristics of the original data – including their quality, the processes used to produce the document to be encoded, and the geometric characteristics of the data (Dowman, Chapter 31). We also need to understand and anticipate the additional problems that can arise in the transfer of analogue information to digital media.

In the second contribution to this section, Ian Dowman presents a review of the simplification, codification, and generalisation that takes place in capturing data from hard copies, together with the checks and balances that can be used to minimise the additional error introduced by the data transfer process. The pitfalls associated with digitising from paper maps are widely known, and the checks that may be used to ensure adequate spatial referencing are straightforward. This is not so clearly the case with regard to aerial photographs, since digital mapping requires specification of the geometry of the image forming system. Moreover, and especially at the scales characteristic of satellite images, relief causes displacement together with changes in scale across the image: these can only be removed by developing a stereoscopic 3-dimensional model or by using ancillary information from a digital elevation model (see Hutchinson and Gallant, Chapter 9). Even digitally corrected orthoimages may still be distorted by tilting of the sensor platform (and will require reconciliation with ground control points), and a range of other transformations may be necessary to reconcile images with spatially referenced systems.

The 'data bottleneck' experienced in early GIS applications stood in stark contrast to the experience of the remote-sensing community, in which the handling and processing of spatially extensive digital images remains a major activity. Satellite imagery has become both more detailed, with developments in satellite technology, and more widely available, particularly from satellites belonging to the former Soviet Union. As a consequence, there has been some convergence of GIS and remote sensing and an opening up of new application areas. Mike Barnsley (Chapter 32) provides an overview of remote sensing as an enabling technology for exploiting the fuller potential of GIS. The availability of more data and the improved positional accuracy associated with them is part of this story. Of more wide-ranging import (at least in applications terms), Barnsley is at pains to emphasise the roles of inference and estimation in moving from surrogate (e.g. land cover) to target (e.g. land use) variables. This provides a reminder that in remote sensing, as elsewhere in GIS, our understanding of the process of inference is far from complete (cf. Getis, Chapter 16) and that irrespective of technique used (cf. Fischer, Chapter 19) we only ever end up with estimates of our chosen

classification. That said, there are a number of important developments in image classification using artificial neural network techniques and contextual information about the apparent configuration of land use: Barnsley sees these as offering the prospect of freeing the classification process from conventional assumptions about statistical distributions and opening up new domains of application, such as urban remote sensing. From a purely technical standpoint, Barnsley also examines the relationship between the design of new sensors and their abilities to estimate the intrinsic properties of the Earth's surface.

The final contribution to this section, by Lange and Gilbert, describes the development of global positioning systems (GPS) as a GIS-enabling technology. They review a range of civilian applications and describe the use of base stations to improve levels of geographical accuracy. The development of GPS vividly illustrates how far technical aspects of GIS data collection have come: Dowman's contribution to this section provided a review of the ways in which digital abstractions can be captured from paper maps which themselves are highly abstract and selective descriptions of geographical reality; by analogy, the GPS user creates an abstraction of the surface of the Earth through the act of moving across it (and perhaps recording additional attribute layers at the same time). This provides a vivid illustration of how much richer GIS data models have become, and how much more control the informed GIS user now has over some aspects of data collection.

30

Spatial referencing and coordinate systems

H SEEGER

The growing use of the Global Positioning System (GPS) offers an apparent panacea for
many GIS-related problems. However, this is partly illusory: to avoid subsequent errors
when matching GPS coordinates to those gathered from pre-existing sources – a normal
requirement – a good understanding of geodesy is essential. This chapter describes the
basic principles of geodetic reference systems, geodetic ellipsoids, and map projections.
It shows the nature of the relationship between WGS84, which is based on geocentric
Cartesian coordinates, and national systems based on an ellipsoid defined locally to suit the
national territory. Most geographical data have hitherto been collected on maps which are
themselves based on nationally defined reference systems and a variety of other factors has
entered into their construction and digital encoding. As a result, coordinate shifts as large
as hundreds of metres may occur in different descriptions of the same place. The
ramifications for the use of GIS to link data together are obvious.

1 INTRODUCTION

Only in a few, very local, applications of GIS can
geodetic reference systems and coordinate systems be
ignored safely. Such situations arise when a particular
study relates only to an area a few tens or hundreds of
metres across and involves no comparisons with
studies elsewhere. In most other circumstances,
however, it is essential that the geographical
descriptions are made using a more global reference
framework (see also Dowman, Chapter 31; Guptill,
Chapter 49; Smith and Rhind, Chapter 47). Typically
in the past this has been achieved through use of
national geodetic frameworks defined and maintained
by national mapping agencies and manifest most
obviously on topographic maps. Now the advent of
the Global Positioning System (GPS: see Lange and
Gilbert, Chapter 33) has offered the possibility of a
uniform reference framework for the whole world and
a means of determining position within that
framework at low cost and instantaneously. It is self-
evident therefore that these developments have major
implications for GIS.

In practice, these developments have even greater
ramifications for GIS users because the apparent

de-skilling of surveying and the globalisation of
reference frameworks conceals a number of major
complexities. It is entirely possible that serious errors
can be made with the new tools, especially when
their use is (necessarily) combined with information
derived from earlier surveying concepts and
methods. In Bosnia, for instance, peace-keeping
troops using GPS to navigate by jeep have been
killed by mines because of disjunctions between
datums underlying the maps of minefields and the
GPS itself. Serious naval incidents have occurred
because of the same factors and blind trust in the
new technology. It follows that serious use of GPS –
and indeed of GIS – requires a basic understanding
of the different geodetic frameworks and the
relationships between them.

This chapter examines the basic concepts
underlying methods of defining location and the
description of that positional information. There is
not space enough to provide comprehensive
guidance on, or the mathematical basis of, a
complex topic. Nor can it give recommendations on
which particular system to use in specific
circumstances. It does, however, give sufficient
information to enable the reader to understand some

of the benefits and pitfalls associated with choosing or coping with different reference systems. Standard texts on geodesy, such as that by Bomford (1980) and Vanicek and Krakovsky (1986), provide detail on background concepts while Lange and Gilbert (Chapter 33) and books by Kennedy (1996), Leick (1990), and Hofmann-Wellenhof et al (1994) give more details on GPS.

2 SOME BASICS ABOUT GEODESY

Geodesy is the science of the shape and size of the Earth, together with its gravity field. It is well-known that the Earth is not a regular geometrically-shaped body, and that variations in its shape occur continuously at the meso and micro scales. In many instances (e.g. in Alpine, Himalayan, or Andean areas), it is simply impossible to make good local approximations by regarding the Earth's level surfaces as flat planes.

In mathematics, there are in general two ways to determine a point position in space – that is, by Cartesian coordinates or by angular coordinates. The coordinates of the point are normally related to a right-handed rectangular coordinate system, or other combinations. The position of the origin of the coordinate system and its orientation is given by the geodetic datum definition.

2.1 Early measurements of the Earth

Eratosthenes was one of the first to measure the size of the Earth with any accuracy but provably accurate definitions of the size of the planet only came about as a result of work in the eighteenth century. French expeditions in 1735 to both Lapland and Peru measured along arcs of meridians using triangulation. The measurements proved that the Earth was almost spherical but flattened at the poles (oblate) and not at the equator (prolate), thus being consistent with Newton's gravitational theories. As a result of triangulation and astronomical observations near Dunkirk and Barcelona, Jean-Baptiste-Joseph Delambre (1749–1822) and Pierre-François-Andre Mechain (1744–1804) proposed the definition of the metre as one ten millionth of a quadrant of the meridian through the Observatory of Paris. This definition was accepted by the French Academy of Sciences in 1795. The use of such universal standards of time and length are crucial to geodesy.

2.2 Geoid and ellipsoid

Defining a single, stable reference surface from which the third dimension is measured was also found to be difficult. The most obvious choice of reference surface is mean sea level but this is not the same everywhere: it is, for instance, affected by ocean currents, wind, and barometric pressure. As a result, a conceptual surface called the *geoid* is used. The geoid is the equipotential surface of the Earth's gravity field that best approximates to mean sea level. An equipotential surface is level in the sense that moving across it results in no work being done against gravity (i.e. the direction of gravity is everywhere perpendicular to it). A simple conceptualisation of the geoid is to imagine the Earth with canals dug across all continents, enabling completely stationary and inert oceans to meet and find their own 'sea level' assuming there were neither tides or weather to affect the result. Even now the geoid is difficult to model: world-wide it is known to a metre or so, although locally it can be found to within 0.01 metres.

The geoid is a smooth surface which can be approximated by an *ellipsoid*. An ellipsoid is defined as an ellipse of rotation about the polar axis. The relationship between these two shapes – the geoid defined by physics and the ellipsoid defined by mathematics – is of prime importance, as will be evident later. This relationship is part of what is termed the geodetic datum (see below).

A further complication is that the Earth is not a rigid and invariant body. Sea tides are quite obvious to observers at the seashore but the same forces also cause the solid Earth to have tides, albeit much smaller ones of only a few centimetres. Because the Earth is elastic, has an atmosphere, is rotating in space, and orbiting the Sun, the poles move. The poles move (polar motion) with both long- and short-term periods. Fortunately, as the instrumentation and methods available to measure the Earth improve, all these temporal movements become measurable but they are time-dependent and thus need to be repeated periodically. They also need to be updated as instrumentation and methods develop.

2.3 Contemporary approaches

From the mid-twentieth century onwards, measurement techniques using light waves and microwaves were considerably improved and the ability to measure long distances (70–100 kilometres) was used by geodesists to refine the nineteenth-century triangulation schemes and re-define the shape of the Earth. The latest methods of obtaining the Earth's size and gravitational field all involve looking outward. Very long baseline interferometry (VLBI) relies on two astronomical telescopes at each end of a baseline (e.g. in Santiago, Chile and in Wetzell, Germany) correlating the signals received from the same quasar. The time delay between one wave-front arriving at the telescope and the same wave-front arriving at the other telescope is measured with a very precise clock. Since the speed of propagation of the wave is known, the distance and direction can be calculated between the two telescopes. The accuracy of these baseline measurements is about one part in 100 million.

In addition, satellites are now used increasingly to determine baselines (distances and directions) and gravitational forces. The most widely used systems come from the USA. The first was TRANSIT, a US Navy navigation satellite which enabled geodesists to measure relative positions over hundreds of kilometres to an accuracy of around 30 centimetres using the Doppler effect of the satellite signals. The second was the GPS which has a constellation of over 24 satellites orbiting the Earth and enables positions between receivers to be deduced to one part in 1000 million using the signal phase differences (see Lange and Gilbert, Chapter 33). Russia has launched a similar system called GLONASS developed initially, like GPS, for military navigation. Satellite laser ranging (SLR) to orbiting satellites from ground stations (e.g. at Herstmonceux, England and Zimmerwald, Switzerland) enables the orbits of such satellites as LAGEOS to be determined and subsequently the effect of the gravity field to be computed. Satellite altimetry using radar to measure from satellites (e.g. ERS-1, ERS-2, and TOPEX POSEIDON) to the ocean surface allows the geoid to be modelled. The latest geodetic methods are also used to determine tectonic plate motion and for earthquake prediction (see, for example, Warita and Nonomura 1997).

3 THE DEFINITION OF POSITION

In this section, a more formal approach is used to define the way in which geographical location or position may be described. Two main types of position information can be distinguished:

- direct positioning – that is, based on coordinates which enable any point in space (one, two, three, or even more dimensions) to be uniquely defined and referred to a (local) coordinate system;
- indirect positioning – that is, based not on coordinates but on values (for instance administrative units, postal addresses, and road numbers) which can be unambiguously mapped – to some defined level of precision and accuracy – to a specific geographical location.

3.1 Direct position

'Direct position' is given by a set of coordinates based on a reference system in which all the physical parameters of the Earth – such as its size, shape, orientation in space, and gravity field – are defined, together with a specific coordinate system. It was pointed out earlier that the Earth itself changes in shape but, in addition and because our knowledge of the Earth is increasing, the values which we gave to some physical constants some years ago are not the 'best' values today. The effect of all this is that reference systems for the Earth, which by their very nature include the physical constants, change periodically.

There are consequently two aspects of direct positioning to be taken into account. The first and essential aspect is that a full coordinate description is required of point, line, area, or volumetric entities. The second aspect is often optional (but worthwhile): it consists of all necessary data to transform these coordinates into coordinates in another reference system.

3.1.1 Geodetic reference systems

All direct positions are defined by a geodetic reference system, including a *geodetic datum*. This contains all the elements necessary to locate a point relative to the surface of the Earth. A geodetic reference system is uniquely identified by an attribute name which is a short acronym or an abbreviation of the full name (e.g. WGS84 for the World Geodetic System 1984). The full name will usually contain the complete name of the coordinate reference system. The optional description attribute contains a more detailed textual description of the coordinate reference system, including recommended usage and possible pitfalls.

The origin, orientation, and rotation of the coordinate system is defined by the datum definition attribute (see section 3.1.4). Most geodetic reference systems have only one datum. However, because of the historic difficulty of surveying and computing horizontal position (bi-dimensional) and vertical position or height (uni-dimensional), the horizontal and vertical components are often separated. Thus a coordinate reference system may contain two datums, the geodetic datum and the vertical datum. However, hybrid datums are here solved as a combination of a bi-dimensional and uni-dimensional geodetic datum within the same datum definition. An optional list of 'anchor', or fiducial, points gives additional positional information which completely defines geodetic reference system. In summary, then, the coordinate system describes the spatial parameters of the 'measurement framework' and how it relates to the Earth as a whole; while coordinates define a 'direct position' within that system.

The easiest way to define a point in space is to use 3-dimensional Cartesian coordinates. Such a geodetic reference system has:

- an origin, O, generally coincident with the centre of mass of the Earth;
- one axis, Z, coincident with the Earth's spin axis;
- one axis, X, lying in the plane of the zero meridian (now close to, rather than exactly at, Greenwich);
- the Y axis completing the mutually perpendicular axes, forming a right-handed system.

For any point P in space, the choice of a geodetic reference system enables Cartesian coordinates X, Y, and Z to be defined as shown in Figure 1.

3.1.2 Geodetic ellipsoids and the zero meridian

As described earlier, geodetic studies in the eighteenth century showed that the mathematical figure which best represents the Earth surface without topography is the geoid and that this can be represented by an ellipsoid of rotation. This is termed a *geodetic ellipsoid*. It is defined by two parameters, for instance the semi-major axis and the 'flattening'.

The simultaneous selection of a geodetic reference system (which defines the 3-dimensional framework) and a geodetic ellipsoid (which defines the geodetic reference surface) is the basic model which distinguishes between horizontal and vertical information through (geodetic) geographical coordinates. When this choice is made, it is assumed that the origin of the ellipsoid coincides with the

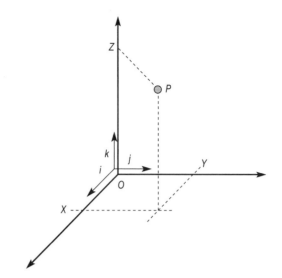

Fig 1. Cartesian coordinates on a geodetic reference system.

origin O of the reference system and that the minor axis (of rotation) of the ellipsoid coincides with the polar axis of the geodetic reference system. We can then define the geographical coordinates Latitude, Longitude, and Height (φ, λ, h) of any point, P, in space (as shown in Figure 2).

In Figure 2 p is the projection of P on the geodetic ellipsoid along the normal to the ellipsoid at P, and h, the ellipsoidal height, is the algebraic

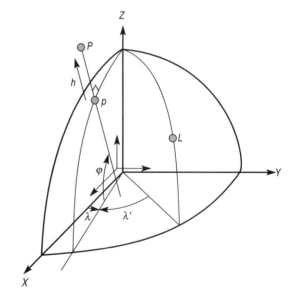

Fig 2. Geographical coordinates.

length of the segment (*pP*) (which is positive outside the ellipsoid). The latitude φ is the angle between the equator of the ellipsoid and the normal to the ellipsoid at *P*. The longitude λ is the angle directed between the Greenwich Meridian plane and the meridian plane of *P*. If another zero meridian is chosen (such as the meridian plane passing through a given point *L*), a different longitude λ' naturally results (as in some French maps before universal adoption of the Greenwich meridian). Following from this, a geodetic reference system, a geodetic ellipsoid, and a zero meridian allow a second type of 3-dimensional coordinates, the *geographical coordinates* (φ, λ, *h*), to be defined for any point *P* in space.

The scientific community has adopted a conventional terrestrial reference system (CTS), which is fixed to the Earth, to be used as reference for scientific and practical works. Its detailed definition and practical implementation have been improved continually. These tasks were the responsibility of the Bureau International de l'Heure (BIH) until 1987 and have been carried out by the International Earth Rotation Service (IERS) since 1988. This system, now named the IERS Terrestrial Reference Frame (ITRF), is truly geocentric and its orientation is the reference orientation of the planet for all Earth rotation studies.

The following relation exists between the ellipsoidal and Cartesian coordinates if the origin of the Cartesian coordinate system coincides with the centre of the rotation ellipsoid, the *X*-axis of the ellipsoid pierces the ellipsoid at the point $\varphi = 0$, $\lambda = 0$ and the *Z*-axis stands perpendicular to the equator:

$$\begin{bmatrix} X \\ Y \\ Z \end{bmatrix} = \begin{bmatrix} [N_E + h]\cos\varphi\,\cos\lambda \\ [N_E + h]\cos\varphi\,\sin\lambda \\ [N_E(1-e^2)+h]\,\sin\varphi \end{bmatrix}$$

With the transverse curvature radius

$$N_E = a\,(1-e^2\sin^2\varphi)^{-\frac{1}{2}}$$

and the first eccentricity

$$e^2 = 1 - \frac{b^2}{a^2}$$

we obtain

$$\lambda = \arctan\frac{y}{x}$$

The inverse problem of φ and *h* can be solved only by iteration, but the solution converges quickly since $h \ll N_E$.

$$h_i = \frac{(x^2 + y^2)^{\frac{1}{2}}}{\cos\varphi_{i-1}} - N_{E(i)}$$

$$\varphi_i = \arctan\left[\frac{z}{(x^2 + y^2)^{-\frac{1}{2}}} \cdot \frac{1}{1 - \dfrac{e^2 \cdot N_{E(i)}}{N_{E(i)} + h_i}} \right]$$

The iteration may be started with

$$\varphi_0 \approx \arctan\left(z \cdot (x^2 + y^2)^{-\frac{1}{2}} \right)$$

3.1.3 Coordinate systems

In formal terminology, the dimension, the coordinate sequence, and the coordinate types are defined by the nature of the particular coordinate system employed. The name, unit scale, unit system, and direction of axis are defined for each axis by the coordinate description entity.

The coordinate sequence is defined according to the axis of the defined datum. If the coordinates are given in a projection, the necessary information is described in the 'entity map projection'. The 'coordinate types' description includes the different height systems which may be used. The range of possible coordinate system types comprises:

Three-dimensional coordinate systems

- Cartesian coordinate system (geocentric);
- geodetic coordinate system with ellipsoidal height (see section 3.1.5);
- geodetic coordinate system with normal height;
- geodetic coordinate system with orthometric height;
- geodetic coordinate system with geopotential number;
- local Cartesian (topocentric);
- astronomical.

Two-dimensional coordinate systems

- geodetic;
- plane;
- local Cartesian;
- astronomical.

One-dimensional systems

- height above ellipsoid;
- orthometric height;
- normal height;
- geopotential number.

A coordinate description always has a name, a definition, and the direction of the axis. An example of a coordinate name is 'E' and of a definition is

'Easting'. A coordinate is necessarily measured according to a system of defined units (e.g. SI units).

3.1.4 Geodetic datum

A tri- or 3-dimensional geodetic datum defines the position of the origin in relation to the Earth mass centre and the orientation of the Z-axis in relation to the conventional Earth rotation axis, the X-axis to the conventional zero meridian, with the Y-axis added in a right-hand system. The parameters of the geodetic ellipsoid are included in the datum definition.

A bi- or 2-dimensional geodetic datum defines the position of a bi-dimensional coordinate system to the Earth body. It is defined by the geodetic coordinates of a main point (the origin of the datum), the deflection of the vertical, and the geoid height at the origin as well as the parameters of the geodetic ellipsoid.

A uni- or 1-dimensional height datum defines the relation of the reference surface of the height system in relation to the mean sea level over the whole ocean (global vertical), the mean sea level as defined by one or more tide gauges (a regional vertical), or to any other preferred reference point.

As indicated earlier, a geodetic datum typically is given a unique name which is a short acronym or meaningful abbreviation of the full name (e.g. ETRS 89 for the European Terrestrial Reference System 1989). The name of the geodetic datum might be the same as the name of the geodetic reference system. The attribute 'full name' should always contain the complete name of the geodetic datum, while the optional description attribute contains a more detailed textual description of the geodetic datum, its recommended usage and possible pitfalls.

To transform coordinates of an older reference system to one of the newer 3-dimensional x, y, z systems is not always a simple task if we do not have the corresponding values for the geoid model. Such uncertainty is surprisingly common: for instance, even in a country as geodetically sophisticated as the United Kingdom, much research was necessary to recover the precise definitions of the 150 or more local, county-related reference systems used in topographic mapping until the 1940s in order to relate historical information to current data. The situation in some other countries is known to be worse with no information being available in certain cases. Two-dimensional ellipsoidal approaches may be easier in such cases though mathematically they tend to be more complex.

3.1.5 The geoid and heights

The Earth gravity field is expressed through its gravity potential which includes both gravitational and centrifugal effects. The gravity g and its spherical directions Φ (astronomical latitude) and Λ (astronomical longitude) are the astronomical coordinates with regard to the CTS.

As defined earlier, the geoid is an equipotential surface of the Earth gravity field which most closely approximates mean sea level. The geoid height N is the positive distance between the ellipsoid and the geoid, defined outside the ellipsoid. In practice, the expression 'closely approximates mean sea level' allows one to consider slightly different equipotential surfaces as geoid models.

Because of multiple geophysical phenomena, the geoid is not a simple mathematical shape although it can be expressed as a sphere with a series of harmonic terms, each one smaller than previously and each one altering the shape to – for instance – a flattened sphere or a pear shape. The larger the number of terms in the expression, the more detail it describes. 'Height' is a number expressing the separation between a point P and a horizontal reference surface. We have already seen the definition of the ellipsoidal height h where the reference surface is a geodetic ellipsoid. But the many types of height normally defined over land typically use the geoid as the reference surface rather than some mathematically defined geodetic ellipsoid. More precisely, they are defined using the geopotential number which is the difference in gravity potential between the geoid and the equipotential surface through the point P and a value for gravity. Geopotential numbers can be converted to linear units. Several options are available and the most commonly used of these are:

- orthometric height H_0, uses a mean value for gravity;
- normal height H_N, uses a more refined gravity value;
- dynamic height H_0, uses an arbitrarily agreed gravity value.

Of these, orthometric heights are most commonly used because they can be determined directly from 'levelling' without first computing geopotential numbers. Orthometric height is simply the linear distance to the geoid, as shown in Figure 3.

The result is that a vertical datum is fully defined by:

- the equipotential surface where the heights are zero or derived by a geoid model. For example, the geoid model used in France for the national geoid

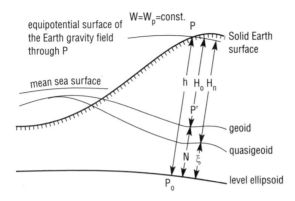

Fig 3. Normal heights and orthometric heights.

system is the equipotential surface passing through the mean level of the Mediterranean sea at Marseilles, determined by tide gauge measurements;
- the type of height (orthometric, normal, dynamic);
- the linear unit in which it is expressed (metre, foot, ...).

The reference system for gravity is given as enumerated values of the gravity system, with the permissible values of IGSN71, Potsdam system and ECS62.

3.1.6 Map projections
Many textbooks provide descriptions of the fundamentals of map projections (e.g. Maling 1973; Snyder 1987, 1993). Put simply, any map projection is a mathematical representation of a geodetic ellipsoid (in some cases, a sphere) as a plane.

Mathematically, if each point P on the geodetic ellipsoid is identified by its geographical coordinates (φ, λ) and each point p' in the plane by its Cartesian coordinates (E, N) then a map projection is defined by two functions f and g such that:

$$E = f(\varphi, \lambda)$$

$$N = g(\varphi, \lambda)$$

From this, one can deduce that, for any point P in space and having chosen a geodetic reference system, a geodetic ellipsoid, a zero meridian, and a map projection, the position of P can be represented by the 3-dimensional coordinates E, N, and h, as shown in Figure 4.

Because of the complexity involved in giving a complete description of all parameters necessary to describe map projections, only well-known projections are recommended here such as Universal Transverse Mercator (UTM) and Gauß-Krüger. For all other projections, the coordinates must be transferred to geographical coordinates if a full description is needed.

3.1.7 Units
As coordinates are given by numerical values (X, Y, Z), (φ, λ, h), (φ, λ), (E, N, h), or (E, N), it is essential to specify the linear and angular units used. Cartesian coordinates (X, Y, Z) are always expressed in metres; heights h and map coordinates (E, N) are in linear units usually given in metres but sometimes

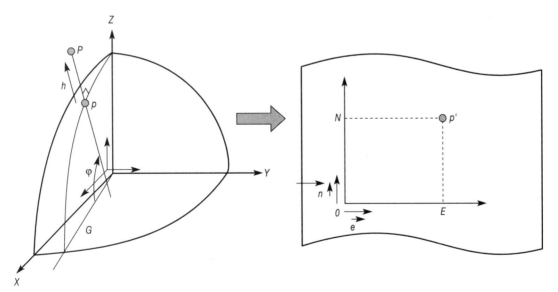

Fig 4. Map projection.

in feet, yards, etc. The geographical coordinates (φ, λ) are usually expressed in angular units (e.g. sexagesimal degrees, minutes of arc, and seconds of arc, but also in gons (grades or centesimal degrees), radians, or decimal degrees.

Local Cartesian coordinates (u, v, w) are defined as Cartesian coordinates of a local reference frame, such as those shown in Figure 5 which define an affine frame around a point P at the Earth's surface. Its orientation with regard to the CTS defines it completely.

3.1.8 National versus global reference systems

The principles and situations described above can be summarised as follows. For reasons of history and the available technology, conventional geodetic coordinates, which are expressed in terms of geodetic latitude and longitude, are related to a national or a continental datum. Thus, for example, the UK Primary Triangulation coordinates are referred to as the OS(SN)80 Datum which is defined by the Airy ellipsoid and the adopted coordinates of the Origin pillar in the grounds of Herstmonceaux Castle in Sussex in southern Britain. Similarly, coordinates in the North Sea prior to the widespread use of GPS were expressed in the ED50 Datum and are related to the Hayford ellipsoid and a point of origin in Germany. Finally, different map projections or different parameters of the same map projections are in use in many countries, each tailored to maximise or minimise some properties held to be of high importance (e.g. distortion; see Maling 1973, Chapter 4). The situation is exacerbated where

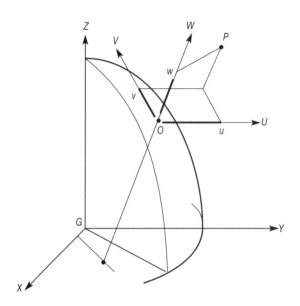

Fig 5. Local Cartesian coordinates.

discontinuities occur in the map projections used, notably in the widely used UTM projection. Britain, for instance, straddles UTM zones and there is therefore an obviously variable relationship between the National Grid coordinates familiar to most of the populace and those of UTM – as illustrated in Figure 6. Where all types of data are held in computer form within one projection system, transformation to another may cause other visual problems; this is most acute where the data are held in raster form, as illustrated in Figure 7 (see also Dowman, Chapter 31).

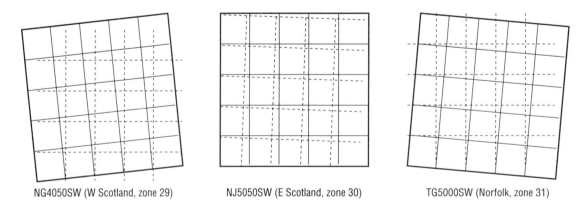

NG4050SW (W Scotland, zone 29)   NJ5050SW (E Scotland, zone 30)   TG5000SW (Norfolk, zone 31)

Fig 6. Ordnance Survey 1:10 000 scale map tiles transformed into WGS coordinates and projected onto the UTM zones in which they fall.

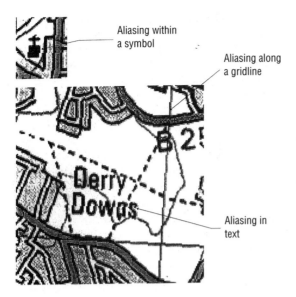

Aliasing within a symbol

Aliasing along a gridline

Aliasing in text

Fig 7. Raster aliasing effect attributable to transforming data encoded in one reference system and map projection combination into WGS coordinates and reprojecting these onto UTM. The effect could be mitigated using smoothing and interpolation techniques.

3.2 Indirect positioning

Common usage of geographical description includes narrative where position is described by place name and relationships to that place (e.g. '25 kilometres north west of Frankfurt'). Alternative indirect ways of describing geography include post (or zip) codes (see Martin, Chapter 6), a variety of zoning systems (such as the French départements), and distances through a network. Some geographical analysis has even been carried out using only the knowledge of the adjacencies between different zones. Although some of these means of describing position are commonplace, they cannot be used to link together different datasets (unless the different datasets each pertain to the same named entities). Spatial relationships cannot be computed nor can the data generally be mapped in geographical space.

It follows that to carry out almost any GIS operation requires that an added value process be carried out: these locations must therefore be described by coordinates of some type, as described in section 3.1. Practical difficulties often occur in this process for the bounds of some entities used in common parlance (e.g. 'the Alps' or 'the Corn Belt') may be fuzzy to some degree (see Mark, Chapter 7, for a general discussion of this topic). Nevertheless,

an increasing number of organisations now maintain and offer mechanisms and data to enable such added value to be achieved. The most common translations of this kind are to relate post (or zip) codes to coordinates or even individual postal addresses within a post (or zip) code to coordinates (as in the British ADDRESS-POINT product).

4 CONVERSION OF DATA REFERENCED TO DIFFERENT POSITIONING SYSTEMS

In recent years positioning based on GPS-observations has proved more and more powerful, almost totally replacing classical triangulation techniques. Thus modern reference systems such as the WGS84 or the ETRS89 are used increasingly in the fields of practical surveying. In so doing, one must be very careful not to generate errors by mixing reference systems and datums. The easiest way to anticipate any difficulties is usually to contact the national mapping organisation for advice in the first instance. Alternatively, it is possible to make calculations of the interrelationship through use of a certain number of stations identical in both the national and the satellite systems (ideally five or more). As positioning through use of navigation satellite systems (GPS, GLONASS, PRARE etc.) will grow rapidly in the future, special attention has to be paid to the different aspects of using various reference systems in daily surveying and, even more important, in the use of data compiled from different sources inside GIS.

The primary reason for this is the historical legacy. Most geographical data in current use have been compiled on the basis of national reference systems which predate the new global system. These reference systems are embedded in national map series at all scales. While some countries (such as Australia and the USA) are committed to converting their maps series to being based on WGS84 or ETRFS, even in these cases the transition period is likely to be long and the cost measured in many millions or billions of dollars. In addition, many users have already expended considerable investments in encoding their own data in relation to these national systems.

This would be less of a problem if everything could simply be converted purely algebraically without loss of precision or accuracy and all those holding data in relation to national frameworks used the same national conversion parameters. In reality, the problem is much greater than this (see Salgé, Chapter 50). In the first instance, there is a huge number of different combinations of geodetic reference system, geodetic ellipsoid, zero meridian, and map projection in use and (surprisingly perhaps) records of which ones have been used are sometimes not kept. The basic work on the geodetic system may, for instance, have been carried out 50 or more years ago, before computers were available to support adjustments in the geodetic networks. Typically, such networks were assembled on a piece-by-piece basis although overall checks were made. As a consequence of this, of generalisations and errors built into maps based on these networks, of the loss of recorded attribute information when digitising the maps and the widespread, free availability of variable quality approximations to the overall algebraic transformations, conversion of existing data into the WGS framework has proved much more troublesome than is generally appreciated.

Ashkenazi (1986) anticipated many of these problems and has demonstrated their significance in subsequent papers. He has shown that – even without errors in the maps or in their digitising – coordinate shifts of the order of several hundred metres can occur between the position of locations as described in WGS84 and pre-existing national frameworks. Even the simple act of transforming from one map projection system to another may cause considerable difficulties, especially in raster topographic datasets (see Figures 6 and 7).

5 CONCLUSIONS

The routine use of GPS receivers or GIS as 'black boxes' can foster a view that 'truth' is being constructed or handled, and that it may be compared directly with other respected and geographically precise sources. This chapter has set out to demonstrate that, even in a relatively well-defined (but crucial) element of GIS, many complexities and gross errors may occur unless some fundamental principles are understood by the user

of the system. Thus care in the selection and use of referencing and coordinate systems is always essential, especially when the need to link different datasets together requires use of information defined and collated in different systems.

The most common difficulties arise when information derived by the use of GPS is combined with that from maps produced using different datums. The combined effects of the use of different geodetic reference systems (including different datums), projection change, grid shifts, and any embedded errors in the maps can cause great practical problems. As indicated earlier, maps are the basis for much current and most historical data in GIS so this problem has to be faced. It can only be faced with a good understanding of the science of geodesy and a good knowledge of the genealogy of the mapping from which data have been derived.

References

Ashkenazi V 1986 Coordinate systems: how to get your position very precise and completely wrong. *Journal of Navigation* 39: 269–78

Bomford G 1980 *Geodesy*, 4th edition. Oxford, Clarendon Press

Hofmann-Wellenhof B, Lichtenegger H, Collins J 1994 *Global positioning systems: theory and practice*, 3rd edition. Berlin, Springer

Kennedy M 1996 *The global positioning system and GIS: an introduction*. Michigan, Ann Arbor Press

Leick A 1990 *GPS satellite surveying*. New York, John Wiley & Sons Inc.

Maling D H 1973 *Coordinate systems and map projections*. London, George Philip

Snyder J P 1987 *Map projections: a working manual*. US Geological Survey Professional Paper 1395. Washington DC, USGS

Snyder J P 1993 *Flattening the Earth: two thousand years of map projections*. Chicago and London, Chicago University Press

Vanicek D, Krakovsky E D 1986 *Geodesy: the concepts*. Amsterdam, New Holland

Warita Y, Nonomura K 1997 The national and global activities of the Japanese national mapping organisation. In Rhind D (ed.) *Framework for the world*. Cambridge (UK), GeoInformation International: 31–47

31

Encoding and validating data from maps and images

I J DOWMAN

This chapter gives an overview of how data for a GIS can be derived from existing graphical information (a map) or from image data such as aerial photographs and satellite images. Two crucial principles underlie this process: (a) the need to understand the characteristics of the original data, including the quality of the source data; the processes used to produce the document to be encoded; and the geometric characteristics of the data; (b) the need to understand the processes used to produce the encoded data and their subsequent quality.

In order to obtain this understanding, this chapter gives some detail of the characteristics of the data sources and of the methods of converting the original data into a form suitable for a GIS. The need to geometrically correct image data leads to the use of elevation data to correct for relief effects and therefore the principles of using stereoscopic data are outlined and methods for deriving digital elevation models. It is assumed that automation will be used in these processes as much as possible, so an outline is given of the automated tools which are now in use. Finally, the question of accuracy is discussed, including the factors which affect accuracy, how accuracy can be assessed, and how this information can be given to and applied by the user of a GIS.

1 INTRODUCTION

Many of the data which are used in GIS come from existing graphical products printed onto paper, or from images derived from photographic cameras or digital sensors (e.g. see Bibby and Shepherd, Chapter 68, Waters, Chapter 59). Some of the latter comes from sensors carried on satellites. In order to use these data in a GIS they must be processed to be in a form which is compatible with the GIS. The first point to consider is whether the GIS requires raster or vector data – assuming that the GIS cannot convert between the two without imposing undesirable constraints. The second point concerns the reference system: all data which are to be overlaid, merged, or compared must be in the same map projection and on the same datum. This implies that any distortions present because of the data capture system must be corrected. The third point is that the quality of the data must be known (see Veregin, Chapter 12). This chapter is concerned with

how vector or raster data can be acquired from existing maps or from images and how they must be processed in order to be used in a GIS. In order to carry out these processes it is essential that the characteristics of the original data and the processes used in the preparation are understood. It is also essential that all of the possible sources of error are recognised, particularly with regard to how such errors affect the quality of the final dataset (see also Fisher, Chapter 13; Heuvelink, Chapter 14).

The term 'map' is used in this chapter to refer to a dataset which contains accurate information about features on the surface of the Earth. In the past there would have been no ambiguity about this because a map user would be handling a piece of paper with the required information printed onto it. Now this information may also be stored in digital form (see Longley et al, Chapter 72, for a discussion of the 'map metaphor').

A knowledge of map projections is essential when handling data from different sources. Any map projection will introduce some distortion into the transformation from the ground to paper and it is important that the sources of this are understood. Data mapped using different projections cannot generally be combined and processes of conversion and transformation must therefore be applied. In addition different datums may be used, even on the same projection, and these too must be understood. A detailed discussion is beyond the scope of this chapter but reference may be made to Snyder (1986; see also Barnsley, Chapter 32; Seeger, Chapter 30).

The first step in using map data in a GIS is to encode it into vector or raster format. This involves either line scanning or raster scanning. Because of the monotony of this task, it is desirable that as much automation as possible is introduced. After encoding, the data must be transformed into the required projection and any known errors removed. A raw image derived from a sensor in an aircraft or on a satellite will contain distortions attributable to the attitude of the sensor and the relief of the ground, and it will not show all the features which a user will need (Barnsley, Chapter 32; Meyers, Chapter 57). Such distortions must be corrected before data derived from the images can be used in a defined reference system. Correction may be possible through 2-dimensional transformation, although 3-dimensional correction using stereoscopic imagery or digital elevation models (DEMs) is frequently also necessary. Such processes are part of the discipline of photogrammetry and this section can be regarded as an introduction to that subject.

A vast amount of useful data already exist in the form of printed maps and an immediate response to the requirement for digital data is to assume that these existing data should form the basis of the digital database. It must, however, be remembered that a great deal of effort, and hence cost, is required to convert printed maps to digital data. A figure of 60–80 per cent is often given, without substantiation, as being the proportion of the cost of establishing a GIS which is attributable to obtaining the data. It must also be remembered that maps may be out of date and may have a number of errors inherent in them, and that the conversion process will itself add further errors. Before embarking on a data acquisition exercise it is sensible to weigh up all the factors involved in data collection, and to consider the available sources of data in the context of the use to which the data will

be put and the methods which will be used to update the data. Logan (1995) provides a useful discussion on these topics.

2 ENCODING MAP DATA

2.1 Principles

Map data have two principle components: geometry and attributes. For use in a GIS every point in a dataset must have unique coordinates (x and y or x, y, and z) and each must be labelled. On a printed map the coordinates usually come from a grid or graticule on the map, provided in such a way that the (x, y) coordinates of any point can be read off a scale. The information about what the point represents will come from the context or the way in which the point is represented, and this representation can be converted to an attribute through a legend. When converting from a printed map to a digital database, both of these components must be recorded.

Digitising will always take place in a rectangular coordinate system which is defined by the digitising system. Conversion from this to a national or regional grid system is therefore relatively straightforward. The procedure is usually to record the corner points of the document to be recorded in the digitising system and then to carry out a transformation (see section 5.3) between the recorded coordinates and the known coordinates in the national system of the corner points. An affine transformation in often used for this to correct for any errors created by paper distortion. Digitising is often carried out in patches (or tiles) which may be the size of a grid square or larger and each tile is transformed to the coordinates of its corners. It is always wise to carry out checks with well-defined points of detail on the map.

Attributes must be assigned by the operator, or automatically where possible, through the use of an attribute menu or key coding.

2.2 Equipment

The most straightforward form of conversion is to digitise manually the information from the printed map into either point, line, or area features and to assign each feature an attribute code. The features are then recorded as coordinates in point or vector form. This can be done on the digitising table on which the map is placed. The operator then selects points or trace lines and the coordinates are recorded as the cursor moves over the map. After digitising, the result needs to be carefully checked and edited to ensure that everything has been recorded and that lines join up where they should and do not overrun. If carried out systematically and carefully this method can result in accurate and complete vector data. The disadvantage is the time which is required and the tedious nature of the task.

A map can be quickly and easily recorded automatically in digital form using scanning techniques. However these methods only record the data in terms of position in a raster format and do not link to separate attribute files, since there is no feature coding or topology. Scanning can be accurately carried out on drum scanners, whereby the printed document is fixed to an accurately rotating drum. The drum is scanned by an optical reading head which determines whether a line or symbol exists over a 'pixel' (picture element) of a given size – which can be as small as 6μm. Colour documents can be recorded by scanning three times, in the red, green, and blue regions of the spectrum. The result is a raster image which reconstructs the original as pixels rather than as lines and symbols and which may look the same as the original but cannot be interpreted by a computer. The accuracy and visual impression is controlled by the pixel size, but a smaller pixel size means more data, more time to scan and display, and increased storage requirements. Current problems with scanning are centred on accurate colour reproduction and the size of files which will be produced for large documents. The former problem can be overcome at the expense of greater data handling if colour separates are available.

Less expensive but less accurate desktop scanners, designed for the publishing trade, are available which will scan at up to 1200 dots per inch (giving a pixel size of 20.16μm). Desktop scanners are subject to geometric distortion but this can mostly be removed by calibrating the scanner and subsequent correction (Sarjakoski 1992).

2.3 Raster-to-vector conversion

The raster image can be converted to vector by a raster-to-vector conversion. This technique is not fully developed as an automatic process and still

cannot add attributes. It also requires extensive editing to remove text and symbols. Raster-to-vector conversion requires salient pixels to be assigned (x, y) coordinates. The actual coordination of a pixel is no problem but deciding which pixel should represent a corner or a junction of two lines is not easy. A junction, for example, may be represented by several pixels and the lines forming the intersection may be several pixels wide; the intersection itself must, however, be represented by a single point. Line thinning will reduce line size but an algorithm is needed to select the optimum point to represent the intersection. This is illustrated in Figure 1, where the node of the intersection could fall at any of the four pixels within the small box. Features such as roads may be represented as single lines or as parallel lines, and the decision as to which to use and how to determine which lines to digitise can also be difficult for an automatic system.

The most efficient solution for raster-to-vector conversion at present may be semi-automatic systems. These are based on line-following techniques which can be done quite well by a computer. The operator is better at starting the process and making decisions about which way to go at intersections of lines. The VTRAK system of Laser-Scan Limited (Cambridge, UK) is a good example of an automatic line-following system: it

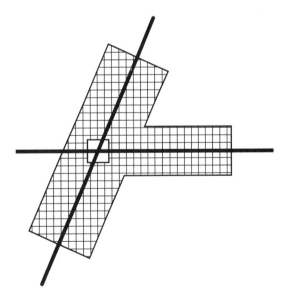

Fig 1. An example of a problem in raster-to-vector conversion.

439

can be used to digitise features into a range of databases, notably object-oriented databases. It also provides editing and quality control facilities and automatic text recognition.

2.4 Editing and error analysis

Editing is an important process in any digitising operation because errors will always be made. The most common error is poor closing of lines at intersections or in polygons. Most digitising packages now include efficient tools for editing data (see Bernhardsen, Chapter 41). Figure 2 shows some examples of common errors.

Errors in content have been discussed above and can be detected by careful checking. Errors in geometry are not so easy to detect. The most important rule in checking accuracy is to have otherwise redundant check points. As explained above a transformation is necessary between the digitising system and the reference system. Usually four points are a minimum requirement to carry out a transformation. A number of additional check points should also be used and the coordinates of transformed points in the digital document should be compared with the original values. Some useful validation statistics are discussed in section 8. A further useful check is possible if other data of the same area exist and the two sets can be superimposed.

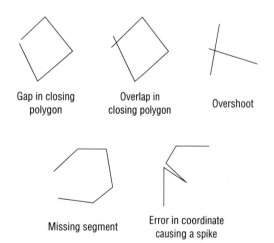

Gap in closing polygon · Overlap in closing polygon · Overshoot

Missing segment · Error in coordinate causing a spike

Fig 2. Common errors in digitising.

3 THE GEOMETRY OF AERIAL PHOTOGRAPHS AND SATELLITE DATA

3.1 Image formation

In order to produce a map from any satellite image it is necessary to first define the geometry of the image forming system, to consider the movement of the platform on which the sensor is mounted, and to define the shape of the ground which is covered by the image. The importance of each of these factors will depend upon the type of sensor used and the path which the sensor is following relative to the ground. This section is concerned with the general effects and these are related to different sensor types. Specific sensors for mapping are discussed in section 4 and by Barnsley (Chapter 32) and Estes and Loveland (Chapter 48).

Images from a particular sensor are usually presented in a standard format covering a defined area of ground. Such an image is called a frame. A frame may be formed in three ways:

1 As a single exposure – that is with no significant movement of the sensor whilst the image is formed as is found in the case of a frame camera. This is used for central projection which is used in all aerial photography and in a number of satellite sensors.
2 As a series of lines almost normal to the track of the sensor. In this case time must be considered in the model for defining the construction of a full frame. The main types of sensor in this category are the push broom scanners, of which SPOT is the best contemporary example.
3 As a series of points each recorded at a separate time. This is the most distorted type of image requiring the most complex mathematical model. The scanner systems such as LANDSAT fall into this category.

The photographic camera, in which an image is formed by light rays from an object being focused onto a focal plane by a lens, may be taken as the standard case. This lens acts as a point through which all light rays pass, so that the process of image formation can be seen as a series of straight lines passing through the perspective centre S onto the focal plane. If the objects all fall on a plane which is parallel to the focal plane, then a direct reduction of the object is found in the focal plane and the amount of reduction is given by the ratio of the object distance to the image distance. This is an approximation to the case of vertical aerial photography when the ground is the object, the object distance is the flying height *H*, and the image distance is the principal distance of the camera *f* (equal to the focal length of the camera when the object distance is effectively infinity). Two points are defined on the vertical line passing through *S* as the nadir points on the image (*n*) and on the ground (*N*). The scale of the photograph in this case is *f/H*. This is illustrated in Figure 3. In practice the ground is not flat and the focal plane is not horizontal.

3.2 Effect of relief

The effect of relief is to cause the scale to change throughout the image and to cause images to be displaced, radially from the nadir, from the position at which they should appear if there were no relief. In Figure 4 a point *A* on the surface of the ground should appear on a map of scale *f/H* at *a'* but actually appears on the photograph at *a*.

The magnitude of this displacement is given by the expression

$$aa' = an \cdot \frac{h}{H} \qquad (1)$$

or in terms of a radial distance from the nadir points (*r*)

$$dr = r \cdot \frac{h}{H} \qquad (2)$$

This expression indicates that the effect of relief displacement is less from greater sensor altitudes. For example on an aerial photograph taken from 10 000 metres, relief of 500 metres will cause a displacement of 5 mm for a point appearing on the edge of a photograph (*r* = 100 mm). This is equivalent to 660 metres on the ground if a wide angle camera is used. On an image taken from a satellite pointing vertically downwards, with an altitude of 700 km the effect of relief on the ground will be 64 metres. For normal relief the effect on this type of satellite image can be ignored, but for photographic cameras and tilted satellite sensors it cannot be ignored and 3-dimensional geometry must be considered. In Plate 19 the effect of relief distortion can be clearly seen at the edge of the photograph where buildings appear to lean outwards. In other words the top of the building is in a different position to the bottom.

The effect of relief can be removed by using a rigorous 3-dimensional model of the geometry of two images or by having a digital elevation model available in order to compute, and hence correct, the effect of relief at every point. An image which has been corrected for the effect of relief is called an *orthoimage*.

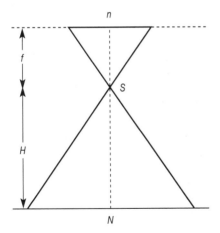

Fig 3. The scale of an image.

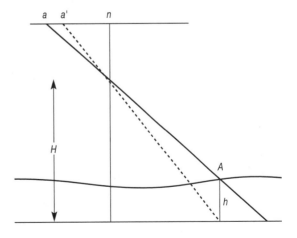

Fig 4. Distortion attributable to relief.

3.3 Distortion because of tilt and other causes

There are other distortions in imagery which must also be considered. These include the tilt of the sensor platform, the movement of the platform and the movement of the Earth. Most of these are small, although they still need to be removed if accurate data are required. A source of distortion which is potentially very serious arises if the sensor is tilted through a large angle – for example SPOT images can be tilted by up to 27 degrees and the new high resolution sensors will tilt by up to 45 degrees. This is an effect which must be removed for accurate mapping. The methods of correcting distortion are described in section 5.

3.4 Acquisition of stereoscopic data

If images from two different positions are available, then the pair of images can be viewed stereoscopically to give a 3-dimensional view of the terrain and measurements can be made to give 3-dimensional coordinates. In order to obtain a stereoscopic image from a pair of photographs certain conditions must be satisfied. The cameras must neither be too close together, nor too far apart. If they are too close together the view will not be different enough to enable a stereoscopic image to be formed; if they are too far apart the views will not be similar enough. Photographs should also be taken with the same or similar cameras and be taken from approximately the same distance from the object. The basic geometric condition can be expressed in terms of base to height ratio.

In Figure 5 the base is shown to be the distance between the cameras, B, and the distance between the base and the object is the height. In aerial photography the distance is equivalent to the flying height, H, and we have a base to height ratio, $B:H$. In order to obtain good stereoscopic viewing $B:H$ should lie between the limits 0.3 and 1.0.

Photographs taken with a camera from an aircraft for the purpose of constructing a map are carefully controlled so that the axis of the camera is pointing almost vertically downwards and each photograph overlaps its neighbour by 60 per cent. This ensures complete coverage of the ground and a

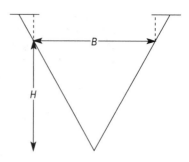

Fig 5. Base to height ratio.

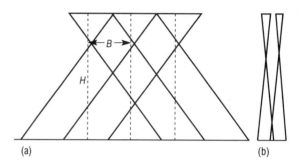

(a)   (b)

Fig 6. Base to height ratios of: (a) aerial photography; (b) vertical satellite images.

convenient base to height ratio. This is shown in Figure 6. Stereoscopic coverage from satellite sensors is more variable and because of the great altitude of the sensor coverage from vertical pointing sensors does not usually give a suitable base to height ratio for mapping.

3.5 Heights from stereoscopic data

As already noted, stereoscopic images can be used for measuring heights. The principle is shown in Figure 7. On two images, two points of different heights, A and B, will produce images at a_1, b_1 on the left hand photograph and at a_2, b_2 on the right photograph. The separation of the images $a_1p_1 + a_2p_2$ and $b_1p_1 + b_2p_2$ is clearly proportional to the heights of A and B. These separations are called parallaxes or disparities.

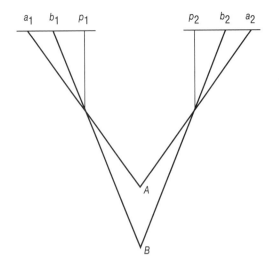

Fig 7. The principle of height determination.

Using the principle as shown in Figure 7 we can derive an expression which relates the height of a point (Z) to the camera geometry and the parallax (p)

$$Z = \frac{fB}{p} \qquad (3)$$

and considering a small change in height (dZ) we can show that:

$$dZ = \frac{Z}{f} \cdot \frac{Z}{B} \, dp \qquad (4)$$

Equation 4 is an expression which shows how a small change in parallax (dp) will give a small change in height (dZ). This equation can be used to give the precision with which height differences can be found. Precision is usually measured by the standard deviation of a set of measurements, or quantities derived from measurement. In this case the standard deviation of height determination, s_z can be found if the precision of observation of parallax s_p is known and the equation becomes:

$$s_z = \frac{Z}{f} \cdot \frac{Z}{B} s_p \qquad (5)$$

which gives an expression showing that the precision with which height can be measured is related to the scale of the photograph (Z/f), the base-to-height ratio (Z/B), and the precision of measurement of parallax.

4 IMAGE ACQUISITION SYSTEMS

Images for mapping may be obtained from sensors placed in aircraft or on satellite platforms. Airborne sensors are mainly film cameras producing high quality images suitable for photogrammetric mapping. Sensors from satellites generally produce images directly in digital form as the satellite moves forwards: the movement enables the necessary large volumes of data to be recorded, but complicates the image geometry. A summary of the main sensors is given in Table 1. The characteristics of satellite sensors are discussed in more detail by Barnsley (Chapter 32) and Estes and Loveland (Chapter 48). Here only a few of the sensors which can be used for mapping will be discussed. Film cameras are used in space and the Russians are the main operators of such cameras. The KFA1000 and KVR1000 are two examples which are included in Table 1. Both of these are designed to give large scale images rather than height information.

Scanners are the most widely used imaging systems from space, and LANDSAT is probably the best known. A scanner records a single pixel (or a group of pixels) at a time; a strip approximately at right angles to the direction of flight is imaged as a mirror rotates about the axis of flight. The raw data obtained from a scanner are subject to many distortions which must be corrected prior to use within a GIS, but because many of the sensors point only directly downwards, rigorous correction is not required.

Pushbroom sensors are the most important for mapping. SPOT was the first of these and was designed as a mapping sensor: it is still dominant in this area, although other systems have been designed and flown. The US high resolution sensors are an important new form of pushbroom scanner.

Stereoscopic overlap can be arranged either fore and aft or laterally so that 3-dimensional models can be formed. Displacement attributable to the relief of the ground is normal to the direction of flight, that is along the arrays, and the amount of displacement is proportional to the distance from the principal point and the altitude. If lateral overlap is used by tilting the sensor than the effect of relief will be large, but a good base to height ratio will be obtained for height measurement.

The SPOT satellite uses a 'lateral overlap' sensor. This gives a good base-to-height ratio but suffers from the disadvantage that stereoscopic images will

Table 1 Characteristics of sensors used for mapping (see also Barnsley, Chapter 32 Table 2).

Platform	Sensor	Launch date	Type	Pixel size (m)	Swath width (km)	B:H	Height accuracy (m)
Aircraft	Film camera		Principal manufacturers are Zeiss and Leica		†	0.3 – 1.0	Depends of flying height: 0.01% overhead
Cosmos	KFA1000	1994	Russian film camera, f = 1000mm	5–10*	†	0.12	30
Cosmos	KVR1000	1987	Russian panoramic film camera, f=1000mm	2*	160	0	N/A
Landsat	Thematic Mapper	1982	Scanning system	29	180	N/A	N/A
SPOT 1-4	HRV	1986–97	Cross track push broom	10,20	60	≤1	10
JERS-1	OPS	1992	2 sensor along track stereo	20	75	0.3	50
Priroda	MOMS	1996	3 sensor along track stereo	6	40	0.9	3–6
IRS-1C	Pan	1995	Cross track stereo	5.8	70	1	5–10§
Earlybird		1997	Along track stereo	3	6	1.2	1.5‡
Quickbird		1997	Along track stereo	1	36	1.2	1.5‡
Space Imaging		1997	Along track stereo	1	11	2	0.4‡
Orbview		1998	Along track and cross track stereo	1–2	8	2	0.4‡

† Varies with altitude   * Originally photographic film but available in digital format.
§ Preliminary result   ‡ Predicted result

be taken a minimum of one day apart – usually longer, as cloud conditions may change from one day to the next. 'Along track' stereo imaging is more efficient as it is very likely to generate stereoscopic data. The optical sensors (OPS) on JERS-1 and MOMS-02 are along track systems: other systems have been designed and will be implemented in the future. Of particular note are the high resolution sensors such as Earlybird and a system from Space Imaging, being developed by commercial organisations in the USA. An example of simulated data is shown in Plate 18, which is part of Space Imaging's CARTERRA San Francisco series of high resolution, high accuracy imagery. It was derived by fusing a one metre resolution panchromatic image and a four metre resolution multispectral image. Laid on top of a United States Geological Survey (USGS) seven-and-a-half minute quad sheet, this image illustrates the value of high resolution imagery in place of traditional line-drawn maps. They are used for a variety of applications, including infrastructure management, urban planning, utilities, and transportation.

Another type of sensor is synthetic aperture radar (SAR). Images from these sensors have a quite

different geometry and appearance to those of other sensors and details can be found in specialist literature on the subject (Schreier 1993). Radar is becoming particularly important at the moment because of its potential to produce high accuracy elevation data through SAR interferometry.

5 CORRECTION OF ERRORS IN IMAGES FOR USE IN GIS

5.1 Requirements

The quality of image data to be used in a GIS must match that of the use of the data. Although many applications require only 2-dimensional data, 3-dimensional information may nevertheless be needed to apply the necessary corrections. There are two main methods of correction. First, a 2-dimensional transformation which is suitable for application to images in which there is very low relief distortion – as, for example, in LANDSAT data. Second, a rigorous 3-dimensional correction which corrects for all the effects of relief, sensor orientation, Earth rotation, and Earth curvature.

For both of these methods ground control points (GCPs) are generally required.

5.2 Ground control

A GCP must be recognisable in both images (or one image if stereo is not being used) and have known coordinates in a suitable ground reference system. The number of points required depends on the method used (Seeger, Chapter 30). GCP coordinates may be obtained directly by survey measurement or from a map. In either case the coordinates will be given in a reference system; this may be geographical (latitude and longitude) or Cartesian (x, y, z) and it may be global, based on the centre of the Earth, or local, based on a national or regional projection. It is always important that the characteristics of the reference system are known and that all coordinates are given in the same system. Further information of reference systems and conversion between systems may be found in Seeger (Chapter 30).

Direct survey measurements may come from surveys based on a local or national coordinate system (e.g. the UK's National Grid), or they may come from the Global Positioning System (GPS) which allows coordinates to be fixed directly from navigation satellites (Lange and Gilbert, Chapter 33). Maps should be used with caution for determining GCPs. Map data at scales of 1:25 000 and smaller are notoriously unreliable because of the many errors which may have accumulated in the map production and map digitising process. These include survey errors (in some parts of the world published maps may be based on topographical sketches), drafting errors, generalisation, paper distortion, and errors in digitising a paper document for use in the validation process. While it is always necessary to take the accuracy of the data into account when using ground control, it is particularly important when using map data.

5.3 Plane transformations

5.3.1 Introduction

For areas of low relief or for low resolution sensors, fairly simple methods of correction may be used. The correction of data in two dimensions may be approached by applying a transformation to the data and resampling the data to produce a corrected image which gives a best fit to the ground control used. The transformation may be based on a theoretical consideration of the errors involved or may be selected on empirical grounds. The method is the one most commonly used to produce an image which is corrected to fit to a given map projection.

A number of transformations are widely used and a brief description of the common ones is given here. An image coordinate system (x, y) and a ground coordinate system (X, Y) is assumed.

5.3.2 Two-dimensional similarity transformation (four parameters)

This transformation is to relate any 2-dimensional rectangular coordinate system to any other 2-dimensional rectangular coordinate system. It preserves the internal geometry of the transformed system, so it is ideal for comparing the geometry of any two systems simply by determining the residuals and the root mean square errors after transformation. For a given control point, the two equations:

$$X = ax - by + c \qquad (6)$$
$$Y = bx + ay + d$$

may define this transformation.

A similarity transformation is performed by applying a scale factor $(m = (a^2 + b^2))$, a rotation angle $(tan\ \alpha = b/a)$ and two translations $(c$ and $d)$. A minimum of two GCPs is needed although it is always desirable to have more for purposes of checking.

5.3.3 Two-dimensional affine transformation (six parameters)

A mathematical relationship for an affine transformation may be expressed in the following:

$$X = a_o + a_1 x + a_2 y \qquad (7)$$
$$Y = b_o + b_1 x + b_2 y$$

An affine transformation enables adjustment to be applied independently in each direction. Thus for scanner images it corrects first-order distortions such as affinity attributable to non-orthogonality and scale difference between scan along track directions which may be caused by earth rotation, map projection, and other geometric distortions. A minimum of three GCPs is required.

5.3.4 Second-order polynomials (twelve-parameters)

Polynomials in the form:

$$X = a_o + a_1x + a_2y + a_3x^2 + a_4y^2 + a_5xy \qquad (8)$$
$$Y = b_o + b_1x + b_2y + b_3x^2 + b_4y^2 + b_5xy$$

are used for correction of scanner data. If polynomials are used, great care must be taken to ensure that a sufficient number of control points is available and that they are distributed over the whole area to be transformed. A minimum of six GCPs is necessarily required to determine the transformation parameters, although it is desirable to have more to build in checks. In addition to first-order distortions, polynomials correct second-order distortions caused by pitch and roll, subsatellite track curvature and scan lines convergence because of Earth rotation and map projection. They may also correct some of the distortions related to the attitude variations along the flight path.

Additional terms may be added to Equation 8 to correct for higher order distortions. The need for care in use of control points is greater for higher orders.

5.3.5 Resampling

After calculating the parameters of the transformations, the transformation must be applied to the image and new Gray level values be computed for each pixel in the operation of resampling. Resampling may introduce some changes to the data and distort the characteristics of some features.

5.4 Determination of orientation elements

The process of extracting 3-dimensional coordinates must include the determination of the orientation and position of the sensor and the application of this information to the measured image coordinates. The information required comprises the position of the sensor given in (X, Y, Z) coordinates in an appropriate reference system, and the attitude of the sensor given as rotations about the reference coordinate system. These six parameters are known as the elements of exterior orientation. They can be found in a number of different ways:

1   By measurement on the platform or sensor. The position can be accurately determined with some satellites, for example ERS-1 where the orbit is accurately determined, or if a GPS receiver is used. There is no suitable system for determining attitude accurately.

2   With the use of GCP. The position and attitude of the sensor can be calculated by relating the image to ground control points. This is relatively straightforward for photographic images, but is more complex when the image is formed over a period of time as with push broom sensors. The sensor geometry and orientation is described mathematically and the unknown parameters determined by a mathematical solution such as a bundle adjustment.

5.5 Principle of correcting height distortion

If the relief is sufficient to cause distortion (section 3.2), then a correction must be applied. The process of distortion due to relief shown in Figure 4 can be reversed to correct for the effect. Figure 8 shows the same geometry as Figure 4 but the terrain is depicted by a series of points, $P_1, P_2, \ldots P_n$ on a regular grid. Each point has a planimetric position (X, Y) and a height (Z), and together they form a DEM. If a corrected image (orthoimage) is to be formed then the correct plan position of the point P_i, that is P_i', will fall at p_i''. To find the correct Gray level value the position p_i in the image is found by projecting P_i into the image. The Gray level value at p_i is placed at p_i'' by resampling. In practice this is done as a mathematical 3-dimensional transformation taking into account the position and attitude of the sensor. The process can take place in a digital photogrammetric workstation or using an image processing package. These are discussed in section 2. The topic of digital elevation models is discussed in section 3.

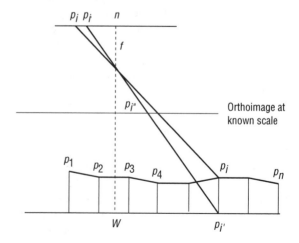

Fig 8. The correction of relief distortion.

6 THE PRODUCTION OF SPATIAL DATA

6.1 Methods

The basic geometry of 3-dimensional imaging systems has been described in sections 3 to 5. This may be incorporated into a computer system in order to allow the extraction of features or height by manual or automatic means. Traditionally, optical mechanical instrumentation has been used by photogrammetrists for map making but this has given way in recent years to analytical instruments in which the operator views the stereoscopic model and follows lines or measures heights under computer control. A degree of automation has also been introduced. Completely digital systems are now being developed in which stereoscopic data acquisition can be carried out as part of a digital image processing system. Such systems allow for stereoscopic viewing and automation is also integral to them.

To use an aerial photograph in a digital system, the photograph first has to be scanned. This is done using a scanner similar to those described in section 2.2, but designed especially for digitising aerial photographs with geometric precision of $1 - 2 \mu m$ and pixel size of down to $7.5 \mu m$.

Digital images can also be viewed with a stereoscope on a split screen display or with the anaglyph principle in two colours. Better quality digital display can be obtained by using polaroid or 'flicker' display on the monitor. In these systems glasses must be worn. For a polaroid display the spectacles have complementary polarisation and in the 'flicker' display the glasses are synchronised with the monitor. Stereo viewing systems are available as an option on workstations.

6.2 Digital workstations

Digital photogrammetric workstations (DPWs) have developed from analogue and analytical plotting instruments designed to produce maps from aerial photographs, and from image processing workstations designed for handling satellite data for interpretation and analysis. They are designed to handle stereoscopic data in digital form and will include software which carries out a number of functions automatically. There will normally be a facility for stereoscopic viewing and for automatic matching of images to produce heights. The main functions are:

- Orientation of stereoscopic images
- Feature extraction and assignment of attributes
- Generation of DEMs
- Creation of orthoimages.

Some digital systems have been developed primarily for photogrammetric work, but others have been developed from GIS or image processing systems designed for handling satellite data. Some, such as Intergraph and ERDAS, are closely linked to GIS and data can be easily transferred between systems. There is also a difference in cost and functionality and accuracy of the products from such systems. Table 2 includes the features of some of these systems. A full discussion of digital workstations with references to detailed descriptions can be found in Dowman et al (1992).

One of the major innovations of digital workstations has been the introduction of automatic processing of data, including the formation of digital elevation models by automatic stereo correlation (image matching). The basic principle

Table 2 Some digital photogrammetric systems.

System	Comment
Leica HAI-750 } Leica HAI-500 }	Two full photogrammetric systems on Unix or PC platforms. Can use SPOT
Zeiss PHODIS	Full photogrammetric system for use with aerial photographs and satellite data.
Intergraph ImageStation	Photogrammetric data capture as part of digital mapping system. Can use SPOT.
ERDAS Imagine with Orthomax	Designed for production of orthoimages with DEM generation. Linked to ARC/INFO. Can use aerial photographs or SPOT.
R-Wel DMS	PC based system for DEM and feature extraction from images as input to and integration with a GIS.
Leica DVP	PC based system for digital mapping from aerial photographics and SPOT.

underpinning stereo correlation is the matching of small patches from two images: the amount of relative displacement to achieve matching is determined and this is used to determine parallax difference which in turn is used to compute heights. The distortion of the patches because of relief and tilt must be taken into account. Errors will occur because of features such as trees and buildings not on the terrain surface, and matching may be incorrect or impossible if the two images are different because of changes on the ground or large discontinuities in the terrain surface.

This can be a particular problem with SPOT when days, weeks, or months may pass between images being obtained. Accuracies of better than 10 metres can be obtained from stereo matching SPOT data. Line maps have been produced from SPOT using traditional photogrammetric methods. Ordnance Survey in the UK and Institut Géographique National (IGN) in Paris have produced maps at 1:100 000 and 1:50 000 scale using analytical plotting instruments (Smith and Rhind, Chapter 47). Image maps are, however, the more usual product, and can be overlaid with vector data.

Plate 19 shows a stereoscopic pair of aerial photographs in which the buildings appear to lean outwards – thereby demonstrating the effect of relief displacement. This effect is most marked on the left image, which appears in the corner of the photograph from which it was taken. The fiducial marks on the top of the images are used to orient the photographs, and they can be viewed stereoscopically using a pocket stereoscope. Part of Plate 19 shows a plot of part of the area produced on a photogrammetric plotter and an example of the output file for two of its features – a lamppost (marked A), which is stored as a symbol along with its coordinates, and a node on a line (marked B) which is part of a (hatched) building and is attributed a line type (solid) and a colour.

6.3 Digital elevation models

6.3.1 Introduction

DEMs are a way of presenting the elevation of the surface of the Earth in numerical form. A DEM will consist of a series of reference heights arranged in regular or irregular form. A typical regular pattern is a rectangular grid. The spacing of the grid will, together with the accuracy with which the heights are given, be a function of the overall accuracy of the DEM. Digital elevation models can be derived from a number of sources and the processes which are required may differ according to the source. A method of derivation will normally be related to the application of the DEM. The most accurate DEM may give heights to an accuracy of a few centimetres with a high density of reference points, but covering a small area. At the other end of the scale, global DEMs may have a spacing of several kilometres and give heights to an accuracy of hundreds of metres. This section will deal with methods of deriving DEMs from aerial and satellite data. Petrie and Kennie (1990) give a full account of DEMs.

6.3.2 Processes

There are three main processes associated with DEMs:

- primary data acquisition;
- resampling to required grid spacing;
- interpolation to extract height of required point.

Each of these stages may introduce errors, the nature of which will depend upon the type of terrain and upon the method used. The data may be acquired from satellites in the form of heights from stereoscopic measurements, from interferometric measurement or from direct measurement with a ranging device. In the case of stereoscopic measurement, automatic correlation will normally be used. In general these methods will produce heights in a quasi-systematic pattern, rows or patches on an approximate grid pattern. These must be resampled to a regular grid. Plate 20 shows a colour coded view of a DEM and a perspective view. The image to the left is a colour coded DEM of the area between Marseilles and Grenoble in southeast France, and is derived from SPOT panchromatic images: the heights were derived on a 50-m grid and are accurate to about 10 metres in height. The image on the right is a shaded view of the same area, derived from the DEM.

DEMs may be generated from stereoscopic aerial photographs or satellite images. The accuracy attainable will depend on the scale of the photograph or on the pixel size. Manually derived DEMs from aerial photography can provide very accurate and reliable information, but the process of recording it is time consuming and tedious. Automatic methods are now widely used but suffer from the problem that a large amount of checking and editing may still be required. The extraction of heights can be done automatically in an off-line mode. Complete scenes of SPOT data can be stereo matched in a few hours in order to obtain a DEM which can be used to produce orthoimages and image maps.

In recent years SAR interferometry has been used to create DEMs. This is still an emerging technology but very good results have been reported in some areas. The method has the advantage of being independent of cloud and also that differential methods can detect shifts in terrain, because of earthquake for example, to a few millimetres.

7 ACCURACY OF DATA

Many aspects of data accuracy have been covered indirectly already, and are also discussed by Fisher (Chapter 13) and Heuvelink (Chapter 14). When encoding map data, the principle source of error is the map itself. Encoding systems can reproduce the features on the map with high apparent precision, but the error present in the original document will remain. These errors are attributable to a number of factors:

- original survey;
- revision;
- generalisation;
- drafting and reproduction.

All available information about the source of the map and the data used to compile and revise it should be collected where possible and stored along with the encoded data. Thus, although no reliable measure of the accuracy is available, at least a sensible estimate of the error can be made.

The errors in images will be dependent on the scale of the image, or the pixel size, and on how rigorously the corrections have been applied. A fully rigorous photogrammetric plot will remove all distortions and the errors will be directly related to the scale of the original imagery. An orthoimage produced with a good quality DEM will also have small errors. However, images covering hilly terrain, corrected with a polynomial, may have large errors present. Errors in this type of product will also be subject to the quality of the ground control which may have been poorly identified, or the coordinates might be in error. When registering two datasets, both sets may contribute to the accuracy of the registration.

8 VALIDATION OF ENCODING PROCESSES

Validation can be carried out by evaluating consistency, precision, and accuracy. Accuracy is defined as the closeness of a measurement to the 'truth'. In practice 'truth' usually means the best available data. Accuracy can be evaluated by the root mean square error which is given as:

$$\text{rmse} = \sqrt{\frac{\Sigma v^2}{n}} \tag{9}$$

where v is the residual error, the difference between computed and reference ground coordinates, and n is the number of observations. Precision is a measure of the closeness of measurements to each other and is defined by the standard deviation of a set of measurements:

$$\sigma = \sqrt{\frac{\Sigma(v-\bar{v})^2}{n-1}} \tag{10}$$

where v is the mean of the residuals. For further discussion of RMSE mesures, see Beard and Buttenfield (Chapter 15) and Fisher (Chapter 13).

Consistency is a more subjective measure, depending upon human interpretation of what looks correct.

9 FUTURE PROSPECTS

The growing use of GIS and the increasing volumes of data from new sensors make it increasingly important that data are corrected and referenced to a common system. These corrections need to be as accurate as possible and need in particular to correct for topographical effects. This increases the need for appropriate DEMs. In the future, systems for image processing and image analysis will almost certainly include software to produce orthoimages and DEMs: this latter requirement means that new data sources to produce DEMs are necessary.

A number of missions are planned which are designed for producing stereoscopic data for mapping. The ideal parameters for a mapping satellite are discussed by Light (1990). The US high resolution satellites are the most exciting of these developments since they will produce very high resolution stereoscopic data to produce high precision (x, y, z) coordinates, orthoimages, and DEMs.

Another development of increasing importance is the automation of processes. Automatic DEM generation has been discussed and is a production process. Automatic registration of images is also possible and work on image-to-map registration is in progress. Extraction of some coverage features such as roads is a problem with considerable potential rewards, but is still at the research stage.

References

Dowman I J, Ebner H, Heipke C 1992 Overview of European developments in digital photogrammetric workstations. *Photogrammetric Engineering and Remote Sensing* 58: 51–6

Light D L 1990 Characteristics of remote sensors for mapping and earth science applications. *Photogrammetric Engineering and Remote Sensing* 56: 1613–23

Logan I T 1995 Cost and benefit considerations in data collection and the application of data collection techniques. Paper presented at Cambridge Survey Officers' Conference

Petrie G, Kennie T J M 1990 *Terrain modelling in surveying and civil engineering*. Caithness, Whittles Publishing

Sarjakoski T 1992 *Suitability of the Sharp JX-600 desktop scanner for the digitisation of aerial colour photographs*. International Archives of Photogrammetry and Remote Sensing 29(B2): 79–86

Schreier G (ed.) 1993 *SAR Geocoding: data and systems*. Karlsruhe, Wichmann

Snyder J 1986 *Map projections – a working manual*. Professional Paper 1395. Washington DC, US Geological Survey

32

Digital remotely-sensed data and their characteristics

M BARNSLEY

This chapter explores the nature and properties of digital remotely-sensed data. Rather than simply summarising the ever-growing range of airborne and satellite sensor systems, together with their technical characteristics, the chapter is divided into three distinct parts, namely: (a) the interaction of electromagnetic radiation with Earth surface materials, focusing on the physical, chemical, and biological properties that control their reflectance, emittance, and scattering characteristics; (b) the impact of sensor and platform design on the ability to record the surface-leaving radiation and the nature of the data that are produced; and (c) the production of data-processing algorithms to translate the recorded signals into estimates of the intrinsic properties of the observed surfaces.

1 PHYSICAL PRINCIPLES

1.1 Remote sensing: inference and estimation

Broadly speaking, the subject matter of terrestrial remote sensing encompasses the set of instruments (sensors), platforms, and data-processing techniques that are used to derive information about the physical, chemical, and biological properties of the Earth's surface (i.e. the land, atmosphere, and oceans) without recourse to direct physical contact. Information is derived from measurements of the amount of electromagnetic radiation reflected, emitted, or scattered from the Earth surface, and its variation as a function of wavelength, angle (direction), wave polarisation, phase, location, and time. A variety of sensors is commonly employed in this context – both passive (i.e. those reliant on reflected solar radiation or emitted terrestrial radiation) and active (i.e. those generating their own source of electromagnetic radiation) – operating throughout the electromagnetic spectrum from visible to microwave wavelengths (see also Dowman, Chapter 31). The platforms on which these instruments are mounted are similarly diverse: although Earth-orbiting satellites and fixed-wing

aircraft are by far the most common, helicopters, balloons, masts, and booms are also used. Finally, a wide range of data-processing techniques has been developed, often in response to advances in sensor technology, but increasingly to meet the demands of a growing set of applications.

The problem with the definition of remote sensing outlined above is that it focuses on the *technology,* as opposed to the *science,* of remote sensing. In doing so, it obscures two fundamental aspects of the remote sensing process, namely *inference* and *estimation*. The role of inference becomes clear when it is understood that very few properties of interest to the environmental scientist can be measured directly by remote sensing. Instead, they must be inferred from measurements of reflected, emitted, or scattered radiation using some form of mathematical model, or via their relationship with a surrogate variable (e.g. land cover) that can be derived more readily from the remotely-sensed data (see also Bibby and Shepherd, Chapter 68; Fisher, (Chapter 13). The accuracy with which a given property can be inferred is therefore dependent on the quality (generality, applicability, reliability, etc.) of the model and algorithms used, or on the degree of correlation between the surrogate and target variables, together with the accuracy and suitability of any land-cover classification scheme involved and the classes that this defines. Unfortunately, our understanding of these relationships is, in many instances, still quite poor (see also Dowman, Chapter 31). Even where our knowledge is well developed, we may be forced to employ models involving a number of approximations or simplifications, perhaps to reduce the computational load in time-critical applications. As a consequence, the values inferred from remotely-sensed data are almost always estimates of the actual quantities of interest.

1.2 Sources of information

There are five main sources of information that can be exploited by remote sensing systems. These relate to variations in the recorded signal as a function of:

- wavelength ('spectral');
- angle ('directional');
- wave polarisation;
- location ('spatial');
- time ('temporal').

1.2.1 Inference and estimation from spectral variations

Most remote sensing studies attempt to exploit spectral (i.e. wavelength dependent) variations in the radiation emanating from the Earth's surface: these are controlled by the physical and chemical properties of Earth surface materials. In the case of healthy green leaves, for example, the principal controlling factors are plant pigments (e.g. chlorophyll, xanthophyl, and the carotenoids), lignin, cellulose, protein, and leaf-water content (Asrar 1989; Jacquemoud and Baret 1990). In the case of soils, the most important factors are the content of moisture, iron oxides, and organic matter, together with surface structure (Price 1990; Jacquemoud et al 1992).

There are two main ways in which the relationship between surface properties and spectral response can be exploited. At one level, the aim may be simply to distinguish different types of surface material. In this case, the objective is to identify those wavelengths at which the contrast between their reflectance, emittance, or scattering characteristics is maximised. Since not all surface materials can be distinguished at a given wavelength, it is common to record data in several parts of the electromagnetic spectrum (i.e. multispectral remote

sensing). A subsequent aim may be to identify the nature of the surface materials; that is, to assign each a label from a set of pre-defined classes, typically expressed in terms of land cover.

The second major use of multispectral data is to estimate values for selected properties of the observed surface materials. For example, many studies have attempted to derive information on the above-ground biomass, leaf area index (LAI), and levels of photosynthetic activity of vegetation canopies. This is commonly based on linear combinations of data recorded in two or more spectral wavebands, generally centred on the visible red and near-infrared wavelengths (Myneni and Williams 1994). Use of this type of empirical model – referred to generically as vegetation indices – is widespread, despite their well-known limitations (Baret and Guyot 1991; Myneni et al 1995); indeed, new indices are continually being developed. The enduring attraction of vegetation indices lies in their conceptual and computational simplicity. This goes some way to explain the enduring popularity of the normalised difference vegetation index (NDVI), most recently for mapping and monitoring vegetation at regional and global scales (Townshend et al 1995).

Recent advances in sensor technology, specifically those relating to improvements in spectral resolution (Vane and Goetz 1993), have prompted more detailed studies of the relationships between spectral response and surface biochemical properties (Wessman et al 1988; Hunt 1991). Many of these studies continue to make use of simple empirical transformations (such as ratios) of data measured in a number of spectral wavebands (Danson et al 1992). Attention has also been focused on locating the so-called 'red edge' (the wavelength of maximum slope in the spectral response of vegetation between 690μm and 740μm; Figure 1), using this as an indicator of photosynthetic activity and leaf biochemistry (Boochs et al 1990; Filella and Peñuelas 1994; Curran et al 1995). More importantly, attempts have also been made to develop physically-based models to account for the optical properties of individual leaves in terms of their chemical and physical characteristics (Jacquemoud and Baret 1990). In principle, these models should be less data-dependent and site-specific than their empirical counterparts. It may also be possible to invert them, so that estimates of their parameters – and, hence, the surface biophysical properties to which they relate – can be derived from multispectral measurements made by remote sensing systems.

Whichever methods are employed, it should be noted that vegetation canopies do not behave simply as 'big leaves', so that problems arise in attempting to apply the techniques described above directly to remotely-sensed images. More specifically, relationships determined *in vitro*, or *in vivo* at the scale of a single leaf, are complicated by differences in, among other things, the spatial and geometric structure of vegetation canopies and variations in the soil substrate (Goel 1989; Asrar 1989). For this reason, attempts to estimate biophysical or biochemical properties at the canopy scale require the use of coupled models of canopy and leaf reflectance (Jacquemoud 1993; Jacquemoud et al 1995).

1.2.2 Inference and estimation from directional variations

The detected reflectance of most Earth surface materials varies, sometimes considerably, as a

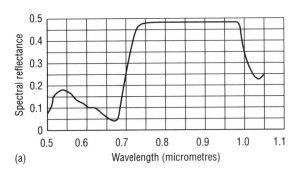

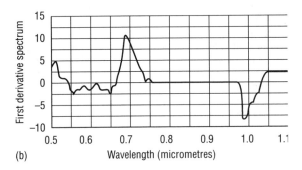

Fig 1. (a) Example leaf reflectance spectrum at visible and near-infrared wavelengths; (b) first derivative spectrum produced from Figure 1(a), showing position of the 'red-edge' (peak in derivative spectrum) at 0.693μm.

function of the angles at which they are illuminated by the Sun and viewed by the sensor. The form and magnitude of this effect are controlled by:

- the optical properties of the component elements of the surface material (e.g. the spectral reflectance and transmittance of plant leaves, stems, tree crowns, and soil facets);
- the spatial and geometric arrangement of these elements;
- the spectral and angular distribution of the incident solar radiation (Goel 1989; Barnsley 1994).

The angular distribution of the reflected radiation is described by the bidirectional reflectance distribution function (BRDF). Research in this area has focused on the development and implementation of various mathematical models (Myneni et al 1990), ranging from simple empirical (e.g. Walthall et al 1985) and semi-empirical functions (e.g. Roujean et al 1992), to models with a more direct foundation in physical principles (e.g. Ahmad and Deering 1992; Kuusk 1994). Interest in these models arises from their potential to derive quantitative estimates of certain biophysical properties of the Earth surface (e.g. the LAI). This can be achieved by inverting the model against measurements of reflected radiation made at a number of different sensor view angles and solar illumination angles with respect to a fixed point on the Earth surface (Plate 21; Goel 1989; Barnsley et al 1994). Estimates of the surface albedo can also be obtained through numerical or analytical integration of the modelled BRDF (Kimes et al 1987; Barnsley et al 1997a).

1.2.3 Inference and estimation from wave polarisation

Electromagnetic radiation considered in wave form has two fields (electric and magnetic) which are perpendicular both to one another and to the direction of propagation (Rees 1990). The orientation of these two fields, known as the wave polarisation of the radiation, has been observed to change as a result of scattering and reflection within the atmosphere and at the Earth surface. The majority of studies in this area have employed data from the microwave region of the electromagnetic spectrum. For example, polarimetric radar data have been used to distinguish different stands in coniferous forests (Grandi et al 1994) and to assess their biophysical characteristics (Baker et al 1994). A smaller number of studies has explored polarisation characteristics of Earth surface materials at visible and infrared wavelengths (e.g. Vanderbilt et al 1991; Ghosh et al 1993). This is partly attributable to the paucity of appropriate airborne and satellite sensors, and partly because the polarisation signal is dominated by the atmosphere at these wavelengths.

1.2.4 Inference and estimation from spatial variations

The amount of radiation reflected, emitted, or scattered from the Earth's surface varies spatially in response to changes in the nature (type) and properties of the surface materials. These variations may be continuous, discrete, linear, or localised, depending on the controlling environmental processes (Davis and Simonett 1991). They may also be manifest at a variety of different spatial scales (Townshend and Justice 1990; Barnsley et al 1997b; Figure 2). The relationships between surface type, surface properties, and spatial variability in land-leaving radiance has been exploited using measures of:

- texture – the statistical variability of the detected signal, typically based on the grey-level co-occurrence matrix, measured at the level of individual pixels (Richards 1993);
- pattern – including the size and shape of discrete spatial entities (regions), typically land-cover parcels, identified within the scene, as well as the spatial relations between them (LaGro 1990; Lam 1990);
- context – referring to the structural and semantic relations between discrete spatial entities identified within the scene (Barr and Barnsley 1997).

1.2.5 Inference and estimation from temporal variations

The reflectance, emittance, and scattering properties of most Earth surface materials vary with respect to time. This may be in response to diurnal effects (e.g. changes in the leaf-angle distribution of vegetation canopies attributable to moisture deficiency or heliotropism), seasonal effects (e.g. phenology), episodic events (e.g. rainfall and fire), anthropogenic influences (e.g. deforestation), or long-term climate change. There are several ways in which these temporal variations can be exploited, namely:

- to assist in distinguishing surface materials, by selecting the time of day or year at which the contrast between their reflectance, emittance, or scattering properties is greatest;
- to detect a change in the dominant land-cover type or biophysical property of an area by measuring

(a)

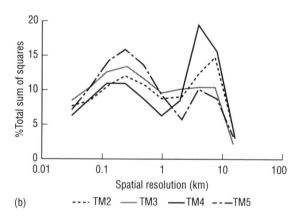

(b) ---- TM2 —— TM3 —— TM4 ---- TM5

Fig 2. (a) LANDSAT-TM sub-scene (2048 by 2048 pixels; near-infrared waveband) covering part of southeast England; (b) Scale variance analysis (Townshend and Justice 1990) applied to the LANDSAT-TM sub-scene of southeast England (a). The diagram shows the different scales of spatial variability that occur in this scene, indicated by the two peaks in variance at approximately 250m and 5km, respectively. Barnsley et al (1977b) suggests that the first peak corresponds to variation in detected reflectance at the scale of individual field parcels, while the second peak relates to broader edaphic and geological differences across the scene.

variations in the amount of surface-leaving radiation over time (known as change detection);
- to determine the physical, chemical, and biological properties of Earth surface materials.

For example, the third approach has been used to produce land cover maps at regional and global scales from coarse spatial resolution satellite sensor images (Lloyd 1990). Lloyd's approach is based on an analysis of the date-of-onset, duration, and amplitude of the 'greening-up' curve, derived from a time-series of NDVI values (Figure 3).

2 MEASURING THE SIGNAL

This section considers the impact of sensor and platform design on the ability to record surface-leaving radiation. While some space is dedicated to specific sensor systems and the characteristics of the data that they produce, the intention is not to provide a summary of current and future remote sensing devices. Rather, the aim is to examine the way in which their general design affects the ability to translate the recorded signals into estimates of the intrinsic properties of the Earth surface. Thus, the sensor system is viewed both as a measurement device and as 'filter' to the surface-leaving signal.

2.1 Spectral resolution and spectral coverage

Since most remote sensing studies – particularly those concerned with the use of optical instruments – exploit spectral variations in surface-leaving radiation, it seems appropriate to begin with a consideration of the spectral characteristics of remote sensing systems, namely:

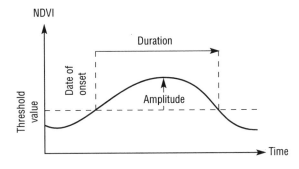

Fig 3. Diagrammatic representation of the variation in the normalised difference vegetation index (NDVI) for a vegetated surface over the growing season. The figure illustrates the concepts of the date-of-onset, the duration, and the amplitude of the 'greening-up' curve used by Lloyd (1990) to map land cover at the regional scale using coarse spatial resolution satellite sensor images.

- the number of spectral wavebands in which the sensor operates (see also Estes and Loveland, Chapter 48);
- the position of these spectral wavebands within the electromagnetic spectrum (spectral coverage);
- the range of wavelengths covered by each waveband (spectral bandwidth or spectral resolution) (Davis and Simonett 1991).

Clearly, the specific configuration adopted for a given sensor is determined by the scientific objectives of the mission, but it is also conditioned by a number of fundamental technical constraints. The latter include the need to locate the wavebands within 'atmospheric windows', the total volume of data that must be handled (including telemetry to Earth, in the case of spaceborne sensors), and the need to achieve an acceptable signal-to-noise ratio (SNR).

2.1.1 Atmospheric windows
The atmosphere scatters and absorbs radiation during its passage from the Sun to the Earth's surface and from the Earth's surface to the sensor. In doing so, it attenuates the amount of radiation reaching the ground and, subsequently, the sensor. It also alters the spectral composition and the angular distribution of this radiation (Diner and Martonchik 1985; Kaufman 1988). The magnitude of these effects varies strongly with wavelength. Sensors designed to study either the land surface or the oceans operate in regions of the electromagnetic spectrum in which the transmission of radiation through the atmosphere is high – known as 'atmospheric windows'. Even so, solar radiation may be scattered within the atmosphere into the path of the sensor without interacting with the Earth surface. Among other things, this component of the signal detected by the sensor, known as path radiance, reduces the apparent contrast between surface materials within the resultant image (Kaufman 1993).

2.1.2 Data volumes and spectral redundancy
Over the last twenty years or so, there has been a continuing trend towards sensors that are able to record data in a greater number of (typically narrower) spectral wavebands, resulting in an increase in the total volume of data acquired. In general, however, the amount of useful information that can be extracted from these data does not increase linearly with the number of available spectral wavebands: there is often a strong statistical

correlation between the data recorded in different parts of the electromagnetic spectrum, particularly those from adjacent spectral wavebands. As a result, the intrinsic dimensionality of a multispectral dataset may be very considerably smaller than the number of available wavebands. This is sometimes referred to as 'spectral redundancy'.

2.1.3 Imaging spectrometers and signal-to-noise ratio

Despite the observations made above, one of the major developments in optical sensor technology in recent years has been the advent of imaging spectrometers and imaging spectroradiometers – instruments capable of recording data in tens, or even hundreds, of very narrow (typically contiguous) spectral channels (Vane and Goetz 1993). The manner in which data from these sensors are generally employed differs from that of conventional multispectral scanners. Instead of focusing on the data simply as a set of 2-dimensional images, emphasis is placed on an analysis of the detailed spectral response recorded for each pixel. These can be compared against spectra for a range of different surface materials, drawn either from an on-line library or from representative pixels sampled within the image itself. In addition to the overall shape of the spectra, comparisons can be made in terms of the presence, depth, and width of absorption features associated with specific biochemical constituents. This may allow the analyst to derive detailed information on the nature, properties, and proportions of the different surface materials present in the corresponding area on the ground.

The Advanced Visible and Infrared Imaging Spectrometer (AVIRIS) instrument operated by NASA is one example of an imaging spectrometer (Vane et al 1993). This airborne sensor records data in approximately 200 narrow spectral channels in the region 0.4µm to 2.5µm. One of the penalties commonly associated with the use of narrow spectral wavebands is a reduction in the SNR of the sensor, because of the smaller number of photons admitted to the detector. This can be compensated for by increasing the sensor dwell-time (at the expense of a reduction in the effective sensor spatial resolution) or by combining images from several successive flights over the same target.

A spaceborne imaging spectrometer, known as HIRIS (High Resolution Imaging Spectrometer), was originally scheduled for launch at the end of the decade as part of NASA's 'Mission to Planet Earth'

programme (Goetz and Herring 1989). The instrument was, however, de-selected at an early stage because of budgetary constraints. A second imaging spectrometer, known as MODIS (Moderate Resolution Imaging Spectrometer) is due to be launched in 1998/9, although this sensor has a much smaller number of spectral wavebands (30, cf. ~200 for HIRIS) and a considerably coarser spatial resolution (250m to 1 km, cf. 30m for HIRIS; Ardanuy et al 1991).

2.2 Radiometric resolution and radiometric calibration

2.2.1 Radiometric resolution

The radiometric resolution of a sensor can be thought of as its ability to distinguish different levels of reflected, emitted, or scattered radiation. Expressed more precisely, radiometric resolution involves three key concepts, namely:

- quantisation;
- signal-to-noise ratio;
- dynamic range.

A digital remote sensing device converts the radiation incident on its detectors initially into an analogue signal (i.e. an electrical voltage) and subsequently into a digital signal. After the analogue-to-digital (A-to-D) conversion, the detected signal is represented as a numerical value, referred to (somewhat tautologously) as a digital number (DN). The set of possible values for the DN is determined by the quantisation level: thus, if a sensor has 8-bit quantisation, it will record values in the range 0 to 255 (i.e. 2^8 or 256 different levels of incident radiation), where a value of 0 indicates the lowest level of detectable radiance and 255 the highest. The sensor designer must also decide how the range of DN are to be used to record incident radiation. It is possible, for example, to design an instrument that is capable of recording the full range of radiances expected under normal illumination conditions from surfaces with reflectances varying between 0 and 1. Alternatively, if the intention is primarily to observe relatively dark targets, such as the oceans, the dynamic range of the instrument might be limited accordingly. Thus, the set of available DNs would be optimised to distinguish surfaces in the desired range of reflectances, although the instrument response would saturate over brighter targets.

2.2.2 Radiometric calibration

One characteristic of some of the most recent and many of the proposed future satellite sensors is the greater attention that is being given to their absolute radiometric calibration. This ensures that the recorded DN can be related accurately to known levels of surface reflectance (or emittance) (Price 1987). This assists in the retrieval of datasets expressed in terms of standard geophysical units and ensures greater consistency between datasets generated by different sensors, or by the same sensor over a prolonged period of time (Hall et al 1991). The lack of accurate radiometric calibration in early satellite sensors has been one of the major hindrances to the use of these data for long-term regional and global-scale environmental monitoring (Hall et al 1995).

2.3 Spatial resolution

In simple terms, the spatial resolution of a sensor determines the level of spatial detail that it provides about features on the Earth's surface. Beyond this, spatial resolution can be defined in a number of different ways (Forshaw et al 1983). For example, the instantaneous field-of-view (IFOV) defines the (nominal) angle, subtended at the sensor, over which the instrument records radiation emanating from the Earth's surface at a given instant in time. The area on the Earth's surface to which this corresponds, known as the ground resolution element (GRE), is therefore controlled by the IFOV and the height of the sensor above the ground. The actual area of ground from which radiation is incident on the detector is, however, larger than this and is determined by the sensor's point spread function (PSF). Finally, 'pixel size' denotes the area of ground covered by a single pixel in the resultant image. This may differ from the GRE because of the effects of over-sampling, variations in the height of the terrain below the sensor, and variations in the attitude and altitude of the platform on which the sensor is mounted (Forshaw et al 1983).

2.3.1 Impacts of sensor spatial resolution

Images produced by digital sensors can be thought of as 2-dimensional grids or arrays of data cells ('picture elements' or pixels). The spatial resolution of the sensor defines the size of these cells, in terms of the area that they represent on the ground (Plate 22). Thus, a remotely-sensed image represents a spatial regularisation of the observed scene (Jupp et

al 1988). One effect of this process is that two or more surface materials may fall within a single pixel, producing a 'mixed pixel' (or 'mixel'). The extent to which this occurs is, of course, dependent on the spatial resolution of the sensor and the spatial variability of the observed surface. The mixed pixel effect has a number of implications for information retrieval from remotely-sensed images. First, the detected spectral response of a mixel will be some composite of the individual spectral signals from the constituent surface materials (Smith et al 1985). Second, the size, shape, and spatial arrangement (pattern) of the major spatial entities present within the scene will be to some extent obscured (Woodcock and Strahler 1987). The first of these two problems has been addressed through the development of a number of techniques designed to 'un-mix' the component spectral responses contained in each pixel (Ichoku and Karnieli 1996). The most widely used of these is linear mixture modelling, in which the composite signal is assumed to be a linear summation of the spectral curves for the component land-cover types, weighted by their relative abundance (i.e. proportion of ground covered) within the pixel. The second problem has received rather less attention in the field of remote sensing, although it is the subject of detailed investigation by landscape ecologists (Barnsley et al 1997b).

2.3.2 Current and future directions

In recent years, there has been an intriguing bifurcation in the spatial resolution of spaceborne optical sensors. One element of this has been the widespread development of 'moderate' (or 'medium') resolution (~1km) devices (as is shown in Table 1 and Figure 4; see also Ardanuy et al 1991; Diner et al 1991; Prata et al 1990). The lineage of these sensors is simple to trace – deriving from the outstanding and, to a certain extent, unanticipated success of the current generation of NOAA's Advanced Very High Resolution (AVHRR) sensors in monitoring the land surface at continental and global scales.

Hand-in-hand with this, there is a continuing trend – initiated with LANDSAT-TM and SPOT-HRV during the mid 1980s – towards sensors with an increasingly fine spatial resolution. This trend has been extended through the availability of data from a range of Russian satellite sensors (e.g. KFA-1000 and KFA-3000), as well as the Panchromatic sensor on-board the Indian satellite IRS-1C, and is set to

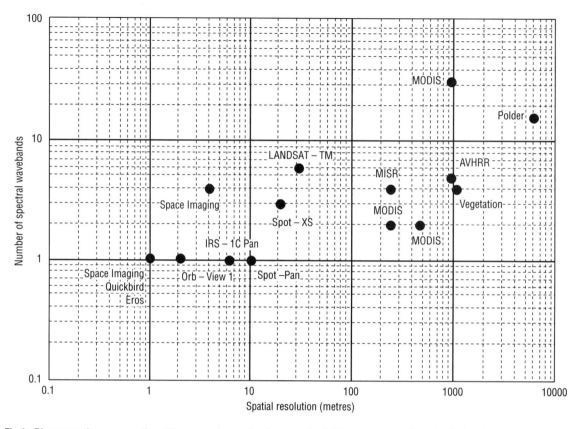

Fig 4. Diagrammatic representation of the range of current and proposed satellite sensors, together with their main spectral and spatial characteristics.

Table 1 Characteristics of a number of 'medium' or 'moderate' spatial resolution satellite sensors currently in operation or scheduled for launch in the next few years. Some other 'standard' remote sensing image sources are set out in Estes and Loveland (Chapter 48) and Dowman (Chapter 31 Table 1).

Sensor	Satellite	Spatial resolution (at nadir)	Number of spectral bands	Year of launch (actual or projected)
ATSR-1	ERS-1	1km	4	1994
ATSR-2	ERS-2	1km	6	1995
POLDER	ADEOS-1	6km by 7km	16	1996
VEGETATION	SPOT-4	1.15km	4	1997/8
MODIS	EOS-1 AM	250m, 500m and 1km	30	1998/9
MERIS	ENVISAT-1	250m (land) and 1km (oceans)	15 (programmable in position and width)	1988
MISR	EOS-1 AM	250m and 1km	4	1998/9

continue with the launch of a number of new, commercially-operated satellite devices (Table 2). Each of these instruments will be capable of producing digital image data with a spatial resolution of between 1 and 5 metres (McDonald 1995; Fritz 1996).

2.4 Angular sampling

Interest in the directional reflectance properties of Earth surface materials has grown considerably in recent years, partly in response to the increasing availability of satellite sensors that can record data at several different angles with respect to the Earth surface. This can be achieved in a number of ways (Barnsley 1994), namely:

- by means of a very wide across-track field-of-view (e.g. NOAA's AVHRR sensors, and the proposed SPOT-4 VEGETATION, MODIS, and MERIS [Medium Resolution Imaging Spectrometer] instruments);
- through the use of a very wide field-of-view in both the along-track and across-track directions

(e.g. the POLDER [polarisation and directionality of the Earth's reflectances] sensor on board the ADEOS-1 satellite);
- by pointing the sensor off-nadir in the across-track direction (e.g. the HRV [high resolution visible] instruments on the SPOT-series of satellites), the along-track direction, or both;
- through the use of multiple sensors pointed forward, nadir and aft of the platform (e.g. the multi-angle imaging Spectroradiometer [MISR] scheduled for launch as part of NASA's Earth Observing System); or
- through the use of a conical scanning motion (e.g. the Along-Track Scanning Radiometer ATSR on the European remote sensing [ERS] satellites).

The range of view angles and solar illumination angles over which a given instrument can acquire data is controlled not only by the geometry of the sensor, but also by the orbital characteristics of the satellite on which it is mounted (Barnsley et al 1994). In most cases, the actual number of angles at which

Table 2 Characteristics of a number of very high spatial resolution satellite sensors currently in operation or planned for launch in the near future (see also Dowman, Chapter 31 Table 1).

Sensor	Year of launch (actual or projected)	Spatial resolution Pan	XS	Swath width Pan	XS	Stereoscopic viewing capability
KFA-1000	1994	6m	—	80km	—	No
KFA-3000	1994	3m	—	27.5km	—	No
IRS-1C	1995	5.8m	—	70km	—	Yes
KVR-1000	1996*	2m	—	40km	—	No
Earlybird	1997	3m	15m	6km	30km	Yes
AVNIR	1997	8m	16m	80km	80km	Yes
EROS	1997	1m	1.5m	15km	15km	Yes
Quickbird	1997	1m	4m	6km	36km	Yes
Space Imaging	1997	1m	4m	11km	11km	Yes
Orbview-1	1998	1–2m	4m	8km	8km	Yes
GDE	1998	1m	—	15km	—	Yes

* digital format

reflectance data can be sampled is quite limited (Figure 5). Appropriate mathematical models are therefore required to interpolate between, and to extrapolate beyond, these sample measurements to describe and account for the full BRDF. If the models are also invertible, it may be possible to retrieve estimates of certain properties of the surface (e.g. LAI) from the sample directional reflectance data. Various BRDF models have been developed for this purpose, ranging from simple empirical formulations through to more complex, physically-based models. While the latter offer significant advantages in principle, inversion of such models typically demands the use of computationally-intensive numerical procedures. For this reason, attention is currently being focused on the use of so-called 'semi-empirical, kernel-driven' BRDF models, which can be inverted analytically (Wanner et al 1995; Barnsley et al 1997c). These models, however, tend not to be specified in terms of measurable biophysical properties of the land surface, so that further work is required to establish the relationships between such properties and the model parameters.

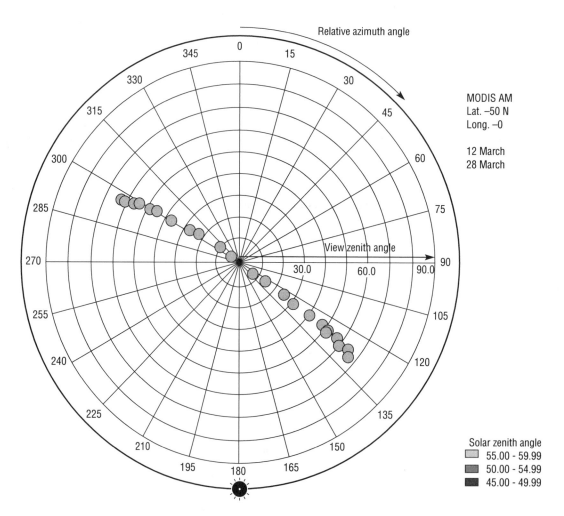

Fig 5. Angular (directional) sampling capability of NASA's proposed MODIS (moderate resolution imaging spectrometer) satellite sensor (in the form of a polar plot) for a fixed site at 50°N over a 16-day period around the vernal equinox (Barnsley et al 1994). Each dot indicates a single occasion on which the sensor is able to observe the target; the position of the dot in the plot indicates the angles at which this was achieved. The figure illustrates the comparatively sparse sample of directional reflectance data that can be acquired using this and other, similar sensors.

2.5 Wave polarisation

The majority of sensors able to measure the polarisation properties of Earth surface materials operate in the microwave region of the electromagnetic spectrum. Sensors such as these can transmit and receive (record) radiation in a given plane polarisation. Where the transmitted and received radiation have the same plane polarisation, the signal is referred to as being like-polarised; where they are of different polarisations, the signal is said to be cross-polarised (Rees 1990). Imaging radar polarimeters can derive a measure of backscattered radiation for any configuration of transmitted and received radiation using a process known as polarisation synthesis (Zyl et al 1987).

The POLDER instrument, launched on board the ADEOS-1 satellite in 1996, offers the capability to measure polarisation properties at visible and near-infrared wavelengths (Deschamps et al 1990). The primary use envisaged for these data is, however, to derive information on atmospheric aerosol properties, rather than the biophysical characteristics of land surface materials.

2.6 Temporal resolution

In simple terms, temporal resolution refers to the frequency with which repeat data can be acquired for a given area on the Earth's surface. This is controlled both by the geometry of the sensor and by the orbital characteristics of the satellite on which it is mounted. In terms of the latter, two main types of orbit are used by Earth-observing sensors, namely: (a) Sun-synchronous, near-polar; and (b) geo-stationary (or geo-synchronous).

Satellites in the first of these two orbits progress in a near-circular path at an altitude of between 500km and 1000km above the Earth's surface. The orbital plane is inclined, so that the satellite passes close to, but not over, the poles (hence 'near-polar'). By taking advantage of precession in the satellite's orbit, it is possible to ensure that the satellite crosses the equatorial plane at approximately the same local solar time on each orbit (hence 'Sun-synchronous'). The rotation of the Earth beneath the satellite means that successive orbits pass over different regions of the surface. Eventually the satellite will complete its sequence of orbits and begin to trace the path of the first orbit again. For a point on the equatorial plane, the period between two such orbits is known as the repeat cycle; LANDSAT-5, for example, has a repeat cycle of 16 days. It is possible to observe a given point on the Earth's surface more frequently than this, depending on the latitude of the site and the configuration of the sensor. Since the satellite's orbital paths converge towards the poles, there is increasing overlap between images acquired on different orbits at higher latitudes; such sites can therefore be imaged more frequently than those at lower latitudes. This effect is also controlled by the field-of-view and, hence, the swath width of the sensor: the wider the swath width, the greater the number of occasions on which a given point can be imaged during the nominal repeat cycle. Even so, key episodic and seasonal events may still be missed because of cloud cover or simply because the event took place while the satellite was tracing another orbit. The former is, of course, less of a problem for active microwave systems, since these can penetrate cloud. The latter can be offset, to a certain extent, by tilting (or pointing) the sensor away from the sub-satellite point (i.e. off-nadir). This allows the sensor to target an area for repeated imaging, even though the satellite is not directly overhead (Barnsley et al 1994).

The second major type of satellite orbit referred to above is the geo-stationary or geo-synchronous orbit. Here, the satellite maintains a fixed position above the Earth's surface, usually at an altitude of around 36 000 km. This orbit is generally reserved for operational meteorological satellites which require frequent coverage (i.e. once every 20 to 30 minutes) of very large areas at a comparatively low spatial resolution (1–5 km) (Kramer 1994).

3 SELECTED DEVELOPMENTS IN DATA PROCESSING

It is not possible, within the scope of this chapter, to provide a comprehensive review of the full range of techniques and algorithms that are used to derive useful information from digital remotely-sensed images; although some have already been mentioned briefly in the preceding sections. Rather, this section attempts to highlight just a few of the most important, recent developments in image data processing.

3.1 Modelling surface-radiation interactions and data assimilation

Perhaps the most significant developments in the processing of remotely-sensed data over recent years have been the increasing focus on converting the detected signals into estimates of key geophysical units and the assimilation of these data into numerical models of various environmental processes (Hall et al 1995; Townshend et al 1995). The first of these two elements is being achieved through the use of increasingly sophisticated, deterministic, radiative-transfer and energy-balance models, some of which have been alluded to in the preceding sections. This general approach is, for example, central to NASA's 'Mission to Planet Earth' programme, embodied in the Earth Observing System (EOS) and its constituent satellite sensors (Running et al 1994). The significance of this development cannot be overstated: it marks the continuing transition of digital remote sensing from being principally an instrument for large-scale (land cover) mapping, to a more comprehensive, robust, and effective scientific tool for environmental monitoring.

The second element – assimilation of remotely-sensed data products into models of, for example, the global carbon cycle, the surface energy balance, and the net primary productivity of the land surface and oceans – is also receiving widespread attention (Running et al 1994; Townshend et al 1995). It reflects the recognition among much of the remote sensing community that, in addition to developing the science and technology to underpin remote sensing *sensu stricto*, there is a need to generate data products that are both appropriate to, and immediately usable by, the broader community of environmental scientists: that the rationale for remote sensing lies not simply in the development of sensors and algorithms, but more importantly in addressing real environmental problems. The scientific challenges that this creates include the requirement (a) to handle very large volumes (i.e. Tera-bytes) of data, often acquired by more than one sensor and/or satellite; (b) to process these data using robust, computationally-efficient, and validated algorithms, based on methods acceptable to most, if not all, of the target community; and (c) to generate usable products at the appropriate spatial and temporal scales, often in near-real time. Ultimately, remote sensing will be measured against how successful it is in meeting these stringent challenges.

3.2 Image classification and segmentation

At a somewhat different level, the production of thematic maps from digital, remotely-sensed images –

commonly referred to as image classification – remains an area of considerable research interest. Attention is, however, shifting from the use of standard, statistical classification algorithms to the wider application of artificial neural network (ANN), fuzzy-set, knowledge-based and evidential reasoning techniques (Fischer, Chapter 19; Bezdek et al 1984; Mesev et al 1995; Schalkoff 1992; Wilkinson 1996). The attraction of ANNs, for instance, lies in their ability to 'learn' by example, as well as their relative freedom from assumptions about the statistical distributions of the candidate classes (cf. conventional statistical classifiers, such as the maximum likelihood algorithm; Foody 1992). Fuzzy-set techniques, on the other hand, move away from the notion that each pixel must be assigned a single label drawn from a set of discrete, mutually exclusive classes. In doing so, they provide another way to account for the mixed pixel ('mixel') effect in remotely-sensed images (Foody 1992; see also Fisher, Chapter 13). Finally, both knowledge-based and evidential reasoning approaches offer ways to incorporate ancillary data (e.g. digital map data exported from a GIS), heuristics, and facts or evidence into the classification process (Wilkinson 1996).

Despite these developments, the overwhelming majority of studies continue to use image-classification algorithms that operate at the level of the individual pixel; that is, algorithms in which each pixel is assigned a label solely on the basis of its multispectral response, without reference to those of neighbouring pixels or the context of that pixel within the scene as a whole. By comparison, relatively limited use has been made of syntactic (or structural) pattern-recognition techniques, which operate on discrete, multi-pixel regions (i.e. meaningful spatial entities or 'objects') to infer further, higher-level information about the scene (Schalkoff 1992). Notable exceptions include the studies by Moller-Jenson (1990) and Nichol (1990) – on the spatial generalisation of thematic maps derived from remotely-sensed data – and, more recently, by Barr and Barnsley (1997) – to infer information on land use in urban areas from satellite sensor images. The comparative lack of attention given to syntactic pattern-recognition techniques in remote sensing to date is probably because of the relatively coarse spatial resolution of the images acquired by the current generation of satellite sensors. This results in uncertainty, not only about the nature (i.e. land cover type) of the principal spatial entities present within the scene, but also their morphological properties (e.g. size, shape, and boundaries) and the spatial (e.g. adjacency, containment, distance, and direction) and structural (e.g. 'forms part of') relations between them. This situation is, however, likely to change with the advent of the new generation of very high spatial resolution (<5m), commercial satellite sensors (but see Smith and Rhind, Chapter 47, for a discussion of some residual limitations). Indeed, data acquired by these new sensors demand alternatives to the conventional, per-pixel classification algorithms, if we are to derive information other than simple land cover about the observed scenes. Syntactic pattern-recognition techniques offer considerable potential in this context.

3.3 Integration of GIS and remote sensing technologies

The relationship between remote sensing and GIS has received considerable attention in the literature and, indeed, remains the subject of continuing discussion (Hinton 1996; Wilkinson 1996). Much of this discussion revolves around the scientific and technical issues relating to 'integration' of the two technologies (Ehlers et al 1989, 1991), so that remotely-sensed images can be used both as a source of spatial data within GIS and to exploit the functionality of GIS in processing remotely-sensed data. Despite this, the actual progress towards the goal of full integration is surprisingly slow. While this is undoubtedly due to the very considerable technical challenge of accessing, manipulating, and visualising vector, raster, and tabular data simultaneously, it seems unlikely that technical constraints have been the sole barrier to achieving full integration. It might be argued that competing imperatives in both remote sensing and GIS have tended to draw attention away from the issue of integration. For instance, from a remote sensing perspective, the recent focus on monitoring global environmental change using coarse (~1km) spatial resolution sensors – and the assimilation of the data that they produce into various environmental simulation models – has deflected some of the attention away from the traditional issues of large-scale mapping, which are more closely allied to the concerns and use of GIS. Similarly, one can see a number of other developments – such as the emergence of GIS functionality (albeit fairly limited) within standard office software, the potential for wider access to GIS software via network/Web

platforms, and the role and application of multimedia technology within GIS – that have consumed much of the research and development effort in the field of GIS.

Nevertheless, there are at least two reasons why the issue of integration is likely to receive fresh impetus in the near future. The first is the increasing availability of data from the very high spatial resolution, commercial satellite sensors that are scheduled for launch over the next few years. These will produce data appropriate to many of the large-scale mapping projects in which GIS have often been used, and are likely to compete directly with the traditional aerial photography market. The second is that these high resolution images require the development of new data processing algorithms, such as syntactic (or structural) pattern-recognition techniques, to extract the maximum amount of information about the observed scene. There is a considerable overlap between the objectives and functionality of these techniques and those used in mainstream GIS, at least in terms of their potential for spatial analysis, and this may also bring the two communities closer together. Estes and Loveland (Chapter 48) provide a more detailed overview of the management of the data products of new remote sensing technologies.

4 CONCLUSIONS

This chapter has attempted to provide a broad overview of the nature of digital remote sensing, including: (a) the physical, chemical, and biological properties that control the interaction of electromagnetic radiation with Earth surface materials; (b) the impact of sensor and platform design on the ability to record these signals and the nature of the data that are produced; and (c) the derivation of useful information from these data. The coverage has necessarily been brief and somewhat partial. It is impossible, within the scope of this chapter, to do justice to all aspects of the subject. For example, little has been mentioned of the development of remote sensing as it relates to the study of the Earth's oceans and atmosphere, or to the exciting advances that have been made in the application of interferometric synthetic aperture radar (SAR) to measure the morphology and deformation of the Earth's crust. Perhaps some of these aspects are of less relevance to the wider GIS community. What should be apparent, however, is the rapid developments taking place in – and the increasing breadth of – digital remote sensing at the present time. Thus, while remote sensing will continue to be an important source of spatial data that can be used within GIS, the nature of these data is set to change in terms of an increase in their diversity and an improvement in their utility, accuracy, and reliability.

References

Ahmad S P, Deering D W 1992 A simple analytical function for bidirectional reflectance. *Journal of Geophysical Research* 97: 18 867–86

Ardanuy A, Han D, Salomonson V V 1991 The Moderate Resolution Imaging Spectrometer (MODIS): science and data system requirements. *IEEE Transactions on Geoscience and Remote Sensing* 29: 75–88

Asrar G (ed.) 1989 *The theory and applications of optical remote sensing*. New York, John Wiley & Sons Inc.

Baker J R, Mitchell P L, Cordey R A, Groom G B, Settle J J, Stileman M R 1994 Relationships between physical characteristics and polarimetric radar backscatter for Corsican pine stands in Thetford Forest, UK. *International Journal of Remote Sensing* 15: 2827–50

Baret F, Guyot G 1991 Potentials and limits of vegetation indices for LAI and APAR assessment. *Remote Sensing of Environment* 35: 161–73

Barnsley M J 1994 Environmental monitoring using multiple view angle (MVA) remotely-sensed images. In Foody G, Curran P J (eds) *Environmental remote sensing from regional to global scales*. London, Taylor and Francis: 181–201

Barnsley M J, Allison D, Lewis P 1997a On the information content of multiple view angle (MVA) images. *International Journal of Remote Sensing* 18: 1937–60

Barnsley M J, Barr S L, Tsang T 1997b Scaling and generalisation issues in land cover mapping, In Gardingen J van (ed.) *Scaling up*. Cambridge (UK), Cambridge University Press

Barnsley M J, Lewis P, Sutherland M, Muller J-P 1997c Estimating land surface albedo through inversion of two BRDF models against ASAS image data for HAPEX-Sahel. *Journal of Hydrology* 188–9; 749–78

Barnsley M J, Strahler A N, Morris K P, Muller J-P 1994 Sampling the surface BRDF: 1. Evaluation of current and future satellite sensors. *Remote Sensing Reviews* 8: 271–311

Barr S L, Barnsley M J 1997 A region-based, graph-theoretic data model for the inference of second-order thematic information from remotely-sensed images. *International Journal of Geographical Information Science*, 11: 555–16

Bezdek J C, Ehrlich R, Full W 1984 FCM: the fuzzy *c*-means clustering algorithm. *Computers and Geosciences* 10: 191–203

Boochs F, Kupfer G, Dockter K, Kuhbauch W 1990 Shape of the red-edge as a vitality indicator for plants. *International Journal of Remote Sensing* 11: 1741–53

Curran P J, Windham W R, Gholz H L 1995 Exploring the relationship between reflectance red, edge and chlorophyll concentration in slash pine leaves. *Tree Physiology* 15: 203–6

Danson F M, Steven M D, Malthus T J, Clark J A 1992 High-spectral resolution data for determining leaf water content. *International Journal of Remote Sensing* 13: 461–70

Davis F W, Simonett D S 1991 GIS and remote sensing. In Maguire D J, Goodchild M F, Rhind D W (eds) *Geographical information systems: principles and applications*. Harlow, Longman/New York, John Wiley & Sons Inc. Vol. 1: 191–213

Deschamps P Y, Herman M, Podaire A, Leroy M, Laporte M, Vermande P 1990 A spatial instrument for the observation of polarisation and directionality of Earth reflectances: POLDER. *Proceedings of International Geoscience and Remote Sensing Symposium (IGARSS90)*. Washington DC, IEEE Publications: 1769–74

Diner D J, Bruegge C J, Martonchik J V, Bothwell G W, Danielson E D, Ford V G, Hovland L E, Jones K LK, White M L 1991 A multi-angle image spectroradiometer for terrestrial remote sensing with the Earth Observing System. *International Journal of Imaging Systems and Technology* 3: 92–107

Diner D J, Martonchik J V 1985 Atmospheric transmittance from spacecraft using multiple view angle imagery. *Applied Optics* 24: 3503–11

Ehlers M, Edwards G, Bédard Y 1989 Integration of remote sensing with geographical information systems: a necessary evolution. *Photogrammetric Engineering and Remote Sensing* 55: 1619–27

Ehlers M, Greenlee D, Smith T, Star T 1991 Integration of remote sensing and GIS: data and data access. *Photogrammetric Engineering and Remote Sensing* 57: 669–75

Filella I, Peñuelas J 1994 The red-edge position and shape as indicators of plant chlorophyll content, biomass, and hydric status. *International Journal of Remote Sensing* 15: 1459–70

Foody G M 1992 A fuzzy sets approach to the representation of vegetation continua from remotely sensed data: an example from lowland heath. *Photogrammetric Engineering and Remote Sensing* 58: 221–5

Forshaw M R B, Haskell A, Miller P F, Stanley D J, Townshend J R G 1983 Spatial resolution of remotely-sensed imagery: a review paper. *International Journal of Remote Sensing* 4: 497–520

Fritz L W 1996 The era of commercial earth observation satellites. *Photogrammetric Engineering and Remote Sensing* 62: 39–45

Ghosh R, Sridhar V, Venkatesh H, Mehta A, Patel K 1993 Linear polarisation measurements of a wheat canopy. *International Journal of Remote Sensing* 14: 2501–8

Goel N S 1989 Inversion of canopy reflectance models for estimation of biophysical parameters from reflectance data. In Asrar G (ed.) *Theory and applications of optical remote sensing*. New York, John Wiley & Sons Inc.: 205–51

Goetz A F H, Herring M 1989 The High Resolution Imaging Spectrometer (HIRIS) for EOS. *IEEE Transactions on Geoscience and Remote Sensing* 27: 136–44

Grandi G de, Lemoine G G, Groof H de, Lavalle C, Siebert A J 1994 Fully polarimetric classification of the Black Forest MAESTRO 1 AIRSAR data. *International Journal of Remote Sensing* 15: 2755–76

Hall F G, Strebel D E, Nickeson J E, Goetz S J 1991 Radiometric rectification: towards a common radiometric response among multidate, multisensor images. *Remote Sensing of Environment* 35: 11–27

Hall F G, Townshend J R G, Engman E T 1995 Status of remote sensing algorithms for estimation of land surface state parameters. *Remote Sensing of Environment* 51: 138–56

Hinton J C 1996 GIS and remote sensing integration for environmental applications. *International Journal of Geographical Information Systems* 10: 877–90

Hunt E R Jr 1991 Airborne remote sensing of canopy water thickness scaled from leaf spectral data. *International Journal of Remote Sensing* 12: 643–9

Ichoku C, Karnieli A 1996 A review of mixture modelling techniques for sub-pixel land cover estimation. *Remote Sensing Reviews* 13: 161–86

Jacquemoud S 1993 Inversion of the PROSPECT + SAIL canopy reflectance model from AVIRIS equivalent spectra: theoretical study. *Remote Sensing of Environment* 44: 281–92

Jacquemoud S, Baret F 1990 PROSPECT: a model of leaf optical properties spectra. *Remote Sensing of Environment* 34: 75–92

Jacquemoud S, Baret F, Andrieu B, Danson F M, Jaggard K 1995 Extraction of vegetation biophysical parameters by inversion of the PROSPECT + SAIL models on sugar beet canopy reflectance data application to TM and AVIRIS sensors. *Remote Sensing of Environment* 52: 163–72

Jacquemoud S, Baret F, Hanocq J F 1992 Modelling spectral and bidirectional soil reflectance. *Remote Sensing of Environment* 41: 123–32

Jupp D L B, Strahler A H, Woodcock C E 1988 Autocorrelation and regularisation in digital images: 1. Basic theory. *IEEE Transactions on Geoscience and Remote Sensing* 26: 463–73

Kaufman Y J 1988 Atmospheric effect on spectral signature – measurement and corrections. *IEEE Transactions on Geoscience and Remote Sensing* 26: 441–50

Kaufman Y J 1993 Aerosol optical thickness and atmospheric path radiance. *Journal of Geophysical Research* 98: 2677–92

Kimes D S, Sellers P J, Diner D J 1987 Extraction of spectral hemispherical reflectance (albedo) of surfaces from nadir and directional reflectance data. *International Journal of Remote Sensing* 8: 1727–46

Kramer H J 1994 *Observation of the Earth and its environment: survey of missions and sensors.* Berlin, Springer

Kuusk A 1994 A multispectral canopy reflectance model. *Remote Sensing of Environment* 50: 75–82

LaGro J 1990 Assessing patch shape in landscape mosaics. *Photogrammetric Engineering and Remote Sensing* 57: 285–93

Lam N S-M 1990 Description and measurement of LANDSAT-TM images using fractals. *Photogrammetric Engineering and Remote Sensing* 56: 187–95

Lloyd D 1990 A phenological classification of terrestrial vegetation cover using shortwave vegetation index imagery. *International Journal of Remote Sensing* 11: 2269–79

McDonald R A 1995 Opening the Cold War sky to the public: declassifying satellite reconnaissance imagery. *Photogrammetric Engineering and Remote Sensing* 61: 385–90

Mesev T V, Longley P A, Batty M, Xie Y 1995 Morphology from imagery – detecting and measuring the density of urban land-use. *Environment and Planning A* 27: 759–80

Moller-Jenson L 1990 Knowledge-based classification of an urban area using texture and context information in LANDSAT-TM imagery. *Photogrammetric Engineering and Remote Sensing* 56: 899–904

Myneni R B, Hall F G, Sellers P J, Marshak A L 1995 The interpretation of spectral vegetation indices. *IEEE Transactions on Geoscience and Remote Sensing* 33: 481–6

Myneni R B, Ross J, Asrar G 1990 A review of the theory of photon transport in leaf canopies. *Agricultural Forestry Meteorology* 45: 1–153

Myneni R B, Williams D L 1994 On the relationship between FAPAR and NDVI. *Remote Sensing of Environment* 49: 200–11

Nichol D G 1990 Region adjacency analysis of remotely-sensed imagery. *International Journal of Remote Sensing* 11: 2089–101

Prata A J, Cechet R P, Barton I J, Llewellyn-Jones D T 1990 The Along-Track Scanning Radiometer for ERS-1: scan geometry and data simulation. *IEEE Transactions on Geoscience and Remote Sensing* 28: 3–13

Price J C 1987 Calibration of satellite radiometers and the comparison of vegetation indices. *Remote Sensing of Environment* 21: 15–27

Price J C 1990 On the information content of soil reflectance spectra. *Remote Sensing of Environment* 33: 113–21

Rees W G 1990 *Physical principles of remote sensing.* Cambridge (UK), Cambridge University Press

Richards J A 1993 *Remote-sensing digital image analysis: an introduction.* Berlin, Springer

Roujean J-L, Leroy M, Deschamps P Y 1992 A bidirectional reflectance model of the Earth's surface for the correction of remote sensing data. *Journal of Geophysical Research* 97: 20 455–68

Running S W, Justice C O, Salomonson V, Hall D, Barker J, Kaufmann Y J, Strahler A H, Huete A R, Muller J-P, Vanderbilt V, Wan Z M, Teillet P, Carneggie D 1994 Terrestrial remote sensing science and algorithms planned for EOS/MODIS. *International Journal of Remote Sensing* 15: 3587–620

465

Schalkoff R J 1992 *Pattern recognition: statistical, structural and neural approaches*. New York, John Wiley & Sons Inc.

Smith M O, Johnson P E, Adams J B 1985 Quantitative determination of mineral types and abundances from reflectance spectra using principal components analysis. *Journal of Geophysical Research* 90: 792–804

Townshend J R G, Justice C O 1990 The spatial variation of vegetation changes at very coarse scales. *International Journal of Remote Sensing* 11: 149–57

Townshend J R G, Justice C O, Skole D, Malingreau J-P, Chilar J, Teillet P, Sadowski F, Ruttenberg S 1995 The 1-km resolution global dataset: needs of the International Geosphere Biosphere Programme. *International Journal of Remote Sensing* 15: 3417–42

Vanderbilt V C, Grant L, Ustin S L 1991 Polarisation of light by vegetation. In Myneni R, Ross J (eds) *Photon-vegetation interactions: applications in optical remote sensing and plant ecology*. Berlin, Springer: 191–228

Vane G, Goetz A F H 1993 Terrestrial imaging spectrometry: current status, future trends. *Remote Sensing of Environment* 44: 117–26

Vane G, Green R O, Chrien T G, Enmark H T, Hansen E G, Porter W M 1993 The Airborne Visible/Infrared Imaging Spectrometer (AVIRIS). *Remote Sensing of Environment* 44: 127–44

Walthall C L, Norman J M, Welles J M, Campbell G, Blad B L 1985 Simple equation to approximate the bi-directional reflectance from vegetative canopies and bare soil surfaces. *Applied Optics* 24: 383–7

Wanner W, Li X, Strahler A H 1995 On the derivation of kernels for kernel-driven models of bi-directional reflectance. *Journal of Geophysical Research* 100: 21 077–89

Wessman C A, Aber J D, Peterson D L, Melillo J M 1988 Remote sensing of canopy chemistry and nitrogen cycling in temperate forest ecosystems. *Nature* 335: 154–6

Wilkinson G G 1996 A review of current issues in the integration of GIS and remote sensing data. *International Journal of Geographical Information Systems* 10: 85–101

Woodcock C E, Strahler A H 1987 The factor of scale in remote sensing. *Remote Sensing of Environment* 21: 311–32

Zyl J J van, Zebker H A, Elachi C 1987 Imaging radar polarisation signatures: theory and observation. *Radio Science* 22: 529–43

33

Using GPS for GIS data capture

A F LANGE AND C GILBERT

Depending on the particular equipment utilised and the techniques used, Global Positioning Systems (GPS) are capable of recording position to a high level of accuracy. Differential GPS is required to obtain the accuracy required by many GIS data capture applications. Differential GPS requires the use of a base station at a known location to remove systematic errors from the GPS signal. Modern GPS/GIS data collection devices have the capability to collect a wide range of GIS attribute data in addition to position of the feature of interest; attributes include point, line, and area features. Since GPS can be used in many diverse GIS data capture applications (including applications as diverse as road centreline mapping, utility pole mapping, and wetland boundary mapping), it is important to understand the limitations of GPS in order to derive full benefits from the technology.

1 FUNDAMENTALS OF GPS

The explosion in interest in GIS as a management tool has been accompanied by the development of a number of enabling technologies, one of the more important of which is the Global Positioning System (GPS). While GIS technology offers tremendous capabilities for more informed management decision-making, rendering competent decisions still depends on having reliable data. In order to realise the benefits of GPS, and not to mis-apply the technology, it is important to understand its limitations. This chapter describes how GPS works and how to obtain reliable data using it. Accuracy (in the surveying sense) of GPS data is an important consideration. The first part of this chapter covers how the GPS accuracy is obtained and the second part discusses how to use GPS for GIS data capture. The term 'GPS' is used interchangeably here to refer to satellite-based navigation systems in general and the ground-based geographical data collection instruments in particular.

1.1 Overview of GPS

The Navigation Satellite Timing And Ranging Global Positioning System, or NAVSTAR GPS, is a satellite-based radio-navigation system that is capable of providing extremely accurate worldwide, 24-hour, 3-dimensional (latitude, longitude, and elevation) location data (Wells 1987). It is one of two such satellite systems, although its Russian counterpart, GLONASS, will not be discussed here. The system was designed and is maintained by the US Department of Defense (DoD) as an accurate, all-weather navigation system. Though designed as a military system, it is available with certain restrictions to civilians. The system has reached its full operational capability, with a complete set of at least 24 satellites orbiting the Earth in a carefully designed pattern.

1.1.1 The fundamental components of GPS

The NAVSTAR GPS has three basic segments: space, control, and user. The space segment consists of the orbiting satellites making up the constellation. This constellation is comprised of 24 satellites, each orbiting at an altitude of approximately 20 000 kilometres, in one of six orbital planes inclined 55 degrees relative to the Earth's equator. Each satellite broadcasts a unique coded signal, known as pseudo random noise (PRN) code, that enables GPS receivers to identify the satellites from which the signals are coming (Hurn 1989). The satellites broadcast the PRN codes as modulation on two carrier frequencies, L1 and L2. The L1 frequency is 1575.5 MHz and the L2 frequency is 1227.6 MHz.

With 24 satellites in the constellation, and the design of the orbits and the spacing of the satellites in the orbital planes, most users will have six or more satellites available at all times. There are a number of different programs used to plot the satellite availability for any geographical location. The universal reference locator (URL) for a World Wide Web (WWW) site with a GPS satellite visibility program on-line is *http://www.trimble.com/satviz*.

The control segment, under the United States Department of Defense's direction, oversees the building, launching, orbital positioning, and monitoring of the system. It provides two classes of GPS service: Standard Positioning Service (SPS) and Precise Positioning Service (PPS). Monitoring and ground control stations, located around the globe near the equator, constantly monitor the performance of each satellite and the constellation as a whole. A master control station updates the information component of the GPS signal with satellite ephemeris data and other messages to the users. This information is then decoded by the receiver and used in the positioning process. Of the two classes of service, the PPS is only available to authorised government users while the SPS is available for civilian use.

The user segment is comprised of all of the users making observations with GPS receivers. The civilian GPS user community has increased

Fig 1. A hand-held 12-channel GPS receiver.

dramatically in recent years, because of the emergence of low-cost portable GPS receivers (Figure 1 and Plate 23) and the ever-expanding areas of applications in which GPS has been found to be useful. Such applications include surveying, mapping, agriculture, navigation, and vehicle tracking (Figure 2; Plate 24). The civilian users of GPS greatly outnumber the military users.

Fig 2. Off-road navigation using GPS.

1.1.2 The limitations of GPS

Although in theory GPS can provide worldwide, 3-dimensional positions, 24 hours a day, in any type of weather, the system does have some limitations. First, there must be a (relatively) clear 'line of sight' between the receiver's antenna and several orbiting satellites. Anything shielding the antenna from a satellite can potentially weaken the satellite's signal to such a degree that it becomes too difficult to make reliable positioning. As a rule of thumb, an obstruction that can block sunlight can effectively block GPS signals. Buildings, trees, overpasses, and other obstructions that block the line of sight between the satellite and the observer (GPS antenna) all make it impossible to work with GPS. Urban areas are especially affected by these types of difficulties. Bouncing of the signal off nearby objects or the ground may create another problem called *multi-path interference*. Multi-path interference is caused by the inability of the receiver to distinguish between the signal coming directly from the satellite and the 'echo' signal that reaches the receiver indirectly. In areas that possess these type of characteristics, as in the urban core of a large city, inertial navigation techniques must be used to complement GPS positioning.

The receiver must receive signals from at least four satellites in order to be able to make reliable position measurements. In addition, this number of satellites must be in a favourable geometrical arrangement, relatively spread out. In areas with an open view of the sky, this will almost always be the case because of the way the satellite orbits were designed. The dilution of precision, or DOP, is a measure of the geometry of the GPS satellites. When the satellites are spaced well apart in the sky, the GPS position data will be more accurate than when the satellites are all in a straight line or grouped closely together.

An additional limitation of GPS is the DoD policy of selective availability (SA). This policy limits the full autonomous accuracy only to official government users. This policy, and methods used to ameliorate its effects, are described in section 1.1.6. A Presidential Directive of May 1996 has declared that the policy of SA will likely be discontinued within a period of more than four years and less than ten years. However, even if SA is eliminated, GPS will not be accurate enough for many mapping and GIS data capture applications.

1.1.3 How a GPS receiver calculates position

The position of a point is determined by measuring the distance from the GPS antenna to three satellites for a 2-dimensional reading (latitude and longitude), and at least four satellites for a 3-dimensional position (latitude, longitude, and altitude). The GPS receiver 'knows' where each of the satellites is at the instant in which the distance was measured. These distances will intersect at only one point, the position of the GPS antenna. How does the receiver 'know' the position of the satellites? Well, this information comes from the broadcast orbit data that are received when the GPS receiver is turned on. The GPS receiver performs the necessary mathematical calculations, then displays and/or stores the position, along with any other descriptive information entered by the operator from the keyboard.

The way in which a GPS receiver determines distances to the satellites depends on the type of GPS receiver. There are two broad classes: *code* based and *carrier phase* based.

1.1.4 Code-based receivers

Although less accurate than their carrier phase counterparts, code-based receivers have gained widespread appeal for applications such as GIS data capture. Their popularity stems mainly from their relatively low cost, portability, and ease of use.

Code-based receivers use the speed of light and the time interval that it takes for a signal to travel from the satellite to the receiver to compute the distance to the satellites. The time interval is determined by comparing the time in which a specific part of the coded signal left the satellite with the time it arrived at the antenna. The time interval is translated to a distance by multiplying the interval by the speed of light. Position fixes are made by the receiver roughly every second, and the GPS receivers designed for use in GIS data capture applications enable the user to store the positions in a file that can be downloaded to a computer for post-processing or analysis.

Under normal circumstances, autonomous position fixes made by code-based receivers would be accurate to within 25 metres. This is the specified accuracy for the SPS. The DoD, however, began imposing its selective availability policy in July 1992, and this limits position fix accuracy to within 100 metres. The purpose of selective availability is to deny potential hostile forces accurate positioning capabilities. Military P(Y)-code receivers are not affected by SA, and are not available to the general public. In order to ameliorate these limits in positioning accuracy, differential GPS (DGPS) techniques have been developed. DGPS enables the user to improve on the SPS, and to remove the effects of SA and some other sources of error. These differential correction techniques can produce positions generally accurate to within a few metres. If and when SA is eliminated the accuracy of GPS will be more accurate than SPS specifications, to about 10 to 15 metres. This is not good enough for most GIS data capture applications and differential GPS will still be required.

There are also now code-based receivers capable of sub-metre differential accuracy. Some sub-metre receivers require longer data collection times (up to 10 minutes). They perform best under very favourable satellite geometry, and with an unobstructed view of the sky. Some newer receivers can provide a sub-metre accurate position each second. These receivers cost more at the outset, but provide a good return on the added investment by facilitating substantially higher productivity.

It is very important that users of code-based receivers understand the positional accuracy limitations of the receiver. Because of SA, each coordinate viewed on a non-differential GPS receiver's display is only accurate to within 100 metres. This accuracy may be improved on by taking an average of 200 or so repeated position observations of the same point. The resulting accuracy would nevertheless still be below what many users would consider acceptable quality. In order to produce acceptable results, GPS data collected in the field *must* be differentially corrected either in real-time, or by post-processing of the data.

1.1.5 Carrier phase receivers

Carrier phase receivers, used extensively in geodetic control and precise survey applications, are capable of sub-centimetre differential accuracy. These receivers calculate distances to visible satellites by determining the number of whole wavelengths and the partial wavelength. Once the number of wavelengths is known, the range may be calculated by multiplying by the wavelength of the carrier signal. It is then a straightforward (but cumbersome) task to compute a baseline distance and azimuth between any pair of receivers operating simultaneously. With one receiver placed on a point

with precisely known latitude, longitude, and elevation, the calculated baseline can be used to determine the coordinates of the other point (Leick 1990).

The relative cost of the carrier phase receivers is high, but technological advances have made the dual frequency (using both L1 and L2) carrier phase receivers much more efficient than the single frequency (using L1 only) receivers that were state-of-the-art only a few years ago. Dual frequency receivers allow for correction of ionosphere delays, and are therefore inherently more accurate at longer baseline lengths over 50 km. With some of the most recent dual frequency receivers very precise measurements (± 1 cm) can be made in real-time. These receivers are used in machine control applications requiring a high degree of accuracy.

1.1.6 DGPS

Differential GPS can be employed to eliminate the error introduced by SA and other systematic errors. DGPS requires the existence of a base station, which is a GPS receiver collecting measurements at a known latitude, longitude, and elevation (Hurn 1993). The base station's antenna location must be known precisely. The base station may store measurements (for post-processed DGPS) or broadcast corrections over a radio frequency (for real-time DGPS), or both.

The assumption made with the base station concept is that errors affecting the measurements of a particular GPS receiver will equally affect other GPS receivers within a radius of several hundred kilometres. If the differences between the base station's known location and the base station's locations as calculated by GPS can be determined, those differences can be applied to data collected simultaneously by receivers in the field (Figure 3). These differences can be applied in real-time (especially applicable for accurate navigation) if the GPS receiver is linked to a radio receiver designed to receive the broadcast corrections. In some GIS mapping applications, these differences are applied in a post-processing step after the collected field data has been downloaded to a computer running a GPS processing software package. GPS processing software is typically integrated with GPS hardware and thus is provided by the receiver manufacturer. As a rule, post-processed DGPS is considered slightly more accurate than real-time DGPS.

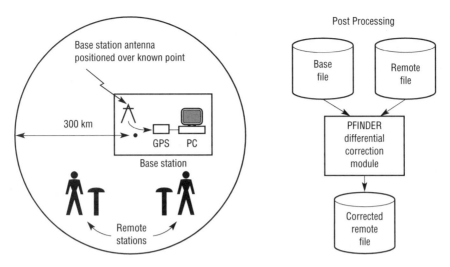

Fig 3. Use of base stations for precise global positioning.

1.1.7 Base stations – the source for reference data for DGPS

There are many permanent GPS base stations currently up and running in the USA that can provide the data necessary for post-processing differentially correcting positions to the users of code-based receivers over an electronic bulletin board.

In many parts of the world a real-time DGPS beacon system is operational. These stations are part of a large network of coastal and river valley stations. These beacons broadcast in the frequency range of 285 to 325 kHz. The range of many of these stations is 300 to 500 kilometres. These stations can provide differential accuracy in the 1-metre range, depending on distance from the station and the quality of the roving receiver.

Carrier phase base differential processing, for centimetre accuracy, does not have the range of code-based differential GPS. To obtain centimetre accuracy, it is now necessary to have the carrier base station within approximately 10 kilometres for real-time carrier differential (sometimes called real-time kinematic) operation. Extending the range of carrier differential beyond 10 kilometres is an active GPS research topic.

1.2 GIS data capture considerations

Several key issues need to be explored when considering whether GPS is an appropriate tool for capturing coordinate data for a GIS database. First and foremost is the need to determine the position accuracy requirements. If the data will be used for site-specific analysis that requires position accuracy to be within a metre, high-quality code-based differential GPS receivers will be necessary. If still higher accuracy is required, of the order of 10 centimetres or better – as for property boundary mapping – then carrier phase differential GPS techniques will be required.

Every GIS database must be referenced to a base map or base data layer, and the reference datum of the various data layers must be the same. Ideally, the database should be referenced to a very accurate base map. If, however, the base map is 1:125 000 or smaller, there could be problems when attempting to view the true spatial relationships between features digitised from such a map and features whose coordinates were captured with GPS. This can be a real problem if the GIS analyst decides to use a particular GIS data layer that was originally generated using small-scale base maps as a base to which all new generated data are referenced (see Weibel and Dutton, Chapter 10). The best way to avoid such incompatibility is to develop an accurate base data layer, based on geodetic control and photogrammetric mapping.

Map datum

Understanding the concept of a map datum is important if useful results are to be obtained from any GPS mapping exercise. A datum is a mathematical model of the Earth over some area. GPS, being a worldwide system, has a datum applicable over the whole earth. This is called the

471

World Geodetic System, 1984 (WGS-84). There are many local datums like European Datum 1950 (ED-50), North America Datum 1927 (NAD-27), and North America Datum 1983 (NAD-83) that have been used to make local maps. A particular point on the surface of the Earth will have different latitude and longitude coordinates depending on the reference datum. For GIS applications, it is recommended that all data be collected and displayed in the most up-to-date datum available. In the US this datum is NAD-83. Fortunately for users in most areas of the world, region-specific datums are quite similar to the GPS natural datum, WGS-84. It is important to use the same datum for all data layers if the data are going to be overlaid in a GIS.

2 USING GPS WITH GIS

There is a range of ways in which GPS is used alongside GIS. First, GPS may be used to identify or refine the geographical coordinates associated with satellite imagery. GPS is used to reduce distortions and to improve the positional accuracy of these images. When three or more distinctive points (the more the better) can be located both on a satellite image and on the ground, GPS receivers can be used to collect accurate geographical coordinates at these locations. The rest of the image can then be adjusted so that it provides a better match to the real-world coordinates. Second, GPS can be used in the ground truthing of satellite images. When a particular satellite image has a region of unusual or unrecognised reflectivity or backscatter, the coordinates of that region can be loaded into a GPS receiver. The GPS receiver will then enable the user to navigate directly to the area of interest so that the nature of the vegetation or the Earth's surface can be examined. With real-time differential GPS, users can even navigate directly to an individual 1m pixel. Third, GPS has developed into a cost effective tool for updating GIS or computer-aided design (CAD) systems. Using GPS to collect data is analogous to digitising a map by moving a mouse or digitising puck over a map. The users of GPS equipment simply move along the surface of the Earth and the geographical coordinates of where they went are computed and stored as a continuous series.

However, GPS technology is better suited to some data collection applications than others. For example, GPS is not particularly well suited to recording the corner stones of skyscrapers because the building itself blocks a large percentage of the sky, and hence the GPS signal. Traditional survey techniques would probably be more efficient. GPS is not necessarily the most efficient tool for evaluating subtle gradational changes in vegetation across a large region. Satellite imagery might be better suited to this task. However, GPS is an excellent tool for data collection in many environments where the user can generally see the sky and is able to get close to the objects to be mapped.

2.1 How is GPS used for GIS attribute collection?

The typical GPS-based data capture tool is a GPS receiver combined with a hand-held computer. These two components may be connected by a cable, or they may be combined as a single, integrated, hand-held unit. The hand-held computer is used for attribute entry and display of data to the user. There are several variations on this theme, including the use of pen computers, laptop computers, and small, hand-held units. The design goal is that the user can focus on observing and entering attribute data, since the position data 'just happens' automatically. Figure 4 illustrates how a bar-coded system may be used to record the locations of utility conduits.

Mapping roads is an important application. By driving the road once, users can record the location of the centreline and both kerbs by noting the offset between the line traversed by their GPS antenna and the linear feature of interest, such as a kerb (see Worboys, Chapter 59). Additionally, they can record a number of attributes such as name, width, condition, materials. The amount of attribute data that can be collected is limited only by how fast the operator (or accompanying colleague) can enter them while driving past. The concepts employed in road mapping can be extended to any type of linear or polygonal feature, such as shorelines, paths, forest fire boundaries, land usage zones, vegetation boundaries, powerlines, and utility right-of-ways: see Plate 24. Presently there are many consulting firms that use GPS to record the attributes and location of thousands of kilometres of road every month.

Recording the location and attributes of point-like features is another very popular application of GPS. To record a point-like feature with GPS, the user will typically walk or drive up to the object to be mapped (such as a pole, storm drain, or soil sample site) and then place the GPS antenna on that location of interest. While the user is busy for a few moments

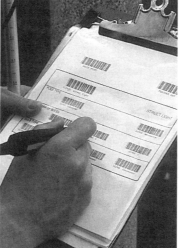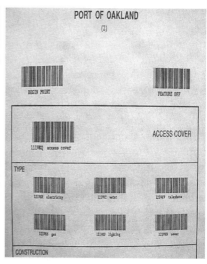

Fig 4. Use of GPS in utilities mapping.

observing and entering attribute data, the GPS receiver is automatically recording position data.

Unlike the hand digitising of a paper map, there is no 'scale' directly associated with this form of 'GPS digitising'. If the user is doing detailed, high accuracy work, it might make sense to record a storm drain as a small polygon (remember the best GPS receivers are accurate to better than 1-centimetre). On the other hand, if the data will be used to populate a crude map, the user might choose to define an entire airport as a point feature. Users are free to define their features as required to best fit their application. GPS is readily used to record the attributes and location of street furniture (such as lamp posts, telephone poles, storm drains, drainage aperture covers, valves, etc.). It is not uncommon that some users see their productivity increase from a typical 100 to 200 points/day before GPS to over 1000 points per day after GPS. The increased productivity allowed by GPS can be stunning.

The use of GPS can be extended to mapping objects which are difficult or dangerous to occupy. By combining the use of a hand-held laser rangefinder with a GPS, users are able to record the location of an object even if they cannot physically occupy that location: see Plate 25. The user can remain in a location that is either convenient or more conducive to good GPS signals, then measure the distance and direction (an offset) to the object of interest with a laser rangefinder. In some GPS, the laser can be connected electronically so that the laser

measurements are transmitted directly to the GPS data collection system; alternatively, the user could enter the offset information by hand. The location of the GPS antenna and the offset measured by the laser are automatically combined to compute an accurate location for the object of interest.

There are many other types of electronic sensing devices that can be used in conjunction with a GPS receiver. Typically when sensor data are coupled with GPS positions, the user has the intention of later presenting these data in the form of a thematic map. The contours of these maps may represent information as diverse as radiation levels (isorad maps: Runyon et al 1994), temperature (isotherm maps), water depth (bathometry: Chisholm and Lenthall 1991), signal strength for cellular telephones, or even biological density.

Some features of modern GPS data collection systems make them extremely effective tools for attribute collection. One common (and quite effective) way that GPS users prepare for data collection is to predefine a customised list of the features and attributes that are to be collected in the field. This list is then transferred into the GPS data collection system so that, in the field, the user will be prompted by the GPS for the appropriate attributes at each location. For example, if the user approaches a telephone pole in the field, the first step is to select the item 'telephone pole' from the screen of the GPS. After having selected the feature name, the GPS would then automatically prompt the user for

473

whichever attribute details are appropriate to telephone poles. These attributes (such as material, condition, identification number, etc.) are taken from a list that was created by the user, are displayed at the appropriate time and in the appropriate sequence, and match the structure of the target database.

When attribute data are collected, they should always be stored in the appropriate form. Alphanumeric and numeric data are stored in the appropriate field types, and numeric values are validated so that they are within predefined maximum and minimum values. Character strings are automatically checked so that they do not exceed the maximum allowable field length of the target database. Whenever appropriate, precise date and time attributes are generated by the GPS, and manually entered date and time values are checked so that they fall within valid ranges. When the attribute is an image, perhaps from an attached digital camera, the data collector is able to store the filename of the image as an attribute value. The attribute values that are collected may be edited at any time after they have been collected – either in the field or during subsequent processing. Collectively, features like the ones described above will greatly reduce field data collection errors.

GPS also feature pause and resume functions which, although apparently simple, are nevertheless invaluable in the field (Gilbert 1994a). Imagine what the path you digitised would look like if you could not momentarily pause the data collection when your hat blew off! When recording roads, many users find that they wish to interrupt data collection every time they stop at a traffic signal or a gate. A problem occurs when a GPS receiver is stationary and continues to record positions once per second, since the resulting road features contain periodic, puzzling clumps of tangled vertices in the roadway.

The ability to repeat previous attribute values is another function which is very simple, yet is often crucial to efficient data collection. After entering 200 attributes for the previous utility pole you will surely appreciate the ability to press one button and have most or all of your attributes transferred to the next pole. The repeat function can also be very useful when recording linear features that contain dynamically changing attribute values. For example, on approaching the end of a block of buildings, a user may wish to end one segment of the road and immediately begin another segment that shares a common node. A well-designed GPS data collector

will allow the user to terminate the first segment and initiate the second while retaining most, or all, of the same attribute values for both nodes – with as little as one keystroke.

Many GPS receivers are unable to collect accurate coordinates effectively while they are moving. This is not a limitation of the satellite GPS themselves, as the best GPS receivers can provide centimetre level results while moving at high speeds. This limitation is simply a by-product of some of the receiver designs and software products on the market. It is sometimes desirable to record point features while moving. One very powerful data collection function allows users to store point features accurately while driving or flying over the point of interest. If it is not possible to travel directly over the point of interest, some GPS will even allow the user to define an offset distance from the GPS antenna to the object of interest. Typically, the direction of the offset is defined as being either to the left or the right, and orthogonal to the direction of travel. Depending on the variability of the attribute data, such a data collection system can allow a skilled user to record thousands of point features per day. It is important to recognise, however, that such rapid recording is often less accurate than even a one second static occupation – most commonly because of user errors in pressing the button at the precise moment the feature is being passed.

2.2 Exporting GPS data to a GIS, CAD system, or database

The final stage of the process of GPS-based data collection for GIS is to transfer the data from the field device to the target database. In the interests of preserving valuable memory, most GPS store data internally in their own proprietary formats. This data transfer is most often accomplished by running a translation program that will convert the data from the compact internal storage format to the database interchange format of the user's choice.

At the most basic level, this task can be accomplished in a very simplistic manner – that is, simply dumping the coordinates and attributes in a comma delimited file. However, most users prefer to employ a translation program that will allow them to modify slightly the output data, so that it matches the characteristics of the target database as closely as possible (Gilbert 1994b). Some of the more important considerations of a translation program are as follows.

2.2.1 Creation of metadata

A useful feature of a data translation program is the ability to generate metadata (Guptill, Chapter 49; Goodchild and Longley, Chapter 40). The resulting output file(s) will contain not only the attribute values that were entered in the field by the user, but also a variety of additional attribute data that are generated by the translation program (Gilbert 1996). Some of these generated attributes are common to the entire dataset, such as the receiver type used to collect the data or the name of the original data file. However, the most useful of the generated attributes are specific to the individual features. Several examples of GPS metadata are described in the following paragraphs.

Quality information about the accuracy of the GPS coordinates is very valuable (see Fisher, Chapter 13). This may take the form of a single quantitative accuracy estimate, or it may be presented as a variety of accuracy indicators that can be used individually or together to infer data quality (Veregin, Chapter 12). Although a single accuracy estimate is simple to use, it may be less reliable. The user must develop an understanding of whether any particular manufacturer's algorithm is reliable enough. Additionally, the user often has no way to tell how the GPS accuracy value was computed and whether it represents a probability of 99 per cent, 95 per cent, 68 per cent, or even as little as 50 per cent. There is also the uncertainty as to whether the reported value refers to horizontal accuracy or 3-dimensional accuracy. It is not uncommon for the horizontal and vertical accuracy of a GPS coordinate to differ by a factor of two. There is no industry standard on to how the GPS report spatial accuracy (see Smith and Rhind, Chapter 47, for a discussion of industry spatial data standards).

Alternatively, the translation software may report a variety of accuracy indicators that can be used individually or together to infer data quality. Such indicators may include occupation time, number of satellites used, maximum dilution of precision, quantity of GPS data collected, or the type of differential processing that was performed, if any. Although this is more complex, it does provide the user with greater understanding and flexibility. When these indicators are available individually as metadata, the users are free to establish whatever data quality rules they desire for their applications. Some agencies will define multiple quality assurance guidelines that vary depending upon the intended use of the data. Storing multiple accuracy indicators has the added benefit that they can be used to verify that the field operator adhered to the appropriate data collection procedures.

2.2.2 Creation of macro files

The transformation of GPS data to an appropriate format can be greatly enhanced by the automatic creation of macro or batch files. While the data are being translated to the desired format, some translation programs will read and compute the information required to create a set of command scripts. These scripts are unique to the translated data and are used to complete the import process quickly and easily. They can automatically create new layers and tables (or append existing ones), automatically associate attribute values with the coordinates, or even create additional data such as labels positioned at the centroid of a polygon feature.

2.2.3 Geodetic datum transformation

As stated in section 1.2, the native geodetic datum of GPS data is WGS-84. Programs that translate GPS data into GIS or CAD compatible formats are usually able to transform the coordinates to the projection, coordinate system, and datum of the user's choice. Most of these translation utilities include a long list of datums and coordinate systems. The more sophisticated programs will also allow the user to create a customised coordinate system.

2.2.4 Configurable ASCII data translation

There are many GIS and CAD compatible import formats. Despite the abundance of formats, GPS manufacturers have discovered that some of the GIS community utilise home-grown, customised import formats. Therefore, even a translation program that supports the dozen most popular formats will seem incomplete to some users. The mainstream GPS data collection systems are all equipped with powerful and flexible configurable ASCII export programs. Such programs give users the ability to translate their GPS data into customised formats that the GPS manufacturers could never have anticipated.

SUMMARY

Depending on the particular equipment utilised and the techniques used, GPS is capable of a wide range of accuracy, from tens of metres to centimetres. Differential GPS techniques are required to obtain the accuracy required by many GIS data capture applications. Differential GPS requires the use of a base station at a known location to remove systematic errors from the GPS signal. Modern GPS/GIS data collection devices have the capability to collect a wide range of GIS attribute data in addition to the position of the feature of interest, including point, line, and area features. GPS can be used in many GIS data capture applications, including applications as diverse as road centreline mapping, utility pole mapping, and wetland boundary mapping. It is nevertheless important to understand the limitations of GPS to derive full benefits from the technology.

References

Chisholm G, Lenthall P 1991 Combining GPS and echo sounders to monitor river siltation. *GISWorld* 5(2): 46–8

Gilbert C 1994a Attribute collection with GPS part 1: Variety is the spice of attributes. *Mapping Awareness* 8(9): 28–30

Gilbert C 1994b Attribute collection with GPS part 2: Assessing ease of use in the field. *Mapping Awareness* 8(10): 30–2

Gilbert C 1996 The software with your GPS Part 1 – data export. *Mapping Awareness* 10(8): 32–4

Hurn J 1989 *GPS: A guide to the next utility*. Sunnyvale, Trimble Navigation Ltd

Hurn J 1993 *Differential GPS explained*. Sunnyvale, Trimble Navigation Ltd

Leick A 1990 *GPS satellite surveying*. New York, John Wiley & Sons Inc.

Runyon T, Hammitt R, Lindquist R 1994 Buried danger: integrating GIS and GPS to identify radiologically contaminated sites. *Geo Info Systems* 8(4):28–36

Wells D 1987 *Guide to GPS positioning*. Fredericton, Canadian GPS Associates

Introduction

THE EDITORS

In this final Technical Issues section, attention turns to the algorithms that allow GIS to transform and link data. Section 2(a) dealt with trends in the computing industry, in computer architectures, programming environments, and communications networks. All of these are likely to influence GIS profoundly over the coming decade. The next section looked at spatial databases, where the general solutions offered by the database industry intersect with the special needs of spatial databases for high performance, multi-dimensional search, and complex structures. The third section dealt with selected technical aspects of spatial data collection. Having addressed architectures, programming environments and data, this final set of chapters examines technical issues in the processing of data in GIS, and focuses on areas where algorithms are particularly challenging and complex, and the subject of ongoing research.

Several disciplines are contributing actively to research in advanced spatial data processing algorithms. The field of computational geometry addresses specifically the computer processing of geometric objects, and has widespread applications in GIS. For example, a fundamental algorithm in computational geometry detects whether two line segments intersect, and if so, where; it finds applications in such GIS processes as polygon overlay. GIS adds a particular context to computational geometry – one of its most distinguishing features is that position can never be exact, and must always be subject to a prescribed tolerance. GIS is an important application field for 'finite precision geometry', the field that attempts to formalise and theorise about such fuzzy spatial problems.

Another distinguishing feature of many GIS applications is the importance of performance, and the ability of algorithms to scale over a very wide range of problem sizes. The field of algorithmic complexity deals formally with performance, and the effects of problem size, and is widely applied in research on advanced GIS algorithms (see De Floriani and Magillo, Chapter 38, for example).

Image processing is yet another cognate field, with many applications to geographical data and GIS. As we saw in the last section, GIS colours these applications with particular characteristics. The surface of the Earth is not flat, unlike a photographic plate, and algorithms are needed: to register images to it accurately; to register images of the same feature to each other; to 'rubber sheet' images for geometric correction; or to change from one projection to another (see Dowman, Chapter 31). Another distinguishing characteristic of image processing when applied to geographical data is the widespread absence of 'truth' (see also Barnsley, Chapter 32) – when imaging the human body, for example, it is possible to regard the label 'liver' as 'true' of a certain part of the image; but suitably precise geographical analogies are much harder to come by – the label 'lake' is confused by many problems of definition (when is a lake a swamp, or a reservoir? See Mark, Chapter 7).

The ability to link data is often cited as the distinguishing feature of GIS. Location on the Earth's surface forms a convenient common key between otherwise disparate datasets and forms of information, allowing data to be linked across the boundaries of disciplines, departments, and agencies. When events occur in the same place, or near to each other, it is easy to believe that they also influence each other, and that both need to be taken into account in making decisions. Contemporary thinking on environmental management urges us to think of all things on the Earth's surface as connected and inter-related – in Tobler's 'first law of geography', 'all things are related, but nearby things are more related than distant things' (Tobler 1970).

Several different forms of linkage can be identified in GIS. Consider two datasets, A and B, covering the same geographical area, and imported to a GIS from different sources. At one extreme, the information provided by B is entirely distinct ('independent' or 'orthogonal' in a statistical sense), and both A and B are necessary in some application. The application will likely require that A and B be overlaid, and the ability to do so will depend on the formats of the two datasets – most GIS will require that they be both raster, or both vector; if both raster, that the rasters be congruent. In another case, the common key between A and B may be a feature identifier, rather than a geographical location. Such cases occur when B provides tabular information to be added to the geographical features in A as additional attributes. Yet another case occurs when B is a source of selective updates for the information in A – for example, B might contain more accurate coordinates for the features shared between it and A. When no common feature identifier is available, the features in B must be matched geometrically to those in A, a process that has been termed 'conflation', and is itself the subject of intensive current research.

Some forms of data transformation and linkage in GIS are straightforward, and do not justify particular attention. Others are made sufficiently special by their geographical context to have emerged as strong subfields for research and development within GIS. In this section the editors have selected several of these, while recognising that the set is by no means exhaustive, and may not survive the test of time – five or ten years from now research and development attention may have turned to quite different problems.

The section contains seven chapters. In the first, Lubos Mitas and Helena Mitasova review the state of the art in spatial interpolation, a vital component in the GIS arsenal because it provides estimates of the value of a variable z at locations (x,y) where it has not been measured. Spatial interpolation is essential in resampling, when data must be shifted from one raster to another; in transformation between representations, such as from a grid to contours; or in dealing with the problems caused by incompatible reporting zones. Mitas and Mitasova review the methods currently available, discuss the bases on which they can be evaluated, and review the applications of the methods in GIS.

Data linkage across GIS layers provides the theme for Ronald Eastman's contribution. Multi-criteria evaluation is concerned with the allocation of land to suit a specific objective on the basis of a variety of attributes that the selected areas should possess. This implies an apparently straightforward

GIS-based overlay exercise, yet this process is complicated on the one hand by differences in data structure (raster versus vector) and, on the other, by ambiguities in the ways in which criteria should be standardised and aggregated into a single summary coverage. Eastman reviews these problems and suggests the use of fuzzy measures as a means of reconciling and developing current practice. In this way, the harshness of using Boolean operators to identify intersection and union of data layers is replaced with an approach which also provides improved standardisation of criteria and better evaluation of decision risk (the likelihood that the decision made will be incorrect).

One aspect that distinguishes GIS from other forms of spatial data processing, notably computer-assisted design (CAD), is its emphasis on representing fields, or variables having a single value at every location on the Earth's surface. Examples of fields are elevation, mean rainfall, or soil type. Because a field is continuous by definition, it must somehow be rendered discrete in order to be represented in the finite space of a digital store. Methods for discretising fields have been discussed at many points in this volume. Among them, two achieve their objective by dividing a plane surface into regular or irregular tiles, forming a 'tessellation'. The mathematics and statistics of tessellations, and their representation and processing in GIS, are important topics for research. The third chapter in this section, by Barry Boots, reviews the state of the art. Some of this research goes well beyond the current state of GIS implementation, particularly in the area of weighted tessellations, but it is easy to see how such methods could be implemented and applied.

GIS has a long history of successful implementation of digital models of the Earth's terrain, collectively known as digital terrain models (DTMs). They include triangulated irregular networks (TINs); and the commonest form of DTM, the digital elevation model (DEM), a rectangular array of spot elevations. DTMs are available in one form or another for much of the Earth's terrestrial surface and for the ocean floor, although the sampling density and conditions of availability vary enormously. One major factor driving the development of DTM technology is its importance in military applications, particularly missile guidance systems ('cruise' missiles navigate largely by recognising the geometric form of the terrain under them).

478

Many sciences, hydrology and geomorphology in particular, have an interest in the form of the Earth's surface, and its influence on the environment. The availability of DTMs, and the facilities in GIS for processing them, have led to an explosion of research on DTM analysis techniques. GIS is now a very significant tool in these fields, and DTMs are also useful in such practical applications as transmission tower location. The fourth and fifth chapters in this section discuss recent DTM research from two perspectives. Lawrence Band reviews the importance of DTMs and related datasets in hydrography and the analysis of landforms; while Leila De Floriani and Paola Magillo discuss representations, and associated algorithms, for the transformation of DTM data into useful information on intervisibility.

The last two chapters in the section move somewhat away from this intensive discussion of theory and algorithms. In Chapter 39, Jorge Nelson Neves and Antonio Câmara look at the role of GIS in the expanding field of virtual reality, or Virtual Environments (VEs). VEs are clearly an important area of application for GIS, particularly if the environment being simulated is in any sense related to the real, geographical world; and even totally artificial environments must be constrained by certain characteristics of the real world if they are to be convincing. VEs require many standard GIS techniques, as well as more generic techniques of image processing and visualisation; and also require much higher performance than traditional GIS applications.

In the final chapter, Michael Goodchild and Paul Longley discuss the issues encountered in using GIS as a data linkage technology. These range from the technical issues of accuracy, compatible data formats and data models, and rules for conflation, to the capabilities of contemporary communication networks for supporting data search and sharing, to institutional issues that are beyond the scope of this section.

Reference

Tobler W R 1970 A computer movie simulating urban growth in the Detroit region. *Economic Geography* 46: 234–40

34

Spatial interpolation

L MITAS AND H MITASOVA

This chapter formulates the problem of spatial interpolation from scattered data as a method for prediction and representation of multivariate fields. The role and specific issues of interpolation for GIS applications are discussed and methods based on locality, geostatistical, and variational concepts are described. Properties of interpolation methods are illustrated by examples of 2-dimensional, 3-dimensional, and 4-dimensional interpolations of elevation, precipitation, and chemical concentrations data. Future directions focus on a robust data analysis with automatic choice of spatially variable interpolation parameters, and model- or process-based interpolation.

1 INTRODUCTION

Spatial and spatio-temporal distributions of both physical and socioeconomic phenomena can be approximated by functions depending on location in a multi-dimensional space, as multivariate scalar, vector, or tensor fields. Typical examples are elevations, climatic phenomena, soil properties, population densities, fluxes of matter, etc. While most of these phenomena are characterised by measured or digitised point data, often irregularly distributed in space and time, visualisation, analysis, and modelling within a GIS are usually based on a raster representation. Moreover, the phenomena can be measured using various methods (remote sensing, site sampling, etc.) leading to heterogeneous datasets with different digital representations and resolutions which need to be combined to create a single spatial model of the phenomenon under study.

Many interpolation and approximation methods were developed to predict values of spatial phenomena in unsampled locations (for reviews see Burrough 1986; Franke 1982a; Franke and Nielson 1991; Lam 1983; McCullagh 1988; Watson 1992; and for a discussion of Kriging and error, see Heuvelink, Chapter 14). In GIS applications, these methods have been designed to support transformations between different discrete and continuous representations of spatial and spatio-temporal fields, typically to transform irregular point or line data to raster representation, or to resample between different raster resolutions.

2 PROBLEM FORMULATION AND CRITERIA FOR SOLUTIONS

The general formulation of the spatial interpolation problem can be defined as follows:

Given the N values of a studied phenomenon z_j, $j = 1, \dots, N$ measured at discrete points $\mathbf{r}_j = (x_j^{[1]}, x_j^{[2]}, \dots, x_j^{[d]})$, $j = 1, \dots, N$ within a certain region of a d-dimensional space, find a d-variate function $F(\mathbf{r})$ which passes through the given points, that means, fulfils the condition

$$F(\mathbf{r}_j) = z_j, \quad j = 1, \dots, N \tag{1}$$

Because there exist an infinite number of functions which fulfil this requirement, additional conditions have to be imposed, defining the character of various interpolation techniques. Typical examples are conditions based on geostatistical concepts (Kriging), locality (nearest neighbour and finite element methods), smoothness and tension (splines), or ad hoc functional forms (polynomials, multi-quadrics). Choice of the additional condition depends on the character of the modelled phenomenon and the type of application.

Finding appropriate interpolation methods for GIS applications poses several challenges. The modelled fields are usually very complex, data are spatially heterogeneous and often based on far from optimal sampling, and significant noise or discontinuities can be present (e.g. see De Floriani and Magillo, Chapter 38). In addition, datasets can be very large ($N \approx 10^3$–10^6), originating from various sources with different accuracies. Reliable interpolation tools, suitable for GIS applications, should therefore satisfy several important demands: accuracy and predictive power, robustness and flexibility in describing various types of phenomena, smoothing for noisy data, d-dimensional formulation, direct estimation of derivatives (gradients, curvatures), applicability to large datasets, computational efficiency, and ease of use.

Currently, it is difficult to find a method which fulfils all of the above-mentioned requirements for a wide range of georeferenced data. Therefore, the selection of an adequate method with appropriate parameters for a particular application is crucial. Different methods can produce quite different spatial representations (Plate 26 (a)–(f)) and in-depth knowledge of the phenomenon is needed to evaluate which one is the closest to reality. The use of an unsuitable method or inappropriate parameters can result in a distorted model of spatial distribution, leading to potentially wrong decisions based on misleading spatial information. An inappropriate interpolation can have even more profound impact if the result is used as an input for simulations, where a small error or distortion can cause models to produce false spatial patterns (Mitasova et al 1996), as illustrated in section 4.2.2 (Plate 27(b)): see also Heuvelink (Chapter 14) for a discussion of error propagation. Quantitative evaluation of interpolation predictive capabilities, for example by cross-validation, is often not sufficient for the selection of an appropriate interpolation method, as the preservation of geometrical properties is in some cases more important than actual accuracy (see Hutchinson and Gallant, Chapter 9). Advanced visualisation and analysis of slope, aspect, and curvature is helpful in detecting geometrical distortions (Brown et al 1995; Mitas et al 1997; Mitasova et al 1995; Nielson 1993; Wood and Fisher 1993).

3 METHODS

In recent years, GIS capabilities for spatial interpolation have improved by integration of advanced methods within GIS, as well as by linking GIS to systems designed for modelling, analysis, and visualisation of continuous fields. Because it is impossible to cover all or even most of the existing interpolation techniques, only methods which are often used in connection with GIS or have the potential to be widely used for GIS applications are included, and references are given to literature for more detailed descriptions.

3.1 Local neighbourhood approach

Local methods are based on the assumption that each point influences the resulting surface only up to a certain finite distance. Values at different unsampled points are computed by functions with different parameters, and the condition of continuity between these functions is defined only for some approaches. The method of point selection used for the computation of the interpolating function differs among the various methods and their concrete implementations.

3.1.1 Inverse distance weighted interpolation (IDW)

This is one of the simplest and most readily available methods. It is based on an assumption that the value at an unsampled point can be approximated as a weighted average of values at points within a certain cut-off distance, or from a given number m of the closest points (typically 10 to 30). Weights are usually inversely proportional to a power of distance (Burrough 1986; Watson 1992) which, at an unsampled location $\mathbf{r}$, leads to an estimator

$$F(\mathbf{r}) = \sum_{i=1}^{m} w_i\, z(\mathbf{r}_i) = \frac{(\sum_{i=1}^{m} z(\mathbf{r}_i) \,/\, |\mathbf{r} - \mathbf{r}_i|^p}{\sum_{j=1}^{m} 1/|\mathbf{r} - \mathbf{r}_j|^p} \qquad (2)$$

where p is a parameter (typically $p=2$; for more details on the influence of this parameter see Watson 1992). While this basic method is easy to implement and is available in almost any GIS, it has some well-known shortcomings that limit its practical applications (Burrough 1986; Franke and Nielson 1991; Watson 1992). The method often does not reproduce the local shape implied by data and produces local extrema at the data points (Plate 26 (c)). A number of enhancements has been suggested, leading to a class of multivariate blended IDW surfaces and volumes (Franke and Nielson 1991; Tobler and Kennedy 1985; Watson 1992). However, most of these modifications are not implemented within GIS.

3.1.2 Natural neighbour interpolation

This uses a weighted average of local data based on the concept of natural neighbour coordinates derived from Thiessen polygons (Boots, Chapter 36) for the bivariate, and Thiessen polyhedra for the trivariate case (Watson 1992). The value in an unsampled location is computed as a weighted average of the nearest neighbour values with weights dependent on areas or volumes rather than distances. The number of given points used for the computation at each unsampled point is variable, dependent on the spatial configuration of data points. Natural neighbour linear interpolation leads to a rubber-sheet character of the resulting surface. The addition of blended gradient information (derived from data points by local 'pre-interpolation') allows the surface to be made smooth everywhere with tautness, analogous to tension, tuned according to the character of the modelled phenomenon. The value of tautness is controlled by two empirically selected parameters which modify the shape of the blending function. The result is a surface with smoothly changing gradients and passing through data points, blended from natural neighbour local trends, with local tunable tautness, and with the capability to calculate derivatives and integrals. The method has been used typically for topographic, bathymetric, geophysical, and soil data (Laslett et al 1987; McCauley and Engel 1997; Watson and Philip 1987).

3.1.3 Interpolation based on a triangulated irregular network (TIN)

This uses a triangular tessellation of the given point data (Boots, Chapter 36) to derive a bivariate function for each triangle which is then used to estimate the values at unsampled locations. Linear interpolation uses planar facets fitted to each triangle (Akima 1978; Krcho 1970; Plate 26(b)). Non-linear blended functions (e.g. polynomials) use additional continuity conditions in first-order, or both first- and second-order derivatives (C^1, C^2), ensuring smooth connection of triangles and differentiability of the resulting surface (Akima 1978; McCullagh 1988; Nielson 1983; Renka and Cline 1984). Because of their local nature, the methods are usually fast, with an easy incorporation of discontinuities and structural features. Appropriate triangulation respecting the surface geometry is crucial (Hutchinson and Gallant, Chapter 9; Weibel and Dutton, Chapter 10; Weibel and Heller 1991). Extension to d-dimensional

problems is more complex than for the distance-based methods (Nielson 1993).

While a TIN provides an effective representation of surfaces useful for various applications, such as dynamic visualisation and visibility analyses (De Floriani and Magillo, Chapter 38), interpolation based on a TIN, especially the simplest, most common linear version, belongs among the least accurate methods (Franke 1982a; Nielson 1993; Renka and Cline 1984).

3.1.4 Rectangle-based methods

These are analogons to a TIN and involve fitting blended polynomial functions to regular or irregular rectangles, such as Hermite, Bezier, or B-spline patches (Chui 1988; Watson 1992), often with locally tunable tension. These methods were developed for computer-aided design and computer graphics and are not very common in GIS applications.

3.2 The geostatistical approach

The principles of geostatistics and interpolation by Kriging are described in a large body of literature (e.g. Burrough 1986; Cressie 1993; Deutsch and Journel 1992; Isaaks and Srivastava 1989; Journel and Huijbregts 1978; Matheron 1971; Oliver and Webster 1990); therefore only the basic notions are outlined here.

Kriging is based on a concept of random functions: the surface or volume is assumed to be one realisation of a random function with a certain spatial covariance (Journel and Huijbregts 1978; Matheron 1971). Using the given data $z(\mathbf{r}_i)$ and an assumption of stationarity one can estimate a semivariogram $\gamma(\mathbf{h})$ defined as

$$\gamma(\mathbf{h}) = \frac{1}{2} Var\left[\{ z(\mathbf{r} + \mathbf{h}) - z(\mathbf{r})\}\right] \approx \frac{1}{(2N_h)} \sum_{(ij)}^{Nh} [z(\mathbf{r}_i) - z(\mathbf{r}_j)]^2 \quad (3)$$

which is related to the spatial covariance $C(\mathbf{h})$ as

$$\gamma(\mathbf{h}) = C(0) - C(\mathbf{h}) \quad (4)$$

where $C(0)$ is the semivariogram value at infinity (sill). The summation in Equation (3) runs over the number N_h of pairs of points which are separated by the vector $\mathbf{h}$ within a small tolerance $\triangle\mathbf{h}$ (size of a histogram bin). For isotropic data, the semivariogram can be simplified into a radial function dependent on $|\mathbf{h}|$. The Kriging literature provides a choice of functions which can be used as theoretical semivariograms (spherical, exponential,

Gaussian, Bessel etc.: Cressie 1993). The parameters of these functions are then optimised for the best fit of the experimental semivariogram.

The interpolated surface is then constructed using statistical conditions of unbiasedness and minimum variance. In its dual form (Hutchinson and Gessler 1993; Matheron 1971) the universal Kriging interpolation function can be written as

$$F(\mathbf{r}) = T(\mathbf{r}) + \sum_{j=1}^{N} \lambda_j \, C(\mathbf{r} - \mathbf{r}_j) \qquad (5)$$

where $T(\mathbf{r})$ represents its non-random component (drift) expressed as a linear combination of low-order monomials. The monomial and $\{\lambda_j\}$ coefficients are found by solving a system of linear equations (Hutchinson and Gessler 1993).

In general, Kriging predicts values at points and blocks in d-dimensional space and enables incorporation of anisotropy. Various extensions enhance its flexibility and range of applicability (Cressie 1993; Deutsch and Journel 1992). Co-Kriging includes information about correlations of two or more attributes to improve the quality of interpolation (Myers 1984), while disjunctive Kriging is used for applications where the probability that the measured values exceed a certain threshold is of interest (Rivoirard 1994). For cases in which the assumption of stationarity is deemed not to be valid, zonal Kriging can be used (Burrough 1986; Wingle and Poeter 1996). Approaches for spatio-temporal Kriging reflect the different behaviour of the modelled phenomenon in the time dimension. Time is treated either as an additional dimension with geometric or zonal anisotropy, or as a combination of the space and time correlation functions with a space-time stationarity hypothesis (Bogaert 1996; Rouhani and Myers 1990).

Recent applications of geostatistics have de-emphasised the use of Kriging as an interpolation and mapping tool while shifting the focus towards models of uncertainty that depend on the data values in addition to the data configuration (Armstrong and Dowd 1993; Deutsch and Journel 1992; Englund 1993; Journel 1996; Rogowski 1996; Yarrington 1996). A stochastic technique of conditional simulation is used to generate alternative, equally probable realisations of a surface, reproducing both data and the estimated covariance. From such a set of statistical samples one can estimate a spatially-dependent picture of the uncertainty which is inherent in the data.

The main strengths of Kriging are in the statistical quality of its predictions (e.g. unbiasedness) and in the ability to predict the spatial distribution of uncertainty. It is often used in the mining and petroleum industries, geochemistry, geology, soil science and ecology, where its statistical properties are of great value (Burrough 1991; Cressie 1993; Isaaks and Srivastava 1989; Oliver and Webster 1990). It has been less successful for applications where local geometry and smoothness are the key issues and other methods prove to be competitive or even better (Deutsch and Journel 1992; Hardy 1990).

3.3 The variational approach

The variational approach to interpolation and approximation is based on the assumption that the interpolation function should pass through (or close to) the data points and, at the same time, should be as smooth as possible. These two requirements are combined into a single condition of minimising the sum of the deviations from the measured points and the smoothness seminorm of the spline function:

$$\sum_{j=1}^{N} |z_j - F(\mathbf{r}_j)|^2 \, w_j + w_0 I(F) = minimum \qquad (6)$$

where w_j, w_0 are positive weights and $I(F)$ denotes the smoothness seminorm (Table 1). The solution of Equation (6) can be expressed as a sum of two components (Talmi and Gilat 1977; Wahba 1990):

$$F(\mathbf{r}) = T(\mathbf{r}) + \sum_{j=1}^{N} \lambda_j \, R(\mathbf{r}, \mathbf{r}_j) \qquad (7)$$

where $T(\mathbf{r})$ is a 'trend' function and $R(\mathbf{r}, \mathbf{r}_j)$ is a basis function which has a form dependent on the choice of $I(F)$. A bivariate smoothness seminorm with squares of second derivatives (Table 1) leads to a *thin plate spline (TPS)* function (Duchon 1976; Harder and Desmarais 1972). The TPS function minimises the surface curvature and imitates a steel sheet forced to pass through the data points: the equilibrium shape of the sheet minimises the bending energy which is closely related to the surface curvature. There are at least two deficiencies of the TPS function: (1) the plate stiffness causes the function to overshoot in regions where data create large gradients; and (2) the second order derivatives diverge in the data points, causing difficulties in surface geometry analysis.

Table 1 Examples of bivariate spline functions, their corresponding smoothness seminorms and Euler–Lagrange equations.

Method	I(F)	Euler–Lagrange Eq.
Membrane	$\int [F_x^2 + F_y^2]\, d\mathbf{r}$	harmonic
Minimum curvature[a]	$\int [F_{xx}^2 + F_{yy}^2]\, d\mathbf{r}$	biharmonic modified
Thin plate spline[b]	$\int [F_{xx}^2 + F_{yy}^2 + 2F_{xy}^2]\, d\mathbf{r}$	biharmonic
Thin plate spline+tension[c]	$\int [\varphi^2 [F_x^2 + F_y^2] + [F_{xx}^2 + ...]\, d\mathbf{r}$	harmonic+biharmonic
Regular thin plate spline[c]	$\int [F_{xx}^2 + ...] + \tau^2 [F_{xxx}^2 + ...]\, d\mathbf{r}$	biharmonic+6th-order
Regular spline with tension[d]	$\sum_{mn} c_{mn}(\varphi) \int [F_{x\,y}^{n\,m}]^2\, d\mathbf{r}$	all even orders

[a] Briggs 1974, Duchon 1975, Hutchinson and Bischof 1983, Wahba 1990
[b] Franke 1985, Hutchinson 1989
[c] Mitas and Mitasova 1988
[d] Mitasova et al 1995; Mitas and Mitasova 1997

The problem of overshoots can be eliminated by adding the first order derivatives into the seminorm *I*(*F*), leading to a *TPS with tension* (Franke 1985; Hutchinson 1989; Mitas and Mitasova 1988). Change of the tension tunes the surface from a stiff plate into an elastic membrane (Plate 30 (a), (b), and (c); Mitas et al 1997). In the limit of an infinite tension the surface resembles a rubber sheet with cusps at the data points. The analytical properties of TPS can be improved by adding higher order derivatives into *I*(*F*), leading to a function with regular second- and possibly higher-order derivatives (Mitas and Mitasova 1988).

To synthesise the desired properties into a single function the *regularised spline with tension (RST)* was proposed and implemented within a GIS (Mitasova et al 1995). The RST function includes the sum of all derivatives up to infinity with rapidly decreasing weights. The resulting surface is of C^∞ class which means that it has regular derivatives of all orders (similar, for example, to a Gaussian) and therefore is suitable for differential analysis and calculations of curvatures (Mitasova and Hofierka 1993; Mitasova et al 1995). Other forms of smoothness seminorm are also possible (e.g. Mitas and Mitasova 1997; Wahba 1990). It is important to note that the splines described in this section are, in general, different from a rich class of piecewise polynomial splines (Chui 1988; Wahba 1990).

RST can be generalised to an arbitrary dimension and the corresponding *d*-variate formula for the basis function is given by

$$R_d(\mathbf{r}, \mathbf{r}_j) = R_d(|\mathbf{r} - \mathbf{r}_j|) = R_d(\mathbf{r}) = \rho^{-\delta}\gamma(\delta, \rho) - (\tfrac{1}{\delta}) \qquad (8)$$

where $r = |\mathbf{r} - \mathbf{r}_j|$, $\delta = (d - 2)/2$, and $\rho = (\varphi\, r/2)^2$.

Further, φ is a generalised tension parameter, and $\gamma(\delta, \rho)$ is the incomplete gamma function, not to be confused with a semivariogram (Abramowitz and Stegun 1964). The somewhat less obvious case for $d = 2$ is given by

$$R_2(\mathbf{r}) = \lim_{d \to 0} [\, \rho^{-\delta}\gamma(\delta, \rho) - (\tfrac{1}{\delta})\,] = -[E_1(\rho) + \ln \rho + C_E] \qquad (9)$$

where $C_E = 0.577215...$ is the Euler constant and $E_1(.)$ is the exponential integral function (Abramowitz and Stegun 1964).

As has been pointed out by several authors (Cressie 1993; Hutchinson and Gessler 1993; Matheron 1981; Wahba 1990), splines are formally equivalent to universal Kriging with the choice of the covariance function determined by the seminorm *I*(*F*). Therefore, many of the geostatistical concepts can be exploited within the spline framework. However, the physical interpretation of splines makes their application easier and more intuitive. The 'thin plate with tension' analogy helps to understand the behaviour of the function also in higher dimensions where the interpolation function models an elastic medium with a tunable tension (Mitas et al 1997; see Plate 30). The RST control parameters such as the tension φ and smoothing weights $\{w_j\}$ proved to be useful and effective for preventing an introduction of artificial features such as waves along contours, artificial peaks, pits or overshoots, often found in the results of less general interpolation techniques, or in spline surfaces with tension set too high (for examples of waves, peaks, and pits: see Plate 26(e) and 29(b)), or too low (overshoots, see Plate 30(c) and (d)) (Mitas et al 1997; Mitasova and Mitas 1993; Mitasova et al 1995; Mitasova et al 1996; Watson 1992). The tension and smoothing parameters can be selected empirically, based on the knowledge of the modelled phenomenon, or automatically by minimisation of the predictive error estimated by a cross-validation procedure (Mitasova et al 1995). Moreover, the tension parameter φ can be generalised to a tensor which enables modelling of anisotropy (Mitas and Mitasova 1997; Mitasova and Mitas 1993).

The interpolation function given by Equation (7) requires solving a system of N linear equations. Therefore processing of large datasets ($N > 10^3$) becomes computationally intractable, as the computer time scales as N^3. Treatment of large datasets is enabled by implementation of an automatic segmentation procedure proposed in various forms (Franke 1982b; Hardy 1990; Mitas and Mitasova 1988; Mitasova and Mitas 1993; Mitasova et al 1995) with computational demands proportional to N. The segmented processing is based on the fact that splines have local behaviour, that is, the impact of data points which are far from a given location diminishes rapidly with increasing distance (Powell 1992). The segmentation uses a decomposition of the studied region into rectangular segments with variable size dependent on the density of data points (Plate 28), using 2^d-trees (Mitasova et al 1995). For a given segment, the interpolation is carried out using the data points within this segment and from its neighbourhood, selected automatically depending on their spatial distribution. For very low tension, this approach requires large neighbourhoods to achieve smooth connection of segments, which makes the segmentation method less efficient for flat, very smooth surfaces with strongly heterogeneous point distributions. Recently, a new, more robust version of RST has been developed which reduces the influence of higher order derivatives and the need for large segment neighbourhoods even for low values of tension (Mitas and Mitasova 1997). Segmentation has allowed users to apply RST to datasets with over a million data points and to interpolate multi-million grid sizes (e.g. Hargrove et al 1995).

Instead of using the explicit solution (7), the minimisation in Equation (6) can be carried out numerically by solving an Euler–Lagrange differential equation corresponding to a given functional (Briggs 1974), for example, by using a finite difference multi-grid iteration method (Hutchinson 1989).

The variational approach offers a wide range of possibilities to incorporate additional conditions such as value constraints, prescribed derivatives at the given or at arbitrary points, and integral constraints (Talmi and Gilat 1977; Wahba 1990). Incorporation of dependence on additional variables, analogous to co-Kriging, leads to partial splines (Hutchinson 1996; Wahba 1990). Numerical solution enables the incorporation of stream enforcement and other topographic features (Hutchinson 1989). Known faults or discontinuities can be handled through appropriate data structures using masking and several independent spline functions (Cox et al 1994). A similar approach can be used to handle regions with spatially variable tension, with blending along their borders, an approach analogous to zonal Kriging. Spatio-temporal interpolation is performed by employing an appropriate anisotropic tension in the temporal dimension (Mitas et al 1997; Mitasova et al 1995).

The spline methods are often used for terrain and bathymetry (Hutchinson and Gallant, Chapter 9; Hargrove et al 1995; Mitasova and Mitas 1993), climatic data (Hutchinson 1996; Hutchinson and Bischof 1983; Wahba 1990), chemical concentrations and soil properties (McCauley and Engel 1997; Mitasova et al 1995), and most recently also for image rectification (Fogel 1996).

While not obtained by a variational approach, similar in formulation and performance are *multiquadrics* (Fogel 1996; Foley 1987; Franke 1982a; Hardy 1990; Kansa and Carlson 1992; Nielson 1993) with $R_d(r) = (r^2 + b)^{1/2}$ or $R_d(r) = (r^2 + b)^{-1/2}$, offering high accuracy, differentiability, d-dimensional formulation, and, with segmentation, applicability to large datasets. Originally ad hoc multiquadrics were later put on more solid theoretical ground. Good performance of multiquadrics, especially in three dimensions, is not surprising, considering that for $d = 3$ in the limit of $b \to 0$ the basis functions $(r^2 + b)^{1/2}$ and $(r^2 + b)^{-1/2}$ are solutions of biharmonic and harmonic equations respectively (Hardy 1990).

3.4 Relationships and differences between the geostatistical and variational approaches

Theoretical and practical issues of the relationship between Kriging and splines have been discussed in several papers (e.g. Cressie 1993; Dubrule 1984; Hutchinson and Gessler 1993; Laslett et al 1987; Matheron 1981; Wahba 1990); therefore only a brief comment is presented here.

Kriging assumes that the spatial distribution of a geographical phenomenon can be modelled by a realisation of a random function and uses statistical techniques to analyse the data (drift, covariance) and statistical criteria (unbiasedness, minimum variance) for predictions. However, subjective decisions are necessary (Journel 1996) such as judgement about stationarity, choice of function for theoretical variogram, etc. In addition, often the data simply lack information about important features of the modelled phenomenon, such as surface analytical properties or physically acceptable local geometries. As mentioned earlier, Kriging is the most successful for phenomena with a very strong random component or for estimation of statistical characteristics (uncertainty).

Splines rely on a physical model with flexibility provided by change of elastic properties of the interpolation function. Often, physical phenomena result from processes which minimise energy, with a typical example of terrain with its balance between gravitation force, soil cohesion, and impact of climate. For these cases, splines have proven to be rather successful. Moreover, splines provide enough flexibility for local geometry analysis which is often used as input to various process-based models.

However, most of the surfaces or volumes are neither stochastic nor elastic media, but are the result of a host of natural (e.g. fluxes, diffusion) or socioeconomic processes. Therefore, each of the mentioned methods has a limited realm of applicability and, depending on the knowledge and experience of the user, proper choice of the method and its parameters can significantly improve the final results. This will be illustrated to some extent in section 4.2 with applications.

3.5 Application-specific methods

There is a large class of methods specially designed for certain applications which use one of the above-mentioned general principles, but they are modified to meet some application-specific conditions. These methods are too numerous to mention, so only a few examples with references related to GIS applications have been selected.

Area to surface interpolations are designed to transform the data assigned to areas (polygons) to a continuous surface, represented by a high-resolution raster. This task is common in socioeconomic applications, for example for transformation of population density data from census units to a raster, while preserving the value for an area (mass preservation condition), and ensuring smooth transition between the area units (Martin, Chapter 6; Dyn and Wahba 1982; Goodchild and Lam 1980; Goodchild et al 1993; Moxey and Allanson 1994; Tobler 1979, 1996).

Voronoi polygons (Boots, Chapter 36) are sometimes used for transformation of qualitative point data to polygons or a raster when the condition of continuity is not appropriate, resulting in a surface with zero gradients and faults (see Plate 26(a)).

Interpolations on sphere are modifications of the methods described in sections 3.1–3.3 for data given in spherical (latitude/longitude) coordinates. The interpolation functions are dependent on angle rather than on distance (Nielson 1993; Tobler 1996; Wahba 1990). These methods are used for applications covering large areas, such as continental and global Earth or other planets datasets.

Contour/isoline data interpolations are modifications of mostly local methods specifically designed to handle isoline data (Auerbach and Schaeben 1990; Weibel and Heller 1991). Modification of splines for contour data (Hutchinson 1989) supports incorporation of topographic features, such as streams and ridges, to improve the quality of the resulting Digital Elevation Model (DEM: Hutchinson and Gallant, Chapter 9).

Raster data resampling and smoothing can be performed by modifications of methods described in sections 3.1–3.3, with increased efficiency achieved by taking advantage of data regularity. Numerous simpler methods are also available, such as bilinear and local least square polynomial functions.

4 GIS APPLICATIONS OF SPATIAL INTERPOLATION

Spatial interpolation is an important component of almost any GIS. While the basic bivariate methods are common, implementation of multivariate tools is restricted to the most advanced systems, because of the lack of data structures and supporting tools for multivariate and temporal data processing and analysis (Peuquet, Chapter 8). In spite of recent advances in the development of methods and algorithms and an exponential increase in computational power, spatial interpolation, especially for large and complex datasets, can still be an iterative, time-consuming task, requiring an adequate knowledge of underlying methods and their implementation.

4.1 Integration of spatial interpolation within a GIS

Depending on application, spatial interpolation can be performed at three levels of integration with a GIS: (1) within a more general program/command; (2) as a specialised command; or (3) using linkage to specialised software.

Interpolation integrated at a 'sub-command' level can be found in many GIS application programs, such as computation of slope and aspect, automatic raster resampling, flow-tracing, hydrological modelling, etc. Mostly simple and fast local interpolations such as IDW, bilinear, or local polynomial methods are used in this case. The interpolation is fully automatic, hidden from the user, and while it is sufficient for most applications, it can result in artefacts in surfaces if an improper method is implemented.

Interpolations integrated at a command level serve as data transformation functions. A limited set of basic and some advanced methods have been integrated within GIS, most often the simpler versions of IDW, TIN, Kriging and splines. Compromises in numerical efficiency, accuracy, and robustness are common and upgrades to improved modifications of methods are slow, especially for commercial systems. Therefore, it is necessary to evaluate the results carefully and, if possible, to use more than one independent interpolation procedure.

Although interpolations performed by specialised software linked to a GIS provide the most advanced and flexible tools, a time-consuming import/export of data, or inconvenient work in a different software environment might be necessary. This approach still can be preferable, especially when data are complex and high accuracy is required. The advanced surface and volume modelling systems with strong interpolation capabilities often also provide some basic spatial data processing, analysis, and graphical tools, thus evolving into specialised GIS.

4.2 Examples of applications

To illustrate the properties of selected interpolation methods as well as their typical GIS applications, a few representative examples and corresponding references are presented. Data were processed by GRASS, ARC/INFO and S-PLUS, and the illustrations were created by SG3d, SG4d and Nviz visualisation programs (Brown et al 1995).

4.2.1 Bivariate interpolation of elevation data

The character of interpolation methods from the simple to the more elaborate ones available in most GIS is illustrated by a common task of interpolation from scattered point elevation data to a raster with 2 m resolution, for an area of approximately 1 km^2. Voronoi polygons (Plate 26(a)), producing a surface with discontinuities, are used to illustrate a relatively homogeneous spatial distribution of the original data. Linear TIN-based interpolation (Plate 26(b)) produces a surface with a typical triangular structure and inadequate description of smaller valleys (triangles creating 'dam' structures across valleys). Application of a non-linear TIN-based method to this dataset resulted in unacceptable overshoots within the triangles. Results of IDW (Plate 26(c)) show a typical pattern with extrema in given points and artificial roughness biased towards the data points. Results of Kriging (Plate 26(d)) and a TPS with tension and stream enforcement (Plate 26(e)) represent a significant improvement; however, subtle discontinuities in Kriging (Plate 26(d)) and small cusps in the data points for both methods are visible at this resolution, although the artefacts are within the data accuracy. The results can be further improved by properly tuning the tension and smoothing, as illustrated by the application of the RST method (Plate 26(f)).

4.2.2 The role of interpolation in modelling

To highlight the impact of subtle interpolation artefacts on the results of a model, an erosion/deposition model (Mitasova et al 1996) was used with DEMs computed from contours (Plate 27(a) white lines) using splines with tension and stream enforcement (Plate 27(b)) and by the RST (Plate 27(c)), and compared with observed depositional areas. Small distortions in the DEM computed by splines with tension set too high lead to an artificial pattern of erosion/deposition biased towards the given contours (Plate 27(b)), because the model is dependent on the highly sensitive surface curvatures.

4.2.3 Interpolation of a large dataset

RST method was used to interpolate a 50 m DEM (400x700 grid) from approximately half a million data points digitised from a contour map of Santa Cruz island in California (Plate 28). A segmentation procedure (Plate 28 inset) was used to make the RST method applicable to a large dataset. The example also illustrates that splines can realistically represent rough surfaces in spite of the smoothness condition, if the roughness is sufficiently described by the input data (Plate 28 inset).

4.2.4 Bivariate and trivariate interpolation of precipitation

Multi-dimensional interpolation is also a valuable tool for incorporating the influence of an additional variable into interpolation, for example for evaluation of precipitation with the influence of topography. Plates 29(a) and (b) illustrate the difference between bivariate spline interpolation of annual precipitation in tropical South America, and interpolation with the influence of topography. The importance of incorporating the terrain data is visible especially in the mountainous areas, where the barrier effect of the Andes is very well represented in the results of trivariate interpolation (Plate 29(b)).

4.2.5 Multi-variate interpolation of scattered spatio-temporal data

Capabilities of RST to model spatial and spatio-temporal distributions of phenomena measured in points scattered in 3-dimensional space and time are illustrated by interpolation of nitrogen concentrations in Chesapeake Bay and their change over the year. The analogy between the tensions for bivariate and trivariate interpolation is illustrated by surfaces and volumes interpolated with increasing tension, changing the surface from a stiff medium with overshoots, through adequate tension, to a highly elastic medium with extrema in data points (Plates 31(a)–(f)). Spatio-temporal interpolation was performed by a quadvariate RST function with time as a fourth independent variable. Anisotropic tension in the third (depth) and fourth (time) dimensions was used to ensure a stable solution, and the appropriate tension and smoothing parameters were found by minimising the cross-validation error. The resulting time series of 3-dimensional rasters (Mitasova et al 1995) was then animated to present the dynamic character of the modelled phenomenon (Plate 30 presents a snapshot; Mitas et al 1997 provide animation).

4.3 Future directions

The following paragraphs identify the tasks considered to be the most important in the future development of spatial interpolation techniques.

4.3.1 Robustness and fully automatic method/parameters selection

For widespread, routine use of GIS by users with little expertise in spatial data processing, fully automatic selection of interpolation methods and their parameters based on the robust analysis of given data or *a priori* information about the modelled phenomenon is crucial. With the fast development of communication technologies and accessibility of a variety of data in different formats, this is becoming one of the most urgent tasks for practical applications.

4.3.2 Increase in accuracy and realism

Improvements in accuracy and realism can be expected by employing spatially-variable adaptive interpolation (Hutchinson 1996; Kansa and Carlson 1992) and by further developments in model-based interpolation. More information can be extracted from field data by using process-based models to extrapolate scattered field observations over both time and space (McLaughlin et al 1993). When field data are combined with model predictions the resulting estimates are able to capture the unique characteristics of a specific area while respecting the general physical principles that control the process influencing the spatial distribution of the studied phenomenon. This can be accomplished by using a stochastic/deterministic model of a process together with the concepts of Bayesian estimation theory.

4.3.3 Synthesis of data from various sources

One of the important developments in geosciences is the increasing availability of data generated by various sources (e.g. local measurements, GPS, satellites, radar) which have diverse character from the point of view of resolution, accuracy, distribution etc. This requires novel approaches to data processing and synthesis so that the extraction of information from all sources of data is properly weighted and optimised (Goodchild and Longley, Chapter 40).

4.3.4 Multi-scale modelling

Currently, new types of simulation methods are being developed which span several spatial or spatio-temporal scales. Such approaches provide new challenges for interpolation focused on the design of a versatile and robust approach applicable across the range of scales. Recent progress in wavelet techniques offers one possibility as scale flexibility is one of the fundamental properties of wavelet construction. However, its potential for general multivariate applications remains to be investigated.

4.3.5 Multi-dimensional representation

Full integration of support for multi-dimensional data, including data structures, analytical, and visualisation tools will stimulate multivariate applications. Although methods have been presented already which are fully capable of treatment of multi-dimensional data, the current GIS computational infrastructure does not effectively support wide application of multi-dimensional and spatio-temporal modelling.

4.3.6 Computational efficiency

High-accuracy interpolation of large datasets is computationally very intensive, and increase in performance is important for both large cutting-edge applications, as well as for routine use. Further development of algorithms and the use of parallel architectures will be one of the options for speeding up calculations.

5 CONCLUSION

This chapter has presented a review of scattered-data spatial interpolation methods which are relevant for GIS applications. It is obvious that there has been substantial development over the past decade from the points of view of accuracy, multivariate frameworks, robustness, variety of applications, and size of problems tackled.

However, the conclusions outlined by Burrough (1986) are still valid: 'It is unwise to throw one's data into the first available interpolation technique without carefully considering how the results will be affected by the assumptions inherent in the method. A good GIS should include a range of interpolation techniques that allow the user to choose the most appropriate method for the job at hand.' Computers will take over a large part of this nontrivial task, but many problems remain to be resolved.

References

Abramowitz M, Stegun I A 1964 *Handbook of mathematical functions*. New York, Dover: 297–300, 228–31

Akima H 1978 A method of bivariate interpolation and smooth surface fitting for irregularly spaced data points. Algorithm 526. *ACM Transactions on Mathematical Software* 4: 148–64

Armstrong M P, Dowd P A (eds) 1993 *Geostatistical simulations*. Quantitative Geology and Geostatistics 7. Dordrecht, Kluwer

Auerbach S, Schaeben H 1990 Surface representation reproducing given digitised contour lines. *Mathematical Geology* 22: 723–42

Bogaert P 1996 Comparison of Kriging techniques in a space–time context. *Mathematical Geology* 28: 73–83

Briggs I C 1974 Machine contouring using minimum curvature. *Geophysics* 39: 39–48

Brown W M, Astley M, Baker T, Mitasova H 1995 GRASS as an integrated GIS and visualization environment for spatio-temporal modeling. In *Proceedings AutoCarto 12, Charlotte*. ASPRS/ASM: 89–99

Burrough P A 1986 *Principles of geographical information systems for land resources assessment*. Oxford, Clarendon Press

Burrough P A 1991a Soil information systems. In Goodchild M F, Maguire D J, Rhind D W (eds) *Geographical information systems: principles and applications*. Harlow, Longman/New York, John Wiley & Sons Inc. Vol. 2: 153–64

Chui C 1988 *Multivariate splines*. Philadelphia, Society for Industrial and Applied Mathematics

Cox S J D, Kohn B P, Gleadow A J W 1994 Toward a fission track tectonic image of Australia: model-based interpolation in the Snowy Mountains using a GIS. Presented at 12th Australian Geological Convention, Perth, W Australia. *http://www.ned.dem.csiro.au/AGCRC/papers/snowys/12agc.cox.html*

Cressie N A C 1993 *Statistics for spatial data*. New York, John Wiley & Sons Inc.

Deutsch C V, Journel A G 1992 *GSLIB geostatistical software library and user's guide*. New York, Oxford University Press

Dubrule O 1984 Comparing splines and Kriging. *Computers and Geosciences* 10: 327–38

Duchon J 1976 Interpolation dès fonctions de deux variables suivant le principe de la flexion des plaques minces. *RAIRO Annales Numériques* 10: 5–12

Dyn N, Wahba G 1982 On the estimation of functions of several variables from aggregated data. *SIAM Journal of Mathematical Analysis* 13: 134–52

Englund E J 1993 Spatial simulation: environmental applications. In Goodchild M F, Parks B O, Steyaert L T (eds) *Environmental modeling with GIS*. New York, Oxford University Press: 432–7

Fogel D N 1996 Image rectification with radial basis functions: Application to RS/GIS data integration. *Proceedings, Third International Conference/Workshop on Integrating GIS and Environmental Modeling. Santa Fe, 21–25 January.* Santa Barbara, NCGIA. CD and WWW

Foley T 1987 Interpolation and approximation of 3-dimensional and 4-dimensional scattered data. *Computers and Mathematics with Applications* 13: 711–40

Franke R 1982a Scattered data interpolation: tests of some methods. *Mathematics of Computation* 38: 181–200

Franke R 1982b Smooth interpolation of scattered data by local thin plate splines. *Computers and Mathematics with Applications* 8: 273–81

Franke R 1985 Thin plate spline with tension. *Computer Aided Geometrical Design* 2: 87–95

Franke R, Nielson G 1991 Scattered data interpolation and applications: a tutorial and survey. In Hagen H, Roller D (eds) *Geometric modelling: methods and applications.* Berlin, Springer: 131–60

Goodchild M F, Anselin L, Deichmann U 1993 A framework for the areal interpolation of socioeconomic data. *Environment and Planning A* 25: 383–97

Goodchild M F, Lam N S-N 1980 Areal interpolation: a variant of the traditional spatial problem. *Geoprocessing* 1: 297–312

Harder R, Desmarais R 1972 Interpolation using surface splines. *Journal of Aircraft* 9: 1989–91

Hardy R L 1990 Theory and applications of the multiquadric-biharmonic method. *Computers and Mathematics with Applications* 19: 163–208

Hargrove W, Levine D, Hoffman F 1995 Clinch River environmental restoration program. *http://www.esd.ornl.gov/programs/CRERP/*

Hutchinson M F 1989 A new procedure for gridding elevation and stream line data with automatic removal of spurious pits. *Journal of Hydrology* 106: 211–32

Hutchinson M F 1996b Thin plate spline interpolation of mean rainfall: getting the temporal statistics correct. In Goodchild M F, Steyaert L T, Parks B O, Johnston C A, Maidment D R, Crane M P, Glendinning S (eds) *GIS and environmental modelling: progress and research issues.* Fort Collins, GIS World Inc.: 85–91

Hutchinson M F, Bischof R J 1983 A new method for estimating the spatial distribution of mean seasonal and annual rainfall applied to the Hunter Valley, New South Wales. *Australian Meteorological Magazine* 31: 179–84

Hutchinson M F, Gessler P E 1993 Splines – more than just a smooth interpolator. *Geoderma* 62: 45–67

Isaaks E H, Srivastava R M 1989 *An introduction to applied geostatistics.* Oxford, Oxford University Press

Journel A G 1996 Modelling uncertainty and spatial dependence: stochastic imaging. *International Journal of Geographical Information Systems* 10: 517–22

Journel A G, Huijbregts C J 1978 *Mining geostatistics.* London, Academic Press

Kansa E J, Carlson R E 1992 Improved accuracy of multiquadric interpolation using variable shape parameters. *Computers and Mathematics with Applications* 24: 99–120

Krcho J 1973 Morphometric analysis of relief on the basis of geometric aspect of field theory. *Acta Geographica Universitae Comenianae, Geographica Physica* Vol. 1. Bratislava, SPN

Lam N S-N 1983 Spatial interpolation methods: a review. *The American Cartographer* 10: 129–49

Laslett G M, McBratney A B, Pahl P J, Hutchinson M F 1987 Comparison of several spatial prediction methods for soil pH. *Journal of Soils Science* 38: 32–50

Matheron G 1971 The theory of regionalised variables and its applications. *Les Cahiers du Centre de Morphologie Mathematique de Fontainebleu* 5. Paris

Matheron G 1981 Splines and Kriging: their formal equivalence. *Syracuse University Geological Contributions*: 77–95

McCauley J D, Engel B A 1997 Approximation of noisy bivariate traverse data for precision mapping. *Transactions of the American Society of Agricultural Engineers* 40: 237–45

McCullagh M J 1988 Terrain and surface modelling systems: theory and practice. *Photogrammetric Record* 12: 747–79

McLaughlin D, Reid L B, Li S G, Hyman J 1993 A stochastic method for characterising groundwater contamination. *Groundwater* 31: 237–49

Mitas L, Mitasova H 1988 General variational approach to the interpolation problem. *Computers and Mathematics with Applications* 16: 983–92

Mitas L, Mitasova H 1997 *Multivariate approximation by regularized spline with tension.* Urbana, National Center for Supercomputing Applications: 1–5

Mitas L, Mitasova H, Brown W M 1997 Role of dynamic cartography in simulations of landscape processes based on multi-variate fields. *Computers and Geosciences* 23: 437–46 includes CD and *http://www.elsevier.nl/locate/cgvis*

Mitasova H, Hofierka J 1993 Interpolation by regularised spline with tension: II. Application to terrain modeling and surface geometry analysis. *Mathematical Geology* 25: 657–69

Mitasova H, Hofierka J, Zlocha M, Iverson L R 1996 Modelling topographic potential for erosion and deposition using GIS. *International Journal of Geographical Information Systems* 10: 629–41

Mitasova H, Mitas L 1993 Interpolation by regularised spline with tension: I. Theory and implementation. *Mathematical Geology* 25: 641–55

Mitasova H, Mitas L, Brown W M, Gerdes D P, Kosinovsky I, Baker T 1995 Modelling spatially and temporally distributed phenomena: new methods and tools for GRASS GIS. *International Journal of Geographical Information Systems* 9: 433–46

Moxey A, Allanson P 1994 Areal interpolation of spatially extensive variables: a comparison of alternative techniques. *International Journal of Geographical Information Systems* 8: 479–87

Myers D 1984 Co-Kriging – new developments. In Verly G, David M, Journel A, Marechal A (eds) *Geostatistics for natural resource characterisation* Part 1. Dordrecht, Reidel: 295–305

Nielson, G 1983 A method for interpolating scattered data based upon a minimum norm network. *Mathematics of Computation* 40: 253–71

Nielson G M 1993 Scattered data modeling. *IEEE Computer Graphics and Applications* 13: 60–70

Oliver M A, Webster R 1990 Kriging: a method of interpolation for GIS. *International Journal of Geographical Information Systems* 4: 313–32

Powell M J D 1992 Tabulation of thin plate splines on a very fine 2-dimensional grid. In Braess D, Schumaker L L (eds) *Numerical methods of approximation theory* 9: 221–44

Renka R J, Cline A K 1984 A triangle-based C^1 interpolation method. *Rocky Mountain Journal of Mathematics* 14: 223–37

Rivoirard J 1994 *Introduction to disjunctive Kriging and non-linear geostatistics.* Oxford, Clarendon Press

Rogowski A S 1996 Quantifying soil variability in GIS applications: II Spatial distribution of soil properties. *International Journal of Geographical Information Systems* 10: 455–75

Rouhani S, Myers D E 1990 Problems in space–time Kriging of geohydrological data. *Mathematical Geology* 22: 611–23

Talmi A, Gilat G 1977 Method for smooth approximation of data. *Journal of Computational Physics* 23: 93–123

Tobler W R 1979b Smooth pycnophylactic interpolation for geographic regions. *Journal of the American Statistical Association* 74: 519–36

Tobler W R 1996 Converting administrative data to a continuous field on a sphere. *Proceedings, Third International Conference/Workshop on Integrating GIS and Environmental Modeling, Santa Fe, 21–25 January.* Santa Barbara, NCGIA. CD and WWW

Tobler W R, Kennedy S 1985 Smooth multi-dimensional interpolation. *Geographical Analysis* 17: 251–7

Wahba G 1990 *Spline models for observational data.* CNMS-NSF Regional conference series in applied mathematics 59. Philadelphia, SIAM

Watson D F, Philip G M 1987 Neighborhood based interpolation. *Geobyte* 2: 12–16

Watson D F 1992 *Contouring: a guide to the analysis and display of spatial data.* Oxford, Pergamon

Weibel R, Heller M 1991 Digital terrain modelling. In Goodchild M F, Maguire D J, Rhind D W (eds) *Geographical information systems: principles and applications.* Harlow, Longman/New York, John Wiley & Sons Inc. Vol. 2: 269–97

Wingle W L, Poeter E P 1996 Site characterization, visualization, and uncertainty assessment using zonal Kriging and conditional simulation (poster). *Global Exploration and Geotechnology, AAPG Annual Meeting San Diego, May*

Wood J D, Fisher P J 1993 Assessing interpolation accuracy in elevation models. *IEEE Computer Graphics and Applications* 13: 48–56

Yarrington L 1996 Sequential Gaussian simulation. Fernald Geostatistical Techniques, Sandia National Laboratories. *http://www.nwer.sandia.gov/fernald/techniq.html*

492

35

Multi-criteria evaluation and GIS

J R EASTMAN

Multi-criteria evaluation in GIS is concerned with the allocation of land to suit a specific objective on the basis of a variety of attributes that the selected areas should possess. Although commonly undertaken in GIS, it is shown that the approaches commonly used in vector and raster systems typically lead to different solutions. In addition, there are ambiguities in the manner in which criteria should be standardised and aggregated to yield a final decision for the land allocation process. These problems are reviewed and the theoretical structure of fuzzy measures is offered as an approach to the reconciliation and extension of the procedures currently in use. Specifically, by considering criteria as expressions of membership in fuzzy sets (a specific instance of fuzzy measures) the weighted linear combination aggregation process common to raster systems is seen to lie along a continuum of operators mid-way between the hard intersection and union operators typically associated with Boolean overlay in vector systems. A procedure for implementing this continuum is reviewed, along with its implications for varying the degrees of 'ANDORness' and trade-off between criteria. In addition, the theoretical structure of fuzzy measures provides a strong logic for the standardisation of criteria and the evaluation of decision risk (the likelihood that the decision made will be incorrect).

1 INTRODUCTION

One of the most important applications of GIS is the display and analysis of data to support the process of environmental decision-making. A decision can be defined as a choice between alternatives, where the alternatives may be different actions, locations, objects, and the like. For example, one might need to choose which is the best location for a hazardous waste facility, or perhaps identify which areas will be best suited for a new development.

Broadly, decisions can be classified into two extensive categories – policy decisions and resource allocation decisions. Resource allocation decisions, as the name suggests, are concerned with control over the direct use of resources to achieve a particular goal. Ultimately, policy decisions have a similar aim. However, they do so by establishing legislative instruments that are intended to influence the resource allocation decisions of others. Thus, for example, a government body might reduce taxes on land allocated to a particular crop as an incentive to its introduction. This is clearly a policy decision; but it is the farmer who makes the decision about whether to allocate land to that crop or not.

To be rational, decisions will be necessarily based on one or more criteria – measurable attributes of the alternatives being considered, that can be combined and evaluated in the form of a decision rule. In some circumstances, allocation decisions can be made on the basis of a single criterion. However, more frequently, a variety of criteria is required. For example, the choice between a set of waste disposal sites might be based upon criteria such as proximity to access roads, distance from residential and protected lands, current land use, and so on.

This chapter focuses on the very specific problems of spatial resource allocation decisions in the context of multiple criteria – a process most commonly known as *multi-criteria evaluation*

(MCE) (Voogd 1983). In some instances, this term has also been used to subsume the concept of *multi-objective* decision-making (e.g. Carver 1991; Janssen and Rietveld 1990). However, it is used here in a more specific sense. An *objective* is understood here to imply a perspective, philosophy, or motive that guides the construction of a specific multi-criteria decision rule. Thus in siting a hazardous waste facility, the objective of a developer might be profit maximisation while that of a community action group might be environmental protection. The criteria they each consider and the weights they assign to them are likely to be quite different. Each is likely to develop a multi-criteria solution – but a different multi-criteria decision. The resolution of these differing perspectives into a single solution is known as multi-objective decision-making – a topic which will not be covered in this chapter (see Campbell et al 1992 and Eastman et al 1995 for two prominent approaches to this problem in GIS).

Almost all of the case study examples in this chapter are based on an analysis of suitability for industrial development for the region of Nakuru, Kenya. Nakuru is a region of strong agricultural potential that has experienced rapid urban development in recent years. It is also the location of one of the more important wildlife parks in Kenya (the large area of restricted development to the south of Plate 32) – one of Kenya's soda lakes in the Great Rift Valley, it is the home of over two million flamingoes as well as a wide variety of other species.

2 TRADITIONAL APPROACHES TO MCE IN GIS

In GIS, multi-criteria evaluation has most typically been approached in one of two ways. In the first, all criteria are converted to Boolean (i.e. logical true/false) statements of suitability for the decision under consideration. (The term *Boolean* is derived from the name of the English mathematician, George Boole, who first abstracted the basic laws of set theory in the mid 1800s. It is used here to denote any crisp spatial mapping in which areas are designated by a simple binary number system as either belonging or not belonging to the designated set.) In many respects, these Boolean variables can be usefully thought of as constraints, since they serve to delineate areas that are not suitable for consideration. These constraints are then combined by some combination of intersection (logical AND) or union (logical OR)

operators. This procedure dominates MCE with vector software systems, but is also commonly used with raster systems. For example, Figure 1 shows how Boolean images, along with their intersection achieved through the characteristic overlay operation of a GIS, may be used here to find all areas suitable for industrial development, subject to the following criteria: suitable areas will be near to a road (within 1 km – upper left), near to a labour force (within 7.5 km of a town – middle left), on low slopes (less than 5 per cent – upper right), and greater that 2.5 km from designated wildlife reserves (middle right). In addition, development is not permitted in wildlife reserves (lower left). These criteria are aggregated by means of an intersection (logical AND) operator, yielding the result on the lower right. Note that the distance to labour force was calculated from a cost distance surface that accounted for road and off-road frictions.

In the second most common procedure for MCE, quantitative criteria are evaluated as fully continuous variables rather than collapsing them to

Fig 1. An example of multi-criteria evaluation using Boolean analysis.

Boolean constraints. Such criteria are typically called *factors*, and express varying degrees of suitability for the decision under consideration. Thus, for example, proximity to roads would be treated not as an all-or-none buffer zone of suitable locations, but rather, as a continuous expression of suitability according to a special numeric scale (e.g. 0–1, 0–100, 0–255, etc.). The process of converting data to such numeric scales is most commonly called standardisation (Voogd 1983).

Traditionally, standardised factors are combined by means of weighted linear combination – that is, each factor is multiplied by a weight, with results being summed to arrive at a multi-criteria solution. In addition, the result may be multiplied (i.e. intersected) by the product of any Boolean constraints that may apply (Eastman et al 1995). For example:

$$\text{suitability} = \sum w_i X_i * \prod C_j$$

where w_i = weight assigned to factor i
X_i = criterion score of factor i
C_j = constraint j

Figure 2 illustrates this approach where a comparable example is developed to that in Figure 1. Again, the intention is to find areas suitable for industrial development, subject to the following criteria: suitable areas will be near to a road (as near as possible – upper left), near to a labour force (as near as possible – middle left), on low slopes (as low as possible – upper right) and as far from the wildlife reserve as possible (middle right). As in Figure 1, development is not permitted in wildlife reserves (lower left) through use of a Boolean constraint. These criteria are aggregated by means of a weighted average of the criterion scores. In this case, all criteria were standardised before weighting to a common numeric range using the most commonly used (but not necessarily recommended) technique – linear scaling between the minimum and maximum values of that criterion. The linear rescaling is to a consistent range (0–255) as follows:

$$X_i = (x_i - \min_i) / (\max_i - \min_i)$$

where X_i = criterion score of factor i
x_i = original value of factor i
$\min_i$ = minimum of factor i
$\max_i$ = maximum of factor i

In addition, to provide the most direct comparison to the results of Figure 1, equal weight (0.25) was assigned to each criterion with the wildlife reserve constraint acting as an absolute barrier to

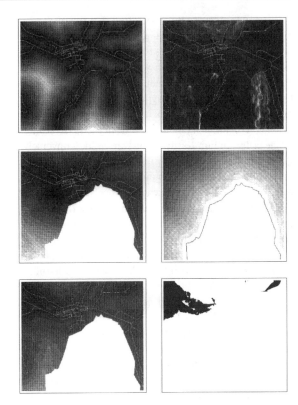

Fig 2. An example of multi-criteria evaluation using weighted linear combination.

development. The result of the averaging process is shown on the lower left. The image on the lower right shows the result of selecting the best areas from this suitability map in order to match the total area of that selected by Boolean analysis in Figure 1. Note that as in Figure 1, the distance to labour force was calculated from a cost distance surface that accounted for road and off-road frictions.

The continuous suitability map shown in Figure 2 has the same numeric range as the standardised factors if the weights that are applied sum to 1.0. A specific decision can then be reached by rank ordering the alternatives (in this case, pixels) and selecting as many of the best ranked areas as is required to meet the objective of the analysis in question. In Figure 2, this has been done in order to select as many of the best areas as were selected by the Boolean analysis in Figure 1.

This procedure of weighted linear combination dominates multi-criteria approaches with raster-based GIS software systems. However, there are a number of problems with both approaches to multi-criteria evaluation.

First, despite a casual expectation that the two procedures should yield similar results, they very often do not. For example, the results of the decision portrayed in the lower right of Figures 1 and 2 are in agreement only by 53 per cent. The reason clearly has to do with the logic of the aggregation operation. For example, Boolean intersection results in a very hard AND – a region will be excluded from the result if any single criterion fails to exceed its threshold. Conversely, the Boolean union operator implements a very liberal mode of aggregation – a region will be included in the result even if only a single criterion meets its threshold. Weighted linear combination is quite unlike these options. Here a low score on one criterion can be compensated by a high score on another – a feature known as *trade-off* or *substitutability*. While human experience is replete with examples of both trade-off and non-substitutability in decision making, the tools for flexibly incorporating this concept are poorly developed in GIS. Furthermore, a theoretical framework that can link the aggregation operators of Boolean overlay and weighted linear combination has, until recently (Eastman and Jiang 1996), been lacking.

The second problem with MCE has to do with the standardisation of factors in weighted linear combination. The most common approach is to rescale the range to a common numerical basis by simple linear transformation (Voogd 1983), as was applied in Figure 2. However, the rationale for doing so is unclear. Indeed, there are many instances where it would seem logical to rescale values within a more limited range. Furthermore, there are cases where a non-linear scaling may seem appropriate. Since the recast criteria really express suitability, there are many cases where it would seem appropriate that criterion scores asymptotically approach the maximum or minimum suitability level.

The third issue concerns the weights that are applied. Clearly they can have a strong effect on the outcome produced. However, not much attention has been focused in GIS on how they should be developed. Commonly they represent the subjective (but no less valid) opinions of one or more experts or local informants. How can consistency and overt validity be established for these weights? Furthermore, how should they be applied in the context of varying trade-off between factors?

A fourth problem concerns decision risk. Decision risk may be considered as the likelihood

that the decision made will be wrong. For both procedures (Boolean analysis and weighted linear combination) it is a fairly simple matter to propagate measurement error through the decision rule and subsequently to determine the risk that a given location will be assigned to the wrong set (i.e. the set of selected alternatives or the set of those not to be included). However, the continuous criteria of weighted linear combination would appear to express a further uncertainty that is not so easily accommodated (see Fisher, Chapter 13). The standardised factors of weighted linear combination each express a perspective of suitability – the higher the score, the more suitable. However, there is no real threshold that can definitively allocate locations to one of the two sets involved (areas to be chosen and areas to be excluded). How are these uncertainties to be accommodated in expressions of decision risk? If these criteria really express uncertainties, why are they combined through an averaging process?

The surprising feature of multi-criteria evaluation is that, despite its ubiquity in environmental management, so little is understood of its character in GIS. In the following sections we survey the issues involved, and offer a perspective on a resolution through the concept of *fuzzy measures*.

3 FUZZY MEASURES

This discussion of fuzzy measures is adapted from Eastman and Jiang (1996). The term fuzzy measure refers to any set function which is monotonic with respect to set membership (Dubois and Prade 1982; see also Fisher, Chapter 13). Notable examples of fuzzy measures include *probabilities*, the *beliefs*, and *plausibilities* of Dempster-Shafer theory, and the *possibilities* of fuzzy sets. Interestingly, if we consider the process of standardisation in MCE to be one of transforming criterion scores into set membership statements (i.e. the set of suitable choices), then standardised criteria *are* fuzzy measures.

A common trait of fuzzy measures is that they follow DeMorgan's Law in the construction of the intersection and union operators (Bonissone and Decker 1986). DeMorgan's Law establishes a triangular relationship between the intersection, union, and negation operators such that:

$$T(a,b) = \sim S(\sim a, \sim b)$$

where T = intersection (AND) = T-Norm
and S = union (OR) = T-CoNorm
and $\sim$ = negation (NOT)

The intersection operators in this context are known as *triangular norms*, or simply T-*Norms*, while the union operators are known as *triangular co-norms*, or T-*CoNorms*.

4 FUZZY MEASURES AND AGGREGATION OPERATORS

A T-Norm can be defined as (Yager 1988):

a mapping T: $[0,1] * [0,1] \rightarrow [0,1]$ such that:

$T(a,b) = T(b,a)$ (commutative)
$T(a,b) \geq T(c,d)$ if $a \geq c$ and $b \geq d$ (monotonic)
$T(a,T(b,c)) = T(T(a,b),c)$ (associative)
$T(1,a) = a$

Some examples of T-norms include:

$\min(a,b)$ (the intersection operator of fuzzy sets)
$a * b$ (the intersection operator of classical sets)
$1 - \min(1,((1-a)^p + (1-b)^p)^{(1/p)})$ (for $p \geq 1$)
$\max(0, a + b - 1)$

Conversely, a T-CoNorm is defined as:

a mapping S: $[0,1] * [0,1] \rightarrow [0,1]$ such that:

$S(a,b) = S(b,a)$ (commutative)
$S(a,b) >= S(c,d)$ if $a \geq c$ and $b \geq d$ (monotonic)
$S(a,S(b,c)) = S(S(a,b),c)$ (associative)
$S(0,a) = a$

Some examples of T-CoNorms include:

$\max(a,b)$ (the union operator of fuzzy sets)
$a + b - a*b$ (the union operator of classical sets)
$\min(1,(a^p + b^p)^{1/p})$ (for $p \geq 1$)
$\min(1, a + b)$

Interestingly, while the intersection ($a*b$) and union ($(a+b) - (a*b)$) operators of Boolean overlay represent a T-Norm/T-CoNorm pair, the averaging operator of weighted linear combination is neither, because it lacks the property of associativity (Bonissone and Decker 1986). Rather, it has been determined (Bonissone and Decker 1986) that the averaging operator falls midway between the extreme cases of the T-Norm (AND) of fuzzy sets (the minimum operator) and its corresponding T-CoNorm

(OR – the maximum operator) – in essence, a perfect ANDOR operator. In fact, Yager (1988) has proposed that weighted linear combination is one of a continuum of aggregation operators that lies between these two extremes. Further, he has proposed the concept of an *ordered weighted average* that can produce the entire continuum. Recently, Eastman and Jiang (1996) have implemented this operator, with modifications, in a raster GIS. In doing so, the traditional aggregation operators of vector and raster GIS have been united into a single theoretical framework.

4.1 The ordered weighted average

With the ordered weighted average, criteria are weighted on the basis of their rank order rather than their inherent qualities. Thus, for example, we might decide to apply order weights of 0.5, 0.3, 0.2 to weight a set of factors *A, B,* and *C* based on their rank order. Thus if at one location the criteria are ranked *BAC* (from lowest to highest), the weighted combination would be $0.5*B + 0.3*A + 0.2*C$. However, if at another location the factors are ranked *CBA*, the weighted combination would be $0.5*C + 0.3*B + 0.2*A$. In the implementation of Yager's concept by Eastman and Jiang (1996), the concept of weights that apply to specific factors has also been incorporated, yielding two sets of weights – *criterion weights* that apply to specific criteria and *order weights* that apply to the ranked criteria, after application of the criterion weights.

The interesting feature of the ordered weighted average is that it is possible to control the degree of ANDORness and trade-off between factors within limits. For example, using order weights of [1 0 0] yields the minimum operator of fuzzy sets, with full ANDness and no trade-off. Using order weights of [0 0 1] yields the maximum operator of fuzzy sets with full ORness and no trade-off. Using weights of [0.33 0.33 0.33] yields the traditional averaging operator of MCE with intermediate ANDness and ORness, and full trade-off. Trade-off is thus controlled by the degree of dispersion in the weights while ANDness or ORness is governed by the amount of skew. For example, order weights of [0 1 0] would yield an operator with intermediate ANDness and ORness, but no trade-off, while the original example with order weights of [0.5 0.3 0.2] would yield an operator with substantial trade-off and a moderate degree of ANDness.

This quality of variable ANDORness has interested some in the decision science field (e.g. Yager 1988) because of the recognition that in human perception of decision logics, it is not uncommon to wish to combine criteria with something less extreme than the hard operations of union or intersection. In the context of GIS, however, it is the property of trade-off that is of special interest. The minimum operator is occasionally used in GIS applications, and represents a form of limiting factor analysis. Here the intent is one of risk aversion, by characterising the suitability of a location in terms of its worst quality. The maximum operator is the opposite, and can thus be thought of as a very optimistic aggregation operator – an area will be suitable to the extent of its best quality. Both of these operations permit no trade-off between the qualities of the criteria considered. Furthermore, in cases where set membership approaches certainty, the results from fuzzy sets will be identical to those of Boolean overlay. However, with weighted linear combination, trade-off is clearly and fully present.

Figure 3 illustrates various degrees of trade-off (and ANDORness) between the same four factors considered in Figures 1 and 2. The same criteria were used as for Figure 2, except that the scaling was changed to facilitate comparison to the results in Figure 1 – a sigmoidal fuzzy membership function was used such that the thresholds used to create the Boolean images in Figure 1 correspond to membership values of 0.5 for the fuzzy criteria in this example (i.e. scaling was asymptotic to membership values of 1.0 and 0.0 at values of 0–2 km for proximity to roads, 0–15 km for proximity to the labour force, 0–10 per cent for slope gradients, and 5–0 km for distance from designated wildlife reserves). In addition, to facilitate comparison, equal criterion weights were applied as in Figure 2. Thus the differences between these aggregations arise solely from the effects of different order weights. The v-shaped sequence, from top to bottom, used order weights of [1 0 0 0], [.60 .20 .15 .05], [.4 .3 .2 .1], [.25 .25 .25 .25], [.1 .2 .3 .4], [.05 .15 .20 .60], and [0 0 0 1]. This sequence progresses from full ANDness and no trade-off for the first (the minimum function), to intermediate ANDORness and full trade-off for the middlemost (equivalent to a standard weighted linear combination), to full ORness and no trade-off with the last (corresponding to the maximum operator). The

image on the middle left is a median operator produced with order weights of [0 .5 .5 0], producing an aggregation with intermediate ANDORness (like weighted linear combination) but almost no trade-off.

The similarity of the result with full ANDness (and thus no trade-off) to the Boolean result in Figure 1 is striking. In fact, when these suitability values are rank ordered and enough of the best pixels are selected to equal the area of the Boolean result, the solution is identical. Thus the reason for the difference in the Boolean and weighted linear combination results is clear – the characteristic Boolean overlay operation of vector GIS produces an aggregation of criteria with full ANDness and no trade-off while the typical weighted linear combination operation of raster GIS produces intermediate ANDness and full trade-off. The results are different because the aggregation operators are different.

Recognising that a full spectrum of aggregation operators exists opens up a much richer set of possibilities for implementing decision rules in GIS.

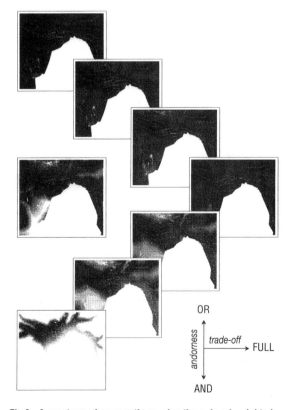

Fig 3. A spectrum of aggregations using the ordered weighted average procedure.

For example, Plate 32 illustrates the effects of combining three of the factors with trade-off and one without. In this case, proximity to roads is given a criterion weight of 0.45, proximity to the labour force is given a weight of 0.12, and the slope factor is given a weight of 0.43. These are combined using a standard weighted linear combination. This result is then combined with the distance from wildlife reserve factor using a minimum operator. The absence of trade-off in this last step is clear – the distance from wildlife reserve factor dominates the result until it no longer represents the limiting factor. The effect is clearly similar to that of a constraint, but lacks the crispness of a traditional constraint. In effect, the minimum operator with a fuzzy measure represents a form of soft constraint. Soft constraints are particularly useful where a specific boundary cannot be reasonably established. Indeed, it might be argued that this is more commonly appropriate than the artificial boundaries of traditional constraints.

5 FUZZY MEASURES AND STANDARDISATION

Clearly, this consideration of fuzzy measures has implications beyond those of the aggregation process alone. It also provides a very strong logic for the process of standardisation. In this context, the process of standardising a criterion can be seen as one of recasting values into a statement of set membership – the degree of membership in the final decision set. Indeed, Eastman and Jiang (1996) argue that such statements of set membership in fact constitute fuzzy sets (a particular form of fuzzy measure), while those of Boolean constraints represent classical sets. This clearly opens the way for a broader family of set membership functions than that of linear rescaling alone. For example, the commonly used sigmoidal (s-shaped) function of fuzzy sets provides a simple logic for cases where a function is required that is asymptotic to 0 and 1. It also suggests that the minimum and maximum raw factor values should not blindly be used as the anchor points for such a function. Rather, anchor points that are consistent with the logic of set membership are clearly superior. For example, in Figure 4, sigmoidal membership functions were created for each factor, with anchor points set at the points where the factor begins to have an effect and where the effect is no longer relevant. The distance

to wildlife reserve factor, for instance, starts to rise above 0.0 immediately at the park boundary, but approaches 1.0 at a distance of 5 kilometres. Further distance does not lead to an increase in the factor score since the distance is far enough.

6 DETERMINATION OF WEIGHTS

Given the consideration of factors as fuzzy sets and the nature of the aggregation process, the criterion weights of weighted linear combination clearly represent trade-off weights – that is, expressions of the manner in which they will trade with other factors when aggregated in multi-criteria evaluation. Rao et al (1991) have suggested that a logical process for the development of such weights is the procedure of pairwise comparisons developed by Saaty (1977). In this process each factor is rated for its importance relative to every other factor using a 9-point reciprocal scale (i.e. if 7 represents substantially more important, 1/7 would indicate substantially less important). This leads to a $n \times n$ matrix of ratings (where n is the number of factors being considered). Saaty (1977) has shown that the principal eigenvector of this matrix represents a best fit set of weights. Figure 5, for example, illustrates this rating scale along with a completed comparison matrix and the best fit weights produced. Eastman et al (1993) have implemented this procedure in a raster GIS with a modification that also allows the degree of consistency to be evaluated as well as the location of inconsistencies to allow for an orderly re-evaluation. The process is thus an iterative one that converges on a consistent set of consensus weights.

A problem still exists, however, in how these weights should be applied in the context of the ordered weighted average discussed above. It seems clear that these weights will have full effect with the weighted linear combination operator (where full trade-off exists), and that they should have no effect when no trade-off is in effect (i.e. with the minimum and maximum operators). It seems logical, therefore, that their effect should be graded between these extremes as the degree of trade-off is manipulated with the ordered weighted average process. However, the logic for this gradation has not been established. In their implementation of the ordered weighted average for GIS, Eastman and Jiang (1996) have used a measure of relative

	1/9	1/7	1/5	1/3	1	3	5	7	9
	extremely	very	strongly	moderately	equally	moderately	strongly	very	extremely

less important ⎯⎯⎯⎯⎯⎯⎯⎯⎯⎯⎯⎯⎯⎯⎯→ more important

(a)

	Proximity to roads	Proximity to labour force	Slope gradient	Distance from wildlife reserves
Proximity to roads	1			
Proximity to labour force	1/3	1		
Slope gradient	1	4	1	
Distance from wildlife reserves	1/3	2	1/2	1

(b)

Factor	Derived weight
Proximity to roads	0.3770
Proximity to labour force	0.0979
Slope gradient	0.3605
Distance from wildlife reserves	0.1647

(c)

Fig 4. Saaty's pairwise comparison procedure for the derivation of factor weights. Using a 9-point rating scale (a) each factor is compared to each other factor for its relative importance in developing the final solution (b). The principal eigenvector of this matrix is then calculated to derive the best-fit set of weights (c).

dispersion (a measure closely related to the entropy measure of information theory) as the basis for this gradation. However, further research is needed on this important aspect of the ordered weighted average procedure.

7 DECISION RISK

Uncertainty in the decision rule, and in the criteria that are considered, implies some risk that the decision made will be wrong. In the case of measurement error, the effects of uncertainty can fairly easily be propagated to the suitability map that is produced in MCE (see Heuvelink, Chapter 14; Heuvelink 1993). Furthermore, Eastman (1993) has developed a simple operator that can convert such an evaluation into a mapping of the probability

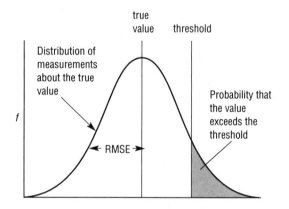

Fig 5. A procedure for calculating decision risk. Assuming a normal distribution of errors, the probability that a data value exceeds or is exceeded by a threshold can be calculated as the area of the normal curve subtended by that threshold. [RMSE = root mean square error]

that locations belong to the decision set (the PCLASS operator of the Idrisi software system). The operator assumes a normal distribution of errors and calculates the area under the normal curve subtended by a threshold that can distinguish the cases that belong in the decision set from those that do not (Figure 5). The result is an expression of decision risk that is directly analogous to the concept of a Type II error in statistical hypothesis testing – that is, the likelihood that the alternative does not belong to the decision set if we assume that it does (Plate 33). The resulting probability map can subsequently be thresholded to see the nature of the decision set at any specified risk level (again see Plate 33).

To the extent that measurement error can be quantified and propagated through an analysis, an expression of decision risk is thus not very difficult to achieve. However, the recognition of factors in MCE as fuzzy sets implies a very different form of uncertainty from that of measurement error. The suitability map that results from weighted linear combination is a clear expression of uncertainty about the suitability of any particular piece of land for the objective under consideration. However, as an expression of uncertainty, it has no relationship to the frequentist notion of probability that underlies the treatment of decision risk in the context of measurement error. Thus a traditional treatment of decision risk as the probability that the decision made will be wrong cannot be developed. Eastman (1996) has therefore suggested that decision risk for such cases be expressed by the concept of *relative risk*.

A mapping of relative risk can be very simply achieved by rank ordering the alternatives and dividing the result by the maximum (i.e. worst) rank that occurs. The outcome is a proportional ranking that can directly be interpreted as relative risk. Then in cases where no specific area requirement for the decision set is being sought (e.g. the best 10 hectares), the final decision set can be established by selecting the alternatives where the relative risk does not exceed a specific threshold (e.g. the best 5 per cent of the areas under consideration). Figure 6 illustrates such a mapping of relative risk for the result of Plate 32 along with a mapping of the best (least risky) 10 per cent of cases outside the wildlife reserve.

Such an expression of relative risk is quite familiar in human experience. By rank ordering the alternatives (on the basis of suitability) and choosing the best ones, we use a procedure that

strives to pick the least risky alternatives (i.e. the ones that are least likely to be poor choices). For example, in screening applicants for employment we may make use of a variety of criteria (e.g. test scores, reference evaluations, years of experience, etc.) that will allow the candidates to be ranked. Then by choosing only the highest ranked candidates we minimise our risk of choosing someone who will perform poorly. However, we do not know the actual degree of risk we are taking; only that the candidates we have chosen are the least risky of the alternatives considered.

From the perspective of considering the criteria of MCE as fuzzy measures, then, it would appear that the expression of decision risk needs to be different from that which arises from a consideration of measurement error. However, in most cases, both

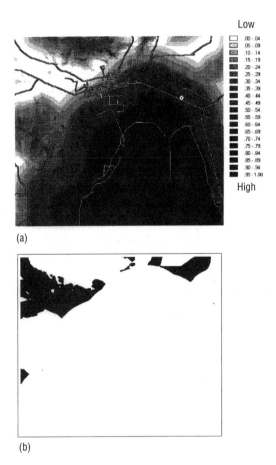

(a)

(b)

Fig 6. (a) A mapping of relative risk for the result in Plate 32. The procedure involves a simple rank ordering of alternatives followed by division by the maximum rank. (b) Areas outside the wildlife reserve with a relative risk of 10 per cent or less.

forms of uncertainty exist. Thus one might anticipate the problem of having to express both forms of risk. For example, given the presence of measurement error in the development of a multi-criteria suitability mapping, and a propagation of those errors to the final suitability map, one realises that this mapping is only one of a large number of possible outcomes that might be produced by randomly introducing the uncertainties in measurement that exist. Thus by Monte Carlo simulation (a capability that unfortunately exists in only a small number of GIS software programs) one could thus tabulate the proportion of simulations in which each location falls within a specific threshold of relative risk, or a specific areal requirement. This then restores the frequentist notion of probability and the usual expression of decision risk.

8 CONCLUSION

In this chapter an attempt has been made to reconcile the differences between the typical approaches to MCE used in vector and raster GIS. By using the theoretical structure of fuzzy measures, both approaches can be seen as special cases of a single family of aggregation operators. In the case of Boolean overlay as very typically used in vector GIS, the decision problem is treated as one of classical set membership, with the intersection and union operations resulting in strict ANDness or ORness with no trade-off. However, it has been shown here that these hard constraints represent no more than the crisp extremes of an underlying logic of fuzzy sets. By considering the more general class of fuzzy measures (of which fuzzy sets are a member) it has been shown that similar operations exist for the continuous factors more commonly associated with raster systems (the minimum and maximum operators). In addition, it has been shown that the weighted linear combination operator commonly used with such factors lies on a continuum with these operators, where it represents the case of intermediate ANDness and ORness, and full trade-off between the factors considered. Furthermore, it has been shown that a more general operator (the ordered weighted average) can produce all of these results along with a continuum of other operators with varying degrees of trade-off and ANDORness. This not only acts as a strong theoretical framework for consideration of the aggregation

operator, but also provides a logic for the standardisation of factors, a rationale for the expression of decision risk, and a high degree of flexibility in the land allocation decision process.

References

Bonissone P P, Decker K 1986 Selecting uncertainty calculi and granularity: an experiment in trading-off precision and complexity. In Kanal L N, Lemmer J F (eds) *Uncertainty in artificial intelligence.* Amsterdam, Elsevier Science

Campbell J C, Radke J, Gless J T, Wirtshafter R M 1992 An application of linear programming and geographic information systems: cropland allocation in Antigua. *Environment and Planning A* 24: 535–49

Carver S J 1991 Integrating multi-criteria evaluation with geographical information systems. *International Journal of Geographical Information Systems* 5: 321–39

Dubois D, Prade H 1982 A class of fuzzy measures based on triangular norms. *International Journal of General Systems* 8: 43–61

Eastman J R 1993 *Idrisi Version 4.1.* Worcester (USA), Clark University

Eastman J R 1996 Uncertainty and decision risk in multi-criteria evaluation: implications for GIS software design. *Proceedings, International Institute for Software Technology Expert Group Workshop on Software Technology for Agenda 21: Decision Support Systems* Section 8

Eastman J R, Jiang H 1996 Fuzzy measures in multi-criteria evaluation. *Proceedings, Second International Symposium on Spatial Accuracy Assessment in Natural Resources and Environmental Studies.* Fort Collins, GIS World Inc.: 527–34

Eastman J R, Jin W, Kyem P A K, Toledano J 1995 Raster procedures for multi-criteria/multi-objective decisions. *Photogrammetric Engineering and Remote Sensing* 61: 539–47

Heuvelink G B M 1993 'Error propagation in quantitative spatial modelling: applications in geographical information systems'. PhD thesis. Utrecht, Netherlands Geographical Studies 163

Janssen R, Rietveld P 1990 Multi-criteria analysis and geographical information systems: an application to agricultural landuse in the Netherlands. In Scholten H J, Stillwell J C H (eds) *Geographical information systems for urban and regional planning.* Amsterdam, Kluwer: 129–39

Rao M, Sastry S V C, Yadar P D, Kharod K, Pathan S K, Dhinwa P S, Majumdar K L, Sampat Kumar D, Patkar V N, Phatak V K 1991 *A weighted index model for urban suitability assessment – a GIS approach.* Bombay, Bombay Metropolitan Regional Development Authority

Voogd H 1983 *Multi-criteria evaluation for urban and regional planning.* London, Pion

Yager R 1988 On ordered weighted averaging aggregation operators in multicriteria decision-making. *IEEE Transactions on Systems, Man, and Cybernetics* 8: 183–90

36

Spatial tessellations

B BOOTS

This chapter describes selected properties of a variety of planar tessellations and examines how these properties influence the use of the tessellations in GIS. Initially, properties common to all planar tessellations are presented. Regular tessellations are considered in the context of image representation and in spatial statistical analysis. Relationships frequently encountered in irregular tessellations are described. A large family of tessellations which arise as the result of data transformations is introduced. Collectively, these are known as generalised Voronoi diagrams. Their use in a variety of contexts, including solving locational problems, defining spatial relationships, and spatial interpolation is illustrated by focusing on two specific types of diagram, the ordinary Voronoi diagram and higher order Voronoi diagrams.

1 PRELIMINARIES

A tessellation of d-dimensional, Euclidean space, $\Re^d$, can be defined from two different yet equivalent perspectives. It may be considered either as a subdivision of $\Re^d$ into d-dimensional, non-overlapping regions or as a set of d-dimensional regions which cover $\Re^d$ without gaps or overlaps. These two perspectives are reflected in the use of various synonyms for tessellation, partitions, and tilings, respectively, being the most common (Grunbaum and Shephard 1986: 16). Although these two perspectives can be resolved into one by adopting a more formal definition of a tessellation, such as that given below, it is useful to maintain the distinction within the context of GIS. If a tessellation is considered as a data model, the two perspectives can be viewed as different data structures for representing the model (Peuquet 1984). The former perspective, emphasising the boundaries of the regions, is consistent with a vector structure that might be used to represent the spatial units in a choropleth map. The latter perspective, focusing on the interiors of the regions, is equivalent to a raster structure such as that formed by the pixels of a remotely-sensed image.

Tessellations have long been the subject of human interest. Indeed, the term is derived from the Latin word 'tessella', a small, square stone used to create mosaics. Despite their antiquity, new forms, such as Penrose tilings, popularly encountered in some screen savers, have only been discovered in the past 25 years (Ammann et al 1992; Grunbaum and Shephard 1986: 10 and 11). Today, tessellations are studied in a wide range of disciplines from astronomy (Zaninetti 1993) to zoology (Perry 1995). An inevitable, but unfortunate, consequence of this widespread use is that many characteristics of tessellations and their component parts are known by a variety of names. Consequently, it is necessary to begin by defining a number of terms. As far as possible usage is consistent with current GIS practice.

To begin with a formal definition of a tessellation: Let S be a closed subset of $\Re^d$, $\Im = \{s_1,...,s_n\}$ where s_i is a closed subset of S, and s_i' the interior of s_i. If the elements of $\Im$ satisfy

$$s_i' \cap s_j' = \emptyset \text{ for } i \neq j \tag{1}$$

$$\bigcup_{i=1}^{n} s_i = S, \tag{2}$$

then the set $\Im$ is called a tessellation of S. Property (1) means that the interiors of the elements of $\Im$ are disjoint and (2) means that collectively the elements of $\Im$ fill the space S. Note that this definition is consistent

503

with practical applications such as those encountered in GIS, where the space under consideration is a bounded region in Euclidean space rather than the unbounded space itself (the usual situation in theoretical treatments). When $d=2$ the tessellation is called a *planar tessellation*. Attention in this chapter is limited to planar tessellations since these are those most commonly encountered in GIS.

Planar tessellations are composed of three elements of d ($d≤2$) dimensions (see Figure 1): cells (2-d), edges (1-d), and vertices (0-d). In GIS these elements are usually referred to as polygons, lines (or arcs), and points respectively. In turn, each of the d ($d>0$)-dimensional elements are composed of elements of (d-1) dimensions. Cells have sides (1-d) and corners (0-d), lines have end points (0-d). A tessellation, such as that in Figure 1, in which the corners and sides of individual cells coincide with the vertices and edges of the tessellation, respectively, is called an *edge-to-edge tessellation*. Individual rectangular cells arranged in a brick wall fashion would not constitute an edge-to-edge tessellation. Only edge-to-edge tessellations are considered here. A d-dimensional tessellation in which every s-dimensional element lies in the boundaries of (d–s+1) cells ($0≤s≤d$–1) is called a *normal tessellation*. Thus, in a normal planar tessellation, each vertex is shared by three cells and each edge is common to two cells.

A *monohedral tessellation* is one in which all the cells are of the same size and shape (e.g. the tessellation in Figure 2). More formally, each cell is congruent (directly or reflectively) to one fixed set S. If r_i denotes the number of edges meeting at the ith corner of a cell in a monohedral tessellation, an *isohedral tessellation* is one in which the ordered sequence of values of r_i is the same for every cell. In short, the cells are completely interchangeable so that, as Bell and Holroyd (1991) note, 'a bug which was put down in one of the [cells] and started to explore the [tessellation] would find exactly the same arrangement of [cells] no matter which [cell] it was originally deposited in'. More formally, all the cells are equivalent under the symmetry group of the tessellation. Thus, the tessellation in Figure 3 is isohedral while that in Figure 2 is not.

A regular polygon is one with equal side length and equal internal angles. Grunbaum and Shephard (1977a) demonstrate that, even if we restrict our attention to tessellations consisting of only one type of regular polygon, the number of such tessellations is infinite. However, by imposing the condition that all the vertices of the tessellation are of the same type, the number reduces dramatically to just 11. These are the so-called *Archimedean tessellations* (see Figure 4), also known as *uniform tessellations*.

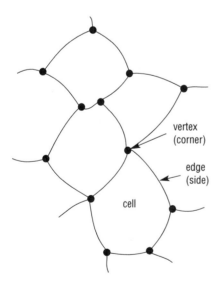

Fig 1. Illustration of terms for tessellations and cells (in parentheses).

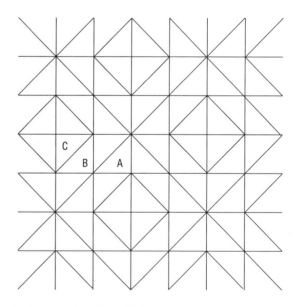

Fig 2. A monohedral tessellation containing three types of cells A, B, and C, each of the same size and shape but arranged differently.

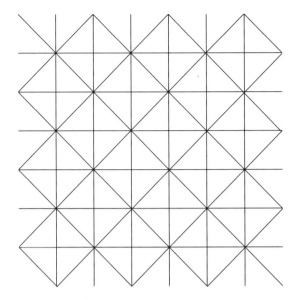

Fig 3. An isohedral tessellation in which all cells are of the same size and shape and are arranged identically.

The three uniform tessellations which are also isohedral (i.e. those consisting of regular triangles, squares, or hexagons) are called *regular tessellations*.

If there is a one-to-one correspondence between the vertices, edges, and cells of one tessellation and the cells, edges, and vertices of another tessellation, the two tessellations are called each other's *dual tessellation* (see Figure 5). A dual tessellation can be generated from a tessellation T in the following way. Select a point q_i in each cell c_i of T. For each pair of cells c_i, c_j which share an edge of T, construct a line segment joining q_i, q_j. Because there is no unique way of selecting q_i, it is possible to generate more than one metric dual.

A d-dimensional tessellation consisting exclusively of cells with $(d+1)$ sides is called a *simplex* (or simplicial graph). The dual tessellation in Figure 5 is a simplex since all its cells are triangles. Another example is the triangulated irregular network (TIN) often encountered in GIS to represent continuous surfaces (Hutchinson and Gallant, Chapter 9).

2 GENERAL PROPERTIES

Here we present properties which hold for all tessellations satisfying a specific condition known as the closure postulate. This postulate requires that at

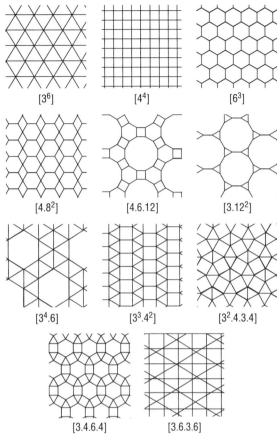

$[3^6]$ $[4^4]$ $[6^3]$

$[4.8^2]$ $[4.6.12]$ $[3.12^2]$

$[3^4.6]$ $[3^3.4^2]$ $[3^2.4.3.4]$

$[3.4.6.4]$ $[3.6.3.6]$

Fig 4. The 11 distinct Archimedean (uniform) tessellations.

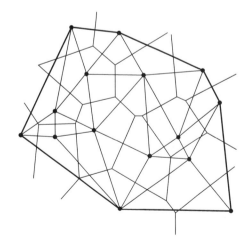

Fig 5. A tessellation (heavier lines) and a dual tessellation (lighter lines).

505

least two edges meet at every vertex and at least two cells meet at every edge. The elements in such tessellations are related to each other through Equation 3 derived by Schlaefli in 1852 as a generalisation of a relationship first formulated by Euler one hundred years earlier (Loeb 1976):

$$\sum_{i=0}^{j} (-1)^i N_i = 1 + (-1)^j \tag{3}$$

where N_i is the number of elements of dimensionality i. For planar tessellations Equation 3 reduces to:

$$N_0 - N_1 + N_2 = 2 \tag{4}$$

where N_0, N_1, N_2 are the number of vertices, edges, and cells respectively. To satisfy the closure postulate, if our tessellation is bounded, we must include in our count of N_2 an outside cell whose number of sides is equal to the number of edges on the boundary of the tessellation.

If we define r=number of edges meeting at a vertex and n=number of corners or sides of a cell, a further equation can be derived from Equation 4, which reveals that we need only three parameters to specify all of the characteristics of the tessellation. This equation is

$$1/\bar{r} - 1/2 + 1/\bar{n} = 1/N_1 \tag{5}$$

where $\bar{r}$ and $\bar{n}$ are the mean values of r and n, respectively. Equation 3 yields expressions for

$$N_0 = 2N_1/\bar{r} \tag{6}$$

and

$$N_2 = 2N_1/\bar{n} \tag{7}$$

The verification of Equations 3–7 is left as an exercise for the reader.

3 REGULAR TESSELLATIONS

This and the next section continue the exploration of properties of tessellations by looking in more detail at specific types of tessellations. However, the emphasis is shifted away from the properties themselves towards an examination of how they might influence the use of the tessellations within specific contexts in GIS. The first example involves the use of tessellations as spatial data models, in particular in image representation.

To be useful in such a role tessellations should ideally possess at least two properties (Ahuja 1983; Samet 1989):

1 be capable of generating an infinitely repetitive pattern, so that they can be used for images of any size;
2 be infinitely (recursively) decomposable into increasingly finer patterns which, collectively, form a hierarchy, to allow for the representation of spatial features of arbitrarily fine resolution.

If attention is restricted to tessellations consisting of only one type of cell, a reasonable starting point is to consider isohedral tessellations, particularly since Grunbaum and Shephard (1977b) demonstrate that there are only 81 such planar tessellations. This number further reduces to 11 if only those tessellations are considered which are topologically distinct (i.e. all edges in the 81 tessellations are straight lines). These 11 (see Figure 6) are known as

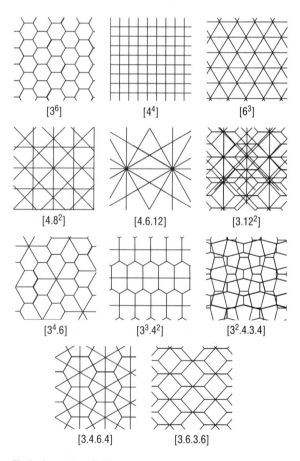

Fig 6. Laves tessellations.

Laves tessellations after the famous crystallographer Fritz Laves. Laves tessellations may also be derived as duals of the uniform (Archimedean) tessellations in Figure 4. Note that three of the Laves tessellations are regular tessellations. To describe Laves tessellations we use a notation based on the number of edges at the vertices of a constituent cell as they are visited in cyclic order (see Figure 6). Thus, the three regular tessellations consisting of triangles, squares, and hexagons are labelled $[6^3]$, $[4^4]$ and $[3^6]$ respectively.

Before pursuing the second property, some additional definitions are required. An *atomic polygon* is an individual cell in a tessellation at the lowest level k ($k=0$) in a hierarchy of tessellations. A *molecular polygon* is an aggregate of atomic polygons used in forming the higher levels ($k>0$) of a hierarchy. The molecular polygon need not be the same shape as the atomic polygon. When the cells at level k of a hierarchy have the same shape as those at level ($k+1$),

the tessellation is said to be unlimited. Alternatively, we can think of unlimited tessellations as lacking a definable atomic polygon since any cell can always be subdivided into cells of the same shape.

The second property suggests that we require unlimited tessellations. Of the 11 Laves nets, only four are unlimited (Bell et al 1983). Two of these are regular tessellations, $[6^3]$ and $[4^4]$, each of which are capable of generating an infinite number of different molecular tessellations where each first level molecular polygon consists of $s=n^2$ ($n>1$) atomic polygons (see Figure 7). The other two unlimited tessellations are $[4.8^2]$ and $[4.6.12]$, each of which gives rise to two types of hierarchy: $[4.8^2]$ has an ordinary ($s=n^2$, $n>1$) and a rotation (135 degrees between levels) hierarchy ($s=2n^2$, $n>1$); while $[4.6.12]$ has an ordinary and a reflection hierarchy ($s=3n^2$, $n>1$) (see Figure 7).

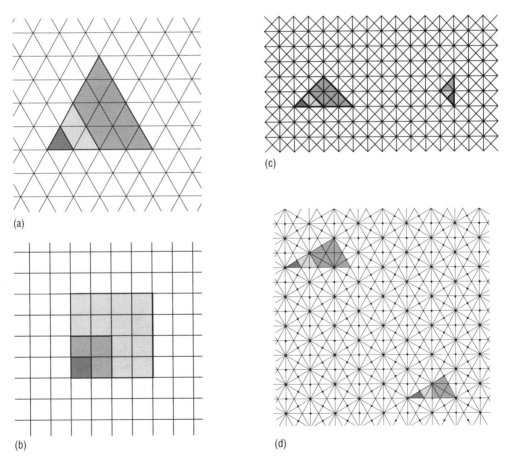

Fig 7. Unlimited tessellations (a) $[6^3]$; (b) $[4^4]$; (c) $[4.8^2]$, ordinary hierarchy (left-hand side), rotation hierarchy (right-hand side); (d) $[4.6.12]$, ordinary hierarchy (left-hand side), reflection hierarchy (right-hand side).

507

Bell et al (1983, 1989) suggest additional tessellation properties which are valuable in image processing and automated cartography applications. Of these, the two most important are uniform adjacency, whereby the distances between the centroid of a given cell and those of neighbouring cells (whether edge or vertex neighbours) are the same, and uniform orientation which means that all cells have the same orientation. However, none of the four unlimited tessellations possess the first property; [4^4] has two adjacency distances, [6^3] three, [4.8^2] eight, and [4.6.12] 16. Further, only [4^4] displays uniform orientation. This situation leads to a reconsideration of the remaining regular tessellation [3^6] which possesses both properties even though it is not unlimited.

Although [3^6] cannot be decomposed beyond the atomic tessellation without changing the shape of the cell, by defining molecular polygons made up of different numbers of atomic hexagons, it is possible to generate a large number of hexagonal hierarchies. While such hierarchies can always be grouped upwards, they cannot necessarily be decomposed downwards. These molecular polygons are referred to as n-*shapes*. More than one *n*-shape is possible for a given value of *n* (see Figure 8) and by using different rotations, the same molecular cell can give rise to more than one hierarchy (Diaz 1986) (compare the 'propeller' in Figure 8(b) with the 'wombat' in Figure 8(c)).

The *n*-shape which has received most attention is the 7-shape whose form most closely resembles that of the atomic hexagon (see Figure 9). Unfortunately, as Figure 9 shows, each level of the hierarchy formed by this 7-shape is rotated by an irrational angle with respect to the previous one.

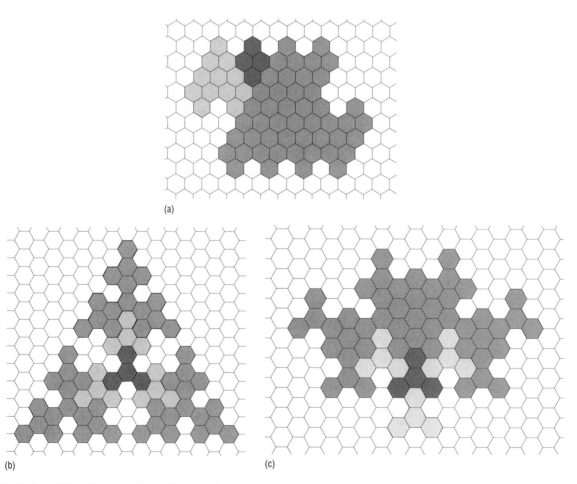

(a)

(b)   (c)

Fig 8. Three different hexagonal hierarchies using 4-shapes.

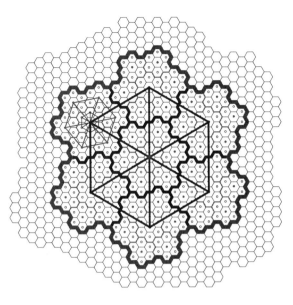

Fig 9. Hexagonal hierarchy using a 7-shape.

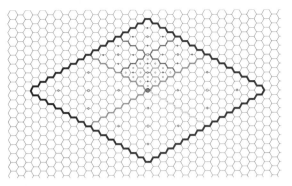

Fig 10. Hexagonal or rhombic (HoR) tessellation.

Even though the cells of a hexagonal tessellation are not infinitely recursively decomposable, by representing each hexagonal cell by its centroid, a triangular lattice is formed in which it is possible to embed similar (triangular), finer grids (Holroyd and Bell 1992) (e.g. the lattices formed by the cell centroids in Figure 9). Recognising this, Bell et al (1989) propose a compromise between [4^4] and [3^6] by presenting a point lattice over which both a hexagonal and a rhombic (4-edge) tessellation can be placed (see Figure 10). At the lowest hierarchical level, the lattice points are simultaneously the vertices of the rhombic tessellation and the centroids of the hexagonal tessellation. The rhombic lattice has the same properties as [4^4] in terms of adjacency and unlimitedness while the hexagonal tessellation can be amalgamated by a 4-shape which is only limited at the atomic level and which, unlike the 7-shape hexagonal tessellation described above, also maintains the same orientation for molecular cells at all levels. Bell et al (1989) call this the hexagonal or rhombic (HoR) tessellation. Despite providing an addressing system for HoR and showing that it has advantages over the addressing system for the 7-shape, they do not appear to have been successful in persuading others to adopt it.

As another illustration of how the properties of regular tessellations influence their use in GIS,

consider a situation from spatial statistical analysis. Recently, there has been considerable interest in integrating spatial data analysis in GIS (see Anselin, Chapter 17; Church, Chapter 20; Fischer, Chapter 19; Getis, Chapter 16; Openshaw and Alvanides, Chapter 18; Ding and Fotheringham 1992; Fotherigham and Rogerson 1993; Goodchild et al 1992). One measure used extensively in spatial analysis which is already incorporated in many GIS is spatial autocorrelation. Although spatial autocorrelation has numerous uses in spatial data analysis (Griffith 1992), one of the most fundamental is as an indicator of the spatial pattern exhibited by the values of a variable x recorded at a set of observations (cells, in the case of tessellations) located in space. The most frequently used measure of spatial autocorrelation is Moran's I given by Equation 8:

$$I = (n \, / \sum_i \sum_j d_{ij}) \, [\sum_i \sum_j d_{ij} \, (x_i - \bar{x}) \, (x_j - \bar{x}) \, / \sum_i (x_i - \bar{x})^2] \quad (8)$$

where n is the number of cells, d_{ij} is a measure of the spatial relationship between cells i and j, x_i is the value of variable x for cell i, and $\bar{x}$ is the mean value of x.

The expected value of I is $E(I) = -[1/(n-1)]$. Positive spatial autocorrelation, $I > E(I)$, occurs when similar values of x are found in spatial juxtaposition while negative spatial autocorrelation, $I < E(I)$, occurs when neighbouring values of x are dissimilar. Since this statistic incorporates a measure of the spatial association between pairs of cells in the tessellation, the calculated value of I reflects the underlying geometry of the tessellation as well as the values of the variable x. In particular, Jong et al (1984) show that the tessellation geometry imposes limits on the feasible values of I. To demonstrate this, first consider an alternative graph theory representation of a tessellation. This involves defining a binary connectivity matrix, **C,** whose elements $c_{ij}=1$ if cells i and j have a common edge, and 0, otherwise (by definition $c_{ii}=0$). Jong et al (1984) demonstrate that the maximum and minimum values of I can be obtained from the largest and smallest eigenvalues of the matrix

$$\mathbf{MCM} = (\mathbf{I}{-}\mathbf{1}\mathbf{1}^T/n)\ \mathbf{C}(\mathbf{I}{-}\mathbf{1}\mathbf{1}^T/n) \qquad (9)$$

The limits on I are investigated here for the three regular tessellations. Since the geometry of a bounded tessellation is not independent of the number of cells n in the tessellation (e.g. as n increases, the proportion of boundary cells decreases), tessellations of three different sizes ($n=64, 256, 1024$) are examined.

Table 1 shows a number of ways in which the limits on I are influenced by the geometry of the tessellation. First, for any given regular tessellation, the range of feasible values of I changes as n changes. Second, for any given value of n, the limits for the tessellations of squares and triangles are quite similar and, further, this similarity increases as n increases. Third, the absolute values for the upper and lower limits for the tessellations of squares and triangles are approximately equal, indicating that similar magnitudes of spatial autocorrelation of both a positive and negative kind can occur. In contrast, for a tessellation of hexagons, extreme positive spatial autocorrelation is approximately twice the magnitude of extreme negative spatial autocorrelation. Clearly, these results suggest that caution should be exercised when interpreting differences in values of I obtained from spatial patterns observed in different tessellations, and that such differences should not be ascribed solely to different spatial characteristics of the variable(s) being studied.

Further results with relevance for GIS can be derived from **C**. Griffith (1996) shows that the

eigenvectors of **MCM** identify the possible mutually exclusive geographical patterns of attribute values with levels of spatial autocorrelation equal in magnitude to the associated eigenvalues. To illustrate this, the three regular tessellations for the three sizes considered above are used. The geographical patterns of spatial autocorrelation remain essentially the same for a given tessellation as n changes. For example, Figure 11 shows the spatial patterns of the first nine eigenvectors of **MCM** for a tessellation of squares. Note that numerical eigenvectors are unique to a multiplicative factor of -1 so that patterns such as those for the third eigenvectors for $n=256$ and $n=1024$ are considered identical even though they are mirror images of each other. Also note that the pattern for eigenvector 1 for $n=64$ is the same as that for eigenvector 2 for $n=256$, while eigenvector 2 for $n=64$ is the same as that for eigenvector 1 for $n=256$ (the same also holds for the eighth and ninth eigenvectors of the two tessellations). This arises because the order of the two eigenvectors is arbitrary since they have the same associated eigenvalue. However, observe that different patterns result for different tessellations. Compare the patterns in Figures 12 and 13 for $n=1024$ for tessellations of triangles and hexagons with the corresponding patterns for the tessellation of squares in Figure 11(c). Only the pattern for the first eigenvector of **MCM** is the same for all three tessellations. Some patterns, such as that displayed by the fourth eigenvector for the tessellation of squares, are unique. Thus, all patterns of spatial autocorrelation are not equally likely to occur for all regular tessellations.

4 IRREGULAR TESSELLATIONS

Many tessellations encountered in GIS, such as those formed by the spatial units of a choropleth map, are highly variable in terms of the characteristics of their constituent cells. Consequently, we might anticipate that it would not be possible to identify any properties of such tessellations beyond the general ones described in Section 2. However, empirical investigation has revealed that two linear relationships hold for a typical (i.e. randomly selected) n-sided cell in many irregular, normal tessellations consisting of only

Table 1 Limits on the value of Moran's *I* for regular tessellations.

n	Squares	Triangles	Hexagons
64	-1.0739 0.9747	-1.0725 1.0330	-0.5519 1.0065
256	-1.0485 1.0216	-1.0477 1.0375	-0.5330 1.0403
1024	-1.0276 1.0206	-1.0273 1.0247	-0.5186 1.0306

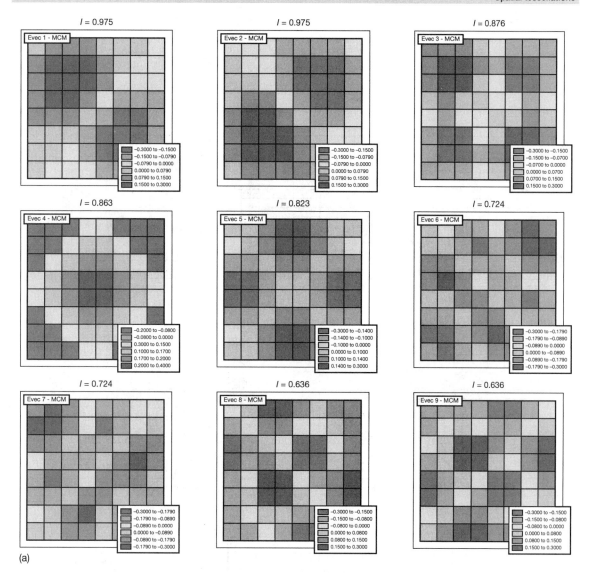

Fig 11. Spatial patterns of the elements of the first 9 eigenvectors of MCM for a tessellation of squares: (a) *n*=64; (b) *n*=256; and (c) *n*=1024. (Continued on pages 512–13)

convex cells. Because of their extensive occurrence, these relationships are usually referred to as 'laws'.

The first is *Lewis' law*, so named because the relationship was first observed by Lewis (1928, 1930, 1931, 1943, 1944) in empirical studies of a variety of biological tessellations. Lewis' law states that the average area of a typical *n*-sided cell increases with *n* in a linear fashion as described by Equation 10:

$$\bar{A}_n = (A_0 / C) + b(A_0 / C)(n–6) \qquad (10)$$

where $\bar{A}_n$ = average area of a *n*-sided cell; A_0 = total area of the tessellation; C = number of cells in the tessellation; and b = a constant.

The other is *Aboav's law*, in recognition that it was first observed by Aboav (1970) in studies of various polycrystalline materials. This law states that the total ñumber of sides of the cells neighbouring a typical *n*-sided cell is linear in *n* as described in Equation 11:

$$nm_n = (6a+\mu_2) + (\bar{n} – a)n \qquad (11)$$

511

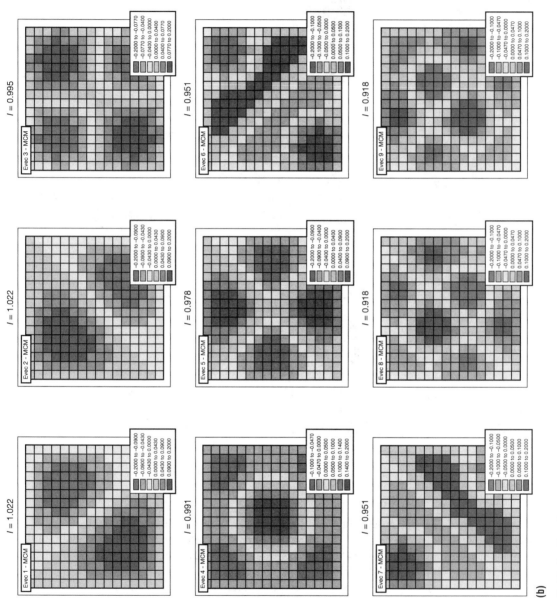

Fig 11. (continued)

(b)

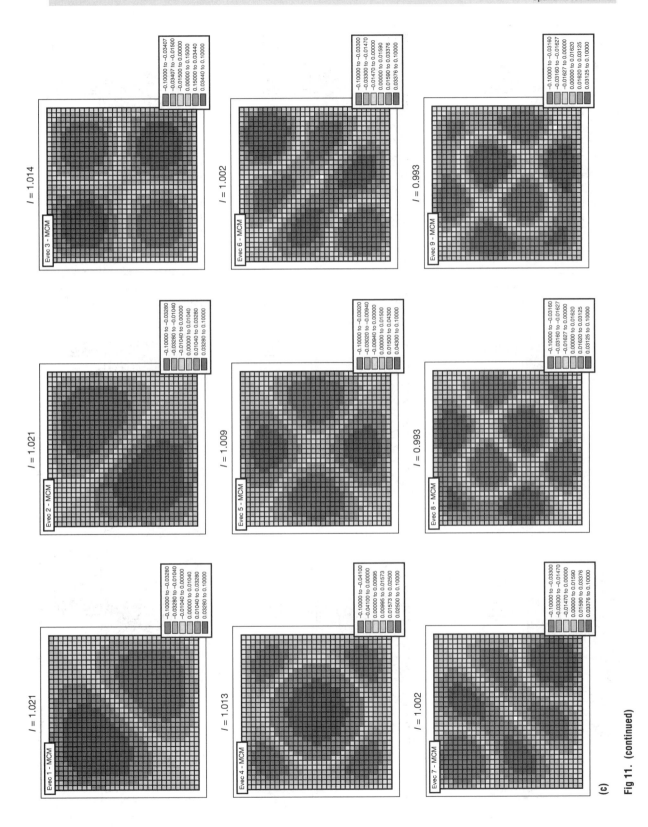

Fig 11. (continued)

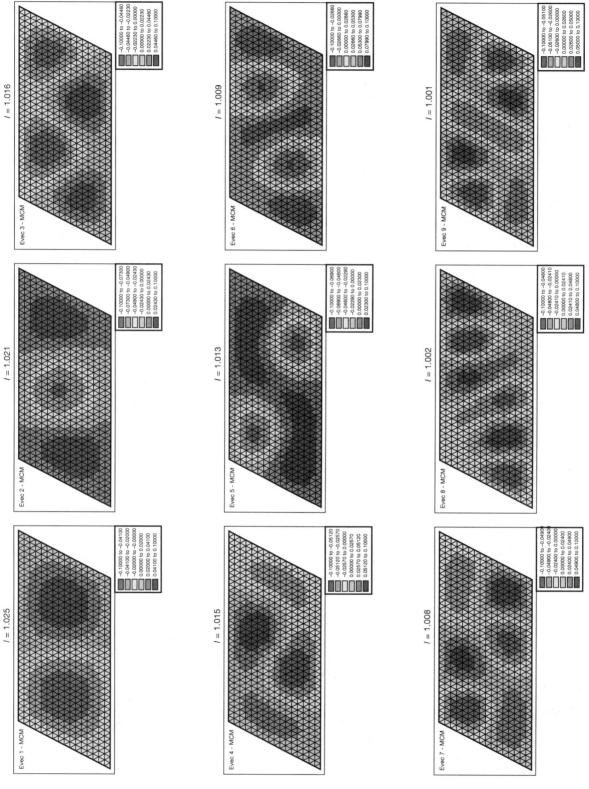

Fig 12. Spatial patterns of the elements of the first 9 eigenvectors of MCM for a tessellation of triangles, n=1024.

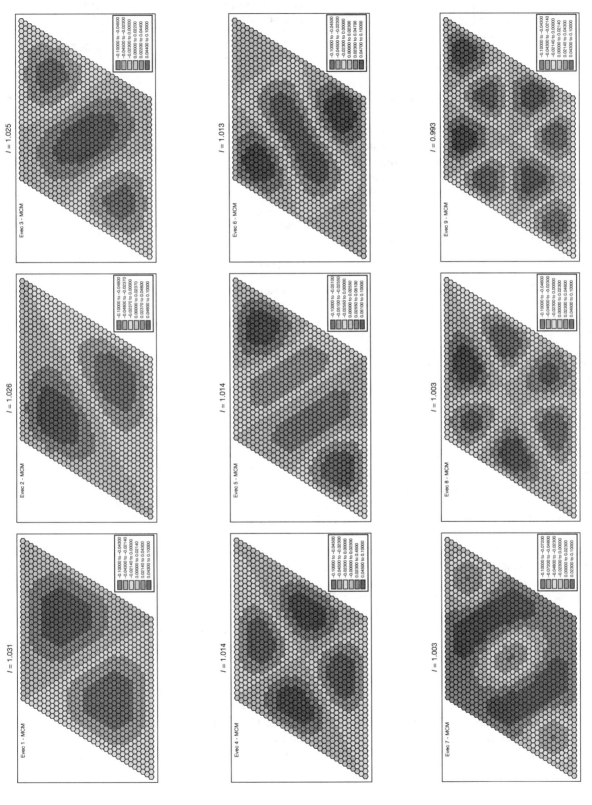

Fig 13. Spatial patterns of the elements of the first 9 eigenvectors of MCM for a tessellation of hexagons, *n*=1024.

515

where m_n = number of sides of a randomly chosen neighbour of a typical n-sided cell; $\bar{n}$ = mean value of n for the tessellation;
$\mu_2 = \overline{n^2} - (\bar{n})^2$;
$\overline{n^2}$ = mean value of n^2 for the tessellation.

This law implies that, on average, cells with few sides are likely to be adjacent to those with many sides, and vice versa.

No satisfactory explanations for the laws have yet been found. Most arguments involve an approach which maximises the entropy of p_n, the distribution of n for the tessellation, subject to a minimal set of constraints (Chiu 1995; Peshkin et al 1991; Rivier 1985, 1986, 1990, 1991, 1993, 1994; Rivier and Lissowski 1982). Both laws are seen as the most likely relationships to arise by chance in the absence of any other constraints.

Given that many kinds of tessellation encountered in GIS (e.g. politico-administrative units at all spatial scales) are likely to contain non-convex cells and may have non-trivalent vertices, it might be anticipated that neither law has much relevance to GIS. However, this conclusion would be incorrect. To illustrate this, consider the counties of the states of the USA. To avoid departing completely from the conditions specified in the laws, states with less than 25 'internal' counties or with more than 10 non-trivalent vertices are not considered. 'Internal' counties are those for which none of the sides are part of the state boundary (including coastlines). These restrictions result in 24 states being examined (see Table 2). Somewhat surprisingly, the counties of 18 states are consistent with either Lewis' law or Aboav's law, while both laws hold for eight states (see Table 2). Figure 14 shows a weighted least squares (WLS) fit of Lewis' law to the data for Pennsylvania (see Table 3), while Figure 15 shows a similar WLS fit of Aboav's law for Georgia (see also Table 4). Both laws also hold for both the 94 interior departments and 222 interior arrondissements of France (Le Caer and Delannay 1993), while Aboav's law holds for the interior administrative subdivisions of Indian states (Boots 1979), parishes of counties in southwest England (Boots 1980), and parishes of Lorraine, France (Pignol et al 1993).

What are the implications of these findings for GIS? The most obvious is that, because of the geometric constraints imposed on the 2-dimensional

Table 2 The incidence of Lewis' law and Aboav's law for counties of selected states in the USA.

State	Lewis' law	Aboav's law
Alabama	*	
Arkansas		*
California		
Colorado	*	*
Georgia	*	*
Illinois	*	*
Indiana	*	*
Kentucky		
Louisiana	*	*
Minnesota		
Mississippi		*
Missouri	*	
Montana	*	*
New York	*	
N. Carolina	*	*
N. Dakota		*
Ohio		*
Oklahoma	*	
Pennsylvania		*
S. Dakota		
Tennessee	*	*
Virginia	*	
W. Virginia		
Wisconsin		

Table 3 Lewis' law for the counties of Georgia, USA.

n	$f(n)$	$\bar{A}_n{}^*$
4	10	0.711
5	32	0.752
6	35	1.030
7	24	1.253
8	6	1.375
9	1	1.093
10	1	1.540

$\bar{A}_n{}^* = \bar{A}_n / (A_0 / C)$

For definition of symbols, see text.

tessellations, there is more order present in irregular tessellations than might be suspected and that other laws may well await discovery. In terms of sampling tessellations for statistical analysis, in order to derive a sample of independent cells, it will be necessary to ensure that cells which are neighbours are not chosen. The findings also have implications for local data structures, storage, and revision.

Table 4 Aboav's law for the countries of Pennsylvania, USA.

n	f(n)	nmn
3	1	23.000
4	2	25.714
5	15	30.588
6	11	37.200
7	5	38.652
8	3	47.304
9	1	51.000

For definition of symbols, see text.

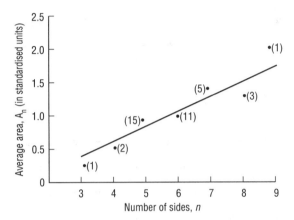

Fig 14. Weighted least squares fit of Lewis' law to the counties of Pennsylvania. Values in brackets are the number of observations used to calculate the mean value.

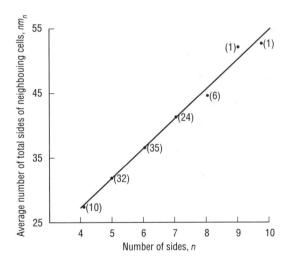

Fig 15. Weighted least squares fit of Aboav's law to the counties of Georgia. Values in brackets are the number of observations used to calculate the mean value.

5 GENERALISED VORONOI DIAGRAMS

As well as occurring directly, tessellations can also arise from data transformations performed in GIS. One such transformation involves an operation which parallels that used to create a dual tessellation (see section 1) and gives rise to a large family of tessellations known collectively as *generalised Voronoi diagrams* (GVDs). In two dimensions such tessellations can be created for any set of s-dimensional ($s \leq 2$) geometric entities (generators) in the plane such as points, line segments, arcs or polygons, or any combination of such elements, by assigning each location in the plane to the 'nearest' generator (Figure 16). By using different distance metrics, different definitions of nearest are possible. The presence of obstacles (non-generator, s-dimensional ($s \leq 2$) entities) can also

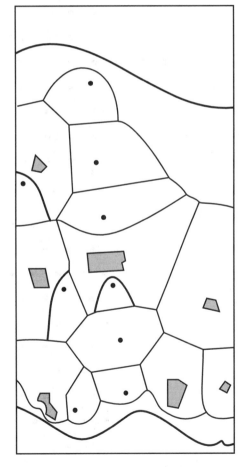

Fig 16. Voronoi diagram of points (stations), lines (rivers), and polygons (parks) in part of the Sumida-Kohto district of Tokyo.

517

be accommodated. By nature of their construction GVDs are necessarily edge-to-edge tessellations. Note that the three regular tessellations (square, triangle, hexagon) described in Section 1 can be generated by defining the Voronoi diagram of a set of points located on a square, hexagonal, and triangular lattice, respectively (see Figure 17). Conceptualised in this way, regular tessellations can be thought of as representing both area and point information simultaneously (Gold 1990).

GVDs are particularly useful for performing a variety of nearest neighbour operations which address locational issues arising in spatial analysis and planning, including solving continuous location problems of both location-allocation and locational optimisation kinds (Okabe et al 1994; Okabe and Suzuki 1995). In addition, individual types of GVD are useful for addressing other issues. To illustrate the potential of GVDs, two types are considered, chosen because of their implications for GIS.

5.1 Ordinary Voronoi diagram

The ordinary Voronoi diagram (OVD) uses individual points and a Euclidean distance metric to define the tessellation. Formally, suppose that we have a set of n

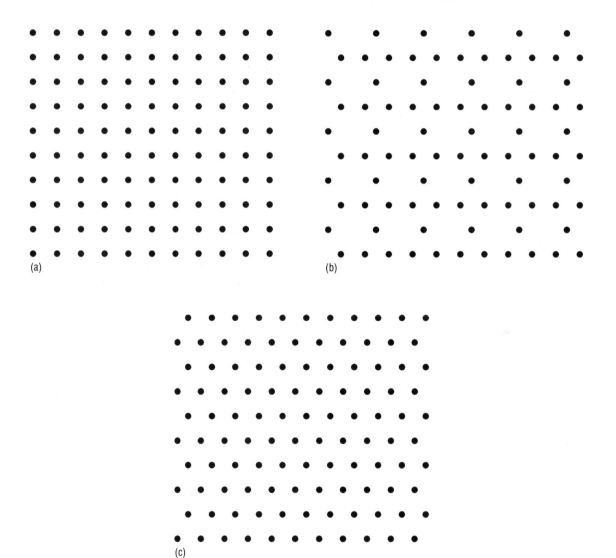

Fig 17. (a) Square, (b) hexagonal, and (c) triangular point lattices.

$(2 \leq n \leq \infty)$ distinct points (generators), $P = \{p_1,...,p_n\}$, located in a finite region S in 2-dimensional space $\Re^2$. To avoid complicated treatment, assume that S is convex. Let $d(p, p_i)$ be Euclidean distance between location p and generator p_i.

We define the region given by:

$$V(p_i) = \{p \mid p \in S; d(p, p_i) \leq d(p, p_j), j \neq i, j=1,...,n\} \quad (12)$$

as the ordinary Voronoi polygon (OVP) associated with p_i and the set given by

$$\upsilon(P) = \{V(p_1),..., V(p_n)\} \quad (13)$$

as the OVD of P.

Thus, the interior of $V(p_i)$ consists of all locations in S which are closer to p_i than to any other generator, while the edges and vertices of $V(p_i)$ represent those locations which are equidistant from two or more generators (see Figure 18). Although Voronoi diagrams can be defined for either raster-based or vector-based data structures, the above definition is consistent with the latter. This situation is maintained for all the GVDs discussed here since it permits a more explicit treatment of their topological relationships.

To show how the OVD can be used for locational decision-making, assume that the set of points P in Figure 18 are fire stations and S is a city and that among the questions to be answered are the following:

1 What is the area of the city for which the nearest fire station is the one at p_i?
2 Which is the nearest fire station to a given location p?
3 Which location in the city is farthest from a fire station?

The answer to 1 is given by the OVP of p_i, while 2 is answered by observing which OVP contains p. For example, in Figure 18, p is closest to the station at p_3. Since $V(p_i)$ is a convex polygon, the location in $V(p_i)$ which is farthest from p_i is found in the vertices of $V(p_i)$. These vertices may include those formed by the set of intersections of cell edges with the edges of S, and the set of vertices of S. Define q_{ij} as the jth vertex of $V(p_i)$, $j=1,..., n_i$ where n_i is the number of vertices of $V(p_i)$. Then the farthest location in $V(p_i)$ from p_i is given by the vertex q_i^* of $V(p_i)$ that satisfies

$$d(p_i, q_i^*) = \max \{d(p_i, q_{ij}) \mid j = 1,...,n_i\}. \quad (14)$$

For example, in Figure 18, for $V(p_1)$ this is q_1^*. (3) may be answered by searching for the longest distance among

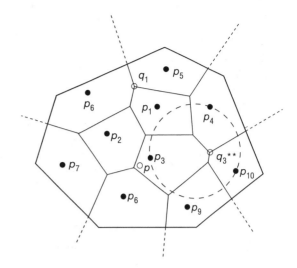

Fig 18. Ordinary Voronoi diagram.

$$\{d(p_i, q_i^*) \mid i = 1,...,n\} \quad (15)$$

or by identifying the vertex q_k^{**} which satisfies

$$d(p_k, q_k^{**}) = \max \{d(p_i, q_i^*) \mid i = 1,...,n\}. \quad (16)$$

In Figure 18, this is the vertex q_3^{**}. Note that q_k^{**} is the centre of a circle radius $d(p_k, q_k^{**})$ (the dashed circle in Figure 18). This circle is the largest circle whose centre is in S and which does not contain any points in P in its interior.

Another fundamental use of the OVD is to define spatial relationships between individual points belonging to a planar point set P such as that defined above. Most work has concentrated on adjacency relationships or the problem of defining a set of neighbours for a given point p_i in P. One solution to this problem is to use the so called 'natural' neighbours of p_i (Sibson 1981) which are those points whose OVPs are adjacent to (p_i) in $V(P)$. For example, in Figure 18 $\upsilon p_1, p_3, p_6, p_7$, and p_8 are the natural neighbours of p_2. This solution has been used in a variety of situations. One involves operationalising spatial models (Besag 1974, 1975; Ord 1975) including spatial autocorrelation models such as Moran's I discussed in section 3. For example, in a study of spatial trends in the grain handling system in the province of Manitoba, Canada from 1943 to 1975, Griffith (1982) set $d_{ij} = 1$ in Equation 8 if two grain handling centres were natural neighbours and $d_{ij} = 0$ otherwise. Another example arises in missing data problems where the unknown value of a variable at a given location must be estimated from known values at other locations. In their study of rainfall data for Kansas and Nebraska, USA, Haining et al (1984) use natural neighbours to identify which weather stations to use in estimating missing values at other stations.

The missing value problem represents a special case of the more general problem of spatial interpolation (Mitas and Mitasova, Chapter 34). Here there exists a set of n data sites $P=\{p_1,...,p_n\}$, located in (a 2-dimensional) space S, at which the values of some variable z are observed. If it is assumed that these values are observations from a surface defined by z over S, spatial interpolation involves finding a function $f(x)$ which best represents the entire surface and which predicts the values of z for locations other than P. Local interpolants represent the value of the surface $f(x)$ at an arbitrary location p in S as a weighted, usually linear, function of values at nearby data sites, $D(p)$ $(D(p) \subset P)$, so that

$$f(x) = \sum_{i=1}^{n_D} w_i z_i, \, p_i \in D(p) \tag{17}$$

where n_D is the number of nearby sites and w_i is the weight attached to p_i.

There are many ways in which $D(p)$ can be selected (Watson 1992) but one, which also has other advantages described in section 5.2.3, is to use the natural neighbours of p (Gold 1991; Sambridge et al 1995; Sibson 1981; Watson and Philip 1987).

That adjacency relationships for generators are uniquely defined in the OVD (and GVDs, in general) has led Gold (1991, 1992) to propose the GVD as an alternative spatial data model to both the raster and vector ones since it possesses desirable properties of both; a known spatial adjacency structure (raster) and a one-to-one mapping with 'real' map objects (vector).

In terms of spatial relationships, however, the OVD (and GVDs, in general) is not limited to considerations of adjacency. Edwards (1993) and Edwards and Moulin (1995) show that a wide range of linguistic concepts of space such as 'near', 'between', 'among', etc. are amenable to such treatment. For example, suppose there exists a set of fixed reference points, $R_1, ...,R_{10}$ as shown in Figure 19(a) and a displaceable query point Q. Realisation of the relative concepts 'near' and 'far' for Q with respect to the pair of reference points R_1, R_2, can be achieved in the

following way. First, define the OVD of the reference points (Figure 19(b)). Next, define the OVD of the reference points plus the query point Q. Figures 19(c) and 19(d) show such OVDs for two different locations of Q. Comparison of Figures 19(c) and 19(d) with Figure 19(b) reveals that the OVP for Q is created by 'stealing' pieces of the OVPs of the reference points (the shaded regions in Figures 19(c) and (d). In Figure 19(c) Q is 'near' to R_1, R_2 and as a result much of its OVP is stolen from the OVPs of these points. In contrast, in Figure 19(d) Q is 'far' from R_1, R_2, so that it steals little of their OVPs. Edwards and Moulin (1995) suggest that the sum of the areas of the regions stolen by Q from R_1, R_2 relative to the area of the OVP of Q may be used as a way of quantifying this notion.

A final use of the OVD is to reconstruct tessellations from incomplete data. For example, in the UK considerable use has been made of the postcode system as a means of georeferencing socio-economic data. However, since no boundaries are defined for unit postcodes, there is a problem reconciling such data with census geography. As a solution, Boyle and Dunn (1991) suggest creating unit postcode zones by defining a Voronoi polygon for each address location contained in the postcode (see Figure 20).

5.2 Higher-order Voronoi diagrams

5.2.1 Order-k Voronoi diagram

As with the ordinary Voronoi diagram, begin with a set of points $P = \{p_1, ..., p_n\}$ but now, instead of dealing with individual points, consider subsets of k points selected from P. Although any value of k ($k<n$) may be considered, for simplicity the situation where $k = 2$ is examined, that is, the focus is on pairs of points. The extension to $k>2$ is described by Okabe et al (1992: 142–158).

Let $A^{(2)}(P) = \{P_1^{(2)}, ..., P_i^{(2)}, ..., P_l^{(2)}\}$, where $P_i^{(2)} = \{p_{i1}, p_{i2}\}, p_{i1}, p_{i2} \in P$ and $l = {}_nC_2$, be all the possible subsets of P which consist of two points. Let p represent an arbitrary location in the plane and $d(p, p_{ij})$ the Euclidean distance from p to p_{ij}. We define the order-2 Voronoi polygon (O2VP) of $P_i^{(2)}$ as

$$V(P_i^{(2)}) = \{p \mid d(p,p_{i1}) \leq d(p,p_j) \tag{18}$$
$$\text{and } d(p,p_{i2}) \leq d(p,p_j) \text{ for } p_j \in P \backslash P_i^{(2)}\}.$$

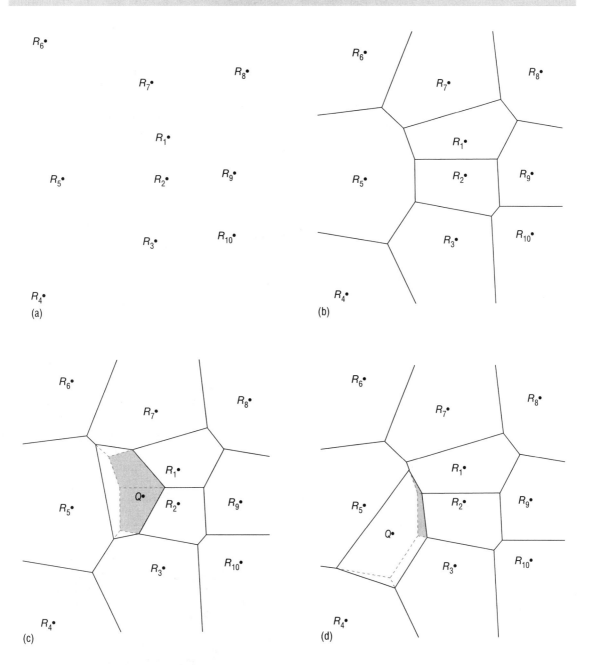

Fig 19. (a) Set of reference points; (b) Ordinary Voronoi diagram of reference points; (c) and (d) Ordinary Voronoi diagram of reference points and query point Q for two different locations of Q.

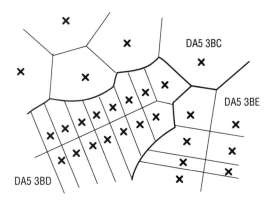

Fig 20. Generating unit postcode boundaries.

Thus, $V(P_i^{(2)})$ consists of all locations for which either p_{i1} or p_{i2} is the first or second nearest point. The set

$$\upsilon(A^{(2)}(P)) = \upsilon^{(2)} = \{V(P_1^{(2)}), ..., V(P_l^{(2)})\} \quad (19)$$

is called the order-2 Voronoi diagram (O2VD) generated by P. Figure 21 shows the order-2 Voronoi diagram for the point set in Figure 18.

5.2.2 Ordered, order-*k* Voronoi diagram

In the order-*k* Voronoi diagram there is no concern with which of the *k* points in $P_i^{(k)}$ is the nearest, second nearest, ..., or *k*th nearest. However, in the ordered, order-*k* Voronoi diagram, this order is

considered explicitly. Again, we examine the situation where $k = 2$, that is, for a pair of points. Let $A^{<2>}(P)$ be the set of all ordered pairs of points obtained from the set of points $P = \{p_1, ..., p_n\}$, that is, $A^{<2>}(P) = \{P_1^{<2>}, ..., P_i^{<2>}, ..., P_l^{<2>}\}$ where $P_i^{<2>} = (p_{i1}, p_{i2}), p_{i1}, p_{i2} \in P$, and $l = n(n-1)$.

Let p be an arbitrary location in the plane and $d(p, p_{ij})$ the Euclidean distance from p to p_{ij}. For a set $P_i^{<2>}$ in $A^{<2>}(P)$, define

$$V(P_i^{<2>}) = \{p \mid d(p,p_{i1}) \le d(p,p_{i2}) \le d(p,p_j), \quad (20)$$

$$p_j \in P\backslash\{p_{i1}, p_{i2}\}\}.$$

The set $V(P_i^{<2>})$ is called the ordered, order-2 Voronoi polygon (OO2VP) associated with $P_i^{<2>}$. $V(P_i^{<2>})$ consists of all locations for which p_{i1} and p_{i2} are the first and second nearest points, respectively.

The set

$$\upsilon(A^{<2>}(P)) = \upsilon^{<2>} = \{V(P_1^{<2>}), ..., V(P_l^{<2>})\} \quad (21)$$

is called the ordered, order-2 Voronoi diagram (OO2VD) of P. The OO2VD corresponding to the O2VD of Figure 21 is shown in Figure 22.

5.2.3 Applications

As with the ordinary Voronoi diagram, higher order Voronoi diagrams can be used to address locational problems. Once more assume that the points in Figure 18 represent the locations of fire stations in a city. However, this time the possibility

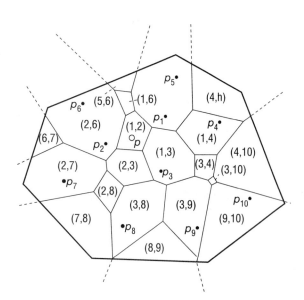

Fig 21. Order-2 Voronoi diagram.

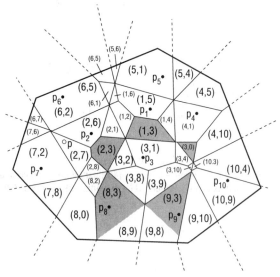

Fig 22. Ordered order-2 Voronoi diagram.

is recognised that, on a given occasion, the equipment at the fire station closest to a given location may already be fully committed, leading to questions such as the following:

1 What are the two closest fire stations to a given location p?
2 What are the first and the second nearest fire stations to a given location p?
3 What is the region of the city for which the station at p_i is the second nearest?

Question 1 may be answered by examining the O2VD of the fire stations and observing in which O2VP the query location p occurs. For example, the two nearest fire stations to p in Figure 21 are those located at p_1, p_2. Questions 2 and 3 require the consideration of the OO2VD of the fire stations. The answer to (2) is found by observing the OO2VP in which p is located (e.g. the first and second nearest fire stations to p in Figure 22 are those at p_2 and p_7, respectively). To answer question 3 we need to find all OO2VPs for which station p_i is the second nearest. The union of these polygons gives the required region (e.g. for the station at p_3 this is the shaded region in Figure 22).

The OO2VD can also be used in spatial interpolation. Recall from the previous section that local interpolants represent the value of a surface $f(x)$ at an arbitrary location p in the plane as a weighted function of the data values at a set of nearby data sites (see Equation 17). It has already seen how natural neighbour relationships can be used to determine the data sites selected. If the ordinary Voronoi diagram is created of the set of data sites $P = \{p_1, ..., p_n\}$ plus p, ϑ $(P \cup p)$, we know that the OVP of p, $V(p)$, can be exhaustively subdivided into OO2VPs $V((p,p_i))$ (see Figure 23). Following Sibson (1981), if we let $| V((p,p_i)) |$ be the area of $V((p,p_i))$, the normalised values $| V((p,p_i)) | / \Sigma | V((p,p_i)) |$ can be used as the weights in Equation 17 (see Okabe et al 1992: 347–51 for details). Note that $V((p,p_i))$ can also be considered as the portion of $V(p_i)$ which is stolen by the OVP $V(p)$ created when p is added to P. Thus, the operationalisations of spatial concepts proposed by Edwards and Moulin (1995) and discussed in section 5.1 can also be thought of in terms of OO2VDs.

The area stealing aspect of OO2VPs is also useful in creating maps of nominal scale variables, such as land use, soil or vegetation types, or surficial geology,

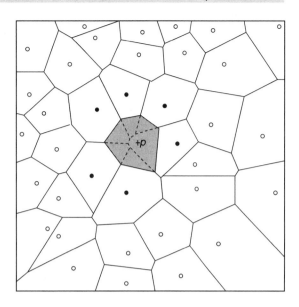

Fig 23. Voronoi diagram ($P \cup p$). Points shown as filled circles are the natural neighbours of p.

from values observed at a set of sampling points $P = \{p_1, ..., p_n\}$, especially if we wish to convey the degree of uncertainty involved in the map content (Lowell 1994). Traditional thematic maps are constructed using a Boolean logic which assigns non-sampled locations to one, and only one, class, even though most locations have the possibility of belonging to several different classes. Lowell argues that such situations are better represented by fuzzy maps in which non-sampled locations are assigned fuzzy membership values (FMVs) reflecting the degree of certainty of their belonging to a given class. To generate the FMVs for a given location q, Lowell suggests generating the OVD of the sample points, P, plus q (see Figure 24) and then examining the OO2VPs $V((q,p_i))$. The FMV for a given class x is obtained as the sum of the areas of the $V((q,p_i))$ for which the value at p_i is x, relative to the area of the OVP of q. Thus, in Figure 24, the FMVs for q for classes A, B and C are 0.25, 0.63, and 0.12 respectively.

6 CONCLUSIONS

In GIS, the attention paid to characteristics of different tessellations depends greatly on the context. Typically, such matters receive explicit consideration only when an appropriate spatial data model is being

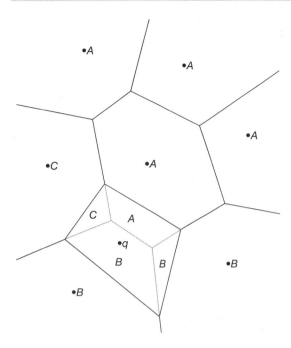

Fig 24. Deriving fuzzy membership values using the ordered order-2 Voronoi diagram.

which such diagrams can perform are the operationalisation of fundamental spatial concepts such as near, adjacent, and between; spatial interpolation of both nominally- and interval/ratio-scaled variables; and solving locational optimisation problems. While the potential for using tessellations in such roles is already considerable, it is expected to grow further as new forms of tessellations continue to be developed in a number of disciplines interested in various aspects of spatial modelling.

References

Aboav D A 1970 The arrangement of grains in a polycrystal. *Metallography* 3: 383–90

Ahuja N 1983 On approaches to polygonal decomposition for hierarchical image representation. *Computer Vision, Graphics, and Image Processing* 24: 200–14

Ammann R, Grunbaum B, Shephard G C 1992 Aperiodic tiles. *Discrete and Computational Geometry* 8: 1–27

Bell S B M, Diaz B M, Holroyd F C 1989 The HOR quadtree: an optimal structure based on a non-square 4-shape. In Brooks S R (ed.) *Mathematics in remote sensing*. Oxford, Clarendon Press: 315–43

Bell S B M, Diaz B M, Holroyd F C, Jackson M J 1983 Spatially referenced methods of processing raster and vector data. *Image and Vision Computing* 1: 211–20

Bell S B M, Holroyd F C 1991 Tesseral amalgamators and hierarchical tessellations. *Image and Vision Computing* 9: 313–28

Besag J 1974 Spatial interaction and the statistical analysis of lattice systems (with discussion). *Journal of the Royal Statistical Society*, Series B 36: 192–236

Besag J 1975 Statistical analysis of non-lattice data. *The Statistician* 24: 179–96

Boots B N 1979 The topological structure of area patterns: Indian administrative divisions. *The National Geographical Journal of India* 25: 149–53

Boots B N 1980 Packing polygons: some empirical evidence. *The Canadian Geographer* 24: 406–11

Boyle P J, Dunn C E 1991 Redefinition of enumeration district centroids: a test of their accuracy using Thiessen polygons. *Environment and Planning A* 23: 1111–19

Chiu S N 1995 A comment on Rivier's maximum entropy method of statistical crystallography. *Journal of Physics A: Mathematical and General* 28: 607–15

Diaz B M 1986 Tesseral addressing and arithmetic – overview and theory. In Diaz B, Bell S (eds) *Spatial data processing using tesseral methods*. Swindon, National Environment Research Council: 1–10

Ding Y, Fotheringham A S 1992 The integration of spatial analysis and GIS. *Computers, Environment, and Urban Systems* 16: 3–19

selected. Once this is accomplished and concerns shift towards analytical issues, tessellation characteristics are rarely considered again. While this chapter reinforces the value of assessing tessellation characteristics in spatial data model selection, it also demonstrates that one should not lose track of them in other contexts. In particular, it is shown by way of examples involving spatial autocorrelation that a tessellation's properties influence the results of analyses of data recorded over the tessellation. Fortunately, staying aware of such properties is not as onerous as might first appear since, as also noted in this chapter, the possible topologic and some metric properties of both regular and irregular tessellations in 2-dimensional space are quite constrained.

Of course, there are many situations in GIS where the objects under consideration do not take the form of tessellations. However, in some of these situations it is possible to generate tessellations from the original objects. Characteristics of these tessellations can then be used to undertake a variety of GIS-related tasks for the original objects. This approach is illustrated by exploring just two of the many members of a family of tessellations known as generalised Voronoi diagrams. Among the tasks

Edwards G 1993 The Voronoi model and cultural space: applications to the social sciences and humanities. In Frank A U, Campari I (eds) *Spatial information theory – a theoretical basis for GIS*. Berlin, Springer: 202–14

Edwards G, Moulin B 1995 Towards the simulation of spatial mental images using the Voronoi model. *Proceedings International Joint Conference on Artificial Intelligence (IJCAI-95), Montreal, Canada, 19–26 August*: 63–73

Fotheringham A S, Rogerson P A 1993 GIS and spatial analytical problems. *International Journal of Geographical Information Systems* 7: 3–19

Gold C M 1990 Spatial data structures – the extension from one to two dimensions. In Pau L F (ed.) *Mapping and spatial modelling for navigation*. Berlin, Springer: 11–39

Gold C M 1991 Problems with handling spatial data – the Voronoi approach. *CISM Journal* 45: 65–80

Gold C M 1992 The meaning of 'neighbour'. In Frank A U, Campari I, Formentini U (eds) *Theories and methods of spatio-temporal reasoning in geographic space*. Berlin, Springer: 221–35

Goodchild M F, Haining R P, Wise S 1992 Integrating GIS and spatial data analysis: problems and possibilities. *International Journal of Geographical Information Systems* 6: 407–23

Griffith D A 1982 Dynamic characteristics of spatial economic systems. *Economic Geography* 58: 177–96

Griffith D A 1992 What is spatial autocorrelation? Reflections on the past 25 years of spatial statistics. *L'Espace Geographique* 21: 265–80

Griffith D A 1996b Spatial autocorrelation and eigenfunctions of the geographic weights matrix accompanying geo-referenced data. *The Canadian Geographer* 40: 351–67

Grunbaum B, Shephard G C 1977a Tilings by regular polygons. *Mathematics Magazine* 50: 227–47

Grunbaum B, Shephard G C 1977b The 81 types of isohedral tilings in the plane. *Mathematical Proceedings of the Cambridge Philosophical Society* 82: 177–96

Grunbaum B, Shephard G C 1986 *Tilings and patterns*. New York, W H Freeman & Co.

Haining R P, Griffith D A, Bennett R 1984 A statistical approach to the problem of missing spatial data using a first-order Markov model. *The Professional Geographer* 36: 338–45

Holroyd F, Bell S B M 1992 Raster GIS: models of raster encoding. *Computers and Geosciences* 18: 419–26

Jong P de, Springer C, Veen F van 1984 On extreme values of Moran's *I* and Geary's *c*. *Geographical Analysis* 16: 17–24

Le Caer G, Delannay R 1993 Correlations in topological models of 2D random cellular structures. *Journal of Physics A: Mathematical and General* 26: 3931–54

Lewis F T 1928 The correlation between cell division and the shapes and sizes of prismatic cells in the epidermis of *Cucumis*. *Anatomical Record* 38: 341–76

Lewis F T 1930 A volumetric study of growth and cell division in two types of epithelium – the logitudinally prismatic cells of Tradescantia and the radially prismatic epidermal cells of *Cucumis*. *Anatomical Record* 47: 59–99

Lewis F T 1931 A comparison between the mosaic of polygons in a film of artificial emulsion and in cucumber epidermis and human amnion. *Anatomical Record* 50: 235–65

Lewis F T 1943 The geometry of growth and cell division in epithelial mosaics. *American Journal of Botany* 30: 766–76

Lewis F T 1944 The geometry of growth and cell division in columnar parenchyma. *American Journal of Botany* 31: 619–29

Loeb A L 1976 *Space structures – their harmony and counterpoint*. Reading (USA), Addison-Wesley

Lowell K 1994 A fuzzy surface cartographic representation for forestry based on Voronoi diagram area stealing. *Canadian Journal of Forest Research* 24: 1970–80

Okabe A, Boots B, Sugihara K 1992 *Spatial tessellations: concepts and applications of Voronoi diagrams*. Chichester, John Wiley & Sons

Okabe A, Boots B, Sugihara K 1994 Nearest neighbourhood operations with generalized Voronoi diagrams. *International Journal of Geographical Information Systems* 8: 43–71

Okabe A, Suzuki A 1995 Using Voronoi diagrams. In Dresner Z (ed.) *Facility location: a survey of applications and methods*. New York, Springer: 103–17

Ord J K 1975 Estimation methods for models of spatial interaction. *Journal of the American Statistical Society* 70: 120–6

Perry J N 1995 Spatial analysis by distance indices. *Journal of Animal Ecology* 64: 303–14

Peshkin M A, Strandburg K J, Rivier N 1991 Entropic predictions for cellular networks. *Physical Review Letters* 67: 1803–6

Peuquet D J 1984 A conceptual framework and comparison of spatial data models. *Cartographica* 18: 34–48

Pignol V, Delannay R, Le Caer G 1993 Characterization of topological properties of 2-dimensional cellular structures by image analysis. *Acta Stereologica* 12: 149–54

Rivier N 1985 Statistical crystallography. Structure of random cellular networks. *Philosophical Magazine B* 52: 795–819

Rivier N 1986 Structure of random cellular networks. In Kato Y, Takaki R, Toriwaki J (eds) *Proceedings, First International Symposium for Science on Form*. Tokyo, KTK Scientific: 451–8

Rivier N 1990 Maximum entropy and equations of state for random cellular structures. In Fougere P F (ed.) *Maximum entropy and Bayesian methods*. Dordrecht, Clair: 297–308

Rivier N 1991 Geometry of random packings and froths. In Bideau D, Dodds J A (eds) *Physics of granular media*. New York, Nova Science: 3–25

Rivier N 1993 Order and disorder in packings and froths. In Bideau D, Hansen A (eds) *Disorder and granular media*. Amsterdam, Elsevier Science: 55–102

Rivier N 1994 Maximum entropy for random cellular structures. In Nadal J P, Grassberger P (eds) *From statistical mechanics to statistical inference and back.* Dordrecht, Clair: 77–93

Rivier N, Lissowski A 1982 On the correlation between sizes and shapes of cells in epithelial mosaics. *Journal of Physics A: Mathematical and General* 15: L143–8

Sambridge M, Braun J, McQueen H 1995 Geophysical parameterization and interpolation of irregular data using natural neighbours. *Geophysical Journal International* 122: 837–57

Samet H 1989 *The design and analysis of spatial data structures.* Reading (USA), Addison-Wesley

Sibson R 1981 A brief description of natural neighbour interpolation. In Barnett V (ed.) *Interpreting multivariate data.* New York, John Wiley & Sons Inc. 21–36

Watson D F 1992 *Contouring: a guide to the analysis and display of spatial data.* Oxford, Pergamon Press

Watson D F, Philip G M 1987 Neighbourhood-based interpolation. *Geobyte* 2: 12–16

Zaninetti L 1993 Dynamical Voronoi tessellation IV. The distribution of the asteroids. *Astronomy and Astrophysics* 276: 255–60

37

Spatial hydrography and landforms

L E BAND

The structure and function of watersheds and landform systems has been incorporated in a range of GIS applications over the past two decades. Initial work focused on methods of automating the extraction of the geomorphic structure of watersheds from a combination of digital terrain data and satellite imagery. More recent work has continued this activity, but has increasingly concentrated on methods of representation within GIS data models that will support spatial analysis of watershed function as well as form. Of necessity, the associated surface properties of soil and vegetation cover have been included in these data models, and new approaches to inferring and representing the covariance of terrain, soil, and vegetation over a range of scales are under construction. The present state-of-the-art includes object data models that incorporate spatial representations of watersheds along with embedded simulation models to handle queries about the static and dynamic properties of watersheds.

1 INTRODUCTION

This chapter reviews the methods that have been developed to represent and process spatial information on watershed geomorphology. First, the pertinent aspects of surface watersheds are outlined from the perspectives of hydrology and geomorphology, including both the topography and the associated patterns of soil cover and vegetation. The algorithmic bases of methods designed to extract and represent watershed-flow path structure using digital terrain models are reviewed. The discussion of topographic skeletonisation (extraction of channel and ridge networks) is then extended to include recovery and representation of nested sub-catchments and hillslopes, and the subdivision of hillslope systems into discrete zones. More recent work is reviewed that seeks to identify a greater range of watershed landforms in an object format, along with formal schema for their organisation. The association of patterns of soils, vegetation, and landforms through watershed systems is discussed and representation within GIS is appraised. Finally, some recent work has been

focused on analysis and representation of watershed form over a range of spatial extents, from small catchments to continental-scale drainage basins. The chapter includes a discussion of the applicability of different approaches, and the form of information that can be extracted and represented, as the focus shifts along this scale range.

Two opposing conceptual models of watersheds have been developed: a model of continuous flow paths and continuous spatial variation of surface attributes; and a model of an exhaustive and mutually exclusive partition of surface area into different discrete land facets. Elements of both continuous and discrete models are often retained within various hydrologic and hydrographic applications. The interaction of these concepts with GIS data models can produce definable bias and inconsistency in spatial representation and analysis. A hybrid model has emerged recently which includes a hierarchical system of embedded sub-catchments, hillslopes, and (potentially) component hillslope positions within which the system of continuous flow paths is defined. Estimation of associated patterns of soil and vegetation by quantification and representation of typical catenary sequences can also be incorporated into a hillslope-based watershed representation and is an interesting current research direction in GIS and spatial environmental modelling. The discussion is extended beyond the spatial domain of individual watersheds to cover methods used to classify and represent a larger range of landforms and systems of landforms. This work builds from the concept of landscape systems or soil-landscapes and embeds knowledge on structure and composition of the surface that has in general not been incorporated into GIS data models in the past.

There have been ample reviews of the techniques developed over the past two decades for channel and ridge network extraction from terrain and image data (e.g. Band 1993; Tribe 1992) and fairly standard methods have been incorporated in commonly-used GIS packages. Discussion is limited here to describing the basic approaches available, the underlying assumptions of the data models, and their shortcomings in different situations. There is a voluminous literature dealing with the extraction of stream networks from digital elevation and remote sensing data that is spread over a range of journals and conference proceedings from different disciplines. A number of these contributions have been motivated by the need to generate watershed structural data for hydrologic or geomorphic research. In addition, stream networks have been very popular systems for the development of machine-vision algorithms. This is partly because of the availability of raster elevation images and partly because of the existence of formal graph models of channel-network patterns in terms of both topology and geometry. This has provided a rich basis for the development and testing of pattern recognition and image analysis techniques. Terrain parameters such as slope, aspect, curvature, and wetness indices are not addressed extensively here (but see Hutchinson and Gallant, Chapter 9).

2 HYDROGRAPHIC NETWORKS AND WATERSHED STRUCTURE

Watersheds can be described as hierarchical systems connected in the direction of flow. Hydrologic and geomorphic stream-, valley- and ridge-networks can be considered to be a topographic skeleton of fluvially eroded landscapes. A rich body of theory

has been developed to describe and explain the form, organisation, and function of watersheds based on the topology and geometry of these networks (summarised by Abrahams 1984). More recent work has concentrated on the scaling behaviour of the hierarchical structure of the topographic form of channel networks and associated drainage areas (e.g. Ijjasz-Vasquez et al 1992). A formal model of watershed form incorporates stream networks as tree graphs with drainage sources and junctions defining the tree nodes, and channel links (unbroken stretches of stream channel between the nodes) defining the edge set as introduced by Shreve (1966). The system of drainage partitions flow-paths between each channel link and forms a complementary graph of divide edges and nodes that intersect the channel system at channel junctions. Figure 1 shows an example of the combined graph structures implemented as a topographic skeleton in a portion of the watershed of Onion Creek, a 25 km^2 headwater catchment in the Sierra Nevada of California. The stream network can be extended to define an exhaustive surface partition by incorporating the set of hillslopes or drainage areas draining into each channel link as the leaves of the tree. This produces a spatially exhaustive and mutually exclusive mapping of channel links to drainage areas, and to left- and right-hand hillslopes (relative to the flow direction), facilitating graph functions to aggregate or distribute surface information through the watershed. The computation and labelling of various stream network and drainage area indices are straightforward recursive graph functions. An alternative partitioning of the surface can be gained with Strahler stream coding (Strahler 1957), in which the labelling of a stream reach does not change until a stream of equal (or higher) order is joined. This does not form a mutually exclusive mapping of all drainage areas draining into uniformly labelled stream reaches, as ordered stream segments form nested sub-catchments.

In this model, it is important to note that drainage divides are not necessarily ridges; they may have no topographic expression other than being the line of steepest descent on a hillslope which terminates at a drainage junction. To define a consistent surface topology, the existence of a channel branch requires the existence of a drainage divide between each set of uniquely labelled tributaries, or a tributary and a mainstream. Warntz and Woldenberg (1967) explored

Fig 1. Perspective view of the components of a simple watershed model consisting of nested channel links, divide segments, link drainage areas, and hillslopes. Channel links are thinned, one-pixel-wide chains and divides are un-thinned segments in this image.

3 HYDROGRAPHIC NETWORK EXTRACTION AND REPRESENTATION

The mass production of digital elevation data has spurred the development of a number of approaches to identifying and organising watershed and other landform information. This section traces the development of the different techniques without delving into the details, which have been summarised elsewhere (Band 1993; Mark 1988; Tribe 1992). The set of image-processing techniques developed over the past three decades to extract watershed-network patterns from digital elevation models (DEMs) and remote-sensing imagery includes:

- classification and pattern-matching algorithms which search for local topographic or image evidence of stream or ridge features in parallel and build the network bottom up;
- global sequential processors that construct space-filling flow patterns over topographic surfaces and generalise the convergent flow system to the connected set of drainage lines;
- knowledge-based techniques that incorporate rules regarding expected drainage-network and landform geometry and topology to infer likely structures from incomplete or noisy data.

The techniques have progressed from simple image-processing algorithms designed to detect the presence of surface-specific points or lines (peaks, pits, valley bottoms, and ridges) to more sophisticated systems that extract the full structure and flow network of a watershed. Object models of watersheds have been implemented recently in which the landforms comprising the watershed are formally defined and instantiated with full object hierarchy and inter-object relations (see Worboys, Chapter 26). Methods to automate the construction of these high-level data models from normally-available DEM and remote-sensing imagery have also been provided.

Topological inconsistencies can exist between the conceptual data model and the implementation in different GIS data structures, which can cascade as a bias into a set of geomorphic and hydrologic applications (see Heuvelink, Chapter 14). The conceptual graph model described above uses a set of surface points, lines, and areas as a well-defined set of discrete areal entities that form an exhaustive partition and connection of the surface. The nodes can be extended to include the set of surface-critical points (peaks, pits, saddles) and stream-channel

the implications of attempting to formalise the complementary nature of divides and channel lines, and Frank et al (1986) repeated the analysis nearly 20 years later. If the drainage lines and divide lines are considered lines of convergent and divergent flow, respectively, and if the drainage lines are extended to the divides at passes, then any ridge junction requires a convergent line, just as any stream junction requires an intermediary drainage divide. Logically, this leads to an infinite sequence of complementary channel and divide lines, unless a stopping criterion is given for the extension of stream sources. However, the conceptual model of drainage basin form that has developed over the past few decades has always incorporated finite extents and terminations of stream sources (Maxwell 1960; Montgomery and Dietrich 1988), negating the need for an infinite hierarchy of ridges and streams. In addition, as mentioned above, drainage divides are not necessarily ridges, such that the assumption of divergent flow in the neighbourhood of a divide is not universally applicable. Defining the upslope terminations of stream channels and valleys is a difficult problem in the field, partially due to the vagueness of landform definition and the continuous variation of surface properties, and a difficult problem in GIS as discussed below (see also Fisher, Chapter 13).

nodes (drainage junctions and sources) defining the set of ridge- or drainage-line segments, which in turn define the set of discrete drainage areas. In keeping with the graph theoretical model, the set of nodes are 0-dimensional and the edges (stream and divide links) are 1-dimensional. This topology is consistent with a vector or Triangulated Irregular Network (TIN) model of a watershed if a sequence of triangle edges can be defined as the channel and divide lines (e.g. Guercio and Soccodato 1996; Jones et al 1990). However, in raster models stream junctions are grid cells with finite areas and perimeters set by the cell resolution. The fixed dimensions of raster cells produce error in the representation of watershed form, particularly in the area immediately around the stream channel. Computations of surface area of wetted channel, or adjacent riparian areas, both of which are critical to the hydrologic behaviour of the catchment, are biased if standard raster algorithms are used. In Figure 1 the stream channels are represented by a one-pixel-wide chain of grid cells. The cells are 30m resolution, while the channels do not exceed about five metres in this watershed and are incised into the valley bottoms in many areas. Commonly, neither the surface area of the channels nor the topography immediately around the channels can be well represented. Run-off dynamics are sensitive to terrain characteristics at this scale, and they are well beyond the resolution of most DEMs. It is probable that properties of the riparian zone, including channel slope, curvature, and width may need to be determined independently and stored as attribute values rather than being computed from raster methods.

Use of a vector or TIN representation of this area may improve geometrical fidelity. Priestal and Downs (1996) have utilised a combination of raster and vector techniques to minimise the problems of raster representation of stream lines, but in general the critical length scales of the near stream environment are so much less than the rest of the watershed that it is often questionable whether topographic information computed for this region by standard GIS algorithms is sufficiently accurate and unbiased. Perkins et al (1996) have resorted to using an 'imposed stream' algorithm in their DEM-based hydrological model in which a vector representation of the stream is registered to the DEM, but this requires adjustment of the stream to maintain its position in the axis of the valley. Stream-channel properties such as width and cross-sectional area are then prescribed rather than computed from the raster data.

While grid-based methods have obvious disadvantages in comparison to more flexible representations for landform representation, including TIN- and contour-based methods (e.g. O'Loughlin 1990; Weibel and Brandli 1995), they are by far the most commonly used in GIS hydrology- and geomorphology-related projects. This is largely dictated by the availability of raster elevation data, the relative compatibility with remote sensing data, and a history of geomorphic and hydrologic models based on finite difference grids.

3.1 Local parallel processors

The first set of approaches to recognise and extract drainage networks from grid DEMs sought to nominate or classify each pixel into a set of discrete surface forms. Greysukh (1967) assigned grid cells to hill, depression, slope, ridge, ravine, or saddle classes based on the pattern of elevations in the surrounding 3×3 kernel. Templates were designed for pattern recognition of each cell. Peucker and Douglas (1975) tested a series of different local classification techniques based on local templates and found most techniques performed poorly, especially in the presence of terrain data noise. A method that was found to be reasonably efficient was to classify grid cells as concave or convex in order to produce potential stream and ridge cells, without distinguishing the richer set of surface forms. In both of these approaches fully connected drainage or divide networks were not produced because of noise or resolution limitations operating on the local processors. In addition, the local classifications of grid cells were not extended to identify connected landform features.

Toriwaki and Fukumura (1978) began with a similar approach, classifying all grid cells into specific surface features on the basis of local terrain patterns, but extended the surface set to distinguish 'top flats' (plateaux) and 'bottom flats' by progressively searching larger local neighbourhoods until the relative position of the flat could be determined. They also went on to form connected-component regions of the discrete landforms, including both linear and areal elements, using a series of post-classification processes which made use of simple adjacency relations between landforms (e.g. slope and stream). One of these methods involved fusing and then thinning streams and ridges into one-pixel-wide chains, while performing complementary operations on adjacent hillslopes. The result is an exhaustive partition of the surface into discrete landforms which could then be associated according to spatial adjacency.

3.2 Global sequential operators

Band (1986b), Douglas (1986), and Smith et al (1990) started with locally nominated stream and ridge pixels that formed incomplete networks (independently classified grid cells). A second step was used to form fully-connected and fully-labelled stream network structures. Band (1986a) did this by labelling segment ends of thinned linear segments as 'upslope' or 'downslope' and then extending the downslope nodes to drain through the DEM by steepest descent until another stream pixel was encountered. Starting at the root of the completed channel network, a recursive algorithm was used to climb and label each channel link according to the topologic code proposed by Shreve (1966, 1967). An extension of the fully-encoded stream network to the set of divides delimiting the drainage area of each link was then gained by performing a grey-weighted thinning procedure from each link upslope to the ridges, with the restriction that divides were connected to all stream junctions. This formed an exhaustive partition of the surface into the drainage areas associated with each channel link in a one-to-one mapping that could reference both the links and drainage areas by position in the drainage system. Smith et al (1990) extended the method of linking thinned channel segments into a fully-connected network to handle conditions of low signal-to-noise DEMs by incorporating variable search windows in a restricted direction forward of an active downstream node.

Marks et al (1984) developed a method of recursively labelling a contributing watershed area by progressively climbing a DEM from a designated outlet to all pixels defined as connected by surface drainage based on local aspect. An important feature is the choice of a minimum local slope below which the pixel is considered flat (because of local noise) and connected to all adjacent pixels. The technique included a simple way of identifying and eliminating pits (grid cells that are local minima and

therefore do not drain into an adjacent cell) by identifying a pour point and raising the elevation of all pit pixels to that level. O'Callaghan and Mark (1984) devised an iterative procedure of locating and starting at local maxima and progressing downslope, accumulating the number of pixels upslope which are stored as the contributing drainage area in each cell (Figure 2). The result is a space-filling tree graph in which each pixel can be a graph node (branch point). The tree must be pruned to a lower level that will define finite stream segments, and this can be done in a number of ways. The most common method is to choose a threshold to the drainage area to set the limit of the stream channel system (Figure 3), adopting a version of Schumm's (1956) constant of channel maintenance.

Band (1986a) produced a recursive version of O'Callaghan and Mark's (1984) method of drainage area accumulation and added a labelling method based on the recursive channel network traverse (Band 1986a) that was extended to include right and left hillslopes draining into each stream link (Band 1989). The basic approach of iterative downslope accumulation or recursive upslope accumulation of drainage area and catchment labelling has been duplicated in various forms by Jenson and Domingue (1988), Martz and Jong (1988), Martz and Garbrecht (1992), Tarboton et al (1988), and a number of others. Jenson and Domingue (1988) popularised the technique by

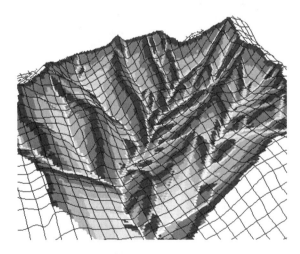

Fig 2. Accumulated drainage area image for a portion of the Onion Creek watershed in California. Each pixel is marked with the number of upslope pixels draining through it using a discrete routing algorithm. Grey scale is stretched for display.

writing and distributing code for personal computer (PC) use. A number of commonly-used GIS currently incorporate similar routines. A novel variant on this approach was developed by Ehlschlaeger (1990) for the Geographic Resource Analysis Support System (GRASS). In this approach a DEM is sorted by increasing elevation and a recursive algorithm branches to all pixels down the sort list (at higher elevation) that are adjacent to the root pixel. This produces a full network of connections from all higher to lower pixels which is then pruned to the drainage pattern using a network optimisation algorithm while accumulating the upslope drainage area (number of pixels). A similar drainage area threshold is used to define a finite stream channel network, along with labelled, associated hillslopes. A similar network optimisation method has been used by Niedda (1996) who points out that the minimisation of cumulative network flow cost is consistent with geomorphic theory of fluvial systems in terms of the dissipation of potential energy.

In the network optimisation all pixels in the DEM are nodes of the graph. The set of edge connections between the nodes must be chosen to minimise $\sum_{ij} c_{ij} x_{ij}$, where the unit cost of a connection (c_{ij}) is proportional to a function of the elevation drop between two adjacent nodes (i and j) and x_{ij} is the quantity of flow between i and j. The x_{ij} are conserved through the accumulation of flow, and in general a constant increment to flow of one is added at each node. This simulates a unit pulse of run-off being added to the full surface. Note that this method allows movement through flat areas or even uphill subject to the global minimisation. The penalty for uphill connections can be set very high, although as shown below this could have important feedback to the full network structure. In many cases the minimum cost algorithm defaults to straight-line segments as minimum-distance solutions if there are not large differences in the gradients along different paths (Plate 34). This solution is not visually intuitive compared to the output of some of the local discrete-drainage-direction algorithms. However, given the data quality of many DEMs this may be a reasonable estimate of flow-paths, and the more irregular nature of the other approaches may simply result from DEM noise.

The method of using a prescribed threshold drainage area to set the stream sources has been criticised as arbitrary, and not reflective of the spatial heterogeneity of the terrain or extent of stream dissection. Tribe (1991, 1992) focused on identifying

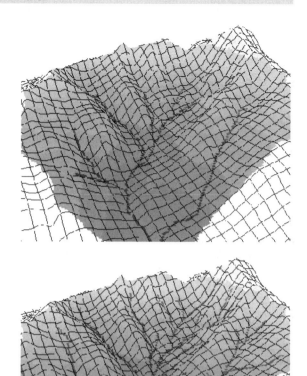

Fig 3. Different extents of the Onion Creek channel network produced by choosing a range of threshold drainage areas to define stream sources.

and extracting valley heads for each valley segment on the basis of local terrain morphology. This approach recognised that what is extracted is actually valley forms as opposed to channel segments, which are probably below the resolution of the DEM. Incorporation of such evidence would improve fluvial network extraction as it is recognised that a spatially-constant drainage area threshold may not be appropriate across a heterogeneous landscape. Evidence of valley head morphology on common high-resolution DEMs (10–100m) appears to be noise sensitive such that the techniques require further development or combination with drainage area information or other ancillary data. Montgomery and Dietrich (1988) proposed a threshold of slope-to-drainage-area gradients for a stream head based on a combination of empirical and theoretical arguments. Whether valley or stream source areas can be discriminated by various methods on normally-available DEMs with sufficient accuracy depends on both the terrain characteristics and the application requirements.

Another approach to forming the scale of the watershed partition from the accumulated drainage area image is to adapt the degree of dissection (or extent of the network) to the local terrain variability in some key surface property. Lammers (1998) has developed a method whereby a drainage area image is first pruned by applying a very low area threshold, forming an extensive network and numerous small hillslopes. A merge procedure is then run which minimises the within-hillslope spherical variance of the surface normal. Adjacent hillslopes are then merged if the surface normal distributions are similar. This forms a more complex partitioning of the terrain and a more detailed extension of the stream network in dissected terrain. Plate 35 shows this adaptability of stream-network and hillslope extraction for the South Platte drainage basin, where the headwaters drain the Front Range of the Rocky Mountains, while the larger part of the watershed drains the high plains of Colorado and Nebraska. This example used 1 km terrain data which generalises much of the terrain variability, but is appropriate for the regional-scale application.

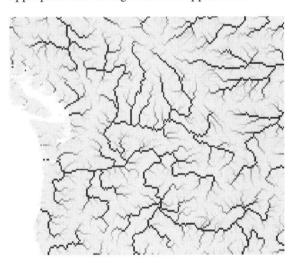

Fig 4. Drainage area accumulation image constructed from ETOPO-5 elevation data (5-minute resolution). Most of the major drainage systems are accurately portrayed with a few exceptions. The low horizontal resolution of the data does not capture accurately the topography of the Fraser River northeast of Vancouver, and the graph optimisation algorithm in GRASS finds a lower cost route for the upper Fraser to drain through the Columbia.

The increasing availability of DEMs covering regional to continental scales (Hutchinson and Gallant, Chapter 9; Hutchinson and Dowling 1991; Jenson 1991) has allowed the application of watershed extraction techniques to continental-scale drainage systems. At this level, a range of physiographic provinces is incorporated, and the performance of the algorithms can be variable, based on local terrain slope and data resolution. A limited amount of work has been published assessing the accuracy of watershed delineation (e.g. Miller and Morrice 1996) and these have generally been limited to areas of moderate to steep relief. In many cases, small errors in local routing where drainage will be connected over a small divide can lead to major topological errors in the connectivity of the network. An example is given in Figure 4 using ETOPO-5 data, a global elevation dataset at five minute resolution (about 10 km in the North–South direction). A portion of the drainage systems extracted for all of North America is shown for the northwest USA and southwestern Canada. While most of the regional drainage is well-represented, portions of the upper watershed of the Fraser River have been connected through the upper Columbia River because of the unresolved narrows of the Fraser Canyon northeast of Puget Sound. In this case, the GRASS4.1 algorithm was used which optimises a flow path. Because of the low resolution of the data and the high penalty associated with traversing the apparent ridge, a major topological error is incurred.

4 LANDFORM FEATURES AND OBJECT-BASED METHODS

The information that can be automatically extracted from low level terrain and image data must be organised into a spatial model commensurate with the watershed representations used by hydrologists and geomorphologists. A number of researchers have added techniques to extract and represent a suite of other topological and geometrical parameters to provide a fuller geomorphometric description of the watershed. Lammers and Band (1990) built a full watershed description from terrain data. Variables were extracted and stored in attribute tables for the set of nodes (stream junctions), lines (stream links), and areas (hillslopes) that constitute the watershed, including stream junction angles, stream link

533

lengths, orientation, slope, drainage area, Strahler order and link magnitude, slope area, spherical mean aspect and gradient, as well as inter-feature relations. The tables were expanded to include information on soils and vegetation from co-registered remote sensing images and soil maps to parameterise distributed hydro-ecological models (Band 1993; Band et al 1991; Nemani et al 1993). In this respect, the hydro-ecological models were distributed on a geomorphic base, given by the hillslope organisation of the watershed, rather than the standard grid-cell approach.

Mackay and Band (1994) and Mackay et al (1992) built an object model of the watershed based on the tree-graph channel-link-based model. A hierarchy of objects representing the nested set of channel links, channel (sub)networks, hillslopes, catchments, sub-catchment, and divide segments was constructed from the information derived by the methods of Lammers and Band (1990). A simple graphical interface was designed to facilitate queries regarding object attributes (e.g. slope, area, orientation, soil, vegetation cover), spatial relations (adjacent to, upslope or downslope of), or drainage aggregation. A hydro-ecological simulation system (RHESSys; Band et al 1993) was tightly coupled to the object system so that queries could be expanded to pose hydrological and ecological questions such as expected run-off production, forest productivity, or soil moisture at specified times in response to meteorological input. The object model is undergoing further development to incorporate temporally defined watershed objects such as clear-cuts or other disturbance features (Mackay 1997) and to formalise its object schema (Robinson and Mackay 1995) as shown in Figure 5.

The *r.watershed* routine in GRASS4.1 includes options to compute a series of terrain and landscape variables in the format required to operate distributed run-off and soil erosion models which are incorporated into the spatial analysis package, as well as drainage information linking the set of slopes and sub-catchments. The incorporation of linked or embedded simulation models of watershed runoff and erosion, or ecological processes within a GIS package with the required spatial data preprocessing techniques, significantly extends the functionality of the spatial analysis system.

5 EXTENSION TO LANDFORM SYSTEMS

The set of landforms comprising a watershed can be extended from the channel, ridge, and slope model described above to include a collection of topographically-distinct regions referred to as bottomland, upslope, midslope, floodplains, valleys, mountains, or ridge areas. Unlike the crisply defined set of graph components that comprise the conceptual model of watershed drainage and ridge networks, these entities are not as precisely defined, nor are their extents or spatial relations. However, they are common terms used in a range of applications to distinguish different functional parts of the watershed in terms of hydrologic, geomorphic, and ecological conditions. As such, it is useful to be able to refer to and analyse 'bottomlands' or 'mid-slope position' within a GIS. Dymond et al (1995) extended the set of watershed features by extracting uniform land facets, which are sections of hillslopes stratified by aspect and distance from the stream or valley bottom. This allows the delineation of areas corresponding to ridge, midslope, or toeslope, and may provide a more detailed subdivision of contributing drainage areas and valley sides. The methods work with region-growing algorithms to agglomerate uniform-aspect pixels into terrain facets. For hilly to mountainous areas these techniques seem to be particularly effective in subdividing the drainage areas according to criteria that would be useful for a variety of scientific or management-oriented activities.

With some exceptions (e.g. Toriwaki and Fukumura 1978; Tribe 1991, 1992) many of the watershed characterisations described above have implicitly assumed steep, incised topography composed of V-shaped valley bottoms and sharp ridge tops surrounding simple slope forms. Many of the techniques for flow-path extraction have severe problems in flat areas due to low signal-to-noise ratios for determining appropriate flow directions. Various methods have been used to route water through these areas by searching larger windows (Smith et al 1990), adjusting elevation patterns on the basis of larger neighbourhood topography (Garbrecht and Martz 1996), or other methods of preprocessing and 'correcting' flow directions in identified flat areas or pits. In most cases, however, the channel lines are still constructed as single-pixel-wide chains through the flat regions even if

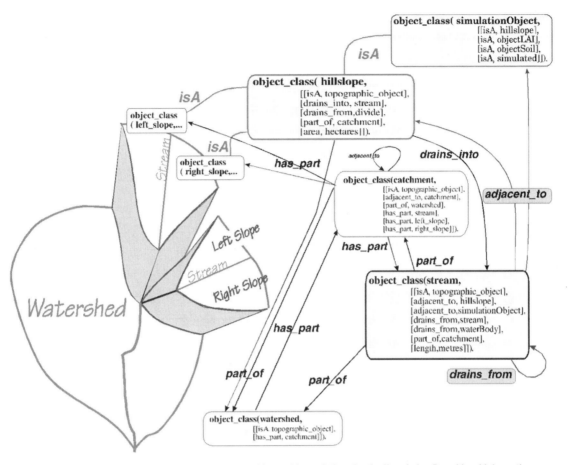

Fig 5. Data model schema defining watershed objects and inter-object relations for the Knowledge-Based Land Information Management System (KBLIMS). (After Robinson and MacKay 1995.)

there are no clearly-defined channels, as in wetlands or open water bodies. It is notable that almost all areas used to develop and demonstrate the channel-network and watershed structure are steeply-sloping fluvial or glacial topographies that are either devoid of lakes or extensive flat bottomland areas or ignore their presence. In areas where these features are ubiquitous, it does not make sense to attempt to ignore their presence as they are dominant features in the landscape. Mackay and Band (1994) addressed the problem of delineating watershed structure in regions dominated by numerous lakes, wetlands, and other flat bottomland features. Rather than attempting to route water through lakes and wetlands, Mackay and Band (1994) and Mackay (1997) first identified all flat regions using a combination of DEM and remote sensing imagery

and then labelled them as lake, wetland, or floodplain objects. A version of the recursive climbing algorithm described by Band (1986b, 1989) was used to extract the flow-path structure and contributing drainage areas for all areas not identified as one of these objects. When the algorithm encounters a flat feature it is designed to traverse its perimeter (the set of boundary pixels of the lake or wetland), recursively climbing and accumulating the surrounding drainage areas (Plate 36). The resulting image shows the flow-path structure which is routed just inside the perimeter of each flat. The flow paths within each flat are then dissolved and the total contributing drainage area to each flat along with its surface area and any other attributes that are available (depth, volume, shape) are stored as part of an object description specific to lakes, wetlands, etc. In this regard the flat areas are explicitly recognised and stored as part of the watershed object hierarchy described above and incorporated into the set of watershed objects (Figure 6).

6 UNCERTAINTY IN LANDFORM LABELLING

A persistent problem is the vagueness of landform concepts in terms of spatial boundaries and identification (labelling). This uncertainty in labelling landforms and landform positions translates into surface attribute uncertainty as attributes are often prescribed on a per-category basis. Note that the simple graph model of watersheds as composed of stream lines, divide lines, and hillslopes obviated some of this difficulty by restricting the range and

detail of surface features. However, it suffers from being both oversimplified and restrictive compared to real watersheds, or the conceptual models of watersheds used by different communities. Varying attempts have been made to incorporate the degree of similarity of a region to one or more of the conceptual zones, or uncertainty regarding local assignment. The adaptation and use of certainty or similarity measures for zonal or landform labels would be an important function for spatial data models to maintain and use the watershed information generated in an efficient manner.

Some attempts to both infer and represent landform position and landform categories as part of a defined landscape system have been developed recently. These techniques use multiple sources of evidence and incorporate measures of uncertainty in both inferring and representing the landscape system

Fig 6. Watershed landform objects incorporating lake, wetland, stream, and slope features into an integrated drainage system.

and components. Skidmore et al (1991) constructed a glossary of the land types characterising different slope positions with characteristic soil/vegetation assemblages for a region of SE Australia. They then used a Bayesian approach to delimit the extents of each land type on the basis of local morphology and terrain position, and were therefore able to infer both the patterns of land components and the characteristic soils expected for the area with known probability. Mackay et al (1992) inferred the presence of types of alpine glacial landforms using mutual evidence theory based on local morphometric characteristics as determined from digital terrain data, and relative position in the landscape. Local evidence of the existence of a cirque within a headwater catchment reinforced the evidence of a glacial trough downslope and vice versa. A certainty measure could then be attached to the landform label for each component as influenced by both local and global information.

7 SOILS AND VEGETATION

The sections above indicate that the extraction of watershed topography and fluvial networks from digital terrain data is reasonably well advanced, at least for hilly to steep terrain, and progress is being made for more gentle topography. For many hydrologic and geomorphic applications, analysis of the composition and structure of a watershed requires information about the patterns of vegetation and soils as well, and their spatial covariance with local topographic form and landscape position. The observation that soils vary in regular patterns within the landscape is old, or even ancient, and it is interesting that this knowledge is just now being built into geographical data models for GIS. The quantification of the relations between topography, vegetation, soils, and climate has been a central concern in physical geography for over four decades. Melton (1958) investigated the correlation structure of a set of variables characterising each of these components using a field- and map-sampling design over a range of watersheds. An important aspect of this and similar research was the collection of field data at similar levels of support. In contrast, a cartographic data layer model typically offers little control over sampling design and the support of the different data layers simply because of the prevalent use of available data, rather than data appropriate to

the analysis at hand (see Weibel and Dutton, Chapter 10). In this sense the technology may act as an inadequate alternative to a good sample design (Goodchild and Longley, Chapter 40), and data-overlay-based systems often may not adequately capture covariance structures that may be significant characteristics of the landscape.

Within a landscape system, each landform component at all levels of the hierarchy can be assigned typical patterns of terrain, types of soils, or soil assemblages, as well as vegetation cover characteristic of the landform category. In this respect, the placement of individual landforms and their expected soil and vegetation cover provides a catenary sequence. The presence of one landform can imply the presence of another related landform either at the same level of the hierarchy or at a level above or below. As an example, in the study of Mackay et al (1992) the presence of a cirque implies a glaciated valley downslope, or a tarn and morainal debris within the cirque. This type of knowledge is essential to understanding a landscape as a system, and is contained within the text and legends accompanying maps of the land systems, but has generally not been translated into GIS which are still dominated by a cartographic data model (but see Mitas and Mitasova, Chapter 34).

Of the set of variables that often needs to be incorporated in a description of a landscape, vegetation and other land cover features (cultural features) can be sampled successfully and mapped with the use of remote sensing or aerial photographic techniques at a similar support to the terrain information extracted from terrain data (see Smith and Rhind, Chapter 47). The weak link in the landscape catenary sequence with respect to GIS applications is the state of soils information, as soil information is rarely available at resolutions commensurate with terrain and vegetation information. As soils generally cannot be mapped using remote sensing techniques (with the possible exception of semi-arid to arid areas), their spatial extent is typically mapped in polygon form. Soil polygons are rarely pure, and contain inclusions of other soils which cannot be spatially located but are simply acknowledged in an accompanying soil report in terms of an areal percentage. This arises because of the coarse scale of soil mapping, the ambiguity of the soil classification schemes in general use, and the spatially continuous variation of soil properties in the field in contradiction to the discrete cartographic polygon model (Fisher, Chapter 13). The mismatch of spatial support for soil information relative to topographic and vegetation information produces error in the covariation of these variables when using the standard overlay technique.

Consequently, a number of techniques are in the process of being developed to attempt to improve the information content of soil data relative to that of terrain and vegetation data. These methods are drawn from the concept of soil–landscape modelling, or the catena concept of Milne (1935). The methods that have recently been explored to quantify and represent soil-landscape patterns range from empirical-statistical correlation through to knowledge-based techniques. Moore et al (1993) investigated the correlations of a set of soil properties with a set of terrain variables for a well-studied area in Colorado. Their working hypothesis was that much of the variation in the soils that lie along toposequences is a response to the movement of water over and through the landscape. Their results indicated significant correlations between soil and terrain variables, particularly the compound terrain index $CTI = \ln (A_s / \tan \beta)$, where A_s is the specific catchment area (see Hutchinson and Gallant, Chapter 9) and β is the slope angle, which serves as an indicator of terrain wetness. In terms of areal prediction of soil variables, the methods used were very data-intensive, requiring a large number of soil measurements to develop an adequate database. Gessler et al (1995) extended this work in Australian field sites by building multivariate models of soil properties as simple functions of terrain variables. They found significant explanations of soil depth and A horizon depth with surface curvature and the Compound Terrain Index (CTI). Once again, the methods are data intensive and it is unknown how generalisable these relations are beyond the specific dataset employed.

Gessler et al (1996) took a different approach to the soil–landscape problem by attempting to synthesise representative slope profiles in terms of both topography and soil properties. This work uses datasets across a climatic gradient and seeks to identify patterns of specific soil properties along convergent and divergent hillslopes in each environment. In this respect, they identify hillslopes as fundamental objects that can be described in terms of soil patterns associated with more easily observable terrain patterns. If a generalisable set of

hillslope soil-terrain behaviour could be identified across observable climatic and geomorphic conditions, this approach may form the basis for extrapolating soil properties across a landscape within a GIS. What is required is a method to encapsulate the knowledge of these typical soil–toposequences in a manner that can be applied to a set of hillslopes extracted by the methods described in the previous section.

Expert systems have also been developed as a method of capturing and representing the knowledge soil scientists have regarding soil-landscape patterns, as an alternative to the demanding sampling procedures required by purely empirical approaches. The Bayesian inference approach of Skidmore et al (1991) was designed to classify landscape elements on the basis of terrain variables in order to infer soil properties. They used schemes developed by local soil scientists and resource managers in southeast Australia to set prior probabilities for each land type, and the typical topographic positions they occupied. Zhu and Band (1994) and Zhu et al (1996) used a fuzzy inference approach to compute membership scores for all grid cells in a raster image for all soil series (or complexes) identified to exist within the area. Figure 7 shows inferred fuzzy memberships for a specific soil complex. The scheme was based on a series of interactive sessions with local soil scientists to attempt to capture and represent their knowledge of soil-landscape patterns by constructing fuzzy membership functions for soil type based on a set of observable landscape variables including terrain, lithology, and canopy cover. The methods worked well in steep, Rocky Mountain topography where terrain-soil relationships are sharp and well defined (and more visualisable to the soil scientists) and less well in more gentle southeast Australian terrain. One of the areas identified as requiring greater development was to incorporate variables that better reflected terrain position, rather than primary terrain variables (e.g. slope, aspect, curvature). These needs were identified by Gessler et al (1996) and Skidmore et al (1991) as being critical to inferring soil variations in gentler topography.

7.1 Extension to larger-area landform systems

A fuller understanding and better categorisation of landforms is gained by placing them within an evolutionary framework such that the formative

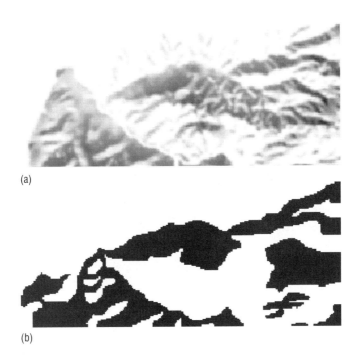

(a)

(b)

Fig 7. (a) Fuzzy and (b) crisp membership in a soil complex in a watershed of western Montana. The fuzzy memberships are computed on the basis of similarity to a central concept for the soil complex in several terrain and forest cover variables.

processes and relations to adjacent and associated landforms are incorporated. In this respect it is important to catalogue not only what can be inventoried within a landscape at a specific period of time, but what its temporal development has been and what the active processes are that maintain or change the system state (Raper, Chapter 5; Raper and Livingstone 1995). This is the basic approach taken in understanding watershed geomorphology and has been incorporated into a number of land systems that have been developed for resource management at national and other jurisdictional levels. The Soil-Landscape Classifications in Canada, US Forest Service Land Systems, and the Australian Soil and Land Survey (McDonald et al 1990) are examples and are well described by Bailey (1996). Each of these systems is hierarchical in nature, progressively subdividing the landscape into regional to local nested landforms, based on the primary geomorphic processes that have shaped the surface and determined the distribution of surface materials. An advantage of these systems is that they seek to embed

knowledge of surficial geomorphology directly within the land-classification scheme. The examples of automating portions of this approach given above were placed in spatial domains of watersheds within which variables such as distance from stream or ridge within hillslopes could be resolved. However, the landscape-systems approach is hierarchical and should extend to much larger regions. Above the level of these small- to medium-size watersheds are large physiographic provinces which combine multiple watersheds and include a mix of dominant geomorphic processes within broader tectonic settings. At these scales, measures pertinent to individual hillslopes are generally well below the resolution of available data, and alternatives must be used. Recognition of landform systems at this level has been attempted by Dikau (1988, 1992), Dikau et al (1995), Fels and Matson (1996), and others. In these cases, initial identification of large-scale physiographic elements is carried out by first identifying geomorphometric parameters at the appropriate scale to distinguish landforms at these levels and the appropriate data resolution for their processing. At global levels an appropriate terrain dataset may be the five-minute elevation data discussed above for recognition of continental-scale features. Classification on the basis of local terrain variables alone is problematic and involves many subjective and interactive decisions regarding thresholds and appropriate variables, which are generally not transferable across different areas or even across different resolutions. Dikau (1992) discusses setting up a formal object hierarchy of landforms that can be aggregated or disaggregated into higher- and lower-level components, respectively. This can aid in identification of individual components, and also ensures consistency with the hierarchical data models used for the land systems. As effective global-scale research and applications using GIS often require a link between continental and local landscape scales, the construction of appropriate object hierarchies for the range of geomorphic systems characteristic of the planet would be an important task.

8 SUMMARY AND DISCUSSION

This chapter has reviewed the methods that have been used to synthesise and represent the geomorphic structure of watersheds and larger landscapes within GIS. Over the past two decades GIS data models have been developed to match more closely the conceptual model of a watershed, progressing from incorporation of the drainage and divide network, to nested sub-catchment and slope systems, to full object models of the watershed. The object models have been integrated with process-based models of hillslope hydrology and ecosystem processes. The catenary system of topographic, soil, and vegetation properties can be combined by standard cartographic overlay, but this method may not preserve significant covariance between these variables. Development and incorporation of soil-landscape models which quantify topographic variation of soils have shown significant progress in the last few years. Like many other areas in GIS, temporal variation in the data model of watersheds has remained rudimentary, although progress is being made in the incorporation of dynamics into surface spatial process and state variables.

There remain certain persistent problems in the recovery and representation of watershed

information. The length scales of significant features are highly variable, with a large gap between the length scales of stream-channel and near-stream-channel environments relative to upland areas. Even with resolution-flexible methods of terrain representation such as the TIN, it is difficult to represent the required information content in these variable regions with layer-based models. Landform definitions are often conceptually vague, specifically regarding spatial extent and coverage. Terrain form and properties are typically spatially-continuous and gradational, and do not conform well to cartographic data models which may require spatially-exhaustive and mutually-exclusive thematic coverages. Both the raster and vector models have topological inconsistencies with watershed features which are most critical in the representation of drainage lines and stream channels.

To a certain extent these problems have been mitigated by resorting to object models of the watershed, although the spatial (landform) objects still need to be instantiated into well-defined classes. However, this problem is not just inherent to GIS representations of the watershed but is endemic to any spatial representation (Martin, Chapter 6; Raper, Chapter 5), such as those developed in distributed hydrologic and geomorphic process models. Continued improvement of GIS data models for watersheds and associated landforms should centre on extending the analytical approaches and level of understanding developed in these applications.

References

Abrahams A D 1984 Channel networks: a geomorphological perspective. *Water Resources Research* 20: 161–88

Bailey R G 1996 *Ecosystem geography*. New York, Springer

Band L E 1986a Analysis and representation of drainage basin structure with digital elevation data. *Proceedings, Second International Symposium on Spatial Data Handling, Seattle*: 437–50

Band L E 1986b Topographic partition of watersheds with digital elevation models. *Water Resources Research* 22: 15–24

Band LE 1989 A terrain-based watershed information system. Hydrological Processes 3: 151–62

Band L E 1993 Extraction of channel networks and topographic parameters from digital elevation data. In Kirkby M J, Beven K (eds) *Channel Network hydrology*. Chichester, John Wiley & Sons: 13–42

Band L E, Patterson P, Nemani R, Running S W 1993 Forest ecosystem processes at the watershed scale: incorporating hillslope hydrology. *Agricultural and Forest Meteorology* 63: 93–126

Band L E, Peterson D L, Running S W, Coughlan J, Lammers R, Dungan J, Nemani R 1991 Forest ecosystem processes at the watershed scale: basis for distributed simulation. *Ecological Modelling* 56: 171–96

Dikau R 1988 Case studies in the development of derived geomorphic maps. *Geologische Jahrbuch A* 104: 329–38

Dikau R 1992 Aspects of constructing a digital geomorphological base map. *Geologische Jahrbuch A* 122: 357–70

Dikau R, Brabb E E, Mark R K, Pike R J 1995 Morphometric landform analysis of New Mexico. *Zeitschrift für Geomorphologie* suppl. 101: 109–26

Douglas D H 1986 Experiments to locate ridges and channels and create a new type of digital elevation model. *Cartographica* 23: 29–61

Dymond J R, Derose R C, Harmsworth G R 1995 Automated mapping of land components from digital elevation data. *Earth Surface Processes and Landforms* 20: 131–7

Ehlschlaeger C 1990 r.watershed manual entry. *GRASS3.1 user's manual*. Champaign-Urbana, US Army Corps of Engineers, Construction Engineering Research Laboratory

Fels J E, Matson K C 1996 A cognitively based approach for hydrogeomorphic land classification using digital terrain models. *Proceedings, Third International Conference on Integrating GIS and Environmental Modeling, Santa Fe, 21–26 January*. Santa Barbara, NCGIA. CD

Frank A U, Palmer B, Robinson V B 1986 Formal methods for the accurate definition of some fundamental terms in physical geography. *Proceedings, Second International Symposium on Spatial Data Handling*: 583–99

Garbrecht J, Martz L W 1996 Digital landscape parameterisation for hydrological applications. *HydroGIS 96: application of geographic information systems in hydrology and water resources management*. IAHS Publ. no. 235: 169–73

Gessler P E, Moore I D, McKenzie N H, Ryan P J 1995 Soil-landscape modelling and spatial prediction of soil attributes. *International Journal of Geographical Information Systems* 9: 421–32

Gessler P E, McKenzie N, Hutchinson M 1996 Progress in soil-landscape modelling and spatial prediction of soil attributes for environmental models. *Proceedings, Third International Conference on Integrating GIS and Environmental Modeling, Santa Fe, 21–26 January*. Santa Barbara, NCGIA. CD

Greysukh V L 1967 The possibility of studying landforms by means of digital computer. *Soviet Geography, Review, and Translations*: 137–49

Guercio R, Soccodato F M 1996 GIS procedure for automatic extraction of geomorphological attributes from TIN-DTM. *HydroGIS 96: application of geographic information systems in hydrology and water resources management*. IAHS Publ. no. 235: 175–82

Hutchinson M F, Dowling T 1991 A continental hydrological assessment of a new grid-based digital elevation model of Australia. *Hydrological Processes* 5: 45–58

Ijjasz-Vasquez E J, Rodriguez-Iturbe I, Bras R L 1992 On the multi-fractal characterisation of river basins. *Geomorphology* 5: 297–310

Jenson S K 1991 Application of hydrologic information automatically extracted from digital elevation models. *Hydrological Processes* 5: 31–44

Jenson S K, Domingue J O 1988 Extracting topographic structure from digital elevation data for geographical information system analysis. *Photogrammetric Engineering and Remote Sensing* 54: 1593–1600

Jones N L, Wright S G, Maidment D R 1990 Watershed delineation with triangle based terrain models. *Journal of Hydraulic Engineering, ASCE* 116: 1232–51

Lammers R B 1998 'Extending hydro-ecological simulation models from local to regional scales'. Unpublished PhD dissertation, University of Toronto

Lammers R B, Band L E 1990 Automating object representation of drainage basins. *Computers and Geosciences* 16: 787–810

Mackay D S 1997 'Representation of heterogeneous topography and forest cover for long-term hydro-ecological modelling'. Unpublished PhD dissertation, Dept. of Civil Engineering, University of Toronto

Mackay D S, Band L E 1994 Extraction and representation of watershed structure including lakes and wetlands from digital terrain and remote sensing information. *EOs, Transactions AGU* 75: 175

Mackay D S, Robinson V B, Band L E 1992 Classification of higher order topographic objects on digital terrain data. *Computers, Environment, and Urban Systems* 16: 473–96

Mark D M 1988 Network models in geomorphology. In Anderson M G (ed.) *Modelling geomorphological systems*. New York, John Wiley & Sons Inc.

Marks D, Dozier J, Frew J 1984 Automated basin delineation from digital elevation data. *Geoprocessing* 2: 299–311

Martz L W, Jong E de 1988 Catch: a FORTRAN program for measuring catchment area from digital elevation models. *Computers and Geosciences* 14: 627–40

Martz L W, Garbrecht J 1992 Numerical definition of drainage network and sub-catchment areas from digital elevation models. *Computers and Geosciences* 18: 747–61

Maxwell J C 1960 Quantitative geomorphology of the San Dimas Experimental Forest. Technical Report 19. Washington DC, Office of Naval Research

McDonald R C, Isbell R F, Speight J G, Walker J, Hopkins M S 1990 *Australian soil and land survey field handbook*, 2nd edition. Melbourne, Inkata Press

Melton M A 1958 Correlation structure of morphometric properties of drainage systems and their controlling agents. *Journal of Geology* 66: 442–60

Miller D R, Morrice J G 1996 An assessment of the uncertainty of delimited catchment boundaries. *HydroGIS 96: application of geographic information systems in hydrology and water resources management*. IAHS Publ. no. 235: 445–52

Milne G 1936 A provisional soil map of East Africa. *Amani Memoirs* 28

Montgomery D R, Dietrich W E 1988 Where do channels begin? *Nature* 336: 232–4

Moore I D, Gessler P E, Nielsen G A, Peterson G A 1993 Soil attribute prediction using terrain analysis. *Soil Science Society of America Journal* 57: 443–52

Nemani R, Running S W, Band L E, Peterson D L 1993 Regional hydroecological simulation system: an illustration of ecosystem models in a GIS. In Goodchild M F, Parks B O, Steyaert L T (eds) *Environmental modeling with GIS*. New York, Oxford University Press: 297–304

Niedda M 1996 Use of network algorithms in spatially distributed models for the study of river basin response. *HydroGIS 96: application of geographic information systems in hydrology and water resources management*. IAHS Publ. no. 235: 207–14

O'Callaghan J F, Mark D M 1984 The extraction of drainage networks from digital elevation data. *Computer Vision, Graphics, and Image Processing* 28: 323–44

O'Loughlin E 1990 Modelling soil water status in complex terrain. *Agricultural and Forest Meteorology* 50: 23–38

Perkins W A, Wigmosta M S, Nijssen B 1996 Development and testing of road and stream drainage network simulation within a distributed hydrologic model. *EOS* 77: 232

Peucker T K, Douglas D H 1975 Detection of surface specific points by local parallel processing of discrete terrain elevation data. *Computer Graphics and Image Processing* 4: 375–87

Priestal G, Downs P W 1996 Automated parameter estimation for catchment scale river channel studies: the benefits of raster–vector integration. *HydroGIS 96: application of geographic information systems in hydrology and water resources management*. IAHS Publ. no. 235: 215–23

Raper J F, Livingstone D 1995 Development of a geomorphological spatial model using object-oriented design. *International Journal of Geographical Information Systems* 9: 359–83

Robinson V B, Mackay D S 1995 Semantic modelling for the integration of geographic information and regional hydro-ecological simulation management. *Computers, Environment, and Urban Systems* 19: 321–40

Schumm S A 1956 Evolution of drainage systems and slopes in badlands at Perth Amboy, New Jersey. *Geological Society of America Bulletin* 67: 597–646

Shreve R L 1966 Statistical law of stream numbers. *Journal of Geology* 74: 17–37

Shreve R L 1967 Infinite topologically random stream networks. *Journal of Geology* 75: 178–86

Skidmore A K, Ryan P J, Dawes W, Short D, O'Loughlin E 1991 Use of an expert system to map forest soils from a geographic information system. *International Journal of Geographical Information Systems* 5: 431–45

Smith T R, Zhan C, Gao P 1990 A knowledge-based, two-step procedure for extracting channel networks from noisy DEM data. *Computers and Geosciences* 16: 777–86

Strahler A N 1957 Quantitative analysis of watershed geomorphology. *Transactions of the American Geophysical Union* 38: 913–20

Tarboton D G, Bras R L, Rodriguez-Iturbe I 1988 The fractal nature of river networks. *Water Resources Research* 24: 1317–22

Toriwaki J, Fukumura T 1978 Extraction of structural information from grey images. *Computer Graphics and Image Processing* 7: 30–51

Tribe A 1991 Automated recognition of valley heads from digital elevation models. *Earth Surface Processes and Landforms* 16: 33–49

Tribe A 1992 Problems in automated recognition of valley features from digital elevation models and a new method toward their resolution. *Earth Surface Processes and Landforms* 17: 437–54

Warntz W, Woldenberg M J 1967 Spatial order – concepts and applications. *Harvard Papers in Theoretical Geography* 1

Weibel R, Brandli M 1995 Adaptive methods for the refinement of digital terrain models for geomorphometric applications. *Zeitschrift für Geomorphologie* suppl. 101: 13–30

Zhu A, Band L E 1994 A knowledge-based approach to data integration for soil mapping. *Canadian Journal of Remote Sensing* 20: 408–18

Zhu A, Band L E, Dutton B, Vertessy R 1996 Soil property derivation using a Soil Land Inference Model (SoLIM). *Ecological Modelling* 90: 123–4

38

Intervisibility on terrains

L DE FLORIANI AND P MAGILLO

Many interesting application problems on terrains involve visibility computations, for example, the search for optimal locations of observation points, line-of-sight communication problems, and the computation of hidden or scenic paths. The solution of such problems requires techniques to answer visibility queries on a terrain efficiently, as well as the development of data structures to encode the visibility of a terrain from one or several viewpoints. In this chapter, visibility structures such as the viewshed (representing the portions of a terrain visible from a single viewpoint) and the horizon (describing the distal boundary of the viewshed) are introduced, as well as visibility structures related to several viewpoints. All these structures admit continuous and discrete encodings (mainly used on triangulated irregular networks and regular square grids, respectively). An overview is provided of algorithms for building visibility structures on triangulated irregular networks and regular square grids as well as for answering visibility queries on a terrain. Finally, some application-specific problems involving visibility computation are illustrated.

1 INTRODUCTION

Many interesting application problems concerning terrains involve visibility computations, for example, the placement of observation points according to suitable constraints, line-of-sight communication problems, the computation of paths with certain visibility properties, etc. Applications include the location of fire towers, radar sites, radio, TV or telephone transmitters, path planning, navigation and orientation, and so on. An exhaustive survey of visibility-related problems is provided by Nagy (1994). The solution of such problems needs methods to answer visibility queries efficiently, through the development of structures for encoding the visibility situation of a terrain. Visibility queries consist of determining whether a given object located on a terrain is visible from a viewpoint and possibly how much of the object is visible. Visibility structures provide information about the visibility of the terrain itself; the knowledge of suitable visibility structures for a terrain helps answer visibility queries.

After recalling some basic notions about digital elevation models (DEMs) in section 2, section 3 defines various kinds of structures encoding visibility information for a terrain. Relevant visibility structures for a single viewpoint are the *viewshed*, which represents the location and extent of the portion of a terrain visible from a given viewpoint, and the *horizon*, which describes the distal boundary of the viewshed. To model the visibility of a terrain with respect to a set of viewpoints, structures based on combinations of the viewsheds of single points are used; depending on the specific problem, it can be useful, for instance, to consider the union or the intersection of viewsheds, or to count how many viewpoints can see each point on a terrain. Visibility structures admit both *continuous* and *discrete* encodings. In general, continuous encodings of the viewshed (called *continuous visibility maps*) are used for triangulated irregular networks (TINs), while discrete encodings (called *discrete visibility maps*) are used on dense regular square grids (RSGs). Visibility information related to multiple viewpoints is commonly represented in a discrete way, because

543

of its size, leading to structures such as *visibility graphs, intervisibility maps* and *visibility counts*; continuous encodings reduce to overlays of continuous visibility maps.

Section 4 provides a survey of algorithms for computing visibility structures (visibility maps, horizons, etc.) on terrain models and for solving visibility queries. RSGs and TINs have very different structures and are usually handled using very different methods. Visibility computation on RSGs is based on line-of-sight processing (Blelloch 1990; Shapira 1990). This is an expensive process because of the size of the grid; parallel algorithms have been defined, which take advantage of the regular spatial structure of an RSG (Mills et al 1992; Teng et al 1993). TINs have deserved strong consideration from both the computational geometry and GIS communities; practical algorithms are reported (Boissonnat and Dobrindt 1992; De Floriani et al 1989; Lee 1991a), as well as algorithms of theoretical interest for their good asymptotic complexity (Edelsbrunner et al 1989; Katz et al 1991; Overmars and Sharir 1992; Preparata and Vitter 1992; Reif and Sen 1988). Finally, *visibility queries* can be efficiently answered based on either visibility maps or on horizons; ad hoc methods have been proposed for solving queries directly, that is, without computation of intermediate data structures (Cole and Sharir 1989).

All visibility computations are sensitive to errors in elevation near the viewpoint, since these are amplified in proportion to the distance. For this reason, various authors suggest that the topography near the viewpoint must be known much more accurately than on the rest of the surface (Cignoni et al 1995; Felleman and Griffin 1990). *Multi-resolution terrain models* play an important role here, since it is possible to obtain from such models terrain representations whose level of resolution in any area of the domain can be specified by the user. Section 5 introduces multi-resolution models of terrains and considers the problem of computing and updating visibility structures on a multi-resolution model, as well as the direct solution of visibility queries on such models.

In section 6 we consider application problems related to visibility, such as viewpoint placement, line-of-sight communication and path problems, and the role of visibility information in solving them. Section 7 contains some concluding remarks.

2 PRELIMINARIES

A *topographic surface* (or *terrain*) can be regarded as the image of a real bivariate function f defined over a domain D in the Euclidean plane. A digital elevation model (DEM) is a model of one such surface built on the basis of a finite set of digital data (see also Band, Chapter 37; Hutchinson and Gallant, Chapter 9). Terrain data consist of elevation measures at a set of points $S \subseteq D$; points in S can either be scattered, or form a regular grid. A DEM built on S represents a surface that interpolates the measured elevations at all points of S. Two classes of DEMs are usually considered in the context of GIS for visibility computation:

- A TIN is defined by a triangulation of the domain D having its vertices at the points of S. Function f is defined piecewise as a linear function over each triangle. Thus, the surface described by a TIN consists of planar patches.
- An RSG is defined by a domain partition into rectangles, induced by a regular grid over D. Functions used on such partitions depend on the degree of continuity desired for the surface; usually, f is either bilinear or constant over each region.

TINs show good capabilities to adapt to terrain features, since they can deal with irregularly-distributed datasets and may include surface-specific points and lines. Often, a Delaunay triangulation is used as a domain subdivision for a TIN, because of its good behaviour in numerical interpolation. A triangulation is a Delaunay triangulation if and only if the circumcircle of each triangle does not contain any other vertex; the Delaunay triangulation can also be characterised as the dual graph of the Voronoi diagram of a point set (Boots, Chapter 36). More recently, data-dependent triangulations have been proposed, which take into account the z values of points in V instead of simply their x, y coordinates; the idea is either to maximise or to minimise some quantity that expresses certain properties of the resulting surface (e.g. the 'roughness' or the 'thin-plate energy', or the maximum jump between adjacent patches; see Dyn et al 1990, for a survey, and Mitas and Mitasova, Chapter 34).

3 VISIBILITY STRUCTURES FOR TERRAINS

Measuring visibility requires computing visibility for (portions of) the surface itself or for objects located on the surface (representing, for example, facilities such as towers, buildings, radio transmitters, etc.). The problem of testing the visibility of an object is essentially a query problem, and must be solved on-line. In contrast, visibility information for the surface itself can be precomputed and stored in appropriate visibility structures; such structures also help answer visibility queries. This section introduces the main structures used to represent the visibility of a terrain with respect to a single or multiple viewpoints; for each structure, first an abstract definition is given, then its encodings on RGSs and on TINs are considered.

Two points V and W on a topographic surface are said to be mutually visible if and only if the interior of the straight-line segment joining them lies strictly above the terrain; such a segment is allowed to touch the surface at most at its endpoints V and W. Any point V lying on or above a topographic surface can be chosen as a viewpoint.

The basic visibility structure for a terrain is the viewshed. Given a viewpoint V on a terrain, the viewshed of V is the set of points of the surface which are visible from V, that is, viewshed $(V)=\{W\in D \mid W$ is visible from $V\}$.

Another relevant form of visibility information is the horizon of a viewpoint V, which corresponds to the 'distal' boundary of the viewshed. Such reduced information can replace the viewshed in some applications, with the advantage of lower storage costs. The horizon determines, for every radial direction around V, the farthest point on the terrain which is visible from V: horizon $(V) = \{W\in D \mid W$ is visible from V and $\forall Q \in D$ if $W\in \overline{VQ}$ then Q is invisible from $V\}$.

Visibility structures for several viewpoints can be defined by combining the viewshed of such points according to some operator; common combination operators are:

- *Overlay*: the terrain is partitioned into regions, in such a way that each region is visible from a given set of viewpoints.
- *Boolean operators* (see Eastman, Chapter 35), such as the intersection of the viewsheds (which gives the portions of a surface visible from all viewpoints), the union (which gives the portions visible from at least one viewpoint), etc.
- *Counting operators*: for example, the surface may be partitioned into regions, such that all the points of a region are visible from the same number of viewpoints.

The set of viewpoints considered is usually restricted to be a subset of the vertices of a TIN, or a subset of the cells of an RSG. In the remainder of the chapter, a visibility structure related to multiple viewpoints will be called a *multi-visibility structure* for brevity; if necessary, the operator used to obtain it will be specified. Note that the overlay of viewsheds contains more information than any other multi-visibility structure.

Any visibility structure can be encoded in either a continuous or a discrete way. For the viewshed (see Figure 1) a continuous encoding subdivides each cell of the DEM; this form is called a *continuous visibility map*, and it is mainly used for

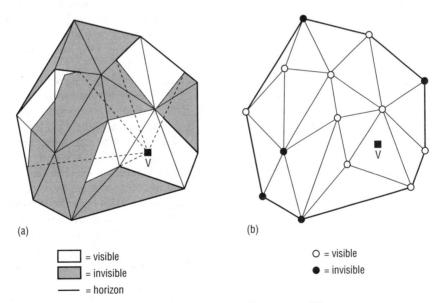

(a)   (b)

☐ = visible   ○ = visible
▨ = invisible   ● = invisible
—— = horizon

Fig 1. (a) The continuous-visibility map and the horizon; (b) the discrete visibility map on a TIN.

TINs. The continuous visibility map of a TIN with n vertices has a worst-case space complexity in $O(n^2)$. On RSGs, the viewshed is usually represented in a discrete way, by marking each grid cell or each vertex as visible or invisible. The resulting array of Boolean values is called a *discrete visibility map*. RSGs are usually dense, so that discrete visibility maps are accurate enough for application needs; on the other hand, for an RSG a continuous visibility map would be difficult to compute (since RSGs do not support linear interpolants) and huge in size. The discrete visibility map has an $O(n)$ worst-case space complexity for a $\sqrt{n}$ by $\sqrt{n}$ regular grid. A discrete visibility map for a TIN is sometimes considered (e.g. Lee 1991); such a map is obtained by marking each triangle or each vertex as visible or invisible; the spatial complexity is $O(n)$ for a TIN with n vertices.

A continuous representation of the horizon of a viewpoint V consists of a sequence of portions of terrain edges, radially sorted around V. As the continuous visibility map, this form is used for TINs. The size of the horizon on a TIN with n vertices is $O(n\,\alpha\,(n))$ (Cole and Sharir 1989), where α is the slowly growing inverse of Ackermann's Function. On an RSG, the horizon can be represented in a discrete way as a collection of grid cells; however, since there is no gain in space complexity with respect to a discrete visibility map, discrete horizon representations are not used in practice.

Visibility structures related to multiple viewpoints can be encoded in a variety of different ways. Discrete encodings are mainly used, due to the huge dimensions of a continuous encoding when many viewpoints are considered. Examples of such encodings are the visibility graph and visibility counts. The visibility graph (Puppo and Marzano 1996) represents the overlay of the discrete visibility maps of several viewpoints. It consists of a graph in which each node corresponds to a vertex or to a cell of a DEM and every pair of mutually visible nodes is joined by an arc. The spatial complexity of a visibility graph is $O(n^2)$ for a DEM with n vertices; visibility graphs are used both for RSGs and TINs, and are fundamental in solving several application problems (see section 6).

Visibility counts are discrete encodings of multi-visibility structures obtained through counting operators. A visibility count is obtained by labelling each vertex or each cell of a DEM with the number of viewpoints from which it is visible. This information is mainly considered on gridded models. A special case is the *intervisibility map* between two regions (Mills et al 1992); given a source region and a destination region (e.g. two rectangular blocks of cells of an RSG), each point of the destination region is labelled with the number of points of the source region from which it is visible. The size of an intervisibility map is determined by the number of cells of the destination region. Discrete representations of multi-visibility structures based on Boolean operators (e.g. union, intersection) reduce to arrays of Booleans.

4 ALGORITHMS FOR VISIBILITY COMPUTATION

In this section, an overview is given of algorithms for visibility computation on terrains. Two subsections are devoted to the computation of visibility structures, in continuous and discrete encodings, respectively; the last subsection deals with visibility queries on a DEM.

4.1 Computation of continuous visibility structures

As mentioned in section 3, continuous encodings of visibility structures are used only for TINs. Thus, the algorithms considered in this subsection are algorithms for TINs; they exploit the fact that a TIN describes a polyhedral surface. In general, polyhedral surfaces have deserved interest both from the GIS community and from computational geometers working on GIS. Some algorithms have a theoretical interest for their good asymptotic complexity, but are difficult to implement; other algorithms have been successfully implemented and show a good practical performance, whereas they exhibit a poor worst-case complexity, or they even lack a precise theoretical analysis.

The problem of computing the continuous visibility map of a TIN is connected with the more general hidden surface removal (HSR) problem for a 3-dimensional scene. Given a viewpoint and an image plane, HSR algorithms build the visible image of a scene, that is, a subdivision of the image plane formed by collecting the projections (images) of the portions visible from the viewpoint of each face of the scene (see Figure 2). Thus some algorithms are reported here for HSR, which can be used to compute the continuous visibility map of a TIN.

A common approach to visibility computation on a TIN consists of processing the faces in front-to-back order from the viewpoint. Given a viewpoint V, a cell c_1 of a DEM is said to be in front of another cell c_2 if a ray emanating from V intersects c_1 before intersecting c_2. A front-to-back order of a DEM is any total order of its cells consistent with the 'in front' relation; if $c_1 < c_2$ then c_1 may be in front of c_2, but not vice versa. A DEM is sortable if a front-to-back order exists. Because of their irregular structure, sortability is not guaranteed for all TINs. Delaunay-based TINs have been shown to be always sortable (De Floriani et al 1991); a non-sortable TIN can always be made sortable by splitting some of its triangles (Cole and Sharir 1989).

The front-to-back approach exploits the fact that a triangle t can be hidden only by triangles in front of it. At each step, a current horizon is maintained and used to determine the visibility of new triangles (see Figure 3). This method was developed and implemented by De Floriani et al

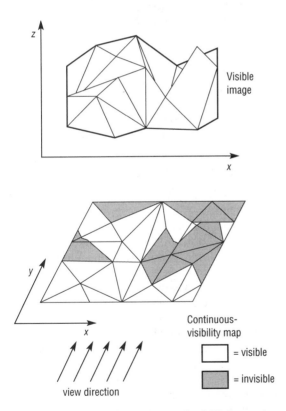

Fig 2. The continuous-visibility map and the visible image of a TIN. The viewpoint lies at infinity in the negative *y* direction, and the *x, z* plane is used as the viewplane.

(1989) and experiments show a nearly-linear time complexity, even though the asymptotic time complexity is $O(n^2 \alpha(n))$ in the worst case. The algorithm was parallelised by De Floriani et al (1994b) by using a data partitioning strategy. The domain is subdivided into radial sectors, which have the viewpoint as their common vertex, and each sector is assigned to a processor. The algorithm has been implemented on a hypercube machine nCUBE-2, a coarse-grained multiple instruction, multiple data (MIMD) architecture.

More sophisticated algorithms, still based on a front-to-back traversal, achieve an output-sensitive time complexity by storing the current horizon into some special data structures (Preparata and Vitter 1992; Reif and Sen 1988). The time is $O((n+d) \log^2 n)$, where n is the TIN size and d is the size of the computed visible image. Interest in these algorithms is mainly theoretical. Reif and Sen (1988) also propose a parallel algorithm, based on a variation of the front-to-back approach.

Still in the realm of theoretically interesting algorithms, the worst-case optimal algorithm by Edelsbrunner et al (1989) can be mentioned. It computes the visible image of a TIN in $O(n^2)$ time, based on a divide-and-conquer strategy. Theoretically efficient output-sensitive HSR algorithms, which can be applied to a TIN, have been proposed by Katz et al (1991) and by Overmars and Sharir (1992). Both algorithms exploit a front-to-back traversal of the scene; a parallel version of the latter algorithm has been proposed by Teng et al (1997).

An on-line algorithm has been proposed by Boissonnat and Dobrindt (1992). This algorithm computes the visible image of a generic set of triangles in the space (and thus the visibility map of a TIN), and is based on an incremental update of the visible image by inserting triangles one at a time. The kernel of the algorithm is a special data structure which provides bounds for the expected time and space complexity when averaging on all possible permutations of the input data. This algorithm was successfully implemented. The algorithm has also been extended to a fully dynamic method, which allows both insertions and deletions of triangles in expected $O(n \log n)$ time (Bruzzone et al 1995; Dobrindt and Yvinec 1993). Dynamic algorithms are of special interest because they are useful to update visibility information for a terrain when the underlying elevation model changes

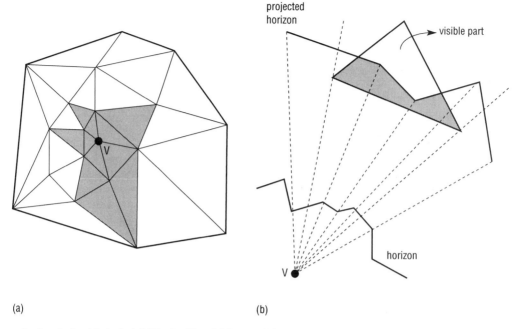

(a)　　　　　　　　　　　　　　　　　　　　　　　(b)

Fig 3. A generic step of a front-to-back visibility algorithm: (a) the set of already-processed triangles (shaded) and the current horizon (thick); (b) determination of the visibility of a triangle by projecting the current horizon on the triangle itself (the portions of the triangle lying above the projected horizon are visible).

(a problem occurring, for instance, with multi-resolution terrain representations, see section 5).

The horizon of a TIN is obtained as a side-product of algorithms for computing visibility maps (for example, front-to-back algorithms build the horizon as an auxiliary structure). The horizon can also be directly computed as the upper envelope of the set of segments obtained by projecting the terrain edges on an image plane. Existing upper-envelope algorithms are of two types: divide-and-conquer methods run in $O(n\ \alpha(n)\ \log n)$ (Atallah 1983) or $O(n \log n)$ (Hershelberg 1989) time (the latter is worst-case optimal since the inherent complexity of the problem is $O(n \log n)$); an incremental approach leads to a sub-optimal $O(n^2\ \alpha(n))$ time in a straightforward implementation, while a randomised version (De Floriani and Magillo 1995) has an expected running time of $O(n\ \alpha(n)\ \log n)$.

Though not very relevant for practical use, continuous encodings of multi-visibility structures for TINs can be obtained starting from the overlay of continuous visibility maps. Efficient algorithms exist which compute the overlay of two maps, the most common of which are based on a sweep-line technique (Bentley and Ottmann 1979; Chazelle and

Edelsbrunner 1992; Clarkson and Shor 1989; Guibas and Seidel 1987; Mairson and Stolfi 1988). The time complexity is sensitive to the size of the final result (which may be quadratic in the size of the two input maps, depending on the number of intersections).

4.2 Computing discrete-visibility structures

Algorithms for discrete-visibility maps are based on the computation of the intersection between lines-of-sight from the viewpoint and edges of the DEM cells. A straightforward approach requires $O(n^2)$ time on a TIN and $O(n\sqrt{n})$ time on an RSG with n vertices. For RSGs, the elevation of each edge is assumed to vary linearly between the two endpoints, and the elevation of the terrain inside each grid cell is usually not considered.

Lee (1991) computes the discrete-visibility map of a TIN by using a front-to-back method, similar to the one used by De Floriani et al (1989) for continuous-visibility maps; here, the discretisation is achieved by considering a triangle as visible if all its three edges are completely visible.

Discrete-visibility maps are mainly used for RSGs. The method of Shapira (1990) traces a line-of-sight from the viewpoint V to any other point P, and starts walking along on the line-of-sight from V to P. The walk terminates when either an intersection between the line-of-sight and a terrain edge is found before reaching P, or when P is reached. This method performs redundant computations, since rays to different points of the grid may overlap partially. Other methods (e.g. Blelloch 1990) consider only rays joining the viewpoint V to a boundary point of the destination region, and determine the visibility of each cell along a line-of-sight by using the current tangent from the viewpoint to the terrain (see Figure 4). The complexity can be further reduced by sampling a subset of line-of-sight directions around a viewpoint, thus leading to approximate discrete-visibility maps.

Visibility computation on RSGs is an expensive process because of the size of the grid, especially when multi-visibility structures (e.g. a visibility graph) are computed. On the other hand, because of their regular topology RSGs are especially suitable to be handled by parallel methods developed for the architecture of massively-parallel computers (see Openshaw and Alvanides, Chapter 18). Parallel algorithms exploit the fact that the regular spatial structure of an RSG can be directly embedded into a parallel single instruction, multiple data (SIMD) architecture, such as the mesh or the hypercube. Mills et al (1992) and Teng et al (1993) have proposed parallel algorithms for computing intervisibility maps on an RSG.

The method of Mills et al (1992) uses a parallel version of the algorithm of Shapira. Every line-of-sight from every viewpoint of the source region is processed in parallel. Elevation data are communicated from one sight-line to an adjacent one, in order to reduce global communication between processors. The algorithm has been implemented on a Connection Machine CM-2.

The method of Teng et al (1993) performs a sweep traversal of the source region, and exploits the coherence at adjacent viewpoints in order to reduce global communications. They consider only lines-of-sight from a point of the source region to a boundary point of the target region, as does Blelloch (1990). If w is the maximum length (in cells) of a line-of-sight, and l is the side of source and destination regions, the time complexity of the algorithm is estimated as $O(l \log w)$, when using $O(l^2 w)$ processors. Experimental results obtained on a Connection Machine CM-2 match this estimate.

4.3 Answering visibility queries

Visibility queries concern the computation of visibility for objects located on the terrain rather than for the surface itself. Given a viewpoint V, a visibility query is defined by providing a query object. The problem consists of determining the visibility of the object from V. A visibility query for a point Q simply requires a test of whether Q is visible from V. As for the visibility of the terrain itself (see section 3), the visibility of a non-point object can be encoded either in a continuous way or in a discrete way. In the continuous approach, the query is answered by computing a partition of the query object into visible and invisible subsets. A discrete answer consists of marking the object with a Boolean value, according to some convention; the answer can be true either when the object is entirely visible, or when it is at least partially visible, or when it is visible for more than a certain percentage of its extent, etc.

Finding the visibility of a query object from a given viewpoint V can be solved by examining the lines-of-sight joining V to a point on the object. The simplest case occurs when the query object is a point P. In this case, a 'brute force' approach that searches for the intersection of segment $\overline{PV}$ with the edges of the DEM takes $O(n)$ time on a TIN and $O(\sqrt{n})$ time on an RSG. If several queries must be solved for the same viewpoint, some kind of preprocessing can reduce the query time.

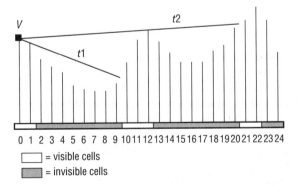

0 1 2 3 4 5 6 7 8 9 10 11 12 13 14 15 16 17 18 19 20 21 22 23 24

☐ = visible cells

▨ = invisible cells

Fig 4. Walking along a line-of-sight in Blelloch's approach: a vertical section of the DEM in the direction of the line-of-sight is shown; $t1$ is the current tangent for cells 2 to 10; $t2$ for cells 13 to 21.

If a visibility map of the terrain is available, then testing the visibility of a query point lying on the surface reduces to a *point location* in such a map. If the query point has a non-null height, then the approach is different depending whether we are considering continuous visibility on TINs or discrete visibility. The latter case reduces to point location in an enriched version of the discrete-visibility map, storing, for each cell, the minimum height which should be added to the cell in order to make it visible (zero if and only if the cell is visible). Such an enriched map can be computed with the same techniques. Point location within the simple or enriched discrete-visibility map can be done in constant time by computing the indices of the row and column containing the query point. As far as continuous visibility on TINs is concerned, testing whether a given point is visible reduces to locating the projection of such a point in the visible image of the TIN. Both the visible image and the continuous-visibility map are plane subdivisions; thus, existing techniques for point location within a plane subdivision can be used (see Kirkpatrick 1983; Lee and Preparata 1977; Preparata and Shamos 1985; Sarnak and Tarjan 1986), typically resulting in logarithmic query times.

Cole and Sharir (1989) proposed a data structure which allows a logarithmic query time by using an amount of memory less than the $O(n^2)$ space required by a continuous-visibility map or a visible image. They reduce a visibility query on a TIN to a ray-shooting query, that is, to the problem of determining the first face of the terrain hit by a ray emanating from the viewpoint and passing through the query point. For answering such queries, Cole and Sharir build a balanced binary tree, which stores a set of partial horizons. The space complexity of the whole tree is only $O(n \, \alpha(n) \log n)$.

5 VISIBILITY ON MULTI-RESOLUTION TERRAIN MODELS

Often, huge sampled datasets are available for a topographic surface. This allows accurate DEMs at the cost of storage space and access times. Since not all tasks require the same level of detail, the use of high-resolution models may affect applications for which many of the details are not relevant. Multi-resolution terrain models have been developed to provide a compact representation of a surface at

different resolutions. A multi-resolution model avoids redundancy of information and supports an easy extraction of terrain representations at any level of detail. This section first introduces multi-resolution terrain models and then briefly reviews the available methods for visibility computation on such models.

5.1 Multi-resolution terrain models

Multi-resolution terrain models are built from a given (huge) set of data by using iterative procedures of two kinds: simplification methods start from the complete dataset and progressively eliminate points, while refinement techniques progressively include points into an initial minimal set. At appropriate steps of the process, fragments of the current DEM are selected to be stored in the model.

A generic multi-resolution model of a terrain is based on a collection of fragments of DEMs, each characterised by a certain degree of accuracy, which can be combined in different ways to provide a description of the whole surface. A multi-resolution model encodes such components together with information needed to combine them into a single DEM satisfying any given level of accuracy. Such additional information typically includes:

- relationships of spatial interference between fragments (two overlapping fragments will provide two representations of the same area at different accuracies);
- information about adjacency and boundary matching between fragments (used to determine whether the union of two fragments forms a proper DEM).

Such information is represented in different ways depending on the structure of the multi-resolution model, which in turn depends on the strategy used for its construction.

A large subclass of multi-resolution models proposed in the literature is characterised by a nested subdivision of the domain; such models are usually termed hierarchical. A hierarchical model can be effectively described by a tree where nodes are the fragments (local DEMs), and arcs correspond to containment of a DEM into a cell of another DEM (see Figure 5). *Quadtrees* (Chen and Tobler 1986; Samet and Sivan 1992) and *quaternary triangulations* (Gomez and Guzman 1979) are regular hierarchical models based on the recursive partition of a square

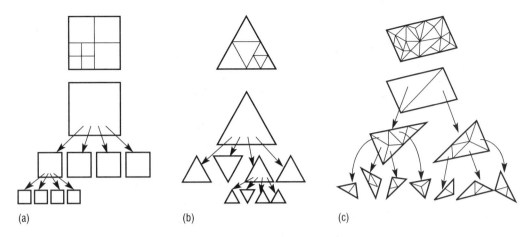

Fig 5. Hierarchical terrain models: (a) quadtree; (b) quaternary triangulation; (c) hierarchical TIN.

or an equilateral triangle, respectively, into four equal parts. In *restricted quadtrees* (Herzen and Barr 1987) the problem of preserving the continuity of the surface is solved by imposing the condition that adjacent squares cannot differ by more than one level in the refinement process, and by triangulating the final cells. Regular hierarchical models can be built both by refinement and by simplification. Other hierarchical terrain models are based on TINs; hierarchical TINs rely on a top-down refinement process, driven by various criteria (e.g. random or accuracy-driven strategies for the insertion of points, Delaunay or heuristic triangulation, etc.). The continuity of the surface is guaranteed through a consistent refinement of edges. Adaptive hierarchical triangulations (Scarlatos and Pavlidis 1990) and hierarchical Delaunay triangulations (HDTs) (De Floriani and Puppo 1995) are the two major examples of such models.

In more general multi-resolution models the spatial interference between two fragments does not necessarily reduce to a containment of one fragment by the other. All existing proposals of non-hierarchical multi-resolution models are based on TINs (see Figure 6). The first proposal is the *Delaunay pyramid* (De Floriani 1989), which encodes fragments from a sequence of Delaunay TINs describing a terrain at a sequence of increasing resolutions. This model does not rely on a special construction technique, and can be built by simplification as well as by refinement. Interference links are stored between pairs of consecutive triangles which have a proper intersection. The model proposed by Berg and Dobrindt (1995) is

built through iterative simplification of a Delaunay TIN; at each step, a set of independent vertices (i.e. vertices that are not endpoints of the same edge) of

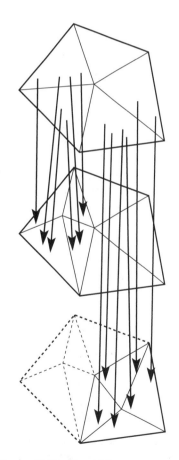

Fig 6. A pyramidal terrain model.

small degree is removed, and the 'holes' left by those vertices are re-triangulated. Interference links are maintained between the triangles incident in a removed vertex and those created to fill the hole.

The multi-triangulation proposed by Puppo (1996) is a general model which includes all previously mentioned models as special cases. The basic idea is a partial order of fragments, which drives their combination. The model can be described as a directed acyclic graph, where the nodes are the fragments, with the properties that two fragments connected by an arc have a spatial interference, and every cut of the graph corresponds to a TIN describing the whole terrain; larger cuts correspond to more detailed representations of the surface (see Figure 7). Cignoni et al (1995) use a different and less complex data structure for encoding a multi-triangulation.

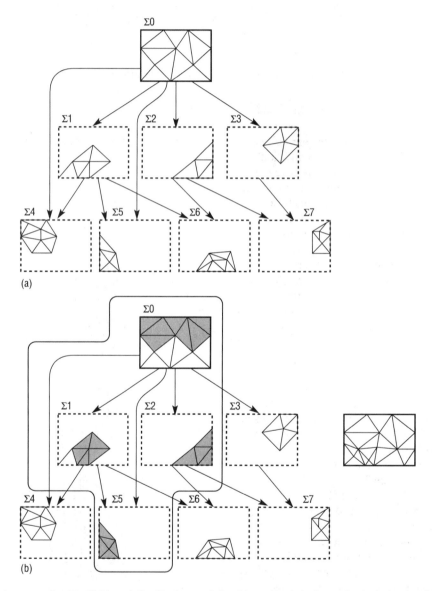

Fig 7. (a) A graph representing a multi-triangulation (the fragment stored in each node is shown, the dashed rectangle is the domain); and (b) a cut and the corresponding TIN defined over the domain (the TIN is obtained by collecting the 'exposed' triangles of the cut, shown shaded).

5.2 Visibility algorithms on multi-resolution terrain models

Since multi-resolution terrain models do not provide an explicit terrain representation, a DEM, describing the surface at a given user-defined level of resolution, must be extracted to be used for visibility computations. The level of resolution may be constant (i.e. a single threshold value is defined for the error over the domain), or variable (i.e. the maximum error over each cell is defined according to a threshold function). In visibility applications, an accuracy that decreases according to the distance from the viewpoint is especially interesting; because errors in elevations near the viewpoint are amplified in proportion to the distance, it is convenient to represent the topography more accurately near the viewpoint than on the rest of the surface (Felleman and Griffin 1990).

A variable-accuracy DEM will be made up of cells from different levels of a multi-resolution model. The main difficulty here is ensuring the continuity of the surface described by the extracted model. Cignoni et al (1995) propose an extraction algorithm for the case of accuracy decreasing with the distance from a viewpoint, which guarantees continuity. The method starts from the triangle having the maximum resolution and enlarges the model by including adjacent triangles; the approach applies to their own model and to any hierarchical TIN. Puppo (1996) proposes a general approach for extracting a DEM at variable resolution, based on interference information only. The method performs a traversal of the graph encoding interference among fragments, and identifies a minimal cut of the graph which satisfies the desired accuracy. This method, with some additional care to avoid cracks in the extracted surfaces, applies also to any hierarchical model (see De Floriani and Magillo 1996).

Once a DEM at the desired resolution has been obtained, any algorithm for visibility computation can be applied to it. On-line algorithms (e.g. Boissonnat and Dobrindt 1992; De Floriani and Magillo 1995) seem to be the most suitable because they can be run in parallel with the construction of the extracted DEM. Magillo and De Floriani (1994) have presented algorithms which compute the visibility map on a hierarchical TIN by navigating the tree-like structure of the model, and thus do not need the explicit construction of a TIN at the given accuracy.

A further problem connected to visibility on multi-resolution terrain models is the update of already-computed visibility structures when changing the resolution in some portion of the domain. This is required in applications such as flight simulation, where the focus of attention is rapidly moving, and thus the maximum resolution is required for different areas at different times. The visibility-update problem can be solved by recomputation or by applying dynamic algorithms. Such algorithms update a visibility structure after the deletion of old DEM cells and the insertion of new ones. Dynamic algorithms have been proposed for horizon computation (De Floriani and Magillo 1995) and for continuous-visibility maps on a TIN (Bruzzone et al 1995; Dobrindt and Yvinec 1993). The core of all these algorithms is a special data structure which helps in locating the parts of the structure affected by an update, providing a good expected performance for a random sequence of updates.

Finally, the brute force approach to visibility queries (see section 4.3) can be combined with a traversal of the tree encoding a hierarchical terrain model to answer visibility queries efficiently at a certain level of resolution. The aim is locating those cells, at the given accuracy, which may hide a query object; the visibility of the object is then tested only against those cells. The problem reduces to an interference query on the extracted DEM, for which the tree-like structure acts as a spatial index (see Figure 8). The major advantages are obtained with regular hierarchical models.

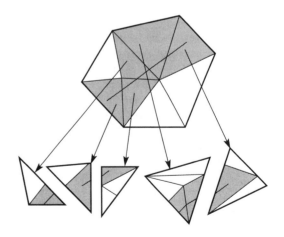

Fig 8. Processing a segment interference query on a hierarchical terrain model. First, the triangles of the root intersected by the segment are found; then the search is repeated on the nodes defining such triangles.

553

6 VISIBILITY-RELATED PROBLEMS AND ALGORITHMS

Interesting application problems on a terrain, which can be solved based on visibility information, can be classified into the following major categories:

- problems which require the placement of observation points on a topographic surface, according to suitable requirements;
- line-of-sight communication problems;
- problems regarding the computation of paths on a terrain, with certain visibility properties.

Viewpoint placement problems require the placement of several observation points on a terrain, in such a way that a large part of the surface is visible from at least one point. Applications include the location of fire towers, artillery observers, radar sites, etc. In general, the aim is either to minimise the number of viewpoints to cover a target area, or, in a dual formulation, to select a fixed number of points in such a way that the visible area is maximised.

For the placement of a single observation point, algorithms running in polynomial time are known. If the height is fixed, a point that can see the whole surface (if one exists) can be determined in $O(n \log n)$ time, while the point with the lowest elevation, from which the entire terrain is visible, can be determined in $O(n \log^2 n)$ time on a TIN with n vertices (Sharir 1988).

The problem of determining an optimal set of viewpoints is usually addressed in a discrete version, by allowing viewpoints to be located only on the vertices of a DEM. Even in this case, the complexity of the problem is exponential in n (Puppo and Marzano 1996), and thus heuristic solutions are used. Several heuristic algorithms are discussed by Lee (1991); Lee considers both the case when the heights of the viewpoints are fixed, and the case when heights are variable and must be minimised. The methods presented are based on a greedy approach; they iteratively add viewpoints to an initially empty set, or eliminate viewpoints from an initial set containing all vertices. Bose et al (1996) address the problem of placing vertex-guards and edge-guards on a terrain as a graph matching problem, and propose simple polynomial-time algorithms to place a worst-case optimal number of guards.

Line-of-sight communication problems consist of finding a visibility network, and connecting two or more sites, such that every two consecutive nodes of the network are mutually visible. Straightforward applications are in the location of microwave transmitters for telephone, FM radio, television, and digital data networks (see Fry, Chapter 58). A typical problem is finding the minimum number of relay towers necessary for line-of-sight communication between a set of sites. The given sites and the relay towers are usually restricted to the vertices of a DEM.

Puppo and Marzano (1996) show that the only visibility information necessary for solving problems of this kind is the visibility graph, and reduce them to classical graph problems. Connecting two sites reduces to a shortest path search on the visibility graph. The construction of a line-of-sight network between several sites reduces to finding a minimum connected sub-graph of the visibility graph, containing the given sites, also known as a *Steiner problem* in graph theory, which is known to be NP-complete. Thus, heuristics developed in the graph literature can be applied. De Floriani et al (1994a) propose a dynamic approach which reduces memory requirements by computing on-line only those portions of the visibility graph that are needed for the computations.

Paths can be defined on a terrain, with application-dependent visibility characteristics. A smuggler's path is the shortest path, connecting two given points, such that no point on the path is visible from a predefined set of viewpoints. Conversely, a path where every point can be seen from all viewpoints is known as a scenic path. Path problems are usually addressed on a DEM by restricting the viewpoints to be vertices, and the path to passing along edges. A solution (if one exists) can be determined by first computing the vertices which are visible or invisible from all the viewpoints, and then applying a standard shortest path algorithm to the edges connecting them. The problem of finding a path 'as hidden as possible' is considered by Puppo and Marzano (1996); a shortest path is computed after having weighted every edge with the number of viewpoints that can (or cannot) see it.

A problem of high practical interest is that of updating visibility for a viewpoint moving along a trajectory. Such a problem has received little attention, perhaps because of its intrinsic difficulty. Existing results are restricted to straight-line trajectories, and, in particular, vertical trajectories (Bern et al 1994; Cole and Sharir 1989); they have mainly a theoretical interest. They are related to the computation of the points along the trajectory at which topological changes occur in the visible image, or to the processing of ray-shooting queries (i.e. finding the terrain face 'seen' by a given visual ray) from arbitrary viewpoints lying on the path.

7 CONCLUSIONS

Many visibility-related problems on terrains still lack practically satisfactory solutions; this group includes such problems as the update of visibility after modifications in the underlying elevation model (e.g. in the context of multi-resolution terrain modelling; see section 5) or for a moving viewpoint, the computation of optimal paths according to given visibility criteria, and several other optimisation problems (see section 6). This lack of efficient solutions is partially attributable to the fact that such problems have received little attention from the research community, both because most of them are intrinsically hard (several optimisation problems have been shown to be NP-complete), and because some of them have come to the attention of the GIS community only recently (for instance, problems on multi-resolution terrain models). In many cases (including path and communication problems), a good definition of the problem is still missing, thus making the work of finding algorithmic methods even more difficult. More research effort should be spent in investigating the problems mentioned above, since they have a high impact on applications and a fundamental importance in the development of information systems of the future.

References

Atallah M 1983 Dynamic computational geometry. *Proceedings, 24th IEEE Symposium on Foundations of Computer Science*. IEEE Computer Society: 92–9

Bentley J L, Ottmann T A 1979 Algorithm for reporting and counting geometric intersections. *IEEE Transactions on Computers* 28: 643–7

Berg M de, Dobrindt K 1995 On levels of detail in terrains. *Proceedings, Eleventh Annual ACM Symposium on Computational Geometry*: C26–C27

Bern M, Dobkin D, Epstein D, Grossman R 1994 Visibility with a moving point of view. *Algorithmica* 11: 360–78

Blelloch G E 1990 *Vector models for data-parallel computing*. Cambridge (USA), MIT Press

Boissonnat J D, Dobrindt K 1992 On-line construction of the upper envelope of triangles in $\mathbf{R}^3$. In Wang C A (ed.) *Proceedings, Fourth Canadian Conference on Computational Geometry*: 311–15

Bose P, Kirkpatrick D, Li Z 1996 Efficient algorithms for guarding or illuminating the surface of a polyhedral terrain. In Fiala F, Kranakis E, Sack J R (eds) *Proceedings, Canadian Conference on Computational Geometry*: 217–22

Bruzzone E, De Floriani L, Magillo P 1995 Updating visibility information on multi-resolution terrain models. In Frank A U, Kuhn W (eds) *Proceedings, Conference on Spatial Information Theory (COSIT 95)*. Lecture Notes in Computer Science 988. Berlin, Springer: 279–96

Chazelle B, Edelsbrunner H 1992 An optimal algorithm for intersecting line segments in the plane. *Journal of the Association for Computing Machinery* 39: 1–54

Chen Z T, Tobler W R 1986 Quadtree representation of digital terrain. *Proceedings AutoCarto*: 475–84

Cignoni P, Puppo E, Scopigno R 1995 Representation and visualization of terrain surfaces at variable resolution. In Scateni R (ed.) *Proceedings, International Symposium on Scientific Visualization*. Singapore, World Scientific. *http://www.disi.unige.it/person/PuppoE*

Clarkson K L, Shor P W 1989 Application of random sampling in computer geometry. *Discrete and Computational Geometry* 4(5): 387–421

Cole R, Sharir M 1989 Visibility problems for polyhedral terrains. *Journal of Symbolic Computation* 17: 11–30

De Floriani L 1989 A pyramidal data structure for triangle-based surface description. *IEEE Computer Graphics and Applications* 9: 67–78

De Floriani L, Falcidieno B, Nagy G, Pienovi C 1989 Polyhedral terrain description using visibility criteria. *Technical Report 17*. Genova, Institute for Applied Mathematics, National Research Council

De Floriani L, Falcidieno B, Nagy G, Pienovi C 1991 On sorting triangles in a Delaunay tessellation. *Algorithmica* 6: 522–32

De Floriani L, Magillo P 1995 Horizon computation on a hierarchical terrain model. *The Visual Computer* 11: 134–49

De Floriani L, Magillo P 1996 A comprehensive framework for spatial operations on hierarchical terrain models. *Technical Report DISI-TR-96-15*. University of Genova, Department of Computer and Information Sciences

De Floriani L, Marzano P, Puppo E 1994a Line-of-sight communication on terrain models. *International Journal of Geographical Information Systems* 8: 329–42

De Floriani L, Montani C, Scopigno R 1994b Parallelizing visibility computations on triangulated terrains. *International Journal of Geographical Information Systems* 8: 515–32

De Floriani L, Puppo E 1995 Hierarchical triangulation for multi-resolution surface description. *ACM Transactions on Graphics* 14: 363–411. *http://www.disi.unige.it/person/PuppoE*

Dobrindt K, Yvinec M 1993 Remembering conflicts in history yields dynamic algorithms. In Ng K W, Raghavan P, Balasubramanian N V, Chin F Y L (eds) *Algorithms and computation. Lecture notes in computer science 762*. Hong Kong, Springer: 21–30

Dyn N, Levin D, Rippa S 1990 Data-dependent triangulations for piecewise linear interpolation. *IMA Journal of Numerical Analysis* 10: 137–54

Edelsbrunner H, Guibas L J, Sharir M 1989 The upper envelope of piecewise linear functions: algorithms and applications. *Discrete and Computational Geometry* 4: 311–36

Felleman J P, Griffin C 1990 The role of error in GIS-based viewshed determination – a problem analysis. *Technical Report EIPP-90-2*. State University of New York, Institute for Environmental Policy and Planning

Gomez D, Guzman A 1979 Digital model for 3-dimensional surface representation. *Geoprocessing* 1: 53–70

Guibas L J, Seidel R 1987 Computing convolutions by reciprocal search. *Discrete and Computational Geometry* 2: 175–93

Hershelberg J 1989 Finding the upper envelope of *n* line segments in O(*n* log *n*) time. *Information Processing Letters* 33: 169–74

Herzen B von, Barr A H 1987 Accurate triangulations of deformed, intersecting surfaces. *Computer Graphics* 21: 103–10

Katz M J, Overmars M H, Sharir M 1991 Efficient hidden surface removal for objects with small union size. *Proceedings, Seventh ACM Symposium on Computational Geometry*. New York, ACM Press: 31–40

Kirkpatrick D G 1983 Optimal search in planar subdivision. *SIAM Journal of Computing* 12: 28–33

Lee J 1991a Analyses of visibility sites on topographic surfaces. *International Journal of Geographical Information Systems* 5: 413–29

Lee J, Preparata F P 1977 Location of a point in a planar subdivision and its applications. *SIAM Journal of Computing* 6: 594–606

Magillo P, De Floriani L 1994 Computing visibility maps on hierarchical terrain models. In Pissinou N, Makki K (eds) *Proceedings, Second ACM Workshop on Advances in Geographic Information Systems (GIS 94)*. New York, ACM Press: 8–15

Mairson H G, Stolfi J 1988 Reporting and counting intersections between two sets of line segments. In Earnshaw R A (ed.) *Theoretical foundations of computer graphics and CAD*. NATO ASI Series F40. Berlin, Springer: 307–25

Mills K, Fox G, Heimbach R 1992 Implementing an intervisibility analysis model on a parallel computing system. *Computers and Geosciences* 18: 1047–54

Nagy G 1994 Terrain visibility. *Computer and Graphics* 18: 763–73

Overmars M, Sharir M 1992 A simple output-sensitive algorithm for hidden surface removal. *ACM Transactions on Graphics* 11: 1–11

Preparata F P, Shamos M I 1985 *Computational geometry: an introduction*. Berlin, Springer

Preparata F P, Vitter J S 1992 A simplified technique for hidden-line elimination in terrains. In Finkel A, Jantzen M (eds) *Lecture notes in computer science 577*. Berlin, Springer: 135–44

Puppo E 1996 Variable resolution terrain surfaces. In Fiala F, Kranakis E, Sack J R (eds) *Proceedings, Canadian Conference on Computational Geometry*: 202–10. Also published in longer version as *Technical Report 6/96*, CNR – Institute for Applied Mathematics, Genova, Italy, 1996. *http://www.disi.unige.it/person/PuppoE*

Puppo E, Marzano P 1996 Discrete visibility problems and graph algorithms. *International Journal of Geographical Information Systems* 11: 139–61

Reif J, Sen S 1988 An efficient output-sensitive hidden surface removal algorithm and its parallelization. *Proceedings, Fourth ACM Symposium on Computational Geometry*. New York, ACM Press: 193–200

Samet H, Sivan R 1992 Algorithms for constructing quadtree surface maps. *Proceedings, Fifth International Symposium on Spatial Data Handling*: 361–70

Sarnak N, Tarjan R E 1986 Planar point location using persistent search trees. *Communications of the Association for Computing Machinery* 29: 669–79

Scarlatos L, Pavlidis T 1990 Hierarchical triangulation using terrain features. *Proceedings IEEE Conference on Visualization*. IEEE Computer Society: 168–75

Shapira A 1990 'Visibility and terrain labeling'. Masters thesis. Troy, Rensselaer Polytechnic Institute

Sharir M 1988 The shortest watchtower and related problems for polyhedral terrains. *Information Processing Letters* 29: 265–70

Teng Y A, Menthon D de, Davis L S 1993 Region-to-region visibility analysis using data parallel machines. *Concurrency: Practice and Experience* 5: 379–406

Teng Y A, Mount D, Puppo E, Davis L S 1997 Parallelizing an algorithm for visibility on polyhedral terrain. *International Journal of Computational Geometry and Applications* 7: 75–8

556

39

Virtual environments and GIS

J N NEVES AND A CÂMARA

The coupling of virtual environments (VEs) and traditional GIS creates what can be termed a virtual GIS. This integration brings to virtual environments the spatial analysis and query functions common to GIS. It also provides GIS users with the visualisation, user interface, and object manipulation features inherent to VEs. Several virtual GIS research projects are discussed in this chapter. Virtual environments provide, in addition, the ground for innovative user interfaces in spatial exploration: bridging miniature with virtual representations of territories; direct data querying; use of sound, time, and space sliders; and logging. These proposals and their relevant geographical applications are also reviewed.

1 INTRODUCTION

A virtual environment (VE) is a broad term that has its roots in notions such as cyberspace, artificial reality, and virtual reality. The notion of cyberspace was introduced by William Gibson in *Neuromancer*, while artificial reality and virtual reality were proposed by their pioneers Myron Krueger and Jaron Lanier, respectively.

In virtual environments, the user is inside a digital world, where he or she can explore and transform objects. The feeling of immersion is achieved by feeding the user's senses with (often 3-dimensional) images and sounds generated in real time. When immersed in VEs, the user can easily scale, rotate, and translate objects and the viewpoint. Virtual environments are thus an appropriate platform for spatial exploration. Indeed, digital terrain models draped with textures are an essential component of many VEs. The digital terrain models are represented, as usual, by regular or hierarchical meshes, or triangulated irregular networks (TINs: see De Floriani and Magillo, Chapter 39; Hutchinson and Gallant, Chapter 9). The textures are derived from aerial photographs, satellite images, or more abstract GIS layers (e.g. soil use, wind, vegetation, aspect).

Thus raster geographical data are well-suited for integration into virtual environments, as manifest in the many civil and military virtual reality applications. Vector data are harder to integrate into virtual environments, particularly when the vectors have to be superimposed onto pre-existing terrain models. However there are reasonably efficient techniques to automate this process.

In section 2, a detailed review of virtual environments is presented, which focuses on object representation, on interfacing within VEs, and also on technical issues relevant to interfacing with the sensory channels. In section 3 the integration of VEs and GIS is analysed in a discussion focusing on representational, performance, design, and interface issues. Section 3 also provides pointers to several projects connecting VEs to GIS, and a short discussion of current technological developments and future research directions.

2 VIRTUAL ENVIRONMENTS

Virtual environments or virtual worlds are those that result from the interaction between the cognitive level of the human being, usually designated as mental maps, and the visual and audible images produced by the computer. A virtual world is a space deliberately designed by humans, to represent real or abstract spaces in which exist objects, which are

557

governed by rules specified for the worlds they inhabit (Jacobson 1994). Virtual worlds can be used to organise, represent, and manipulate data with multi-dimensional characterisitcs. They can be seen as planar images, 2.5-dimensional models in a conventional monitor, or as a true 3-dimensional space in an immersive environment. Virtual worlds can also include information with more than three dimensions, including the temporal dimension (Jacobson 1994).

The use of virtual reality (VR) techniques allows high-fidelity spatio-temporal representations, because the decision-maker is able to interact directly with the elements which form the basis of the user's decision. This ability to feel, as opposed to qualify, the results of an action makes the understanding clearer and universal. Virtual environments facilitate human–computer interaction by the use of a 3-dimensional representation and direct manipulation of virtual objects (Burdea and Coiffet 1994). In traditional VEs the user is immersed in a 3-dimensional world generated by a computer using a head-mounted display, and position and orientation sensors.

2.1 Defining terms

Several concepts are associated with VEs, and a detailed introduction to the subject must necessarily begin with their identification. The field of VE has been strongly influenced by the pre-existing fields of teleoperation and simulation. After the emergence of VE, a new field with a lot of similarities with VE appeared, called augmented reality.

In *teleoperation systems* a real environment is sensed by special-purpose sensors capable of giving visual and audible records of the environment. This information is sensed by a robot, which is called a telerobot because it is being manipulated by a human being. The human operator has access to the information sent by the robot using a specially-designed human–machine interface which conveys such information in multi-sensory display devices. The human takes actions in the form of motor actions that, being sensed by the interface, are transformed into signals to control the telerobot. The degree of control that the human operator has on the telerobot varies. When the telerobot is completely autonomous it is called simply a robot. The purpose of teleoperation

is to extend the sensorimotor system of the human organism (Durlach and Mavor 1994), making the human operator more able to sense an environment remotely or locally.

VEs are directed not to sense a real environment but to interface with a simulated environment. The human operator is called a user, who can interface with a computer using a human–computer interface in order to solve some kind of problem. What makes VEs unique is that this interface immerses the user in a digital environment by the use of multi-sensory displays and controls: displays that impress a multitude of senses, and controls that react to inputs coming from multiple human language constructs (speech, body language).

Virtual worlds are populated by 3-dimensional objects (with locations and orientations in 3-dimensional space) that have encapsulated behaviours capable of enriching the human experience. The user can move independently of these objects, manipulating them in real time. Special input/output devices give the user a way to interact with the digital environment using a variety of motor output channels, directly connected to the senses. This multi-sensory environment serves vividly to convey the feeling of presence in the simulated environment, whether realistic or abstract.

VEs are used to present and interact with information stored in a computer. There is a transfer of information from the computer to the human such that the human's knowledge, both physical and psychological, is increased. This makes VEs especially attractive for education and training and for scientific visualisation. Virtual environment systems are also being used in the areas of medicine, marketing, telecommunication, information visualisation, product design, manufacturing, and entertainment.

In *augmented-reality (AR) systems* virtual and real environments are combined to form a unique environment which is shown to the user. Usually it consists of a see-through display where the information from the real world (coming from video cameras) and from the digital world (coming from the computer) are overlaid upon one another.

The human–machine interface is very important to all of these systems (teleoperation, VEs, and AR) because it strongly affects the performance and subjective experience of the human operator in the real or virtual environment. The interaction must be transparent, that is, the electronic mediation must be perceptually neutral. This is particularly important for the subjective sense of presence experienced by the user. The sense of virtual presence is dependent on the amount of stimulation coming from the immediate environment and also on the variety and simplicity of the interaction.

VEs are unique mainly because of their multi-sensory, immersive, interactive, and 3-dimensional nature. VEs are computer objects with behaviours and appearance resembling real-world objects, and the user's interaction with them has strong similarities with interaction in the real world. This is the reason why VE interfaces are known as 'no metaphor' interfaces.

In recent years the senses most researched have been, in descending order, vision, audition, and haptics. Olfactory and gustatory displays are not common. The basic piece of technology used to display visual and auditory displays is the head-mounted display (HMD: see Figure 1); gloves are used to display haptic signals. HMDs always involve the tracking of the user's head position and orientation, giving the computer the parameters to render the visual and auditory signals in real time. Gloves measure hand position and finger joint angles. Feedback is usually presented by a virtual hand, although this arrangement does not give haptic sensory feedback. Devices capable of giving force and tactile feedback, or texture and temperature displays, are still not available commercially.

2.2 Sensory channels

As noted above, the most-used channels in the field of VE are the visual, the auditory and the haptic. The

Fig 1. User wearing a head mounted display (HMD) in a typical immersive VE system.

other channels are rarely used and are not as suitable for the spatial information systems considered here. Since the projects reviewed in this survey do not use the haptic channel, only the visual and auditory channels will be reviewed in this chapter.

2.2.1 Visual channel

The field of VE has always been associated with HMDs. These displays include the capacity to present 3-dimensional images and sounds, through the use of tiny screens close to the human eye, with additional optics to make these images perceptible; earphones to convey audio; and tracking systems to measure head position and orientation in order to generate such displays. These displays still have strong limitations in the resolution and field of view of the visual images, in addition to certain ergonomic problems.

A different type of display, intended to deal with the limitations of HMDs, is called an 'off the head' display (OHD). A good example is the Virtual Environment Theatre (VET) described below. The Boom from Fake Space Labs, lightweight stereographic glasses, and desktop stereo display screens are other kinds of display systems currently available. See-through displays are basically HMD where digital and environmental images (from video) are mixed and presented to the user in the same display. They still suffer from registration problems (arising out of misalignment of digital and environmental images), which have been the subject of extensive research (State et al 1996).

2.2.2 Auditory channel

There are two approaches to the presentation of auditory signals to the user: earphones, which are usually built into the HMD, and loudspeakers. Loudspeakers are preferable if the goal is to generate loud sounds and also to provide a physical impression. Earphones are more suited to individual experiences, where the information is presented to a single user or multiple users inhabiting the same virtual world but receiving different signals.

The measurement of head-related transfer functions (HRTF) and their incorporation into 3-dimensional audio-processing boards are still inadequate, particularly in the simulation of sounds located in front of or behind the user. Moreover, recording and playback of sounds requires large amounts of disk storage, and also the real-time generation of sounds still does not include environmental sounds. Only methods to generate music and speech exist, and their application in VE remains limited. The use of auditory signals for sensory substitution or to present information in parallel with the visual signals is an important field of study, mainly in the design of effective user interfaces for VEs.

2.3 Objects and operations

Objects that can be incorporated in VE have a hierarchical structure: nodes group several objects, objects group several polygons, and polygons group several points or vertices (usually three). The atomic element is a vertex with three coordinates x,y,z that represent its location in 3-dimensional space.

Objects in VE can be realistic or abstract. They can have a 1:1 relation with their physical counterparts (when they represent real-world objects), or may be scaled in time or space. In the case of abstract objects they can represent a different sensory modality (sound icons, for instance) or they can be represented through a visual metaphor (Stuart 1996).

The objects that populate a VE can be produced by keying the vertices' and polygons' definitions directly into a text editor, using a 3-dimensional modelling package or directly from real objects using a 3-dimensional digitising system. They can also be produced in real-time in the virtual environment as a result of user modelling or as an output of a simulation process (Stuart 1996).

Objects have several attributes that characterise them or that make explicit their links to multimedia information: position and orientation, scale, colour, feel, texture and shape, visibility, interaction with other objects, attached sound, weight, and inertia. Each of these attributes can be changed, thus defining the set of operations available on objects. Intelligent behaviour, like that of robots, is a further characteristic of objects that is being studied extensively.

2.4 Interacting with VEs

The most basic interaction mode in VE is navigation: walking (terrain following) or flying over the virtual space. Other possibilities include (Stuart 1996, following Mine 1995):

- selection;
- manipulation;
- scaling;
- menu and widget interaction.

These interaction modes are related to our usual modes of interaction with real-world objects like touching, grabbing, and moving. Touching objects can also activate some kind of simulated behaviour, such as the effects of gravity, or a pre-recorded animation, such as opening a door. The approach used to represent the interactions with these objects also can vary, and may entail (Stuart 1996, following Mine 1995):

- direct manipulation (body tracking and gesture recognition);
- use of physical controls (buttons, joysticks, and other haptic devices);
- use of virtual controls, such as controls that resemble physical controls or that follow the WIMP (windows, icons, menus, pointer) model.

Virtual space- and time-sliders, along with the possibility of taking notes (logging), can increase the usability of the virtual environment. When users are exploring virtual spaces they often need to have information about a particular location, whether static or dynamic (resulting from a simulation process). Details on this subject can be found in the work of Dias et al (1995).

Spatial sound and generation of non-spoken audio are critical areas of research in virtual environments, as they can dramatically increase their level of realism. Buxton et al (1993) present the main questions and the answers currently available. The generation of non-spoken audio assumes that the audible world is composed of a set of acoustic objects. Features such as tone, intensity, and rhythm may specify the identity of objects and render significant events or actions in the virtual world. These features may be manipulated systematically, creating a sound symbology extending from common sounds to the more abstract mapping of statistical data into sound parameters (Buxton et al 1993).

The nature of sound makes it the best choice to complement the role of images in portraying views of the state of the system. Sound has strengths that seem to solve some of the problems inherent in the visual representation of scientific data (Shepherd 1994):

- it can be used to interpret multi-dimensional data because many parameters can be used to denote change;
- it can be used to represent parameters that change over time;
- sound is omni-directional and is not affected to the same extent as light by physical barriers.

Above all, sound is very useful because it works in parallel with vision and so does not overload the user with information.

3 VIRTUAL GIS

The current generation of GIS have graphical user interfaces (GUIs) with numerous pull-down menus and widgets, following the WIMP model (see Egenhofer and Kuhn, Chapter 28). In such systems interaction is predominantly with the mouse in a point-and-click fashion. Additionally, working with geographical information means working with several layers of information simultaneously, which is a task inappropriate to common GUIs. VEs provide a very natural way of interacting with and visualising data, and can be used effectively to interpret multi-layer data and to analyse the interactions between layers (Faust 1995). Immersion (total or partial) within geographical databases facilitiates very intuitive means of navigation; it can also give the user new opportunities to understand the spatial relationships between GIS variables, and allow the user to perform vertical or horizontal analysis (Faust 1995) while visualising data. Simulation models could be integrated into the same exploration-based interface (Neves et al 1995), making results directly perceptible.

The way-finding metaphor provided by VEs can be considered a human–problem interaction metaphor instead of a human–computer interface metaphor (Edwards 1992). As Edwards points out, the formation of contextual mental models of problems is only possible if the human has experienced the problem first-hand. Virtual environments provide a rich environment to position the user in the space and time of the problem, making possible the formation of mental maps which are already rich in contextual information.

3.1 Bridging VE and GIS

Two opposite approaches have been used in the past to integrate GIS and VE:

- develop a GIS inside a VE, in order to implement GIS functions; or
- integrate both the GIS and the VE, allowing them to exchange objects.

The first approach was until recently the only one available because commercial GIS designs did not follow object-oriented principles (see Worboys, Chapter 26). With the advent of object orientation in GIS the second alternative has become more feasible, and some projects have already followed this approach (Neves et al 1995). The continuing decrease of price/performance ratios in personal computers, making computers with several processors relatively inexpensive, will make the second alternative even more attractive. Strong market pressures are leading GIS vendors in the direction of open object formats, making the development of GIS functionality much easier than before. The leading GIS vendors are already following this path.

3.2 Level of detail management

In order to explore large terrains, level of detail management algorithms (LODs) are essential for real-time rendering. The goal is to maximise image quality while maintaining a sufficiently rapid frame rate for an immersive walk-through, knowing that triangle size (in TINs) and texture resolution should vary inversely with the distance to the viewpoint. LODs are often based on quadtree representations for both the terrain and the textures. The different levels of detail may be pre-computed and stored on disk, or may be dynamically computed using a wavelet decomposition of the information (images and terrain). This enables not only the storage of different levels of detail but also its update on a local and multi-resolution basis.

Implementations using several levels of detail for terrain data fall in two categories. The first aims to represent data with high resolution but covering relatively small areas, while the second aims to represent data defined on the sphere (i.e. a simplified Earth's surface). In the former category the original data are generally arranged in an elevation matrix and a hierarchical tree is built subdividing the space in rectangular patches (Hughes 1993; Lindstrom et al 1996; see Plate 37). The area covered is relatively small so the function domain can be considered to be a plane region. Otherwise, the data need to be mapped to the sphere using parameterisation. However, mapping errors always occur since there is no globally smooth parameterisation for the sphere. The second category starts on a base shape of triangles, resulting in several trees, each one subdividing a triangle of the base shape recursively (Fekete 1990). This way, each triangle can have a corrected parameterisation.

The triangulations used on both categories are generally regular triangulations. As wisely stated by Scarlatos and Pavlidis (1992), terrains form continuous linear patterns which generate critical lines like ridges or channels. A regular triangulation generates edges that do not conform with these features. For this reason, when using a regular tessellation, a large number of triangles must be used to create a consistent approximation. But irregular triangulations can be built according to terrain features and can be arranged hierarchically, in order to provide different levels of detail across space. However, they cannot be updated at multiple resolutions. Scarlatos and Pavlidis (1992) describe such a representation, but it is rather complex since it considers several triangle-split strategies at the same time (see also De Floriani and Magillo, Chapter 38; Hutchinson and Gallant, Chapter 9).

The real-time rendering of terrains with their level of detail defined on a local basis involves the correct determination of the minimum level of detail to use for each area. Lindstrom et al (1996) present rules for this purpose, by reducing topography and texture data based on both distance and orientation.

3.3 Multi-resolution representation

Multi-scale simulation and visualisation has the advantage of being more efficient, because detail is used only when and where needed. The resolution can be increased or decreased because the detail information allows changing scales. However, none of the representations cited above allows multi-resolution updates. Recently, Lounsbery at al (1994) presented a multi-resolution representation for surfaces of arbitrary topological type. The base shape used is an octahedron (eight faces). However, classical wavelets are used, which are defined for the real line or the plane. Hence, the method relies on parameterisation and consequently it is not the actual data that are represented but an approximation of them. This problem has been overcome by Schröder and Sweldens (1995), who describe a representation that uses second-generation wavelets defined on the sphere.

In the representation used by Muchaxo et al (1996) the terrain data are considered to be planar. Multiple resolution is achieved using wavelets defined on the Haar basis, which was chosen because it has compact support and is fast to compute. Other wavelet bases can be used but have inferior performance. In order for wavelets to be used, the size of each data matrix must be a power of two in both dimensions; however it is not required that the powers be equal. A previous implementation was recently presented that does not impose this restriction, but does not use multiple resolutions (Muchaxo 1995).

In order to represent information pertaining to an extensive terrain in several levels of detail, the terrain could be recursively subdivided using a quadtree and several levels of detail computed and saved. However, redundant information would exist among the several levels of detail and the update of data would be nearly impossible to perform. To avoid this, the quadtree must be created in multi-resolution form: a low resolution version of the data should be used for the root of the quadtree, and detail coefficients should be organised at increasing resolutions in the hierarchy. However, to arrange the quadtree according to level of detail, the detail coefficients need also to be organised according to their spatial influence. A method to accomplish this is presented by Muchaxo et al (1996).

A continuous resolution increase across the terrain requires a level-of-detail management in terms of data points. While this is appropriate for multi-resolution painting of images, it is not appropriate for 3-dimensional systems since the rendering is done on a polygon basis and the texture assigned to it is considered to have a homogeneous resolution. Hence the levels of detail are stored and managed on a quad cell basis using small sub-matrices. The size of these is chosen based on the Haar system described by Muchaxo et al (1996).

Most raster geographical data can be converted to multi-resolution form by the process described by Muchaxo et al (1996). No distinction is made between different data types when creating the quadtree, except for the number of independent components of each. As an example, an aerial photograph may have three independent colour components, while topographic data have only one component. A quadtree is built for each component. Subtitles can also be used to map the values of the layer components to colours shown to the user. Since the Haar wavelet transformation averages values to obtain a lower resolution, only layer types for which the interpolation of values makes sense can be considered. Since all the updates performed in real time on the layer components are considered to be linear operations, other kinds of operation are not possible.

3.4 Design of virtual geographical systems

The purpose of designing virtual geographical information systems (VGIS) is to provide environments in which efficiency in the interaction with spatial problems is maximised. Performance in VEs is influenced by several factors that are related to the characteristics of the task, the user, and the medium (Stanney 1995). Bennett et al (1993) suggest that 3-dimensional displays should be used in tasks that require integration of information and that tasks requiring focused attention should use 2-dimensional displays. It has also been found (Stanney 1995) that texturing virtual objects can be effective to represent additional data dimensions.

User characteristics also affect efficiency in exploring VEs and should be considered when designing VGIS. The level of experience of the user influences the design of human–computer interfaces (Stanney 1995) and technical aptitudes like spatial visualisation and orientation are important issues to consider, particularly in VGIS interface design. Empirical observation shows that the great majority of users get lost in large VEs, so interfaces should be designed to assist users in maintaining spatial orientation within virtual worlds (Stanney 1995).

The integration of multi-modal interaction is also an important issue in the design of virtual worlds. Studies indicate that sensory redundancy improves performance in VE (Burdea and Coiffet 1994; Larijani 1994). This sensory integration can be addressed by two different approaches (Stanney 1995):

- command coordination, with the input being unimodal and the output multi-modal;
- sensory transposition or sensory substitution, used in particular when there is no way to display sensory information in the channel being substituted (Massimino and Sheridan 1993).

The implications of the factors described above in the design of VGIS interfaces should be the object of further study, as the acceptance of such systems on a day-to-day basis depends on it. The related area of intelligent user interfaces has produced good results which should be incorporated in the design of VGIS.

3.5 Interfaces

The previous section presented issues that should be considered in designing VGIS interfaces. Several previous interface designs have introduced interesting concepts that could be applied to VGIS interfaces. The Worlds In Miniature concept (WIM) (Pausch and Burnette 1995; Stoakley et al 1995) is one such interesting development. In it, interface navigation and object manipulation in the VE were integrated in the same interface using a tablet and a hand-held track-ball. The tablet presented a miniature representation of the simulated environment, and whenever the user wanted to manipulate or scale an object the operation was performed in that miniature world. Also any change in viewpoint was achieved by manipulating an object that represented the user in the WIM.

In the Virtual GIS Room project described below (Neves et al 1995) a similar approach is followed. The user has a virtual camera (i.e. a position- and orientation-tracked small object resembling a camera) in one hand, a wireless pen in the other hand, and a sensing tablet. The tablet is the WIM of the VGIS. Visual and audible display is presented in the computer monitor and speakers and in the HMD.

Usually navigation and interaction do not coexist: the user has two different modes of interaction and changes between them by touching a button on a 3-dimensional mouse or joystick. In these types of system large-scale navigation or large-scale object manipulation is performed by pointing to the destination. The user then has the option to choose from an instantaneous position-orientation transformation, or to draw a path between the initial and the destination point. Additional fish-eye views give some cues to avoid the user becoming lost in the virtual world (Koller et al 1995).

In VGISs the information that usually resides in a GIS is reachable directly because the physical objects have a 3-dimensional representation in the virtual world. So actions such as touching, grabbing, and moving can always display information about the object being manipulated, in the form of 2-dimensional or 3-dimensional text or in the form of sound.

GISs have always had visually-dominated interfaces, and this fact has influenced the design of VGISs. No VGIS is known that uses speech processing for querying data. Usually sound is used only as an output modality and not as an input modality. The ability to make voice annotations would dramatically improve the efficiency of VGISs mainly in their educational component (see Forer and Unwin, Chapter 54; Longley et al, Chapter72).

3.6 Applications

Systems allowing real-time visualisation of terrain datasets have been the object of attention in the past few years. NASA's Virtual Planetary Exploration Project (Hitchner 1992), the Terra Vision System (T-Vision Project, *http://www.artcom.de/projects/terra*; Leclerc and Lau 1994), and the MAGIC project (*http://www.magic.net*) are three good examples of implemented systems that fall into this category. Another system, the NPSNET (Macedonia and Zyda 1994) is a large-scale distributed virtual environment for military simulation and training. It is one of the first systems to implement distributed virtual environments with large datasets.

The Virtual GIS project (Koller et al 1995; Lindstrom et al 1996) is an integrated real-time 3-dimensional GIS. It is truly immersive, allowing the user to navigate and understand complex and dynamic terrain databases. The interface to the GIS data is based on the direct manipulation metaphor in both the window-based version and the virtual-reality version. The emphasis is on the high-resolution rendering of geographical features based on fast access to large databases on remote servers. Very efficient rendering algorithms (using LOD) and tightly-coupled distributed systems with high-speed processing units for parallel computation and input/output, together with a specialised graphics system for real-time rendering and interaction, provide the means to make such systems possible (Plate 38).

The work of Worldesign (Jacobson 1994) is one of the few examples of a company that is actually selling the idea of joining virtual worlds to GIS. This company has developed the VET which is an immersive projected environment, consisting of a set of screens surrounding the users (270 degree field of view) where images are projected. Worldesign has built a strong lead in the application of VE to planning and environmental impact assessment (EIA), and has made available tools to convert data from GIS to VE systems. AEPD-3D is a software engine that automatically converts output from AEPD, a GIS-like module for the planning and management of electric power networks, into a 3-dimensional virtual world.

The work being conducted by Jonathan Raper at Birkbeck College in London (Raper et al 1993; and see Raper, Chapter 5) has contributed to the issues of interfacing GIS with virtual reality technology. Work at Delft University in the Netherlands (Kraak et al 1995) is oriented to the use of immersive virtual worlds to access GIS databases and perform GIS queries. The authors and their colleagues at the New University of Lisbon are developing the Virtual GIS Room project (Neves et al 1995). Extensive work has been done on LOD algorithms (Muchaxo 1995), and extensions to dynamic LOD by the use of wavelets (Muchaxo et al 1996; Figure 4). The user interface issues have recently been the object of detailed study in the context of the above project (Figure 5).

4 CONCLUSIONS AND FUTURE DEVELOPMENTS

This chapter has discussed the advantages to the field of creating virtual GIS, in terms of the consequences to the interaction of GIS users with geographical data. New VE interfaces dealing specifically with geographical problems have been presented, and been reviewed, with emphasis on concepts relevant to VGIS applications.

The future success of VGIS will depend on the commercial adoption of interoperability standards such as CORBA (Common Object Request Broker Architecture) and OLE (Object Linking and Embedding). Interoperable environments will, in addition, allow VGIS to run spatial simulation and other analytical tools uncommon in commercial GIS (Sondheim et al, Chapter 24).

The 'moving worlds' specification from Silicon Graphics established the Virtual Reality Markup Language (VRML) 2.0 format, adding some features that made VRML effectively a good option to deliver geographical information on the WWW (Coleman, Chapter 22). This specification is supposed to evolve to integrate multiple users, making collaborative work on the Web a reality.

Will the interoperable objects wave reach the Web? If so, will GIS vendors then follow that path, and how quickly will that happen? The evolution of these trends will determine the way GIS will be used in the future. Immersive sessions of experts on the Web could make human–expert–problem communication a reality.

References

Bennett K B, Toms M L, Woods D D 1993 Emergent features and graphical elements: designing more effective configurational displays. *Human Factors* 35: 71–97

Burdea G, Coiffet P 1994 *Virtual reality technology*. New York, John Wiley & Sons Inc.

Buxton W, Gaver W, Bly S 1993 *Auditory interfaces: the use of non-speech audio at the interface*. Cambridge (UK), Cambridge University Press

Dias A E, Silva J P, Câmara A S 1995 BITS: browsing in time and space. *Conference Companion CHI 1995*: 248–9

Durlach N, Mavor A S (eds) 1994 *Virtual reality: scientific and technological challenges*. National Research Council, National Academy Press

Edwards T M 1992 Virtual worlds technology as a means for human interaction with spatial problems. *Proceedings of GIS/LIS San José, 10–12 November*: 208–20

Faust N L 1995 The virtual reality of GIS. *Environment and Planning B: Planning and Design* 22: 257–68

Fekete G 1990 Rendering and managing spherical data with sphere quadtrees. *Proceedings of Visualisation 90 (First IEEE conference on visualisation, San Francisco, 23–26 October 1990)*. Los Alamitos, IEEE Computer Society Press.

Gonçalves P, Diogo P 1994 Geographic information systems and cellular automata: a new approach to forest fire simulation. *Proceedings, European Conference on Geographical Information Systems, Paris*

Hitchner L E 1992 Virtual planetary exploration: a very large virtual environment. *Proceedings SIGGRAPH – tutorial on implementing immersive virtual environments*

Hughes P 1993 Terrain renderer for Mars Navigator. In Wolff R S, Yaeger L (eds) *Visualization of natural phenomena*. Santa Clara, TELOS – The Electronic Library of Science

Jacobson R 1994 Virtual worlds capture spatial reality. *GIS World* (December): 36–9

Koller D, Lindstrom P, Ribarsky W, Hodges L F, Faust N, Turner G 1995 Virtual GIS: a real-time 3D interface for geographical information system. *Georgia Institute of Technology technical report*. *ftp://ftp.gvu.gatech.edu/pub/gvu/tech-reports/95–14.ps.Z*

Kraak M-J, Smets G, Sidjanin P 1995 Virtual reality, the new 3-dimensional interface for geographical information systems. Presented at *First Conference on Spatial Multimedia and Virtual Reality, Lisbon, Portugal, 18–20 October*

Larijani L C 1994 *The virtual reality primer*. New York, McGraw-Hill

Leclerc Y G, Lau S Q Jr 1994 TerraVision: a terrain visualization system. *SRI International* 540 (April)

Lindstrom P, Koller D, Ribarsky W, Hodges L F, Faust N, Turner G 1996 Real-time, continuous level of detail rendering of height fields. *Proceedings of SIGGRAPH, New Orleans, 4–9 August*: 109–18

Lounsbery M, DeRose T, Warren J 1994 Multiresolution analysis for surfaces of arbitrary topological type. *Technical Report*. Department of Computer Science and Engineering, University of Washington. *ftp://cs.washington.edu/pub/graphics/TR931005b.ps.Z*

Macedonia M R, Zyda M J 1994 NPSNET: a network software architecture for large scale virtual environments. *PRESENCE: Teleoperators and Virtual Environments* 3: 265–87

MAGIC project, MAGIC Consortium. *http://www.magic.net/*

Massimino M J, Sheridan T B 1993 Sensory substitution for force feedback in teleoperation. *PRESENCE: Teleoperators and Virtual Environments* 2: 145–57

Mine M 1995 Virtual environment interaction techniques. *Technical Report TR95-018*. Department of Computer Science, University of North Carolina, Chapel Hill

Muchaxo J 1995 Multi-scale representation for large territories. *Proceedings First Conference on Spatial Multimedia and Virtual Reality, Lisbon, Portugal, 18–20 October*

Muchaxo J, Neves J N, Câmara A 1996 Wavelets and level- of-detail management: real time and multiresolution for geographical data. Internal report

Neves J N, Gonçalves P, Muchaxo J, Jordão L, Silva J P 1995 Virtual GIS Room: interfacing spatial information in virtual environments. *Proceedings, First Conference on Spatial Multimedia and Virtual Reality, Lisbon, Portugal, 18–20 October*

Pausch R, Burnette T 1995 Navigation and locomotion in virtual worlds via flight into hand-held miniatures. *Proceedings SIGGRAPH 1995, Los Angeles*: 399–400

Raper J F, McCarthy T, Livingstone D 1993 Interfacing GIS with virtual reality technology. *Proceedings Association for Geographic Information Conference Birmingham, 16–18 November* 3: 1–4

Scarlatos L, Pavlidis T 1992 Hierarchical triangulation using cartographic coherence. *CVGIP: Graphical Models and Image Processing* 54: 147–61

Schröder P, Sweldens W 1995 Spherical wavelets: efficiently representing functions on the sphere. *Proceedings SIGGRAPH 1995 6–11 August, Los Angeles*

Shepherd I 1994 Multi-sensory GIS : mapping out the research frontier. *Proceedings of SDH94 Edinburgh*: 356–90

Stanney K 1995 Realizing the full potential of virtual reality: human factors issues that could stand in the way. *Proceedings VRAIS 1995 Research Triangle Park, North Carolina, 11–15 March*

State A, Hirota G, Chen D T, Garret W F, Livingston M A 1996 Superior augmented reality registration by integrating landmark tracking and magnetic tracking. *Proceedings SIGGRAPH 1996 New Orleans, 4–9 August*: 429–38

Stoakley R, Conway M, Pausch R 1995 Virtual reality on a WIM: interactive worlds in miniature. *Proceedings CHI 1995 Denver*: 265–72

Stuart R 1996 *The design of virtual environments*. New York, McGraw-Hill: 175–80

40

The future of GIS and spatial analysis

M F GOODCHILD AND P A LONGLEY

The chapter explores factors affecting spatial analysis, in theory and practice, and their likely impacts. A model is presented of the traditional role of spatial analysis, and is examined from the perspectives of increasing costs of data, the increased sharing of data between investigators across a wide range of disciplines, the emergence of new techniques for analysis, and new computer architectures. Practical problems are identified that continue to face investigators using GIS to support spatial analysis, including accuracy, the technical problems of integration, and the averaging of different feature geometries. The chapter ends with a speculation on spatial analysis of the future.

1 INTRODUCTION

GIS and spatial analysis have enjoyed a long and productive relationship over the past decades (for reviews see Fotheringham and Rogerson 1994; Goodchild 1988; Goodchild et al 1992). GIS has been seen as the key to implementing methods of spatial analysis, making them more accessible to a broader range of users, and hopefully more widely used in making effective decisions and in supporting scientific research. It has been argued (e.g. Goodchild 1988) that in this sense the relationship between spatial analysis and GIS is analogous to that between statistics and the statistical packages. Much has been written about the need to extend the range of spatial analytic functions available in GIS, and about the competition for the attention of GIS developers between spatial analysis and other GIS uses, many of which are more powerful and better able to command funding. Specialised GIS packages directed specifically at spatial analysis have emerged (e.g. Idrisi; see also Bailey and Gatrell 1995). Openshaw and Alvanides (Chapter 18) have set out some of the ways in which developments in computation may feed through to enhanced GIS-based spatial analysis. Finally, Anselin (Chapter 17), Getis (Chapter 16) and others have discussed the ways in which implementation of spatial analysis

methods in GIS is leading to a new, exploratory emphasis.

The purpose of this chapter is to explore new directions that have emerged recently, or are currently emerging, in the general area of GIS and spatial analysis, and to take a broad perspective on their practical implications for GIS-based spatial analysis. In the next section, we argue that in the past the interaction between GIS and spatial analysis has followed a very clearly and narrowly defined path, one that has more to do with the world of spatial analysis prior to the advent of GIS than with making the most of both fields – the path is, in other words, a legacy of prior conditions and an earlier era (see also Openshaw and Alvanides, Chapter 18). The following section expands on some of the themes of the introduction to this volume by identifying a number of trends, some related to GIS but some more broadly based, that have changed the context of GIS and spatial analysis over the past few years, and continue to do so at an increasing rate. The third section identifies some of the consequences of these trends, and the problems that are arising in the development of a new approach to spatial analysis. The chapter concludes with some comments about the complexity of the interactions between analysis, data and tools, and speculation on what the future may hold, and what forms of spatial analysis it is likely to favour.

2 TRADITIONS IN SPATIAL ANALYSIS

2.1 The linear project design

In the best of all possible worlds, a scientific research project (the term 'research' will be interpreted broadly to include both scientific and decision-making activities) begins with clearly stated objectives. Some decision must be made, some question of scientific or social concern must be resolved by resorting to experiment or real-world evidence. A research design is developed to resolve the problem, data are collected, analyses are performed, and the results are interpreted and reported. Although this implies a strictly linear sequence of events, the most robust research designs also include feedbacks and checks in order to ensure that the principles of good scientific research are not overly compromised in practical implementation. This simple, essentially linear, structure with recursive feedbacks underlies generations of student dissertations, government reports, and research papers. It is exemplified by the classic social survey research design illustrated in Figure 1. The sequential events in this design together constitute a holistic research project, and the feedbacks are all internal to the research design. Thus once the project has been initiated, the availability of existing data has no further influence upon problem definition; methods of analysis that are consistent with the type, quality, and amount of data to be collected are identified at the design stage (and the data collection method changed if no suitable analytical method exists); the sample design is not guided by considerations and priorities that lie outside the remit of the research; and so on.

In this simple, sequential world, the selection of methods of analysis can be reduced to a few simple rules (in the context of statistical analysis, see for example Levine 1981: chapter 17; Marascuilo and Levin 1983: inside cover; Siegel 1956: inside cover). Choice of analytic method depends on the type of inference to be drawn (e.g. whether two samples are drawn from the same, unknown population, or whether two variables are correlated), and on the characteristics of the available data (e.g. scale of measurement – nominal, ordinal, interval, or ratio: Wrigley 1985). Inference about, and exploration of, the research problem will take place in what is loosely described as the confirmatory (hypothesis testing and

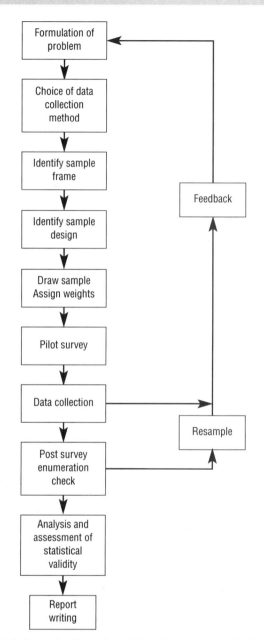

Fig 1. Sequential stages of a typical social survey research design.

inference seeking) and exploratory (pattern or anomaly seeking) stages of the research.

In contrast to this coherent research design, the terms 'data-driven' and 'technique-driven' are highly pejorative in research generally, as are such phrases as 'a technique in search of a problem' – in this ideal world, the statement of the problem strictly precedes the collection of data and the performance of analysis.

2.2 Spatial analysis

Spatial analysis, or spatial data analysis, is a well-defined subset of the methods of analysis available to a project. One might define spatial analysis as a set of methods useful when the data are spatial, in other words when the data are referenced to a 2-dimensional frame. More narrowly, the Earth's surface provides a particular instance of such a frame, the geographical frame, with its peculiar properties of curvature. This definition of spatial analysis is arguably too broad, because in basing the definition on the properties of data it does not address the question of whether the 2-dimensional frame actually matters – could the same results have been obtained if the frame were distorted in some way, or if objects were repositioned in the frame? More precisely, then, spatial analysis can be defined as that subset of analytic techniques whose results depend on the frame, or will change if the frame changes, or if objects are repositioned within it. To distinguish analytic methods from more mundane operations, they might be defined as methods for processing data with the objective of solving some scientific or decision-making problem.

Methods of spatial analysis have accumulated in a literature that spans many decades, indeed centuries (see Getis, Chapter 16). They have been invented in many disciplines, including mathematics, particularly geometry; statistics, particularly spatial statistics and statistical geometry; and in geography and other Earth sciences. Compendia have been published (among others, see Bailey and Gatrell 1995; Berry and Marble 1968; Haining 1990; Taylor 1977; Unwin 1981), and various approaches proposed for structuring this body of technique. Spatial analytic techniques may also be classified into those which are confirmatory and those which are exploratory. Choice of analytical method also relates to data characteristics – documented since Chorley and Haggett's (1965) analogies between (respectively) nominal, ordinal, interval and ratio data and point, line, area, and surface objects (see also Chrisman 1997; Martin 1996).

2.3 The well-informed analyst

Traditionally, the responsibilities of the inventor of a technique ended when the technique had been tested and described. Even the testing of a technique can be suspect in an academic world that often values theory over empiricism, and is suspicious of

empirical results that cannot be demonstrated to be generally true. The advent of the digital computer changed this world fundamentally because it became possible for a scientist to perform a method of analysis automatically, without taking personal responsibility for every aspect of the performance. It was now possible using the 'black box' of the computer to perform an analysis that one did not know everything about – that one could not perform by hand. Methods emerged, beginning in the 1970s and particularly in the area of multivariate statistics, that would be impossibly impractical to perform by hand. Pedagogically, a fundamental shift became possible in how analysis was taught – that one might learn about a technique by studying the nature of its response to particular inputs, rather than by studying the procedure which generated the response. But there is a fundamental difference between these two positions: between whether one understands the results of a principal components analysis, for example, as the extraction of eigenvalues from a specific matrix, or the generation of statistics that broadly indicate some concept of 'relative importance' without presuming any understanding of what eigenvalues are and *how* they formalise the structure in data.

Exactly where this change occurred is open to debate, of course. It may have occurred when students were no longer required to perform statistical analyses by hand before being let loose on computer packages; or when FORTRAN appeared, making it necessary to understand less about how instructions were actually carried out; or when the growth of the scientific enterprise had reached such a level that potential replication of every result was a practical impossibility.

Of course the digital computers that were introduced to the scientific community beginning in the late 1950s produced rapid change in the labour demands of many statistical methods. The intricate calculations of factor analysis (Harman 1976) could be performed by a fully automatic machine, provided the researcher could command sufficient computer time, and provided labour was available to punch the necessary cards. Computers and the brains of young humans are in some ways similar: both begin 'hard wired' with the primitive elements of reasoning (e.g. binary processing in computers, linguistic abilities in infants) and both can build enormously complex structures out of simpler ones, apparently ad infinitum (see Fischer, Chapter 19, for a broader discussion of computer 'reasoning'). What began in the 1960s as a set of uncoordinated efforts by individual scientists writing their own programs had developed by the 1990s into a complex of enormously sophisticated tools, each integrating a large number of methods into an easy-to-use whole.

If software packages and user-friendly computer environments have made aspiring spatial analysts less aware of the computational and statistical context to inference, then the opposite is true to some extent of exploratory analysis, where the innovation of computer graphics and windows, icons, mice, and pointers (WIMPs) has created a more intuitive context to the interrogation of spatial data (see Anselin, Chapter 17; Kraak, Chapter 11). Indeed one of the criticisms of GIS-based graphics developments from the spatial analysis community has been that the computer graphics medium has been allowed to dominate the spatial analysis message, by analogy to 'data-led' thinking as described in section 2.1.

2.4 Extending the functions of analytic software

Although they show clear evidence of their roots, the packages used by the scientists of the 1990s are different in fundamental respects from the programs of the 1960s. Besides implementing large numbers of statistical methods, today's packages also provide support for the maintenance of data and the creation of information. There will be tools for documenting datasets, and describing their properties, such as accuracy and history. Other tools will support the sharing of data, in the form of format converters or interfaces to the Internet. In short, the functions of today's digital computers in supporting research go far beyond those of a simple calculating machine, carrying out well-defined methods of analysis. The same digital computer may now be involved in:

- the selection and formulation of a problem, by providing access to automated library catalogues and on-line literature;
- the collection of data through support for real-time data acquisition;
- management of data, performance of analysis, visualisation of results, writing of conclusions;
- even publication through access to the Internet and the World Wide Web.

The computer is no longer part of the research environment – we are rapidly approaching a world in which the computer *is* the research environment.

These trends are all echoed strongly in GIS. Although a particular scientist might use a GIS in ways that are more analogous to the early days of statistical computing, by performing a single buffering operation, for example, scientific applications are much more likely to include integration of many GIS functions. Today's scientist or decision-maker is likely to see a GIS as an environment for research, rather than as a means of automating analysis. The GIS is likely to be involved in the project from beginning to end, and to be integrated with other tools and environments when these are needed. GIS will be used for collecting, assembling, verifying, and editing the data; performing some of the analyses required by the project; and presenting and interpreting the results. Moreover, much GIS use may not be tied to a specific project – GIS finds extensive use in the collection of data for purposes that may be generic, or not well-defined, or may be justified in anticipation of future demand. Even though these may not be 'spatial analysis' in the sense of the earlier discussion, analysis may still be necessary as part of the data production process – for example, when a soil scientist must analyse data to produce a soil map.

2.5 When to choose GIS

A related issue is the extent to which GIS remains a separately identifiable technology, and in what senses the 'GIS environment' is distinctive. The general drift of many of the chapters in this 'Technical issues' Part of this book is that GIS is increasingly becoming both a background technology (more akin to wordprocessing than, say, spatial interaction modelling), and a technology that can be broken up and packaged as niche products (Elshaw Thrall and Thrall, Chapter 23). And yet the various discussions in the 'Principles' Part of this book document the important agenda for spatial analysis set in the environment of GIS, and why GIS-based spatial analysis is likely to remain a distinctive area of activity for the foreseeable future.

If GIS has multiple roles in support of science and problem-solving, then one might not be surprised to find that the choice between GIS alternatives is complex and often daunting. The many GIS packages offer a wide range of combinations of analysis functions, housekeeping support, different ways of representing the same phenomena, variable levels of sophistication in visual display, and performance. In addition, choice is often driven by: the available hardware, since not all GIS run on all platforms; the format in which the necessary data have been supplied; the personal preferences and background of the user; and so forth. Even the extensive and frequently updated comparative surveys published by groups such as GIS World Inc. can be of little help to the uninitiated user.

The existence of other classes of analytic software complicates the scene still further. Under what circumstances is a problem better solved using a package that identifies itself as a GIS, or using a statistical package, or a mathematical package, or a scientific visualisation package? Under what circumstances is it better to fit the square peg of a real problem into the round GIS hole? GIS are distinguished by their ability to handle data referenced to a 2-dimensional frame, but such capabilities also exist to a more limited extent in many other types of software environment. For example, it is possible to store a map in a spreadsheet array, and with a little ingenuity to produce a passable 'map' output; and many statistical packages support data in the form of images.

Under what circumstances, then, is an analyst likely to choose a GIS? The following conditions are suggested, although the list is certainly not complete, and the items are not intended to be mutually exclusive:

- when the data are geographically referenced;
- when geographical references are essential to the analysis;
- when the data include a range of vector data types (support for vector analysis among non-GIS packages appears to be much less common than support for raster analysis);
- when topology – representation of the connections between objects – is important to the analysis;
- when the curvature of the Earth's surface is important to the analysis, requiring support for projections and for methods of spatial analysis on curved surfaces;
- when the volume of data is large, since alternatives like spreadsheets tend to work only for small datasets;

- when data must be integrated from a variety of sources, requiring extensive support for reformatting, resampling, and other forms of format change;
- when geographical objects under analysis have large numbers of attributes, requiring support from integrated database management systems, since many alternatives lack such integration;
- when the background of the investigator is in geography, or a discipline with strong interest in geographical data;
- when the project involves several disciplines, and must therefore transcend the software traditions and preferences of each;
- when visual display is important, and when the results must be presented to varied audiences;
- when the results of the analysis are likely to be used as input by other projects, or when the data are being extensively shared.

3 ELEMENTS OF A NEW PERSPECTIVE

This section reviews some of the changes that are altering the context and face of spatial analysis using GIS. Some are driven by technological change, and others by larger trends affecting society at the turn of the millennium.

3.1 The costs of data creation

The collection of geographical data can be extremely labour-intensive. Early topographic mapping required the map-maker to walk large parts of the ground being mapped; soil mapping requires the exhausting work of digging soil pits, followed often by laborious chemical analysis; census data collection requires personal visits to a substantial proportion of (sometimes all) household respondents; and forest mapping requires 'operational cruise', the intensive observation of conditions along transects. Although many new methods of geographical data creation have replaced the human observer on the ground with various forms of automated sensing, there is no alternative in those areas that require the presence of expert interpreters in the field.

Many of the remaining stages of geographical data creation are also highly labour-intensive. There is still no alternative to manual digitising in cases where the source document is complex,

571

compromised, or difficult to interpret. The processes of error detection and correction are difficult if not impossible to automate, and the methods of cartographic generalisation used by expert cartographers have proven very difficult to formalise and replace. In short, despite much technical progress over the past few decades, geographical data creation remains an expensive process that is far from fully automated.

Labour costs continue to rise at a time when the resources available to government, the traditional source of geographical data, continue to shrink (see Elshaw Thrall and Thrall, Chapter 23). Many geographical datasets are collected for purposes which may be far from immediate, and it is difficult therefore to convince taxpayers that they represent an essential investment of public funds, especially in peacetime. Governments in financial straits call for evidence of need: for example, census organisations are under continual pressure to demonstrate that their costly operations do not replicate information that is available elsewhere; and many governments have moved their mapping operations onto a semi-commercial basis in order to allow demand to be expressed through willingness to pay (see Rhind, Chapter 56). To date, the US Federal mapping agencies have resisted the trend, but internationally there is more and more evidence of the emergence of a market in geographical information.

Within the domain of geographical data the pressures of increased labour costs favour data that can be collected and processed automatically. Given a choice between the labour-intensive production of vector topographic data, and the semi-automated generation of such raster products as digital elevation models and digital orthophotos, economic pressures can lead only in one direction. It is easy to imagine a user trading off the ability to identify features by name against the order of magnitude lower cost, and thus greater potential update frequency, of raster data.

The broader context to these changes is that we now live in a digital world, in which far more data are collected about us, in computer readable form, than ever before. This is what has been termed the 'information economy', in which government no longer has a monopoly in the supply of geographical information, and in which information has become both a tradeable commodity and a strategic organisational resource. Global trends such as deregulation and privatisation, allied to the

increasing competitive edge of consumer-led markets, are multiplying the potential number of sources of information, yet at the expense of system-wide standardisation (Rhind, Chapter 56). Such data are not ideally suited to the linear research design set out in section 2.1, yet (in socioeconomic research at least) they frequently are far richer in detail than anything that has been collected hitherto. A clear challenge to spatial analysis is therefore to reconcile diverse datasets with different data structures or spatial referencing systems, and to gauge how representative they are with reference to existing (more limited or less frequently updated) public sources. A good example of this concerns the development of geodemographic indicators which have traditionally been derived from census data. These are typically updated only every ten years and are frequently reliant upon very indirect indicators of likely consumer behaviours (Longley and Clarke 1995). 'Lifestyles' approaches based upon questionnaire returns from a range of self-selecting respondents (Birkin 1995) offer the prospect of 'freshening up' and in time replacing conventional census-based geodemographics, although thorny issues of representativeness and bias must be grappled with before credible 'data fusion' can be deemed to have taken place.

Of course, the principle of information commerce is alien to the scientific community, which is likely to resist strongly any attempt to charge for data that is of interest to science, even peripherally. But here too there are pressures to make better use of the resources invested in scientific data collection. Research funding agencies increasingly require evidence that data collected for a project have been disseminated, or made accessible to others, while recognising the need to protect the interests of the collector.

Trends such as these, while they may be eminently rational to dispensers of public funds, nevertheless fly directly in the face of the traditional model of science presented earlier. For example, the best-known definition of the discipline of geography is that it is 'concerned to provide accurate, orderly, and rational description and interpretation of the variable character of the Earth surface' (Hartshorne 1959). As a general rule, commercial datasets are not accurate (they provide little indication of the sources of unknown errors in data collection or the ways in which they are likely to operate in analysis); they are orderly only in a minimalist sense (for example, satellites provide frequently-updated *coverage* information yet cannot comprehensively measure land *use*; 'lifestyles' data do not provide information about all groups in society); and they are not rational in that they separate still further the analyst from the context to the research problem and lead to data- or machine-led thinking. How can projects fail to be driven by data, if data are forced to obey the economic laws of supply and demand? Where in traditional science are the rules and standards that allow scientists to trade off economic cost against scientific truth? It seems that economic necessity has forced the practice of science to move well beyond the traditions that are reflected in accepted scientific methodologies and philosophies of science.

3.2 The life of a dataset

In the traditional model presented earlier data were collected or created to solve a particular problem, and had no use afterwards except perhaps to historians of science. But many types of geographical data are collected and maintained for generic purposes, and may be used many times by completely unrelated projects. For other types, the creation of data is itself a form of science, involving the field skills of a soil scientist, for example, or a biologist. Thus a dataset can be simultaneously the output of one person's science, and the input to another's. This is to conceive of spatial analysis within GIS as the process of building 'models of models' – whereby the outcome of a 'higher level' spatial analysis is dependent upon data inputs which are themselves a previously modelled version of reality. These relationships have become further complicated by the rise of multidisciplinary science, which combines the strengths and expertise of many different sciences, and partitions the work among them. Once again, the linear model of science is found wanting, unable to reflect the complex relationships between projects, datasets, and analytic techniques that exist in modern science. The notion that data are somehow subsidiary to problems, methods and results is challenged, and traditional dicta about not including technical detail in scientific reports may be counterproductive.

In truth, of course, this is nothing new in the sense that most spatial analysis in the socioeconomic realm has been based upon crude surrogate data, obtained for inappropriate areal units in obsolete time periods. Thirty-five years ago we were all 'information poor' and the limited data-handling capabilities of early

spatial analysis methods reflected this. Arguably, it was this impoverishment that was the root cause of the failure of many such methods to generate detailed insights into the functioning of social systems (see Openshaw and Alvanides, Chapter 18). The potential to build detailed data-rich depictions of reality within GIS will make some problems more transparent, yet others will likely be further obscured. From a pessimist's standpoint, data-rich modelling within GIS represents a return to the shifting sands of naive empiricism. For the optimist, sensitive honing of such data to context allows data-rich models to shed light upon a wider range of social and economic research problems.

In this new world, a given set of data is likely to fall into many different hands during its life. It may be assembled from a mixture of field and remote sensing sources, interpreted by a specialist, catalogued by an archivist or librarian, used by scientists and problem-solvers, and passed between its custodians using a range of technologies (Figure 2). It is quite possible in today's world that the various creators and users of data share little in the way of common disciplinary background, leaving the dataset open to misunderstanding and misinterpretation. Recent interest in metadata, or ways of describing the contents of datasets, is directed at reducing some of these problems, but the easy access to data provided by the Internet and various geographical data archives has tended to make the problem of inappropriate use or application worse.

These issues are particularly prominent in the case of data quality, and the ability of the user of a dataset to understand its limitations, and the uncertainty that exists about the real phenomena the data are intended to represent. To take a simple example, suppose information on the geodetic datum underlying a particular dataset – potentially a very significant component of its metadata – were lost in transmission between source and user; alternatively, suppose that the user simply assumed the wrong datum, or was unaware of its significance. This loss of metadata, or specification of the data content, is equivalent in every respect to an actual loss of accuracy equal to the difference between the true datum and the datum assumed by the user, which can be several hundreds of metres. This is perhaps an obvious example, but what, say, is the magnitude of error associated with soil profile delineation? What is the magnitude of likely ecological fallacy associated

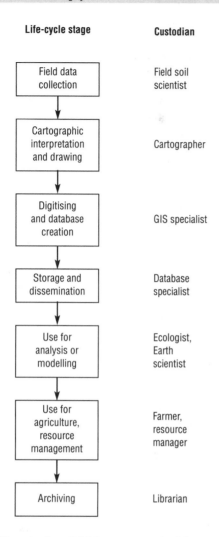

Life-cycle stage	Custodian
Field data collection	Field soil scientist
Cartographic interpretation and drawing	Cartographer
Digitising and database creation	GIS specialist
Storage and dissemination	Database specialist
Use for analysis or modelling	Ecologist, Earth scientist
Use for agriculture, resource management	Farmer, resource manager
Archiving	Librarian

Fig 2. The life-cycle of a soil database: an example of the complex patterns of custodianship and transfer now common for many types of spatial data.

with comparison of a geodemographic classifier with the results of a survey? In short, the quality of a dataset to a user is a function of the difference between its contents and the user's understanding of its meaning, not the creator's.

3.3 Data sharing

In this new world of shared data, the term metadata has come to function as the equivalent of documentation, cataloguing, handling instructions, and production control. The US Federal Geographic Data Committee's Content Standard for Digital

Geospatial Metadata (FGDC 1994) has been very influential in providing a standard, which has been emulated frequently (see Salgé, Chapter 50). If the custodian of a large collection of geographical datasets provides metadata in this form, it is possible for others to search its records for those that match their needs. The FGDC's National Geospatial Data Clearinghouse (*http://www.fgdc.gov*) is one such directory (see also the Alexandria Digital Library project to provide distributed library services for geographically referenced datasets: Smith et al 1996; and see *http://alexandria.sdc.ucsb.edu*).

The user of a traditional library will rarely know the exact subject of a search – instead, library search has an essential fuzziness, which is supported by the traditional library in several essential ways. By assigning similar call numbers to books on similar subjects, and shelving by call number, the traditional library is able to provide an environment that allows the user to browse the collection in a chosen area. But this support is missing when the records of a metadata file are searched using simple Boolean methods. It would make better sense to model the search process as one of finding the best fit between a metadata record representing the user's ideal, and metadata records representing the datasets available. It is very unlikely, after all, that data exist that perfectly match the needs of a given problem, especially in the ideal world of problem-solving represented earlier. This is especially true when the object of search is to find data covering a particular location, as it often is in the GIS context. In such cases, it seems very unlikely that there will be an exact fit between the area requested by the user, and the area covered by a data set in an archive.

3.4 New techniques for analysis

Several chapters in this section have focused on new methods of spatial analysis, particularly new methods that have emerged in the data-rich computational environment now available to scientists. These include neural nets (Fischer, Chapter 19), new methods of optimisation such as simulated annealing and genetic techniques, and computationally intensive simulation. The term *geocomputation* (Openshaw and Alvanides, Chapter 18) has been suggested. Anselin (Chapter 17) and others have extended the principles of exploratory data analysis (Tukey 1970) to spatial data.

In science generally, the combination of vast new sources of data and high-speed computation have led to an interest in methods of *data mining*, which implies the ability to dredge data at very high speed in a search for patterns of scientific interest. In a geographical context, the vague notion of 'scientific interest' might suggest the need for methods to detect features or measurements that are inconsistent with their surroundings, in apparent violation of Tobler's 'first law of geography' (Tobler 1970; see also Johnston, Chapter 3). Linearities in images are of potential interest in geological prospecting; and one can imagine circumstances in which atmospheric scientists might want to search large numbers of images for patterns consistent with weather events. Such techniques of pattern recognition were pioneered many years ago in particle physics, to search vast numbers of bubble-chamber photographs for the tracks characteristic of rare new particles.

One might argue that such techniques represent a renewal of interest in inductive science – the search for regularities or patterns in the world that would then stimulate new explanatory theories (see Fischer, Chapter 19). Inductivism has fallen out of fashion in recent decades, at least in disciplines that focus on geographical data, leading one to ask whether a renewal of interest represents a fundamental shift in science, or merely a response to the opportunities for data-led thinking offered by more powerful technology. On this issue the jury is clearly still 'out' – geocomputation has not yet provided the kinds of new insights that might support a broad shift to inductivism.

3.5 New computer architectures

The communication technologies that have emerged in the past decade have allowed a fundamental change in the architecture of computing systems. Instead of the early mainframes and later stand-alone desktop systems, today's computers are linked with high-speed networks that allow data, software, and storage capacity located in widely scattered systems to be integrated into functioning wholes. Data can now be 'served' from central sites on demand, avoiding the need to disseminate many copies, with subsequent confusion when updates are needed. Coleman (Chapter 22) reviews the architectural alternatives now common in computing systems, and their technical impacts on GIS.

The new approaches to computing that are possible in this interconnected environment are having a profound effect on spatial analysis. Because it is no longer possible to assume a lifetime association between a user and a particular system design, there are mounting pressures for standards and interoperability between systems to counter the high costs of retraining of staff and reformatting of data.

The proprietary GIS that once dominated the industry attempted to provide a full range of GIS services in one homogeneous environment. Data were stored in proprietary formats, often kept secret by vendors to maintain market position, but making it difficult for others to expand the capabilities of the system by programming extra modules. The 'open GIS' movement (Buehler and McKee 1996 and see *http://www.ogis.org*) mirrors efforts in other areas of the electronic data processing world to promote interoperability, open standards and formats, and easy exchange from one system to another. While such ideas were often regarded as counter to the commercial interests of vendors, there is now widespread acceptance in the industry that they represent the way of the future.

The implications of open systems for spatial analysis are likely to be profound. First, they offer the potential of a uniform working environment, in which knowledge of one system is readily transferable to another. To make this work, however, it will be necessary to achieve a uniform view, and its acceptance across a heterogeneous user community. There is no prospect of interoperability and open systems without agreement on the fundamental data models, terminology, and objectives of GIS-based analysis. Thus much effort will be needed on the part of the inventors and implementors of spatial analysis to develop this uniform view.

Second, the possibility of easy sharing of data across systems gives even greater momentum to efforts to make geographical information more shareable, and even greater demands on the existence and effectiveness of metadata.

Third, interoperability is likely to create an environment in which it is much easier to implement methods of spatial analysis in GIS. Traditionally, vendors of monolithic systems have added functions when market demand appears to justify the development costs. It has been impossible, in a world of proprietary systems, for third parties to add significant functionality. Thus expansion of spatial analytic capabilities has been slow, and has tended to reflect the needs of the commercial market, rather than those of science and problem-solving, when these diverge. In a world of open systems it will be much easier to add functions, and the new environment will encourage the emergence of small companies offering specialised functionality in niche markets.

Finally, new interoperable approaches to software will encourage the modularisation of code (Sondheim et al, Chapter 24). It is already possible in some mainstream software environments to launch one specialised application within another – for example, to apply spreadsheet functions to information in a word processing package. This 'plug and play' environment offers enormous scope to GIS, since it will lead ultimately to a greater integration of GIS functions, and map and imagery data in general, into mainstream electronic data processing applications.

The scientific world has grown used to a more or less complete separation between data, and the functions that operate on and manipulate data. Functions are part of 'analysis', which plays a role in the traditional approach to problem-solving outlined earlier that is clearly distinct from that of data. But it has already been argued that in a world of extensive data sharing and interaction between disciplines it is impossible to think of data in isolation from its description, or metadata, which allows the meaning of information to be shared.

In the abstract world of object-oriented methods, it is argued that the meaning of data lies ultimately in the operations that can be performed. If datasets exist in two systems, and pairs of functions exist in both systems that produce the same answers, then the two datasets are the same in information content, irrespective of their specific formats and arrangements of bits. It makes sense, then, to *encapsulate* methods with data. When more than one method is available to perform a given function, it makes sense for the choice to be made by the person best able to do so, and for the method thereafter to travel with the data. For example, a climatologist might encapsulate an appropriate method for spatial interpolation with a set of point weather records, because the climatologist is arguably better able to select the best method of spatial interpolation, given his or her knowledge of atmospheric processes.

In future, and especially given the current trend in computing to object-oriented methods, it is likely that the distinction between data and methods will become increasingly blurred. Commonly used techniques of spatial analysis, such as spatial interpolation, may become encapsulated with data in an extension of the concept of metadata to include methods. Of course this assumes that methods are capable of running in a wide variety of host systems, which takes the discussion back to the issue of interoperability introduced earlier.

4 SPATIAL ANALYSIS IN PRACTICE

At this stage, it seems useful to introduce a discussion of the practical problems which face the users of today's GIS. While it is now possible to undertake a wide range of forms of spatial analysis, and to integrate data from a range of sources that would have seemed inconceivable as little as five years ago, there continue to be abundant limitations that impede the complete fulfilment of the technology's promise. The following subsections discuss several of these enduring impediments.

4.1 Absolute and relative position

First, and perhaps foremost, are problems of varying data quality. In science generally it is common to express quality in terms such as 'accurate to plus or minus one per cent'. But while such methods are useful for many types of data, they are less so when the data are geographical. The individual items of information in a geographical dataset are typically the result of a long and complex series of processing and interpretation steps. They bear little relationship to the independent measurements of traditional error analysis. Section 2 dealt at length with the data quality issue, and the theme is taken up again in the context of decision-making by Hunter (Chapter 45), and those discussions will not be repeated here. Instead, the following discussion is limited to the particular problems encountered when merging datasets.

While projections and geodetic datums are commonly well-documented for the datasets produced by government agencies, the individual scientist digitising a map may well not be in a position to identify either. The idea that lack of specification could contribute to uncertainty was discussed earlier, and its effects will be immediately apparent if a dataset is merged with one based on another projection or datum. In practice, therefore, users of GIS frequently encounter the need for methods of *conflation*, a topic discussed in detail below.

The individual items of information in a geographical dataset often share lineage, in the sense that more than one item is affected by the same error. This happens, for example, when a map or photograph is registered poorly – all of the data derived from it will have the same error. One indicator of shared lineage, then, is the persistence of error – all points derived from or dependent on the same misregistration will be displaced by the same or a similar amount. Because neighbouring points are more likely to share lineage than distant points, errors tend to show strong positive spatial autocorrelation (Goodchild and Gopal 1989).

Rubber-sheeting is the term used to describe methods for removing such errors on the assumption that strong spatial autocorrelations exist. If errors tend to be spatially autocorrelated up to a distance of x, say, then rubber-sheeting will be successful at removing them, at least partially, provided control points can be found that are spaced less than x apart. For the same reason, the shapes of features that are less than x across will tend to have little distortion, while very large shapes may be badly distorted. The results of calculating areas, or other geometric operations that rely only on relative position, will be accurate as long as the areas are small, but will grow rapidly with feature size. Thus it is important for the user of a GIS to know which operations depend on *relative* position, and over what distance; and where *absolute* position is important (of course the term absolute simply means relative to the Earth frame, defined by the Equator and the Greenwich meridian, or relative over a very long distance).

When two datasets are merged that share no common lineage (for example, they have not been subject to the same misregistration), then the relative positions of objects inherit the absolute positional errors of both, even over the shortest distances. While the shapes of objects in each dataset may be accurate, the relative locations of pairs of neighbouring objects may be wildly inaccurate when drawn from different datasets. The anecdotal history of GIS is full of such examples – datasets which were perfectly adequate for one application, but failed completely when an application required that they be merged with some new dataset that had no common lineage. For example, merging GPS measurements of point positions with streets derived from the US Bureau of the Census TIGER (Topologically Integrated Geographic Encoding and Referencing) files may

lead to surprises where points appear on the wrong sides of streets. If the absolute positional accuracy of a dataset is 50 metres, as it is with parts of TIGER, then such surprises will be common for points located less than 50 metres from the nearest street. In a similar vein but in the context of the fragmented data holdings of UK local authorities, Martin et al (1994) describe the problems and mismatches inherent in matching individual and household information with property gazetteers.

4.2 Semantic integration

Some of the most challenging problems in GIS practice occur in the area of semantic integration, where integration relies on an understanding of meaning. Such problems can occur between geographical jurisdictions, if definitions of feature types, or classifications, or methods of measurement vary between them. It is common, for example, for schemes of vegetation classification to vary from one country to another, making it difficult to produce horizontally merged data (Mounsey 1991). 'Vertical' integration can also be problematic, as for example in merging the information on land classification maps produced by different agencies, or different individuals (Edwards and Lowell 1996).

While some of these problems may disappear with more enlightened standards, others merely reflect positions that are eminently reasonable. The problems of management of ecosystems in Florida are clearly different from those of Montana, and it is reasonable that standards adopted by the two states should be different (see Fisher, Chapter 13). Even if it were possible to standardise for the entire US, one would be no further ahead in standardising between the US and other countries. Instead, it seems a more reasonable approach is to achieve interoperability without standardisation, by more intelligent approaches to system design.

4.3 Conflation

Conflation appears to be the term of choice in the GIS community for functions that attempt to overcome differences between datasets, or to merge their contents. Conflation attempts to replace two or more versions of the same information with a single version that reflects the pooling of the sources; it may help to think of it as a process of weighted averaging. The complementary term *'concatenation'* refers to the integration of the sources, so that the contents of both are accessible in the product. The polygon overlay operation familiar to many GIS users is thus a form of concatenation.

Two distinct forms of conflation can be identified, depending on the context:

1 conflation of feature geometry and topology, and concatenation of feature attributes;
2 conflation of geometry, topology, and attributes.

As an example of the first case, suppose information is available on the railroad network at two scales, 1:100 000 and 1:2 million. The set of attributes available is richer at the 1:2 million scale, but the geometry and topology are more accurate at 1:100 000. Thus it would be desirable to combine the two, thereby discarding the coarser geometry and topology.

As an example of the second case, consider a situation in which soils have been mapped for two adjacent counties, by two different teams of scientists. At the common border there is an obvious problem, because although the county boundary was defined by a process that was in no way dependent on soils, the border nevertheless appears in the combined map. Thus it would be desirable to 'average' the data at and near the boundary by combining the information from both maps in compatible fashion. As these two examples illustrate, the need for conflation occurs both horizontally, in the form of edge matching, and 'vertically'. A further example of the second case is provided by spatially extensive property valuation exercises, such as that which accompanied the introduction of the UK Council Tax (Longley et al 1994): surveyors were each individually responsible for allotted areas and conflation of estimates around the area boundaries was used to enhance consistency.

4.4 Perfect positioning

Some of the problems of conflation, and of relative and absolute positional accuracy, might be expected to dissipate as measurement of position becomes more and more accurate, leading eventually to 'perfect' positioning. Unfortunately, there are good reasons to anticipate that this happy state will never be reached. Although the positions of the Greenwich meridian and various geodetic control points have been established by fixing monuments, fundamental uncertainty will continue to be created

by seismic motions, continental drift, and the wobbling of the Earth's axis. Any mathematical representation of the Earth's shape must be an approximation, and different approximations have been adopted for different purposes. Moreover, there will always be a legacy of earlier, less accurate measurements to deal with. Thus it seems GIS will always have to deal with uncertainty of position, and with the distinctions between relative and absolute accuracy, and their complex implications for analysis.

Instead, strategies must be found for overcoming the inevitable differences between databases, either prior to analysis or in some cases 'on the fly'. Consider, for example, the problems caused by use of different map databases for vehicle routing. Systems are already available on an experimental basis that broadcast information on street congestion and road maintenance to vehicles, which are equipped with map databases and systems to display such information for the driver. In a world of many competing vendors, such systems will have to overcome problems of mismatch between different databases, in terms both of position and of attributes. For example, two databases may disagree over the exact location of 100 Main Street, or whether there *is* a 100 Main Street, with potentially disastrous consequences for emergency vehicles, and expensive consequences for deliveries (see Cova, Chapter 60). Recent trends suggest that the prospects for central standardisation of street naming by a single authority are diminishing, rather than growing.

5 CONCLUSION

The prospects for spatial analysis have never been better. Data are available in unprecedented volume, and are easily accessed over today's communication networks. More methods of spatial analysis are implemented in today's GIS than ever before, and GIS has made methods of analysis that were previously locked in obscure journals easy and straightforward to use. Nevertheless, today's environment for spatial analysis raises many issues, not the least of which is the ability of users to understand and to interpret correctly. Questions are being raised about the deeper implications of spatial analysis, and the development of databases that verge on invasion of individual privacy (Curry, Chapter 55). Our expectations may be unreasonable given the inevitable problems of spatial data quality.

Postmodern scientific discourse has fragmented, and with regard to GIS there is diversity not just in the sources of digital geographical information, but also increasingly (and especially with respect to human systems) in its interpretive meaning. There is a need to communicate clear interpretive conceptions of the rich but widely-distributed and piecemeal data holdings of networked GIS. We need now to think about spatial analysis not just in terms of outcomes, but also in terms of inputs. Metadata will come to fill a crucial role in the comparative assessment *between* different datasets just as, in a previous era, exploratory data analysis allowed *within* dataset assessment to take place. Such assessment and interpretation will become essential in an era in which the relative importance of conventional, governmental data providers is set to diminish. It seems clear that tomorrow's science will be increasingly driven by complex interactions, as data become increasingly commodified, technology increasingly indispensible to science, and conclusions increasingly consensual. New philosophies of science that reflect today's realities are already overdue.

These changes are profound and far-reaching, but they provide grounds for cautious optimism about the future of GIS-based spatial analysis. The established self-perception of rigour among spatial analysts has hitherto been to some extent misplaced, in that data quality, resolution, and richness have not always been commensurate with the sophistication of spatial analytic methods. However nostalgically we may at times now view it, the linear project design was by no means a panacea in practice.

If science and problem-solving are to be constrained by these new realities, then what kinds of spatial analysis are most likely to dominate in the coming years? The points raised in this chapter's discussion suggest that the future environment will favour the following:

1 data whose meanings are clearly understood, making it easier for multidisciplinary teams to collaborate;
2 data which are routinely collected in the day-to-day functioning of society and the everyday interactions between humans and computers;
3 data with widespread use, generating demands that can justify the costs of creation and maintenance;
4 data with commercial as well as scientific and problem-solving value, allowing costs to be shared across many sectors;
5 methods of analysis with commercial applications, making it more likely that such methods will be implemented in widely available form;
6 methods implemented using general standards, allowing them to be linked to other methods using common standards and protocols.

References

Bailey T C, Gatrell A C 1995 *Interactive spatial data analysis*. New York, John Wiley & Sons Inc.

Berry B J L, Marble D F (eds) 1968 *Spatial analysis: a reader in statistical geography*. Englewood Cliffs, Prentice-Hall

Birkin M 1995 Customer targeting, geodemographics and lifestyle approaches. In Longley P, Clarke G (eds) *GIS for business and service planning*. Cambridge (UK), GeoInformation International

Buehler K, McKee L (eds) 1996 *The Open GIS guide*. Wayland, The Open GIS Consortium Inc.

Chorley R J, Haggett P 1965 Trend surface models in geographical research. *Transactions of the Institute of British Geographers* 37: 47–67

Chrisman N R 1997 *Exploring geographic information systems*. New York, John Wiley & Sons Inc.

Edwards G, Lowell K E 1996 Modeling uncertainty in photointerpreted boundaries. *Photogrammetric Engineering and Remote Sensing* 62: 377–91

FGDC 1994 *Content standards for digital geospatial metadata*. Washington DC, Federal Geographic Data Committee, Department of the Interior. *http://www.fgdc.gov*

Fotheringham A S, Rogerson P A (eds) 1994 *Spatial analysis and GIS*. London, Taylor and Francis

Goodchild M F 1988 A spatial analytic perspective on geographical information systems. *International Journal of Geographical Information Systems* 1: 327–34

Goodchild M F, Gopal S 1989 *Accuracy of spatial databases*. London, Taylor and Francis

Goodchild M F, Haining R P, Wise S 1992 Integrating GIS and spatial analysis: problems and possibilities. *International Journal of Geographical Information Systems* 6: 407–23

Haining R P 1990 *Spatial data analysis in the social and environmental sciences*. Cambridge (UK), Cambridge University Press

Harman H H 1976 *Modern factor analysis*. Chicago, University of Chicago Press

Hartshorne R 1959 *Perspective on the nature of geography*. Chicago, Rand-McNally/London, John Murray

Levine G 1981 *Introductory statistics for psychology: the logic and the methods*. New York, Academic Press

Longley P, Clarke G 1995 Applied geographical information systems: developments and prospects. In Longley P, Clarke G (eds) *GIS for business and service planning.* Cambridge (UK), GeoInformation International: 3–9

Longley P, Higgs G, Martin D 1994 The predictive use of GIS to model property valuations. *International Journal of Geographical Information Systems* 8: 217–35

Marascuilo L A, Levin J R 1983 *Multivariate statistics in the social sciences: a researcher's guide.* Monterey, Brooks/Cole

Martin D 1996 *Geographic information systems: socioeconomic applications.* London, Routledge

Martin D, Longley P, Higgs G 1994 The use of GIS in the analysis of diverse urban databases. *Computers, Environment, and Urban Systems* 18: 55–66

Mounsey H 1991 Multisource, multinational environmental GIS: lessons learned from CORINE. In Maguire D J, Goodchild M F, Rhind D W (eds) *Geographical information systems: principles and applications.* Harlow, Longman/New York, John Wiley & Sons Inc. Vol. 2: 185–200

Siegel S 1956 *Nonparametric statistics for the behavioral sciences.* New York, McGraw-Hill

Smith T R, Andresen D, Carver L, Dolin R, and others 1996 A digital library for geographically referenced materials. *Computer* 29(5): 54, 29(7): 14

Taylor P J 1977 *Quantitative methods in geography: an introduction to spatial analysis.* Boston, Houghton Mifflin

Tobler W R 1970 A computer movie simulating urban growth in the Detroit region. *Economic Geography* supplement 46: 234–40

Tukey J W 1970 *Exploratory data analysis.* Reading (USA), Addison-Wesley

Unwin D J 1981a *Introductory spatial analysis.* London, Methuen

Wrigley N 1985 *Categorical data analysis for geographers and environmental scientists.* Harlow and New York, Longman

APPENDIX: GLOSSARIES AND ACRONYMS

GLOSSARIES

Extensive glossaries of GIS terminology can be found at the following sources:

Books

Arlinghaus S L 1994 *Practical handbook of digital mapping terms and concepts*. Boca Raton, CRC Press
Burrough P A, McDonnell R A 1998 *Principles of geographical information systems*, 2nd edition. Oxford, Oxford University Press
Clarke K C 1997 *Getting started with geographic information systems*. Upper Saddle River, Prentice-Hall
DeMers M N 1996 *Fundamentals of geographic information systems*. New York, John Wiley & Sons Inc.
McDonnell R A, Kemp K 1995 *International GIS dictionary*. Cambridge (UK), GeoInformation International
Padmanbhan G, Leipnik M R, Yoon J 1992 *A glossary of GIS terminology*. NCGIA Technical Papers Series 92–13, Santa Barbara, NCGIA
Worboys M F 1995 *GIS: a computing perspective*. London, Taylor and Francis

World Wide Web sites

There is a huge range of online GIS glossaries, including:
http://www.geo.ed.ac.uk/agidict/welcome.html
http://www.esri.com/base/users/glossary.glossary.html
http://www.lib.berkeley.edu/EART/abbrev.html

ACRONYMS

AACR	Anglo-American Cataloguing Rules
AAG	Association of American Cartographers
ABS	Australian Bureau of Statistics
ACIC	Automated Cartographic Information Center
ACID	Atomicity; Consistency; Isolation; Durability
ACM	Association of Computing Machinery
ACMLA	Association of Canadian Map Libraries and Archives
ACSM	American Congress on Surveying and Mapping
ADL	Alexandria Digital Library
ADT	abstract data type
AEC	Australian Electoral Commission
AGI	Association for Geographic Information
AGILE	Association of Geographic Information Laboratories in Europe
AGNPS	agricultural non-point source
AI	artificial intelligence
ALA	American Library Association
ALE	Association Liègeoise d'Electricité
AM/FM	automated mapping/facilities management
AML	Arc Macro Language
ANC	African National Congress
ANN	artificial neural network
ANSI	American National Standards Institute
ANZLIC	Australia and New Zealand Land Information Council
API	application programming interface
AR	augmented-reality/arc routing
ARMGS	autonomous route management and guidance system
ASAP	All-hazard Situation Assessment Programme
ASPRS	American Society of Photogrammetry and Remote Sensing
ATCS	advanced travel conditions system
ATKIS	Amtliches Topographisch-Kartographisches Informations-system
ATM	asynchronous transfer mode
ATOS	advanced travel orientation system
ATPS	advanced trip planning system
ATSR	Along-Track Scanning Radiometer
AUSLIG	Australian Surveying and Land Information Group
AVDS	Automatic Vehicle Dispatch System
AVHRR	Advanced Very High Resolution Radiometer
AVI	automatic vehicle identification
AVIRIS	advanced visible and infrared imaging spectrometer
AVLS	automatic vehicle location system
AVMS	automatic vehicle monitoring system
AVNS	automatic vehicle navigation system

AZP	automated zone design problem		CI	computational intelligence
BCS	British Cartographic Society		CIS	Customer Information System
BEAR	base-wide environmental analysis and restoration		CLI	Canadian Libraries Initiative
			CMLS	chemical movement through layered soils
BEV	Bundesant für Eich- und Vermessungswesen (Austria)		CNES	Centre National d'Etudes Spatiales
BFKG	Bundesamst für Kartographie und Geodäsie		CNIG	Comité de l'Information Géographique
BIH	Bureau International de l'Heure		COBOL	common business-oriented language
BLG	binary line generalisation (tree)		CODASYL	Conference on Data Systems and Language
BLPU	(British Standard) basic land and property unit		COGO	coordinate geometry
BMDP	BioMedical Data Processor		COM	common object model
BMS	Battlefield Management System		CORBA	common object request broker architecture
BPI	bits per inch			
BRDF	bidirectional reflectance distribution function		COTS	commercial-off-the-shelf
			CPD	Central Postcode Directory
BSI	British Standards Institution		CPGIS	Chinese Professionals in GIS
BSP	binary space partitioning		CPR	Continuing Property Records
BT	British Telecommunications		CPS	Collaborative Planning System
CAD	computer-aided design		CPU	central processing unit
CADD	computer-aided design and drafting		CRC	class, responsibilities, and collaborators
CAG	Canadian Association of Cartographers		CRESP	crisis and response prototype (NATO)
CAM	computer-aided mapping		CRT	cathode ray tube
CAPDU	Canadian Association of Public Data Users		CSCW	computer-supported cooperative work
			CSEP	cumulative seasonal erosion potential
CARL	Canadian Association of Research Libraries		CSIR	Council for Scientific and Industrial Research (S Africa)
CASE	computer-assisted software engineering		CSIRO	Commonwealth Scientific and Industrial Research Organisation
CBRED	Central Board for Real Estate Data (Sweden)		CSR	complete spatial randomness
			CTI	compound terrain index
CCD	census collector district		C3I	command control and communication information
CD-ROM	compact disk, read-only memory			
CDV	cartographic data visualiser		CTS	conventional terrestrial reference system
CEM	comprehensive emergency management		DBTG	database task group
CEN	Comité Européen de Normalisation		DBMS	database management system
CEOS	Committee on Earth Observation Systems		DCDB	digital cadastral database
			DCM	digital cartographic model
CERCO	Comité Européen des Responsables de la Cartographie Officielle		DCP	distributed computing platform
			DCW	digital chart of the world
CGA	colour graphics adapter		DED	district electoral division
CGIA	California Geographic Information Association; also Center for Geographic Information and Analysis		DEM	digital elevation model
			DEMON	digital elevation model networks
			DGIWG	Digital Geographic Information Working Group
CGIS	Canadian Geographic Information System		DGPS	differential global positioning system
CHEST	Combined Higher Education Software Team		DIF	directory interchange format
			DIGEST	digital geographic information exchange standard
CHGIS	critical history of GIS			

DIME	dual independent map encoding
DLG	digital line graph
DLI	Data Liberation Initiative
DLL	dynamic link library
DLM	digital landscape model
DN	digital number
DoD	Department of Defense (US)
DOE	Department of Environment (UK)
DOP	dilution of precision
DOS	disk operating system
DOT	Department of Transportation
DPI	dots per inch
DPW	digital photogrammetric workstation
DSP	depository services programme
DSS	decision support system
DTM	digital terrain model
DXF	digital exchange format
EASD	Empowerment for African Sustainable Development
ECDIS	electronic chart display
ECU	European currency unit
ED	enumeration district
EDA	exploratory data analysis
EDAMS	Electoral Distribution and Mapping System
ED-50	European Datum 1950
EDIS	economic development information system
EEA	European economic area
EGA	enhanced graphics adapter
EIA	environmental impact assessment
EIS	executive information system
ENV	(European standards)
EOS	Earth observing system
EPIC	erosion productivity impact calculator
EPZ	emergency planning zone
ERA	entity–relationship–attribute
ERS	European resource satellites (of the European Space Agency and of Japan)
ERTS	Earth resources technology satellite
ESDA	exploratory spatial data analysis
ESRI	Environmental Systems Research Institute Inc.
ESTDM	event-based spatio-temporal data model
ETRS	European terrestrial reference system
EU	European Union
EUROGI	European UmbRella Organisation for Geographic Information
FACC	feature and attribute coding catalogue

FAO	Food and Agriculture Organisation (United Nations)
FDDI	fibre distributed data interface
FDLP	Federal Deposit Library Program
FEMA	Federal Emergency Management Agency (US)
FEPD	Forum for Effective Planning and Development (S Africa)
FGDC	Federal Geographic Data Committee (US)
FHWA	Federal HighWay Administration
FINDAR	facility for interrogating the National Directory of Australian Resources
FIPS	Federal Information Processing Standard
FMV	fuzzy membership value
FOIA	Freedom of Information Act
FORTRAN	formula translation
FRAMME	facilities rule-based application model management environment
GADS	geodata analysis and display system
GAM	geographical analysis machines
GAP-tree	generalised area partitioning
GATT (1994)	General Agreement on Tariffs and Trade, 1994
GBF/DIME	geographic base file/dual independent map encoding
GCEM	geographical correlates exploration machine
GCM	global circulation model
GCP	ground control point
GDF	geographic data files
GDIS	geodemographic information system
GENIE	Global Environmental Network for Information Exchange (UK)
GFS	geo-facilities information system
GGI&S	Global Geospatial Information and Services
GIC	geospatial information community
GICC	Geographic Information Coordinating Council (North Carolina)
GILS	Government Information Locator Service
GISDK	GIS developers kit
GISRUK	GIS Research UK
GISy/GISc	subdivisions of GIS: systems/science
GKS	graphics kernal system
GLEAMS	groundwater loading effects of agricultural management systems
GLIS	global land information system

GNU	Government of National Unity (South Africa)	IERS	International Earth Rotation Service
GOTS	government-off-the-shelf	IETF	Internet Engineering Task Force
GPAC	Government Performance Audit Committee (US, North Carolina)	IFLA	International Federation of Library Associations
GPO	Government Printing Office (US)	IFOV	instantaneous field-of-view
GPS	global positioning system	IGBP	International Geosphere Biosphere Programme
GRASP	greedy randomised adaptive search procedure	IGES	Initial Graphic Exchange Standard
GRASS	geographic resource analysis support system	IGN	Institut Géographique National
		IHO	International Hydrographic Organisation
GRE	ground resolution element	IMap	intelligent mapping, accounting, and provisioning
GRIA	global regional interchange algorithm		
GRIDS	geographic roadway information display system	IP	Internet Protocol
		IRMC	Information Resources Management Commission (US; formerly the Information technology Council)
GRSA	GIS-relevant spatial analysis		
GSDI	global spatial data infrastructure		
GUI	graphical user interface	ISBD	international standards for bibliographic description
GVD	generalised Voronoi diagram		
HARN	statewide high accuracy reference network	ISBN	international standard book number
		ISDN	integrated services digital network
HAS	highway analysis system	ISM	industry structure model
HCI	human computer interaction	ISO	International Standardisation Organisation
HDT	hierarchical Delaunay triangulation		
HIPSS	health information for purchaser planning system	ISPRS	International Society for Photogrammetry and Remote Sensing
HIRIS	High Resolution Imaging Spectrometer	ITA	interactive territory assignment
		ITS	intelligent transportation system
HMD	head-mounted display	ITU	International Telecommunications Union
HMLR	Her Majesty's Land Registry		
HPC	high performance computing	JERS	Japanese Earth Resources Satellite
HPF	highly parallel FORTRAN	KA	knowledge acquisition
HRTF	head-related transfer functions	KB	kilobyte
HRV	high resolution visible	KBLIMS	knowledge-based land information management system
HSR	hidden surface removal		
HTML	hypertext markup language	KHz	kilohertz
http	hypertext transmission protocol	LAI	leaf area index
ICA	International Cartographic Association	LAN	local area network
		LANDSAT	LAND resources assessment SATellite system (US)
ICT	information and communication technology		
		LAPC	Land and Agriculture Policy Centre
IDGIS	interactive distributed geographical information system	LDBS	land data bank system
		LEACHM	leaching and chemistry estimation
IDNDR	International Decade for Natural Disaster Reduction	LGMB	Local Government Management Board
IDW	inverse distance weighted interpolation	LIDARS	light detection and ranging system
IEC	International Electrotechnical Commission	LIS	land information system
		LISA	local indicators of spatial analysis
IEEE	Institute of Electrical and Electronics Engineers	LISS	linear imaging self-scanning system

LOD	level of detail management algorithm
LP	linear programming
LP/IP	linear programming/integer programming
LUNR	land use and natural resource inventory (State of New York)
LWIR	long wavelength infrared
MAGI	Maryland automated geographic information
MAN	metropolitan area network
MAP	map analysis package
MAPS	Montana agricultural potential system
MARBI	machine-readable bibliographic information
MARC	machine-readable cataloguing
MAUP	modifiable areal unit problem
MB	megabyte, or mbyte
MBR	minimum bounding rectangle
MBS	minimal bounding sphere
MC&G	mapping, charting, and geodesy
MCE	multicriteria evaluation
MCNC	Microelectronics Center of North Carolina
MEL	Master Environmental Library
MERIS	medium resolution imaging spectrometer
MESSR	multispectral self-scanning radiometer
MHZ	megahertz
MIMD	multiple instruction, multiple data
MIPS	million instructions per second
MIS	management information system
MISR	multi-angle imaging spectrometer
MIT	Massachusetts Institute of Technology
MLRA	major land resource area
MODIS	moderate resolution imaging spectrometer
MOPTT	Ministry of Post and Telecommunications (Saudi Arabia)
MOS	marine observation satellite
MPI	message passing interface
MPO	multi product operations
MPP	massively parallel processor
MSC	Mapping Science Committee
MSR	Microwave Scanning Radiometer
MSS	multispectral scanner
MST	minimum spanning tree
MVA	multiple view angle
NAD-27	North American Datum 1927
NAD-83	North American Datum 1983
NAFTA	North American Free Trade Agreement
NAPA	National Academy of Public Administration
NRSA	National Remote Sensing Agency
NASA	National Aeronautical and Space Administration (US)
NASDA	National Space Development Agency (Japan)
NATO	North Atlantic Treaty Organisation
NATSGO	national soil geographic database
NAVSTAR	navigation satellite timing and ranging system
NCGA	National Computer Graphics Association
NCGIA	National Center for Geographic Information and Analysis
NCIH	North Carolina Information Highway
NC-REN	North Carolina Research and Education Network
NCSA	National Centre for Supercomputing Applications
NDVI	normalised difference vegetation index
NESPAL	National Environmentally Sound Production Agriculture Laboratory (US)
NFM	network flow model
NGDF	national geospatial data framework
NGS	National Geodetic Survey
NISO	National Information Standards Organisation (US)
NLIS	national land information system
NLS	National Land Survey
NMA	National Mapping Agency
NMAS	national map accuracy standards
NMCA	National Mapping and Cadastre Agency
NMO	National Mapping Organisation (Netherlands Standardisation Organisation)
NNI	
NNSDP	National Nutrition and Social Development Programme (S Africa)
NNWG	Natal Nutrition Working Group
NOAA	National Oceanographic and Atmospheric Administration (US)
NP-Hard	non-deterministic polynomial-hard
NPR	national performance review
NRIS	national resource information system
NRSA	National Remote-Sensing Agency (India)
NSF	National Science Foundation (US)
NSDI	national spatial data infrastructure
NTF	national transfer format
OCLC	online computer library centre

ODMG	Object Database Management Group	PR	proportional representation
OECD	Organisation for Economic Cooperation and Development	PRN	pseudo random noise
		PROM	programmable read-only memory
OEEPE	(European Organisation for Experimental Photogrammetric Research)	PSNP	primary school nutrition programme
		PSS	planning support system
		PSTN	public standard telephone network
OGC	Open GIS Consortium	QA/QC	quality assurance/quality control
OGDC	Office of Geographic Data Coordination	QTM	quaternary triangular mesh
		RAM	random access memory
OGIS	open geodata interoperability specification	RDBMS	relational database management system
OGM	open geodata model	RDP	Reconstruction and Development Programme (S Africa)
OHD	'off the head' display		
OID	object identifier	RFP	request for proposals
OLE	object linking and embedding	RICS	Royal Institution of Chartered Surveyors
OLE/COM	object linking and embedding/ component object model	RIN	realtors information network
		RINEX	receiver independent exchange
OMB	Office of Management and Budget (US)	RMSE	root mean square error
		ROM	read-only memory
OMG	Object Management Group	RS	remote sensing
OMT	object modelling technique	RSG	regular square grid
ONS	Office of National Statistics (UK)	RST	regulated spline with tension
OODB	object-oriented database	RUSLE	revised universal soil loss equation
OODBMS	object-oriented database management system	RVIS	reliability visualisation tool
		SA	selective availability
OOP	object-oriented programming	SAIF	spatial archive and interchange format (Canadian)
OOPL	object-oriented programming language		
		SAR	synthetic aperture radar
OO2VD	ordered order-2 Voronoi diagram	SAS	statistical analysis system
OpenGIS	Open Geodata Interoperability Specification	SBC	Santa Barbara climate
		SCADA	supervisory control and data acquisition
OPS	optical sensor		
ORNL	Oak Ridge National Laboratory	SCDF	spatial cumulative distribution function
OS	operating system		
OS (GB)	Ordnance Survey (Great Britain)	SDA	spatial data analysis
OS (NI)	Ordnance Survey (Northern Ireland)	SDE	spatial data engine
OSM	OGIS Services Model	SDSS	spatial decision support system
O2VD	order-2 Voronoi diagram	SDTS	spatial data transfer standard
OVD	ordinary Voronoi diagram	SERDP	strategic environmental research and development program
OVP	ordinary Voronoi polygon		
PACE	prediction and coverage estimation	SFWMD	South Florida Water Management District
PAIS	Prototype Allied Command Europe Intelligence System		
		SIGMOD	special interest group on the management of data
PC	personal computer		
PCT	personal construct theory	SIMD	single instruction, multiple data
POES	polar orbiting environmental satellite	SINES	spatial information enquiry service
POLDER	polarisation and directionality of the Earth's reflectances	SLC	spatial location code
		SLM	structure line model
POST	Parliamentary Office for Science and Technology	SLOSH	sea, lake, and overland surge from hurricane
PPS	precise positioning service		

SLR	satellite laser ranging
SMI	Strategic Mapping Incorporated
SMMT	Society of Motor Manufacturers and Traders
SMSA	standard metropolitan statistical area (US)
SNR	signal-to-noise ratio
SOC	soil organic carbon
SOHO	small office/home office
SoLIM	soil land inference model
SPA	shortest path analysis
SPOT	Satellite Pour l'Observation de la Terre
SPS	standard positioning service
SPSS	statistical package for the social sciences
SQL	structured query language
SRAM	static random access memory
SSC	Swiss Society of Cartography
SSM	system status management
STAC	space–time attribute creature
STATSGO	State soil geographic database (US)
STEP	standard for the exchange of product model data
STV	single tranferable voting
SWIR	short wavelength infrared
TAZ	traffic analysis zone
TIGER	topologically integrated geographic encoding and referencing
TIMS	transportation information management system
TIN	triangulated irregular network
TM	thematic mapper
TLP	Telefones de Lisboa e Porto
TPS	thin plate spline
TRIPS	trade-related aspects of intellectual property rights
UCC	uniform commercial code
UCGIS	University Consortium for GIS
UDMS	Urban Data Management Society (Europe)
UGISA	University Geographic Information System Alliance
UKMARC	United Kingdom machine-readable cataloguing
UNCRD	United Nations Centre for Regional Development

UNDRO	Office of the United Nations Disaster Relief Coordinator
UN ECE	United Nations Economic Commission for Europe
UNESCO	United Nations Economic, Social, and Cultural Organisation
UNIDROIT	International Institute for the Unification of Private Law
UNITAR	United Nations Institute for Training and Research
UPRN	unique property reference numbers
URISA	Urban and Regional Information Systems Association
URL	uniform resource locator
USBC	United States Bureau of Census
USGS	United States Geological Survey
USMARC	United States machine-readable cataloguing
UCT	universal time coordinated
UTM	universal transverse mercator
UTMS	urban transportation model system
VAR	value added reseller
VE	virtual environment
VET	virtual environment theatre
VGIS	virtual geographical information system
ViSC	visualisation in scientific computing
VLBI	very long baseline interferometry
VNIR	visible and near-infrared radiometer
VOA	Valuation Office Agency
VR	virtual reality
Vro	vehicle routing
VRT	variable rate treatment
VTIR	visible and thermal infrared radiometer
WAN	wide area network
WCED	World Commission on Environment and Development
WGS-84	World Geometric System 1984
WIM	worlds in miniature
WIMP	windows, icons, mice, and pointers
WLS	weighted least squares
WOFOST	World Food Studies
WTO	World Trade Organisation
XGA	extended graphics adapter
ZDES	zone design system
ZIP	zone improvement plan

AUTHOR INDEX

Abakuks A 958, 965, 1089
Abbott T 903, 908, 1023
Abdelmoty A 383–4, 1080
Abel D J 388, 399, 1000, 1007, 1023
Aber J D 452, 466, 1098
Abery J C G 968, 980, 1093
Abler R F 50, 59, 1023
Aboav D A 511, 524, 1023
Abrahams A D 528, 540, 1023
Abrahamson F 87–8, 1023
Abram G D 387, 399, 1049
Abramowitz M 485, 490, 1023
Acevedo W 841–2, 1035, 1038, 1071
Ackermann F 109, 121, 1023
Acquilla S D 931, 936, 1043
Acredolo L 89, 1093
Adams J B 457, 466, 1090
Adams T 829, 841, 844, 1023, 1096
Adams W M 968, 975, 1023
Addicott J F 971, 975, 1023
Adelson B 410–12, 1023, 1056
Adler P 681, 799, 901–8, 1019, 1036
Affholder J-G 140, 144, 154, 1082
Agarwal N 325–6, 328, 1046
Agee J K 970, 976, 1023
Agre P E 762, 765, 1023
Agterberg F P 970, 976, 1030
Aguilar R 763–5, 1023
Agumya A 638, 640–1, 1023, 1045
Ahearn S C 968, 975, 1023
Ahlberg C 409–10, 1023
Ahmad S P 453, 464, 1023
Ahnert F 983, 994, 1040
Aho J M 971, 975, 1023
Ahonen P 220, 233, 1087
Ahuja N 506, 524, 1023
Aitken S C 57, 59, 1023
Akima H 110, 121, 483, 490, 1023
Akom A A 886–7, 1023
Alaniz M A 984, 992, 994, 1045
Albrecht J 405, 412, 1101
Albrektsen G 931, 938, 1085
Aldrich R C 956, 964, 1023
Alexander D 847–8, 856, 1023
Alexander M 968, 975, 1023
Alexander S 838, 841, 1023

Alexandroff P 37, 1023
Ali Khan N 320, 328, 1038
Allanson P 487, 490, 1075
Allen D 200, 204, 1040
Allen J F 98, 101, 1023
Allen K 265–6, 1045
Allen L H Jr 987, 994, 1038
Allison D (Plate 21) 453, 461, 1027
Allison L J 982–3, 989, 996, 1078
Al-Taha K K 101, 1023
Al-Thani Sheik Ahmed Bin Hamad 16
Altman D 197, 203, 1023
Alvanides S 8, 44–5, 64, 73, 77–8, 137, 165, 235–7,
        243–4, 267–83, 287, 298, 303, 310, 404, 509,
        549, 567, 573–4, 658, 856, 930, 940, 958, 970,
        992, 1010, 1018, 1023, 1079
Alvares L 138, 152, 1026
American Library Association 906, 908, 1024
American Society for Photogrammetry 180, 187, 1024
American Society of Civil Engineers 180, 187, 1024
Ammann R 503, 524, 1024
Amrhein C G 243, 248, 1024
Amundson S E 878, 888, 1070
ANC 914, 924, 1024
Anderson G W 992, 994, 1032
Anderson H 968, 979, 1089
Anderson J R 651, 956, 964, 1024
Anderson K 205, 1101
Anderson M G 541, 1070
Andes N 933, 936, 1024
Andresen D 574, 580, 1090
Andrewartha H G 970, 975, 1024
Andrews D F 254, 264, 1024
Andrienko G 264, 1024
Andrienko N 264, 1024
Andrieu B 453, 465, 1060
Andrieux P 195, 198, 204, 1064
Angelstam P 970, 977, 1055
Annitto R 320, 324, 1024
Anon 922, 924, 1024
Anselin L 7, 20, 77, 80, 128, 157, 167, 221, 225, 229,
        235–7, 242–5, 248–50, 253–66, 284, 291, 402,
        409, 487, 491, 509, 570, 574, 928, 930, 936,
        971–2, 1000, 1010, 1024–5, 1042, 1047, 1050,
        1052, 1055, 1062, 1066, 1077
ANSI 378, 384, 415, 423, 681, 692, 1025

Antenucci J 2, 362, 369, 802, 818, 1025
Anthes R A 99, 101, 1025
Antolin M F 971, 975, 1023
ANZLIC 754–5, 1025
Appleton J H 974–5, 1025
Arbia G 208, 216, 243–4, 248–9, 286, 291, 1025, 1054–5
Archer H N 779, 785, 1025
ArcView 245, 248, 1025
Ardanuy A 456–7, 464, 1025
Arentze T A 298, 302, 1025
Argentati C 903, 908, 1023
Arlinghaus S L 1025, xxvii
Armstrong M P 147, 152, 163, 167, 172–3, 264, 298, 302, 484, 490, 723, 727, 731–2, 849, 856, 881–2, 885, 887, 930, 932, 938, 1025, 1034, 1068, 1087, 1089–90
Armstrong P 726, 732, 1084
Arnold G W 967, 973, 977, 1057
Aronoff S 2, 181, 187, 598, 600–3, 606, 609, 620, 1025
Aronson P B 114, 123, 1070
Arrow K 772, 785, 1025
Arseneau B 325, 327, 1025
Arthur R M 830, 838–9, 841, 1025
Ashkenazi V 436, 1025
Askov D (Plate 13) 221–2, 225, 232, 1069
Aspinall R J 42, 242, 244, 262, 910, 954, 967–80, 1025–6, 1073
Asrar G 452–3, 464, 1026, 1051, 1076
Association of Research Libraries 902, 908, 1026
Astley M 482, 490, 1032
Astrahan M 374, 384, 1026
Atallah M 548, 555, 1026
Atkinson M 403, 410, 1026
Attewell P 626, 631, 1087
Audit Commission England and Wales 621, 629, 1026
Auerbach S 110, 121, 487, 490, 1026
Aufhauser E 286, 291, 1026
Aumann G 112, 121, 1026
Aurenhammer F 110, 121, 1026
AUSLIG 634, 641, 1026
Austin M P 970, 976, 984, 996, 1026, 1072
Austin R F 820, 826, 1075
Avery B W 198–9, 203, 1026
Babolola A 832, 844, 1097
Bacastow T 621, 631, 894, 899, 1081
Bach H K 1004, 1007, 1097
Baeijs C 138, 152, 1026
Baella B 151–2, 1026
Baier M 761, 765, 1026
Bailey R G 539–40, 1026

Bailey T C 77–9, 243, 247, 249, 254, 264, 567–9, 579, 926, 928, 930–2, 936–7, 1026, 1050
Bain G D 763, 765, 1026
Baker A M 992–3, 1026, 1029
Baker B E 852, 858, 1064
Baker J R 453, 464, 1026
Baker T (Plates 10 and 12) 99, 102, 221, 223, 225, 233, 482, 485–6, 488–91, 1032, 1074
Baker W L 970, 972, 976, 978, 1026, 1058
Balasubramanian N V 555, 1042
Balch J 924, 1072
Baldwin R A 866, 875, 1076
Balisky A C 968, 976, 1033
Ball J 927, 936, 1026
Ballou N 374, 384, 1063
Bancilhon F 403, 410, 1026
Band L E 107, 112–13, 116, 118, 121, 197, 479, 527–42, 1026, 1065, 1069, 1076, 1101
Bao S 242, 248, 254, 256, 258, 261–4, 1024, 1027
Bär H R 145, 152, 1027
Barata P 822, 826, 1027
Barbic F 101–2, 1070
Bardon K S 877, 887, 1027
Baret F 452–3, 464–5, 1027, 1060
Barholdi J J III 388, 399, 1076
Barker J 462, 465, 1087
Barker J S F 971, 979, 1091
Barkhi R 297, 303, 1088
Barkley D 263–4, 1027
Barnes K B 928, 935, 937, 1050
Barnes T J 42–3, 46, 1027
Barnett C 852, 857, 1049
Barnett V 123, 526, 936, 1041, 1089
Barney T L 985, 995, 1055
Barnsley M J (Plate 21) 3, 92, 105, 108, 202–3, 244, 248, 426, 438, 440, 443, 451–66, 477, 661, 663, 665, 667–9, 672, 770, 780–1, 806, 850, 880, 887, 896, 931, 954–5, 971, 981–3, 1010, 1013, 1027
Barr A H 551, 556, 1057
Barr C J 955, 964, 1032
Barr R 73, 79, 626, 629, 754–5, 768, 785, 1027
Barr S L 202–3, 244, 248, 454, 457, 462, 464, 880, 887, 1027
Barrera R 394, 399, 1049
Barrett G W 972, 976, 1027
Barrett-Connor E 931, 938, 1085
Bart P 301, 303, 1093
Barthel K M 638, 641, 1027
Barton G S 684, 692, 1027
Barton I J 457, 465, 1082
Basoz N I 849, 853, 858, 1063

Bassa L 291, 1046

Bassoglu U 636, 641, 1027

Batey P 45–6, 1027

Batey P W J 933, 936, 1031

Batini C 415, 423, 1027

Battista C 853, 856, 1027

Batty M 74, 153, 195, 203, 242, 247–51, 263–4, 268, 282, 284, 292, 307, 309–17, 359–60, 407, 462, 465, 614–15, 699, 708, 721, 723–6, 730–1, 830, 840–2, 865, 883–6, 888, 1014, 1020, 1027–8, 1030, 1035, 1040, 1051, 1053, 1056–7, 1068, 1073, 1079, 1100

Baumann J 821, 826, 1028

Bayer R 385, 388, 395, 399, 1028

BCS 754–5, 1028

Beard M K 130, 147, 149, 152–4, 163–4, 172, 176, 179, 186, 188, 196, 219–33, 258, 339, 449, 588, 633, 635, 639–41, 989, 1010, 1028, 1030, 1033, 1069, 1080

Beasley J 300, 302, 1028

Beaumont J R 297, 302, 714–15, 721, 1028

Beck M B 123, 1074

Becker R A 221, 232, 254–5, 264, 749–50, 756, 1028, 1075

Beckett P 196, 203, 1035

Beckmann N 395, 399, 1028

Bédard Y 352–3, 372, 378, 384, 405, 413–24, 463–4, 637–9, 641, 829, 897, 1028, 1033, 1044

Bednarz S W 752, 755, 1028

Beech D 402, 412, 1092

Belbin L 968, 978, 1072

Bell D 362, 369, 1028

Bell J C 984–6, 993, 1028

Bell M 827, 841–2, 844, 1030, 1060

Bell S B M 68–9, 504, 507–9, 524, 1028, 1041, 1058, 1076

Ben Akiva M 834, 841, 1029

Benedetti R 243, 248, 1025

Bengtsson L 91, 103, 1090

Beniger J R 759, 765, 1029

Bennett C J 760, 765, 1029

Bennett K B 563, 565, 1029

Bennett R J 42–3, 46–7, 80, 251, 291, 520, 525, 1029, 1055, 1080

Benoit D 716, 721, 1029

Bentham C G 934, 937, 1056

Bentham G 934–5, 937, 1061, 1065

Bentley J L 386, 399, 548, 555, 1029

Beresford J 927, 937, 1044

Berg M de 132, 147, 150–2, 551, 555, 1029

Bern M 554–5, 1029

Bernado D J 984, 988, 995, 1050

Bernard M 749, 755, 1029

Bernhardsen T 2, 361, 425, 440, 583, 586–7, 589–600, 602, 606, 613, 673, 739, 1029

Bernknopf R L 608–9, 1029

Berry B J L 43, 47, 178, 188, 248, 285, 291, 569, 579, 992–3, 1029, 1039, 1100

Berry J K 749, 755, 877, 887, 1029

Bertalanffy L von 973, 976, 1029

Bertazzon S 840–1, 1029

Bertin J 161, 170–2, 221, 225, 227, 230, 232, 1029

Bertuglia C S 716, 721, 1029–30, 1035

Besag J 243, 248, 272, 282, 519, 524, 932, 936, 1029

Best R G 852, 857, 1054

Best R H 957, 964, 1029

BEV 860, 866, 875, 1029

Beven K J 108, 120–3, 212, 216, 540, 986, 996, 1026, 1029–30, 1083

Bezdek J C 197, 203, 270, 282, 287, 291, 462, 464, 1029

Bhaskar R 62, 69, 1029

Bhopal R S 931, 936, 1029, 1043

Bibby P R 71, 437, 451, 885, 909–10, 953–65, 971, 1029–30

Bicking B 226, 232, 1030

Bideau D 525, 1085

Bier E 409, 412, 1092

Biggar J W 990, 997, 1096

Biggeri A 930, 936, 1035

Biljon W R van 914, 924, 1061

Binley A 120–1, 1030

Birch L C 970, 975, 1024

Birkin M 45–6, 71, 240, 268, 282, 286, 291, 293, 334, 337, 343, 572, 579, 585, 601, 612, 620, 640, 707, 709–22, 748, 758, 762, 767, 792, 801, 805, 815–16, 820, 832–3, 841, 884, 892, 928, 935, 1017, 1030

Birks H J B 968, 978, 1087

Birnie R 955, 964, 1030

Bischof R J 485–6, 491, 1059

Bishop G D 198, 203, 1033

Bishop I 167, 173, 1090

Bithell J 930, 936, 1030

Black W 827, 844, 1097

Blad B L 453, 466, 1097

Blaha M 353, 358, 363, 369, 380, 382, 384, 1087

Blais A 940, 944, 950, 1030

Blakemore M 80, 180, 188, 233, 291, 399, 636, 641, 857, 1030, 1047, 1050, 1063, 1079

Blalock M 73, 79, 1030

Blanchard G 837, 841, 1030

Blasgen M 374, 384, 1026

Blelloch G E 544, 549, 555, 1030
Bliss N B 987–9, 994, 1030
Bloksma R J 203–4, 1039
Blöschl G 108, 117, 120–1, 1030, 1053
Bly S 560, 565, 1033
Board C 159, 172, 1030
Bober M L 984, 994, 1030
Bocco G 849, 856, 1030
Bogaert P 484, 490, 1030
Bogardi I 208, 217, 1100
Bogdanor V 950, 1036
Bognar J A 983, 995, 1065
Bohme R 656, 666, 1030
Boissonnat J D 544, 547, 553, 555, 1030
Bolstad P V 112, 121, 1030
Bomford G 428, 436, 1030
Bonham-Carter G F 2, 158, 172, 970, 976, 1030
Bonissone P P 496–7, 502, 1030
Bonsall P 827, 841–2, 844, 1030, 1060
Booch G 95, 101, 353, 357, 414, 418, 420, 423, 1030
Boochs F 452, 464, 1030
Boote K J 987, 994, 997, 1038, 1099
Booth J G 984, 995, 1058
Boots B N 66, 77, 79, 110, 132, 140, 242, 248, 262,
    285–6, 292, 478, 483, 503–26, 544, 854, 933,
    941, 987, 1000, 1031, 1050, 1058, 1076–8, 1094
Borgers W J 298, 302, 1025
Born S M 973, 978, 1070
Borning A 407, 410, 1031
Bos J van den 396, 399, 1078
Bose P 554–5, 1031
Bossler J 180, 188, 1093
Bothwell G W 457, 464, 1041
Botkin D A 972, 978, 1058
Bougé L 282, 1095
Bouma J 992, 994, 1031
Bouwman W 993, 1030
Bouzigues R 195, 198, 204, 1064
Bovy P H L 832, 841, 1031
Bowyer J K 116, 120, 123, 1067
Boyce R 374, 384, 1034
Boydell B 981, 983, 990–2, 997, 1095
Boyle P J 520, 524, 929, 938, 1031, 1090
Boyle S L 185, 188, 1092
Brabb E E 539, 541, 1041
Bracken I 77, 79, 1031
Bradley R 247, 250, 255, 260, 264–5, 928, 937, 1031,
    1056
Brady S J 974, 977, 1047
Braess D 492, 1082
Braga M 849, 856, 934, 937, 1031, 1035

Brainard J S 934, 937, 1031, 1056
Brand M 698, 705, 1031
Brand S 726, 731, 1031
Brandeau M L 297, 302, 1031
Brandeis L D 758–9, 763, 766, 1097
Brandenberger A J 656, 666
Brandenberger C 151–2, 1031
Brandli M 530, 542, 1098
Branki N E 725, 731, 1061
Branscomb A W 585–6, 768, 785, 1031
Bras R L 528, 531, 541–2, 1059, 1093
Brassel K 133, 136, 143, 152, 179, 183, 188, 394,
    399, 411, 1031, 1063–4
Braun J 110, 123, 520, 526, 1087
Breckling B 974, 976, 1031
Breemersch H 825–6, 1048
Breheny M 957, 964, 1031
Breininger D R 970, 976, 1031
Brewer C A 225, 232, 1031
Bridgewater P B 968, 976, 1031
Briggs D 927, 932, 936, 1036
Briggs I C 112, 121, 485–6, 490, 1031
Brill E D Jr 301, 303, 1031
Brinkmann W L 44, 47, 1085
Briuer F 257, 266, 1099
Broadwin R 932, 938, 1068
Brodahl M K 984, 998, 1100
Brody H 194, 203, 1031
Bromley R 625, 629, 722, 1031, 1100
Brooks F 362, 369, 1031
Brooks K 263–4, 1027
Brooks R 325, 327, 1036
Brooks S R 524, 1028
Brookshire D L 608–9, 1029
Broome F R 79, 1031
Broschart M 970, 979, 1080
Brouer B H 992–4, 1033
Browder J A 972, 978, 1058
Brown J F 938, 988, 994–7, 1031, 1068, 1092
Brown J R 219, 232, 1050
Brown K 2, 362, 369, 802, 818, 1025
Brown M D 158, 173, 1072
Brown P J B 45–6, 282, 933, 936–7, 1027, 1031,
    1057, 1078
Brown R 792, 795, 1032
Brown R B 984, 992, 994–5, 1032, 1048
Brown R S 836, 844, 1097
Brown W M (Plates 10–12) 99, 102, 221, 223, 225,
    233, 482, 485–91, 1032, 1073–4
Bruegge C J 457, 464, 1041
Bruns T 405, 410, 1032

Brunsdon C 621, 630, 928, 936, 1032, 1079
Brunt J W 975, 1026, 1034, 1069
Brussard P F 968, 976, 1032
Bruzewicz A J 845, 856, 1032
Bruzzone E 547, 553, 555, 1032
Bryant P 850, 852, 858, 1067
Bucher F 179, 183, 188, 1031
Buckland S T 970, 976, 1032
Budge A 184, 188, 1075
Buechner M 970, 979, 1091
Buehler K 4, 10, 308, 322, 347–59, 369, 382, 402,
    410, 415, 564, 575, 579, 584, 681, 683, 694,
    698–9, 705, 725, 802, 850, 865, 1010, 1032
Bufton J L 109–10, 122, 1055
Buja A 254, 256, 264–5, 1032, 1037
Bullen N 933, 936, 1032
Bunce R G H 955, 964, 1032
Bundred P 933, 936–7, 1031, 1057
Bundy G L 131–2, 138–45, 152–3, 155, 1032, 1061,
    1097
Burch G J 112, 123, 1074
Burdea G 558, 563, 565, 1032
Bureau of the Census 193, 200, 203, 1032
Burges S J 986, 994, 1037
Burgess K 362–3, 369, 1047
Burgess T M 214, 216, 1071
Burke I C 972, 976, 1032
Burke L M 849, 956, 1032
Burley T M 954, 964, 1032
Burnette T 563, 565, 1081
Burns T 749, 755, 1032
Burril A 975, 1026
Burrough P A 2, 5, 20, 34–7, 67, 69, 83, 88, 114,
    121, 186, 188, 195–7, 203–4, 207–8, 210,
    215–16, 227, 232, 481–4, 490, 637, 639, 641,
    698, 705, 785, 988, 994, 1012, 1020, 1031–2,
    1037, 1040, 1057, 1064, xxvii
Burton I 847, 856, 1033
Burton L L 636, 641, 1033
Burton P J 968, 976, 1033
Bush J 934, 938, 1094
Butler D 940, 942, 945, 950–1, 1033, 1069, 1072
Buttenfield B P 126, 129, 141, 144, 149, 152–5,
    163–4, 170, 173, 176, 179, 186, 196, 219–33,
    258, 339, 449, 588, 633, 635, 640–1, 968, 979,
    989, 1010, 1028, 1033, 1043, 1052, 1066, 1074,
    1076, 1086, 1089, 1098
Buttimer A 42–3, 46, 1033
Buxton W 560, 565, 1033
Bydler R 828–30, 841, 1033
Cabrera J 254, 265, 1037

Cadoux-Hudson J 721, 756, 818, 978, 1044, 1052,
    1058, 1073, 1076, 1095
Cahn M D 922–4, 1033
Cai G 639, 641, 1044
Caicco S 968, 979, 1089
Cain B E 944, 946, 950, 1033
Calhoun C 621, 629, 1033
Caliper Corporation 833–4, 841, 1033
Calkins H W 602, 609, 618, 620, 1041
Callahan S 680, 692, 1061
Callihan R H 984, 995–6, 1065, 1082
Calvert C 656, 665–6, 770, 779, 785, 787, 1033
Câmara A S 105, 107, 110, 310, 402, 405, 479, 557–65,
    725, 727, 731, 1017, 1033, 1041, 1048, 1075
Cameron A H 928, 938, 1078
Campari I 37–8, 88, 247, 249, 525, 1037, 1049–51,
    1070, 1081, 1090
Campbell C S 170, 172, 1033
Campbell G 453, 466, 1097
Campbell H J 410, 583, 587, 599, 602–3, 611,
    621–31, 877, 885–8, 923, 953, 964–5, 1033,
    1071, 1098
Campbell J B 194, 196, 202–3, 1033
Campbell J C 83, 88, 494, 502, 1033
Campbell R H 608–9, 1029
Campbell W G 198, 203, 1033
Camus J-P 837, 841, 1033
Canters F 208, 216, 1048
Capper B 754, 756, 1096
Caprio J M 987, 996, 1076
Carbone G J 984, 987, 994, 1033
Carbonell J G 147, 152, 1033
Cardenas A 407, 411, 1082
Carey H C 240, 248, 1033
Carey M J 374, 382, 384, 402, 412, 1054, 1092
Carlo Rota G 282, 1057
Carlson G R 990, 992, 994–5, 1033, 1068
Carlson R E 486, 489, 491, 1062
Carmichael A 414, 423, 1033
Carneggie D 462, 465, 1087
Carnes S A 854, 858, 1091
Caron C 416–17, 419, 423, 1028, 1033
Carr P M 990, 994, 1033
Carrara A 856–8, 1037, 1062, 1065
Carreras E M 849, 858, 1092
Carroll J B 89, 1099
Carron J 167, 173, 1090
Carstairs A McL 944, 950, 1033
Carstairs V 933, 936, 1034
Carter J R 112, 121, 210, 216, 1033
Carter S E (Plates 62 and 63) 984, 987, 994, 1037

Cartwright J 852, 856, 1034
Carvalho M S 928, 936, 1076
Carver L 574, 580, 1090
Carver S J 227, 233, 494, 502, 881, 887, 1034, 1079
Cascio J L 83, 88, 184, 188, 1046
Casetti E 245, 248, 1034
Cashwell J W 99, 101, 1049
Caspary W 180, 188, 1034
Cassettari S 2, 184, 188, 1034
Casti J L 67, 69, 1034
Castle G H 818, 1084
Castner H W 131, 155, 1087
Catling I 837, 841, 1034
Cavill M 310, 329, 1076, 1082
CBRED 860, 863, 866, 875, 1034
CCGISE/IGISE 749, 755, 1034
Cechet R P 457, 465, 1082
CEN 179, 183, 188, 350, 358, 1034
Centro Nacional de Epidemiologia 930, 936, 1034
Ceri S 383–4, 411, 415, 423, 1027, 1034, 1054
Chakraborty J 849, 856, 1034
Chamberlin D 374, 384, 1026, 1034
Chambers J M 221, 232, 1034
Chamoux J P 769, 785, 1034
Champine G A 318, 327, 1034
Chandler J H 124, 1099
Chang H 147–9, 152, 1034
Chang K C 849, 858, 1063
Chang W 374, 382, 384, 1054
Chapal S E 1007, 1034
Charlton M 45, 47, 227, 233, 254, 257, 265, 272,
    282, 621, 630, 928, 930–1, 936, 938, 1032,
    1048, 1079–80
Chatfield T A 852, 857, 1054
Chazelle B 548, 555, 1034
Checkland P 973, 976, 1034
Chen D T 559, 565, 1091
Chen P P-S 384, 1034
Chen Z T 550, 555, 1034
Chetwynd A D 930, 936, 1041
Chevallier P 108, 123, 986, 996, 1083
Cheylan J 282, 1079
Chi-Chang L 202, 204, 1052
Chilar J 452, 462, 466, 1094
Chin F Y L 555, 1042
Chisholm G 473, 476, 1034
Chiu S N 516, 524, 1034
Chiu S S 297, 302, 1031
Choo Y K 325, 327, 1034
Chorafas D 319, 327, 1034
Chorley R J 43, 46–7, 569, 579, 974, 976, 1029,
    1034, 1054

Chou C-L 829, 844, 1096
Chou Y H 849, 856, 1034
Chow M H 881, 888, 1101
Chrien T G 456, 466, 1096
Chrisman N R 2, 35, 38, 58–9, 73, 80, 94, 101, 178,
    180, 188, 192, 203, 311, 316, 569, 579, 748, 756,
    956, 964, 1034–5, 1065, 1076, 1081
Christaller W 240, 248, 1035
Christensen T 74, 80, 1067
Christerson M 414, 410, 423–4, 1060
Christian E 683, 692, 1035
Chronoles M J 418, 424, 1069
Chui C K 140, 152, 483, 485, 490, 1035
Church M R 198, 203, 1033
Church R L 185, 235, 237, 293–303, 509, 710, 719,
    741, 827–8, 832–3, 841, 849, 852–4, 857, 946,
    1010, 1035, 1038, 1050, 1058, 1068
Churcher C 325, 327, 1035
Churcher M 325, 327, 1035
Chuvieco E 881, 887, 1035
Cignoni P 544, 552–3, 555, 1035
Cislaghi C 929–30, 936, 1031, 1035
Claassen E 396, 399, 1078
Clapham S B 633, 641, 1028
Clark G E 848–9, 857, 1044
Clark J A 452, 464, 1039
Clark J D 99, 102, 970, 976, 1035, 1060
Clark P J 241–2, 248, 662, 666, 1035, 1084
Clark S C 841–2, 1035
Clark W A V 241, 248, 1035
Clarke A L 108, 112, 121, 1035
Clarke D 710–11, 721, 1065
Clarke G P 45–6, 71, 80, 196, 203, 240, 268, 282, 293,
    334, 337–8, 343, 345, 572, 579–80, 585, 601, 612,
    620, 640, 707, 709–22, 748, 758, 762, 767, 792,
    795, 801, 805, 815–16, 820, 832–3, 841, 884, 892,
    928, 935, 1017, 1027, 1029–30, 1035, 1065, 1070
Clarke K C 2, 76, 79, 108, 110, 121, 1035, 1068–9,
    xxvii
Clarke M 45–6, 71, 240, 268, 282, 288, 291, 293,
    334, 337, 343, 585, 601, 612, 620, 640, 707,
    709–22, 748, 758, 762, 767, 792, 801, 805,
    815–16, 820, 832–3, 841, 884, 892, 928, 935,
    1017, 1030, 1035
Clarke R 762, 765, 1035
Clarkson K L 548, 555, 1035
Claval P 47, 1061
Clayton A M H 973–4, 976, 1035
Clayton M K 992, 997, 1100
Cleveland W S 221, 224–5, 232, 258, 262, 264–5,
    1024, 1028, 1034–6

Cliff A D 42, 46–7, 62, 69, 241, 248, 258, 260, 265, 284–5, 291–2, 925, 927, 936, 1036, 1054, 1061, 1100
Clifford E 948, 950, 1036
Clifford J 35, 37, 93, 100–2, 1036, 1060
Cline A K 483, 492, 1084
Cline N 904, 908, 1036
Clinton W J 782, 785, 1036
ClockWork Software 168, 172, 1036
Clogg R 945, 950, 1036
Coad P 414, 423, 1036
Cocks K D 970, 980, 1096
Codd E 374, 376, 384, 1036
Coelho J D 719, 722, 1099
Cofinas M 970, 978, 1076
Cogan C B 970, 979, 1092
Cohen R B 735, 743, 1036
Cohn A G 65, 69, 1036
Coiffet P 558, 563, 565, 1032
Cole G 112, 121, 1036
Cole R 544–7, 550, 554–5, 1036
Coleman A 954, 964, 1036
Coleman D J 4, 152, 271, 307, 310, 314, 317–29, 337, 345, 401, 564, 575, 625, 630, 726, 730, 748, 770, 779, 785, 803, 840, 851, 862, 865, 885, 1007, 1010, 1014, 1020, 1025, 1036, 1045
Coleman P R 970, 978, 1064
Collar N J 968, 976, 1036
Collier A 62–3, 69, 1036
Collins J 428, 436, 1058
Collins M 20, 663, 666, 769, 785, 1036
Collins S 928, 932, 936, 1036
Colomer J L 151–2, 1026
Coltelli M 110, 113, 123, 1065
Columbia Human Rights Law Review 759, 765, 1036
Comas D 87–8, 1071
Commission of the European Communities 760, 765, 1036
Congalton R G 196, 203, 992, 994, 1036, 1075
Connor P 824, 826, 1036
Connor S J 931–2, 938, 1093
Conquest J 323, 327, 1037
Conry T 892, 899, 1037
Conway M 563, 565, 1092
Cook D 254, 256, 258–9, 264–6, 1037, 1070, 1093
Cook P 984, 997, 1097
Cook S 353, 357, 414–15, 423, 1037
Cooksey D 984, 987–8, 997, 1099
Coombes M 954, 964, 1037
Coop R D 318, 327, 1034
Cooper A 853, 857, 1061

Cooper L 833, 842, 1037
Cooper R H 318, 328, 625, 630, 1045
Coopers and Lybrand 657, 660, 769, 772–3, 779, 785, 1037
Copeland G 91, 101, 1037
Coplien J O 420, 423, 1037
Coppin P 962, 964, 1029
Coppock J T 20, 55, 58–9, 767, 785, 802, 818, 845, 856, 953, 955, 964, 1037
Corbett J D (Plates 62 and 63) 984, 987, 994, 1037
Corbley K P 829, 841, 851–3, 857, 1023
Cordey R A 452, 464, 1026
Cormack R M 248, 1041
Cornin C 836, 843, 1067
Cortese A 840, 842, 1037
Costa-Cabral M 986, 994, 1037
Costanza R 99, 103, 1090
Costello B 322, 328, 1037
Cotting C C 817–18, 1037
Couclelis H 11, 23–4, 29–38, 70, 77–8, 81, 88, 99, 101, 204, 355, 723, 731, 968, 974, 1037, 1047, 1083, 1093
Coughlan J 534, 541, 1026
Coulson M R C 134, 154, 625, 629, 749, 755, 1031, 1037, 1085
Coulson R N 970, 979, 1087
Coulter M C 971, 977, 1057
Council of Europe 760, 765, 1037
Cousins S 972, 976–7, 979, 1031, 1054
Cova T J 78, 237, 578, 723, 742, 798, 828, 830, 845–58, 893, 946, 1038
Coward L P 968, 976, 1033
Cowen D J 6, 20, 50, 59, 411, 1038, 1070
Cox D R 220, 232, 1038
Cox N J 983, 995, 1063
Cox S J D 486, 490, 1038
Coyne R D 885, 887, 1038
Craft A W 44, 47, 272, 282, 930, 938, 1079–80
Craglia M 70, 204, 628–31, 785, 953, 965, 1012, 1020, 1032–3, 1038, 1047, 1071, 1083, 1098
Craig C S 719, 721, 1051
Craig P 247, 250, 255, 260, 265, 928, 937, 1056
Craig W J 320, 328, 1009, 1020, 1038
Crane M P 67, 69, 123–4, 491, 969, 977, 994, 997, 1039, 1052, 1059, 1087, 1092
Crane R 99, 101, 1053
Crang P 46, 1038
Crawford-Tilley J 841–2, 1038
Cressie N A C 209–10, 216, 220–1, 226, 231–2, 243–4, 248, 254–60, 265, 286, 291, 483–5, 487, 490, 1037–8, 1070, 1093

Crilly A J 124, 1097
Crist T O 971, 978, 1061
Crockett M 13, 20, 1038
Cromley R G 130, 152, 1038
Cross A 272, 282, 621, 630, 931, 938, 1079
Crossley D C 323, 328, 970, 978, 1038, 1076
Croswell P L 2, 362, 369, 626, 630, 802, 818, 1025, 1038
Crouch G 841, 1029
Croze H 928, 937, 1066
CRU/OPCS/GRO(S) 73, 79, 1038
CSDC 778, 785, 1038
Csuti B 968, 979, 1089
Cullinan V I 971, 976, 979, 1038, 1090
Cullingworth J B 953, 956, 964, 1038
Cullis B J 626, 628, 630, 1038
Culnan M J 763, 765, 1038
Cumming S G 968, 976, 1033
Cunningham R L 984–6, 993, 1028
Curran P J 452, 464, 1027, 1038
Current J R 298, 303, 1038
Curry D J 761, 765, 1038
Curry M R 50, 54–7, 59, 79, 237, 578, 585, 607, 647,
    651, 663, 714, 746, 755, 757–66, 772, 775, 839,
    962, 1010, 1038
Curry R B 987, 994
Curtis A 243, 249, 1048
Cushnie J 74, 79, 1038
Custer S G 982, 994, 1039
Cuthbertson J 728, 731, 1045
Cutter S L 847, 857, 1039
Cuzick J 928, 930, 932, 936–7, 1039, 1044
Cyert R M 603, 609, 1039
Czaplewski R L 216–17, 1048
D'Erchia F 968, 979, 1089
Daamen R A 212, 216, 1060
Dacey M F 241, 248, 285, 291, 1039
Dahab M 208, 217, 1100
Dair B 821, 826, 1039
Dale A 79, 1027
Dale P F 2, 193, 203, 314, 657, 751, 754–5, 769, 798,
    859–75, 919, 956–7, 964, 1039
Dale V H 970, 976, 980, 1039, 1095
Daly C 120–1, 1039
Dando L P 779, 785, 1039
Dangermond J 52, 795, 845, 857, 1039
Daniel D L 244, 250, 1073
Daniel R 682, 684, 692, 1098
Daniels J 353, 357, 414–16, 423, 1037
Daniels M J 816–18, 1037, 1073
Danielsen D K 829, 836, 843, 1075
Danielson E D 457, 464, 1041

Danko D 685, 692, 1039
Dansby B 603, 609, 1039
Dansby H B 644, 651, 1039, 1061
Danson F M 452–3, 464, 1039, 1060
Date C J 363, 369, 377, 384, 1039
Davey S M 970, 976, 978, 1039, 1074
David M 491, 1075
Davidson D A 203–4, 974, 976, 1039
Davies C 402, 410, 622, 630, 1039
Davies R L 715, 721, 1039
Davis B E 2, 1039
Davis E 98, 100–1, 1039
Davis F W 244, 250, 295, 301, 303, 454–5, 464, 667,
    675, 954, 964, 968–70, 976, 979, 1035, 1039,
    1050, 1073, 1089, 1091
Davis J E 933, 936, 1024
Davis J R 1000, 1007, 1023
Davis L S 544, 547, 549, 556, 1093
Davis M R 984, 992, 994, 1045
Davis S L 194, 204, 1039
Davis W S 983, 996, 1078
Dawe P 323, 328, 1039
Dawes W 117, 121, 537–8, 542, 1040, 1090
Day A 315–16, 1040
Day T 109, 121, 1040
De Floriani L 107, 132, 152, 200, 204, 477, 479,
    482–3, 543–57, 562, 974, 976, 1032, 1040, 1069
De Ploey J 983, 994, 1040
De Roo A P J 210, 216, 988, 994, 1040
Deal T E 626, 630, 1040
Deane M 932, 938, 1068
Decker K 496–7, 502, 1030
Deering D W 453, 464, 1023
DeFanti T A 158, 173, 1072
DeGloria S D 984, 990–1, 996, 1068, 1081
Deichmann U 77, 80, 487, 491, 984, 994, 1040, 1052
Delanny R 516, 525, 1066, 1082
Delcourt H R 972, 976, 1040
Delcourt P A 972, 976, 1040
DeLotto J S 129, 152, 1033
Demazeau Y 138, 152, 1026
DeMers M N 2, 1040, xxvii
Demetrius-Kleanthis D 167, 173, 1090
Dennis P 970, 979, 1090
Densham P J 163, 172, 243, 249, 298, 301–3,
    315–16, 716, 721, 723, 731, 830–3, 842, 881–2,
    885, 887, 935–6, 1025, 1028, 1040, 1048
Department of Land Affairs 919, 924, 1040
Department of National Health and Population
    Development 914, 924, 1040
Department of the Environment 955–7, 959, 962,
    964, 1040

Derose R C 534, 541, 1043
DeRose T 562, 565, 1068
Deschamps P Y 453, 461, 464–5, 1041, 1086
Desmarais R 484, 491, 1055
Desmet P J J 984, 986–7, 994, 1041
Despotakis V K 881, 887, 1041
Deursen W van 195, 203, 1032
Deutsch C V 211, 216, 483–4, 490, 1041
Deux O 374, 384, 1041
Devogele T 129–30, 152, 1041
Dewdney J G 200, 204, 1041
Dewitt B A 208, 216, 1091
DeWitt D 403, 410, 1026
DeWitt W J 841–2, 1041
Dhinwa P S 499, 502, 1083
Dias A E 560, 565, 1041
Diaz B M 507–9, 524, 1028, 1041
DiBiase D 158, 167, 170–2, 225, 232, 1041
Dickinson G C 954, 956, 964, 1041
Dickinson H J 602, 609, 618, 620, 1041
Didier M 772, 786, 1041
Diepen C A van 212, 216, 1042
Dietrich J 1003, 1007, 1093
Dietrich W E 529, 532, 542, 1074
Difani C 853, 857, 1041
Diggle P J 242–3, 248, 285, 291, 926–32, 936–7,
     1029, 1041, 1050
Digital Geographic Information Working Group
     349, 356, 358, 1041
Dikau R 127, 131, 152, 539–41, 1041
Dillard A 125, 153, 1041
Diner D J 453, 455, 457, 464–5, 1041, 1063
Ding Y 263, 265, 286, 291, 509, 524, 1042
Diogo P (Plate 39) 565, 1051
Dittnick K 403, 410, 1026
Dively D D 601–7, 610, 1101
Dixon R G 942–3, 945–6, 950, 1042
Dixon R K 983, 997, 1095
Dixon T H 108–10, 122, 1042
Dobkin D 554–5, 1029
Dobrindt K T G 132, 152, 544, 547, 551, 553, 555,
     1029–30, 1042
Dobson J E 294, 297, 303, 970, 976, 1042
Docherty M 841–2, 1042
Dockter K 452, 464, 1030
Dodds J A 525, 1085
Dodson R T 257, 261–2, 264, 286, 291, 753, 755, 1042
Dolan T 928, 937, 1066
Dolin R 574, 580, 1090
Dolton L 853, 857, 1041
Domingue J O 107, 113, 116, 122, 531, 541, 1061

Donnay J P 416, 424, 1080
Donnelly K P 242, 248, 1042
Doralswamy P 984, 997, 1097
Dorling D 78–9, 98–9, 101, 929, 936, 1042
Douglas D H 112, 122, 129, 153, 396, 399, 530–1,
     541–2, 752, 755, 1042, 1081
Douglas M 603, 609, 1042
Douglas R 770, 786, 1042
Dougligeris C 852, 858, 1101
Douven W 884, 888, 1042
Dow K 848–9, 857, 1044
Dowd P A 484, 490, 1025
Dowers S 68, 70, 95, 102, 327–8, 802, 818, 1051, 1083
Dowling T I 113, 117, 121–2, 533, 541, 1059
Dowman I J 105, 108–9, 127, 160, 182, 311, 341,
     425, 427, 434, 437–52, 458, 477, 593–4, 663–8,
     804, 806, 830, 880, 956, 971, 1042
Downs A 603, 609, 1042
Downs P W 530, 542, 1083
Dozier J 531, 541, 969, 976, 1039, 1071
Drabek T E 845, 857, 1042, 1051
Draeger W C 760, 780–2, 786, 1042
Draper D 841, 1029
Draper G 282, 1079
Draper S W 732, 1076
Dresner Z 525, 833, 842, 1042, 1078
Drummond S T 991, 997, 1092
Drummond W J 887–8, 1033, 1042
Duan N 68, 70, 96, 102, 1081
Dubayah R C 244, 250, 1073
Dubbink D 728, 731, 1042
Dubin R 245, 248, 1042
Dubois D 496, 502, 984, 998, 1042, 1100
Dubrule O 487, 490, 1042
Duchon J 484–5, 490, 1042
Dueker K J 835, 843, 1076
Duggin M J 955, 964, 1042
Dumais S 410–12, 1023, 1056
Dunbar W S 841–2, 1042
Duncan J S 42, 46, 1027
Dungan J 167, 173, 534, 541, 1026, 1090
Dunlop C 621, 630, 1043, 1046, 1087
Dunn C E 520, 524, 852–4, 857, 931, 935–6, 1031,
     1043
Dunn J E 970, 976, 1035
Dunn R 954, 956–7, 964, 1043
Durlach N 558, 565, 1043
Dutra L 110, 113, 123, 1065
Dutton B 538, 542, 1101
Dutton G 3, 26, 65, 105, 110, 125–55, 181, 188, 196,
     246, 396, 471, 483, 537, 658, 776, 810, 886, 891,
     988, 1002, 1010, 1043, 1090

Dyke P T 988, 997, 1099
Dykes J 78, 80, 158, 167, 172, 1043, 1047, 1100
Dymon U J 851, 853, 857, 1043
Dymond J R 534, 541, 1043
Dyn N 487, 490, 544, 556, 1043
Earnshaw R A 124, 555, 1070, 1097
Eason K D 621, 625, 627, 630, 1043
Eastman J R 197, 245, 248, 312, 405, 478, 493–502,
      545, 636, 749, 756, 831, 835, 841, 881, 888, 923,
      970, 987, 992, 994, 1043
Ebner H 112, 121–2, 447, 450, 1026, 1042–3
Echoud M van 775–7, 786, 1043
Eckert J W 849, 858, 1092
Eddy F 353, 358, 363, 369, 380, 382, 384, 1087
Edelsbrunner H 544, 547–8, 555–6, 1034, 1043
Edmonds E A 725, 731, 1061
Edwards D 1007, 1034
Edwards G 195, 204, 463–4, 520, 523, 577, 579,
      1043–4
Edwards R 930, 936, 1039
Edwards T C 968, 971, 978–9, 1058, 1089
Edwards T M 561, 565, 1044
Egbert S L 167, 170, 172–3, 1033, 1044, 1090
Egenhofer M J 31, 38, 85–8, 312, 325, 328, 337, 343,
      352, 372, 374, 383–4, 401–12, 420, 423, 561,
      588, 725, 730, 928, 1007, 1010, 1032, 1039,
      1044, 1047, 1062–4, 1071, 1076, 1089, 1098
Eglese R 852–4, 858, 1089
Ehlers M 463, 1044
Ehlschlaeger C 523, 541, 1044
Ehrlich R 462, 464, 1029
Ehrliholzer R 151, 153, 155, 1044, 1098
Einstein A 32, 38, 1044
Eiselt H A 297, 303, 1044
Eknert B 971, 979, 1085
Ekstrund S 120, 122, 1044
Elachi C 461, 466, 1101
Elliot R L 984, 988, 995, 997, 1050, 1087
Elliot S 850, 857, 1044
Elliot W 321, 328, 1044
Elliott C 714, 721, 1044
Elliott C C H 971, 980, 1097
Elliott C J 877, 887, 1027
Elliott P 927–8, 932, 937, 1044
Elmes G A 639, 641, 970, 978, 1044, 1067
Elshaw Thrall S 5, 10, 73, 78, 162, 168, 254, 307–8,
      310, 312, 322–3, 331–45, 360–1, 402, 407, 570,
      572, 614, 620, 666, 694, 708–9, 714, 730, 750,
      771, 792–3, 802, 815–16, 827, 829, 840, 862,
      884, 903, 993, 1010, 1016, 1044, 1094
Elston D A 970, 976, 1032

Elzinga J 396, 399, 1044
Emani S 845, 848–9, 857, 1044
Emanuel W R 982, 989, 996, 998, 1082, 1101
Emmi P C 849, 857, 1044
Engel B A 483, 486, 491, 988, 994, 1072
Engel H 109, 123, 1044, 1063
England J 953, 964, 1045
Englebart D 325, 328, 1045
English R 928, 932, 937, 1044
Englund E J 225, 227, 232, 484, 490, 1045
Engman E T 457, 462, 465, 1055
Enmark H T 456, 466, 1096
ENV 704–5, 1045
Epstein D 554–5, 1029
Epstein E 638, 641, 1045
Escobar D E 984, 992, 994, 1045
Espa G 243, 248, 1025
ESRI 151, 153, 256, 262, 265, 749, 756, 835, 842, 1045
Estes J E 2–3, 20, 92, 339, 345, 440, 443, 455, 458,
      463, 584, 594, 653, 656, 661, 663, 665–75, 703,
      705, 770, 780–1, 806, 849, 857, 896, 931, 954,
      968, 979, 983, 1010, 1013, 1045, 1091
Eswaran K 374, 384, 1026
EUROGI 704–5, 1045
European Commission 356, 358, 770, 786, 1045
European Conference of Ministers of Transport
      837, 842, 1045
European Parliament and Council Directive 776,
      786, 1045
European Space Agency 674–5, 1045
Evans A S 926, 937, 1062
Evans B J 892, 899, 1065
Evans B M 110, 113, 123, 1088
Evans F C 241–2, 248, 1035
Evans P A 199, 204, 1083
Everitt J H 984, 992, 994, 1045
Ezigbalike I 318, 328, 625, 630, 1045
Faber B G 728, 731, 1045
Fabos J 638, 641, 1066
Fahrig L 970–1, 976, 979, 1045, 1093
Fairfax A E 949–50, 1054
Fairfield J 986, 994, 1045
Faith D P 968, 980, 1097
Falcidieno B 132, 152, 200, 204, 544, 547–8, 555, 1040
Faloutsos C 388–91, 395, 399, 1045, 1062
FAO/UNESCO 198–9, 204, 1045
Farley J A 257, 265, 1045
Farnes P 982, 994, 1039
Farrell D M 944, 950, 1045
Faulhaber G R 607, 609, 1045
Faulkenberry G D 244, 249, 1045

Faust N L 561–5, 885, 888, 1045, 1063, 1067
Fayyad U 409, 411, 1046
Feast K 923–4, 1076
Federa K 1000, 1007, 1046
Fegeas R G 83, 88, 184, 188, 1046
Feigenbaum E 607, 609, 1046
Fekete G 131, 141, 152, 561, 565, 1046
Felleman J P 544, 553, 556, 1046
Fels J E 539, 541, 1046
Ferguson R B 992, 995, 1053
Ferrari D 326, 328, 1046
Ferraz de Abreu P 726, 731, 1046
Ferreira F C 725, 731, 1048
Fetterer A 325, 328, 1046
Feuchtwanger M 637, 641, 1073
Feyerabend P K 62, 65, 69, 1046
FGDC 184, 188, 220, 228, 232, 574, 579, 656, 666,
    679, 683, 692, 782–3, 786, 1046
Fiala F 555–6, 1031, 1083
Field B C 604–6, 609, 1046
FieldWorks Inc. 321, 328, 1046
Fifer S 841, 843, 1071
FIG 859, 861, 875, 1046
Filella I 452, 464, 1046
Fincham R J 160, 767, 909, 913–24, 1046, 1064,
    1072, 1082
Fingleton B 244, 251, 258, 260, 266, 286, 292, 1097
Finke P A 208, 210, 216, 984, 994, 1046
Finkel A 555, 1082
Fischer M M 57, 77, 99, 147, 235–6, 241–4, 247–9,
    262, 264, 272, 282–92, 426, 462, 509, 570,
    574–5, 887, 928, 955, 970, 984, 1010, 1024,
    1026, 1046, 1052–3, 1066, 1079
Fisher P F 57, 67, 75, 78, 80, 83, 138, 153, 158, 164,
    167, 176–8, 191–205, 208–9, 216, 219–22, 226,
    232, 239, 354–5, 383, 406, 437, 449, 451, 462,
    475, 482, 492, 496, 529, 538, 577, 606, 633, 640,
    749, 756, 856, 917, 927, 936, 974–6, 989, 1010,
    1026–7, 1041, 1047, 1097, 1100
Fishkin K 409, 412, 1092
Fites D V 710–11, 721, 1047
Fitzgibbon M 852, 857, 1047
Flaatten P O 362–3, 369, 1047
Flaherty D H 758, 760, 765, 1047
Flamm R O 973, 977, 1047
Flanagan C 806, 818, 1047
Flanagan N 806, 818, 1047
Flasse S 931–2, 938, 1093
Flather C H 970, 972, 974, 977, 1047, 1057
Fletcher D 829, 842, 1047
Fletcher G A 849, 852, 1045

Fletcher R 274, 282, 1047
Fleury B 749–50, 756, 1075
Florax R J G M 245, 248–50, 1024, 1042, 1047,
    1050, 1055, 1062, 1066
Florence J 402, 411, 1047
Florencio C A 916, 924, 1047
Flowerdew R 77, 80, 243, 249, 283, 291, 855, 857,
    1047, 1053
Flynn L J 763, 765, 1047
Fochler-Hauke G 969, 977, 1047
Fogel D N 484, 490, 1047
Foley T 486, 490, 1047
Folke C 973, 979, 1081
Folmer H 245, 249, 1047
Folse L J 970, 977, 1048, 1087
Folving S 975, 1026
Fonseca A 725, 731, 885, 888, 1033, 1048
Food and Agriculture Organisation 983, 994, 1048
Foody G M 194, 197, 202, 204, 462, 464, 1027, 1048
Foote K 753, 756, 1048
Forbes H 934, 937, 1048
Ford V G 457, 464, 1041
Forer P 7, 49, 66, 563, 585, 599, 620, 745, 747–56,
    792, 969, 1048
Foresman T W 2–3, 20, 841–2, 849, 857, 1038, 1045,
    1048, 1071, 1084
Forester J 727, 731, 1048
Forier F 208, 216, 1048
Forman R T T 970, 977–8, 1048, 1073
Formentini U 37–8, 525, 1037, 1051, 1090
Fornaro G 100, 113, 123, 1065
Forrester Research Inc. 326, 328, 1048
Forsberg B R 244, 250, 1073
Forshaw M R B 457, 464, 1048
Forsyth G G 160, 767, 909, 913–24
Fortin J P 244, 250, 1033, 1070
Fortin M 837, 841, 1020
Foster C D 583, 585–6, 601, 609, 770, 773, 786, 1048
Foster G R 986, 996, 1084
Foster M J 1020, 1069
Foster S A 245, 249, 1048
Fotheringham A S 8, 20, 243, 247, 249, 254, 257,
    263–5, 272–3, 282–3, 291–2, 316, 228, 345, 509,
    524–5, 567, 579, 936–7, 1026, 1028, 1042, 1048,
    1050, 1055–6, 1079
Foucault M 62, 65, 69, 1048
Fougere P F 525, 1085
Foulger F 711, 721, 1030
Foulon F 825–6, 1048
Fournier R A 244, 250, 1070
Foussereau X 984, 995, 1048

Fowler M 420, 423, 1048
Fowlkes E B 254, 264, 1024
Fox C J 953, 964, 1048
Fox G 544, 546, 549, 556, 1073
Fraigniaud P 282, 1095
Franceschetti G 110, 113, 123, 1065
Frank A U 5, 20, 33–8, 66–9, 81–5, 88, 97, 101, 131,
        153–5, 186, 188, 197, 203, 247, 249, 383–4, 394,
        399–403, 410–12, 420, 423, 525, 529, 541, 555,
        748, 756, 1032, 1037, 1044, 1048–51, 1057, 1062,
        1064, 1071, 1081, 1090, 1094, 1097–8, 1101
Frank I 282, 1079
Frank S 901, 908, 1049
Frank T D 971, 977, 1049
Franke R 111, 122, 481–6, 490–1, 1049
Franklin J 242, 249, 971, 977, 1049
Franklin S E 113, 122, 1051
Franz E H 986, 995, 1086
Franzosa R D 31, 38, 1044
Frawley J 109–10, 122, 1055
Frederiksen P 112, 122, 1049
Freeman G T 986, 995, 1049
French S P 877, 884, 887–8, 1049
Freundschuh S M 87–8, 1071
Frew J 531, 541, 1071
Frey A E 42, 46, 285, 292, 1054
Friedl M A 244, 250, 1073
Friedman J H 386, 399, 1029
Friedrich C J 32, 38, 1049
Friel C 852, 857, 1049
Fritsch E 140, 144, 153–4, 1049, 1082
Fritz L W 459, 464, 1049
Fry C 298, 554, 721, 791, 798, 819–26, 892
Fuchs H 387, 399, 1049
Fujimoto R M 94, 101, 1049
Fukumura T 530, 534, 542, 1094
Fulcher C 852, 857, 1049
Full W 462, 464, 1029
Fuller N 822, 826, 1049
Fuller R M 955, 964, 1032, 1049
Furse E 138–42, 152, 1032
Gadia S K 93, 100–2, 1060
Gaebler T 770, 786, 1080
Galbraith J M 984, 991, 996, 1068
Gale F 205, 1101
Gale S H 251, 292, 934, 937, 1056, 1094
Gallant J C 24–6, 44, 66, 83, 105–24, 168, 186, 207,
        226, 338, 426, 482–7, 505, 528, 533, 538, 544,
        557, 562, 982–4, 986–7, 995, 997, 1049, 1059,
        1074, 1099
Gallion C 674–5, 1053

Gamma E 420, 423, 1049
Ganter J 99, 101, 1049
Gao P 531, 534, 542, 1090
Garbrecht J 109, 122, 531, 534, 541, 1049, 1071
Gardels K 4, 10, 308, 322, 347–59, 369, 382, 402,
        410, 415, 564, 576, 584, 681, 683, 694, 698, 725,
        802, 850, 865, 1010
Gardingen J van 454, 457, 464, 1027
Gardner R H 103, 970, 978–9, 1064, 1077, 1082,
        1090, 1095
Garey M R 298, 303, 1050
Gargantini I 68–9, 1050
Garman S L 967, 977, 1055
Garner B 948, 950, 1058
Garnett S 934, 937, 1050
Garoui A 244, 249, 1045
Garret W F 559, 565, 1091
Gartner C 771, 787, 1086
Garza J 374, 384, 1063
Gaston G G 850, 858, 1086
Gaston K J 968, 977, 1050
Gatrell A C 5, 32, 38, 71, 75, 77–80, 237, 240, 243,
        247, 249, 254, 264, 567, 569, 579, 709, 794, 845,
        854, 857, 909, 919, 925–38, 1026, 1031, 1041,
        1050, 1055, 1064, 1068
Gault I 625, 630, 1050
Gaus M P 849, 858, 1063
Gauthier H L 831, 833, 841, 843, 1093
Gaver W 560, 565, 1033
Geary R 241, 249, 1050
Geiger C 928, 935, 937, 1050
Geleta S 984, 988, 995, 997, 1050, 1087
Genard M 975, 978, 1066
Genderen J L van 181, 188, 1050
GeoData Institute 749, 756, 1050
GEO-EAS 244, 249, 1050
Geographic Data BC 350, 358, 1050
GEOHUB 353, 358, 1050
George F 288, 291, 719, 721, 1030
George J 945, 950, 1055
Gerdes D P (Plates 10–12) 99, 102, 221, 223, 225,
        233, 482, 485–6, 489, 491, 1074
Gerrard R A 301, 303, 1050
Gershon N 219, 232, 1050
Gerth H H 610, 1098
Gessler P E 108, 113–14, 120, 122, 484–5, 491, 538,
        542, 984–5, 995, 1049–50, 1059, 1074
Getis A 8, 20, 77, 235, 239–51, 254, 257–62, 264,
        267, 284–7, 291–2, 404, 426, 509, 569, 707, 710,
        928, 936, 970–2, 977, 992, 1010, 1018, 1024,
        1031, 1046, 1050–2, 1066, 1075, 1080

Gholz H L 454, 464, 1038
Ghosh A 719, 721, 1051
Ghosh R 454, 464, 1051
Ghosh S K 656, 666, 1031
GI2000 706, 1051
Giaoutzi M 881, 887, 1041
Gibbons G 837, 842, 1051
Giddens A 62, 69, 627, 630, 1051
Giffin T L C 163, 172, 1051
Gilat G 484, 486, 492, 1093
Gilb T 362, 369, 1051
Gilbert C 11, 26, 66, 105–8, 321, 333, 349, 425–9,
    445, 467–76, 615, 650, 782, 804, 806, 830, 897,
    981, 990, 1010, 1051
Gilbert D 45–6, 50, 59, 1051
Giles P T 113, 122, 1051
Gillies C F 901, 908, 1064
Gittings B M 327–8, 802, 818, 1051
Glaister S 682, 684, 692, 1066–7, 1089, 1092
Glanz J 986, 995, 1051
Glasmeier A K 46, 1051
Gleadow A J W 486, 490, 1038
Glendinning S 67, 69, 123–4, 491, 969, 977, 994,
    997, 1039, 1052, 1059, 1087, 1092
Gless J T 494, 502, 1033
Gluck M 901, 908, 1091
Gober P A 39, 44, 46, 1051
Godby J 682, 684, 692, 1098
Goderya F 208, 217, 1100
Godron M 970, 977, 1048
Godschalk D R 847, 857, 1051
Goebel J J 983, 995, 1062
Goel N S 453, 464, 1051
Goetz A F H 456, 465–6, 1095
Goetz S J 452, 457, 465, 1051, 1055
Golay F 416, 423, 1051
Gold C M 141, 155, 518, 520, 525, 752, 756, 1051, 1101
Gold J R 752, 756, 1051
Goldberg J 892, 899, 1037
Golledge R 88, 412, 721, 887, 1037, 1040, 1098
Gomes A L 725, 731, 1033
Gomez D 550, 556, 1051
Gonçalves P (Plate 39) 561, 563–5, 1051, 1076
Gong P 244, 249, 1051, 1097
Good I J 254, 265, 1051
Goodchild M F (Plate 9) 1–20, 25, 27, 34, 36, 38, 45,
    49, 52–6, 58–60, 62, 67, 69, 77–80, 88, 107,
    112–13, 122–4, 127, 131, 141, 153–4, 177–80,
    185–6, 188, 202–4, 216, 221–3, 231–2, 226–7, 236,
    244–51, 262, 265, 268, 282–7, 292, 295, 298, 303,
    308, 311, 316, 327–8, 334, 355, 357, 384, 388–90,
    399, 405, 411, 425, 437, 464, 475, 479, 487,
    489–92, 509, 525, 537, 542, 563, 567–80, 601, 609,
    615, 635, 637, 641, 677, 707, 714, 721, 739, 747–9,
    751–2, 756, 769, 785–6, 792–3, 795, 818, 828, 833,
    842–4, 856–8, 875, 880, 887, 897, 902, 906, 926–8,
    935–8, 964–5, 969, 971, 977–8, 994, 997, 1000,
    1007, 1009–21, 1032, 1035, 1037, 1039–40, 1043,
    1045, 1047, 1049–53, 1056, 1058–61, 1063, 1066,
    1070, 1075–80, 1084, 1087, 1091–100
Goodman J M 46, 1051
Goodman P S 626, 631, 1091
Goodwin C W H 829, 837, 842, 1052
Gopal S 36, 38, 80, 152, 216, 244, 249, 251, 283,
    288–92, 577, 579, 641, 1043, 1046, 1052–3,
    1079, 1094
Goper P 46, 1052
Gordon S R 829, 842, 1052, 1078
Gordon T 326, 328, 1052
Gore A 770, 782, 786, 1052
Gorr W L 245, 249, 1048
Goss J 50, 56, 59, 1052
Gotelli N J 970, 977, 1053
Gottlob G 383–4, 1034
Gotway C A 992, 995, 1053
Gould M D 87–8, 409, 411, 1071
Gould P R 47, 88, 92, 99, 101, 1037, 1053, 1061, 1100
Goulstone-Grunland A G 852, 857, 1053
Gouveia C 725, 731, 855, 858, 1048
Gouveia M C 726, 731, 1048, 1053
Govers G 984, 986–7, 994, 1041
Goyal B 325–8, 1046
Graff L H 118, 122, 899, 1053
Graham L A 674–5, 1053
Graham S 56, 59, 310, 315–16, 1053
Grandfield A W 388, 399, 1052
Grandi G de 453, 456, 1053
Granger K 849, 857, 1053
Grant D M 660, 666, 768, 778, 786, 1074
Grant E D 387, 399, 1049
Grant L 454, 465, 1095
Grant W E 970, 977, 1048, 1087
Grasman J 216, 1060
Grassberger P 526, 1085
Gratton D J 244, 250, 1070
Gray J 402, 412, 1026, 1092
Grayson R B 105–8, 112, 115–17, 123, 212, 216,
    1053, 1074
Grayson T H 828, 842, 1053
Graziani G 323, 328, 1067
Green D R 795, 818, 836, 843, 972, 976–7, 979,
    1031, 1054, 1067, 1076, 1093
Green F 265–6, 1045
Green M 77, 80, 243, 249, 283, 291, 855, 857,
    1047, 1053

Green R E 212, 216, 1067
Green R O 456, 466, 1096
Greenlee D 463–4, 1044
Gregoire T G 208, 216, 1062
Gregory D A 42–7, 50, 52, 59, 65, 69, 264, 1053, 1061
Gregory K J 42, 46, 1053
Gregory S 43, 47, 1061
Grehan J R 968, 977, 1053
Grelot J-P 764–5, 769, 776, 786, 1053
Greysukh V L 530, 541, 1053
Griffin C 544, 553, 556, 1046
Griffith D A 208, 216, 242–4, 248–9, 284–6, 291–2, 509–10, 519, 525, 852, 857, 1024, 1053–5
Griffiths P 374, 384, 1026
Griggs R H 988, 997, 1087
Grimshaw D J 602–3, 608–9, 795, 1054
Grimson R 928, 937, 1060
Griswold v. Connecticut 763, 765, 1054
Grofman B 950–1, 1042, 1067, 1076
Gronning M 931, 938, 1085
Groof H de 453, 465, 1053
Groom G B 453, 464, 955, 964, 1026, 1049
Gross M 638, 641, 1066
Grossman R 554–5, 1029
Grothe M 884, 888, 1042
Groves C 968, 979, 1089
Grubb T G 970, 977, 1054
Grün A 108, 112, 121, 1035
Grunbaum B 503–4, 506, 525, 1024, 1054
Grünreich D 127, 145–6, 151, 153, 1054
GS+ 244, 250, 1054
Guber A L 852, 857, 1054
Gucinski H 983, 997, 1095
Guercio R 530, 541, 1054
Guibas L J 544, 547–8, 556, 1043, 1054
Gunnink J L 256, 265, 1054, 1056
Günther O 412, 420, 423, 1092
Guptill S C 95, 100–1, 158, 173, 184, 188–9, 219, 226, 356–7, 427, 475, 584, 615, 633–4, 653, 677–92, 739, 783, 1031, 1054, 1061, 1085, 1088
Gurak L J 763, 765, 1054
Gustafson E J 970, 977, 1054, 1057
Gustafson G 930, 937, 1054
Güting R 403, 411, 1054
Guttman A 394–7, 399, 1054
Guy C 717, 721, 1054
Guyot G 452, 464, 1027
Guzman A 550, 556, 1051
Guzzetti F 856–8, 1037, 1062, 1065
Haarslev V 406, 411, 1054

Haas L M 374, 382, 384, 1054
Habermas J 62, 65, 69, 627, 630, 1054
Hacking I 62, 69, 1054
Hagen H 491, 1049
Hagens W W 949–50, 1054
Hägerstrand T 44, 46, 91, 99, 101, 926, 1054
Haggett P 42–3, 46–7, 285, 292, 569, 579, 925, 927, 936, 973–4, 977, 1034, 1036, 1054
Haines-Young R 42, 47, 972, 976–7, 979, 1031, 1054, 1093
Haining R P 208, 216, 226, 232, 244, 249–50, 257–8, 260, 262, 265, 284–7, 292, 509–10, 525, 567, 569, 579, 793, 795, 928, 932, 937, 1000, 1007, 1052, 1054–5
Haithcoate T W 985, 995, 1055
Hakala J 684, 692, 1055
Hake G 134, 153, 1055
Hakimi L S 833, 842, 1055
Halcrow Fox/SERRL 962, 964, 1055
Hall A D 973, 977, 1055
Hall D 462, 465, 1087
Hall F G 452, 456, 462, 465, 1055, 1075
Hall G B 197, 202–3, 205, 749, 756, 1055, 1097
Hall P 957, 965, 1055
Halliday S L 984, 995, 1055
Halverson J A 970, 978, 1067
Hamid A 880, 887, 1027
Hamilton A C 329, 1089
Hamlett J M 990, 997, 1096
Hammer R D 985, 995, 1055
Hammersley J M 211, 216, 1055
Hammitt R 473, 476, 1087
Hamre R H 216–17, 1048
Han D 457, 464, 1025
Han J 264–5, 1076
Han S Y 877, 888, 1055
Hancock J R 221, 232, 1055
Hand G 945, 950, 1055
Handscomb D C 211, 216, 1055
Handy C B 583, 626, 630, 1055
Hanley T A 970, 977, 1057
Hanocq J F 452, 465, 1060
Hansen A J 525, 967, 977, 1055, 1085
Hansen E G 452, 466, 1096
Hanski I 972, 977, 1055
Hanson S 841–3, 1055, 1058, 1076, 1080
Hansson L 970, 972, 1055
Hao L V 387, 399, 1071
Harbison-Briggs K 148, 154, 1072
Harche F 719, 721, 1051
Harder R 484, 491, 1055

Hardie P 957, 965, 1055
Harding D J 109–10, 122, 1055
Hardy E F 956, 964, 1024
Hardy P 151, 153, 1055
Hardy R L 484, 486, 491, 1055
Hargitai P 182, 189, 1096
Hargrove W 486, 491, 1055
Harley J B 186–8, 1055
Harman H H 569, 579, 1055
Harmsworth G R 534, 541, 1043
Harp E 853, 857, 1055
Harris B 723, 725, 731, 883, 888, 1055–6
Harris R 768, 780, 786, 1013, 1020, 1056
Harris T M 47, 50, 54, 57–60, 1056, 1079, 1098
Harrison A 954, 956–7, 964, 1043
Harrison C M 853, 857, 1056
Harrison J 968, 977, 1056
Harslem E 408, 410, 412, 1090
Hartgen D T 828, 836, 843, 1065
Hartman R 1013, 1020, 1056
Harts J 217, 938, 1068, 1098
Hartshorne R 39–40, 47, 572, 579, 968, 977, 1056
Harvey D 31, 38, 43, 47, 241, 250, 1056
Harvey R 974, 977, 1056
Harvison P E 942, 950, 1056
Haskell A 457, 464, 1048
Haslett J 167, 172, 243, 247, 250, 255–7, 260, 264–5,
    928, 937, 1031, 1056
Hatton T J 117, 124, 1096
Haunold P 402, 411, 1056
Haussler D 409, 411, 1046
Havens M W 984–6, 993, 1028
Hawke D 327–8, 1056
Hawkins G 255–6, 266, 1095
Haynes D 885, 888, 1038
Haynes K 337, 345, 1056
Haynes R M 934, 937, 1056
Hayward S 171–2, 1056
Hazelhoff L 210, 216, 256, 265, 988, 994, 1040, 1056
Hazelton B 94, 102, 1082
Hazelton N W J 94–5, 101, 1056
Healey P 956, 965, 1056
Healey R G 68, 70, 153–4, 172, 188, 204, 327–8,
    377, 384, 802, 818, 828, 842, 1043, 1051, 1056,
    1063–5, 1089, 1096
Health Systems Trust 914, 924, 1056
Hearn D W 396, 399, 1044
Hearnshaw H M 79–80, 158, 172, 204, 232, 383–4,
    402, 411, 417, 424, 600, 936, 1033, 1042, 1044,
    1047, 1049–50, 1052, 1056, 1073, 1100
Hecht-Nielsen R 288, 292, 1056

Heikkila E J 838, 842, 1056
Heimbach R 544, 546, 549, 556, 1073
Heinselman R C 318, 327, 1034
Heipke C 447, 450, 1042
Heit M 641, 1083
Held G D 374, 384, 1056
Heller M 64, 69, 105–10, 122, 124, 147, 153, 483,
    487, 492, 1056, 1098
Helm R 420, 423, 1049
Helstrom C W 210, 216, 1057
Hendersen J 749, 755, 1032
Hendersen-Sellers A 753, 756, 1072
Henein K 970, 979, 1093
Hennell T 933, 937, 1057
Henry M 242, 248, 263–4, 1027
Hensley S 109–10, 124, 1101
Hepner G F 849, 858, 1068
Hepple L W 245, 250, 1057
Hergert G W 992, 995, 1053
Herman M 461, 464, 1041
Herr A M 971, 977, 1057
Herring J 66, 69, 85–8, 401, 403, 411, 1044, 1057
Herring M 456, 465, 1051
Hershelberg J 548, 555, 1057
Herskovits A 83, 88, 1057
Herzen B 551, 556, 1057
Hess G 970, 977, 1057
Hesse W J 866, 875, 1057
Hester v. United States 759, 765, 1057
Heuvelink G B M 105, 164, 176, 196, 199, 204,
    207–17, 219, 227, 231–2, 298, 303, 437, 449,
    481–2, 500, 529, 640, 989, 1010, 1057, 1098
Hewitson B 99, 101, 1053
Hewitt M J III 185, 188, 1092
Heywood D I 721, 756, 978, 1044, 1052, 1058, 1073,
    1095
Hickman C E 829, 842, 1057
Hick S 827, 844, 1097
Higgs G 74, 80, 577–8, 580, 880, 888, 1068, 1071
Hill M O 970, 977, 1057
Hillary N M 984, 994, 1063
Hillis W D 268, 282, 1057
Hills M 927, 937, 1044
Hillsman E L 298, 300–3, 1057, 1086
Hinterberger H 393, 399, 1076
Hinton J C 120, 122, 463, 465, 1057
Hirota G 559, 565, 1091
Hirschfield A F G 933, 936–7, 1031, 1057
Hirschheim R A 622, 627–8, 630, 1057, 1069
Hitchner L E 564–5, 1057
Hjalmars U 930, 937, 1057

HM Land Registry 860, 875, 1057
Hoare A G 47, 1061, 1100
Hobbs F 108, 122, 1057
Hobbs N T 970, 977, 1057
Hobbs R J 967, 973, 979, 1057, 1076
Hodder I 248, 1042
Hodges L F 561–5, 1063, 1067
Hodgson M E 849, 857, 971, 977, 1057
Hodgson M J 934, 937, 1057
Hoekstra T W 972, 977, 1057
Hoetmer G J 845, 857, 1042, 1051
Hof R 840, 842, 1037
Hoffman F 486, 491, 1055
Hofierka J 111, 123, 482, 485, 488, 491, 986, 996, 1074
Hofman H 255–6, 266, 1095
Hofmann-Wellenhof B 428, 436, 1058
Hofstadter D R 309, 316, 1058
Hogweg P 99, 102, 1058
Holdridge L R 982, 995, 1058
Holland A 837, 842, 1058
Holland J 973, 977, 1058
Holm T M 769, 780–2, 786, 1042
Holroyd F C 504, 507–9, 524, 1028, 1058
Holt D 243, 250–1, 1058, 1100
Holt-Jensen A 968–9, 973, 977–8, 1058
Homer C G 971, 978, 1058
Honeycutt D M 180, 188, 1058
Hooker C A 32, 38, 1058
Hoop S de 27, 1074
Hoover J R 985, 996, 1086
Hopkins L D 878, 888, 1058
Hopkins M S 539, 541, 1072
Horn M E T 198, 293, 832, 909–10, 939–51, 1058
Horne B van 972, 980, 1099
Hornsby A G 984, 995–6, 1048, 1058, 1076
Hornsby K 402, 411, 1047
Horowitz J L 834, 842, 1058
Horton C A 849, 857, 1044
Hossain H 888, 1063
Hössler R 112, 122, 1043
Houlding S 68–9, 1058
Hovland L E 457, 464, 1041
Howard D (Plate 13) 221–2, 225, 232, 1069
Howarth P J 244, 249, 1051, 1097
Howe A 714, 721, 1058
Howerton P 328, 1045
Hoyle B S 841–2, 1058
Hsie E Y 99, 101, 1025
Htun N 663, 666, 675, 767, 770, 786, 1058
Huber D L 295, 303, 1058
Hubert L J 241, 250, 1058

Hudak S 257, 261–4, 286, 291, 1025
Hudson D 637, 641, 1058
Hudson K 953, 964, 1045
Hudson R 954, 965, 1085
Huete A R 462, 565, 1087
Hugenholtz P B 775–7, 786, 1043, 1058
Hughes C A 945, 950, 1061
Hughes D F 991, 997, 1092
Hughes J R 947, 950, 1058
Hughes P 561, 565, 1058
Huijbregts C J 62, 69, 209, 213, 216, 483, 491, 1061
Hummel J W 992–4, 1033
Hunsaker C T 972, 978, 1058
Hunt E R Jr 452, 465, 1058
Hunt L 243, 250, 1058
Hunter G J (Plate 9) 91, 101–2, 113, 122, 176, 199,
    203, 219, 221–3, 231–2, 576, 583, 588, 633–41,
    1023, 1045, 1058
Hunting Technical Services 956, 965, 1058
Hurcom S 821, 826, 1058
Hurley C 254, 265, 1037
Hurn J 468, 470, 476, 1059
Husby O 684, 692, 1055
Hutchinson C F 935, 937, 1059
Hutchinson J 983, 996, 1068
Hutchinson M F (Plates 60 and 61) 24–6, 44, 66, 83,
    105–24, 168, 186, 207, 226, 338, 426, 482–7, 489,
    491, 505, 528, 533, 538, 541, 544, 557, 562, 970,
    978, 982–4, 988, 995, 1049–50, 1059, 1065, 1067
Hutflesz A 387, 399, 1059
Hutson J L 990, 995, 997, 1096
Huttenlocher D P 130, 153, 1059
Huxhold W E 2, 601–3, 606–9, 620, 626, 630, 1059
Hyman J 489, 491, 1072
Ibaraki T 942, 950, 1059
ICA 126, 134, 153, 1059
Ichoku C 457, 465, 1059
IDON Corporation 326, 328, 1059
Idrisi 245, 248, 295, 303, 1059
Ihse M 971, 979, 1085
Ijjasz-Vasquez E J 528, 541, 1059
Imhof E 221, 232, 1059
Ims R A 972, 978, 980, 1059, 1099
Ingoldsby T R 319, 328, 1059
Ingram B 979, 1088
Ingram K T 987, 997, 1099
Inhelder B 84, 88, 1081
Inkley D B 974, 977, 1047
Innes J E 624, 630, 1059
Inskeep W P 984, 987–90, 995, 997, 1060, 1099
Irby C 408, 410, 412, 1090

Ireland P J 706, 714, 721, 1060, 1087
Irvin A 928, 937, 1066
Isaaks E H 8, 20, 258, 265, 283, 292, 483–4, 491, 928, 937, 1060
Isbell R F 198–9, 204, 539, 541, 1060, 1072
ISO 350, 352, 358, 415–16, 423, 681, 692, 1060
Itami R M 99, 102, 970, 979, 1060, 1081
Iverson L 111, 123, 482, 488, 491, 986, 996, 1074
Jackson M J 507–8, 524, 806, 818, 1028, 1060
Jackson P 47, 1086
Jacobi O 112, 122, 1049
Jacobsen J S 984, 987–94, 997, 1033, 1091, 1099
Jacobson B 402, 411, 1060
Jacobson I 353, 357, 414, 418–20, 423–4, 1030, 1060
Jacobson R 558, 564–5, 1060
Jacquemoud S 452–3, 464–5, 1060
Jacquez G M 243, 250, 926–28, 937, 1060
Jagadish H V 388, 399, 1060
Jaggard K 452, 465, 1060
Jain R 326, 328, 1060
Jakeman A J 123, 1074
Jammer M 30, 38, 1060
Jankowski P 884, 888, 1060
Jansen M J W 208–12, 216, 1046, 1060
Janssen R 494, 502, 1060
Jantzen M 555, 1082
Jasper J D 122, 1059
Jayaraman V 297, 303, 1088
Jayasuroya M D A 117, 124, 1096
Jeffrey D 837, 842, 1060
Jenkins A 748, 752, 756, 1051, 1060
Jennings C 806, 818, 1047
Jensen C S 93, 100–2, 1060
Jensen J R 970–1, 997–8, 1057, 1060
Jenson S K 107–8, 113, 116, 121–2, 531–3, 541, 1061, 1096
Jepson W H 78–9, 1067
Jessup R E 984, 995, 1058
Ji W 999, 1007, 1061
Jiang H 496–9, 502, 1043
Jibson R 853, 857, 1055
Jin W G 494–5, 499, 502, 881, 888, 1043
João E 65, 69, 1061
Jobson T A 914, 924, 1061
Johannesson P 959, 965, 1061
John S 621, 630, 1068
Johns R C 837, 842, 1061
Johnson A R 972, 978, 1061
Johnson D S 298, 303, 680, 692, 1050, 1061
Johnson G O 845, 849, 857, 1061
Johnson J M 836, 844, 1097

Johnson J P 644, 651, 1061
Johnson L B 972, 978, 1061
Johnson M E 33, 211, 216, 1061, 1065
Johnson P E 457, 466, 924, 1072, 1090
Johnson R 420, 423, 1049
Johnston C A 67, 69, 123–4, 491, 969, 972, 977–8, 994, 997, 1039, 1052, 1059, 1061, 1087, 1092
Johnston K 970, 978, 1073
Johnston R J 11, 23–4, 41–3, 47, 50, 53, 62, 575, 848, 943–6, 950–1, 968–9, 974, 1018, 1061, 1072, 1093
Jolley D 927, 937, 1044
Jones A P 323, 328, 853, 857, 934, 937, 1061, 1067
Jones A R 955, 964, 1049
Jones C A 988, 997, 1061, 1099
Jones C B 2, 129, 138–45, 152–3, 155, 1032, 1061, 1063, 1097
Jones D 680, 692, 1062
Jones H 124, 1097
Jones J G 11, 119, 124, 1097
Jones J W 987, 994, 997, 1038, 1099
Jones K L 457, 464, 925, 933, 936–7, 1032, 1041, 1061
Jones N L 530, 541, 1061
Jones R M 725, 731, 1061
Jones S 827–9, 843, 1070
Jong P de 509–10, 525, 531, 541, 1061, 1071
Jonsson P 414, 424, 1060
Jordan L 405, 412, 877, 888, 1092
Jordan T 50, 53, 59, 1061
Jordão L 561–5, 1076
Jørgensen S E 67, 69, 974, 978
Journel A G 62, 69, 200, 204, 209–13, 216, 483, 487, 490–1, 971, 979, 1041, 1061, 1075, 1086
Joyce L A 970, 972, 976, 979, 1032, 1088
Jupp D L B 457, 465, 1062
Justice C O 452, 454, 462, 465–6, 1087, 1094
Kainz W 95, 102, 182, 188, 1062, 1086
Kaiser M 256–9, 265, 1070
Kalaris T M 984, 995, 1063
Kamel I 395, 399, 1062
Kanakubo T 172, 1030
Kanal L N 502, 1030
Kanaroglou P S 242, 248, 1031
Kangas J 968, 973, 978, 1062
Kansa E J 486, 489, 491, 1062
Kaplan A 197, 204, 1062
Kapuscinski J 639, 641 1084
Karacapilidis N 325–28, 1052, 1062
Karimi H A 637, 641, 852, 857, 1062, 1073
Karjala D S 776, 786, 1062
Karnieli A 457, 465, 1059

Kates R W 847, 856, 1033
Kato Y 525, 1085
Katoh N 942, 950, 1059
Katz M J 544, 547, 556, 1062
Katz R 317, 319, 325, 328, 1062
Katz v. United States 759, 765, 1062
Kauahikaua J 849, 857, 1062
Kaufman Y J 455, 462, 465, 1062, 1087
Kay W 717, 721, 1062
Kearney A 325, 327, 1025
Kedem Z M 387, 399, 1049
Keefer B J 208, 216, 1062
Keen P G W 627, 630, 828, 842, 1062
Keighery G J 968, 978, 1072
Kelejian H H 242, 245, 250, 1062
Keller M M 166, 172, 1062
Keller P R 166, 172, 1062
Keller S F 138–9, 146–50, 152, 154, 1062, 1099
Keller T 680, 692, 1062
Kellogg R L 983, 995, 1062
Kelly G A 974, 978, 1062
Kelly R E 109, 122, 1062
Kelly-Bootle S 191, 204, 1062
Kelmelis J 35, 38, 94–5, 102, 1062
Kelsall J L 930, 937, 1062
Kelsey J L 926, 930, 937, 1062
Kemp K K 67, 69, 84, 88, 748–9, 756, 1062–3, 1072,
    xxvii
Kemp W P 984, 995, 1063, 1067
Kemp Z 38, 785, 1037
Kennedy A A 626, 630, 1040
Kennedy B A 43, 46, 974, 976, 1034
Kennedy M 423, 436, 1063
Kennedy S 482, 492, 1094
Kennie T J M 448, 450, 1081
Kern J S 982–4, 988–9, 994, 1063
Kessel S R 99, 102, 849, 858, 1063
Keulen H van 212, 216, 1042
Kevany M 2, 362, 369, 802, 818, 1025
Khanna R 325, 328, 1036
Kharod K 499, 502, 1083
Kharrazi M 932, 938, 1068
Kidner D B 129–32, 145, 153, 1061, 1063
Kilby P J 1000, 1007, 1023
Killpack C C 816–18, 1073
Kim S H 849, 858, 1063
Kim T J 838, 842, 877, 888, 1055–6
Kim W 103, 374, 381, 384, 1063, 1090
Kimball R 408, 410, 412, 1090
Kimerling A J 158, 173, 1085
Kimes D S 453, 465, 1063

Kindleberger C 726, 731, 1063
King L J 241, 250, 285, 292, 1063
King M 984, 987, 994, 1033
King R M 970, 977, 1047, 1054
King S A 849, 853, 858, 1063
King W 374, 384, 1026
Kingham S 935–7, 1042, 1063
Kirby K C 405, 411, 1063
Kirby R F 832, 842, 1063
Kirby R P 955, 964, 1037
Kiremidjian A S 849, 853, 858, 1063
Kirkby M J 540, 983, 994–5, 1026, 1040, 1063
Kirkpatrick D G 550, 554–6, 1031, 1063
Kirwan M 934, 937, 1050
Kishimoto H 411, 1063–4
Kitchen N R 991, 997, 1092
Kite G W 120, 123, 1063
Klaassen L 244, 251, 286, 292, 1080
Klanderman G A 130, 152, 1059
Klauber M R 931, 938, 1085
Kleiner A 394, 399, 1063
Kleiner B 221, 232, 1034
Kleinschmidt I 927, 937, 1044
Klesh L 837, 842, 1063
Klijn F 969, 978, 1063
Kling R 621, 626, 630, 1043, 1046, 1063, 1087
Klinger A 407, 411, 1082
Klir G J 192, 204, 1063
Kloster S 403, 411, 1068
Klosterman R E 723, 731, 883, 888, 1063
Knapp L 167, 173, 1090
Kneeshaw D D 968, 976, 1033
Knight D 933, 937, 1057
Knight R L 837, 843, 1057, 1089
Knisel W G 988, 995, 1066
Knowles R D 841–2, 1058
Knuth E 411, 1044
Koch T 684, 692, 1055
Koelsch F 735, 743, 1063
Koerper G 983, 997, 1095
Kohn B P 486, 490, 1038
Kolejka J 968, 979, 1081
Koller D 561–5, 1063, 1067
Konecny G 109, 123, 1063
Korporal K D 984, 994, 1063
Korte G B 362, 369, 1063–4
Kosinovsky I (Plates 10–12) 99, 102, 221–5, 233,
    482, 485–6, 489, 491, 1074
Kosko B 197, 204, 1064
Kousser J M 945, 950, 1064
Koussoulakou A 170, 172, 1064

Kowalski R 101–2, 1064
Kraak M-J (Plate 7) 26–7, 70, 78, 98–9, 102, 105,
    128, 152–5, 157–73, 224, 258, 336–9, 407,
    410–11, 776, 922, 1002, 1041, 1043–4, 1047,
    1064, 1073, 1083–4, 1087, 1094, 1097–8, 1100
Kragt J E 852, 858, 1064
Krakovsky E D 428, 436, 1096
Kramer H J 461, 465, 1064
Kramer L A 986, 988, 996, 1080
Kranakis E 555–6, 1031, 1083
Kranzler G 996, 1075
Krcho J 483, 491, 1064
Kreps P 374, 399, 1092
Kreveld M van 147, 150–2, 155, 1029, 1098
Kriegel H-P 395, 399, 1028
Krige D 913, 924, 1064
Krishna Iyer P V A 241, 250, 1064
Krishnan V V 970, 979, 1091
Kroes J G 984, 994, 1046
Krohn W B 970, 978, 1067
Kruck E 109, 123, 1063
Kruess A 970, 978, 1064
Krummel J R 970, 978, 1064
Krygier J B 167, 170–2, 225, 232, 1041, 1064
Kubat M 1000, 1007, 1046
Kubik K 112, 122, 1049
Kucera G L 98, 100, 102, 1064
Kucera H A 95–6, 102, 1064
Kuhbauch W 452, 464, 1030
Kuhn T S 42, 47, 62–5, 69, 85, 1064
Kuhn W 37–8, 88, 153, 155, 312, 337, 343, 352, 372,
    374, 401–12, 555, 561, 588, 725, 730, 928, 1007,
    1010, 1044, 1049, 1056, 1064, 1071, 1081,
    1097–8, 1101
Kulldorff M 930, 932, 937, 1057, 1064
Kundert K 928, 937, 1066
Kunze A E 818, 1037, 1073
Kuo Y H 99, 101, 1025
Kupfer G 452, 464, 1030
Kupfer J A 452, 464, 971, 975, 978, 1064
Kuusipalo J 968, 973, 978, 1062
Kuusk A 453, 465, 1064
Kyem P A K 494–5, 499, 502, 881, 888, 1043
L'Eplattenier R 928, 937, 1066
Lachapelle G 992, 995, 1068
Ladson A R 107, 112, 115–17, 123
Laffey S C 830, 842, 1064
Lagacherie P 195, 198, 204, 1064
Lagazio C 930, 936, 1035
Lagrange J-P 126, 134, 140, 152–5, 1049, 1072,
    1075, 1087, 1091, 1098

LaGro J 454, 465, 1064
Lahiri S 256, 258–9, 265, 1070
Lai P 901, 908, 1064
Lakatos I 62, 65, 69, 1065
Lake R W 51–2, 57, 59, 356, 358, 1065
Lakoff G 33, 38, 81–3, 88, 1065
Lalanne L 240, 250, 1065
Lam D C L 972, 978, 1058
Lam N S-N 454, 465, 481, 487, 491, 1052, 1065
Lamb R 986, 996, 1083
Lamb W R 892, 899, 1065
Lamberts J 420, 423, 1054
Lammers R B 533–4, 541, 1026, 1065
Lanari R 110, 113, 123, 1065
Landis J D 878, 881, 884, 888, 1067
Lane J S 828, 836, 843, 1065
Lang L 726, 731, 1065
Lang T 693, 706, 1065
Lange A 11, 26, 66, 105–8, 321, 333, 349, 425–9,
    445, 467–76, 615, 650, 782, 804, 806, 830, 897,
    981, 990, 1010
Langford I 930, 935, 937, 1065
Langford M 749, 756, 1065
Langran G 35, 38, 68–9, 91–5, 100, 170, 172, 183,
    188, 1065
Langston P 710–11, 721, 1065
Langton C G 288, 292, 1065
Lanter D P 185, 188, 208, 216, 1065, 1096
Lanza L 849, 858, 1065
Lapis G 374, 382, 384, 1054
Laporte M 461, 464, 1041
Larijani L C 563, 565, 1065
Larrivée S 421, 423, 1028
Larsen L C 78, 794, 909, 911, 999–1007, 1094
Larson E 761, 765, 1065
Larsgaard M 5, 11, 681, 799, 901–8, 1019
Larsson G 859, 975, 1065
Laslett G M 483, 487, 491, 1065
Lass L W 984, 995, 1065
Lathrop R G 983, 995, 1065
Latour B 54, 58–9, 1065
Lattuada R 68–9, 1065
Lau S Q Jr 564–5, 1066
Lauenroth W K 972, 976, 1032
Lauer D T 769, 780–2, 786, 1042
Laurini R 2, 319, 328, 410–11, 725, 732, 1039,
    1064–6, 1076, 1089
Lavalle C 453, 465, 1053
Laviolette M 197, 204, 1066
Law K H 849, 853, 858, 1063
Lawless J 656, 663, 666, 1045

Lawrence V V 1015, 1020, 1066
Laws D 638, 641, 1066
Lawson A B 932, 937–8, 1066, 1097
Layard R 603–4, 609–10, 1066–7, 1089, 1092
Lazar R A 83, 88, 184, 188, 1046
Le Caer G 516, 525, 1066, 1082
Le Sueur D 919, 924–6, 932, 938, 1066, 1089
Leahy F J 95, 102, 1056
Leary T 852, 857, 1049
Leberl F W 138, 153, 1066
Leclerc Y G 564–5, 1066
Lecordix F 151, 154, 1086
LeDuc L 950, 1030
Lee B D 831, 843, 1066
Lee B S 849, 858, 1096
Lee D 150–1, 153, 1066
Lee G 849, 858, 1063
Lee J 108, 110, 123, 544–50, 554, 556, 1066
Lee R 752, 756, 1051
Leenaers H 209, 216, 1066
Leenhardt D 208, 216, 1066
Lees B G 970, 978, 1066, 1074
Legates D R 112, 123, 1066
Leick A 428, 436, 470, 476, 1066
Leifker D 395, 400, 1086
Leigh C M 833, 844, 1099
Leipnik M R 1080, xxvii
Leitner M 149, 153, 1066
Lemmens M J P M 108–9, 123, 1066
Lemmer J F 502, 1030
Lemoine G G 453, 465, 1053
Lenczowski R 663, 666, 770, 786, 1066
Lenthall P 473, 476, 1034
Leonard R A 988, 995, 1066
Lepper M J C de 925, 937, 1066
Lerman S 834, 841, 1029
Leroy M 453, 461, 464–5, 1041, 1086
LeSage J P 245, 250, 1066
Lescourret F 975, 978, 1066
Lessard P 928, 937, 1066
Lester M 93, 97, 102, 1066
Lettenmaier D P 120, 124, 1099
Leung Y C 197, 202, 20 227, 232, 288–9, 292,
    1052, 1066
Levin D 544, 556, 1043
Levin J R 568, 580, 1070
Levin R M 50, 57, 59, 1056, 1098
Levin S A 972, 978, 1066
Levine D 486, 491, 1055
Levine G 568, 579, 1066–7
Levine J 878, 884, 888, 1066

Levinsohn A G 2, 601, 608–9, 620, 626, 630, 1059
Levy R M 78–9, 1067
Lewis A 116, 120, 123, 1067, 1074
Lewis F T 511, 525, 1067
Lewis P A W (Plate 21) 211, 216, 453, 460, 464,
    1027, 1067
Leymarie P 986, 994, 1045
LGMB 797, 799, 1067
Li S G 489, 491, 1072
Li X 460, 466, 880, 882, 888, 1097, 1101
Li Z 554–5, 1031
Liang T C 212, 216, 1067
Library of Congress 682, 692, 1067
Lichtenegger H 428, 436, 1058
Lichtner W 138, 154, 1067
Liebhold A M 970, 978, 984, 995, 1067
Liederkerke M van 323, 328, 853, 857, 1061, 1067
Liggett R S 78–9, 1067
Light D L 449–50, 1067
Lijphart A 945, 950–1, 1042, 1067, 1076
Lillie J 621, 630, 1079
Lillywhite J 678, 692, 1067
Limp W F 257, 265–6, 1045, 1099
Lin C C 227, 232, 1066
Lind M 74, 80, 1067
Lindberg J B 44, 47, 1085
Lindenmayer D B 970, 978, 1067
Lindquist P 934, 938, 1068
Lindquist R 473, 476, 1087
Lindsay B 374, 382–4, 402, 412, 1054, 1092
Lindstrom P 561–5, 1063, 1067
Linz A 850, 852, 858, 1067
Lissowski A 516, 526, 1085
Lithgow T 836, 843, 1067
Little R 604, 609, 1067
Litwin W 348, 358, 1067
Livingston M A 559, 565, 1091
Livingstone D N 43–5, 47, 55, 60, 64, 68, 70, 539,
    542, 564–5, 1067, 1083
Livingstone S A 970, 978, 1067
Llewellyn-Jones D T 457, 465, 1082
Lloyd B B 88, 1086
Lloyd D 455, 465, 1067
Loague K 212, 216, 1067
Loban S R 302–3, 1035
Lobley J 19
Lochovsky F H 66, 70, 1095
Lock B F 181, 188, 1050
Lockhart J 257, 265, 1045
Loeb A L 506, 525, 1067
Logan I T 438, 450, 1067

Lohman G M 374, 382, 384, 1054
Lohmann P 109, 123, 1063
Lolonis P 163, 172, 298, 384, 881, 885–7, 926,
    930–1, 938, 1025, 1087
Lomas T 932, 938, 1068
Lombard K 302–3, 1035, 1068
London Planning Advisory Committee 960, 965, 1068
Long D S 984, 991–2, 995–6, 1068
Longley P A 1–20, 34, 45–6, 56, 62, 71, 78, 107–8,
    112, 121, 127, 177, 195, 203, 226, 236, 242,
    247–51, 262, 268, 282–4, 292, 311, 334, 338,
    345, 355–7, 405, 425, 437, 462, 465, 475, 479,
    489, 537, 563, 567–80, 601, 615, 677, 707, 711,
    714, 721, 739, 748, 769, 793–5, 841–2, 856, 880,
    888, 897, 902, 906, 1007, 1009–21, 1028, 1030,
    1035, 1040, 1051, 1053, 1057, 1069–71, 1073,
    1079, 1100
Longmore R 108, 116, 123, 1076
Loon J C 108, 112, 121, 1035
Lopez X R 621, 630, 1068
Lopez-Abente G 932, 938, 1068
Lorensen W 353, 358, 363, 369, 380, 382, 384, 1087
Lorie R 374, 384, 1026
Lösch A 240, 250, 1068
Louis A 198, 205, 1093
Loukes D K 829–30, 843, 1068
Lounsbery M 562, 565, 1068
Louviere J 834, 843, 1068
Love B 830, 843, 1068
Love D 934, 938, 1068
Love J 772, 786, 1068
Love R F 297, 303, 1098
Loveland T R 3, 92, 440, 443, 455, 458, 463, 584,
    594, 653, 656, 661, 665–75, 703, 705, 770,
    780–1, 806, 896, 931, 954, 983, 988, 991, 996–7,
    1010, 1013, 1031, 1068
Lovett A A 849, 856, 927–8, 932, 936–7, 1031, 1041,
    1056, 1068, 1092
Lowe S N 124, 1099
Lowell K 523, 525, 1068
Lowell K E 577, 579, 1043
Lownsbery B 680, 692, 1068
Lowry J H 849, 858, 1068
Löytönen M 925, 936–8, 1031, 1050, 1055, 1064, 1068
Lu X 852, 858, 1068
Lucena e Vale M J 725, 731, 1033
Ludwig G 752, 756, 1028
Luk W S 403, 411, 1068
Lummaux J C 769, 786, 1068
Lundberg G 85, 88, 1086
Luo L Q 131–2, 152, 1061

Luppi G 929–30, 936, 1031
Lyman P 901, 908, 1068
Lynch C 930, 938, 1087
Lyon D 839, 843, 1068
Lyon J D 999, 1007, 1061
Lyytinen K 622, 630, 1069
Maamar Z 416–19, 423, 1028
MacDevette D R 160, 767, 909, 913–24, 1046, 1068
Macdonald Holmes J M 943, 950, 1069
MacDougall E B 167, 172, 255, 265, 1069
MacEachren A M (Plate 13) 98–9, 102, 158–9,
    166–7, 170–3, 220–2, 225, 230, 232, 1031, 1041,
    1064, 1069, 1090, 1093
Macedonia M R 564–5, 1069
MacGill S M 719, 722, 1099
Machlis G E 938, 978, 1072
MacInnes S 112, 121, 1036
Mackaness W A 92, 97, 102, 138, 145, 149, 152–4,
    1028, 1047, 1069
Mackay D S 534–5, 537, 541–2, 1069, 1086
Mackenzie D 762, 765, 1069
Mackey B G 108, 116, 120, 123, 970, 978, 1067, 1069
Mackey H E 971, 977, 1057
Mackinlay J 227, 232, 1069
MacLennan M H 749, 756, 1055
MacLeod F 68, 70, 95, 102, 1083
MacMillan R A 195, 203, 1032
MacMillan W D 46–7, 946–7, 950, 1069, 1079, 1084
Macur R E 989–90, 995, 1060
Maddocks A 934, 937, 1050
Maffini G 75, 80, 1069
Magalhaes G C 820, 826, 1069
MAGIC project 565, 1069
Magillo P 107, 131, 200, 477, 479, 482–3, 543–57,
    562, 974, 1032, 1040, 1069
Magrogan P J 418, 424, 1069
Maguire D J 1–20, 45, 80, 88, 124, 153, 188, 203,
    265, 268, 270, 282, 308, 311–12, 316, 334,
    359–69, 383–4, 402, 411, 417, 424, 437, 464,
    490–2, 563, 580, 590–1, 596, 601, 612, 622, 707,
    716, 721, 739, 748, 753, 756, 785–6, 792–5,
    802–3, 816–18, 827–9, 835, 840–4, 858, 875,
    887, 891, 935–8, 964–5, 1000, 1007, 1009–21,
    1028, 1032, 1035, 1037, 1039–40, 1047,
    1049–50, 1056, 1060, 1069, 1075, 1079–80,
    1084–5, 1095, 1099–100
Maguire S A 362, 369, 1070
Mahmassani H S 832, 842, 1101
Mahoney R P 625, 630, 1070
Maidment D R 67–9, 123–4, 491, 530, 541, 969, 977,
    994, 997, 1039, 1052, 1059, 1061, 1087, 1092

Maier D 403, 410, 1026
Maiocchi R 101–2, 1070
Mairson H G 548, 556, 1070
Majid D A 784, 786, 1070
Majumdar K L 499, 502, 1083
Majure J 254–6, 258–60, 265–6, 1037, 1070, 1093
Makarovic B 108, 123, 1070
Makela M E 970, 979, 1087
Makki K 555, 1069
Mäler K-G 973, 979, 1081
Maling D H 221, 233, 433–4, 436, 1070
Malingreau J-P 452, 462, 466, 1094
Mallet J-L 68–9, 1070
Malthus T J 452, 464, 1039
Mandelbrot B B 8, 20, 114, 123, 1070
Mann J R 928, 938, 1078
Manola F A 388, 399, 1080
Mantei M 410, 1031
Mantel N 241, 250, 1070
MapGuide 325, 328, 1070
Mapp H P 984, 988, 995, 1050
Mapping Science Committee 615, 620, 782, 784,
    786, 1070
Marascuilo L A 568, 580, 1070
Marble D F 2, 248, 285, 291, 327–8, 569, 579, 878,
    888, 893, 1029, 1039, 1070, 1081
Marceau D J 244, 250, 1070
March J G 603, 609, 1039
Marchi M 930, 936, 1035
Marciano R 264, 1025
Marcus M G 43, 47, 1070
Marechal A 491, 1075
Margerum R D 973, 978, 1070
Margriter S 849, 857, 1062
Margules C R 968, 970, 976, 978, 1026, 1070, 1072
Marion K E D 832, 844, 1097
Mark D M 5, 11, 25, 31–3, 37–8, 75, 81–9, 91, 108,
    112–14, 118, 122–3, 149, 154, 291, 353–4,
    402–3, 406, 409–11, 435, 477, 529, 531, 541–2,
    658, 724, 959, 975, 986, 996, 1010, 1023, 1039,
    1044, 1049, 1052, 1057, 1064, 1070–2, 1076–8,
    1089, 1101
Mark L 348, 358, 1067
Mark R K 539, 541, 1041
Markham R 754, 756, 1085
Marks B 967, 977, 1055
Marks D 531, 541, 1071
Marks M 338, 345, 1094
Marks R J 282, 1029
Markus L 626, 630, 1071
Marmie A 323, 328, 1071

Marsh C 79, 1027
Marshak A L 452, 465, 1075
Marshall K 836, 844, 1097
Marshall R J 928, 932, 938, 1071
Marshall W F 817–18, 1071
Marston S A 857, 1053
Martin C 919, 924, 1066
Martin D J 2–3, 5, 11, 25–7, 49, 54, 71–80, 160–1,
    173, 179, 187, 193, 203, 283–5, 311–12, 336,
    341–3, 345, 353, 374, 388, 399, 414, 424, 435,
    487, 540, 569, 577–80, 638, 658, 674, 802, 820,
    855, 880, 888, 919, 935, 938, 954–6, 962, 971,
    1000, 1010, 1031, 1068, 1071
Martin J 620, 803, 818, 1071
Martinelli L A 244, 250, 1073
Martonchik J V 455–7, 464, 1041
Martz L W 531, 534, 541, 1049, 1071
Marvin S 310, 315–16, 1053
Marx R W 79, 1031
Marzano P 546, 554–6, 974–6, 1040, 1083
Maskarinec G 932, 938, 1071
Mason D C 68–9, 1076
Masser I 80, 233, 282, 291, 621–31, 706, 711, 721,
    768, 776–9, 783–6, 857, 885–8, 953, 964–5,
    1012, 1020, 1027, 1032–3, 1035, 1042, 1047,
    1050, 1059, 1071, 1074, 1078–9, 1086, 1098
Massicotte L 940, 944, 950, 1030
Massimino M J 563, 565, 1071
Masters R 953, 964, 1044
Masuoka P 841, 843, 1071
Mather P M 69, 328, 1027, 1051, 1076
Matheron G 483–5, 487, 491, 1071
Matheson W 984, 996, 1085
MathSoft 260, 263, 265, 1071
Matson K C 539, 541, 1046
Matson P A 971, 979, 1086
Matsuyama T 387, 399, 1071
Matti J C 608–9, 1029
Mavor A S 558, 565, 1043
Maxwell J C 529, 541, 1072
Mayhew L 928, 938, 1072
McAleer M 123, 1074
McAusland S 625, 630, 1072
McBratney A B 214, 216, 483, 487, 491, 992, 996,
    1065, 1072
McBride R A 984, 994, 1030
McCabe G J 986, 998, 1100
McCarthy J 999, 1007, 1061
McCarthy T 564–5, 1083
McCauley J D 483, 486, 491, 1072
McConnell P R H 109, 122, 1062

McCool D K 986, 996, 1084
McCorduck P 625, 630, 1046
McCormick B 158, 173, 1072
McCreight E 385, 388, 395, 399, 1028
McCubbrey D J 362–3, 369, 1047
McCullagh M J 110, 123, 481–3, 491, 1072
McDaniel P A 987, 996, 1076
McDermott D 98, 103, 1089
McDonald G 754, 756, 1089
McDonald J A 254, 256, 264, 1032
McDonald R A 459, 465, 1072
McDonald R C 539, 541, 1072
McDonnell R A 2, 84, 88, 1032, 1072, xxvii
McDowell L 40, 47, 1072
McGill M E 254–5, 264–5, 1024, 1036
McGill R 224, 232, 1036
McGlade M S 44, 47, 1074
McGranaghan M 221, 224–5, 233, 1072
McGrath M 914, 924, 1072
McGraw K L 148, 154, 1072
McGuffie K 753, 756, 1072
McGuire C B 785, 1025
McGwire K C 3, 20, 849, 857, 1045, 1091
McHaffie P H 69, 1072
McHarg I L 3, 20, 294, 303, 878, 888, 1072
McJones P 374, 384, 1026
McKay L 323, 328, 1072
McKay R B 950–1, 1042, 1067, 1076
McKee L 575, 579, 698–9, 705, 1032
McKee M J 608–9, 1029
McKendry J E 749, 756, 968, 978, 1043, 1072
McKenzie N H 538, 541, 1050
McKenzie N J 108, 120–2, 984–5, 995–6, 1050, 1072
McKenzie N L 968, 978, 1072
McKerrow N 916, 924, 1072
McKimley M I 892, 899, 1065
McLaren M 848–9, 858, 1072
McLaren R A 314, 625, 639, 657, 769, 798, 859–75,
    956–7, 964, 1019–20, 1070, 1072
McLaughlin D 489, 491, 1072
McLaughlin J D 2, 193, 203, 318, 323, 327–9, 625,
    630, 770, 785, 859, 875, 1014, 1020, 1036, 1039,
    1045, 1089
McLean I 940–2, 945, 950–1, 1033, 1069, 1072
McLean M 335, 345, 1094
McMahon J P 970, 978, 982–3, 995, 1059, 1067
McMahon T A 121–2, 212, 216, 1053
McMaster R B 126, 133–6, 143–5, 147–55, 732, 849,
    858, 1028, 1033–4, 1072, 1074, 1076, 1086,
    1089, 1098
McMichael A J 969, 978, 1072

McMillen D P 245, 250, 1073
McPherson J 374, 382, 384, 1054
McQueen H 110, 123, 520, 526, 1087
McRae H 621, 630, 1072
Mead R A 196, 203, 1036
Means G D 984, 995, 1058
Medyckyj-Scott D 402, 410–11, 600, 622, 630, 692,
    1039, 1044, 1049, 1061, 1067, 1073
Meehan W J 808, 816, 818, 1073
Megiddo N 396, 399, 1073
Megretskaia I 256, 266, 1093
Mehl J 374, 384, 1026
Mehta A 454, 464, 1051
Meixler D B 79, 1031
Mejia-Navarro M 849, 858, 1073
Melack J M 244, 250, 1073
Melillo J M 452, 466, 1098
Mellor S 414, 424, 1090
Melton J 403, 411, 1073
Melton M A 537, 541, 1073
Mendeleev D I 240, 250, 1073
Menthon D de 544, 549, 556, 1093
Merchant J W 982–3, 988, 994, 996, 1031, 1068, 1073
Merriam G 970, 979, 1093
Merrill D W 929, 938, 1089
Mertes L K 244, 250, 1073
Mesev T V 462, 465, 1073
Metropolis N 282, 309, 316, 1057, 1073
Metz J A J 970, 980, 1096
Meyburg A H 828, 843, 1092
Meyer-Roux J 975, 1026
Meyers J R 361, 438, 591, 801–18, 1073
Miao Q 829, 842, 1078
Michaelsen J 244, 250, 1073
Michalak J 254–6, 264, 1032
Michel S M 57, 59, 1023
Michener W K 975, 1026, 1034, 1069
Miellet P 749, 756, 1029
Migliaccio M 110, 113, 123, 1065
Mignotte A 282, 1095
Mildenberger S J 109, 122, 1062
Miller A 38, 328, 1073
Miller B 51, 57, 60, 1073
Miller D R 535, 542, 974, 978, 1073
Miller E 682–4, 692, 1098
Miller H J 297, 303, 828, 833–4, 843, 849, 858, 926,
    938, 1068, 1073
Miller H T 953, 964, 1048
Miller J 112, 121, 1036
Miller K A 852, 857, 1054
Miller P F 457, 464, 1048

Miller R G 637, 641, 1073
Milleret-Raffort F 725, 732, 1066
Milligan P 931–2, 938, 1093
Mills E S 303, 1084
Mills K 544–6, 549, 556, 1073
Milne A K 122, 1059
Milne B T 970–2, 978, 1061, 1073
Milne G 538, 542, 1073
Mimno P 321–3, 328, 1073
Mine M 560, 565, 1073
Minghi J V 198, 204, 1085
Mirless J 604, 609, 1067
Mishoe J W 987, 997, 1099
Missikoff M 411, 1054
Misund G 132, 154, 1073
Mitas L (Plates 10–12) 99, 102, 105–7, 110–11, 123, 210, 221–5, 233, 478, 481–92, 520, 537, 544, 986, 1073–4
Mitasova H (Plates 10–12) 99, 102, 105–7, 110–11, 123, 210, 221–5, 233, 478, 481–92, 520, 537, 544, 986, 996, 1032, 1073–4
Mitchell L C 999, 1007, 1061
Mitchell P L 453, 464, 1026
Mitchell W 729, 732, 1074
Mitrani J E 845, 858, 1076
Miyares I M 44, 47, 1074
Moellering H 170, 173, 184, 188, 633, 641, 696–7, 706, 1074
Molenaar M 27, 70, 126, 131, 147, 153–5, 410–12, 1041, 1043–4, 1047, 1064, 1073–4, 1083–4, 1087, 1094, 1097–8, 1101
Moller-Jenson L 462, 465, 1074
Monckton C G 196, 204, 1074
Mondschein L G 845, 858, 1074
Monk J 752, 756, 1051
Monmonier M 98, 102, 131, 154, 158, 167–8, 173, 220, 224–5, 232, 255, 265, 776, 786, 1074
Montagne C 987, 996, 1076
Montani C 547, 555, 1040
Montgomery D R 120, 124, 529, 532, 542, 986, 998, 1074, 1101
Moody D 423, 1074
Moon G 925, 933, 936–7, 1032, 1061
Mooney D J 660, 666, 768, 778, 786, 1074
Mooney H A 973, 978, 1074
Mooneyhan D W 656, 663, 666, 1045
Moore D M 970, 978, 980, 1074, 1097
Moore G C 621, 630, 1074
Moore I D 105–8, 112–17, 123, 212, 216, 538, 542, 984–8, 995–6, 1049–50, 1053, 1074, 1080
Moore J E 838, 842, 1056

Moore R 849, 857, 1062
Moore T E 88, 1086
Moraczewski I R 194, 204, 1075
Morain S 184, 188, 1075
Morais Arnaud A 706, 1071
Moran P A P 241, 250, 1075
Morehouse S 151, 154, 1075
Moreira J 110, 113, 123, 1065
Morgan G 624–6, 630, 638, 641, 1027, 1075
Morgan J M 749–50, 756, 1075
Morgan P H 953, 965, 1075
Morgan R 924, 1072
Morrey I 362, 369, 1028
Morrice J G 533, 542, 974, 978, 1073
Morrill R L 44, 47, 942–6, 950–1, 1075
Morris K P 459–61, 464, 1027
Morris K S 970, 978, 1060
Morris R 933, 936, 1034
Morrison A C 243, 250, 1075
Morrison J L 158, 173, 18–9, 220, 233, 615, 620, 636, 641, 1027, 1031, 1061, 1075, 1085, 1087
Mortenson D C 198, 203, 1033
Mortenson M E 68–9, 1075
Mortimore R 942, 951, 1072
Morton M S S 828, 842, 1062
Moulin B 416–19, 423, 520, 523, 525, 1028, 1043
Mounsey H M 188, 577, 580, 666, 768, 786, 1052, 1075, 1084
Mount D 542, 556, 1093
Mowrer H T 208, 216–17, 1048, 1075, 1100
Mowshowitz A 759, 765, 1075
Moxey A 487, 491, 1075
Moy C C 820, 826, 1075
Moyer D D 829, 836, 843, 1075
Muchaxo J 561–5, 1075–6
Muehrcke P C 158, 173, 220, 233, 1085
Mueller R 984, 997, 1097
Muir C S 929, 938, 1090
Muir K R 928, 938, 1078
Mulla D J 114, 123, 971, 979, 992, 996, 1075, 1086
Müller F 974, 976, 1031
Müller J-C 126, 128, 145, 152, 154–5, 164, 172, 188, 1052, 1064, 1072, 1075, 1087, 1091, 1098
Muller J-P A L 109, 121, 459–62, 464–5, 880, 887, 1027, 1040, 1087
Murray K 656, 665–6, 770, 779, 785, 1033
Musgrave A 62–3, 69, 1075
Mustiére S 150, 154, 1075
Mwalyosi R B B 968, 978, 1075
Myers D E 484, 491–2, 1075, 1086
Myneni R B 452–3, 464–6, 1075–6, 1095

Nadal J P 526, 1085
Nadin V 953, 956, 964, 1038
Nagao M 387, 399, 1071
Nagy G 132, 152, 200, 204, 543–4, 547–8, 555–6, 1040, 1076
Naiman R J 972, 978, 1061
Naisbitt J 623, 630, 1076
Narumalani S 970, 978, 984, 987, 994, 1033, 1060
NASA 673, 675, 1076
Natal Nutrition Working Group 917, 924, 1076
Naude A 923–4, 1076
Navathe S B 415, 423, 1027
Naylor B F 387, 399, 1049
NCGIA 1009, 1020, 1076
Nebert D D 323, 328, 1076
Neft D S 240, 250–1, 1076, 1097
Negroponte N 310, 316, 409, 411, 1076
Neilson R P 112, 120–1, 982, 994, 1039
Nelder J A 286, 292, 1076
Nelder V J 970, 978, 1076
Nelson B 244, 250, 1073
Nemani R 534, 541–2, 1026, 1076
Neves J N 105–7, 110, 310, 402, 405, 479, 557–65, 727, 897, 1017, 1075–6
Newell J 272, 282, 932, 936, 1029
Newell R G 350, 369, 802, 818, 1076
Newman I 692, 1061, 1067
Newsom D E 845, 858, 1076
Newton H 562, 565, 1068
Newton J 680, 685, 692, 1076
Newton P W 320, 329, 879, 888, 1076, 1082
Ng K W 555, 1042, 1076
Ng R 264–5, 1076
Ngxongo S 919, 924, 1066
Nichol D G 462, 465, 1076
Nicholls A O 970, 976, 1026
Nickerson B G 138, 147–9, 154, 1076
Nickeson J E 457, 465, 1055
Nicola G de 824, 826, 1076
Niedda M 532, 542, 1076
Nielsen A 1003, 1007, 1093
Nielsen D R 992, 996, 1076
Nielsen G A 987, 990–2, 994–6, 1033, 1068, 1074, 1076, 1091
Nielson G M 264, 480–3, 486–7, 491, 1049, 1076
Niemi R G 944, 950–1, 1030
Nievergelt J 155, 393, 399, 1076, 1098
Nijkamp P 287, 291–2, 639, 641, 881, 884, 887–8, 936, 1024, 1041–2, 1046, 1053, 1076, 1079
Nijssen B 530, 542, 1081
Niklasz L 860, 866, 875, 1076

Nikula A 970, 979, 1087
Nilsson A 828–30, 841, 1033
Nisbet R A 972, 978, 1058
Nix H A 108, 116, 120–3, 968, 970, 978–9, 982–3, 995, 1059, 1067, 1076
Nobre F F 928, 938, 1076
Nofziger D L 984, 996, 1076
Nonomura K 429, 436, 663, 666, 770, 787, 1097
Nordbeck S 77, 80, 1076
Norman D A 727, 732, 1076
Norman J M 453, 466, 1097
Norris H 852, 857, 1049
Norris P 950, 1030
North Carolina General Statutes 737, 743, 1076
Norton D J 970, 980, 1099
Norton T W 968, 979, 1076
Norval R 928, 937, 1066
Noss R 968, 979, 1089
Nott S 953, 965, 1074
Nulty W G 388, 399, 1076
Nunes J 87–8, 1071
Nyerges T L 410–11, 748, 756, 828–35, 841–3, 1039, 1064, 1076, 1089
Nyland H 931, 938, 1085
Nyquist M 970, 975, 1023
O'Callaghan J F 17, 118, 123, 531, 542, 948, 950, 986, 996, 1058, 1076
O'Conaill M A 68–9, 1076
O'Kelly M E 831–3, 841–3, 1093
O'Loughlin E 530, 537–8, 542, 1077, 1090
O'Loughlin E M 112, 123, 1074
O'Loughlin J 242, 250, 1077
O'Neill R V 970, 978–9, 1064, 1078, 1082, 1095
O'Riordan P D 362–3, 369, 1047
O'Shaugnessy P J 117, 124, 1096
Oaks S D 849, 858, 1073
Obermeyer N J 583, 587, 590–2, 601–10, 620, 626, 657, 792–4, 797, 816, 837, 862, 1077
Odell J J 414, 424, 803, 818, 1071
Oden N L 971, 979, 1091
OECD 760, 765, 771, 786, 1078
Office of National Statistics 193, 204, 1078
OGC 352–8, 410, 415–16, 424, 1078
Ohlen D O 983, 988, 994, 996, 1031, 1068
Okabe A 242–3, 250, 518, 520, 523–5, 1078
Okazaki D 167, 173, 1090
Okunieff P 829, 842, 1078
Oldfield B 821, 826, 1039
Oliver M A 114, 123, 198, 205, 244, 250, 483–4, 491, 928, 938, 1078, 1098
Olle T W 414, 424, 1078

Olmstead v. United States 759, 765, 1078
Olsen E R 970, 979, 1078
Olsen L 1007, 1089
Olson J S 410–12, 982–3, 989, 996, 998, 1023, 1056, 1078, 1101
Olson P 983, 996, 1068
Olsson G 251, 292, 1094
OMB 779, 786, 1078
Omernik J M 983, 996, 1078
Onsrud H J 56, 60, 177, 188, 583, 586–8, 607, 621, 626, 630–1, 634, 643–51, 666, 763, 768, 772, 775, 785–6, 861, 888, 1025, 1039, 1042, 1052, 1059, 1071, 1074, 1078, 1086
Oosterom P van 11, 129–31, 154, 343, 372, 385–400, 415, 802, 1078
Opdam P 970, 980, 1096
Openshaw C A 237, 270, 282, 1079
Openshaw S 8, 20, 39, 44–52, 59–60, 64, 70, 73, 77–80, 99, 101, 137, 165, 193, 199, 204, 227, 233–7, 243–4, 251, 254, 264, 267–83, 287–8, 292, 298, 310, 401, 404, 411, 509, 549, 567, 573–4, 621, 630, 637, 641, 658, 726, 732, 752, 756, 856, 930–1, 938, 946, 958, 964–5, 970, 972, 1010, 1018, 1042, 1078–80, 1084, 1095
Oppong J R 833, 843, 1080
Oracle Corporation 353, 358, 1080
Orav E J 211, 216, 1067
Orbeton P 410, 1031
Ord J K 8, 20, 62, 69, 121–2, 241–3, 248, 258, 260, 262, 265, 285–6, 291, 519, 525, 971, 977, 982–3, 995, 1036, 1041, 1051, 1059, 1080
Ordnance Survey 956, 959, 962, 965, 1080
Orenstein J A 388, 399, 1080
Orlob G T 99, 102, 1080
Ormeling F J (Plate 7) 158, 163–5, 170, 172, 1064
Osborne D 770, 786, 1080
Ostresh L M 828, 833, 843, 1080, 1087
Oswald H 112, 123, 1080
Oswald R 323, 329, 1080
Ottens H F L 217, 1098
Ottmann T A 548, 555, 1029
Ottoson L 73, 80, 1080
Overgaard G 414, 424, 1060
Overmars M H 544, 547, 556, 1062, 1080
Owen R B 970, 978, 1067
Ozsoyoglu G 100, 102, 1080
Packard J M 970, 978, 1048, 1087
Padilla D K 971, 975, 1023
Padmanbhan G 1080, xxvii
Paelinck J 244, 251, 286, 292, 1080
Pahl P J 483, 487, 491, 1065

Painho M 237, 265, 291, 1024
Pallage M 825–6, 1048
Palm R 847, 849, 857–8, 1057, 1080
Palmer B 529, 541, 1049
Palmer C 325, 329, 1082
Paloheimo J 971, 976, 1045
Pannatier Y 209, 216, 1080
Pantazis D 416, 424, 1080
Panuska J C 986–8, 996, 1080
Papadias D 325, 328, 1062
Papageorgiou G J 607, 610, 1080
Papathanassiou K 110, 113, 123, 1065
Paradis J 221–3, 233, 639–41, 1080
Parfitt J 849, 856, 1031
Parker D 80, 282, 384, 936, 1071, 1079–80
Parker G R 970, 977, 1054
Parker P 91, 102, 180, 188, 878, 888, 1091
Parkes D N 91, 102, 180, 188, 1080
Parkes S E 928, 938, 1078
Parks B O 67, 69, 123–4, 490–1, 542, 858, 880, 969, 977–8, 994, 997, 1039, 1045, 1052, 1058–9, 1076, 1084, 1087, 1091, 1096
Parlange M B 992, 996, 1076
Parrot R 855, 858, 1080
Parton W J 972, 976, 1032
Paruelo J M 969, 979, 1091
Pas E I 833, 843, 1080
Pascoe R T 327, 1035
Pastor J 970, 979, 1080
Patel K 454, 464, 1051
Paterson M S 388, 399, 1080
Pathan S K 499, 502, 1083
Pathirana S 202, 204, 1047
Patkar V N 499, 502, 1083
Paton N 383–4, 1080
Patten B 974, 978, 1061
Pattenden S 927, 937, 1044
Patterson B 320, 327, 1024
Patterson J 734–5, 743, 1080
Patterson P 534, 540, 1026
Pau L F 525, 1051
Paulsson B 880, 888, 1080
Pausch R 563, 565, 1081, 1092
Pavlidis T 132, 155, 551, 556, 562, 565, 1088
Pawlak Z 197, 203–4, 1081
Pazner M 405, 411, 1063
Pearman H 957, 965, 1081
Pearson D M 969–71, 976, 1026
Pearson E J 845, 848, 858, 1096
Pearson M L 850, 858, 1086
Peart R M 987, 994, 1038

Pederson E 85, 88, 1081
Pennell K D 984, 995, 1058
Peñuelas J 452, 464, 1046
Peplies R W 970, 976, 1042
Pereira J M C 970, 979, 1081
Perkins W A 530, 542, 1081
Pernici B 101–2, 1070
Perrée T 273, 281–2, 1079
Perrings C 973, 979, 1081
Perritt H H 648, 651–2, 772–9, 786, 1081
Perry B 928, 937, 1066
Perry J N 503, 525, 1081
Perry M I 892, 899, 1065
Perry R 854, 858, 1081
Perttunen J 970, 979, 1087
Peshkin M A 516, 525, 1081
Petach M C 990, 996, 1081
Petch J R 42, 47, 968, 979, 1054, 1081
Petchenik B B 54, 60, 1081
Petersen J K 411, 1081
Peterson C 983, 990, 996–7, 1081, 1095
Peterson D L 452, 466, 534, 541–2, 1026, 1076, 1098
Peterson G A 538, 542, 1074
Peterson G W 110, 113, 123, 1088
Peterson M P 159, 170, 173, 1081
Peterson T A 992, 995, 1053
Petrie G M 448, 450, 892, 899, 1065, 1081
Pettigrew A M 628, 630, 1081
Petrochemical Open Software 356, 358, 1081
Peucker T K 73, 80, 129, 153, 396, 399, 530, 542, 1042, 1081
Peuquet D J 2, 5, 25–6, 35, 38, 68, 70, 85, 88, 91–103, 353–5, 358, 384, 405, 487, 503, 525, 621, 631, 850, 894, 899, 926, 1000, 1010, 1081
Peutherer D 625, 630, 1050
Pfeffer J 627, 630, 1081
Phatak V K 499, 502, 1083
Philip G M 483, 491, 520, 526, 1097
Phillips D L 112, 120–1, 982, 994, 1039
Piaget J 84, 88, 1082
Pick H 89, 1093
Pickering A 763, 765, 1081
Pickles J 24, 27, 33, 37–8, 45–7, 49–60, 63–5, 70–1, 79–80, 186, 237, 663, 752, 756, 1010–11, 1017, 1038, 1052–3, 1056, 1072, 1082, 1085
Pidd M 852–4, 858, 1089
Pienovi C 131, 152, 200, 204, 544, 547–8, 555, 1040
Pierce T 946–7, 950, 1069
Pierson S M 198, 203, 1033
Pignol V 516, 525, 1082
Pigot S 94, 102, 1082

Pike R J 114, 123, 539, 541, 1041, 1082
Pile S 45, 47, 1082
Pilon D 325, 329, 1082
Pinker S 82, 88, 1082
Pinto J K 620, 626, 630, 1077
Piper S 913, 924, 1046, 1082
Piransesh H 374, 382, 384, 1054
Pissinou N 555, 1069
Pizano A 407, 411, 1082
Pla M 151–2, 1026
Planchon O 108, 123, 986, 996, 1083
Plane D A 46, 1051
Planning 961, 965, 1082
Platt R H 853, 857, 1043
Plazanet C 136, 138–44, 149, 154–5, 1082, 1087
Pleuwe B 4, 20, 323–5, 329, 1082
Plotnick R E 970, 979, 1082
Plowden F J 583–6, 601, 609, 770, 773, 786, 1048
Pocknee S 981–3, 990–2, 997, 1095
Podaire A 461, 464, 1041
Podolcsák Á 866, 875, 1076
Poeter E P 484, 492, 1100
Poiker T K 53, 57, 60, 748, 756, 1082
Polyorides N 725, 732, 1082
Poon J 245, 248, 1034
Popko E 321, 329, 1082
Popper K 65, 70, 1082
Porter W M 456, 466, 583, 1096
Portugali J 36, 38, 1082
POST 773, 786, 1082
Post W M 982, 989, 996, 998, 1082, 1101
Potts R B 832, 842, 1063
Potvin C 242, 251, 1082
Powell K 953, 964, 1045
Powell M J D 486, 492, 1082
Power G M 255, 260, 265, 1056
Powitz B M 145–6, 151, 153, 1054
Poxon J 953, 964, 1033
Prade H 496, 502, 1042
Prata A J 457, 465, 1082
Pratchett T 191, 204, 1082
Prather T S 984, 996, 1082
Prato T 852, 857, 1049
Premerlani W 353, 358, 363, 369, 380–2, 384, 1087
Preparata F P 544, 547, 550, 556, 1066, 1082
Prescott J R V 194–5, 198, 204, 1039, 1082
Pretty R W 319, 329, 1082
Price D J de S 763, 765, 1083
Price J C 452, 457, 465, 1083
Price K P 971, 978, 1058
Pries R A 110, 123, 1083

Priestal G 530, 542, 1083
Primavesi A L 199, 204, 1083
Prince W 841–2, 1038
Prisley S P 637, 641, 1083, 1090
Probst J R 972, 979, 1083
Proctor J D 6, 20, 49, 55, 60, 1100
Provancha M J 970, 976, 1031
Pugh D 957, 965, 1083
Pugh J 362, 369, 1028
Puglisi G 110, 113, 123, 1065
Pullar D 754, 756, 1089
Puppo E 132, 154, 544–7, 551–6, 974, 976, 1035, 1040, 1083, 1093
Putzolu G 374, 384, 1026
Pyramid Research 821, 826, 1083
Qi Y 971, 979, 1083
Qian L 94–6, 102, 1081
Quang P X 244, 251, 1083
Queen L P 971, 977, 1057
Quek S 325, 327, 1025
Quimby W F 984, 995, 1063
Quinn P F 108, 123, 986, 996, 1083
Radcliffe N J 973–6, 1035
Radke J 494, 502, 948, 858, 1033, 1083
Radner R 785, 1025
Raetzsch H 112, 123, 1080
Raghavan P 555, 1042
Ralphs M P 957, 965, 1087
Ralston B A 841–2, 1041
Ramachandran B 68, 70, 95, 102, 1083
Ramsey R D 970–1, 978–9, 1058, 1078
Rao L 78, 80, 274, 282, 1079
Rao M 499, 502, 1083
Rao P S C 984, 995, 1058
Rao P V 984, 995, 1058
Raper J F 5, 11, 24–5, 45, 54, 61–71, 74, 80, 105, 187, 203, 282, 353, 374, 407, 539–42, 564–5, 658, 725, 731, 749, 756, 850, 956, 959, 962–4, 974, 1010, 1029, 1048, 1065, 1079, 1083–4
Rappoldt C 212, 216, 1042
Ratcliffe D A 969, 979, 1083
Ratick S J 848–9, 857, 1044
Raubal M 404, 412, 1094
Rauscher H M 970, 976, 1039
Ravenhill W L D 44, 47, 1083
Ravenstein E G 240, 251, 1083
Raynal L 129–30, 152, 1041
Rebello K 840, 842, 1037
Rector J M 816–18, 1083
Redfern P 73, 80, 1084
Redman T C 180–2, 188, 1084

Reed B C 983, 988, 994, 996–7, 1031, 1068, 1092
Rees P H 833, 844, 1099
Rees W G 453, 461, 465, 1084
Reeves C 167, 170–2, 225, 232, 1041
Regnauld N 144, 154, 1084
Reichenbacher T 138–9, 146–9, 154–5, 1084, 1099
Reid H G 714, 722, 1084
Reid L B 489, 491, 1072
Reif J 544, 547, 556, 1084
Reilly W J 240, 251, 1084
Rein M 727, 732, 1089
Reinhardt A 112, 122, 1084
Reinhardt W 321, 329, 1043
Reisner P 403, 411, 1084
Reiter P 243, 250, 1075
Rejeski D 639, 641, 845, 858, 1084
Remetey-F‚ll^pp G 866, 875, 1076
Renard K G 986, 996, 1084
Renka R J 483, 492, 1084
Renner S A 684, 692, 1084
Replogle M 843, 1084
ReVelle C S 297–300, 303, 1035, 1084, 1094
Rewerts C 988, 994, 1044
Rey S 242, 245, 248–9, 1025, 1047
Reybold W U 987–9, 994, 996, 1030, 1084
Reynecke C D 914, 924, 1061
Reynolds H 243, 248, 1024
Reynolds J 714, 722, 1084
Rhind D W 1–20, 45–7, 55–9, 62, 72–4, 80, 83, 88, 124, 152–4, 177, 188, 203–4, 265, 268, 282, 308, 311, 316, 334, 338, 362, 384, 401, 411, 425–7, 437, 448, 463–4, 475, 490–2, 537, 563, 572, 580, 584–5, 601–3, 607, 610, 615, 620, 669, 672–4, 694–6, 703, 706, 721, 726, 732, 745–8, 755–6, 765–87, 792, 802–3, 807, 818, 829, 842–4, 856–9, 861–5, 875, 885–7, 893, 897, 905, 935–8, 953–4, 957–8, 964–5, 974, 979, 1009–21, 1028–9, 1032–41, 1045–9, 1053, 1056–60, 1069–70, 1074–5, 1079–80, 1083–7, 1095–9
Rhoda R 298, 303, 1057
Rhyne T 167, 173, 1090
Ribarsky W 561–5, 1063, 1067
Riccio D 110, 113, 123, 1065
Richards J A 454, 465, 1085
Richards K S 124, 1099
Richardson C W 988, 996, 1085
Richardson D E 131, 145, 154–5, 1085, 1093
Richardson J S 971, 975, 1023
Ridley M 825–6, 1085
Rieger M 134, 154, 1085
Riemann H 849, 956, 1030

Rietveld P 494, 502, 1060
Rigau-Peres J G 243, 250, 1075
Riise T 931, 938, 1085
Riley J 752, 756, 1051
Ringrose S 984, 996, 1085
Ripley B D 243–4, 251, 285, 292, 1085
Rippa S 544, 556, 1085
Ritman K 970, 978, 1066
Rivier N 516, 525–6, 1081, 1085
Rivoirard J 484, 492, 1085
Rix D 754, 756, 836, 843, 1067, 1076, 1085
Rizzo B 327–8, 1052
Robert P C 994–7, 1031, 1068, 1076, 1085, 1091–2
Roberts S M 45, 47, 1085
Roberts Y 282, 1095
Robertson G P 971, 979, 1085
Robertson J G M 971, 979, 1085
Robertson P K 224, 227, 230, 233, 1085
Robey D 626, 630, 1071
Robinson A H 44, 47, 83, 88, 159, 172, 220, 233,
    241, 251, 1085
Robinson C J 281, 291, 1029
Robinson D P 242, 245, 250, 1062
Robinson G J 149, 154, 1085
Robinson G K 8, 20, 1085
Robinson J T 386, 400, 1085
Robinson V B 44, 47, 85, 88, 197, 204, 529, 534–7,
    541–2, 1049, 1069, 1085–6
Robinson W A 771, 787, 1086
Rodcay G K 835, 843, 1086, 1097
Rodrigues L 927, 937, 1044
Rodriguez-Iturbe I 528, 531, 541–2, 1059, 1093
Roessel J W van 377, 384, 1086
Rogers A 241, 251, 285, 292, 1086
Rogers C 806, 816, 818, 1086
Rogers D S 715, 721, 1039
Rogers E H 626, 631, 1086
Rogers G O 854, 858, 1091
Rogerson P A 247–9, 264, 282, 292, 316, 509, 525,
    567, 579, 936–7, 1026, 1028, 1048, 1055, 1079
Rogowski A S 484, 492, 984–5, 996–7, 1086
Rohrer J 930, 938, 1087
Roller D 490, 1049
Rolstad J 972, 978, 1059
Romme W H 970, 979, 1077
Ronai M 769, 785, 1034
Roos T 155, 1098
Rooyen M M van 914, 924, 1061
Root R 970, 975, 1023
Rooyen M M van 914, 924
Rosch E 83, 88, 1086

Rose G 45–7, 927, 939, 1044, 1082
Roseman S 390–1, 399, 1045
Rosen P A 109–10, 124, 1101
Rosenberg J B 387, 393, 400, 1086
Rosenblueth E 211, 216, 1086
Rosenblum L 264, 1032
Rosenfeld C L 850, 858, 1086
Rosenthal A S 684, 692, 1084
Roshannejad A A 95, 102, 1086
Rosing K E 301, 303, 1086
Rosing-Vogelaar H 301, 303, 1086
Ross J 453, 464, 466, 1076, 1095
Ross S M 211, 216, 1086
Rossi R E 971, 979, 984, 995, 1067, 1086
Rossing W A H 212, 216, 1060
Rossiter D J 943, 946, 950, 1061
Rota G C 316, 1073
Roughgarden J 971, 979, 1086
Rouhani S 484, 492, 1086
Roujean J-L 453, 465, 1086
Rourke T J 852, 858, 1064
Rousseau D 151, 154, 1086
Rousseau T 151, 154, 1086
Roussopoulos N 348, 358, 395, 399–400, 1045,
    1067, 1086
Routh K 852, 858, 1087
Roux F G 421, 424, 1087
Rowe L A 382, 384, 401, 412, 1092
Rowe P 933, 937, 1057
Rowell R C 829–30, 843, 1086
Rowlingson B S 243, 249, 926, 930–2, 936–7, 1029,
    1041, 1050
Royal Town Planning Institute 878, 888, 1086
Rozema W J 114, 123, 1082
Ruas A 131, 136, 138–43, 150, 154–5, 1086–7
Rubright P R 984, 987–8, 997, 1100
Rucklidge W J 130, 153, 1059
Ruggles A 167, 173, 1090
Ruggles C 692, 1061, 1067
Rule J B 626, 631, 759, 765, 1087
Rumbaugh J 353, 357–8, 363, 369, 380–2, 384,
    414–18, 420, 424, 1030, 1087
Rumley D 198, 204, 1087
Rundstrom R 57, 60, 1087
Running S W 105, 123, 462, 465, 534, 541–2, 971,
    979, 982, 997, 1026, 1076, 1086–7
Runyon T 473, 475, 1087
Rural Development Commission/Housing
    Corporation 959, 965, 1087
Rusak Mazur E 131, 155, 1087
Rush R M 970, 976, 1042

Rushton G 172, 188, 298, 301, 303, 609, 621, 630, 716, 721, 768, 786, 828, 833, 843, 881–2, 885–7, 926, 930–1, 938, 1025, 1040, 1052, 1076–8, 1080, 1087

Rust R H 994–7, 1031, 1068, 1076, 1085, 1091–2

Ruttenberg S 452, 462, 466, 1094

Ryan P J 537–8, 541–2, 984–5, 995, 1050, 1090

Rydin Y 953, 958, 965, 1087

Rystedt B 73, 77, 80, 1076, 1080

S+Spatial Stats 244, 247, 251, 1087

Saarenmaa H 970, 979, 1087

Sabbagh G J 984, 988, 995, 997, 1050, 1087

Sabel C E 957, 965, 1087

Sack J R 555–6, 1031, 1083

Sack R 102, 1087

Sacks S 929, 938, 1089

Sadler G J 880, 887, 1027

Sadowski F 452, 462, 466, 1094

Saeed S E O 825–6, 1087

Sétersdal M 968, 979, 1087

Sahay S 621, 625, 628, 631, 1087

Sainsbury R M 196, 204, 1087

Sale R D 220, 233, 1085

Salgé F 10, 13, 62, 151, 179, 183–4, 188, 220, 226, 332, 349, 352, 356, 436, 574, 584, 615, 653, 660, 684–5, 693–706, 738, 770, 783, 862, 891–3, 897, 906, 1010–12, 1020, 1031–2, 1087

Salomonson V V 457, 462–5, 1025, 1087

Salski A 970, 979, 1087

Sambridge M 110, 123, 520, 526, 1087

Samet H 131, 140, 155, 388, 391, 393, 506, 526, 550, 556, 1088

Sampat Kumar D 499, 502, 1083

Samuelson P 776, 787, 1088

Sanchez R 849, 856, 1030

Sandgren U 769, 787, 1088

Santiago M 243, 250, 1075

Sargent B 852, 857, 1049

Sarjakoski T 439, 450, 1088

Sarnak N 550, 556, 1088

Sasowsky K C 110, 113, 123, 1088

Sasse C 945, 950, 1055

Sastry S V C 499, 502, 1083

Saunders D A 967, 973, 977, 979, 1057, 1076

Savigny D de 925, 938, 1076, 1088

Sawyer J E 992, 997, 10878

Sayer A 43, 47, 1088

Scarano J G 684, 692, 1084

Scarlatos L 132, 155, 551, 556, 562, 565, 1088

Scarponcini P 829, 843, 1088

Scarrow H A 950–1, 1042, 1067, 1076

Scateni R 555, 1035

Schaeben H 110, 121, 487, 490, 1026

Schaerstrom A 926–7, 938, 1088

Schalkoff R J 462, 466, 1088

Schardt J A 418, 424, 1069

Scheepers C F 914, 924, 1061

Schein E H 626, 631, 1088

Schein R H 45, 47, 1085

Schek H-J 412, 1092

Schenkelaars V 129, 131, 154, 398–9, 1078

Scheuring R 180, 188, 1034

Schilling D A 297–8, 303, 1038, 1088

Schimel D S 244, 250, 972, 976, 1032, 1073

Schirra S 147, 150–2, 1029

Schlegel A 150–1, 155, 1088

Schlesinger M 103, 1090

Schmidt C 145–6, 151, 155, 1054

Schmidt J W 269, 274, 282, 411, 837, 843, 1054, 1080, 1089

Schmiegelow F K A 970, 979, 1089

Schneider J B 828, 843, 1089

Schneider R 395, 399, 1028

Scholten H J 216, 247–9, 264, 287–9, 291, 502, 639, 641, 706, 884, 888, 925, 937, 1024, 1042, 1046, 1060, 1066, 1076, 1079, 1098

Schon D A 727, 732, 1089

Schotman A 970, 980, 1096

Schott H F 197, 204, 1062

Schrage M 326, 329, 1089

Schreier G 444, 450, 1089

Schröder P 140, 155, 562, 565, 1089

Schroeter H 1007, 1089

Schüller J K 698, 705, 880–2, 888, 990–2, 997, 1031, 1088

Schulz T T 970, 979, 1088

Schumaker L L 492, 1082

Schumm S A 531, 542, 1088

Schwabisch M 110, 113, 123, 1065

Schylberg L 145, 155, 1088

Scogley E O 990, 994, 1033

Scopigno R 544, 547, 552–3, 555, 1035, 1040

Scott G 770, 787, 1088

Scott J M 968, 979, 1089

Scovell L M 841–2, 1042

Seaman J W 197, 204, 1066

Sedunary M E 323, 329, 1089

Seeger B 395, 399, 1028

Seeger H 83, 425–36, 438, 445, 655, 663, 679, 778

Segev A 93, 100–2, 1060

Seidel R 548, 556, 1054

Sellers P J 452–3, 465, 1063, 1075

Sellis T 395, 399, 1045
Sellwood R 957, 965, 1089
Selman P H 974, 976, 1039
Selvin S 929, 938, 1089
Sen A K 604, 610, 1089
Sen L 327–8, 1070
Sen S 544, 547, 556, 1084
Senior M L 5, 75, 237, 240, 709, 794, 909, 919, 925–38, 1089
Sennett R 766, 1099
Serazzi G 326, 328, 1046
Settle J J 453, 464, 1026
Sevcik K C 393, 399, 1076
Seymour J 763, 765, 1089
Shaffer M J 984, 998, 1100
Shamos M I 550, 556, 1082
Shand P J 706, 1020, 1069, 1087
Shapira A 544, 548–9, 556, 1089
Sharif A R B M 86, 88, 1089
Sharir M 544–7, 550, 554–6, 1036, 1043, 1062, 1080, 1089
Sharma P 754, 756, 1089
Sharp B L 919, 924–6, 932, 938, 1089
Sharp E 953, 964, 1033
Sharpe B 919, 924, 1066
Sharpe R 879, 888, 1076
Shaw M A 954–6, 964, 1041
Shea K S 126, 133–6, 143–5, 153–4, 1072, 1089
Shekita E 374, 382, 384, 1054
Shelley P 680, 692, 1061
Shephard G C 503–4, 506, 525, 1024, 1054
Shepherd I 560, 565, 752, 756, 1051, 1089
Shepherd J R 71, 1083
Shepherd J W 74, 80, 437, 451, 880, 885–7, 909–10, 953–65, 971, 1027, 1029–30, 1089
Sheppard E S 42–3, 47, 50–1, 54, 57, 60, 1089
Sheridan T B 563, 565, 1071
Shiffer M J 45–7, 69, 83, 312, 407, 412, 585, 667, 707–8, 716, 723–32, 856, 862, 881, 885, 888, 922, 947, 1019, 1089, 1099
Shlaer S 419, 424, 1089
Shneiderman B 408–10, 412, 1023, 1089
Shoham Y 98, 103, 1089
Shor P W 548, 555, 1035
Short D 117, 121, 537–8, 542, 1040, 1090
Shortread A 641, 1083
Shortridge J 953, 964, 1045
Show I T 99, 103, 1089
Shrestha R L 208, 216, 1091
Shreve R L 528, 531, 542, 1089
Shugart H H 971, 990, 1097

Shu-Quiang W 849, 858, 1089
Shyu M-J 254, 264, 1028
Sibson R 110, 123, 519–20, 523, 526, 1089
Siccardi F 849, 858, 1065
Siderelis K 585, 708, 733–43, 767
Sidjanin P 564–5, 1064
Siebert A J 453, 465, 1053
Siegel D 829, 842, 1052, 1078
Siegel S 568, 580, 1089
Siegl B 255–6, 266, 1095
Sikos T T 291, 1046
Silva F N de 852–4, 858, 1089
Silva J P 560–5, 885, 888, 1041, 1048, 1076
Silverman B W 283, 292, 1090
Simberloff D 970, 979, 1053
Simkowitz H J 827–8, 835–6, 843, 1089, 1097
Simmel G 758, 766, 1090
Simmons A J 91, 103, 1090
Simmons M A 971, 979, 1090
Simon D 841, 843, 1090
Simon H A 973, 979, 1090
Simon T P 983, 996, 1078
Simonett D S 454–5, 464, 1039
Simpson D W 624, 630, 1059
Simpson I A 970, 979, 1090
Singh V P 122, 1053, 1063
Sinton D F 178, 188, 877, 888, 1090
Sivan R 550, 556, 1088
Sivaplan M 108, 120–1, 1030
Six H-W 387, 399, 1059
Skidmore A K 537–8, 542, 1090
Sklar F H 99, 103, 1090
Skole D 452, 462, 466, 1094
Sloan T M 327–8, 802, 818, 1051, 1090
Slocum T A 167, 172–3, 1044, 1090
Smallbone K 928, 932, 936, 1036
Smans M 929, 938, 1090
Smart J J 38, 1060
Smets G 564–5, 1064
Smith B 65, 70, 957, 965, 1090
Smith D A 362, 369, 602–8, 610, 1090
Smith D C 408, 410, 412, 1090
Smith D M 44, 47, 1090
Smith J L 208, 216, 637, 641, 968, 975, 1023, 1062, 1083, 1090
Smith K G 847, 858, 970, 976, 1035, 1090
Smith L C 901, 908, 1091
Smith M O 457, 466, 778, 787, 932, 938, 1068, 1090
Smith N S 52, 60–2, 72–4, 83, 151, 220, 233, 338, 427, 448, 463, 475, 537, 584, 607, 615, 653–66, 674, 703, 768–70, 779, 783–5, 829, 859, 862, 885–6, 893, 897, 957, 1010, 1033, 1087, 1091

Smith R 827–9, 843, 1070
Smith R B 970, 976, 1031
Smith S J 47, 1086
Smith T 401, 412, 463–4, 1044, 1090
Smith T R 531, 534, 542, 574, 580, 1090
Smith W H F 111, 123, 1090
Smith W L 734, 743, 1080
Smith Patterson J 585, 708, 733–43, 767
Smyth C S 1012–13, 1021, 1090
Snodgrass T R 35, 38, 93, 100–2, 1060, 1080, 1090
Snyder J P 433, 436–8, 450, 1090–1
Snyder R D 982–4, 987–90, 994–7, 1039, 1060,
    1076, 1099
Soccodato F M 530, 541, 1054
Soil Classification Working Group 198, 205, 1091
Soil Conservation Service 989, 997, 1091
Soil Survey Staff 198–9, 205, 989, 997, 1091
Sokal R R 971, 979, 1091
Soller D R 608–9, 1029
Solomon C 841–2, 1035
Soluk D A 971, 975, 1023
Sondheim M 4, 10, 95–6, 102, 308, 322, 347–59, 369,
    382, 402, 410, 415, 418, 423, 564, 576, 584, 681–3,
    694, 698, 725, 802, 850, 865, 1010, 1028, 1064
Sonis M 36, 38, 1082
Sonnen D 335, 345, 1091
Sorensen J H 854, 858, 1091
Sorensen J W 849, 858, 1092
Sorensen P 301, 303, 833, 841, 1035
Soriano A 969, 979, 1091
South West Thames RHA 621, 631, 1091
Sox J 852, 857, 1053
Spangrud D J 992, 997, 1091
Sparks L 710, 717, 722, 1091
Speaker R C 942, 950, 1056
Speer E 323, 327, 1037
Speight J G 115, 118, 124, 539, 541, 1072, 1091
Spiess E 128, 155, 1091
Springer C 509–10, 525, 1061
Sproull L S 626, 631, 1091
Sridhar V 454, 464, 1051
Srinivasan R 988, 994, 1044
Srivastava R M 8, 20, 258, 265, 283, 292, 483–4,
    491, 928, 937, 1060
SSC 127, 155, 1091
Stafford H A 46, 1051
Stafford S G 975, 1026, 1034, 1069
Stamp D 954, 964, 1091
Stamps J A 970, 979, 1091
Stangenberger A G 982, 989, 996, 998, 1082, 1101
Stanislawski L V 208, 216, 1091
Stanley D J 457, 464, 1048

Stanney K 563, 565, 1091
Star J L 2–3, 20, 339, 345, 667, 675, 1091
Star T 463–4, 1044
Starks P 109, 122, 1049
Starr J 841–2, 1035
State A 559, 565, 1091
Staufer P 244, 249, 1046
Stearns F 182, 188, 1091
Steckler G A 992, 994, 1032
Steel C 866, 875, 1091
Steel D G 243, 250–1, 1058, 1100
Stegun I A 485, 490, 1023
Stein A 196, 204, 210, 216, 1057
Stein J A 116, 120, 123, 1067
Stein J L 116, 120, 123, 1067
Steinitz C 405, 412, 878, 888, 1091
Steinnocher K 244, 249, 1046
Stenseth N C 972, 980, 1099
Stephan E-M 179, 183, 188, 1031
Stern E 838, 841, 1031
Stern R M 925, 928, 932, 937, 1044, 1066
Steven M D 452, 464, 1039
Stevens M 928, 938, 1078
Steward H J 134, 155, 1091
Stewart A R 651, 1024
Stewart J Q 240, 251, 1091
Steyaert L T 67, 69, 108, 119, 121, 123–4, 490–1,
    542, 858, 880, 969, 977–8, 983, 994, 997, 1000,
    1007, 1039, 1045, 1052, 1058–9, 1076, 1084,
    1087, 1091–2, 1096
Stiglitz D 604, 610, 1092
Stileman M R 453, 464, 1026
Still D A 988, 995, 1066
Stillman S T 982, 997, 1092
Stillwell J C H 502, 1060, 1079
Stitt S C F 970, 975, 1023
Stoakley R 563, 565, 1092
Stockwell D R B 970, 976, 1039
Stockwell J R 849, 858, 1092
Stolfi J 548, 556, 1070
Stolorz P 409, 411, 1046
Stoms D A 301, 303, 1050
Stoms D M 295, 639, 641, 968, 970–1, 979, 1035, 1092
Stone H F 185, 188, 1092
Stone J F 985, 988, 995, 1050
Stone M 409, 412, 1092
Stone N D 970, 979, 1087
Stonebraker M R 100–1, 374, 382–4, 402, 412, 1054,
    1056, 1092
Stopher P R 828, 843, 1092
Storm J D 828, 834, 843, 1073

Storrier A L G 934, 937, 1057
Stothers N 877, 887, 1027
Stout J 296, 303, 1092
Stow D A 972, 979, 1092
Stowe T 113, 121, 1030
Stra-kraba M 974, 978, 1061
Strahler A H 457, 460–1, 465–6, 1062, 1087, 1097, 1100
Strahler A N 457, 460–2, 464–6, 528, 542, 1027, 1092
Strand E J 321–2, 325, 329, 1092
Strandburg K J 516, 525, 1081
Strasser T 904, 908, 1092
Straten G van 216, 1060
Streb D 320, 329, 1092
Strebel D E 457, 465, 1055
Stuart R 560, 565, 1092
Stuetzle W 254–6, 264, 1032, 1092
Stutz F P 855, 858, 1080
Stynes K 78, 80, 1100
Subaryono 197, 202–3, 205, 1097
Sudduth K A 991, 997, 1092
Sudweeks F 885, 888, 1038
Sugarbaker L J 314, 337, 583, 587, 605–6, 611–20, 622, 663, 709
Sugihara K 242, 250, 518–20, 523–5, 1078
Suguhara G 970, 977, 1064
Sui D Z 50–2, 60, 1092
Summerside A 625, 630, 1072
Sun F 829, 844, 1096
Sun G 208, 216, 222, 227, 232, 969–71, 977, 1052
Sunny K R 984, 992, 994, 1045
Surbey C 185, 188, 1065
Sutherland I 407, 412, 1092
Sutherland M 460–1, 1027
Sutter J F 608–9, 1029
Suzuki A 518, 525, 1078
Svensson P 405, 412, 1092
Svorou S 85, 88, 1071
Swain R 299–303, 1084, 1092
Swann D 338, 660, 665, 672, 697, 718, 799, 889–99, 1017
Swanson T M 968, 979, 1092
Swanwick C 956, 964, 1043
Swayne D F 254–6, 264, 1032
Sweldens W 140, 155, 562, 565, 1089
Sylvester J J 396, 400, 1092
Symanzik J 256–9, 265, 1037, 1070, 1093
Taaffe E J 831, 833, 841, 843, 1093
Tagashira N 243, 250, 1078
Takaki R 525, 1085
Takeyama M 36, 38, 1093
Talen E 258, 266, 1093
Talmi A 484–6, 492, 1093

Talmy L 83, 89, 1093
Tamminen M 386, 400, 1093
Tanca L 383–4, 1034
Tang L 112, 121, 1026
Tanton M T 970, 978, 1067
Taormino T (Plate 13) 221–2, 225, 232
Tarboton D G 531, 542, 1093
Tarjan R E 550, 556, 1088
Tasco C 928–30, 936, 1031
Tate N J 928, 938, 1068
Tavernier R 198, 205, 1093
Taylor D R F 158, 172–3, 232, 1031, 1064, 1069, 1090, 1093
Taylor F W 625, 631, 1093
Taylor G H 982, 994, 1039
Taylor J R 191, 196, 205, 209–10, 217, 1093
Taylor M 680, 692, 1061
Taylor M A P 838, 843, 879, 888, 1076, 1093
Taylor P 833, 843, 970, 979, 1093
Taylor P J 43, 46–7, 51, 60, 73, 80, 243, 251, 569, 580, 751–2, 756, 945, 951, 1072, 1080, 1093
Teillet P 452, 462, 465–6, 1087, 1094
Teitz M B 301, 303, 1093
Teng Y A 544, 547–9, 556, 1093
TeSelle G W 983, 987–9, 995–6, 1062, 1084
Tessar P 320, 328, 1038
Thapa K 180, 188, 1093
Theocharopoulos S P 203–4, 1039
Thoen B 321, 329, 1093
Thomas C D 938, 972, 977, 980, 1055, 1093
Thomas C J 722, 1100
Thomas D H L 968, 975, 1023
Thomas E N 241, 251, 778, 787, 1090, 1093
Thomas J 410, 412, 1100
Thomas J M 971, 976, 979, 1038, 1090
Thomas R 936, 1041, 1093
Thomas R G 984, 997, 1096
Thompson D 2, 753, 756, 1093
Thompson R C 145, 155, 1093
Thompson R J 769, 780–2, 786, 1042
Thompson S K 244, 251, 1093
Thompson W D 926, 937, 1062
Thomson M C 931–2, 938, 1093
Thorkilsen M 1003, 1007, 1093
Thorne C R 115, 124, 1101
Thornton P E 105, 123, 982, 997, 1087
Thrall G 5, 10, 73, 78, 162, 168, 254, 307–10, 312, 322–3, 331–45, 360–1, 402, 407, 570, 572, 614, 620, 666, 694, 708–9, 714, 730, 750, 771, 792–3, 802, 815–16, 827, 830, 840, 862, 884, 903, 993, 1010, 1016, 1093–4

Thrift N J 47, 91, 102, 180, 188, 251, 1061, 1080, 1100
Thrower N 170, 173, 768, 787, 1094
Thünen J H von 240, 251, 1094
Tiefelsdorf M 242, 251, 1094
Tilman D 972, 980, 1094
Tim U S 984, 997, 1094
Timmermans H J P 298, 302, 721, 887, 1025, 1040
Timpf S 131, 155, 404, 412, 1094
Tinline R 927, 938, 1094
Tissot A 221, 223, 1094
Tobler W R 7, 20, 77, 80, 170, 173, 239–40, 251,
        284, 292, 477, 482, 487, 492, 550, 555, 575, 580,
        1034, 1094
Todd C S 970, 978, 1067
Todd P 933–4, 936–7, 1031, 1048
Toledano J 494–5, 499, 502, 881, 888, 935, 937,
        1043, 1059
Tolley R 841, 844, 1094
Tomlin C D 3, 20, 36, 38, 99, 103, 207, 217, 405,
        412, 831, 843, 877, 888, 1066, 1094
Tomlinson R F 17, 362, 369, 602–8, 610, 666, 953,
        965, 1085, 1090, 1094
Toms M L 563, 565, 1029
Toppen F J 752, 756, 1094
Toregas C 300, 303, 1094
Toriwaki J 525, 530, 534, 542, 1085, 1094
Tortona J 411, 1054
Tosta N 18, 666, 768, 782, 786, 1094
Townshend J R G 452–4, 457, 462–6, 1048, 1055,
        1095
Traiger I 374, 384, 1026
Tranmer M 243, 250–1, 1058, 1100
Transportation Research Board 837–8, 844, 1094
Travis J 242, 251, 1082
Trevisan J 129–30, 152, 1041
Tribe A 118, 124, 528–9, 532–4, 542, 1094
Trudeau M 331, 345, 1094
Tsai V 110, 124, 1095
Tsang T 454, 457, 464, 1027
Tscharntke T 970, 978
Tsitchrizis T C 66, 70, 1095
Tucker C J 971, 980, 1097
Tufte E R 224–6, 230, 233, 1095
Tukey J W 220, 233, 254, 266, 574, 580, 1095
Tukey P A 221, 232, 254, 264, 1024, 1034
Turkman K 936, 1041
Turnbull D 187–8, 1095
Turner A K 751, 756, 1095
Turner D P 983, 997, 1095
Turner G 561–5, 1063, 1067
Turner K 68–9, 1050, 1083

Turner M G 103, 970–3, 978–80, 1047, 1058, 1077,
        1090, 1095
Turner M L Jr 942, 950, 1056
Turton B 841, 844, 1094
Turton I 268, 282, 1094
Tuttle M 841, 843, 1071
Tuzhilin A 35, 37, 1036
Tyler D A 992, 997, 1091
UCGIS 348, 358, 1010, 1021, 1095
Udo de Haes H A 968, 978, 1063
Ulliman J 968, 979, 1089
UN ECE 859, 875, 1095
UNEP 968, 980, 1095
Unwin A 167, 172, 243, 247, 250, 255–6, 260, 265–6,
        287–9, 291–2, 928, 937, 1056, 1095
Unwin D J 7, 49, 66, 72, 78–80, 158, 172, 204, 232,
        247–9, 287, 291, 563, 569, 580, 585, 599, 620,
        747–56, 792, 849, 858, 936, 956, 964, 969, 1024,
        1033, 1042, 1046, 1050–2, 1056, 1090, 1095,
        1100
Unwin T 968–9, 973–5, 980, 1095
Upton G J 244, 251, 258–60, 266, 286, 292, 1095
Urban D L 967, 977, 1055
US 760–1, 766, 1095
US Defense Mapping Agency 661, 666, 1095
US Geological Survey 149, 155, 349, 358, 1095
US Senate 760, 766, 1096
Usery E L 118, 122, 981–3, 990–2, 997, 1053, 1095
Ustin S L 454, 466, 1095
Vail L W 120, 124, 1099
Vail S G 968, 980, 1095
Väkevä J 970, 979, 1087
Valle J del 337, 345, 1094
Valliére D 415–19, 423, 1028
Vance S 852, 857, 1049
Vanderbilt V C 454, 462–5, 1087, 1095
Vanderpost C 984, 996, 1085
Vane G 452, 456, 466, 1095–6
Vanicek D 428, 436, 1096
Vaughn B M 845, 858, 1096
Vckovski A 179, 183, 188, 1031
Veen F van 509–10, 525, 1061
Vega-Garcia C 849, 858, 1096
Venkatesh H 454, 464, 1051
Ventura S J 638, 641, 990, 997, 1027, 1096
Verboom J 970, 980, 1096
Verdin K L 108, 124, 1096
Veregin H 57, 68, 72, 91, 107, 164, 177–89, 195–6,
        207–8, 216–19, 225, 298, 303, 356, 437, 475,
        681, 1010, 1065, 1096

Verity J 840, 842, 1037
Verly G 491, 1075
Vermande P 461, 464, 1041
Verplank B 408–10, 412, 1090
Vertessy R A 117, 124, 538, 542, 1096, 1101
Vijlbrief T 398–9, 1078
Vincent P 845, 854, 857, 1050
Vincent V 984, 997, 1096
Vinken R 153, 1054
Viridians 199, 205, 1096
Visone D L 899, 1053
Vitter J S 544, 547, 556, 1083
Vlissides J 420, 423, 1049
Voelcker J 319, 329, 1096
Vonderohe A 829, 844, 1096
Voogd H 494–6, 502, 1096
Vorhauer C F 990, 997, 1096
Voss H 326, 328, 1052
Vrana R 91, 103, 1096
Wachowicz M 68, 70, 1096
Wade B 374, 384, 1026
Wade G 984, 997, 1096
Wadge G 845, 848–9, 858, 1096
Wagenet R J 990, 995–7, 1081, 1096
Wagner D F 327, 329, 1096
Wagner D G 984, 998, 1100
Wahba G 484–5, 490, 492, 1043, 1096
Wakeford R 953, 965, 1096
Waldo D 953, 965, 1097
Walford R 46–7, 1053, 1061
Walker B 928, 937, 1066
Walker D A 692, 1061, 1067
Walker J 539, 541, 1072
Walker M D 972, 980, 1097
Walker P A 968–72, 980, 1096–7
Wallace W 728, 731, 1045
Waller L A 928, 932, 937–8, 1060, 1097
Wallin D O 971, 980, 1097
Walmsley D J 943, 951, 1097
Walsby J C 196, 205, 1097
Walsh W J P 829–30, 843, 1068
Walsham G 621, 625, 628, 631, 1087
Walter S D 928, 938, 1097
Walthall C L 453, 466, 1097
Wan Z M 462, 465, 1087
Wang C A 555, 1030
Wang F 197, 202–3, 205, 1097
Wang M W 990, 997, 1089, 1097
Wang Z 145, 154, 1075
Wanner W 460, 466, 1097
Ware J M 131–2, 138, 141–5, 153, 155, 1061, 1097

Warford R 852, 857, 1049
Warita Y 429, 436, 663, 666, 770, 787, 1097
Warnecke L 836, 844, 1097
Warner T A 50, 57, 59, 1056, 1098
Warntz W 240, 251, 528, 542, 1091, 1097
Warren I R 1004, 1007, 1097
Warren J 562, 565, 1068
Warren S 758–9, 763, 766, 1097
Wartenberg D 242, 251, 928, 937, 1060, 1098
Waters N M 293, 310, 333, 337, 402, 406, 437, 620,
    694, 749, 755, 770, 798, 827–44, 856, 934, 1023,
    1025, 1029, 1037, 1097, 1099
Watkins K B 984, 988, 995, 1050
Watson A I 974, 976, 1039
Watson C C Jr 849, 858, 1097
Watson D F 481–2, 486, 492, 520, 526, 1097
Watson G H 110, 119, 124, 1097
Watson V 374, 384, 1026
Watts J A 982–3, 989, 996, 1078
Watts M J 47, 1072
Waugh T C 70, 153–4, 172, 188, 204, 327–8, 802,
    818, 1043, 1051, 1063–5, 1089
Weatherbee O 970, 978, 1060
Weaver J R 300, 303, 1035
Webb T A III 972, 976, 1040
Webcrawler Survey 321, 329, 1097
Weber A 240, 251, 1098
Weber C 167, 173, 407, 412, 1090, 1098
Weber M 604, 608, 610, 625, 631, 1098
Weber R 170, 173, 928, 938, 1098
Webster C J 318, 329, 878–80, 888, 1098
Webster R 114, 123, 198, 205, 214–15, 244, 250,
    483–4, 491, 1071, 1078, 1098
Wedderburn R W M 286, 292, 1076
Wee C 968, 975, 1023
Weesies G A 986, 996, 1084
Wegener M 622, 631, 1098
Wegge P 972, 978, 1059
Wegner L 411, 1044
Weibel R 3, 26, 65, 105–6, 110, 125–55, 181, 196,
    246, 658, 776, 810, 886, 891, 988, 1002, 1010,
    1031, 1072, 1075, 1087–8, 1091, 1098
Weibel S 682, 684, 692, 1098
Weih R C 637, 641, 1090
Weiner D 47, 50, 54, 57–60, 1056, 1079, 1098
Wei-Ning X 852, 857, 1053
Weinrich J 972, 979, 1083
Weiringa J 121, 124, 1098
Weisberg P J 970, 976, 1026
Weiss M J 761, 766, 1098
Welch R 756, 1048

Welles J M 453, 466, 1097
Wells D 467, 476, 1098
Wendroth O 992, 996, 1076
Wentz E 96, 102, 1081
Werner C 109–10, 124, 1101
Werschlein T 140, 155, 1098
Wesolowsky G O 297, 303, 1098
Wessel M 406, 411, 1054
Wessel P 111, 123, 1090
Wesseling C G 208, 217, 1098
Wessels D P 892, 899, 1065
Wessman C A 452, 466, 1098
Westin A F 759, 765, 1098
Westlake A 927, 937, 1044
Whalen T 325, 329, 1082
Wheeler D J 969, 980, 1098
Whelan B M 992, 996, 1072
Whitby M C 964, 1032
White D 399, 411, 1064, 1078
White G F 847, 856, 1033
White H 291–2, 1099
White M 830, 837, 844, 1099
White M L 457, 464, 1041
White R M 221, 224–5, 233, 1072, 1101
Whiteford A 914, 924, 1072
Whittington D 621, 629, 1033
Whorf B L 82, 89, 1099
Wickham J D 970, 980, 1099
Widmayer P 155, 387, 399, 1059, 1098
Wiens J A 970–2, 978, 980, 1061, 1099
Wiggins L L 726, 732, 1049, 1056, 1099
Wigmosta M S 120, 124, 530, 542, 1081, 1099
Wijeyaratne P 925, 938, 1076, 1088
Wild R 935, 951, 1099
Wildschut A 919, 924, 1099, 1101
Wilhelmi F 971, 980, 1097
Wilkerson G G 987, 997, 1099
Wilks A R 221, 232, 255, 264, 1028
Wilkinson G G 462–3, 466, 1099
Wilkinson N 420, 424, 1099
Williams D L 452, 464, 1076
Williams H C W L 719, 722, 935, 938, 1071, 1099
Williams I P 257, 266, 859, 866, 875, 1099
Williams J C Jr 946, 951, 1099
Williams J R 988, 997, 1087, 1099
Williams M 383–4, 1080
Williamson I P 91, 95, 101–2, 1056–7, 1058, 1099
Williamson R A 780–2, 787, 1099
Williamson T 193, 196, 205, 1099
Willis G 167, 172, 1056
Willmott C J 112, 123, 1066

Wills G 243, 247, 250, 255, 260, 265, 928, 937, 1056
Wilms P F 374, 382, 384, 1054
Wilson A G 240, 251, 268, 282, 709, 712, 716, 719, 721–2, 828, 833–4, 841, 844, 935, 1029–30, 1035, 1099
Wilson J P 115, 121, 124, 244, 909–10, 981–98, 1007, 1017, 1039, 1049, 1060, 1074, 1091, 1099
Windham W R 452, 464, 1038
Wingle W L 484, 492, 1099
Winn D S 970, 979, 1078
Winner L 762, 766, 1099
Winston P H 81, 89, 1099
Wirasinghe S C 832, 844, 1097, 1099
Wirth L 758, 766, 1099
Wirtshafter R M 494, 502, 1033
Wise J A 410, 412, 1100
Wise S 113, 124, 262, 265, 284–7, 292, 509, 524, 567, 579, 793–5, 928, 937, 1000, 1017, 1052
Wislocki A P 845, 848, 858, 1096
Wittie L D 319, 329, 1100
Woelk D 374, 384, 1063
Wohl E E 849, 858, 1073
Wojtowicz J 315–16, 1100
Woldenberg M J 43, 47, 528, 542, 1097, 1100
Woldt W 208, 217, 1100
Wolf J K 212, 216, 984, 997, 1042, 1086
Wolfe M L 984, 995, 1055
Wolff R S 565, 1058
Wolfram S 970, 980, 1100
Wolkowski R P 992, 997, 1100
Wollenhaupt N C 985, 992, 995, 997, 1055, 1100
Wolock D M 986, 998, 1100
Wong D W S 73, 80, 243, 249, 273, 282–3, 292, 1048
Wong E 374, 384, 1092
Wong R S 85, 88, 1086
Wood D 79–80, 186, 189, 984, 994, 1030, 1100
Wood J D 78, 80, 158, 167, 172, 221, 226, 232, 482, 492, 1047, 1052, 1100
Wood J S 46, 1051
Wood M 158, 173, 1101
Wood W B 1013, 1021, 1100
Woodard P M 849, 858, 1096
Woodcock C E 457, 465–6, 1062, 1100
Woodhouse J 931, 936, 1043
Woods D D 563, 565, 1029
Woodsford P A 150, 155, 806, 818, 1060, 1100
Woollam R 823, 826, 1100
Worboys M F 2, 11, 31, 35, 38, 66, 95, 100–3, 204, 270, 282, 312, 351–4, 358, 363, 372–85, 403, 414–18, 424, 472, 529, 561, 693, 802–3, 818, 828, 877, 891, 975–6, 1025, 1047, 1074, 1079, 1096, 1100, xxvii

World Commission on Environment and Development (WCED) 968, 980, 1100
World Intellectual Property Organisation 764, 766, 1100
Worrall L 600, 884, 887–8, 1100
Wortman K 696–7, 706, 1074
Wösten J H M 208–10, 216, 984, 994, 1046
Wraith J M 984, 988–90, 995, 997, 1060, 1099
Wren A 837, 843–4, 1088, 1100
Wright D A 988, 996, 1085
Wright D F 970, 976, 1030
Wright D J 6, 20, 49, 55, 60, 1100
Wright J K 221, 233, 1100
Wright J R 842, 1056
Wright R G 968, 979, 1089
Wright S G 530, 541, 1061
Wright Mills C 610, 1098
Wrigley N 44, 47, 80, 243, 251, 286, 291–2, 568, 580, 711, 722, 1080, 1100
Wu J 971, 979, 1083
Wulfsohn T 625, 631, 1100
Wyatt R 888, 1063
Wylie B K 984, 998, 1100
Wymer C 44–7, 272, 282, 930, 938, 1079–80
Wyss M von (Plate 13) 221–2, 225, 232, 1069
Xiang W N 852, 858, 1068
Xie Y 263–4, 312, 316, 462, 465, 723–5, 731, 884, 888, 1028, 1073
Xu T 113, 122, 1059
Yadar P D 499, 502, 1083
Yaeger L 564, 1058
Yager R 497–8, 502, 1100
Yang L 983, 996, 1073
Yang S 131, 140, 153, 208, 216, 222, 227, 232, 969–71, 977, 1052
Yang W 141, 155, 983, 996, 1073, 1101
Yao F F 388, 399, 1080
Yapa L 57, 60, 1101
Yarrington L 484, 492, 1100
Yeh A G-O 723, 794, 799, 877–88, 1101
Yoder D C 986, 996, 1084

Yonker C M 972, 976, 1032
Yoon J 1080, xxvii
Yost R S 212, 216, 1067
Young E 194, 205, 1101
Young F J 985, 995, 1055
Young H P 943, 951, 1101
Yourdon E 414, 423, 1036
Yuan B 192, 204, 1063
Yuan M 97, 103, 192, 204, 405, 412, 1101
Yvinec M 547, 553, 555, 1042
Zadeh L A 197, 205, 1101
Zaninetti L 503, 526, 1101
Zdonik S 403, 410, 1026
Zebker H A 109–10, 124, 461, 466, 1101
Zeigner A 326, 328, 1046
Zerbe R O Jr 601–7, 610, 1101
Zevenbergen L W 115, 124, 1101
Zhan C 531, 534, 542, 1090
Zhan C-X 85, 88, 1081
Zhan F B 272, 282, 831, 844, 1048, 1101
Zhang W 120, 124, 986, 998, 1101
Zhdanov N 769, 787, 1101
Zhexue H 405, 412, 1092
Zhou Y 852, 857, 1049
Zhu A 538, 542, 1101
Ziegler B 67, 70, 326, 329, 1101
Zietsman H L 919, 924, 1101
Ziliaskopoulos A K 832, 843, 1101
Zimmerman H-J 291, 1046
Zinke P J 982, 989, 996, 998, 1082, 1101
Zipf G K 240, 251, 1101
Zlocha M 111, 123, 482, 488, 491, 986, 996, 1074
Zografos K G 852, 858, 1101
Zubin D 64, 70, 85, 88, 1071, 1101
Zubrow E 265–6, 1045
Zurada J M 282, 291, 1029
Zureik E 839, 843, 1068
Zwart P R 320, 329, 1082
Zyda M J 564–5, 1069
Zyl J J van 461, 466, 1101

SUBJECT INDEX

0-dimensional 419–20
1-dimensional 117, 377, 385, 388–90, 393, 419–20,431–2, 987
2-dimensional 6, 33, 67–8, 85–6, 126, 166, 194, 224, 284, 385–90, 393–4, 419, 431–2, 438, 444–5, 987
3-dimensional (Plates 3 and 4) 6, 32, 67–8, 105, 166, 220–1, 224, 314–15, 385, 419, 430–2, 438, 442–4, 467, 557–64, 846, 974, 987
4-dimensional 68, 387
9-intersection model 85–7
Aboav's law 511, 516–17
Absolute position 576–7
Abstract data types 352–3
Abstraction 130, 829
Access to data and information 585–6
Access to health services 933–6
Accessibility 934, 943, 960
Accident analysis 836
Accuracy 178–8l, 191, 439, 449, 467, 489, 570, 762, 804, 846
Ackermann's Function 546
Acyclic graph 394
Adaptive hierarchical triangulation 551
Address
    geocoding 830, 962–3
    matching 959
Address Manager 959–60, 962–3
ADDRESS-POINT 74, 435
Adjacency 84
Advanced travel systems 837
Advanced Very High Resolution 457–9, 671, 983
Advanced Visible and Infrared Imaging
        Spectrometer 456
Aerial photograph (Plate 19) 109, 440–6, 668, 861, 991
Affine transformation 434, 438, 445–6
Aggregation 139–40, 419, 957–8, 1007
        aggregate objects 380
        effects 298
        operators 145, 494–502
AGI see Association for Geographic Information
Agriculture 909–10, 915
    agricultural applications 981–93
Air quality 663
Airy ellipsoid 434
Akima's method 10
Alexandria Digital Library Project 574, 902, 906–7

Algorithm 134, 137–9, 147, 439, 546–50, 832, 891, 986
Allocation Model see Location–Allocation Model
Altimetry 429
AM/FM see Automated Mapping and Facilities
        Management
Ambiguity 192, 197–9
Amplified intelligence 138
Analysing 404–5, 880
Analysis
    model 414
    spatial see Spatial analysis
    tool 1002
Analytic
    hierarchy process 295
    software 570
Analytical techniques 984–7
Anglo-American Cataloguing Rules 905
Animation see Visualisation
Anisotropy 110–11, 484, 486, 975
Annotation mechanisms 725–6
Anti-trust law 644, 775
Apple 312
Applet 10, 840
Applications
    analysts 613
    development 364–8, 615–16, 816, 923
    domain 289
    viewpoint 701
Application-specific methods 487
Apportionment 941–2, 946
Approximation 5 78
ARC/INFO (Plate 15) 150, 312–13, 337, 360, 364–6, 371, 488, 674, 716, 828, 831, 835, 883–4, 928, 947–8, 959–60, 987–8, 990, 992
Arc Macro Language 312, 360, 366
Archimedean tessellations 504–5
Arc routing 832
ArcView(Plate 49) 312–13, 337, 360, 366–7, 674
    Avenue 366–7, 853
Area 943
    data 927, 932
    to surface interpolation 487
Areal
    interpolation 855
    interpretation 283
    phenomena 77

Artificial intelligence 12, 98, 970, 1010–11, 1017
Artificial neural network *see* Neural net
Artillery 894
ASCII 348, 350, 365, 406, 475
Aspatial 100
Assessment 999–1007
Assignment 196
Association 380, 419
Association for Geographic Information 12, 184, 754
Association Liègeoise d'Electricité: case study 825
Astronomy 503
Asynchronous Transfer Mode 320
Atlas 337, 915
    digital 332
Atmospheric windows 455
Atomic polygon 507
Attribute 76, 284, 375, 438–9, 473, 578, 804
    accuracy 184, 594, 636
    completeness 183
    data 100, 473–4, 594, 829, 831, 916, 926
    transformation 134
Auditory channel 559–60
Augmented-reality systems 558
Australia
    and electoral districting 939–50
    case study 778–9
Australian Electoral Act 941
Australian experience 947–9
Autocorrelation 245, 257–8
    spatial 72, 126, 241, 509–10, 577, 971
Automated
    mapping and facilities management 797, 802–18,
        820, 823, 851
    zone design 274
Automatic vehicle
    dispatch system 827, 831
    identification 837
    location system 827, 830, 837, 852
    navigation system 827, 830, 837
Automation techniques 803–5
Availability of digital data 678
AVDS *see* Automatic Vehicle Dispatch System
AVHRR *see* Advanced Very High Resolution
Balance of errors 211–12
Bandwidth 898, 930
Barrack 892–3
Base-maps 804–7, 886–9
Base-plant 890, 896
Base stations 471
Bathometry 473
Battlefield 893–4, 897–8, 1017
Bayes estimation 930

Bayesian 489, 537–8, 970
BBC Domesday 45, 753
Behaviour 82
Beliefs 496
Bell Canada: case study 823–4
Benchmark 598
Benefit–cost analysis *see* Cost-benefit analysis
Bettmann Archive 777
Bidirectional reflectance distribution function 453
BigBook 3D browser 315
Binary 9, 299, 569, 672, 802
    logit 833–4
    tree 550
Bintree 386
Biodiversity 120, 910, 968–9
Bird statistics 1006
Bivariate interpretation 488–99
    map 226, 336, 341
    spline 484–5
BLG-tree 129–31, 397–8
Boolean
    classification 196
    logic 196–7, 494–5, 523, 970
    modelling 86, 196, 200–1, 388–9
    objects 194–6
    operators 98, 478, 495–8, 545–6, 992
    searches 323, 574, 686
    values 382, 549
Bootstrapping 984
Boundary 65, 246
Box map 258
Branching 93, 97
Brightness 167–8, 225, 254–5, 262
British Telecommunications: case study 824–5
Browsing 403–4
Bruntland Report 968
BSP-tree 387–8
B-tree 385
Budgeting 618–19
Buffer zone 714–15, 726, 927, 932
Buffering 344, 830, 877–8, 992
Burden sharing 890–1
Business 709–21, 959–60
Business Week 840
CAD *see* Computer-aided design
CAD/CAM 378
Cadastral
    information 658–9, 783, 822
    mapping 10, 71, 656, 660, 804, 859
Cadastre 859–60
    and organisations 862–3

California Urban Features Model 884
Canonical correspondence 970
Cantor sets *see* Boolean
Cartesian 32, 35
    coordinates 33, 95, 428–31, 433–4, 445
    science 52, 974
Cartogram 74, 78, 168, 929
Cartographer 135, 148, 158
Cartographic
    bias 185–6
    data 167
    modelling 971
    representation 957
    visualisation 221, 310–11
Cartography 33, 76, 99, 126, 166, 220–1, 769 and
    GIS 14
CASE *see* Computer-assisted software
    engineering
Catalogue *see* Data catalogue
Cataloguing 902, 906–8
Catena concept 538
Cathode Ray Tube 128
Causality 99
Cause-and-effect 98
Cell database
Cell *see* Pixel
Cellular automata 99, 970
Census 3, 94, 638, 707, 714, 720, 783, 905, 917, 948
    collector districts 948
    data processing 274–5, 855
    geographies 131, 193
Census of Population (UK) 73, 78, 713
    enumeration district 73, 193, 658, 947, 959
Census of Population (US) 73, 901
Censuses 72, 274, 572
Centre for Geographic Information and Analysis
    (North Carolina) 538
Central Postcode Directory 74
Centralist model 864
Certification 704
Change 92–3
CHANGE system 145–6, 151
Chernoff faces 257
Cholesky decomposition 211
Chorochromatic maps 163–5
Choropleth 10, 78, 165, 167, 258, 336, 339, 510,
    928–9
Chronon 93
CIF quadtree 393
Circumferential compactness 946–7
Cirque 537

Classes 378
Classification 134, 181, 192, 244, 419, 540, 586,
    889, 1002
    of geographical data 192
    of GIS 5
    of landscape 969
Clearinghouse 356, 653, 679, 683, 685–7, 783–4, 903
Client-server model 319, 802, 894
Climate model (Plate 63)
Climatology 576, 663
Clip Art 335–6
Closure postulate 505
Cluster analysis 244, 832, 930–1
Coastal morphology 1006
Code-based receivers 469–70
Cognition 1010
Cognitive science 4, 33, 81–7, 410, 417
Collaborative planning system (Plate 42) 325–6,
    726–8
Combination electoral system 940
Combined Higher Education Software Team 904
Command-line input 408
Command Control and Communication
    Information 892–5
Commerce 585–6
Common Object Request Broker Architecture 351,
    381, 564, 699
Common Transfer Format 349–50
Commonwealth Scientific and Industrial Research
    Organisation 948
Communications 943, 1014–17
Communications Act (US) 823
Communication technology *see* Data transfer
Community of interest 942–3
Compactness 946
Comparison
    of spatial data components 165–6
    of error propagation techniques 211
Compatibility of data structures 298
Competition, change, and GIS 817–18
Completeness 183–4, 594–5
Complete spatial randomness 242
Compound Terrain Index 538
Comprehensive emergency management 845, 853
Computation 546–8
Computational efficiency 490
    intelligence 287–9
    technology 283
Computational zone design 273–4
Computer 569, 759
    architecture 575–6, 1015–16
    as research environment 570

cartography 757
graphics *see* Visualisation
networking 318–21
Computer-aided design 33, 413, 472, 474–5, 478, 695, 802, 836
Computer-assisted software engineering 414–23, 615, 753
Computer-supported cooperative work *see* Collaborative planning and decision-making
Computers and Geosciences 14
Computers, Environment and Urban Systems 14, 827
Computer science – discipline 755, 828
Concatenation 577–8
Concept 95, 413–14
Conceptualisation 71, 105, 627
Conceptual models 132–6, 179, 848
Confidentiality 947
Configuration 947
Conflation 478, 576–8
Conflict 136
Confusion Matrix 196
Connections 288
Connectivity 877–8
Conservation 838, 919
Consistency 136, 182–3, 449
Constituency 940
Containment 84
Content 679–80
Content Standard for Digital Geospatial Metadata 574
Context 454
    characteristics 228
Continental assessment 981–3
Continuous 82, 92
Continuous-visibility
    maps 543, 545–7
    structures 546–8
Contour 108–9
    flowline networks 117–18
    isoline data interpolation 487
Contract law 588, 599, 618, 644–51
Contract liability concepts 645–7
Control monitoring 1004
Controlled uncertainty 199–200
Convention on Biological Diversity 968
Conversion 435–6
Cartesian *see* Cartesian coordinates
    geometry 808
    systems 427–36, 829
Copyright 7, 607, 775–7, 861, 902

Copyright Act (US) 779
CORBA *see* Common Object Request Broker Architecture
Core Curriculum 748
Core standards 703–4
CORINE project 768
Corporate Geographic Database 739–48
Corporate growth model 710–11
Correction or error in images 444–6
Correlation 258, 348–9
Correspondence 130
Cost–benefit analysis 590–3, 597, 601–9, 792, 837–8
Cost-effectiveness and evaluation 603–4
Cost recovery 773–5
Costs and benefits of AM/FM/GIS 816–18
Counting operators 545
Coverage 354–5, 595
CPD *see* Central Postcode Directory
Criterion weights 497
Critical
    applications of AM/FM/GIS 806–16
    history of GIS group 55
    science 41
    social history 55
Crop 910
Cross-validation 227, 229, 982
Crowding 136
Crown copyright 780
Customer support 612–13
Customisation 359–69
    costs 362–2
    programming *see* Arc Macro Language (ARC/INFO), Avenue (ArcView)
Customised Zone Design *see* Zone design
Custom map creation 323
Cyberspace 36, 59, 323, 557
Cycle 99
Data 134–7, 589–90, 885–6, 895, 898
    analysis and display 2
    access 323–5, 678, 764, 906–7
    accuracy 178–9, 191, 449, 593–4
    acquisition 592–5, 702, 954
    admimstration 615
    and commercial interests 572
    archiving 672–3
    assimilation 461–2
    attribute data *see* Attribute
    automation 816
    availability 893
    capture 807, 919
    catalogue 357, 677–91

characteristics 183, 569, 655–6
collection 11, 76
completeness 183, 594–5
conversion *see* Rasterisation, Vectorisation
creation costs 571–3
currency 595
definition 374, 860
delivery 323–5
description 239
dimensions 227
distribution 323
domain 1012–13
encapsulation 1013
exchange 350, 415
exploratory *see* Exploratory
exporting GPS data 474–5
facilitation 1013–14
'fitness-for-use' 678, 691
fusion *see* Data integration
handling 702
integration ('data fusion') 572, 597
inventory 493
life of a dataset 573–4
management 583, 861, 897–9, 919, 973
manipulation 76, 247, 374, 377
matching 760
mediation 684
metadata *see* Metadata
mining 575
modelling 66, 139–42, 353–6, 374, 378–83, 592
output 76
ownership 862
presentation 702
pricing 862
processing 461–3, 625
profiling 761–2
protection 757–8, 775, 862
quality *see* Data quality
relationship 803
representation 139–42
resolution 594–5
sharing 574, 862, 919
sources 655–66, 971–2
spatial 191–203, 253–64, 447–9
spatially continuous 927–8
standards 693–705, 767–85
status 228
stereoscopic 442–3
structure 891

suppression *see* Confidentiality
transformation 31, 133
versus metadata 678
view 701
visualisation *see* Visualisation
volumes 455–6
Data-centric standardisation 694
Data Liberation Initiative (Canada) 905
Data models, object-based 378–83, 592
Data quality 177–87, 223, 593–5, 639–40, 862, 886,
      972–3
  components 178–83, 593–5
  metadata *see* Metadata
  parameters 177–87
  standards 183–5
Database administration 374, 614
  design 595
  developments 352
  of real estates (Austria) 867–9
  schema 376
  systems 373–4
  technology 831
  validation 988–90
Database management 374–5, 878–9
  manipulation 719, 721
Database management system 318, 371–84, 413,
      571, 614–15, 797, 825, 862, 877, 1015–16
  and SQL 372, 374, 376–8, 382
  object-oriented 129, 372, 378–82
  relational 9, 373–8, 382, 805, 828–9
  spatial 25, 413, 417–23
DBMS *see* Database management system
Decision risk 496, 500
Decision support systems 138, 877
Decision-making 635–9, 708, 748, 795, 969
Decision tree 284, 970
Defence 889–96
Definition 5, 192–5, 208–9, 413–14
Delambre, Jean-Baptiste-Joseph 428
Delaunay triangulation 110, 132, 141–3, 544, 551
Delegation 409
DEM *see* Digital elevation model
Demand representation 298
Democracy 708, 723, 760–2, 913, 939
Demography 3, 663, 914, 942
De Morgan's Law 496
Dempster–Shafer theory 496
Density 930
Deregulation 817, 819–23
Description 99, 126, 724–5, 880
Descriptive tree 144

Design 562, 812
  model 414
  requirements 596–7
Desktop GIS *see* GIS
Detecting errors *see* Error
Deterministic 99
Development
  control 953
  strategy 914
Differential rectification 470
Different place/different time (Plate 45) 730
Different place/same time (Plates 43 and 44) 727–9
DIGEST 183, 349–50, 352, 356
Digital 10–11
  individuals/personae 762
  line graph 830
  terrain data 131–2
Digital atlas *see* Atlas
Digital Chart of the World 662, 665
Digital cadastral database 860, 870–3
  map 874
Digital cartography 159–60, 447
Digital Elevation Model (Plates 4, 5, 58 and 59)
  10, 26, 83, 108–20, 179, 200, 207, 438, 448–9,
  529–32, 544, 550, 553–4, 572, 672, 974, 982,
  985–6
  applications 108, 119–21, 985
  error/quality assessment 109–14, 179, 200, 207,
    448–9, 985
  generation 106–13
  interpretation 107–8, 114–19
  visualisation 107–8, 531
Digital
  generalisation 150–1
  geographical information 890
  landscape model 127, 159–60
  libraries 902–6
  line graph 95
  map 728, 778, 802
  orthoimagery 656, 659
  orthophotography 806
  photogrammetric systems 447–8
  terrain model 26, 53, 105–8, 478, 557
Digital Geographical Information Exchange
  Standard 695–7, 893, 895–6
Digital Geographical Information Working
  Group 891
Digital Geospatial Data Framework 783–4
Digitising 3, 438, 806
DIME *see* Dual Independent Map Encoding
Dimension 228

Dimensionality 83
Dimensionality of representations of spatial
  data 168
Direct
  manipulation 408–9, 727
  positioning 429–35
Direction 85
Directory Interchange Format 681–2
Dirichlet tessellation 110
Disaggregation 974
Disaster management *see* Emergency management
Disaster recovery and mitigation 742
Discord 192, 198
Discounting 604–6
Discrete 82, 543, 833
  SDA 286
  visibility maps 543, 545–6
  visibility structures 548–9
Disease *see* Epidemiology
Disparity 442
Displacement 134, 139–40, 142–4
Display *see* Visualisation
Display time 171
Displacement vectors 142–3
Distance 84–5
Distortion 442, 944–5
Distributed computing 318–21, 1010
Distributed network *see* Network
District 940–4
Districting (Plate 57) 940–6
Documentation 164
Domain specific standards 703–4
Domesday Project *see* BBC Domesday
Doppler effect 429, 991
Dot density map 339–40
Douglas-Peucker algorithm 129–30, 396
Downsizing 601
Drainage 112–15
Dual Independent Map Encoding 3–4, 73–4, 636, 761
Dual tessellation 505
Dublin Core 683, 684, 907
DTM *see* Digital terrain model
Dutch National Cadastre 864–5
Duration 173
Dynamic 99, 158, 166, 383, 409, 456, 798, 829
EAGLE system (Plate 65) 1003–4, 1006
Earth 40, 428, 569
Earthquake 849–53
Earth surface materials 859
Earth observation *see* Remote sensing
Earth Observation Science 14

Earth Resources Technology Satellite 669
Earth sciences 569, 907
Ecological 99
    class generalisation 196
    fallacy 73, 573–4
    MAUP *see* Modifiable areal unit problem
        modelling 99, 973
Ecology 484, 968, 974
Economic
    dependency ratio 917
    development 733–5, 742
Economics 974
    of geospatial data 772–5
Ecoregion maps 983
Economy 663, 733–43
Ecosystems science 972
EDAMS *see* Electoral Distribution and Mapping
    System
Edge matching 578
Edge-to-edge tessellation 504
Edinburgh Cray T3D supercomputer 268–9
Editing 440
Education 585, 741, 747–55, 902, 913–14
Eelgrass growth 1004
Eigenvector 246, 449, 510–11
El Niño 92
Electoral Distribution and Mapping System (Plate
    57) 948
Electoral districting (Plate 57) 910, 941–9
Electromagnetic radiation 932, 955
Electronic atlas 905
Elements of a GIS *see* GIS
Elevation *see* Digital elevation model
Elevation
    data 106, 656, 659
    extraction 897
    matrix (Plate 55) 896
Ellipsoid 428, 431–2
Embedded Systems 269
Emergency
    management 742, 798, 845–56
    planning zone 854
    service accessibility (Plate 56) 926–7, 934
Empirical science 40–2
Enabling model 864
Encapsulation 95, 378–80, 403, 576
Encoding 437–49
    map data 438–40
    processes 449
Engineers 894
Enhancement 973

Enterprise computing implementations 321–3
Entity 75, 82–4, 94–5, 143, 378, 657
Entity–attribute–value model 178
Entity relationship modelling 374, 417
Entity–relationship–attribute approach 378
Entity – geographical 431
Entry 196
Enumeration district *see* Census of Population (UK)
Environment 878
    and GIS 39–46
    and natural resources 740–1
Environmental
    applications 909–11
    data 878, 928
    hazards 847
    impact assessment 564, 880, 999–1007
    modelling 105–20, 846, 884, 969–70, 1001–2
    monitoring 818, 911, 999–1002
    programme 890
    science 451, 968
    studies 848
Epidemiology 909, 925–32
    disease – incidence 75, 925
Episodes 92, 97
Epistemology 50–3, 62–5, 187
Equality of representation 941–2
Equipment 438–9
Erastosthenes 428
ERDAS 447
Ergon 43–4
Erosion productivity impact calculator 988
Error 178–83, 195–6, 207–15, 219–32
    analysis 226–7, 229–31, 440, 449, 973
    and accuracy 178–80, 440, 449
    and consistency 182–3, 449
    and completeness 183
    and precision 181–2, 449
    and vagueness 200–2
    graphical detection 219–31
    propagation 176, 207–22, 298, 637
ESDA *see* Exploratory data analysis
Estimation 245, 451–5, 834, 930
Euclidean (space) 31–6, 84, 179, 503–4, 518, 520,
    544, 928
Euler–Lagrange 485–6
European Database Directive 745
European Journal of Operational Research 833
European Resource Satellite 672
European Umbrella Organisation for Geographic
    Information 784
Evacuation (Plates 51–54) 849–54

Evacuation vulnerability mapping 854
Evaluation 211, 219–31, 493–502
Examples of graphical methods 220–3, 255–63
Executive Agency 779–80
Exogenous variable 660
Experiencing 405–6
Experientialism 81–2
Expert system 838, 877, 884, 1010
Exploratory 287
    data analysis 220–1, 235–7, 254, 257
    spatial data analysis 254, 256, 258–62, 287, 930–2
Externalities 607–8
Extrapolation 98
Facilities 807, 810–12, 890
Facilitating model 864
Factors 495, 569, 885–7
Fair Credit Reporting Act (US) 761
Fair information principles 760, 762–3
Farming see Agriculture
Feature 116–18, 353–4, 654
    and Attribute Coding Catalogue 356
    displacement 128
    domain 130
    extraction 891, 897
    locking 802
    name 809
Federal Geographic Data Committee 183–4, 220,
        228, 574, 656, 679, 683, 686–7, 782–3, 906
Federal Depository Library Program 901, 905
Emergency Management Agency 845, 850
Feedback 1004–6
Field 82–4, 927
Field-based systems 321
Field-tree 394
Filtering 409
FINDAR 679–80
FIPS 182
Fire 848–9, 852
Flood control 848–9
Food and Agriculture Organisation 983
Forest 852
Forestry 914
Formal
    language 417
    methods for spatial database analysis and design
        417–21
Formalisation 164
FORTRAN 285, 308, 312, 569
Forward star 855
Fourier Power Spectrum 114
Four-step model 828
Fractal 107, 114

Frame 287, 440
Framework 356, 614–15
    for graphical methods 227–31
    nature of 655–6
Framework data 10, 655–6, 772, 859, 886
Characteristics of 657–8
Fratar balancing 834
Fraud in the inducement 648–9
Freedom of Information Act 775, 779
Frequency 99, 114, 171
Functions 140, 576
Fuzzy 78, 222, 354
    classifiers 227, 230–1
    inference 538
    c Means algorithm 197
    measures 496–9
    membership 523–4, 538–9
    set theory 83–5, 176, 192, 197, 202–3, 462, 497–8
    viewshed 200–2
Gap of understanding 727
GAP-tree 130–1, 398
GATT see General Agreement on Tariffs and Trade
Gaussian 209, 211, 243, 290, 484–5
Gauß Krüger 423
Geary's c 241–2, 258, 285, 971
General Agreement on Tariffs and Trade 777
Generalisation 125–51, 419
    algorithms 26, 134, 139, 144–6
    and multiple representations 129–32
    by example 150
    objectives 134, 138
    operators 134–7, 143, 147
    process-oriented 129
    quality assessment 150
    representation-oriented 129
Generalised Voronoi diagram 517–18
Generic standards 703
Geochemistry 484
Geocode 757–64, 866, 959
Geocomputation 6, 12, 267–81, 574–5, 1018
Geocomputational paradigm 270
Geodata Analysis and Display System 828
Geodemographics 78, 572, 707, 757, 820, 841
    and lifestyles 572, 707
    and privacy 757, 761–2
    targeting by 712, 822, 933–4
Geodesy 427–36, 769
Geodetic
    control 656
    coordinates 431
    datum 428–9, 432, 475, 576
    ellipsoid 430–1

framework 659
reference systems 429–30
surveying 639
Geo-Facilities Information System 802
Geographic Information Coordinating Council
(North Carolina) 738–9
Geographic Information Science 752, 1011
Geographic Information Systems *see* GIS
Geographic Information Systems 827
Geographic Resource Analysis Support Systems 5,
223, 532–3
Geographic Roadway Information Display System 835
Geographical Analysis 14
Geographical
Analysis Machine 272, 930
coordinates 431
Correlates Exploration Machine 272
coverage 595
data modelling 353–6
data policies 767–85
database 91–101
datasets 643–51
Information Community 356
Information Exchange 772–7
Information System *see* GIS
objects 193–5
relevance 725–6
Geographical Systems 827
Geographically Enabled Programming Language
344–5
Geographically Enabled spreadsheets 334–5
Geography
– discipline 30, 39–40, 42, 49–53, 569, 572, 694,
748, 753, 828, 968, 1018
of governmental units 659
of health 925
of politics 939–50
of service provision 711–14
Geoid 428, 432
Geoinformatica 14
Geo Info Systems 14
Geolineus 185
Geological Survey (US) 836
Geology 484, 914, 1011
Geomatics 6
Geometric integration 1013
Geometrical 133
Geometry 30–1, 438, 569, 578, 943
of aerial photographs and satellite data 440–3
Geomorphology 479, 527, 539
Geomorphometric 539

Geoprocessing 318, 698, 702
Georeference 74
Georeferencing 72–4, 593–4, 679, 702
ADDRESS-POINT 74, 435
CPD 74
using postcodes 435, 520, 713, 715, 822
Geoscience 751
Geospatial *see* Geographical
Geostatistics 8, 243–4, 258, 483, 486–7, 900
Gerrymandering *see* Modifiable areal unit problem
Gestalt 751, 957
Getis and Ord *G* statistics 971
Gibson, William 557
GIM International: Geomatics Info Magazine
Gini index 945
GIS
and agriculture 981–93
and geography 50–2, 939–50
and GPS 472–5
and Internet *see* Internet
and landscape conservation 967–75
and land use analysis 714–19
and models in retail analysis 714–19
and multimedia 723–31
and politics 939–51
and privacy 585, 760–4
and remote sensing 667
and society 585, 1010
and spatial statistics 246–7
and virtual environments 557–64
and WWW *see* Internet
application development tools 364–8
applications of 6, 14–15, 119–20, 212–14, 285,
387–90, 488–9, 564, 890–6, 957–9, 981–93
application trends 792–3
as a spatial science 44–5
as an analysis tool 1002
as management tool 584–5
as social theory 45–6
as urban infrastructure 315–16
choosing 570–1, 589–600
competence 599
contribution to the whole economy 733–43
current trends 8–11, 570
customisation 359–69
data 803–6
data layers 984–7
definition and classification 5, 1019
desktop 312–12, 331–45, 815
diffusion 792–3
ease-of-use 337

efficiency and effectiveness 583, 587–8
elements of 137–50
future 567–79
global applications 981–93
hardware/software revolutions 310–12
in business and service planning 709–21
in cartography 133, 438
in emergency management 845–56
in business and service planning 709–21
in emergency management 845–56
in environmental monitoring 1000–1
in land administration/reform 959–74, 919–22
in land use analysis 955
in libraries 901–8
in network planning 820–1
in North Carolina's economic and governmental
    decision-making 739–42
in South Africa 913–24
in telecommunications 819–25
in the utilities 801–18
in urban planning 877–8, 922–3
institutional consequences of adoption 621–9
interacting with 401–10
liability 643–51
literature 569, 602–3, 622, 625, 753, 847–9,
    852–3, 928
management 611–20
national applications 981–3
national assessment 983–90
networked environments 317–27
new technology 309–16
principles of GIS 6, 14–15, 254–6, 413–23, 438,
    446
progress 747–55
resources for 14
safe to use 583, 587–8
science and 14
service policies 902–4
social implications 49–59
software engineering 362
spatial analysis tools 270–1
technologies 897, 914, 1014–17
trends 884
'turnkey' products 606
virtual 561–4
GIS Africa 14
GIS Asia Pacific 14
GISc 747–55
GIS Europe 14
GISt 747–55
GIS–T (Plate 50) 827–41

GIS World 14
GISy 747–55
Glimpse 686
Global
    assessment 981–3
    circulationa models 91
    trends 823
Global Environmental Network for Information
    Exchange 679
Global Geospatial Information and Services 665
Global Information Infrastructure 700
Global Land Information System 687–8
Global Positioning System 11, 26, 66, 83, 106, 333,
    349, 425, 427–46, 467–75, 650, 757, 762, 804,
    830, 836, 853, 909, 926, 932, 981, 990–1, 1015,
    1017
    with GIS see GIS
Global Spatial Data Infrastructure 770, 1014
Government
    as business 585–6
    role 770–3
Government Information Locator Service 682–3
GPS see Global Positioning System
GPS World 837
Graduated symbol map 340–1
Granularity 181
Graph optimisation 533
Graphical methods of detection 219–31
    variables (Plate 7), 161
Graphical user interface 9, 139, 149, 361, 374, 401,
    408–9, 561, 730, 802–3, 948
Graphic
    design issues 224–5
    display options 229–31
Graphics 407, 570
GRASS see Geographic Resource Analysis Support
    System
Graticule 438
Gravity model 833
Gray
    curve 389
    ordering 388
Grid file 393–4
Grid-based spatial access 393–4
Ground
    control 445–6
    resolution element 457
    truthing 955
Groundwater modelling 987
Group Areas Act 919
Group decision support system 884
GUI see Graphical user interface

Haar 562
Habitat suitability 970
Hagiography 55
Hairsine-Rose 987
Hardware 310–12, 589, 595–9, 816, 877, 894–5, 898, 1001, 1014–17
Harsono, Soni 1013
Hausdorff distance 130
Hayford ellipsoid 434
Hazard 847–52
    maps 988
Hazardous waste 849–51
Head-mounted displays 559, 563
Health 922, 925–36
    care 925–36
Information for Purchaser Planning System 935
Hermeneutics 40–2
Heterogeneity 285–6, 769, 928
Heuristics 85, 145, 300–1, 554
Hexagonal tessellation 509
Hidden line/surface removal 546
Hierarchical 550
    neutral models 970
    triangulation 551
High performance computing 268–70
High resolution satellite imagery 456, 669
Higher-order Voronoi diagram 520–3
Highway Analysis System 836
Hilbert 388–9, 395–6
Holdridge life zones 982
Homogeneity 144, 942
Horizon 543
Host-based system 317–18
Human–computer interaction 149–50, 401–10, 558–63, 885, 975
Human
    database interaction 374
    expert 148
    resource 894–5
Hurricanes (Plates 46 and 47) 742, 849–52
Hybrid 377
Hydrodynamic modelling 1004
Hydro-ecological model 534
Hydrographic data 1003–4
Hydrography 656, 659, 985
Hydrology 105, 479, 890, 985
Hypercard 315
Hypermaps 159
Hypermedia 45
Hypermedia systems 884
Hypertext 9, 684
Idealism 63, 137

Identification error 209–10
Identity 378–9, 403
Idrisi 185, 312, 501, 567, 636, 835, 987, 992
IDW (Plate 26c), 488
Image 437–49
    acquisition systems 443–4
    exploitation 897
    formation 440–1
    rectification 486
    segmentation 462–3
ImageNet 323
Imaging spectrometer 456
Impact assessment 878
Implementation 363, 561, 627–30
Implications 628–9
Index map 919–21
Indexes 382
Indirect positioning 429, 435
Induction 329
Inductive science 575
Industry 741–2, 754
Inference 451–4
Information 33–4
    communities 356
    economy 572
    kiosks (Plate 44) 730
    management 825
    processing system 661
    providers 777
    society 771, 1011
    superhighway 735
    system 705
    technology 620, 699, 735–6, 753, 889
Information Resources Management Group (US) 736
Infrastructure 663, 739
Inheritance 96, 380, 402–3
Input error 212
Instance variables 378
Instantaneous Field of View 457
Institut Geegraphique National 448
Institute of Terrestrial Ecology (GB) 955
Institutional
    consequences 621–9
    constraints 861–2
    mandate 187
Instrumental rationality 625
Intangible benefits 608–9, 656–7
Integrated Services Digital Network 825
Integration 463, 488, 838–9, 881, 884, 1010
Intellectual property rights 644, 768, 775–7, 902, 1012

Intelligence 661
  agents 755
  staff 894
Intelligent
  GIS 716
  mapping 823
  network model 797, 817
  transportation systems 837–8, 856
Intelligent Vehicle/Highway System 827, 837
Interaction 158, 163, 402, 561, 716, 727–30, 973–4
  modalities 406–8
  paradigms 403–6
  styles 408–9
Interactive 332
Interactive modelling 948
Interactive techniques and spatial data
  analysis 253–64, 254–6
Iterative approach 591–2
Interface 350–2, 363, 683, 1013
Interferometry 449, 672
Intergraph 151, 337, 417, 422, 447, 674, 778, 820,
  835, 1015
International Cartographic Association 698
International data standards *see* Data
International dimensions 662–5
International Geosphere Biosphere Programme 674
International Graphic Exchange Standard 693
International Hydrographic Organisation 184, 695,
  697–8
*International Journal of Geographical Information
  Science* 14, 827
International Map of the World 768
International Society for Photogrammetry and
  Remote Sensing 698
International Standards Organisation 684–5, 695,
  770
Internet 9–10, 12, 17, 29, 219, 237, 264, 269, 307–8,
  320–7, 336, 570, 573, 677, 686, 699, 701, 729,
  779, 783, 793, 840, 851–2, 856, 866, 873, 885,
  905, 1007, 1009, 1015
  and GIS 4, 9–10, 17, 58, 237, 264, 269, 308,
    314–15, 322–7, 333–4, 336, 401–4, 564, 681,
    691, 702, 726, 730–1, 748, 751–3, 793, 840,
    851–2, 856, 885, 905, 1007, 1009
  and mapping 314–15, 333–4, 575, 589, 683–4,
    701, 865
  intranet 413, 866, 885
  Java 10, 325
WWW 4, 12, 58, 308, 314–15, 321–3, 325, 333,
  401–4, 468, 564, 570, 661–2, 679, 681, 683–91,
  701, 726, 730–1, 748, 768, 840, 885, 907, 1009

Interoperability 12–13, 347–57, 761, 850, 1010, 1012
Interpolation 106, 110–13, 487–8, 878, 932, 992
IDW (Plate 26c) 488
  Kriging *see* Kriging
  spatial 448, 481–90, 520, 576
  spatio-temporal 486, 489
TIN-based 483
Interpretation 107–8, 114–19
Intersection 98, 494–7
Inter-urban freight 837–8
Interviewing 404
Intervisibility 429, 543–55
  graphs 544
  map 546
Intranet 413, 866, 885
Inverse distance weighted
  interpolation *see* Interpolation
Invitation to tender 597–8
Irregular tessellation 510–17
Isohedral tessellation 504
Isoline 78, 165, 336, 341, 487
Isorad 473
Isotherm maps 473
Iteration 225, 431, 584
IVHS *see* Intelligent Vehicle/Highway System
IVHS Journal 827
Japanese Earth Resources satellite 671
Jarman Index 199
Java *see* Internet
Journal of Advanced Transportation 827
*Journal of the Urban and Regional Information
  Systems Association* 14
Journal of Transportation Planning and Technology
  827
K-tunction 242–3, 246, 284–5, 930–1
Kaiser–Nagel 946
Kalman filter 113
KD-tree 386
Kernel 244, 283, 930
Key
  management issues 674
  multimedia applications 725–7
Knowledge-Based Land Information Management
  Group 585
Knowledge-based method 147–9
Kriging (Plates 26c and 26d) 38, 110, 210, 212–13,
  221, 229, 243, 259, 283, 354, 481, 483–8, 928,
  932, 971, 992–3
Krueger, Myron 557
Labelling 536–7
Lagrangian relaxation 300

Lambert Azimuthal Projection (Plates 62 and 63)
Land
  administration 798, 859–74
  classification 577
  cover 663, 885, 910, 919, 953–63
  Data Bank System (Sweden) 864–9
  development 878
  information management 861–4
  information system 860
  ownership 193
  parcels 956–7
  reform 919–22
  registration 859, 962
  suitability map 881
  taxation 860
Landform
  analysis 527–40
  features 533–4
  labelling 536–7
  systems 534–6
Land Records Management Programme (North
  Carolina) 738
LANDSAT 7, 194, 443–4, 457, 461, 665, 668–70,
  780–1, 955
Landscape 969–70
  architecture 33
  as scenery 974–5
  chronology 969
  classification 969
  conservation 910
  ecology 968–71
  geometry 974
  morphology 969
  visualisation 105–20
Landschaft 969
Land Status Automated Systems 836
Land use 30, 663, 878–9, 884–5, 910, 919, 922,
  953–63
  analysis 957–8
  policy areas 958–60
Language 82
Lanier, Jaron 557
Latitude 431, 687
Lattice 258, 260–2
Laves tessellation 506–7
Layer *see* Coverage
Leaf area index 452
Level of detail 561–4
Lewis' law 511, 516
Lexis-Nexis 763–4
Liability 643–51, 775, 861

Libraries 799, 901–8
Life of a dataset *see* Data
Lifestyles *see* Geodemographics
Limited intervention model 864
Limitation 220, 664
Line
  data 927
  graph 830, 836
Lineage 184, 594, 902
Linear
  imaging 670–1
  mixture modelling 457
  modelling 970
  project design 247, 568
  programming 828
  referencing 798, 829
Linguistics 98
Linked views 168–9
LIS *see* Land Information System
Local applications of GIS *see* GIS
Local area
  monitoring 1003–6
  networks 318–20
Local expert 313
  indicators of spatial association 262, 286
  neighbourhood approach 482–3
  parallel processors 530–1
  surface patches 111
Locality 481, 932–4
Locally adaptive gridding 111–12
Location 284, 300, 679, 829
Location modelling 293–302, 741
  impediments to 297–8
  integration within GIS 298–300
Location Science 833
Locational
  data 926
  planning 934
Location–allocation model 45, 300, 719, 828, 831–3,
  883–5, 934
Logical consistency 184, 594
Logos 43–4
Longitude 421, 687
Lotus 763
Low cost housing 922–3
Machine-readable cataloguing 681–2, 906
Macro 361, 402, 475
MAGE 138, 143
Main memory access methods 386–8
Maintenance 811
Majorative change 92

Malapportionment 910, 944–5
Management 6, 14–15, 583, 845
    information systems 828, 882
    oversight 619
    support system 1003
    tool 584–5
Managerial rationalism 588, 624–6
Managing
    an operational system 612–16
    a whole economy 733–43
    uncertainty 633–40
Manhattan 390
Manual interpretation 821
Map
    algebra 36, 207, 405
    analysis 99
    catalogue production 890
    continuous visibility 543
    datum 471–2
    design 922
    digitising 402
    editing 808
    generalisation operators 135–6
    generaliser 150–1
    overlay 877–8
    production 3, 809–10
    projection 433
    roads 604
    scale 810
    thematic 636, 725
    thermal 668
MapBASIC 344
MapInfo 312, 948, 992
MapObjects 367–8
Mapping Awareness 14
Mapping 878–9
    ad hoc 810
    applications 808–10
    systems 333–45
Maputaland Malaria Programme 919
Maps 2–3, 6, 8, 17, 65, 67, 86–7, 128–9, 149, 157–71,
    401, 407, 436–49, 578, 714–20, 724–5, 804–18,
    822, 825
MARC *see* Machine-readable cataloguing
Marine Observation Satellite 671
Marketing 714–17
Markov techniques 99
Master Environmental Library 686, 688–91
Matching 130
Materialism 63
Mathematics – discipline 30, 98, 569

Matrix 831, 891
MAUP *see* Modifiable areal unit problem
Maximal covering 299–300
McTrans programs 828
Mean 930
Measuring the signal 455–61
Mechain, Pierre-François-Andre 428
Mechanical perception 955–6
Medora project 821
MEGRIN 220, 663–4, 666, 770
Mental maps 561
Mergers and acquisitions 717–18
Mesoscale 93, 120
Metadata 13, 164, 184, 219–20, 225–6, 357, 404,
    475, 573–6, 615, 653, 658, 677–91, 778, 783,
    973, 1019
    in libraries 905–7
    standards 184, 220, 782–3
Metaphysics 62–5
Method 417–21, 447, 482–7
Methodology 51–3
Metropolitan area networks 320
Microcomputer 311
Migration 831, 926
Military applications 889–99
Minimum
    bounding rectangle 130
    enclosing rectangle 129
    environmental cost route 831
    spanning tree 144
Misrepresentation 648–9
Mitigation 742, 845–6, 848–50, 856
Modalities 406–8
Mobile phones 821
Modal split analysis 834
Model
    completeness 183
    data 803–5
    errors 212
    generalisation 127
    of uncertainty 191–203
    validation 988–90
Modelling 164, 207–15, 831–3, 880, 1004
    applications 987–8
    data-rich 573
Models 414
Moderate Resolution Imaging Spectrometer 45–60
Modifiable areal unit problem 8, 44, 64, 73, 241,
    243, 273–4, 1017
    gerrymandering 910, 941, 944–5
Molecular polygon 507

Monitoring
  data 1002
  land cover and land use 953–63
  Landscape Change Project 956
  programmes 1004
  system 999–1006
Monohedral tessellation 504
Monopoly trading 778
Monotonic functions 970
Monte Carlo method 211, 243, 502
Moore's law 748, 794, 1014
Moran's *I* 241–2, 258, 262, 285, 509–10, 519, 971
Moran scatterplot 261–2
Morphology 532, 537, 969
Mortality 930
Motion key 388, 392
MSS *see* Multispectral scanner
Multi-criteria evaluation 493–502
Multimedia 171–2, 402, 866, 1017
Multi-objective 494
Multi-path interference 469
Multiple instruction, multiple data 547
Multiple representations 125–50
  views 108–9
Multiquadrics 486
Multi-resolution
  representation 562
  terrain models 544, 550–3
Multi-scale analysis 118, 489
Multiscale databases 130
Multiscale spatial access methods 396–8
Multispectral
  classification 821
  scanner 669–71
Multivariate 538, 970
Multi-visibility structure 545
Nadir 441
NASA *see* National Aeronautics and Space
    Administration
National Academy of Public Administration 779
National Aeronautics and Space Administration
    669, 673, 681–2, 769–70, 780
National
  applications of GIS *see* GIS
  assessment of GIS *see* GIS
  coordination 1013–14
National Center for Geographic Information and
    Analysis 14, 1010–1
National Environmentally Sound Production
    Agriculture Laboratory 990–2

National Geospatial Data Clearinghouse 574,
    582–3
National Geospatial Data Framework (UK) 584–5
National Land Information Service (UK) 860–6,
    870–3, 957
National Land Registration System (Hungary)
    867–9
National Land Use Stock System/Classification
    (UK) 954, 956
National Mapping Agencies/Services 149, 660–5,
    679, 703, 770, 774–5, 863, 1013
National Nutrition and Social Development
    Programme (South Africa) 916–19
National Science Foundation 1010
National Soil Geographic Database 989
National Spatial Data Infrastructure 5, 184, 219,
    746, 767, 769, 773, 779, 782–5, 903, 1010, 1013
National Spatial Development Framework (South
    Africa) 915–16
National Survey of Vacant Land in Urban Areas
    (England) 958
National Transfer Format 184, 695–6
NATO 693, 697, 891, 894
Natural hazard 849–50
Natural neighbour interpolation 483
Natural resources 846, 890
Nature
  conservation 915
  reserves 968–9
Nautical chart 697, 836
Navigation 560
Navigational aids 726
NCGIA – Initiative-19 49, 54–8
Nearest neighbour 113, 242, 246, 518, 930
Needs assessment 932–3
Negligence 644, 647–8
Negotiation 98
Neighbourhood classification *see* Geodemographics
Neophytes 52
Nested relations 382
Network 312–16, 326–7, 1017
  architecture 319
  computer 317–27
  data model 854
  management 836
  planning 820–1
  records management 820
  sampling 244
  systems 820
  technology 1007

Network-centric approach 1015
Network modelling 812–14, 832
Neural net 99, 288–9, 462
New technology 309–16
NMAS 185
NOAA 682, 780
Nodal digital video (Plate 45) 726, 730
Node 828–9, 854–5
Non-interventionist model 864
Non-specificity 192, 198
Normalisation 420
Normalised difference vegetation index 452–5, 932
NSDI *see* National Spatial Data Infrastructure
NTF *see* National Transfer Format
Nyquist 673
Object 75, 130, 322, 356, 378–83
    and operations 560
    class of 378–9
    composition 380
    database management group 382
    generalisation 127
    management group 381
    modelling technique 363
Object-based methods 533–4
Object DBMS 380–2
Object link embedding 351, 367–8, 564, 699, 1016
Object-oriented 66, 95–6, 131, 312, 363, 378, 414,
    417–18, 576, 803
Object paradigm 547, 553
Octree 68
OECD *see* Organisation for Economic Cooperation
    and Development
OLE *see* Object link embedding
On-line
    algorithm 547, 553
    metadata 10–11
    see also Internet
Ontology 26, 32, 52, 62–5
Open Geodata Model 351, 699
OpenGIS
    movement 575, 770, 1017
    specification 351–5, 382–3, 698–9
Operational
    criteria 1004
    GIS 611–20
    staff 894
Operations 207, 828
    research 828, 833
    support 613–14
Operators 97–100, 134–7
Operator polymorphism 380

Optical sensor 670–1
Optimisation 719, 946
Optimising retail networks 718–19
Order 171
Ordered weighted average 497–9
Ordering 388
Ordinary Voronoi diagram 518–20
Ordnance Survey 143, 149, 177, 434, 448, 607, 656,
    659, 686, 774, 780, 829, 956–8
Organisation 599–603, 885
Organisation for Economic Cooperation and
    Development 771
Organisational issues 862–3
Orientation 446
Originality 776
Orthoimage 441, 446, 449
Orthographic display techniques (Plate 45) 726, 730
Orthophoto 10
OS *see* Ordnance Survey
Oscilloscope 311
Overlap 84
Overlay 545, 726, 894
p-median 299
Pach algorithm 395
Paideia 43
Panchromatic data 669
Paper maps 67, 159, 186, 762, 893, 954
Paperwork Reduction Act (US) 779
Paradigm interaction 403–6
Parallax 442–3, 448
Parallel algorithm 547–9
Parameterisation 561
Participation and education 1017–18
Partitioning 211–12, 832
Partnerships 904–5
Pascal 312
Path analysis 852
Pattern 454, 973, 1003
    analysis 242–3
    recognition 289
Patterns of association 286
PC environments 317–18
Peano key 388–91
Pedology 984
Penrose tilings 503
Perceived reality 178
Perceptibility 136
Perception 82
Percolation models 970
Perfect positioning 578
'Personality'/'moral right' model 764

Perspectives 622–8
Phenomena 82–4
Philosophy – discipline 30, 98
Photogrammetric Engineering and Remote Sensing
     14
Photogrammetry 10, 193, 438, 639, 668, 673, 830
Photolog technology 836
Physics – discipline 30, 32
Physical principles 451–5
Physical and behavioural criteria 943
Physical geography 943
Pilot project 596
Pixel 18, 181–2, 194, 439, 457, 531–2, 955
Plan, urban 880–2
Planar 353, 504–24, 544, 562, 829, 832, 969
Plane transformations 445–6
Planning 723–31, 845, 922–3, 932–5, 968
     application processing 879
     business 713–21
     fiat 958
     metaphor 1017
     models 714–16
     options 881
     support system 882–4
Plausibility 496
Plotter 916
Poesis 43
Point
     data 926–7
     location 550
     quadtree 391–2
Pointing 408–9
Point-in-polygon 636
Poisson 242, 285
Polarisation 461
Political
     criteria 943–4
     dimension 1013–14
Pollution 927, 932
Polygon 130, 297, 343, 387, 440, 504, 519, 971
     atomic 507
     molecular 507
Polymorphism 402
Polynomial 449
Population 8, 72, 831, 851, 919, 922, 934, 941–2, 970
     biology 972
     census 769, 836, 927, 933–4
     data 274–80, 403
     density 76, 820
     models 846
     trends 820

Position
     absolute 576–7
     definition of 429–33
     perfect 578
     relative 576–7
Positional accuracy 184
Positioning 897
Possibility 496
Postal
     sectors 716
     service 761
Postcode
     address 957, 962–3
     as georeference *see* Georeferencing; *see also* Zipcode
Powell-Fletcher method 274
Practice 73–4
Precision 181–2, 443, 449
Prediction 880
Preparedness and response 845, 850–3
Preprocessing 129–30
Prescription 880
Presentation 902
Primitive 130, 140
Principles of GIS *see* GIS
Prism maps 336, 341–2
Privacy 585, 607, 746, 757–64, 902, 947, 1012
     legislation 760–1
     rights 644, 758–60
Private
     sector perspectives 1012–13
     visual thinking 158
Privatisation 801, 823, 825
Probability 99, 176, 195–6, 496, 500
Probable viewshed 202
Problem 245–6, 481–2
     of land parcels 956–7
Process-centric standardisation 694
Process tracing 149
Processing 196, 288
Procurement 591
Product
     launch 717
     quality 636–7
Production
     cartography 192
     line view 701
Professional development 754
Project management 616, 816
Projection 221, 880
Propagation 207–15
Property

ownership 922
    reference number 861
Proportional representation 940
Protection 776–7, 1012
Prototyping 363–4
Provenance 777, 902
PR-quadtree 391–2
Pseudo random noise code 467–8
Psychology
Public 58
Public
    access to data 906–7
    discourse 723–31
    procurement laws 575
Public–private sector partnership 585, 705, 708,
    735–6, 771, 778, 831, 862, 866–7
Public visual communication 158
Publicly Available Specification 701
Pyramids 141
Quadtree 68, 131, 141, 391–3, 550–1, 562
Qualitative 98, 160
Quality control 185, 999
Quantification 224
Quantisation 456
Quantitative 24, 98, 160, 181
Quaternary
    triangular mesh 141–2
    triangulations 550–1
Query 168–9
    language 403–6
    tool 311
Querying 403
Radar 444
RADARSAT 670, 672
Radial compactness 946
Radiation 454–61, 668, 854
Radio location 821
Radiometric resolution/calibration 456–7
Random variable 208–9
Raster 4, 25, 66–8, 83, 186, 311, 438, 520, 530, 572,
    667, 806, 830, 880, 891
    data resampling 487
    representation (Plate 2)
Raster-to-vector conversion 312, 439–40
Rasterisation 933
Rate of change 171
Rate of time preference 605
Rationale
    for graphical methods 219–20
    for formal analysis and design 414
RDBMS *see* Database management system
    r. watershed 534

Reactive tree 130, 396–8
Readymade maps 335–6
Real-time monitoring 1002–3
Real world 75
Realism 81–2, 489
Realtors Information Network (US) 866, 870–3
Rebuilding a country: South Africa 913–24
Receiver in dependent exchange
    format 349
Recollection 724
Reconstruction and development programme 914
Records 375–8
Recovery 845, 853–4
Rectangle-based methods 483
'Red edge' 452–3
'Red Routes' 959
Reduction of uncertainty 637
Redundancy 182
Re-expression 168–9
Reference
    frames 85
    model 702–4
Referencing *see* Georeferencing
Reflectance 668
Reflection 388
Regard 167, 247, 255, 259, 287, 928
Region quadtree 392–3
Regional
    assessment of GIS *see* GIS
    evaluation vulnerability 854–5
    market shares 711–12
    planning 953–63
Regional Research Laboratory 14
Regionalisation 832, 969
Regionalised variables 62
Regression 212, 286, 833–4, 934
Regular
    square grid 543–9
    tessellation 505–10
Regularised spline with tension (Plates 26c and 27c)
    485, 488–9
Relation 375
    schema 375, 377
Relational
    databases 373–84, 417, 802, 823, 828
    DBMS *see* Database management system
    table 403
    technology 377–8
Relationship 100
Relative
    position 577

risk 501
space 32, 36, 64
Relativism 36
Reliability
    diagrams 221
    visualisation tool 221, 225
Relief displacement 448
Remote sensing 33, 202–3, 451–61, 665–74, 780–2,
        880, 981, 990
    and GIS 3
Remotely-sensed data 667–75, 955
    facilities 673–4
    in operational use 109–10, 451–61, 668
    sources 668–72
Remotely-sensed imagery 72
Remote surveillance 757, 762
Reporting 830
Representation 49–89, 105–55
    extensions 1010
    issues 36–7, 92–3
    multiple 125, 128–32
    of socioeconomic phenomena 71–9
    of terrain (Plate 55) 105–21, 131–2, 535–7, 890,
        895
    spatial 61–9, 71–88
    see also Visualisation
Representation-oriented view 129
Requirements analysis 892, 999
Research Initiative-19 54–8
Resolution 181–2, 594–5, 898
    spectral 455
Resource inventory 880
Response 845, 850–3
Responsibility for frameworks 658–61
Restricted quadtree 551
Retail site location 712–14
Retrospective forecast 1004
Revised universal soil loss equation 986–7
Risk 847–51
Risk estimation 640
Road 130, 637, 778, 828, 855
Robotics 558
Rogers, Everett 792
Root mean square error 114, 179, 196, 200, 209,
        223, 226, 228, 445, 449
Route finding 332–3, 852
Routing 578, 828, 836
Row-prime ordering 388
Royal Institution of Chartered Surveyors 753
RST see Regularised spline with tension
R-Tree 394–6

Rubber-sheeting 483, 577
Rural
    settlements 958–9
    transportation 837–8
SABE dataset 663–6
SAIF see Spatial archive and interchange format
Sales and marketing 822
Same place/different time (Plate 45) 727–30
Same place/same time 727–9
Sampling 244, 973
Sapir-Whorf hypothesis 82
SAR see Synthetic aperture radar
Satellite 194, 429, 457–61, 667–72, 931
    data 440–3, 465, 665–75
    imagery 931, 971, 991, 1006, 1013
    LANDSAT see LANDSAT
    remote sensing 454–60, 780
    SPOT see SPOT
Scale 107, 114, 119–20, 246, 954, 1010
    and zoning 243
    variance analysis 454
Scanning 438–9
Scatter diagram 257, 262
Scene analysis 974–5
Schlaefi 506
Schema 130, 356–7, 377, 405
School districting 828–30
Science 845
    approaches to 40–2
    cognitive 81–7
    conventions 573
SDA see Spatial analysis
SDS 142
SDTS see Spatial data transfer standard
Search
    engines 404
    unction 597
    mechanisms 686
Second-order
    methods 285
    polynomials 446
Security 614, 902
Sedimentation 1006
Sediment transport 987, 1004
Segmentation 144
Segregation 486
Seismic 578, 849
Selected library programmes 907–8
Selective availability 469
Selling data 607
Semantic 147, 183, 348–9, 403

data model 374
    import model 197, 200
    integration 577, 1013
Sensor spatial resolution 457
Sensory channels 559–60
Service architecture 351
Service planning 298–300, 709–21
Setting a vision 701–2
Settlement patterns 241
Settlements 959
Shape recognition 143–4
Sharing information 356–7
Shortest path analysis 828, 831
Siepinski curve 388–90
Signal-to-noise ratio 455–6, 534
Similarity Relation Model 197
Similarity transformation 445
Simplex 505
Simplification 134, 139, 145
Simulated annealer 274–5, 574
Simulation 229–31, 553–64, 636–7, 890
Simulation modelling 67, 852
Single instruction, multiple data 549
Single transferable voting 540
Sinuosity 143
Site identification 298
Sketching 407–8
Slope 922, 987
SmallWorld Systems 313, 361–2
Smoothing 134, 139, 145
Social
    applications 909–11
    implications of GIS see GIS
    interactionalism 588, 626–8
    theory 45–6, 49–59
Social science benchmark code 269
Social sciences 974
Social survey research design 568
Societal issues 585
Society 186–7
Sociocultural 33
Socioeconomic data 72–8, 572, 694, 831, 846, 878,
        885, 919, 922, 927, 933, 962, 1012
Software 421–2, 571, 589, 595–9, 816, 877, 884,
        894–5, 898, 928, 947, 970, 1014–17
    for GIS-T 834–5, 928
Soil 197–9, 453, 537–40, 663
    drainage 986
    map 570, 679, 982–5, 989
    organic carbon 989
    pedon 982–4, 989

properties 989
    science 484, 573–4
    survey 988–9
Soil landscape
    classification 539
    map 985–6
Sorites Paradox 197
Sound 406–7, 560
Source 211–12, 803–5
Sources
    of framework data 661–2
    of information 452–5
Space 64, 178–9
Space-filling curve 388–90
SpaceStat 245, 256
Space-time
    analysis 91, 931
    attribute creatures 272
    data 29–37
    stationarity 484
Spatial 7, 132–3
Spatial
    access methods 385–98
    accuracy 179–80
    archive and interchange format 96, 349, 352
    autocorrelation 72, 126, 241, 509–10, 577, 971
    database 477–579, 699, 829–30
    decision support system 846, 882–3, 926, 935
    dependence 284–5, 971
    distributions 157–71
    econometrics 244–5
    epidemiology 928
    evacuation vulnerability 854–5
    generalisation 196
    geometric view 285
    heterogeneity 284–5, 971
    hydrography 527–40
    information (Plate 7) 852, 903
    infrastructure 768
    interactions 240, 833, 883–4, 926–8
    lag 259–61
    location code 398
    metadata 679 (see also Metadata)
    modelling 207–15, 245, 623, 878–80
    querying 415
    relations 84–7, 1010
    referencing 427–36
    representation (Plate 1) 24, 61–9, 71–9, 81–7,
        132–3, 956
    resolution 181–2, 457–9
    sampling 182, 244

schematic 948
science 44–5
smoothing 928
statistics 239–47, 569, 931, 1010
structuring 971
tessellations 503–24
traditions 568–71
variations 454
visualising 259
weights 246, 258, 260
Spatial analysis 52, 235–7, 253–64, 267–91, 413–23, 569–79, 702, 710, 879–80, 970–3, 1010
and GIS 567–79, 878
and holistic measures 133
correlation 241
cumulative distribution function 258
discrete SDA 286
exploratory SDA 254, 256, 258–62, 287, 930–2
techniques 569, 574–5
traditions 568–71
typology of technologies 272–3, 716, 926
Spatial Archive and Interchange Format 695
Spatial Association 241–2
Spatial data 125–50, 569, 677–91, 755
analysis *see* Spatial analysis
infrastructure 660, 726
integration 846
model 66–7
transfer standard 82, 84, 183–5, 220, 349
Spatial database
analysis 413–23
design 413–23
engine 360
for health research 926–8
Spatial interpolation 448, 481–90, 520, 576
Spatialisation 36
Spatially continuous data 927–8
Spatio-temporal
data handling 93–7
interpolation 486, 489
perspectives 30–3
representation 29, 34
Spectral
analysis 114
redundancy 455–6
variation 452, 455
Speculation 725
Speech 406–7
Sphere-tree 396
Spider diagram 831
Spill monitoring 1004, 1006

Spillovers 697–8
Spiral ordering 388–9
Spline 110–12, 481–7 (*see also* Regularised spline with tension, Thin plate spline)
Splitting primitive 387
Sporadic change 92
SPOT 109, 113, 440, 442–4, 447–8, 665, 668–70, 781, 987, 992
Spreadsheet 313, 831
array 571
SQL *see* Structured query language
Staffing 886–7, 904
Stakeholders 604
Standard for digital data 782–3
Standardisation 164, 381, 495, 499, 572, 577, 681–5, 694–5, 1012–13
Standards 657, 862
Statistical
analysis *see* Analysis, Spatial analysis
data *see* Data
geometry 569
interpolation *see* Interpolation
modelling *see* Model, Modelling
Statistics 220–1, 239–47, 569
spatial analysis *see* Spatial–analysis
spatial statistics *see* Spatial statistics
visualising *see* Visualisation
Steiner problem 554
Stereo correlation 448
Stochastic 208–10, 484, 834
Storage 374
Strategic functions 1012
Strahler order 534
Stratège 138
Strategic
data 703–4
planning 953–63
Strategic issues for GIS educators 751–4
Strategy for managing uncertainty 635–6
Street
addresses 829
centreline maps 829
Strict products liability 649
Structure 132, 143–4
line model 144–5
of a GIS-T 828–30
Structured Query Language 7, 100, 313, 337, 343, 352, 372, 374, 376–7, 403, 415, 693, 698, 828
Substitutability 496
Supervisory control and data acquisition 814, 1000
Support staff 816

Surface
    albedo 453
    modelling *see* Continuous surface model, Terrain
        representation, Terrain Visualisation
Surface-radiation interaction 461–2
Surface-specific point elevation 108
Surveying 31
Sustainability 910, 968–9
Symbolisation 134, 1002
Symbology 808, 922
Synchronisation 171
Syntax 403
Synthesis 489
Synthesising
    cartographic data 959–60
    urban areas 962
Synthetic aperture radar 109, 113, 670–2
Syntropy 353
System 326–7
    evaluation 598
    interoperability 415
    specification 596–7
    status management 296
    viewpoint 701
Tangible benefits 602, 656–7, 816–17
Target management structure 894–5
Targeting funding 916–19
Taxicab distance 31
Taylor Series 210–11, 214
Tcl/Tk 371
Techniques 14–15, 806
Technological
    determinism 588, 623–4
    hazard 850
Technology 6, 845
    supporting GIS 865
Telecom Italia: case study 824
Telecommunications 735–7, 819–25, 914–15
    infrastructures 822, 860
    networks 797–8, 822–5
Teleconferencing 729
Telemetrics 673, 821
Teleoperation system 558
Template 658
Temporal 97–8, 178, 196, 229, 383
    accuracy 180
    cycles 97
    distance 97
    resolution 182, 461
Terrain 544
    analysis 890, 895, 897, 986

data 537, 544–54
generalisation 148
intervisibility 543–54
nominal 178–9
parameters 115–16
representation 107, 131–2, 535–7
visualisation (Plate 55) 890, 895
Tessellation 140
    Archimedean 504–5
    dual 505
    edge-to-edge 504, 518
    hexagonal 509
    irregular 510–17
    isohedral 504–5
    Laves 506–7
    monohedral 504
    normal 504
    regnlar 505–10
    spatial/planar 503–24
Tetrahedron (Plate 3) 68
Texture 454
Thematic 178
    accuracy 181
    environmental data 725
    mapping 161, 336, 338, 407, 473, 609, 855, 879
    resolution 182
    view (Plate 6b)
Theory 72–3, 210–12
Thermal sensor 668
Thiessen
    polygon 66, 77, 483, 854–5, 933, 941, 987 *see also*
        Voronoi polygon
    polyhedra 483
    tessellations 1000
Thin plate spline 354, 484, 488, 982
Tiered electoral system 940
TIGER *see* Topologically Integrated Geographic
        Encoding and Referencing
TIGER/Line 835, 901, 908
Tile indexing 388
Time 68, 91–101, 356, 484, 604–6, 926, 1000
'Time geography' 926–7
TIN *see* Triangnlated Irregular Network
Tissot's indicatrix 227
Toolbox 405, 877, 1017
Toolkit 947
Tools for protecting databases 777
TOPOG 117
Topographic 107
    mapping 660, 882, 897

surface 544
template framework 768
Topography 26, 114, 117, 489, 528, 534, 663, 820,
    850, 919, 971, 983
Topological
    relationship 803
    view (Plate 6a)
Topology 4, 30–1, 74, 96–7, 571, 578, 594, 804
Topologically Integrated Geographic Encoding and
    Referencing 74, 577, 829
Toponymy 658–9
Toposcale 120
Tort 588, 643–4, 647–51
Tract geography 908
Trade-off 496–8
Traditions in spatial analysis 568–71
Traffic
    analysis zones 831
    assignment 834
Training 816, 904
Transaction management 803
Transactions in GIS 14, 827
Transfer 678
Transformation 31, 133
    in cartography 133, 438
    in G1S 348, 445
Transforming objects 77
Translation 348, 351, 475
Transport 697, 735, 739, 851, 878, 943
Transport Departments 835–6
Transportation 656, 827–41, 846, 852, 884
Triangulated data model 143
Triangulated Irregular Network 26, 67, 77, 83,
    106–7, 110–11, 132, 147, 243, 478, 483, 488,
    505, 530, 543–7, 553–4, 557
Triangular norms (T-Norm) 497
Triangulation 110–11, 131, 142, 428, 562
    adaptive hierarchical 551
Trip generation–trip attraction 833–4
Trip Planner 840
'Tripquest' 324
Trivariate interpretation 489
Trouble call management 813–14
Tuple 95, 375–7
Turing 54
UK: case study 779
Uncertainty 27, 199–200, 208, 576–8, 633–9, 846,
    1010
    absorption 637–9
    and risk 606–7

in landform labelling 536–7
management of 199–200, 635–6
reduction 637
Unified Modelling Language 353, 417–18
Uniform
    resource locator 9
    tessellations 504
Uniform Commercial Code 644–52
Union 98, 494
Unique change 92
Unique property reference number 74
United Nations 845, 848, 859, 1014
Universal function approximator 289
Universal reference
    locator 333
    time coordinate 356
    Transverse Mercator 433–5
University Consortium for GIS 1010
Unlimited tessellation 507–8
Updating 405
UPRN *see* Unique property reference number
Urban 3, 960–2
    growth 961–2
    planning 723–31, 799, 877–87, 953–63
    transportation 837–8
Urban and Regional Information System
    Association 3, 624, 827
Urban transportation model system 833–4
URISA *see* Urban and Regional Information
    System Association
USA: case study 779
User
    interface 882
    requirements 591
    satisfaction issues 227
    types 228–9
User-centred design 627
Utilities 801–18
    and new technology 803–5
    and privatisation/deregulation 816–17
    applications 846
    mapping 473, 806–10
    pole mapping
Vagueness 192–3, 196–7, 200–3
Validation 437–49
Value added
    analysis 603–4
    resellers 707, 865
    services 11
Value completeness 183
Varenius project 1011–12

Variable rate transport 990
Variance 257
Variational 484–6
Variogram 114, 213–14, 243, 258
   box plot 259
   cloud 259
Vector 4, 25, 66, 68, 95, 311, 520, 806
   data 438, 571, 667, 830, 880, 891
   displacement 142
   representation (Plate 3)
Vector-raster conversion 312
Vegetation (Plate 62) 537–40, 577, 919
Vehicle routing 578, 832
Very long baseline interferometry 429
Video 407
   conferencing (Plate 43)
   Privacy Act (US) 760
Videodisk 726
Viewshed 200–2, 534, 545, 910, 974
Virtual
   environment 310, 314–15, 402, 405, 727
   Environment Theatre 559–64
   GIS 557–64
   households 762
   Private Network 825
   reality *see* Virtual environment
   Reality Makeup Language 564
Visibility
   algorithm 548, 553
   computation 546–50
   counts 544, 546
   graphs 544, 546
   queries 544, 549–50
   structures 544–6
Visual 98, 559
   analysis 162–6
   imagery 725
   languages 406
Visual Basic 322, 344, 363, 367–8, 422
Visual C++ 344, 367
Visual exploration 166–71
Visual variables 161, 221, 224
Visualisation 27, 76–8, 107–8, 138, 157–71,
     258–9, 298, 402, 878–80, 897, 983, 1002, 1010
   and cartography 221, 310–11
   and multimedia 157, 159, 171
   and spatial analysis 157–72, 253–64
   animation 159, 166, 168, 170–1, 281, 407
   design issues 36–7, 92–3, 99, 224–5
   graphical display options 229–31
   graphical methods for detecting errors 219–32
   graphical methods in statistics 220–1

graphical methods related to GIS 221–3
graphical user interface 9, 149, 361, 374, 401,
     408–9, 561, 730, 802–3, 811, 948
graphical variables (Plate 7) 161
   limititations 220
   of epidemiological data 928–9
   representation of socioeconomic
   phenomena 71–9
   representation of terrain (Plate 55) 105–21,
     131–2, 535–7, 890, 895
Voice mail 730
Volygon 68
Voronoi
   diagram 141, 517–23, 544
   polygon (Plate 26) 242
Voxel (Plate 2) 68
Vulnerability 847–51, 854–5
Warranties 730
Water
   and sewer infrastructure 740
   quality 663
   supply 915
Waterfall Model 362–3
Watershed
   delineation 533
   structure 528–9
Wavelets 140, 561–2, 564
Weighted linear combination 495–6
Weights 499
   determination of 499–500
   spatial 246, 248, 260
Weights of evidence 970
Wetland classification 783
What You See Is What You Get (WYSIWYG) 312
Wide area network 9, 318, 861–2
Wildlife conservation 967–74, 1004
WIMP *see* Windows, icons, mice, and pointers
Windows 9, 312, 361, 364
Windows, icons, mice, and pointers 401, 408–9,
     560–1, 570
Wire frame maps 341–2
WOFOST 212–14
Word query 343
Workload planning 161–18
Work orders 811–12, 823
World Conservation Monitoring Centre 663
World Intellectual Property Organisation 777
World Meteorological Organisation 772
World Trade Agreement 770
World Trade Organisation 745, 769
World Wide Web 4, 12, 58, 308, 314–15, 321–3, 325,

333, 401–4, 468, 564, 570, 661–2, 679, 681,
683–91, 701, 726, 730–1, 748, 751–3, 768, 840,
885, 907, 1009, 1015
Worlds in Miniature 563
XDISP (Plate 64) 1002
X Windows 794
Yellow Pages 315, 956, 959
Zero meridian 438–41

Zipcode 715–16, 761 *see also* Postcode
Zonal 73, 831
Zone 77
    for urban modelling 946
Zone design 273–82
    automated/computational 273–4
Zoning maps 960
Zoology 503, 791